Springer Wien New York

T0181720

CISM COURSES AND LECTURES

Series Editors:

The Rectors
Giulio Maier - Milan
Jean Salençon, Palaiseau-Paris
Wilhelm Schneider - Wien

The Secretary General
Bernhard Schrefler - Padua

Executive Editor
Carlo Tasso - Udine

This series presents lecture notes, monographs, edited works and proceedings in the field of the Mechanics, Engineering, Computer Science and Applied Mathematics.
Purpose of the series is to make known in the international scientific and technical community results obtained in some of the activities organized by CISM, the International Centre for Mechanical Sciences.

INTERNATIONAL CENTRE FOR MECHANICAL SCIENCES

COURSES AND LECTURES – No. 486

AMST'05

ADVANCED MANUFACTURING SYSTEMS AND TECHNOLOGY

PROCEEDINGS OF THE SEVENTH INTERNATIONAL CONFERENCE

EDITED BY

ELSO KULJANIC
UNIVERSITY OF UDINE

SpringerWien NewYork

This book was printed with the contribution of Consorzio
Universitario del Friuli and of Regione Friuli Venezia Giulia

This volume contains 650 illustrations

This work is subjected to copyright.
All rights are reserved,
whether the whole or part of the material is concerned
specifically those of translation, reprinting, re-use of illustrations,
broadcasting, reproduction by photocopying machine
or similar means, and storage in data banks.

© 2005 by CISM, Udine

In order to make this volume available as economically and
as rapidly as possible the authors' typescripts have been
reproduced in the original forms. This method unfortunately
has its typographical limitations but it is hoped that they in
no way distract the reader.

ISBN-10 3-211-26537-6
ISBN-13 978-3-211-26537-6
SpringerWienNewYork

MATERIALS SCIENCE AND THE SCIENCE
OF MANUFACTURING, INCREASING PRODUCTIVITY
MAKING PRODUCTS MORE RELIABLE AND LESS EXPENSIVE

ORGANIZERS

University of Udine – College of Engineering – Department of
Electrical, Management and Mechanical Engineering – Italy
Centre International des Sciences Mecaniques, CISM – Udine – Italy
University of Rijeka – College of Engineering – Croatia

CONFERENCE VENUE

University of Udine – PALAZZO ANTONINI
Via Tarcisio Petracco 8, Udine, Italy

PREFACE

*Manufacturing a product is not difficult, the
difficulty consists in manufacturing a product
of high quality, at a low cost and rapidly.*

*Drastic technological advances are changing global markets very rapidly. In such
conditions the ability to compete successfully must be based on innovative ideas and new
products which has to be of high quality yet low in price. One way to achieve these objectives would be through massive investments in research of computer based technology and
by applying the approaches presented in this book.*

*The First International Conference on Advanced Manufacturing Systems and
Technology AMST'87 was held in Opatija (Croatia) in October 1987. The Second
International Conference on Advanced Manufacturing Systems and Technology AMST'90
was held in Trento (Italy) in June 1990. The Third, Fourth, Fifth and Sixth Conferences on
Advanced Manufacturing Systems and Technology were all held in Udine (Italy) as follows:
AMST'93 in April 1993, AMST'96 in September 1996, AMST'99 in June 1999 and
AMST'02 in June 2002.*

*The Seventh International Conference on Advanced Manufacturing Systems and
Technology - AMST'05, which was held in Udine in June 2005, aimed at presenting up-to-
date information on the latest developments – generated by research activities as well as
industrial experience – in the field of machining of conventional and advanced materials,
high speed machining, hard and dry machining, CIM, forming, modelling, simulation, non-
conventional machining processes, new tool materials, tool systems, tool condition and process monitoring, rapid prototyping, rapid tooling and rapid manufacturing, ecodesign -
assembly and disassembly, and quality assurance, thus providing an international forum for
a beneficial exchange of ideas and the furthering of favorable and productive cooperation
between research and industry.*

Elso Kuljanic

FOREWORD

It is the duty of men of science and institutions who work at fostering development, to promote those territorial aspects which – for their intrinsic interest, but also for indications of a general nature that can be drawn – deserve careful consideration.

Working in this direction has reaffirmed the strength of a concept whose affirmation dates back to Charles Bally and as far as 1909: the existence of a homogeneous European civilization, of a true "European mentality", fruit of a century-long process of cultural convergence, where powerful elements have come together, such as our Greek-Latin heritage, the spread of Christianity, the role of the languages of culture, and a civilization which over time has spread even beyond Europe.

A case in point is the 1987 establishment of the "Work Group for the study of multilingualism in the territory of Alpe-Adria", under the patronage of The Conference of University Rectors of Alpe-Adria. As a result of the work in this direction, "The International Centre on Multilingualism" (CIP) was established at the University of Udine, which is similar to different Centres in Europe in this field such as the Brussels Research Centre on Multilingualism (Centre de Recherche sur le Pluringuisme, the Uppsala University's Centre for Multi-ethnic Research (Centre for Multiethnic Research), and the Mannheim University's Eurolinguistischer Arbeitskreis Mannheim.

The International Conference on Advanced Manufacturing Systems and Technology has an important role within this process, in that they allow knowledge to converge on Udine where theory and practice continually confront.

The social world we live in has become more and more complex, thus requiring pluralistic, targeted and serious action. This action requires to develop constant innovation and to promote an integration among professions in order to be able to build, design and develop together. With this aim in mind and hoping in a future where exact science and the science of man will be able to work together more closely – in a wider and further reaching concept of universitas and in conformity with institutional expectations whose hearts are set on promoting this Conference – the Proceedings of the AMST'05 Conference will certainly help in spreading this knowledge locally, across the territory of Alpe Adria and globally.

<table>
<tr><td align="center">*Roberto Gusmani*
Rectoral Delegate to Alpe-Adria
University of Udine</td><td align="center">*Giovanni Frau*
President
Consorzio Universitario del Friuli</td></tr>
</table>

HONOUR COMMITTEE

R. ILLY, President of Giunta Regione Autonoma Friuli-Venezia Giulia
M. STRASSOLDO DI GRAFFEMBERGO, President of Provincia di Udine
F. HONSELL, Rector of the University of Udine
D. RUKAVINA, Rector of the University of Rijeka
A. STELLA, Dean of the College of Engineering, University of Udine
S.CECOTTI, Mayor of Udine
G. FANTONI, President of the Associazione Industriali della Provincia di Udine
G. FRAU, President of Consorzio Universitario del Friuli
B. SCHREFLER, General Secretary of CISM
N. AMENDUNI, Acciaierie Valbruna Spa
G. BENEDETTI, Danieli & C.
M. FANTONI, Gruppo Fantoni
A. PITTINI, Gruppo Pittini
E. SNAIDERO, Gruppo Snaidero

SCIENTIFIC COMMITTEE

E. KULJANIC, (Chairman), University of Udine, Italy
E. ABELE, T.U. Darmstadt, Germany
N. ALBERTI, University of Palermo, Italy
P. BARIANI, University of Padova, Italy
A. BUGINI, University of Bergamo, Italy
R. CEBALO, University of Zagreb, Croatia
G. CHRISSOLOURIS, University of Patras, Greece
N.L. COPPINI, University UNIMEP, Brasil
M.F. DE VRIES, University of Wisconsin Madison, U.S.A.
R. IPPOLITO, Polytechnic of Torino, Italy
F. JOVANE, Polytechnic of Milano, Italy
H.J.J. KALS, University of Twente, The Netherlands
R. KEGG, TechSolve, Ohio, U.S.A.
F. KLOCKE, T.H. Aachen, Germany
R. LEVI, Polytechnic of Torino, Italy
B. LINDSTROM, Royal Institute of Technology, Sweden
J.A. Mc GEOUGH, University of Edimburg, UK
M.E. MERCHANT, TechSolve, Ohio, U.S.A.
T. MIKAC, University of Rijeka, Croatia
B. MILCIC, INAS, Zagreb, Croatia
A. MOISAN, ENSAM, France
T. NAKAGAWA, University of Tokyo, Japan
S. NOTO LA DIEGA, University of Palermo, Italy
R. PASQUINO, University of Salerno, Italy
J. PEKLENIK, University of Ljubljana, Slovenia

H. SCHULZ, T.U. Darmstadt, Germany
K. SCHUTZER, University UNIMEP, Brasil
Q. SEMERARO, Polytechnic of Milano, Italy
G. SPUR, T.U. Berlin, Germany
N.P. SUH, MIT, Mass., U.S.A.
H.K. TOENSHOFF, University of Hannover, Germany
B.F. von TURKOVICH, University of Vermont, U.S.A.
K. UEDA, Kobe University, Japan
A. VILLA, Polytechnic of Torino, Italy
H.-J. WARNECKE, IPA Stuttgart, Germany
R. WERTHEIM, Iscar Ltd, Israel
J. ZHU, University of Nanjing, China

ORGANIZING COMMITEE

E. KULJANIC (Chairman)
M. SORTINO, A. PUPPATTI, G. TOTIS, A. KULJANIC, G. CUKOR

SUPPORTING ORGANIZATIONS

Giunta Regione Autonoma Friuli-Venezia Giulia, Provincia di Udine, Comune di Udine, University of Udine, Consorzio Universitario del Friuli, Danieli & C., Gruppo Fantoni, Gruppo Pittini, Gruppo Snaidero, Fondazione Cassa di Risparmio di Udine e Pordenone, Friulcassa, Comitato per la promozione degli studi tecnico scientifici, Centro Convegni ed Accoglienza, ONRG

CONTENTS

Machine Tools and Flexible Manufacturing Systems

Non-Conventional Processes

Tool Condition Monitoring and New Materials

Rapid Prototyping, Rapid Tooling and Rapid Manufacturing

Ecodesign Assembly and Disassembly

Quality Assurance

SCIENCE, TECHNOLOGY AND SOCIAL INNOVATION

H.-J. Warnecke

Fraunhofer-Institute for Manufacturing and Automation (IPA), Stuttgart, Germany

KEYWORDS: Paradigms of Manufacturing, Fractal Factory, Leadership.

ABSTRACT. In the last twenty years the increased international competition in the wake of globalization has changed the methods of manufacturing, especially the structure and organization of a factory. More self-similarity among the enterprise, the groups and the employees, self-organization and self-optimization have arisen. For this structure the name Fractal Company has been used. Many companies are following this perception, but it is a long and difficult way to go which needs vision and leadership. Some examples are given here.

1 MANUFACTURING IN A TURBULENT MARKET

Around 1990 a study was published which was performed by the Massachusetts Institute for Technology (MIT) and ordered by the automotive industry comparing European, US-American and Japanese plants. The outcome of this study named "The Machine That Changed the World" was terrible, especially for the investigated plants in Germany. Roughly speaking one could say: the Japanese engineers produce with half time consumed, half space and half number of employees. The German plant was said to use one third of manpower in order to repair what two thirds used to produce in poor quality before. There was still the attitude quality costs money. This shock initiated the change of organization, responsibilities and processes. The known rules of manufacturing were questioned. It was realized: manufacturing is service to the market, to the customer. Rather simultaneously arose another impact: markets and competition became global and turbulent. Planning and foresight have become more and more uncertain. Monthly quantities of production and a lot of different sizes were kept stable, and surplus over the market demand was put in store binding capital. That is no longer the right way: manufacturing must follow the market demand. Especially larger companies with their hierarchical and bureaucratic structures became too slow and lost competitiveness. The paradigms of manufacturing changed. In the last fifteen years many developments in technology and organization have appeared to increase the learning and adaptive capabilities of the manufacturing processes.

Whereas a few decades ago the survival of a company was seen in diversification, what we see now is focusing on using better the limited resources of qualified, competent employees and more targeted and efficient capital, communicating and networking with other specialists.

2 THE FRACTAL COMPANY

In 1992 I personally contributed to a proposal for a Fractal Company which was published in the book "The Fractal Company" to raise the speed of reaction and show that it is structured out of many small control loops with high autonomy and responsibility for their range of work. The

Published in: E. Kuljanic (Ed.) *Advanced Manufacturing Systems and Technology*,
CISM Courses and Lectures No. 486, Springer Wien New York, 2005.

main feature of a fractal structure in mathematics is self-similarity, self-organization and self-optimizing. The name fractal was introduced by Mandelbrot in the U.S.A. in the 1960s, because in describing algorisms broken (fractal) orders, like $x^{2,8}$ are eminent, as there are many structures between an exact geometrical plain and a cube. Self-similar means that the total structure is repeated down to the smallest unit. In a company that means, that you will find the company in the thinking and acting of the smallest element, i.e. the human being, the employee. With this aim the enterprise is organized in small, fast, partly autonomous control loops. This fundamental idea was taken by many enterprises that formed autonomous and responsible groups with budgets and targets. Each company according to its situation must find its own solution and adapt continuously. The leading management must realize that qualified and motivated co-workers are the most valuable asset of a company.

The other, already mentioned feature of a fractal structure is its ability of self-organization and self-optimization. It is a must for interdisciplinary project teams of today. Only they themselves can reorganize a company, when and how long a specialist should be integrated to solve a complex task. This flexibility is also expressed in the design of new buildings. Therefore, the new project house (Projekthaus) of BMW in Munich is very transparent and of high flexibility in its inner layout.

3   FUTURE DEVELOPMENT OF A FRACTAL STRUCTURE

In my opinion these ideas need further development to increase social innovation in society and enterprises. We would like that the citizen identifies himself with his state and the employee with his enterprise. But why would he do that when he knows that politicians are not fair and do not tell the truth and the enterprise will fire him very fast when the situation requires that and the management has no feeling of responsibility for him.

We must find a new design of working contracts, which must lead to a higher flexibility, adaptability and survival of the enterprise and accordingly to a higher motivation of the employees, when they are more dependent on the success and misfortune of their company in their income. The gap between independent and dependent employment is too big. Even if more and more young people are willing to take the risk of independent activity, we must realize, that in the future many qualified persons will also be dependent employed, think alone of the capital-intensive business. These employees must and can be given more autonomy, responsibility and risk. We must begin to think about possible ways not only of new working contracts, but also about new working culture and mentality. It will be a long and difficult way.

4   MANAGEMENT AND LEADERSHIP

This requires politicians and leaders in the economy who have visions. In Germany a politician said several years ago: who has visions should go to a medical doctor but not to government. I think vision makes the difference between politicians and a statesman, between a manager and leadership. A leader must incorporate values and give an example.

Each organization must define its values and aims. That must be internally communicated in a simple, clear and understandable way, periodically confirmed and complemented. It is strange that in an organization you can only have a very limited number of values and aims to communicate and follow up. The quality of an organization can only be seen by the results outward, internally there are only costs. A leader has always the dilemma to follow the set targets stubbornly and with reliability, on the other hand he has to correct the target in time when it is becoming illusionary without getting the image of being opportunistic and unreliable. To lead a group or company means also to give the employees freedom and to take risks. People who act make failures, but they must be sure to have the back-up of their leader. In such a culture innovations will take place. We know that only about 10% of all innovation-processes become a success. It is not worthwhile and poisonous to look for the guilty. Innovation-processes are uncertain and unplannable in the result.

5 THE NEW WORLD OF WORKING

Back to the new world of work and here is an example of a company where I am a member of the supervisory board: it is the Brose, an automotive supplier in Germany, Coburg. Thanks to the leading owner, they have created new working conditions: high flexibility in the design and lay-out of the offices, your work place, your working hours and your income depend strongly on your efficiency, while the company offers many possibilities for recreation, the difference between working and leisure time is no longer sharp. The result is that the Brose has become an appreciated employer, especially for young engineers, it has grown considerably and is one of the leading innovators.

Leadership is different from Management, which is – only – the continuous optimization of the networking of humans, machines and influences. Continuously significant figures are compared and influenced. It is mainly an administration. Whereas leadership develops the strength of the co-works it makes their weaknesses insignificant. The culture in the company is important and it also takes into account the culture of the region.

6 A LEADER: ERNEST SHACKLETON

Therefore, for states and enterprises besides structure and organization, the leadership is decisive. In history there are several examples or guides. One of the most known may be mentioned here: it is the Englishman Ernest Shackleton (1874-1922). Shortly before the outbreak of the First World War he and 28 men started on an expedition to the South Pole. The ice enclosed their ship, the Endurance, before reaching the shore of the Antarctic. The ship did not stand the pressure of the ice and sank. The crew, dogs, provision, tents and life saving boats were rescued on a large ice floe. In this hopeless situation Shackleton became an eminent leader, a guide who demanded more from himself than from his men. He was hard, consequent, reliable and foreseeable in his attitudes, but simultaneously motivating and human. A team-leader. This interesting and teaching story cannot be told here in detail. After many months he sailed in one of the open boats to the Falkland Islands to ask for help. Most men waited in the icy cold. The happy end is that the dogs had to die to give food to the crew and all men, without any loss, arrived in England in 1917. Shackleton was awarded "Sir" because of his outstanding leadership.

Today many politicians and managers are far from this example and are not appreciated by society or employees. On the other hand I am sure that there are many people in a leading position but they are not recognized in public regardless of their fulfilling tasks and excellent duties, results. The success of a company will come when further development of its technology is embedded in a culture created by good leadership.

CHARACTERISTICS OF MODERN MANUFACTURING TECHNIQUES

H. Schulz

Institute of Production Management, Technology and Machine Tools (PTW)
Darmstadt University of Technology, Germany

KEYWORDS: Manufacturing, Machining Technologies, Future Trends.

ABSTRACT: Today 5 production paradigms can be identified as actual and with future potential. The advanced machining technologies and the developments of machine tools and tools are already initiated, but must be implemented to a growing extend.

1 INTRODUCTION

Today, the manufacturing industry is on the threshold to the 5th manufacturing revolution.
1)   Substitution of hand work by machines
2)   Automation of the factory
3)   Integration of computers into the production process
4)   Integration of information and communication techniques enables globalization
5)   New production techniques by penetration of nano- and bio-technologies and application of new materials

Every revolutionary step demands for new production paradigms.

Today, technology doesn't dominate anymore because the new mind of thinking also demands for the consideration of nature, economy, society and technology (NEST).

However, in every period of paradigm changing, the production technology has been subject to a fundamental and continuous change. This process of steady changing will persist without important leaps in innovation having to be expected, but certain developments, such as the application of linear motors or parallel kinematics on machine tools, undoubtedly will speed up the overall progress in manufacturing.

In general, it will be a reduction of process chains and an increased safe performance of processes, for instance by increased use of simulation technologies.

Manufacturing becomes faster and more flexible but only when the corresponding production techniques, the manufacturing technologies and digital manufacturing etc. are available.

Therefore, to be faster and more flexible are the fundamentals of advanced manufacturing today.

2 TODAY'S AND FUTURE PRODUCTION OBJECTIVES

Significant drivers of innovation are actually the ability of innovation to follow the requests of the market:
-      lower prices

Published in: E. Kuljanic (Ed.) *Advanced Manufacturing Systems and Technology*,
CISM Courses and Lectures No. 486, Springer Wien New York, 2005.

- high quality
- large variety
- short delivery times, i.e. immediate availability
- innovative products
- environment friendly products

Today, 5 production paradigms can be identified in the consumer goods manufacturing (Figure 1).

	Craft and Small Batch Production	Mass Production	Flexible Production	Customized Production	Sustainable Production
Paradigm-Start	~ 1850	~ 1910	~ 1980	~ 2000	~ 2020
Market Characteristic	small volume per product	steady demand	small and medium volume per product	globalization, fluctuating demand	environment
Technology Enabler	electricity	interchangeable parts	computer	information technology	nano-, bio-, material technology
Process Enabler	machine tools	moving assembly, linked production lines	flexible manufacturing systems, robots	reconstructable machining systems	new manufacturing processes to add atom to atom

FIGURE 1. Paradigms of Production Technology

- Small Production or Craft Production
This means to make exactly the product that the customer asks for, usually one product at a time. Highly skilled workers and the use of very flexible machines are the characteristics.

- Mass Production
The products are manufactured in big series and produced with extremely high quantities, e.g. cars, telephones, photo equipment etc.
Because of the large quantities, products can be produced at lower costs which, in turn, enable a number of people to buy the products.

- Flexible Production
This was introduced in the late 1970s in order to respond to the change of the market, that started to be saturated by mass produced goods and the request for more diversified products.

- Customized Production
The goal is to produce a variety of almost-customized products at mass production prices. It is a society-driven paradigm, as customers are asking for a larger variety in consumer products. This mass-customization and personalization paradigm is driven by globalization, intended as the creation of a single worldwide market. Globalization creates a huge excess of global production capacity of high quality products that can be produced in several countries.

- Sustainable Production
Based on society's needs for a better environment and therefore "clean products".

The new emerging technologies of nano, bio and material technology alone or combined with each other will provide the possibility to achieve this future requirements and the goals of society in 2020 and later.

3 ADVANCED MACHINING TECHNOLOGIES

Considering the significant drivers of innovation and the mentioned general trends in innovative production technologies there are the following general trends of manufacturing technologies, machine tools and tools:
- Reduction of manufacturing time
- Reduction of planning and manufacturing costs
- Realization of high accuracy standards
- Miniaturization of products
- Application of new high performance materials, e.g. metal matrix composites, ceramics, fiber-reinforced material etc.
- Ultra-precise surfaces

3.1 MANUFACTURING TECHNOLOGIES

- High Speed Machining (HSC/HSM)
Today, HSM technology has established itself in the industrial countries as state of the art. This is also due to the fact that standard machine tools have become faster and thus allow application of machining processes at higher speeds. HSC means an increase of the cutting speed by a factor of 5 to 8 (Figure 2)
This results in a reduction of the manufacturing time and costs and an improvement of quality.

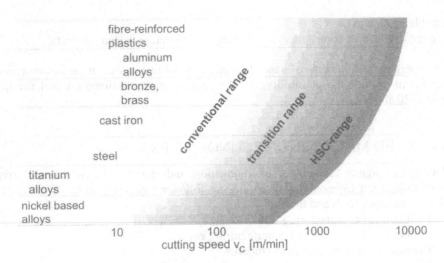

FIGURE 2. Cutting Speed Characteristics Depend on Material

- Dry and Minimal Lubrication Machining

Economical and ecological reasons and the legislation force to reduce the application of coolants. When dry machining, the most important functions of the coolant, as cooling, lubrication and chip removal must be substituted. The lack of cooling causes a rise of temperature, and as a consequence internal stress within the workpiece as well as dimensional and shape deviations, fringe layer effects, chip melting and chip build-up on tools and workpieces. Possible solutions can be seen in vacuuming the chips or blowing them away by means of compressed air.

Modification of the geometry and special coatings can improve the suitability of tools for dry machining. But a minimum lubrication will often be better. Very small quantities of lubricant (below 100 ml/h) have to be exactly applied to the cutting zone (Figure 3)

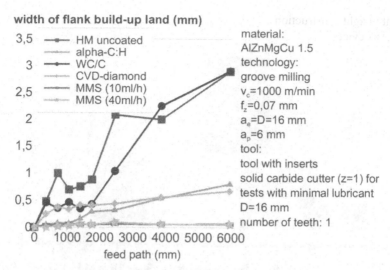

FIGURE 3. Minimal Lubrication when Milling Aluminum

- Hard Machining

Highly tempered steels expect a great deal from both the heat hardness as well as the tenacity of the used tools. For hard machining (> 58 HRC) TiAlN-coated finest-grain carbides have especially proved true. In every case hard machining decisively shortens the process chain: Hard turning substitutes all following operations. In die and mold making erosion processes can be substituted by hard milling. So, an optimum combination of hard milling and erosion can lead to a higher efficiency.

- Micro-Cutting

It is a well-known fact that the miniaturization of components keeps persisting. Good examples for this development in particular are high standard products for everyday life such as mobile phones, video recorders etc. An increasing integration of functions in small and miniature components is demanded. These micro-systems are produced by etching, primarily in such cases where injection-moulding or electro-forming dies, etch masks or even miniature components have to be made directly. Micro-cutting of dies normally involves machining of very difficult materials. This means that monocrystalline cutting materials have to be used for tools to ensure cutting edge sharpness and stability. Also only materials having a homogeneous structure can be machined adequately.

3.2 MACHINE TOOLS

- HSC-Machine Tools

The requirements to a hsc machine demand for a close interaction between the manufacturing technological process, the components of the machine and the tools. HSC-machines differ from normal nc-machines by the following essential points:
- high frequency motor spindles
- fast cnc-controls
- highly dynamic feed drives (Figure 4)

- light weight construction
- safety devices

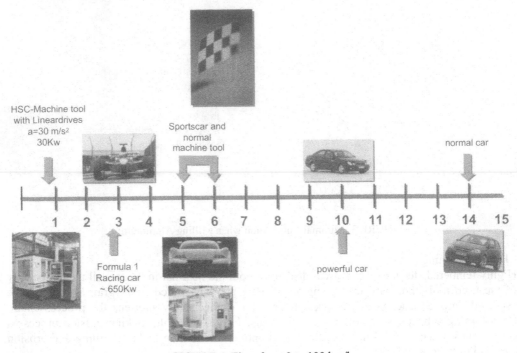

FIGURE 4. Time from 0 to 100 km/h

Higher purchase and operating costs of linear motor machines are compensated by an extreme productivity increase. As an example Figure 5 shows machining of a graphite electrode with machining time having been tremendously reduced.

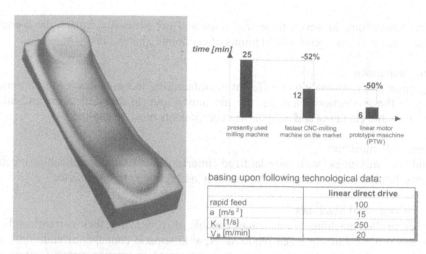

FIGURE 5. Example for the Efficiency of a Linear Motor Machine

basing upon following technological data:

	linear direct drive
rapid feed	
a [m/s^2]	100
	15
K$_v$ [1/s]	250
V$_B$ [m/min]	20

- Machines for Micro-Cutting

Of course it is also necessary to develop appropriate new machine tools for micro-cutting. The ideal machine tool provides extremely high spindle speeds up to 500,000 rpm for achieving the required cutting speed at very small tool diameters. The moving components all move without friction and must be mechanically uncoupled from each other.

- Parallel Kinematic Structures

Machine tools with parallel kinematics (Figure 6) have aroused much discussion. Their benefits are:

- inherent stiffness
- simple frame components
- reduced moved masses
- identical feed drives in each axis

FIGURE 6. Machines with Parallel Kinematics

Like other innovations, however, these machines will not generally replace the conventional machine tool but will have certain focal points of economic application.

- One setup machines
The integration and combination of different manufacturing technologies lead to a tremendous shortening of the production chain, e.g. pre-machining and finishing, cleaning and measuring processes. Flexible clamping and peripherical components must be applied.

- Reconfigurable machines
New kind of machines with standardized interfaces, reusable applicable mechatronic components, integrated sensors and actors result in a big range of performance.

- Accuracy-Controlled Machines
Accuracy-controlled machines are machines capable of recognizing their current condition and of correction it automatically when required. New kinematic concepts or additional actuator equipment must allow adaptive movements to compensate orientation errors. Current position measuring must be removed from the axes and must be relocated at the real point of process action. This means that the measuring system must be capable of making three-dimensional tool position and orientation identification. For this reason, speed and position determination of the machine must permit transmission directly at the point of machining (Figure 7).

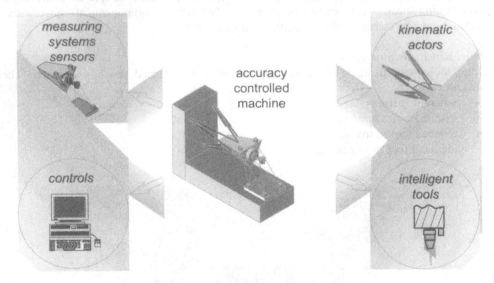

FIGURE 7. Accuracy Controlled Machine

- Lifecycle oriented machines
The simulation of the production process enables to recognize the virtual life time of the machine, beginning at development and design, regarding the manufacturing use and ending in the recycling phase (Figure 8). The best selection of economical production based on specific application parameters will be possible.

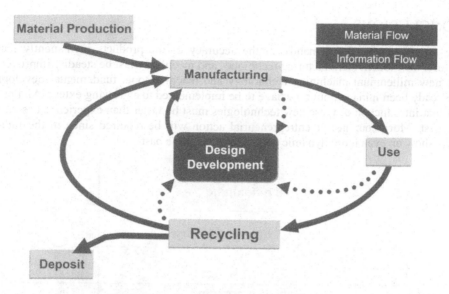

FIGURE 8. Lifecycle Oriented Machine Design

3.3 INTELLIGENT TOOLS

Today's tools will be replaced by intelligent tools equipped with sensors for recognition of accuracy-relevant factors (such as wear, process forces, chatter) located as near as possible to the point of machining, as well as with integrated actuator devices, based on small and fast elements for error compensation (displacement, vibration, edge wear) (Figure 9).

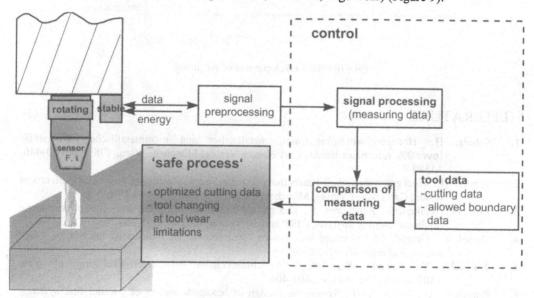

FIGURE 9. Intelligent Tool

4 CONCLUSIONS

Figure 10 shows that the demands on the accuracy of the products permanently increase. Therefore, the modern manufacturing techniques and machines must be steadily improved.
In the new millennium machining technology and machine tools fundamental developments have already been initiated, but now have to be implemented to a growing extent. An important fact is that introduction of these new technologies must be faster than experience has taught us in the past. More courage for entrepreneurial action will be required since in the future the markets show an even more dynamic behaviour than in the past.

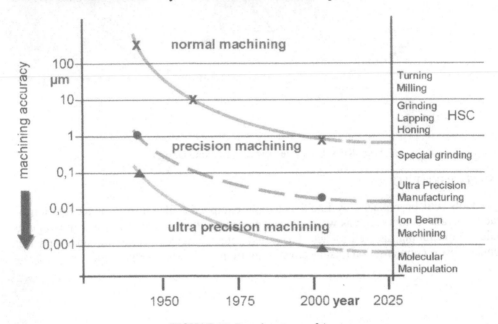

FIGURE 10. Developments of Accuracy

5 LITERATURE

1. Schulz, H.: Hochgeschwindigkeitsfräsen metallischer und nichtmetallischer Werkstoffe (over 200 references inside) Carl Hanser Verlag, München, Wien, 1989 ISBN 3-446-15589-9
2. Schulz, H.: Hochgeschwindigkeitsbearbeitung - High-Speed Machining (over 300 references inside) Carl Hanser Verlag, München,Wien, 1996, ISBN 3-446-18796-0
3. Zirn, O., Weikert, S., Rehsteiner, F.: Design and optimization of fast axis feed drives using nonlinear stability analysis, CIRP annals, Volume 46/1/97, 363-366
4. Heisel, U., Gringel, M.: Machine tool design requirements for high-speed machining, CIRP annals, Volume 46/1/97, 389-392
5. Takeuchi, Y., Sawada, K., Sata, T.: Ultraprecision 3-D micromachining of glass CIRP annals, Vol. 46/1/97, 401-404
6. Pritschow, G., Wurst, K.-H.: Systematic Design of hexapods and other parallel link systems, CIRP annals, Vol. 46/1/97, 291-296

7. Shamoto, E., Moriwaki T.: Rigid XYO table for ultraprecision machine tool driven by means of walking drive, CIRP annals, Vol. 46/1/97, 301-304

8. Schulz, H.: High-Speed Machining in Die and Mold Manufacturing, proceedings of, AMST, 4th international conference on advanced manufacturing systems and technology, Udine, Italy, 1996

9. Schulz, H.: Der Werkzeug- und Formenbau braucht schnelle Maschinen, proceedings of 3D-Erfahrungsforum Werkzeug- und Formenbau, Darmstadt,d 1998

10. Zielasko, W.: Kühlschmierstoffmengenreduzierung, Trockenbearbeitung und Sprühnebelkühlschmierung in der spanenden Fertigung, Schriftenreihe Praxis-Forum, 1993

11. Weck, M., Wahner, U.: Linear magnetic bearing and levitation system for machine tools, CIRP annals, Vol. 47/1/98, 311-314

12. Brussel Van, H., Braembussche Van Den: Robust Control of feed drives with linear motors, CIRP annals, Vol. 47/1/98, 325-328

13. Warnecke, H. J., Neugebauer, R., Wieland, F.: Development of Hexapod based machine tool, CIRP annals, Vol. 47/1/98, 337-340

14. Molinari-Tosatti, L., Bianchi, G., Fassi, I., Boer, C. R., Jovane, F.: An integrated methodology for the design of parallel kinematic machines (PKMs), CIRP annals, Vol. 47/1/98, 341-346

15. Pritschow, G.: A comparison of Linear and Conventional Electromechanical Drives, CIRP annals, Vol. 47/2/98, 541-548

16. Schulz, H.: High speed machining needs very fast and accurate machine tools, Proceedings of the International Seminar on Improving Machine Tool Performance, San Sebastian, Vol. I/98, 57-64

17. Tönshoff, H.K., Karpuschewski, B., Lapp, C., Andrae, P.: New machine techniques for high-speed machining, Proceedings of the International Seminar on Improving Machine Tool Performance, San Sebastian, Vol. I/98, 65-74

18. Neugebauer, R., Schwaar, M., Wieland, F.: Accuracy of parallel-structured machine tools, Proceedings of the International Seminar on Improving Machine Tool Performance, San Sebastian, Vol. II/98, 521-532

19. Kalhöfer, E.: Dry machining - Technology and requirements to the machine tool, Proceedings of the International Seminar on Improving Machine Tool Performance, San Sebastian, Vol. II/98, 633-642

20. Minekawa, H., Inasaki, I., Makoto, N., Suzuki, S., Kamima, T., Yokota, H.: Cutting with minimal quantity lubricant, Proceedings of the International Seminar on Improving Machine Tool Performance, San Sebastian, Vol. II/98, 655-664

21. Schulz, H., Würz, T., Huerkamp, W.: Tools for high speed machining - Safety concepts, Proceedings of the International Seminar on Improving Machine Tool Performance, San Sebastian, Vol. II/98, 715-726

22. Schulz, H.: New Cutting Strategies and Machine Tools for High Speed Cutting in Die and Mold Machining Applications, Proceedings of 31. CIRP International Seminar on Manufacturing Systems, University of California at Berkeley, USA

23. Schulz, H.: Efficiency and Accuracy of Linear Motor Driven Milling Machine, Proceedings of the 1998 SAE Conference, Long Beach, 1998

COMPETITION AND COLLABORATION IN PRODUCTION SCIENCE

H. K. Toenshoff

Institute for Production Engineering and Machine Tools, University of Hannover, Germany

KEYWORDS: Competition and Collaboration, Process Chain, Gear Production.

ABSTRACT. Competition and collaboration are the driving forces for science in general and for production science in specific. Competition is an appropriate mean to focus rare sources. Collaboration enables a holistic approach to production and can be seen as a new paradigm for production research. An example for the new paradigm of process chain development is given: the decreased process chain for the production of gears for the automotive industry. Material scientists, metal forming engineers, grinding experts and automation specialists work together with industrial partners.

1   TWO SIDES OF A COIN

Competition and collaboration are two sides of one coin. We see that in the national and international economic scale. Globalization means the internationalization of markets, of trade, of capital, of goods and services. What is true in economy is also true in science. As a matter of fact, globalization is not the invention of economists and business executives. Globalization and competition on global scales has existed for decades and centuries in the scientific world – at least in the upper level of scientific communities. So competition seems to be self-evident on the first glimpse – nevertheless, from the point of view of the German university system, there seems to be several deficits, which I would like to speak on.

Competition

♦ sharpens the profile of a university, of a faculty

♦ allocates rare resources to the best groups/labs

♦ overcomes overcapacities

♦ attracts the best students

♦ attracts the best professors

♦ makes potential transparent to industry

FIGURE 1. Why do we need competition in science?

The question is why do we need competition in science (Figure 1)? Actually, the answers are not too far away, as industrialists could give them for their own field. Competition sharpens the profile, makes clear the strong and weak points of an academic unit, of a university, of a faculty or

Published in: E. Kuljanic (Ed.) *Advanced Manufacturing Systems and Technology*,
CISM Courses and Lectures No. 486, Springer Wien New York, 2005.

even of a smaller unit like a laboratory or a seminar. This profile is the best scale for allocating resources. Resources are always rare. Right now, this is especially true for German universities and research institutes. Although there are many complaints, I think such a situation has some advantages. It gives the chance to give the money, the funds and the personnel to the best groups or to the best laboratories instead of distributing the resources by principle of giving everyone a slice of the cake.

From market economy we know that competition is a very effective means to overcome overcapacities. This steering role is of course especially valuable in times of rare resources. But in the scientific world, brains are the most important currency. Competition provides the necessary basis for attracting the best students, under the condition that students can identify the competition. This calls for university fees. The problem is, that in Germany there are no fees. There is a vivid discussion about this fact, and I am convinced it is only a matter of time before we introduce them. But right now, there are some influential opinion leaders who are against the introduction of fees. We'll have to wait.

Along the same line, is necessary to attract the best professors to the best faculties. This has to be reached by resources as well as by wages. And here again, there is very little leeway in German universities. This has to be changed.

Finally – and this seems to me especially important – competition makes the potential of a scientific group, of the human and fund resources transparent for customers, i.e. the partners in industry and the partners in society, that are especially the students.

Collaboration

♦ facilitates spezialisation and a broader approach

♦ gives different views on complex processes

♦ offers a better usage of machines and equipment

♦ brings faster results

♦ corresponds to the new pardigm: product instead of function

FIGURE 2. Why do we need collaboration in science?

We come now to the second side of the coin, to collaboration (Figure 2). In the academic world, I would like to distinguish between two types of disciplines or scientists. One type knows everything about nothing. This means that knowledge is extremely deep, but is only spread over a very small domain, almost nothing. The other type knows nothing about everything. They have extremely wide, but also very shallow knowledge. Collaboration between disciplines provides the very valuable opportunity to combine specialization with deep knowledge with a broader approach.

This also means that under the condition of collaboration different views on complex systems and processes are possible. Often, such views provide new ideas and valuable innovations. The use of machines and equipment can be improved, results are gained faster, and finally a new paradigm may be introduced in science, especially production science, which is already common in industry: If production science with its applied research claims partnership with production industry, it has to see this paradigm change which took place one or two decades ago. The organizational structure in many companies and very often in innovative companies has changed from a functional orientation to a process orientation. The product and its path through the company come into the focus of managing strategies.

My colleagues in Hannover and I have concluded from this deep change in structure and process-oriented organization, that it is not the single and isolated process in research projects that must be looked at, but the manufacturing of a whole product. For our partners, a specific process is not of interest, but the total process chain which a product runs through [1,2].

One example of this approach in production science shall show you how my colleagues and I react to this new challenge. I have chosen an example of new process chains of high performance components from the automotive industry. I'll speak about the production of gears.

2 PROCESS CHAIN FOR GEARS

Gears belong to the most important machine elements in machines, cars, machine tools and energy transformation systems. Although we live in a century with a wide range of applications in electronics and photonics, with micro-devices like micro-electronic or photonic elements, still, mechanical elements like gears are indispensable for the transformation of speed or torque. The history of gears goes far back, more than 2000 years (Figure 3). The first evidence on the existence of gears was given by Philon 230 B.C. who wrote the μηχανικη συνταξις, the compilation of inventions. Philon shows a water hoisting device in Luxor, an old Egyptian city, where important testimonies of culture were built more than 3000 years ago. The picture also shows a modern, extremely compact power shift transmission for a heavy truck. The development in gears, especially for the automotive industry, is mainly concerned with the transmission of higher power, the reduction of weight and construction volume, an increase in comfort and in reliability. These achievements have been reached through materials of higher strength, through higher macro and micro-geometric accuracy and through a considerable improvement in surface integrity. Besides that, of course, productivity is a main criterion of gear manufacturing.

My group works in this field together with other groups of material science, metal forming and automation, in cooperation with the automotive industry and with machine tool manufacturers. The approach of our project is an entirely new process chain, as can be seen in Figure 4[3]. Conventional production comprises of at least 14 steps, starting with a sawing or shearing process from bar stock. After storage, the steel block has to be heated, then forged and deblurred by shearing. Cooling and temporary storage again follows until the forged part is intermediately machined in the soft state by turning and hobbing. Since heat treatment is done in batch operation, storage is necessary again. Finally, the hardened and tempered part is finished by abrasive processing.

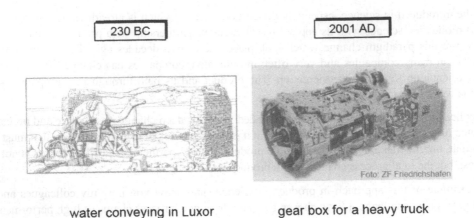

water conveying in Luxor   gear box for a heavy truck

FIGURE 3. Yesterday's and today's gears

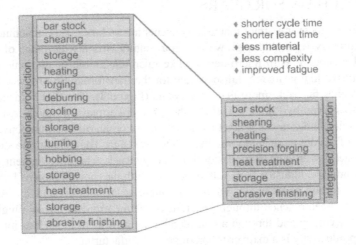

FIGURE 4. Gear manufacturing

The integrated production chain is drastically shorter, and takes place mainly in a flow line, so that only one storage step is necessary. Production starts again by shearing from tubing or a rod. Heating, forging and heat treating stations are integrated into the production line. Forming is done by a so-called precision forging process, which takes place in a closed die. Therefore, the forging result has no burr, and thus does not need deburring by shearing. The main point is, that the teeth of the gear are already shaped with a tolerance in the range of 0.2 mm to 0.4 mm. In Figure 5 precision-forged helical gears can be seen. It is especially interesting that additional form elements can be formed, which do not need further machining operations. Heat treatment is done using the forging heat, with integrated, controlled cooling by gas, or with a two-phase fluid. Integration of the hardening and tempering process into the forging line saves a considerable amount of energy, because the material is heated only once.

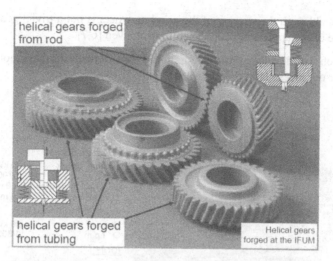

FIGURE 5. Precision forging process with integrated heat treatment

The precision forging is of a quality that needs only hard machining by grinding. The part is finished in two clampings, in a universal grinding machine, and in a gear grinding process. There are several possibilities, as shown in Figure 6. Principally, generating (cinematic) and profile grinding can be distinguished. In profile grinding, the gear tooth geometry is determined by the shape of the grinding wheel, whereas in cinematic grinding, tooth geometry is generated by cinematic coupling of the tool and the work piece movement [4, 5]. Each group of processes can be further subdivided into discontinuous and continuous processes. Because of quality and productivity reasons, in our project we concentrated on discontinuous profile grinding and on continuous grinding with a cylindrical warm-threaded wheel or involutes screw.

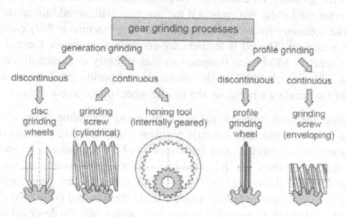

FIGURE 6. Gear grinding processes according to German DIN 3960 classification

Discontinuous profile grinding means that each tooth space is ground in a single path. The grinding wheel has to be shaped according to the profile of the two adjacent teeth. The process is highly productive, although the shaping of the grinding wheel is not a trivial geometric problem. I'll report on some of our approaches and the results achieved [6,7].

The machine which was developed in collaboration with a German machine tool builder and my research group for our investigations is shown in Figure 7.

processes:
- profile grinding
- cont. screw grinding
- external honing

tool spindle:
- nom. power   20 kW
- r.p.m.   0-7000 1/min
- wheel diam. max.   210 mm
- max. speed   75 m/s

work spindle:
- nom. power   20 kW
- r.p.m.   0-3000 1/min

tools:
- dressable vitrified wheels
- non dressable galvanic wheels

gear data:
- modul   0,5 -10 mm
- head diam.   50 - 250 mm
- max. helix angle $\pm$ 35°

control:
- Siemens   840 D

FIGURE 7. 6-axes gear grinding center Kapp KX1

The machine is versatile – which is very advantageous for R & D purposes. It was designed for discontinuous profile grinding, for continuous screw grinding and for external tooth honing. It has six numerically-controlled axes. We equipped the machine with an additional dressing unit and, of course, with the necessary force measuring devices. The machine is fully encapsulated, in order to use oil as a coolant. The oil is temperature-controlled, to ensure thermal stability of the machine. The NC-control 840 D was designed so that arbitrary electronically-linked gear trains could be defined. This was essential for the cinematic and profile grinding processes as well as for the dressing of the grinding wheels, i.e. the profile wheel or the screw wheel.

For profile grinding, the main problem was calculation and generation of the tool or wheel profile. It is firstly a three-dimensional geometric problem where the given contact line of the tooth flank has to be transformed into the axial cross-section of the tool, as can be seen in Figure 8. I will not go into the mathematics, which is based on vector algebraic calculations. The theory has been state-of-the-art since the end of the sixties. But since modern gears have been profile-corrected several times, to improve fatigue strength and vibration and noise behavior, the involute equations had to be extended, to provide for arbitrary mathematically describable modifications. This was not trivial. Besides that, for precision forged blanks, the tooth foot and the tooth head roundness had to be incorporated into the wheel profile. Especially the latter is important for the fatigue life of the gear, because in forgings, in comparison to pre-machining by milling, a protuberance cannot be generated. Thus in profile grinding, the foot of the tooth is ground, too.

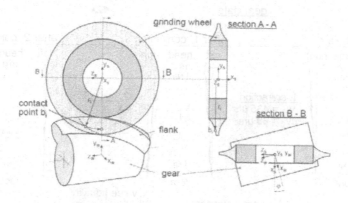

FIGURE 8. Transformation of the contact line into axis cross section of the tool

The quality of the manufactured gear is dependent on several geometric deviations of the machine and the tool. This might be due to thermal effects as well as to elastic deformations or deviations of the theoretical machine positions and movements. Because we set the goal to achieve at least quality 5 according to ISO standards, we developed a system to reconsider different deviations of the geometric system. The Figure 9 shows six influences, displacements of the tool in x_s and y_s direction, tilting of the tool around the y_s and the z_s axis, and of the workpiece around x_w, and finally the rotation of the workpiece around z_w. The effects for these erroneous influences can be calculated, so that preventive correction is possible, or reasoning backwards, the influences can be determined by the different effects. The Figure 10 shows how this correction model works. The profile lines for the left and right flanks are improved by two corrections from the initial state, where the quality was in the range of quality 9 to 10. The quality value of 5 to 4 could only be reached by applying the correction model. Quality 4 and 5 is beyond the superior industrial demands which are now in the range of 6.

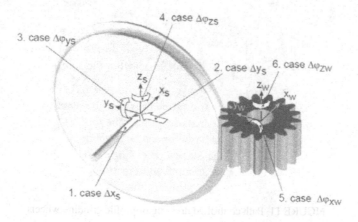

FIGURE 9. Possibilities of erroneous positioning between grinding wheel and gear

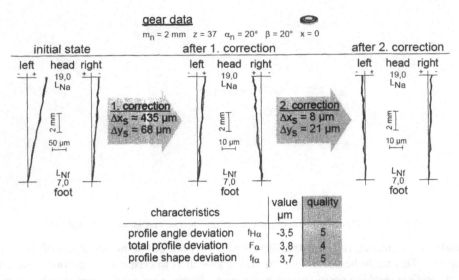

FIGURE 10. Correction of grinding wheel position with the correction model

Let us assume that the tool profile has been correctly determined. It then has to be achieved at the real grinding wheel. For the wheel made of sintered alumina, which was used, we developed a special dressing procedure, working together with a dressing wheel manufacturer. The rotating dressing tool is equipped with hand-set diamonds of a special strength [8]. The simplest possibility to generate the right and left half of the profile is to dress it in one path. But that would implicate that one side is dressed by push and the other by pull. This would lead to different loads, and thus to different elastic deformations of the grinding wheel. This is the reason that a twin edged dressing wheel was developed (Figure 11).

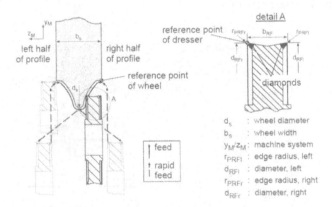

FIGURE 11. Path controlled dressing of profile grinding wheels

This example of developing a new process chain for gear manufacturing shows the essential collaboration of different disciplines, and of research institutes and industrial companies (Figure 12).

Integrated heating, forging, hardening and tempering made it necessary to think about the material. Material scientists had to enter the project. Because in precision forging, die wear resistance is of crucial importance, the material scientists also had to give their advice on how to improve this feature. Presently, thick coating of dies at the critical parts of the surface seems to be the best solution.

	Industry	Research Institutes
Material	Thyssen Krupp, Remscheid Hanomag, Heat Treatment, Hannover	Institute for Material Science, U Hannover
Forming	Müller Weingarten, Weingarten Hirschvogel Automotive Group, Decklingen Hammerwerk Fridingen GmbH, Fridingen	Institute for Metal Forming, U Hannover
Grinding	KAPP Group, Coburg Rappold, Winterthur Dr.Kaiser, Diamantwerkzeuge, Celle	Institute for Production Engineering and Machine Tools, U Hannover Institute for Transport Systems and Automation, U Hannover
Component	Ford Motor Company, Cologne	Center of Production Technology, Hannover

FIGURE 12. Project consortium

For the development of the precision forging process itself, metal forming engineers were responsible. The optimal design of the die, the material flow considerations and the definition of the process parameters was their task. Together with the material scientists, they developed the hardening and tempering process which was integrated into the forging process. The Institute for Transport Systems and Automation had to be consulted to ensure the time-critical material flow from the shearing device via heating and forging to heat treating.

My group, together with the metal forming engineers, investigated the optimal shape of the forged blank. If the forged blank is too near to the final geometry the forging process is complicated, pressure in the die increases and adds to the wear of the forging tool. On the other hand, the allowance of the forged blank must not be too high, so the flanks can be ground in only two paths. The optimal geometric and technological interface between the size and the shape of the forging and the subsequent finishing operation had to be determined. The strong and weak points of the adjacent processes had to be investigated.

Besides that, the industrial partners played an essential role in the project. The forging press manufacturer developed a machine with sufficient power and press force and an extremely short pressure contact time, since the duration of contact while the pressure is held in the die dominantly determines the heat flux from the heated workpiece, approximately 1200° C, to the die. If the die is overheated, the die material looses strength, and due to this, wear increases.

Cooperation with the gear grinding machine manufacturer enabled my group to obtain a specific grinding machine for the given purpose. Development of the dressing system and its integration into the gear grinder was the result of collaboration with a small, but innovative dressing roll manufacturer. Finally, demands were made and input given by the automotive parts supplier, in connection with the car manufacturer.

The process-chain-approach needs inevitably collaboration between partners. Several academic groups as well as industrial partners have to cooperate simultaneously and in an open mind manner. As a matter of fact the new paradigm brings industry and academia closer together. Thus the other customer of academia, the students gain also to a great extent. In any case the participating worlds, industry – research groups – students need driving forces: These are competition and collaboration.

REFERENCES

1. Toenshoff, H.K. (1992): Survey of Manufacturing Processes. In Dubbel, Handbook of Mechanical Engineering, Springer, K1.
2. Toenshoff, H.K., Kramer, N. (2004): Technologische Prozessketten – ein strategischer Ansatz in der Forschung. Tagungsband Hannover Kolloquium 2004, PZH Produktionstechnisches Zentrum GmbH, 9-30.
3. Behrens, B.-A., Doege, E, Reinsch, S.,.Telkamp, K., Daehndel, H., Specker, A. (2004): Precision Forging Processes for High-Duty Automotive Components, International Conference on Advanced Manufacturing Technology, 828-834.
4. Bausch, T (1994).: Moderne Zahnradfertigung. 2.Auflage, expert Vertlag.
5. Boucke,T. (1994): Zahnradprofilschleifen mit keramisch - gebundenen CBN - Schleifscheiben. Dr.-Ing. Diss. RWTH Aachen.
6. Toenshoff, H.K., Marzenell, C (1999).: Comparison of Surfaces Generated by Different Finishing Processes for Gears. Proc. 32nd Int. Symp. On Autom. Techn. & Autom., Wien, pp. 51-60.
7. Tuerich, A. (2002): Werkzeug-Profilerzeugung für das Verzahnungsschleifen. Dr.-Ing.Diss. Univ. Hannover.
8. NN: Product Catalog of Dr. Kaiser Diamantwerkzeuge, Celle, Germany, www.drkaiser.com.

A NEW STRUCTURE OF AN ADAPTABLE MANUFACTURING SYSTEM BASED ON ELEMENTARY WORK UNITS AND NETWORK INTEGRATION

J. Peklenik

University of Ljubljana, Slovenia

KEYWORDS: Elementary Work System (EWS), Adaptation, Network, Integration, Information, Logistics.

ABSTRACT. The existing industrial production structures for various products are based on A.W. Taylor principles of the division of work and operations, developed almost a hundred years ago. The previous century introduced into manufacturing technology, systems and their control a large number of innovations, based on fundamental and applied research in various fields of interest. The contribution presents a new approach to the structuring of an Adaptable Network Manufacturing System (ANMS) for the production of various HT-products. The ANMS is based on a competent selection of Elementary Work Systems (EWS). Some structures and characteristics of their elements and units are explained in detail. The second part of the contribution presents the procedure for the structuring of an optimal ANMS. The role of innovation management and marketing research is analyzed. The active cooperation with a team of researchers, dealing with advanced investigations related to the new product, and the team of product designers is explained. Special attention is focused on the subjects S and their competence, as well as on the methods of how to increase the necessary knowledge, needed by the subjects operating an ANMS.

1 A BRIEF HISTORY OF THE DEVELOPMENT OF MANUFACTURING SYSTEMS

At the beginning of the second half of the previous century, the manufacturing technology became a field of significant changes. In 1952, the CINCINNATI – MILACRON [1] developed the first computer-controlled machine tool, an NC-milling machine, creating a revolution in the development of production technologies.

The computer capacity and the physical size of the process computers were being improved fast. These achievements enabled more flexible, faster and reliable production activities. The introduction of NC into manufacturing technologies became very intensive, indicating significant changes of the philosophy of manufacturing industrial products.

About fifteen years later, in 1967, a second revolutionary step changed significantly the technology of new products. The first manufacturing system »MOLINS 24« was developed in England, by the physicist Williamson [2]. The M-System consisted of six NC-machine tools with loading units, a transportation system Molac, connecting the individual MT with the pallet rack and the work setting station, Figure 1. A process computer for integration and coordination of various activities on six MT controlled various machining and other processes, necessary for manufacturing geometrically and technologically similar parts, based on the group technology principles.

Published in: E. Kuljanic (Ed.) *Advanced Manufacturing Systems and Technology*,
CISM Courses and Lectures No. 486, Springer Wien New York, 2005.

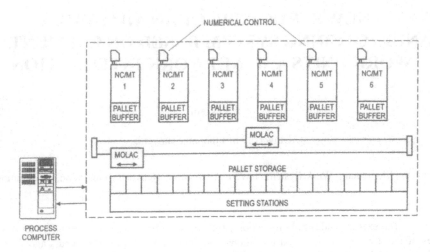

FIGURE 1. Integrated manufacturing system (MOLINS 24)

A year later, two important developments of manufacturing systems were realized: the Cincinnati Variable Mission {3} and the IKEGAI Turning System COMPTURN [4]. In the next four decades, the principle of integration of NC machine tools and other computer controlled equipment into the manufacturing systems was widely applied in the production of HT- products.

2 THE WORKABILITY AND CAPACITY OF A MANUFACTURING SYSTEM

The development of the first manufacturing system »MOLINS 24« focused on two types of integration:
- the mechanical integration of classical work units, work setting stations and storage places by means of various transportations and handling devices, and
- the computer integration of the functions for running, control and supervision of the manufacturing system.

The manufacturing of parts in a MS with various processes is, however, limited in terms of
- dimensions of parts to be produced,
- accuracy and surface requirements,
- number of available processes and tools,
- ability of the process to form a required shape in a given material, etc.

These limitations cannot be changed if the parts to be produced have different requirements (dimensions, processes etc.), which the available MS does not cover. This means that a new M-system must become available and put in operation. The limited flexibility of the MS becomes an important cost problem, affecting the competition of the firm. This situation can be solved effectively by large companies which have sufficient financial means and intellectual capacities to meet the manufacturing requirements. When the dimensions of parts, technological processes, accuracy etc. require a new manufacturing system, a new investment will be necessary. The SME (Small and Medium Size Enterprises) can not easily solve this type of problems.

Another observation, related to the percentage of time employed by the MS, is of great importance since it affects the production costs. A full day employment is 100 %. Let us assume that the normal production time is 8 hours or 33 % of the full day. Active employment of the MS might be perhaps 4 hours or 16.5 %. This means that the time in which a MS generates parts is only ~ 1/6 of the full day. The low usefulness of this distribution is not desirable.

These are the reasons why it is necessary to search for a new solution related to manufacturing, which may provide better results and much lower production costs.

In order to explain the proposed solution, let us consider the following topics in the development of a new approach to form the future structure of manufacturing systems with the existing SMEs:
- the structure and properties of the EWS,
- the selection of the necessary EWS for manufacturing an innovative HT product,
- the network integration of EWS by information communication and logistics in order to structure an ANMS for the production of a HT-product.

3   MOTIVATION FOR THE DEVELOPMENT OF A NEW SOLUTION

The last decade brought a number of new ideas and solutions into the field of industrial production. The most important among them is, perhaps, the application of the communication networks into the manufacturing activities.

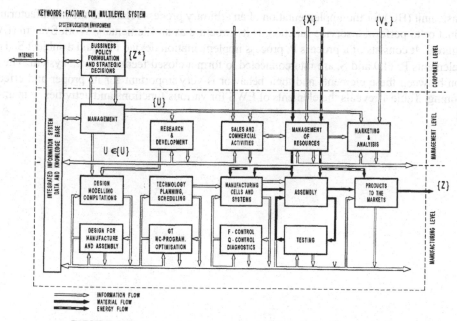

FIGURE 2. Factory as a large complex multilevel adaptive system

The classical production system, (or factory), represents a fixed structure of various working units on 1. the cooperative level (for business policy formulations and strategic decisions), 2. the management level (management, marketing, research and development, sales etc.), and 3. the manufacturing level (design, technology planning, manufacturing cells & systems, assembly, etc.). Figure 2 [7] These units are located in various buildings, forming a fixed structure of the factory. Any changes in the production programme usually involve considerable expenses.

To improve this situation, Large Manufacturing Corporations (LMC) introduced a new approach when organizing an enlargement of their production capabilities. Based on the principle of »Manufacturing in Networks«, developed by researchers in the USA, Germany, Japan, UK etc., they introduced the Network Structures. Selected Small and Medium Size Enterprises (SME) are integrated into an information communication network with the major LMC directing, managing, controlling and also selling the products on the global markets. The individual SME-members of the clusters usually produce various components of the products which have been developed and designed by the LMC. This type of production is still based on F.W. Taylor principles which, today, may be questionable [9]. The SMEs working in this type of networks are limited in their development, decision making, independence and financial means.

In the EU, the SME employ about 66 % of the workforce and generate ca. 56 % GNP. [10] For this reason, it is important to find new ways of significantly increasing the innovativeness, productivity and production output of the SMEs in order to decisively increase the GNP of nations.

4   THE MORPHOLOGY OF AN ELEMENTARY WORK SYSTEM (EWS)

The basic unit (BU) for the implementation of an arbitrary process related to the manufacturing of a product or to perform a service function, is its structure as an Elementary Work System (EWS) [7] Figure 3. It consists of a process P, process implementation device PID and a subject S. These three elements P, PID and S, are interconnected to form a closed feedback work system. The correlation between these elements and their behavior is very important for its proper and effective functioning. Table 1 reveals the elements of EWS for various functions and activities in manufacturing.

TABLE 1. Elements of the EWS

PROCESS P	PID	SUBJECT S	INFORMATION, COMMUNICATION
DESIGN	COMPUTER	DESIGNER	Software System
PLANNING & SCHEDULING	COMPUTER	PLANNING PERSON	Software System
TURNING	LATHE	OPERATOR	-
GRINDING	GRINDING MACHINE	OPERATOR	-
MEASURING	M-INSTRUMENT	MEASURER	-
MANAGEMENT	COMPUTER	MANAGER	Software System
ASSEMBLY	ASSEMBLY DEVICE	OPERATOR	-
WELDING	WELDING DEVICE	WELDER	-
DECISION MAKING	(COMPUTER)	MANAGER	
MARKETING	COMPUTER	MARKETING MANAGER	Software System
TRANSPORTATION	TRANSPORTATION DEVICE	DRIVER	LOGISTICS
DEVELOPMENT	(COMPUTER)	RESEARCHER	SUPPORTING DEVICE

Let us identify these three elements:
- what are their characteristics,
- how can they be described analytically,
- what are the possibilities to determine the functioning of the EWS under selected conditions.

In order to structure properly an EWS for a certain process in the development, design, production, assembly etc. of a product, it is necessary not only to describe these three elements of the EWS in words, but to define these elements by analytical methods.

A manufacturing process, for instance, turning, grinding, assembling etc. must be described analytically in such a way that the structure of the description indicates the factors, affecting the output in terms of process performance and quality. The analytical tools for process and PID description are the transfer functions or the describing functions. These functions define the relation between the output and input, which expresses the conditions affecting the process, dynamics, costs and quality. The describing function is used for the description of nonlinear systems. Tools and methods have been developed for the estimation of various influences related to the conditions, materials and dynamics, affecting the process output. Research into the identification and control of manufacturing processes, machine tools and other devices brought a number of descriptions of

P and PID by transfer and describing functions. Our future task will be to collect and standardize this information for future use.

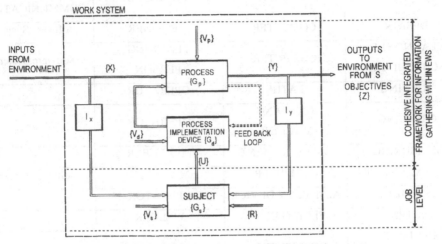

FIGURE 3. Cybernetic model of the elementary work system

The third element of the EWS is the Subject S (Operator, Designer, Manager, Team ...). The assessment and selection of an appropriate person for the implementation of a certain process or function depends upon his competence, trust, ability to adapt at a proper level, to communicate, to work in a team, and the ability for self-organization. Creativity is a highly desirable subject characteristic as well, it can be activated under certain conditions, created by the management of the project development [6].

Competence is defined as the sum of knowledge and experience

$$C = K + E \tag{1}$$

The estimates of competence should, of course, be properly documented and available for persons active in an enterprise.

5 THE INTEGRATION OF THE EWS INTO AN ADAPTABLE NETWORK MANUFACTURING SYSTEM (ANMS)

An analysis of the capabilities of SMEs to create new innovative HT-products and to manufacture it for global markets shows that the available knowledge and financial means of these enterprises usually cannot meet these problems [5]. On the other hand, the SMEs employ a number of creative and intelligent subjects who would be capable of generating new ideas, leading to new innovative HT-products. However, the present structures and organization of the factories and the transfer of knowledge, research and advanced working methods, are not the best solutions for the improvements of production activities.

Several years ago the author proposed a new structure of an Adaptable Network Manufacturing System which was developed and successfully tested, [6].

5.1 THE ORGANISATIONAL STRUCTURE OF AN INNOVATIVE HT-PRODUCT SPECIFICATION

An unidentified subject, employed perhaps in an SME, research institute or university gets an innovative idea about a new product. Let us assume that this idea is presented to a clever manager of an SME. He becomes enthusiastic and decides that the idea should be tried out, Figure 4. For this reason the manager organizes a small group of selected product developers or designers with creative ideas, who start to search for various solutions, together with the originator of the innovative idea. The aim of this effort is a well developed HT-product specification, considering the functions, performance, loading, quality and reliability, price, delivery time, etc.

However, the knowledge of these subjects in various fields is usually not sufficient. They are, for instance, not informed about the research accomplishments in the areas which can contribute a great deal to the innovation of the product.

For this reason, a carefully selected research team should be employed in order to provide the newest and most relevant knowledge for the solution. This integration of two teams working together is highly important and the only way to transfer effectively the scientific research results into new products.

An important question to which the innovation management must get a reliable answer is »whether the new product has the chance to be sold on the global markets.« The specification of the HT-product enables the marketing research to obtain an appropriate answer. The decision whether detailed design and manufacturing should be initiated and financially supported, depends on the marketing research results, making expectation assessments more real.

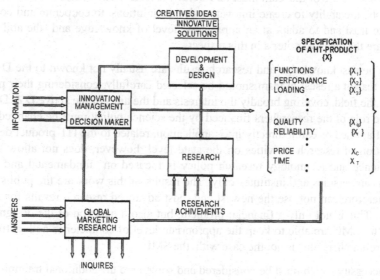

FIGURE 4. Process of high-tech product specification

5.2 DEVELOPMENT AND DESIGN OF AN HT-PRODUCT RESULTING IN THE PROTO-TYPE

Figure 5 reveals a corresponding work structure to accomplish this task.

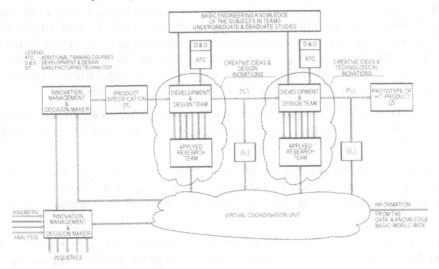

FIGURE 5. Structure of the "Innovative" development & design process cycle

The product specification $\{X_i\}$ describes the input data which the D & D team has to consider. The intellectual structure of the team must be carefully selected. The competence of the individual D & D subjects, the ability to create innovative design solutions, to cooperate and communicate in the team, to trust and to adapt at an appropriate level of knowledge and the ability for self-organization are the decisive criteria in this respect.

However, the newest knowledge and research results are usually not known to the D & D team. For this very reason a research team should be selected carefully, considering their previous investigation in the field, covering broadly the interests and the aims of the active D & D team. This means that the role of the researchers financed by the science policy must be changed to support the transfer of applied research directly into application, related to the HT-product development. The organization of research activities on the state level, however, does not allow this type of approach. Science and technology research policy is focused on fundamental and applied research within universities and institutes only. The results of this work are the published papers. The SMEs therefore can not use the newest and most advanced research results for their development work. This is not only a financial problem, but also a question of the designers' knowledge. While the LMC are able to keep the appropriate level of competence and active cooperation with the researchers, this is not the case with the SME.

The second necessity which must be considered and solved are the additional training courses for the D & D. The volume and quality of knowledge in various fields of working technologies have grown tremendously. Information and communication knowledge, design methodology, control and systems theory, materials, etc., in particular, are spheres that subjects working in various fields are often not familiar with. For this reason their work productivity and quality are rather limited. With selected training courses on various topics, the subjects will get a lot of useful knowledge which will significantly contribute to the improvement of their work performance and productivity.

The next step in the process is the manufacturing of the prototype of the HT-product, Figure 5. A manufacturing team develops the technology for the prototype, entering into this process by the information $\{Y_D\}$. The activities of the team, their knowledge of the most advanced technologies, etc. can be specified in the same way as for the D & D team. The role of the manufacturing research team and the importance of the additional training courses have the same objectives and procedures as earlier discussed.

The output $\{Y_P\}$ of these activities is the prototype of the developed HT-product, which will be tested and improved before the final decision is made, which means that this product will be planned for production on an appropriate large scale for global markets.

The role of the Virtual Coordination Unit (VCU). The organizational structuring of working teams, as well as the responsible decision making concerning their activities, are in the hands of the innovation management. For this type of activities, the manager must have a large amount of well- organized and quickly accessible information at his disposal.

The solution of this requirement was developed and realized as Virtual Coordination Unit (VCU) shown in Figure 6 and Table 2 [6]. This new working unit provides all the information needed for the activities: from the decision making to the manufacturing of the prototype.

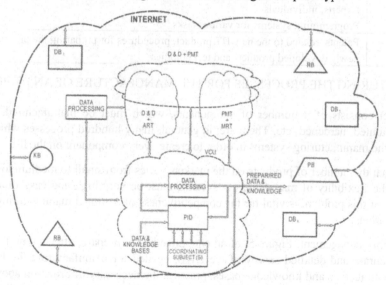

FIGURE 6. Structure of the virtual coordination unit

The VCU is connected via Internet and various local networks with the data and knowledge bases worldwide. The problem of obtaining systematically organized information, that the production needs, from the data and knowledge bases, was realized by many research institutions and industries. This was the reason for the organization of the European Network of Excellence, with the objective to solve these important problems within the EU-project VRL-CiP (Virtual Research Laboratory – Communication in Production) [7]. It will take quite a long time for this problem to be solved successfully.

TABLE 2. Data and knowledge bases of the VCU

Data Bases	- products, available on the markets - materials - components, elements . .	(mechanical electrical biotechnical . hydraulic)

Bases on Production Capacities	Enterprise A
	                  B
	                  .
	                  K
		(employer, available machinery, EWS's capacities, financial means etc.

Knowledge Bases	Research, focused in certain field Research institution, individuals Training courses, relevant for work improvement, teachers, etc. Experts, individuals Programming systems for various tasks	
Patents ·	Patents, related to the new HT product, procedures for preparing the patents of newly developed products and technologies	

5.3 STRUCTURING THE PROCESSES FOR THE MANUFACTURE OF AN HT-PRODUCT

Each product consists of a number of components, which must be manufactured, measured, sometimes painted, hardened, etc. There are, in general, many hundred processes which must be available in the manufacturing systems in order to create every component on the list.

The fact is that the number of products on the markets varies from small to medium to large. This means that the flexibility of manufacturing systems must be very high and easy to accomplish. The solution of this problem is vital for the competitiveness of national manufacturing industries on global markets.

The innovation management, Figure 5, of an enterprise has to prepare, after the prototype is finished, an accurate and detailed survey of the EWS, needed for manufacturing the HT-product. The VCU, with its data and knowledge bases, provides the necessary information about the EWS in various SMEs. Its structure is shown in Figure 6, and indicates the information about the data and knowledge available on the website.

The manager should select the EWS, situated in various SMEs, which are able to manufacture the proposed product. The interest of the contacted SME will decide about the cooperation between interested firms.

A great advantage of this approach is the structuring of an adaptable network manufacturing system (ANMS) by selecting the appropriate EMS via the local data bases with the procedures related to the VCU. This means that a team of experts led by the innovation manager selects the available and proper EMS for the jobs. The selectors must not only take into account the technological properties, but also the daily free- of- work time in which the SMEs can be employed. This type of approach may contribute to higher employment of the SME in comparison with the classical organization of the factory. No new high financial costs are required to implement the production of the HT-product.

6 INTEGRATION OF THE EWSs INTO AN ANMS

The innovation manager S1 of the SME1 selects from various SME2 ... SMEk ... SMEn, via VCU, a set of the EWS1,2,...t, able to produce the planned HT-product. He invites his colleagues to form a team in order to discuss the production of the proposed product. Figure 7 indicates the basic structure of an Adaptable Network Manufacturing System (ANMS), with the objectives as described above, [6].

The selection of the EWSs and their locations, as well as the VCU, is finally established by a competent team which also solves the logistics, based on the principle of »just in time«. Figure 8 reveals an arbitrary distribution of the EWS, integrated via Internet, LAN and the material transportation logistics (MTL) into an Adaptable Network Manufacturing System (ANMS).

For the implementation of various production functions the SMEs have a number of EWSs working together. For instance, for the manufacture of an axis with two types of gears, the following processes are required: 1 – turning; 2 – milling; 3 – hardening; 4 – grinding. The limitations are the maximal dimensions and the accuracy of the processes. Therefore, a SME, cooperating in the ANMS, employs more than one EWS, called production unit PU, as shown in Figure 9. This type of manufacturing structure represents an optimal solution, related to much better exploitation of the machinery than in the case of a classical factory.

The adaptability of the manufacturing system (ANMS) structure depends upon the innovation team of managers who are determined to achieve as low as possible investment costs in the process implementation devices (PID), by effectively employing the existing PID. This is a very effective way of increasing the competitiveness of the SME on the global markets.

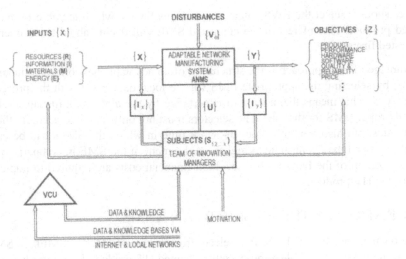

FIGURE 7. Basic structure of an ANMS

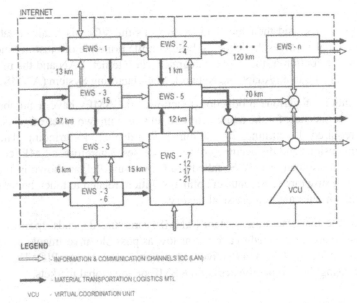

FIGURE 8. Network structure of a complex adaptive factory system with distributed EWSs

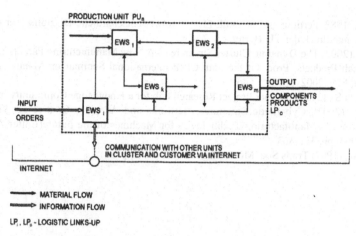

FIGURE 9. Structure of production unit PUn in an SME

7 CONCLUSIONS

The existing industrial production structure, based on A.W. Taylor principles developed almost a hundred years ago, were analyzed. It was shown that the SMEs, which contribute more than 60 % to the GNP and employ more than 50 % of the work force, are not in the financial and intellectual position to develop and manufacture innovative HT-products for global markets.

This research paper, supported by experiments, shows how an innovation manager of a SME initiates the development of an innovative HT-product. The procedure of the product specification, the role of the development and research team, as well as the marketing for the decision making of the management, are discussed in detail. The product specification presents a carefully prepared and tested input into the D & D and the manufacturing of prototype activities. In order to accomplish innovative products, active support of research transfer for the design and production technology is necessary. To increase the knowledge for the work improvement of the cooperative subjects, various training courses are planned and executed. A virtual coordination unit (VCU) is established for the selection and control of the information related to the processes involved.

The Elementary Work System (EWS) necessary for the implementation of various processes which are not at the disposal of various SMEs. These elements are integrated by communication and logistic means in order to formulate an Adaptable Network Manufacturing System (ANMS) for manufacturing an HT-product, competitive on global markets.

REFERENCES

1. Cincinnati Milacron Report 1952.
2. Williamson D.T.N. (1967): A New Concept of Manufacture Proc. 8th International MTDR-Conference Manchester 1967 pp.372-376 Pergamon.
3. Cininnati Milling Machine Co: Company Documentation 1967/68.
4. IKEAI Iron Works Tokyo Documentation 1968.

5.   Peklenik J,. 1988, Fertigungskybernetik-Eine neue wissentschaftliche Disziplin fuer die Produktion-
     stechnik, Sonderdruck der TU Berlin, pp. 23-24.
6.   Peklenik J.,(2002) The Dynamic Cluster Structures – A New Manufacturing Paradigm for Production
     of High-Tech Products, Proc. Of the 35th CIRP-International Seminar on Manufacturing Systems,
     Seul/Korea, May 2002, pp. 539-546.
7.   Tichkiewitch S., (2004), Virtual Project Research Lab for a Knowledge Community in Production.
8.   Peklenik J., (2003), Cybernetic Structures, Networks and Adaptive Control of Work Systems in
     Manufacturing: In Manufacturing Technologies for Machines of the Future(Editor A. Dashchenko)
     Springer 2003, pp. 331-363.
9.   Taylor F.W., (1907) Trans Soc. Mech. Eng. Vol 28, 1907 pp. 31.

SOME APPROACHES IN THE MACHINING RESEARCH

E. Kuljanic, M. Sortino

Department of Electrical, Management and Mechanical Engineering, University of Udine, Italy

KEYWORDS: Machining, Modeling, Experiments

ABSTRACT The paper discusses some approaches in machining research. Development of empirical technology, as well as of science-based (predictive) technology, and development of computer-based technology are presented. The application of mathematics of statistics and design of experiments, simulation of machining processes such as analytical simulation, geometrical simulation, finite element simulation, and supervision systems in machining are discussed. Also, the importance of machining research for computer integrated manufacturing enterprise in global market conditions is discussed.

1 INTRODUCTION

Initially, machining was an art. There was no engineering basis for determining proper machining parameters such as cutting speed, feed, depth of cut and cutting tool characteristics for obtaining higher productivity in machining operations. According to E. Merchant [1], basic technology of the machining process began to evolve in the 20th century, going through three main stages in that timeframe. These were:

1. Development of empirical technology, beginning in the early 1900s.

2. Development of science-based (predictive) technology, beginning in the 1940s.

3. Development of computer-based technology, beginning in the 1970s.

Each of these stages was triggered by a key event and, interestingly enough, all three of these stages today co-exist and synergize each other. The aim of the research in machining is to find a solution which will make a product of high quality, at a low cost and rapidly. These objectives are summarized in the first author's maxim "Manufacturing a product is not difficult, the difficulty consists in manufacturing a product of high quality at a low cost and in a short time." [2].
This aim cannot be achieved without a better understanding of machining processes - new approaches, such as stochastic approach, modeling, simulation, supervision systems, new methodologies, etc. The paper will present some approaches in machining research and their related examples.

2 EMPIRICAL RESEARCH

The empirical research for engineering of the machining process was initiated by F.W. Taylor. In 1880 he started a massive, factory-based research program that lasted 26 years without any publication. In 1906 he published the results of his research in his classic book - that is actually a huge paper, with more than 300 pages "On the Art of Cutting Metals" [3], which was presented at the Winter Annual Meeting of the American Society of Mechanical Engineering in New York.

Published in: E. Kuljanic (Ed.) *Advanced Manufacturing Systems and Technology*,
CISM Courses and Lectures No. 486, Springer Wien New York, 2005.

That research produced an empirical understanding of machining process and empirical equations such as Taylor equation:

$$v_c T^n = C \tag{1}$$

where: v_c is the cutting speed, m/min, T is tool life, min, n is exponent whose value depends on the specific workpiece material, tool material and some other factors of the particular operation such as machine tool, and C is empirical constant whose value depends on the specific workpiece material, tool material and some other factors of the particular operation. This relationship is the basic equation for the identification of machining process. The application of Taylor's results had generated an increase of approximately 200-300 percent in the productivity of the machine tools. After the publication of Taylor's results a strong effort emerged to continue the research in the same direction. Different tool life equations were developed. An equation which explains tool wear and tool life phenomenon better is for example the "New Tool Life Equation" proposed by E.Kuljanic [4] - the first of this kind,

$$T = C' v^{k'_v} f_z^{k'_f} S^{k'_S} (vz)^{k'_{vz}} (vS)^{k'_{vS}} (zS)^{k'_{zS}} (vf_z S)^{k'_{vf_z S}} (vzS)^{k'_{vzS}} \tag{2}$$

Where: C' is empirical constant, v is cutting speed, m/min, T is tool life, min, z is number of teeth, f_z is feed per tooth, mm, S is stiffness of the machining system, N/mm, k'_v, k'_f, k'_z, k'_S, k'_{vz}, k'_{vS}, k'_{zS}, $k'_{vf_z S}$, k'_{vzS} are exponents. This equation includes interactions between different variables. The interaction is present when the effect of one variable depends on the level of the other variable or variables. The exponents k'_v, k'_f, etc. and the constant C' given in the equation (2) are obtained from the experiments. New Tool Life Equation includes only statistically significant factors and interactions. Tool life equations are needed for the identification of the machining process and its optimization.

3   SCIENCE BASED RESEARCH

In the late 1930s a new approach to machining began to evolve the science based research. The basic characteristic of science-based technology for engineering of machining is that it draws on the science of physics. Therefore, it is independent of empirical information. Hans Ernst investigated the mechanism by which a cutting tool removes metal from a workpiece, i.e. the process of chip formation. He published his main findings in the paper "Physics of Metal Cutting" [5]. He proposed the concept of the "shear plane" in chip formation, i.e. the very narrow plastic zone-"plane" between the body of the workpiece and the body of the chip.

M.E. Merchant applied the science of the mechanics of solids bodies to the "shear plane" concept. This resulted in the model of the equilibrium force system acting in the chip-tool-workpiece system, Figure 1, [6]. The result was a science-based-predictive model of the basic process of chip formation - the first of this kind, i.e. the first science-based predictive model in engineering of the machining process.

With the application of unmanned machining system: machine tool, workpiece, tool, fixture and control unit, chip formation, research has become more important, for example, the form of the chip is more important. The long chip can stop the machining process and damage the tool and the workpiece. Even in milling there can be a long chip. For example, E.Kuljanic, in his research

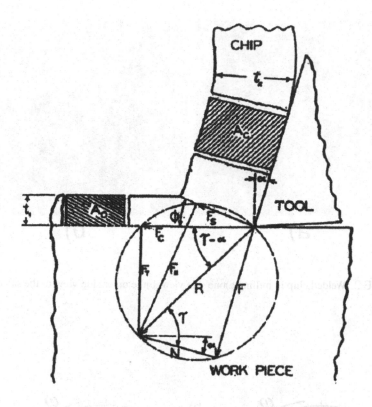

FIGURE 1. Condensed form of the M.E. Merchant orthogonal cutting force system (1944)

for the Ph.D. thesis in milling at the University of Cincinnati (USA), obtained the curled long chip in milling stainless steel, Figure 2.

The width of the workpiece was only 50 mm, and a carbide cutter was used. The chips were more than 500 mm long and welded on each cutting edge. They would brake when hitting the body of the milling machine. Such "impossible" chips were obtained due to the welding joints of many short, approximate 8 mm long, chips, Figure 2. There were two welding joints. First the chip No.1 was welded to the tooth at the exit of the tooth from the workpiece, Figure 3a). At the next entrance of the tooth, the chip formation of the chip No.2 started and at the same time when this chip was welded to the chip No.1. At the exit of the tooth the chip No.2 was welded to the tooth, but the actual chip consisted of chip No.1 and chip No.2, Figure 3b). Since the cutter was rotating, after the third exit of the tooth from the workpiece, the chip was three times longer and so on. In this way a long chip was welded from a number of short chips No.1, No.2, No.3, etc. Each tooth produced a long chip which was welded onto the cutting edge. Therefore, the long chips were rotating with the cutter and would break only after hitting the body of the machine tool.

This phenomenon happened during two hours in milling testing over a period of more than four months. That means that it is very difficult to have the right welding conditions - such as tem-

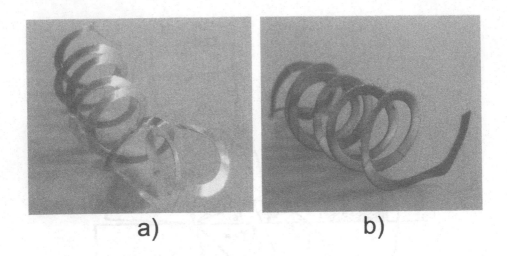

FIGURE 2. Welded chip in milling a)one side view, b)the other side view of the same chip

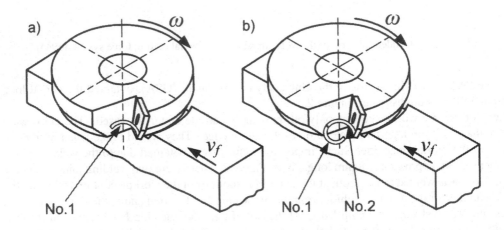

FIGURE 3. Welded chip in milling, a)chip No.1 welded on the cutting edge, b)chip No.1 welded to the chip No.2

perature, pressure, etc. - at two different places, i.e. at the entrance: chip to chip, and at the exit: chip to cutting edge. Max Kronenberg's comment was "Such a long chip in milling can be obtained only in Cincinnati".

Merchant's discovery of the science-based predictive model in engineering of the machining process, based on Ernst's shear plane, opened a new era in chip formation, i.e. in machining.

4 COMPUTER BASED TECHNOLOGY

The computer based technology has a significant influence on approaches in machining research. The numerical control brought a radical change in the field of machining in the 1950s. We were convinced that the research of numerical control started at Massachusetts Institute of Technology - MIT. However, the inventor of the numerical control of machine tools was John T. Parson [7]. He filed his initial patent on numerical control in 1952. He started to work on the numerical control concept in the latter part of the 1940s. In 1949 he signed a con-tract with the US Air Force "to design and construct a milling machine using servo-mechanisms actuated by punch cards on tape to produce wing sections" for supersonic air planes. However, he was well aware of the inability of his company to carry out such an under-taking alone. He took as subcontractor the Servomechanism Laboratory at MIT. However, MIT made a prototype of the numerically-controlled servo system and filed a patent on the proto-type in 1952. It is interesting to note that the machine tool industry and the manufacturing industry did not want to apply the numerical controlled servo-system for machine tools. The time required to manually prepare the program for NC machining of a part could be, in some cases, more than 50 times the time required to machine that part. MIT solved the problem by automating the programming process which was the beginning of the computer based technology.

The computer technology and the numerical control radically changed the industrial conditions in production. One of the most important strengths of the computer technology, as applied to machining, was its capability to combine both empirical and science-based technology in the engineering of machining operations. In particular, it proved to be able to simulate the actual ingoing performance of such an operation; i.e. it could create "dynamic" models. However, of even broader significance was the fact that it provided power capability to integrate these dynamic models of machining performance with the performance of all the other components of the overall system of manufacturing, as first envisioned by Merchant in 1961 [8].

After more than forty years the industrial conditions in production completely changed. The operator's precious experience and knowledge cannot be directly used in automated machining operations. This can be compensated with the results obtained from machining research.

Furthermore, the facilities and techniques to do the research completely changed as well. For example, in machining research we can use: design of experiments, mathematics of statistics, modeling, computer and new software, new methodologies, etc.

5 MATHEMATICS OF STATISTICS AND DESIGN OF EXPERIMENTS

The application of statistical methods and design of experiments in machining research is very important. It is of such an importance that the mathematics of statistics affects the way of thinking which yields a better understanding of many processes.

Billions of euros are spent for experiments around the world every year. However, it is amazing how little the design of experiments is used. It is obvious that without the application of the design of experiments and mathematics of statistics the experimental results could be poor, less reliable, and the experiment could be time consuming and more expensive. Therefore, the aim of this section is to emphasize the need of the application of mathematics of statistics and design of experiments in machining research.

Some examples of the application of mathematics of statistics and design of experiments in machining research are given as follows: comparison of tools from different manufacturers, regression analysis-effect of tool life data analysis on tool life equation and some examples of design of experiments.

5.1 COMPARISON OF TWO TOOLS FROM DIFFERENT MANUFACTURERS

In manufacturing production there is a need to find out which tool is better for machining a given material. This can be done by tool wear testing and by applying statistical analysis [9]. Different tool wear data in turning were obtained for two different tools, Table 1.

TABLE 1. Tool flank wear land after 30 minutes turning steel C45 with tools from two different manufacturers

Test	VB Tool A [mm]	VB Tool B [mm]
1	0,43	0,32
2	0,39	0,43
3	0,40	0,35
4	0,45	0,26
5	0,34	0,36
6	0,45	0,42
7	0,36	0,31
8	0,35	0,37
Mean	0,40	0,35

The arithmetical mean of tool wear for the tool A was $VB_A = 0,40$ mm and for the tool B was $VB_B = 0,35$ mm. From the t-test, we obtain:

$$|t| = 1,72 \text{ and } t_{\alpha=0,05} = 2,145.$$

We may conclude that the arithmetical means of the tool A and the tool B are not significantly different since $|t| < t_{\alpha=0,05}$.

More reliable comparison can be done by testing at different cutting speeds and with the application of regression analysis to determine tool life equations. These equations are then compared statistically by using the methodology given in [10].

5.2 APPLICATION OF REGRESSION ANALYSIS

Regression analysis is, for example, used to determine the association between two variables in experiments, where one of the variables is a non-stochastic variable, the values of which are predetermined when the experiment is designed-planned.

Regression analysis is a useful mathematical tool which can be used in research particularly for identification of machining process needed for optimization of the process. To show how important it is to determine which variable is non-stochastic variable an example in machining is presented. One of the reasons, why the so called Taylor exponents in (1) are usually different when the tool life experiments and the regression analysis are done at the same conditions in USA and in Europe, is the wrong determination of the non-stochastic variable. Hundred years ago F.W. Taylor (USA) proposed the tool life equation (1). To determine the exponents and the constant C, Taylor plotted cutting speed v_c against tool life T without applying regression analysis.

Mostly the way, cutting speed v_c versus tool life T is used in USA today when regression analysis is applied. In Europe the tool data analysis is done in opposite way, i.e. tool life T versus cutting speed v_c. In order to convince Max Kronenberg that it was possible to obtain different tool life equations from the same data, when the analysis of data was done v_c versus T and T versus v_c the first author made the following statistical analysis in 1976 [11].

The tool life data for turning steel with HSS tool from the PhD thesis [12] were used for regression analysis done in two different ways: v_c versus T and T versus v_c. In Figure 4, two different tool life equations determined from the same tool life data are given. The equations a) are obtained with the regression analysis T versus vc, and equations b) are obtained with regression analysis v_c versus T. It can be seen that the Taylor exponent $m = 0,108$ (T versus v_c) is ap-proximately 40% greater than the $m = 0,077$ obtained with the regression analysis v_c versus T. The reason why the exponent m is different can be seen from Figure 4. In this case, T versus v_c, the regression analysis is done with the values y for each point, and in the case of v_c versus T the analysis is done with values x.

The difference of the exponent m affects the optimal tool life very strongly

$$T_e = (\frac{1}{m} - 1)(t_c + \frac{C_g}{C_1 + C_o})\tag{3}$$

For the criterion of optimization-minimum machining cost, where T_e is tool life for minimum machining cost, min, t_c is tool change time, min, C_g is the tool cost between two resharpening, euro/cutting period, C_1 is direct labor cost on the machining system, euro/min, C_0 is overhead cost (rate) of the machining system, euro/min.

For example, for turning steel with HSS tool the tool life for minimum machining cost is $T_e = 100$ min for $m = 0,077$, and $T_e = 69$ min for $m = 0,108$ [11],[13] due to and respectively. The tool life T_e for identical turning operation is 45% greater when regression analysis is done v_c versus T in comparison to T versus v_c.

According to these results E.Kuljanic proposed to ISO (International Standard Organization), in 1979, to standardize the testing procedure in which the regression analysis should be done - tool life versus cutting speed, since the tool life physically depends on the cutting speed. The ISO accepted the proposal in the ISO standards 8688/1 and 8688/2.

5.3   DESIGN OF EXPERIMENTS

The design of experiments is a very useful technique to increase the reliability of experimental results and to decrease test time, experimental expenses and required time. The best results are obtained when the design of experiments is combined with appropriate statistical analysis and with good understanding of the process examined.

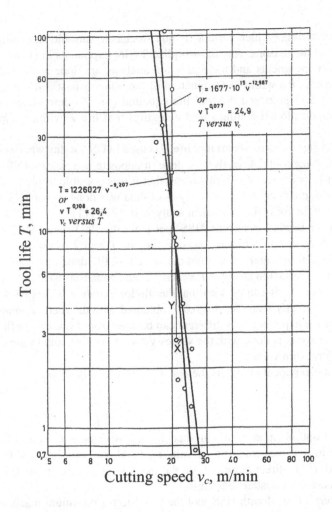

FIGURE 4. Different tool life curves obtained from the same data

DESIGN OF EXPERIMENTS AND REGRESSION ANALYSIS - SURFACE ROUGHNESS

This is an example where a complete 3^4 design of experiments was used to determine the relationship between the surface roughness R_a in turning and the following non-stochastic variables: cutting speed v_c, depth of cut a_p, feed f and radius of the insert r_e . Regression analysis was used, and the Student's t-test was applied to determine the significant effect of the variables. The experiment was turning a brass bar: cutting speed v_c = 99-196-393 m/min; feed f = 0,040 - 0,125 - 0,315 mm; depth of cut a_p = 0,2 - 1,0 - 2,0 mm; radius of the insert r_ϵ = 0,4 - 0,8 - 1,2 mm. After dropping the non-significant variables: cutting speed and depth of cut, we obtain:

$$R_a = \frac{1,533 f^{0,444}}{r_\epsilon^{0,66}} \tag{4}$$

It can be seen that the Equation (4) is highly significant according to the analysis of variance of the regression:

$$F = 393 > F_0 = 4{,}92$$

This is an example when testing procedure is short, that is not very usual in this area. There-fore, it was possible to use a complete 3^4 design of experiments.

DESIGN OF EXPERIMENTS - ANALYSIS OF VARIANCE AND EFFECT OF SIGNIFICANT FACTORS

In order to find out the effect of four variables: number of teeth in the cutter z, stiffness of the machining system S, cutting speed v_c and feed per tooth f_z on tool life T in face milling, a 2^4 design of experiments was applied, milling tests were carried out and analysis of variance and effects of significant factors were determined. The result of the research was the proposed new tool life equation (2), for face milling stainless steel with carbide cutter E.Kuljanic [2],[4]

$$T = 211,79 \cdot 10^5 v_c^{-4,023} f_z^{-1,454} z^{-10,267} S^{-1.329} \cdot$$
$$\cdot (v_c z)^{2,3913} (v_c S)^{0,3880} (zS)^{0,8384} (v_c f_z S)^{0,0190} (v_c zS)^{-0,1972} \tag{5}$$

It was proved, for the first time, that the number of teeth in the cutter, stiffness of the machining system and some interactions have a significant effect on tool life in milling stainless steel with carbide cutter.

The application of mathematics of statistics in manufacturing was very inadequate 25 years ago. Only a few researchers in CIRP (College International pour la Recherche en Productique - Paris) such as R. Levi, G. Lorenz, E. Kuljanic and J. Peklenik were using and promoting the statistical methods and design of experiments.

6 SIMULATION OF MACHINING PROCESSES

Simulation of machining process is becoming more and more useful in machining research. The main advantage provided by machining simulators is to enhance the comprehension of the machining results, forecast the machining outputs to an acceptable degree and thus reduce the cost and the number of machining experiments. Obviously, simulation of machining is a very complex task. It can be performed at different levels. The main procedures of a machining simulation are given in table 2.

TABLE 2. Classification of simulation systems

SIMULATION TASK	SIMULATION SYSTEM		
	ANALYTICAL	GEOMETRIC	FINITE ELEMENT
Workpiece / tool interaction - Uncut chip thickness estimation	Analytical	Finite volumes method, geometric intersection	Finite element
Cutting forces estimation	Empirical	Empirical	Constitutive equation
Temperatures estimation	Empirical, not common	Empirical, not common	Constitutive equation
Workpiece surface characteristics estimation	Analytical	Geometric	Deformed mesh

6.1   ANALYTICAL SIMULATION OF MACHINING

Analytical simulation is the simplest approach of machining simulation. It can be used to estimate the uncut chip thickness when the geometry of the operation is very simple. For example,
an analytical simulation of machining was used together with experimental data to evaluate the
influence of tilt angle on cutting forces and surface finish in ball-nose end milling by Schulz et
al. [14]. The cutting forces can be estimated from the uncut chip thickness by using and empirical approach such as the Kienzle force model, equation (6), or by shearing and ploughing force
model, equation (7), as follows:

$$F_c = k_{s0} h(t)^{1-a_h} L \tag{6}$$
$$F_c = k_{sc} h(t) L + k_{pc} L g(t) \tag{7}$$

Where k_{s0} is the unit cutting force, N/mm^2, $h(t)$ is the instantaneous uncut chip thickness, a_h
is the uncut chip thickness exponent, L is the width of cut, k_{sc} is the shearing coefficient, k_{pc}
is the ploughing coefficient and $g(t)$ is the contact function, whose value is 1 when the cutter is
in contact with the workpiece and 0 elsewhere. Usually, cutting temperatures are not estimated
by applying this approach. The main advantage of this approach is its simplicity. Also, it is
not time consuming method. The main disadvantage is that it can not be applied for complex
geometries such as hobbing. A common use of this approach is to provide quick data for cutting
force coefficient estimation in very simple machining cases, such as turning and milling.

6.2   GEOMETRICAL SIMULATION OF MACHINING

In geometrical simulation of machining the volume removed from the workpiece is determined
numerically by geometrical intersection of the tool and the workpiece.
In this approach the tool is represented by a set of basic solids such as tetrahedrons, while the
workpiece can be represented in several ways. The most common workpiece models are Z-map
models, first developed by Anderson in 1978 [15], sliced-sections models [16] and solid models.
In 2003, a new approach was presented by E.Kuljanic and M.Sortino, where the workpiece is
represented by an octree [17]. Different machining operations can be simulated with the same
model of the workpiece without modification.This is the most important characteristic of this
methodology. An example of simulation of hobbing, turning and drilling of a gear is given in
Figure 5. Also, this method is suitable for simulation of complex machined parts and tools.
The cutting forces can be estimated in the same way as they are estimated in the analytical
simulation of machining.
The geometrical simulation of machining can be useful in several applications such as development of new tools, simulation of complex machining operations, both for supervision and
experimental data analysis - for example cutting forces, simulation of machining very expensive workpieces, etc. The main disadvantage of this approach is that it is based on empirical
equations. Therefore, there is a need to have reliable equations.

6.3   FINITE ELEMENT SIMULATION OF MACHINING

The finite element simulation of machining is mostly used by researchers in this field. The aim is
to predict all variables quantitatively with acceptable accuracy such as flow of the chip, the stress

FIGURE 5. Simulation of gear manufacturing by hobbing

and the temperature.

To accomplish this task, a relationship between forces, friction, strain, strain rate and temperature is needed. This equation is, in general, quite complex and its reliability depends on the parameters that are difficult to determine since the conditions of workpiece material tests are different from the real cutting conditions.

The Johnson-Cook constitutive equation [18],[19] is widely used to model the workpiece material behavior under high strain, high strain rate and at high temperatures:

$$\sigma = (C_1 + C_2 \epsilon_{pl}^n)[1 + C_3 log(\frac{\dot{\epsilon}}{\dot{\epsilon_0}})][1 - (\frac{T - 20}{T_m - 20})^m] \tag{8}$$

Where σ is the stress, ϵ_{pl} is the plastic strain, n is the strain index, $\dot{\epsilon}$ is the strain rate, $\dot{\epsilon_0}$ is the reference strain rate, T is the temperature, T_m is the reference temperature, m is the temperature sensitivity index and C_1, C_2 and C_3 are the material constants, see Table 3.

In 2002, Hamann et al. presented a paper on the determination of constitutive equation by direct methods such as the split Hopkinson's pressure bar bench - SHPB test. The paper provided guidelines for the application of experimentally determined constitutive equations by using the finite element approach [20].

Another approach to determine the stress response of workpieces in orthogonal machining was presented by Batzer et al. [21]. Two commercial steels, AISI L6 and AISI O1 were analyzed by using quasi-static tensile, split Hopkinson's bar and orthogonal machining testing. The coefficients of Johnson-Cook model were determined by using test data. The predictions of machining simulation were high correlated to real dry-cutting machining tests.

TABLE 3. Johnson-Cook equation parameters for AISI4340 [19]

Parameter	Value
C_1	950 MPa
C_2	725 MPa
C_3	0,015
n	0,375
m	0,625
$\dot{\epsilon}_0$	$3.500 s^{-1}$

The most common way to determine the workpiece material properties at high strain rates is the Hopkinson's bar test.

Finite element theory relies on the hypothesis of linearity, therefore, it applies to small displacements. This theory has been extended to large displacements by introducing iterative methods instead of linear methods. The most common non-linear iterative method is the updated Lagrangean method with Euler's stress-strain relation, where the change of shape is taken into account and incompressibility constraint in the deformation zone is relaxed.

$$\{\dot{F}\} + \{F_t\} = \{[K_M] + [K_G] + [K_F]\}\{v\} \qquad (9)$$

Where $\{\dot{F}\}$ is the nodal force rate, $\{v\}$ is the velocity of nodes, $[K_M]$ is the stiffness matrix, $[K_G]$ is the geometrical stiffness matrix, $[K_F]$ is the correction matrix for load, $\{F_t\}$ is the thermal force load caused by volumetric change.

The simulation of chip formation using finite element is quite difficult. A parting criterion should be introduced such as maximum stress/strain or a geometrical criterion. A geometrical separation method that is considered to be the most effective for ductile materials was presented by Obikawa in 1996 [22]. Obviously, the parting process changes the stiffness matrix of the workpiece, thus it adds a non-linearity in the iterative method. A common way to include it in the model is to add fake elements to the workpiece mesh. The behavior of the fake elements is non-linear, and when the plastic strain reaches a critical value, the element is deleted. This process is very complex and time-consuming. Therefore, the application of finite-element method to practical cases is rare. In 1996, CIRP proposed a round-robin on FEM simulation. The aim of the round robin was to simulate the cutting forces, the chip contact length and the shear plane angle, the temperatures and friction of a turning operation of a low-carbon steel. Several research groups participated in this work, in which each group proposed a different finite-element approach. The results obtained by simulation were compared to the experimental test results. The differences between the obtained simulation results where from 50% to 400%.

The simulation of high speed machining by applying a physics base modeling technology which includes the change in the constitutive equation and friction characterization at cutting speeds up to 400 m/min, was presented in 2002 by Ng et al.[19]. 3D finite element modeling of orthogonal and oblique cutting operations with both continuous and segmental chip formations was also part of this work.

Modeling of machining operations of a low machinability material TiAl6V4 was presented by F. Klocke et al.[23], in which a 5-axes milling operation of a turbine blade was modeled. The simulation results helped find the influence of cutting parameters on the process. The estimated results obtained by simulation were in good accordance to the test results.

7 SUPERVISION SYSTEMS

In machining, direct and indirect supervision systems could be used [24]. Direct system measures, for example, the actual flank wear land VB or the crater wear. The tool wear, tool breakage and some important process parameters can be determined by applying indirect measuring processes. This methodology is, in general, less accurate than direct methods.

Tool wear is determined by vision systems from a picture. In 2003, an algorithm for automatic measurement of the flank wear land was proposed by M. Sortino and implemented in the Wearmon software (University of Udine) [25]. The algorithm was able to identify the shape of the worn area and measure the average and maximum tool wear in many practical cases. Similar approaches were presented by M. Lanzetta [26], by T.Pfeifer and L.Weigers [27]. The vision system can not be applied during cutting. In order to improve machining, a flow of information about tool condition is necessary. Indirect measuring processes are suitable for on-line tool condition monitoring. They are based on the application of sensors such as dynamometers, acoustic emission sensors, accelerometers, etc.

For tool condition monitoring, it is necessary to obtain the information required from the sensor signal, and the correlation between the signal characteristic and the tool condition. In many cases, there are external disturbances and limited computer power. One approach to solve this problem could be to use faster computers. It would make possible to use the original acoustic emission signal instead of the root mean square acoustic emission signal (AERMS), which is now broadly used.

In 1995, the state of the art on the application of sensors for tool condition monitoring in research and industrial conditions had been presented by Byrne et al. [28]. According to the authors, the tool condition monitoring systems are not applied in industrial conditions due to their limited reliability. A method for estimating the tool wear in milling by applying the cutting forces measured with a rotating dynamometer was presented by the authors in 2005 [29]. This method is based on the ratio between actual cutting forces and the cutting forces when the tool is sharp. A good correlation between the experimental results and the estimated tool wear was found.

Multiple sensor systems are more reliable from an extended research. It is obvious that, the architecture of the system is more complex and requires advanced artificial intelligence techniques to take into consideration all information. In 2003 Scheffer et al.[30] investigated the application of a multiple sensor tool condition monitoring system for hard turning. Their approach was based on the application of dynamometers, acoustic emission, vibrations and temperature sensors. The monitoring system was able to estimate the tool wear with a good reliability. Generally, the question is which combination of sensors will provide the required information at the minimum cost.

8 CONCLUSIONS

The conditions in global market are changing very rapidly. In such conditions the competitiveness has to be based on new ideas and new products. It is important to have computer integrated manufacturing enterprise in which machining systems must be able to autonomously increase the effectiveness of the machining system and avoid or correct machining errors during cutting in order to manufacture a product of high quality at low cost and in a short time. Thus, there is a further need of research in computer based technology by applying the approaches discussed in the paper such as empirical approach, science based approach, stochastic approach, modeling, simulation and supervision. Also, there is a need to develop new methodologies and suitable decision making strategies.

REFERENCES

1. Merchant, M.E., 1999, 20th Century Evolution of Basic Machining Technology - an Interpretive Review, Proceedings of the 5th International Conference on Advanced Manufacturing Systems and Technology - AMST'99, Springer Wien New York, 1-10

2. Kuljanic, E., 1996, Machinability Testing in the 21st Century - Integrated Machinability Testing Concept, Proceedings of the 4th International Conference on Advanced Manufacturing Systems and Technology - AMST'96, Springer Wien New York, 23-36

3. Taylor, F.W., 1906, On the Art of Cutting Metals, Transactions, American Society of Mechanical Engineers, 28

4. Kuljanic, E., 1975, Effect of Stiffness on Tool Wear and New Tool Life Equation, Journal of Engineering for Industry, Transactions of the ASME, Ser. B, 939-944

5. Ernst, H., 1938, Physics of Metal Cutting, Machining of Metals, ASM

6. Merchant, M.E., 1944, Basic Mechanics of the Metal Cutting Process, Journal of Applied Mechanics 11, 168-175

7. Parsons, J.T., Stulen, F.L., 1958, Motor Controlled Apparatus for Positioning Machine Tool, U.S. Patent No. 2,821,187.

8. Merchant, M.E., 1961, The Manufacturing System Concept in Production Engineering Research, Annals of the CIRP 10, 77-83, 1-9

9. Kuljanic, E., 1999, Mathematics of Statistics and Design of Experiments in Engineering, Forme e contenuti per una statistica di base nella formazione universitaria, FORUM, 89-95

10. Hald,A., 1967, Statistical Theory with Engineering Applications, John Wiley and Sons Inc., New York, London, Sidney

11. Kuljanic, E., 1976, Effect of Tool Life Data Analysis on Tool Life Equation, Annals of the CIRP, 25/1, 105-100

12. Henkin, A., 1962, The Influence of Some Physical Properties on Machinability of Metals, PhD Thesis, University of Michigan

13. Kuljanic, E., 1983, Different Taylor Equations Obtained from the Same Data - Effect on Optimal Tool Life, KOMEOS, University of Rijeka, College of Engineering, Rijeka, 1-14

14. Schulz, H., Hock, S., 1995, High speed milling of dies and moulds. Cutting conditions and technology, Annals of the CIRP, vol. 44/1, 35-38

15. Anderson, R.O., 1978. Detecting and Eliminating Collisions in NC Machining, Computer Aided Des., vol. 10, 231-237

16. Bouzakis, K.-D., Aichouh, P., Efstathiou, K., 2002. Determination of the Chip Geometry, Cutting Force and Roughness in Free Form Surfaces Finishing Milling, whit Ball-Nose End Tools, International Journal of Machine Tools and Manufacture, vol. 43, 499-514

17. Kuljanic, E., Sortino, M., 2003, A New Method for Tool Engagement Estimation in Machining, AITEM2003

18. Johnson, G.R., Cook, W.R., 1985, Fracture Characteristics of the Metals Subjected to Various Strain, Strain Rates, Temperature and Pressure, Engineering Fracture Mechanics, 21, 31-38

19. Ng, E.-G., El-Wardany, T. I., Dumitrescu, M., Elbestawi, M. A., 2002, Physics Based Simulation of High Speed Machining, Proc. 5th CIRP International Workshop on Modeling of Machining Operations, 1-20

20. Hamann, J.C., Meslin, F., Sartkulvanich, J., 2002, Criteria for the Quality Assessment of Constitutive Equations Dedicated to Cutting Models, Proc. 5th CIRP International Workshop on Modeling of Machining Operations, 21-30

21. Batzer, S. A., Subhash, G., Olson John, W., Sutherland, W., 2002, Large Strain Constitutive Model Development for Application to Orthogonal Machining, Proc. 5th CIRP International Workshop on Modeling of Machining Operations, 31-40

22. Obikawa, T., Usui, E., 1996, Computational Machining of Titanium Alloy - Finite Element Modeling and a Few Results, Transactions, ASME, Journal of Manufacturing Science and Engineering, 118, 208-215

23. Klocke, F., Markworth, L., Messner, G., 2002, Modeling of TiAl6V4 Machining Operations, Proc. 5th CIRP International Workshop on Modeling of Machining Operations, 63-70

24. Kuljanic, E., Sortino, M., 2002, Recent Development and Trends in Tool Condition Monitoring, Proceedings of the 6th International Conference on Advanced Manufacturing Systems and Technology - AMST'02, Springer Wien New York, 15-36

25. Sortino, M., 2003, Application of Statistical Filtering for Optical Detection of Tool Wear, Int. Journal of Machine Tools and Manufacture, Vol. 43/5, 493-497

26. Lanzetta, M., A New Flexible High-Resolution Vision Sensor for Tool Condition Monitoring, Journal of Materials Processing Technology, 119, 73-82

27. Pfeifer, T., Wiegers, L., 2000: Reliable tool wear monitoring by optimised image and illumination control in machine vision. Measurement 28, 209-218

28. Byrne G., Dornfeld D., Inasaki I., Ketteler G., Koenig W., Teti R., 1995, Tool Condition Monitoring (TCM) - The Status if Research and Industrial Application, Keynote Paper, Annals of the CIRP, Vol. 44/2/1995, 541-567

29. Kuljanic, E., Sortino, M., 2005, TWEM, a method based on cutting forcesÛmonitoring tool wear in face milling, International Journal of Machine Tools and Manufacture, Vol. 45/1, 29-34

30. Scheffer, C., Kratz, H., Heyns, P.S., Klocke, F., 2003, Development of a tool wear monitoring system for hard turning, International Journal of Machine Tools & Manufacture, Vol. 43/10, 973-985

NUMERICAL & EXPERIMENTAL METAL CUTTING ANALYSIS: AN APPRAISAL

L. Settineri, A. Zompì, R. Levi

Department of Production Systems and Economics, Polytechnic of Turin, Italy

KEYWORDS: Metal Cutting, Turning Experiments, Numerical Simulation.

ABSTRACT. Numerical analysis of metal cutting operations is increasingly relied upon in computer experiments, in order to clarify mechanical, thermal and tribological aspects in view of process optimization. The acid question is, however, whether impressive color displays and detailed figures describe what is actually taking place, or if some results might more properly belong to virtual reality. A classic experimental design for second order response surface work, concerning a simple cutting operation on mild steel, was run both in the metal cutting laboratory and on dedicated software, and results compared also in the light of classic models. Agreement concerning responses such as forces, temperature, and cutting ratio, was found to range from definitely fair to quite poor, underlining that reliance on numerical model may not always be fully justified. Analysis of deviations throws some light on a complex situation.

1 INTRODUCTION

Detailed numerical models of metal removal processes, currently implemented in commercial software with a wealth of graphic displays, provide fast, detailed description of mechanical and thermal aspects of operation. Users are expected to assume models and finite element discretization to be adequate, numerical analysis flawless, and therefore provided proper material parameters are fed into the program correct results are invariably obtained. Laboratory tests being expensive and time consuming, numerical results are often taken at face value, and experimental verification is conveniently dispensed with.

In a number of instances one may put up with numerical results, even of questionable value, if only because seldom people bothers checking, and some inconsistencies with experimental evidence may be blamed upon scatter in machining tests. However in some cases predictions fly on the face of established facts, and the situation must be addressed squarely. In the case at hand, failure of some predicted results to satisfy elementary checks spilled the beans. Doubts did first arise when some numerical results were found inconsistent with equilibrium conditions in simple, two-dimensional steady state cutting operations, and force components on rake face did not match friction coefficient, an user selected parameter. Failure to find satisfactory explanations prompted a specific investigation involving a set of laboratory tests, covering a comprehensive range of machining parameters. Existing doubts were confirmed by independent, published experimental evidence.

Published in: E. Kuljanic (Ed.) *Advanced Manufacturing Systems and Technology*,
CISM Courses and Lectures No. 486, Springer Wien New York, 2005.

2 LABORATORY TESTS

Free orthogonal cutting tests were performed turning dry C45 (AISI/SAE 1045), BHN 190 steel billets, 190 mm dia. by 350 mm long, using uncoated carbide tool inserts HM NOVATEA MT 6 ISO P30 with a nominal cutting edge radius of 0.02 mm (a fresh tool was used for every treatment combination; individually measured radii ranged from 0.02 to 0.025 mm). Relief angle was 11°, and a constant width of cut of 5 mm was achieved in free orthogonal plunge cutting by previously machining into the workpiece radial grooves 5 mm apart, and using a tool 6 mm wide, see Figure 1. Turning experiments were carried out on a UTITA CNC lathe with a peak power of 30 kW, and reduction in workpiece diameter due to infeed was compensated for automatically by increasing rotational speed, thus keeping cutting speed close to nominal value.

Cutting forces were measured with a KISTLER Type 9263 piezoelectric tool-post dynamometer, and rake face temperatures were read with embedded Chromel/Alumel thermocouples inserted into two 0.5 mm dia. holes, sunk by EDM into tools 3.1 mm apart, with the hot junction 0.8 mm behind cutting edge and 0.2 mm below rake face, see Figure 1. All signals were sampled at 1 kHz rate and fed via an A/D converter into a PC for storage and analysis.

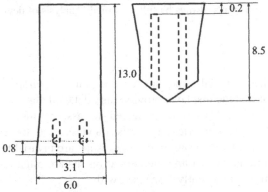

FIGURE 1. Top and front view of tool, showing layout of holes for embedded thermocouples

Factors (and levels) were machining parameters, namely cutting speed v_t (50, 125, 200 m/min), rake angle γ (- 10°, 0, 10°), and feed f (0.05, 0.125, 0.20 mm/rev). Responses were main (tangential) and normal cutting forces F_t and F_n, chip thickness h_c (enabling computation of cutting ratio r_c), and rake face temperature T at a selected location. A classic RSM second order design was selected, in the shape of a face centered cube in sample space, three replications at central point catering for error estimation; treatment combinations are listed in Table 1. Numerical simulation with a commercial software product was also performed for all treatment combination listed in the above mentioned table, introducing in every instance the cutting edge radius pertaining to the corresponding tool used in actual cutting tests. Material properties were specified for both tool and workpiece in terms of supplied data base; introduction of experimental tool-chip friction coefficient, attempted first, had to be dispensed with later in favour of default values, as it led to inconclusive results. A typical result (temperature distribution, with mesh) is shown in Figure 2.

TABLE 1. Experimental design, showing treatment combinations

Test no.	Rake angle γ	Cutting speed v_t, m/min	Feed f, mm/rev
1	-10	50	0.05
2	-10	200	0.05
3	-10	125	0.125
4	-10	50	0.20
5	-10	200	0.20
6	0	125	0.05
7	0	50	0.125
8	0	125	0.125
9	0	125	0.125
10	0	125	0.125
11	0	125	0.125
12	0	200	0.125
13	0	125	0.20
14	10	50	0.05
15	10	200	0.05
16	10	125	0.125
17	10	50	0.20
18	10	200	0.20

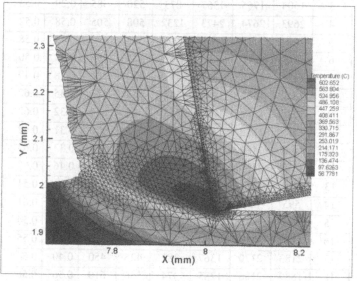

FIGURE 2. Typical temperature pattern over zone of deformation and tool, showing mesh

Experimental and numerical results (computed temperature corresponding to actual thermocouple location) are shown in Table 2. Test results appear rather consistent with predictions based upon classic metal cutting mechanics; thus unit cutting force k_s is found to range between 1.9 and 3.2 GPa, that is 1 ÷ 1.6 times BHN according mainly to feed and rake angle. Average coefficient of friction µ on rake face (estimated in terms of force components and rake angle, according to classic metal cutting mechanics) ranges between 0.5 and 1.2, agreeing with typical representative values reported in classic literature [1, 2].

A second order polynomial equation for unit cutting force k_s is found to explain some 96% of variation, with feed, rake angle, and product of feed by cutting speed accounting for over 85% and a handful of second order terms making up the remainder, residual mean square being consistent with estimate of pure error from replications. Tool temperature and cutting ratio, both within machining parameter volume considered, are also adequately described in terms of parsimonious second order equations.

3   ANALYSIS OF RESULTS

The degree of agreement between experimental and numerical values shown in Table 2 may be appreciated at a glance considering absolute $D=Exp.-Num.$ and percent deviations $D\% = 100(Exp. - Num.)/Exp.$, see Table 3.

TABLE 2. Experimental and numerical results

Test no.	Cutting force F_t, N		Normal force F_n, N		Temperature T, °C		Cutting ratio r_c	
	Exp.	Num.	Exp	Num.	Exp.	Num.	Exp.	Num.
1	805	833	603	467	383	400	0.14	0.41
2	751	765	724	469	474	570	0.17	0.53
3	1504	1765	1357	873	520	585	0.29	0.53
4	2693	2674	2473	1232	508	505	0.58	0.52
5	2004	2738	1610	1222	592	695	0.40	0.58
6	705	688	599	361	425	495	0.33	0.50
7	1616	1588	1313	543	439	450	0.28	0.49
8	1542	1582	1186	558	514	550	0.35	0.62
9	1513	1582	1205	558	518	550	0.32	0.62
10	1465	1582	1186	558	515	550	0.31	0.62
11	1487	1582	1208	558	520	550	0.33	0.62
12	1398	1600	996	589	552	600	0.46	0.62
13	2186	2427	1587	747	534	580	0.42	0.51
14	595	651	340	288	251	365	0.50	0.61
15	635	644	532	315	407	540	0.54	0.54
16	1324	1438	926	316	447	515	0.49	0.55
17	2183	2170	1367	381	425	450	0.40	0.54
18	1889	2186	1285	374	566	610	0.56	0.56

TABLE 3. Absolute and percent deviations between experimental and numerical results

Test no.	Cutting force F_t		Normal force F_n		Temp. T	Cutting ratio r_c	
	D, N	$D\%$	D, N	$D\%$	D, °C	D	$D\%$
1	-28	-3	136	23	-17	-0.27	-193
2	-14	-2	255	35	-96	-0.37	-218
3	-261	-17	484	36	-65	-0.24	-83
4	19	1	1241	50	3	-0.18	-31
5	-734	-37	388	24	-103	-0.17	-42
6	17	2	238	40	-70	-0.17	-52
7	28	2	770	59	-11	-0.21	-75
8	-40	-3	628	53	-36	-0.27	-77
9	-69	-5	647	54	-32	-0.30	-94
10	-117	-8	628	53	-35	-0.31	-100
11	-95	-6	650	54	-30	-0.29	-88
12	-202	-14	407	41	-48	-0.17	-37
13	-241	-11	840	53	-46	-0.09	-21
14	-56	-9	52	15	-114	-0.10	-20
15	-9	-1	217	41	-133	0.00	0
16	-114	-9	610	66	-68	-0.06	-12
17	13	1	986	72	-25	-0.14	-35
18	-297	-16	911	71	-44	0.00	0

Computed main cutting forces appear to match reasonably experimental values, percentage deviation averaging – 8 with a standard deviation of about 10; disregarding however an apparent outlier, these figures would drop respectively to – 6, and 6 (normally distributed), quite compatible with extended uncertainty of experimental estimates.

Lack of agreement between experimental and computed normal force appears on the other hand substantial, relative differences averaging 47%, and ranging between 15 and 72%, experimental values always exceeding computed estimates. Those of cutting ratio r_c, on the contrary, are found to exceed experimental values by some 65% on the average, with a maximum over 200%; excluding however the first two treatment combinations, with small uncut chip thickness and negative rake angle, average relative difference would fall to a still substantial - 48%, and lower limit to – 100%.

These findings appear to agree with published results [3] obtained at NIST on the same type of steel in a comprehensive series of experimental tests, and numerical simulations, see the probability plots of Figures 3 and 4, showing the distributions of percentage deviations between experimental and computed values for normal force and cutting ratio in both instances. But for two instances in the latter case, either corresponding to a feed substantially lower than those tested at NIST, both sets of data appear to belong to the same distributions, definitely suggesting existence of sizeable, systematic deviations in both instances between predicted and computed responses. The agreement among both sets of experimental results appears remarkable, also in view of substantial differences in cutting edge radius, and of cutting parameter range covered in either test program. Experimental values of cutting ratio r_c too fall far short of computed estimates, also in definite agreement with results referred to above, two out of the three largest deviations at either end corresponding to feed well below the lower limit of the range covered at NIST.

Computed tool temperatures appear on the other hand in reasonable agreement with experimental values, overestimating the latter by an average of some 50 °C, by no means an excessive amount in view of the rather steep gradients near rake face, catering for substantial uncertainties of experimental estimates. A marked association is observed between DF_n and DT, with a coefficient of correlation falling just short of 0.7 (highly significant given the relevant degrees of freedom), possibly hinting at some clues towards inherent model shortcomings, see Figure 5.

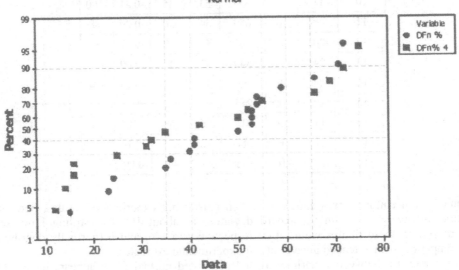

FIGURE 3. Probability plots pertaining to percentage deviations between experimental and computed values of normal force F_n, pertaining to data listed in Table 3, $DF_n\%$, and obtained at NIST, $DF_n\% 4$. Distributions are seen to overlap substantially

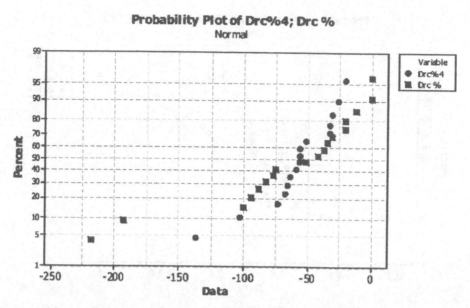

FIGURE 4. Probability plots pertaining to percentage deviations between experimental and computed values of cutting ratio r_c, pertaining to data listed in Table 3, $Dr_c\%$, and obtained at NIST, $Dr_c\% 4$. But for the case of some tests performed with very small feed, distributions appear to overlap to a substantial extent

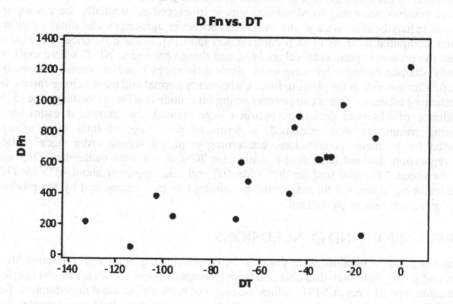

FIGURE 5. Scatterplot of deviations DFn, N, vs. DT, °C, showing substantial association

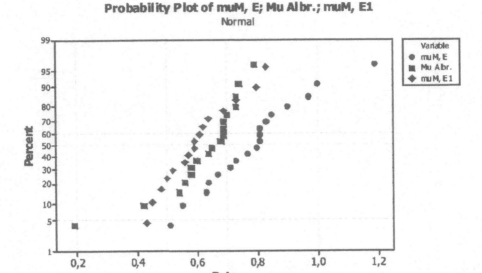

FIGURE 6. Cumulative distributions of experimental estimates of average coefficient of friction μ on rake face, computed according to Merchant from raw experimental data of Table 2, *muM E*, from NIST data, *muM E1*, and from data of Table 2 corrected for plowing force according to Albrecht, *muM Albr*, showing closeness among the last two (first and second distributions from the left)

Experimental estimates of average coefficient of friction μ on rake face, computed in terms of force components according to Merchant, appear to exceed substantially the corresponding numerical values both in average and in scatter; no better agreement was obtained either for estimates computed in terms of chip compression ratio [4]. Some differences were observed between the present experimental values of μ, and those obtained at NIST, where tools with a definitely sharper cutting edge were used. Such differences however almost disappear, see Figure 6, after correction for plowing force components normal and main cutting forces, that is by subtracting estimated forces components acting on rounded surface of cutting edge [5, 6].

Dependence of observed deviations between experimental and numerical estimates from machining parameters were examined, in terms of percentage of total sums of squares accounted for by major contributions, concerning empirical second order models fitted by linear regression. By and large feed accounts for 70% of variation concerning DF_n, cutting speed for about 30% (and feed for 20%) for *DT*, and rake angle for about 45% for *DT*, the remainder being accounted for mainly the remaining two parameters, and by the products of cutting speed and rake angle, or feed.

4 DISCUSSION AND CONCLUSIONS

Experimental results obtained were found to be consistent in general with reference literature values, and particularly with data obtained from a comprehensive set of cutting tests performed on the same type of steel at NIST, within the range of relevant extended uncertainties. Results obtained from numerical simulation were in fairly good agreement with experimental data, as far as main cutting force and rake face temperature estimates are concerned; evolution in time of main cutting force appears to be modeled in realistic terms too.

Substantial systematic deviations were however observed in terms of normal force, chip compression and average friction coefficient on rake face, all being seriously underestimated by numerical values. The latter failed also to meet equilibrium conditions in terms of classic metal cutting mechanics, friction coefficient assumed by default differing substantially from corresponding nominal value computed in terms of rake angle and force components; nor did introduction of user selected values for friction coefficient make the situation any better. Accounting for plowing forces did not bring computed values any closer to agreement with experimental evidence, nor did estimation of friction coefficient according to Kronenberg (1958) in terms of chip compression ratio, since the relevant numerical estimates obtained for the latter appeared also to be substantially flawed.

Bias in these numerical estimates observed in the current investigation appears to be both substantial and systematic, definitely matching comparable results obtained at NIST. No satisfactory explanation was found for the fair agreement on tool face temperature estimates, observed in spite of substantial bias for numerical values of normal force and friction coefficient, both seriously underestimated. Apparently some kind of compensation took place, which may not however be relied upon to occur in other instances; in fact, quoting recent findings [7], "Results are highly dependent on the material model and friction behaviour, indicating that caution be used when finite element analysis is to predict rather than interpret machining temperatures". In the light of observed results, it would be wise to entertain some doubts on the accuracy of numerical estimates of responses in metal cutting analysis, unless corroborated by substantial experimental evidence.

ACKNOWLEDGEMENTS

Collaboration given by Mr. Roberto Calzavarini of ISTEC – CNR in carrying out accurately laboratory tests is gratefully acknowledged.

REFERENCES

1. Merchant, M.E., (1950), Metal Cutting Research – Theory and Application, in: Machining – Theory and Practice, American Society for Metals, 5.
2. Shaw, M.C., (1960), Metal Cutting Principles, Technology Press, 3rd ed., MIT, Cambridge, U.S.A.
3. Ivester, R.W., Kennedy, M., (2004), Comparison of Machining Simulations for 1045 Steel to Experimental Measurements, AvantEdge Users Conference.
4. Kronenberg, M., (1958), A new approach to some relationships in the theory of metal cutting, ASTME Paper 86/58, Philadelphia.
5. Albrecht, P., (1960), New Developments in the Theory of the Metal-Cutting Process. Part 1: The Ploughing Process in Metal Cutting, Trans. ASME, J. Eng. Ind., Vol. 82, 348.
6. Masuko, M., (1953), Fundamental Researches on Metal Cutting. A New Analysis of Cutting Forces, Trans. Japan Society of Mechanical Engineers, Vol. 19, 32.
7. Davies, M.A., Cao, Q., Cooke, A.L., Ivester, R.W., (2003), On the measurement and prediction of temperature fields in machining AISI 1045 steel, Annals of the CIRP, Vol. 52/1, 77.

FRICTION STIR WELDING:
A SOLID STATE JOINING PROCESS

N. Alberti, L. Fratini

Dipartimento di Tecnologia Meccanica, Produzione e Ingegneria Gestionale,
University of Palermo, Italy

KEYWORDS: Joining, Welding, Friction Stir.

ABSTRACT. A wide experimental investigation on the friction stir welding of AA6082-T6 sheets is presented. In particular the influence of some of the most relevant process parameters was taken into account, with the aim to maximize the joint strength; furthermore the joint fatigue behavior was considered too. An accurate analysis on the obtained parts was carried out, with particular attention to the material microstructure evolution. The obtained results permitted to develop a proper FSW process engineering.

1 INTRODUCTION

Sheet metal joining processes have always played a relevant role in mechanical industries in particular as regards the building of vessels and the automotive and aeronautic industries. The evolution of the joining technology has allowed to pass from the riveted joints (both as cold processes and what is more as hot ones) to the weldings which are simpler and more economical processes. Unfortunately such processes are not always utilizable depending on the materials to be welded. Actually, the reached temperatures, necessary to melt the base material, normally determine strong variations in the material microstructure.

In this way, still today the joining of aluminum alloys sheets represents an open border in the research activity of the scientific community. The so called un-weldable or difficult to be welded alloys are common materials in automotive or aerospace industries. Actually quite weak joints are obtained with classic welding processes, i.e. TIG or laser, due to a strong increase of the average grain size in the melted material and to quite large thermally affected zone. Furthermore strong precautions have to be considered during classic weldings in order to avoid inclusions and other typical defects in the joint core.

On the other hand, friction stir welding (FSW) is a solid state welding process, which was first proposed in the early nineties. The process uses a specially designed rotating tool characterized by a properly shaped pin-end which is inserted into the adjoining edges of the sheets to be welded with a proper nuting angle and then moved all along the joint line. Actually an heat flux is generated by frictional forces work and plastic deformation work, but no material melting is observed during the process [1]. Furthermore, as the tool moves, material is forced to flow around the tool in a quite complex flow pattern: assuming for sake of simplicity a null nuting angle, a single point of the material at the tool contact surface moves along a cycloid curve.

Published in: E. Kuljanic (Ed.) *Advanced Manufacturing Systems and Technology*,
CISM Courses and Lectures No. 486, Springer Wien New York, 2005.

Several researched were aimed to investigate what happens at a microstructural scale in the joint section [2-3]. Actually, considering a section of the joint normal to the tool movement direction, an asymmetric metal flow is obtained. In particular an advancing side and a retreating one are observed: the former is characterized by the "positive" composition of the tool feed rate and of the peripheral tool velocity; on the contrary, in the latter the two velocity vectors are opposite. Overall the tool action determines the increase of temperature and the material softening and, what is more, the metal flow which generates the blanks welding.

A detailed observation of the material microstructure in the joint section allows also to individuate a few different areas: the so called parent material in which no material deformation has occurred, the heat affected zone (HAZ) in which material has undergone a thermal cycle which has modified the microstructure and then the mechanical properties, a thermo-mechanically affected zone (TMAZ) in which the material has been plastically deformed by the tool, and the heat flux has also exerted influence on the material. Finally, at the core of the welding zone the so called nugget is found out; it is a recrystallized area in which the original grain and subgrain boundaries appear to be replaced with fine, equiaxed recrystallized grains characterized by a nominal dimension of few mm. It has been showed that the joint resistance is strongly related to the nugget area extension [4].

Actually the effectiveness of the obtained joint is strongly affected by several operating parameters; in particular geometrical parameters such as the height and the shape of the pin (cylindrical, conical, screwed, etc) and the shoulder surface of the tool have got a relevant influence both on the metal flow and on the heat generation due to friction forces. What is more, probably the most important technological parameter determining the joining process is the force superimposed on the rotating tool during the process itself. It should be observed that the generated pressure on the tool shoulder surface and under the pin end determines the heat generation during the friction stir welding process. Furthermore both rotating speed and feed rate have to be properly chosen in order to obtain effective joints, since the determine the power conferred during the process. Finally, material physical and thermomechanical characteristics have to be considered in order to control the microstructure evolution during the joining process [5-7]. Furthermore a few researches on the joints fatigue resistance have been developed taking into account several different aluminum alloys [7-11]. In such investigations the fracture mechanics in fatigue tests is highlighted and what is more some indications are given on the fracture positioning along the welded joint transverse section. The obtained results have been also compared with those obtained by conventional TIG and MIG welding processes showing the effectiveness of the FSW process [11]. Furthermore the correlation between corrosion evolution and fatigue behavior of FSW joints is pointed out showing the insurgence of galvanic couples determined by the electrochemical behavior of the different zones along the joint section. The different behaviors of the material zones along the joint section strongly reduce the corrosion resistance of the joint.

Strictly connected to fatigue behavior of welded joints is the residual stress state occurring in the joint after welding. A few authors [12-15] have highlighted the residual stress state in the friction stir welded joint.

In the recent past a few research activities have been developed on the numerical simulation of FSW processes, in order to develop a proper computer aided engineering of the process. Two main approaches have been followed: first of all thermal models, taking into account the heat generated by both friction forces work and the material deformation work, have been proposed by Song et al. [16-17], Schmidt et al. [18] and Chao et al. [19], trying to highlight the temperature distributions nearby the rotating pin. On the other hand, finite element thermo-mechanical models have been presented by Chen et al. [20] and Lockwood et al. [21] with the aim to investigate the stress and strain distribution during the FSW process. Furthermore Xu et al. [22-23] and Deng et al. [24] presented the results of both 2D and 3D thermo-mechanical analyses based on the finite element method. In particular the researches were aimed to the investigation of the material flow, in terms of material circumferential strain rate all around the rotating pin; what is more a correlation was developed between the material final microstructure and the plastic strain distribution. The authors have recently proposed a 3D FEM model [25] aimed to highlight the FSW process mechanics and to show the distribution of the most important field variables, namely temperature, strain and strain rate. What is more the model was able to detect the insurgence of the tunnel defect, a typical shape defect occurring in friction stir welding processes for an insufficient value of the conferred thermal power. Such numerical tool has been already improved implementing a microstructure evolution model depending on the local values of strain, strain rate and temperature in order to follow the recrystallization phenomena occurring during FSW [26].

It should be observed that the above reported numerical models have to be properly tuned in order to became effective devices in the design of FSW processes. Actually correct and detailed material data have to be introduced, i.e. variations of the material thermo-mechanical characteristics with the temperature; what is more, effective contact algorithms must be utilized in the thermo-mechanical FEM analysis in order to take into account the right heat generation due to friction forces work.

In this paper, a wide experimental investigation on the friction stir welding of AA6082-T6 sheets is presented. In particular the influence of some of the most relevant process parameters was taken into account, with the aim to identify those values which maximize the joint strength and to describe the joint fatigue behavior. An accurate analysis on the obtained parts was carried out, with particular attention to the material microstructure evolution, In order to permitted to highlight the process mechanics and to develop a FSW process engineering.

2 FSW PROCESS MECHANICS

As briefly described before, FSW of butt joints is obtained inserting a specially designed rotating pin into the adjoining edges of the sheets to be welded and then moving it all along the joint. The pin is characterized by a rather small nuting angle (θ) limiting the contact between the tool shoulder and the sheets to be welded just to about one half of the shoulder surface. As the pin is inserted into the sheets, the blanks material undergoes to a local backward extrusion process up to reach the tool shoulder contact. Initially, the tool rotation determines an increase of the material temperature due to the friction forces work. As a consequence the material mechanical characteristics locally decrease and the blanks material reaches a sort of "soft" state but no melting is observed, while a circumferential metal flow is obtained all around the tool pin and close to the tool shoulder contact surface.

As local material softening is obtained, the tool can be moved along the joint avoiding the pin fracture due to excessive material reaction. The tool feed movement determines heat generation due to both friction forces work and material deformation one along the welding line. Furthermore the composition of the tool spin vector and of the feed rate vector determines a peculiar metal flow all around the tool contact surface (Figure 1).

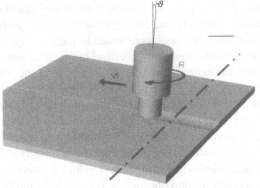

FIGURE 1. A sketch of the FWS butt joint

It should be observed, in fact, that the tool composed movement with respect to a fixed referring system is such that, assuming a null nuting angle and on the basis of the actual values of the tool pin rotation speed (R) and of the tool feed rate (V_f), a single point of the tool contact surface moves along a cycloid curve. As a consequence, considering a section of the joint normal to the tool movement direction (Figure 2), an asymmetric metal flow is obtained. An advancing side and a retreating one are observed in the joint section: the former is characterized by the "positive" composition of the tool feed rate and of the peripheral tool velocity; on the contrary, in the latter the two velocity vectors are opposite. Overall, the tool action determines the material softening and, what is more, the metal flux which allows the blanks welding.

A detailed observation of the material microstructure in the joint transverse section of two AA6082-T6 sheets allows also to discern a few different areas (Figure 2) as described below.

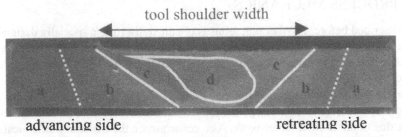

FIGURE 2. Material microstructures in a typical joint section (AA 6082-T6)

a – Parent material: no material deformation has occurred; such remote material has not been affected by the heat flux in terms of microstructure or mechanical properties and shows an average grain size of 80μm.

FIGURE 2a. Parent material

b – Heat affected zone (HAZ): in this region the material has undergone a thermal cycle which has modified the microstructure and consequently the mechanical properties. However, no plastic deformation occurred in this area (Figure 2b).

FIGURE 2b. Heat affected zone

c – Thermo-mechanically affected zone (TMAZ): in this area, the material has been plastically deformed by the tool, and also thermally stressed. In the case of aluminum alloys no recrystallization is observed in this zone; on the contrary, extensive deformation is present (Figure 2c).

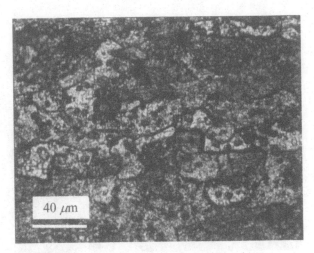

FIGURE 2c. Thermo-mechanically affected zone

d – Nugget: the recrystallised area in the TMAZ in aluminum alloys is generally called the nugget. In such zone the original grain and subgrain boundaries appear to be replaced with fine, equiaxed recrystallized grains characterized by a nominal dimension of few μm (Figures 2d).

FIGURE 2d. The nugget transition zone

A few considerations can be developed about the recrystallization phenomena occurring in the nugget zone [26-27]. Several authors suggest that the microstructure observed in the nugget does not result from a discontinuous dynamic recrystallization phenomenon characterized by recrystallization nuclei formation and gross grain-boundary migration occurrence (DRX). In turn, the microstructural evidence suggests that a "continuous" dynamic recrystallization (CDRX) process, analogous to that which gives rise to subgrain formation during hot rolling, occurs. Apart from specific microstructural considerations, it should be observed that the "continuous" dynamic recrystallization phenomenon is due to the tool pin disruptive mechanical action. Actually, no time is given for grain-boundary recrystallization phenomena, even if dynamic. On the other hand, the thermo-mechanical action of the tool pin rather determines a grains demolition in the blanks material up to a microstructure characterized by very fine, equiaxed grains. CDRX process, as the more known DRX, is affected by a few variables; in other words, the final dimension of the continuously recrystallized grain is influenced by the local value of a few field variables, such as the strain, the strain rate and the temperature levels, and, of course, by the mechanical properties of the considered material. These last consideration fully explain the final microstructure observed in the nugget zone of a FSW joint section.

3   PERFORMANCES OF AA6082-T6 BUTT JOINTS

3.1   EXPERIMENTAL PROCEDURES

A properly designed clamping fixture was utilized in order to fix the specimens to be welded on a milling machine (Figure 3a). The steel plates composing the fixture where finished at the grinding machine in order to assure an uniform pressure distribution on the fixed specimens.

As far as the utilized tool is regarded, it was made in H13 steel quenched at 1020 °C, characterized by a 52 HRc hardness; a cylindrical pin was used with the following geometrical characteristics: pin diameter equal to 3.00 mm and pin height equal to 2.90 mm (Figure 3b). Finally the shoulder diameter was equal to 10 mm. Such values were chosen on the basis of the technical literature; no investigation was developed on the influence of such geometrical parameters (i.e. the pin shape and its dimensions) on the process effectiveness.

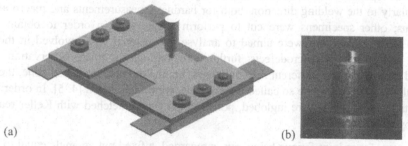

(a) (b)

FIGURE 3. (a)The clamping fixture, (b) The cylindrical pin

The utilized base material was an AA 6082 T6 aluminium alloy in 3 mm thick sheets, characterized by a yield stress of 280 MPa and an ultimate tensile stress (UTS$_b$) of 319 MPa. The material showed a microhardness equal to 120 HV and grain dimensions between 20 and 30 μm. The sheets were reduced in square specimens of 100 mm edges.

As mentioned before a few process variables were taken into account in this paper, namely the tool pin rotation speed (R), the tool feed rate (V$_f$), the nuting angle (θ), and the tool sinking into the specimen during the process (Δh). The influence of each variable on the process mechanics and on its effectiveness was investigated.

In particular a reference testing condition was fixed characterized by the following process parameters: R=715 rmp, V_f=143 mm/min, θ=2° and Δh=2.80 mm. Each variable was then varied in a proper range assuming different levels and the effects on the joining process in terms of joint resistance and of local microstructure were observed.

In the following Table 1 the investigated levels, for each considered variable, are reported.

TABLE 1.

Process Variable	Investigated values
R [rmp]	340, 490, 715, 1040, 1500
Vf [mm/min]	71, 100, 143, 215
θ	1°, 1.5°, 2°, 2.5°, 3°
Δh [mm]	2.70, 2.75, 2.80, 2.85, 2.90

In order to take into account the combined effects of the investigated variables on the process mechanics the influence of a summary parameter was considered. In particular, for each test, the Specific Thermal Contribution (STC) was calculated as the ratio between the conferred power and the tool feed rate [J/mm]. The former results from the energy due to the friction forces and to material distortion dissipated into heat. Since a direct evaluation of the conferred power was not possible, an indirect and average measurement was adopted: the power was measured by the difference between the electrical power absorbed by the milling machine during the test and the one absorbed in no tool penetration conditions.

Each test was repeated three times and from each joint different specimens were cross-sectioned perpendicularly to the welding direction, both for hardness measurements and macro and micro observations; other specimens were cut to perform tensile tests in order to obtain the joint strenght. Macro observations were aimed to analyse the material area involved in the process mechanics and eventually macrodefects; furthermore, through the micro observations of grain shapes and dimensions, the different material zones, i.e. the thermally affected zone, the thermo-mechanically affected ones, the so called nugget zone, were highlighted [4, 5]. In order to obtain such results the specimens were inglobed, polished and finally hetched with Keller reagent and observed by a LM.

Furthremore, as far as joint fatigue behaviour is regarded, a fixed nuting angle equal to 2° otherwise the following process variables levels were investigated:

- tool pin rotation speed (R): 490, 715, 1040, 1500 rmp;

- tool feed rate (Vf): 71, 104, 143, 215 mm/min;

- tool sinking into the specimen during the process (Δh): 2.85, 2.90 and 2.95 mm.

Each test was repeated three times and from each joint three different specimens were cross-sectioned perpendicularly to the welding direction; in this way nine tests were utilized in order to develop the joints fatigue characterization.

In the next Table 2 the results of the developed tests are summarized and for each combination of the operative parameters an identification index (ID) is introduced.

TABLE 2. The developed tests

σ_r [MPa]	Δh [mm]	R [rpm]	V_f [mm/min]	ID
172,2	2,9	715	143	A
125	2,9	715	71	B
167,5	2,9	715	100	C
128,3	2,9	715	215	D
138	2,85	715	143	E
156,5	2,95	715	143	F
88	2,9	490	143	G
188	2,9	1040	143	H
194	2,9	1500	143	I

The fatigue characterization of the joints was carried out on an Instron resonance test machine. The oscillation frequency used was in the interval of 77 – 103 Hz. Test specimen dimension used were 100 x 25 x 3 mm (length, width, thickness) as shown in the next Figure 4.

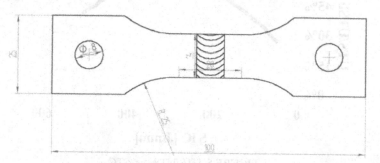

FIGURE 4. The specimen dimensions (mm)

The tests have been conduct both on welded and parent material specimens (sound specimens). As far as the latter specimen is regarded the fatigue curve and endurance limit have been carried out. Specimen dimension admit a variation of the cross section area that has been considered in the definition of the applied tension. The utilized stress ratio of the sinusoidal load-time function was R = P_{max}/P_{min} = 0,2. Such large ratio was chosen since aluminum alloys are generally less sensitive to the effects of the load ratio itself and furthermore a minimum preload value was aimed to.

The imposed load upper limit has been varied in the 8 - 6 kN range for the sound specimen and in the 4 - 2 kN range welded one, with steps of 1 kN. The fatigue limit that has been considered was for 106 cycle, even if in literature it comes also considered for 107 cycles. This assumption has been accepted because at this cycle (106) a fatigue curve deflection is already observed.

A preliminary inspection of the specimens has been led before test execution in order to discard joints for which macro-defect were highlighted (for instance bottom defect); in this way two specimens were eliminated since they showed an unwelded part on the bottom face.

3.2 JOINT RESISTANCE

In the present paragraph the obtained results will be presented: firstly the results in terms of joint resistance will be shown, then the micro-hardness tests will be reported.

Starting from the above reported reference testing condition the influence of each variable on the process mechanics was investigated. What is more, the STC value was utilized to summarize the effect of the investigated variables on the FSW process; it should be observed that a large value of the tool pin rotation speed (R), a low tool feed rate (V_f), and a large tool sinking determine a large STC value. In this way, in the next Figure 5 all the obtained results are reported in order to highlight the influence of the STC on the UTS of the obtained joints in terms of percentage of the base material. In particular the quadratic regression of the obtained results is reported. In this way the overall effect of the investigated variables on the joint resistance is shown.

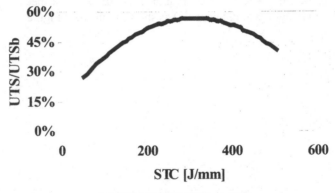

FIGURE 5. UTS/UTS$_b$ vs. STC

The obtained trend shows a maximum: it should be observed that for low STC values, an insufficient heat flux is conferred to the welding area. In particular an insufficient temperature level is reached and the material flux results ineffective giving rise to typical defects named tunnel defects; in other words, an incomplete material filling is observed under the tool pin head. In Figure 6 a typical tunnel defect is shown: such defect was obtained with the following operative parameters: R=715 rmp, V_f=143mm/min, θ=2° and Δh=2.70mm; the calculated STC value was 102 J/mm.

FIGURE 6. The tunnel defect

Furthermore, for too large values of STC a reduction in the joint resistance is observed; an accurate micro and macro analysis of the obtained joints permitted to interpret the occurring phenomena. In fact, for large values of STC the following observation are derived: an enlargement of both the thermo-mechanically affected zone (TMAZ) and of the heat affected zone (HAZ) is observed (Figure 7); a subsequent enlargement of the average value of the grain size is obtained; local micro-fusions are noticed. All the mentioned effects concur to reduce the joint strength.

FIGURE 7. Cross section of a welded joint – STC=435 J/mm (R=1500rpm, V_f=71mm/min, θ=2° and Δh=2.85mm)

In Figure 8, in turn, the cross section of one of the "best" results is reported, the test was developed with the following operative parameters: R=1500 rmp, V_f=143mm/min, θ=2° and Δh=2.80mm; the calculated STC value was about 350 J/mm.

FIGURE 8. Cross section of an effective joint

The different material zones are quite discernible in Figure 8 as described in paragraph 2. Furthermore in the next Figure 9 the temperature distribution obtained at the bottom of the joint utilizing the same "best" process conditions is reported [28].

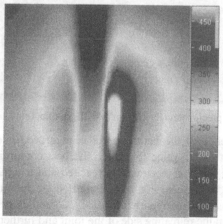

FIGURE 9. The temperature distribution at the bottom of the "best" joint

Another correlation was then investigated: the joint resistance vs. the nugget area (Figure 10); again the quadratic regression of the obtained results is reported. A strict relation is highlighted between the joint resistance and the measured nugget area in which the tool pin action determines the largest deformation energy and the smallest grain dimensions.

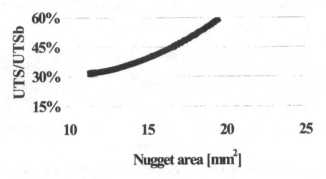

FIGURE 10. Joint resistance vs. nugget area

Other interesting considerations can be developed on Si precipitations (Figure 11) during the welding process. It has been observed that the Si particles density in the nugget area decreases at increasing V_f, i.e. at decreasing STC; furthermore the measured average size of Si particles varies from 7 μm to 12 μm, and this value increases with the increasing of the R value, i.e. at the STC increasing.

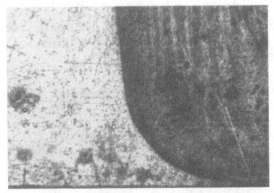

FIGURE 11. Si particles precipitates (R=715rpm, V_f=71mm/min, θ=2° and Δh=2.80mm) (62.5X)

Finally as far as the influence of the nuting angle is regarded, a low sensitivity of the joint strength was shown, except for 2° value for which the best compromise between the shoulder contact area and its sinking into the specimens is reached and a significant improvement in the joint resistance is observed. In the next Figure 12 the effect of the nuting angle on the joint surface finish and on the flash shape are reported for θ=1°, 2° and 3° respectively. In particular for θ=3° flash is observed just in the retreating side of the joint and furthermore the stronger action of the shoulder edge fragments the flash which becomes discontinuous.

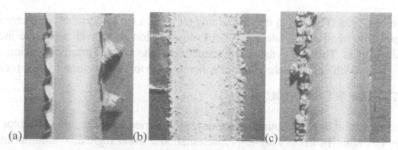

FIGURE 12. Joint surface finish for $\theta=1°$ (a), $2°$ (b) and $3°$ (c)

For each of the developed joints hardness tests were developed with the aim to investigate the mechanical properties of the welding zone. In particular different hardness tests were developed on the mean plane of the joints at different distances from the symmetry plane towards both the retreating side of the joint and the advancing one.

First of all the micro-hardness in the transverse section of the welded joints was investigated [29]. In Figure 13 the obtained micro-hardness values are reported at different depths from the specimen surface (y variable). It should be observed that a quite strong local softening of the material occurs because of the thermal action of the welding process. In particular the maximum reduction in hardness is observed nearby the heat affected zones (HAZ). In order to understand such material softening phenomenon, the Mg_2Si precipitates density after welding has to be investigated. Actually an image analysis of the transverse section of the welded joint led to different values of precipitates density all along the section itself [29].

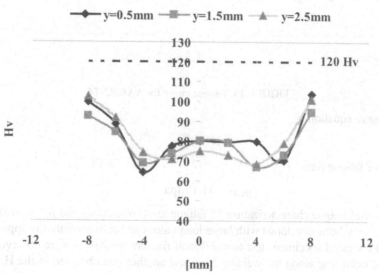

FIGURE 13. Micro-hardness values (R=1040rpm, V_f=104mm/min, $\theta=2°$ and Δh=2.90mm)

It should be observed (Figure 13) that in the nugget zone, i.e. close to the symmetry axis of the joint, a contrasting trend of the micro-hardness values is found out with a relative increasing of the material hardness. This due to the strong reduction of the average grain dimension obtained in the nugget zone, which contrasts the material softening due to the reduction of the precipitates.

3.3 JOINT FATIGUE BEHAVIOUR

First fatigue curve and limit for sound specimens were carried out using nine different alternative stress value. In Figure 14 using a log-log representation a linear regression straight line has been plotted.

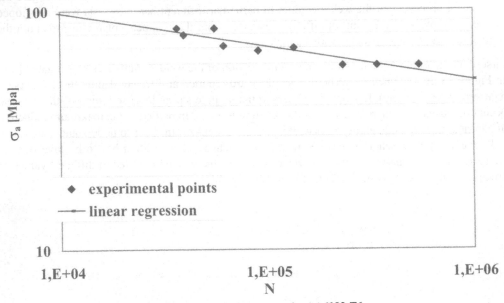

FIGURE 14. Fatigue curve for AA6082-T6

The fatigue curve equation is:

$$\sigma^7 \cdot N = 10^{18} \tag{1}$$

and the relative fatigue limit:

$$\sigma_{L,Rf} = 51{,}1 \text{ Mpa} \tag{2}$$

After this material fatigue characterization 56 fatigue tests were conducted for welded specimens. The first tests have been developed with large load values and subsequently the applied load was reduced. In the welded specimens test two different fracture modalities were observed, namely a fracture mode occurring along the welding line, and another one observed in the HAZ (heat affected zone).

Table 3 summarizes the fatigue parameters determined from all the developed experimental tests.

TABLE 3. Fatigue parameters for all the developed tests

Specimens ID	Fatigue curve equation	Fatigue limit
A	$\sigma^{20} \cdot N = 10^{36}$	$\sigma_{L,Rf} = 32,15$ MPa
B	$\sigma^{38} \cdot N = 10^{65}$	$\sigma_{L,Rf} = 31,8$ MPa
C	$\sigma^{38} \cdot N = 10^{65}$	$\sigma_{L,Rf} = 26,8$ MPa
D	$\sigma^{15} \cdot N = 10^{28}$	$\sigma_{L,Rf} = 27,8$ MPa
E	$\sigma^{10} \cdot N = 10^{21}$	$\sigma_{L,Rf} = 27,56$ MPa
F	$\sigma^{29} \cdot N = 10^{50}$	$\sigma_{L,Rf} = 29,8$ MPa
G	$\sigma^{11} \cdot N = 10^{23}$	$\sigma_{L,Rf} = 29,9$ MPa
H	$\sigma^{15} \cdot N = 10^{27}$	$\sigma_{L,Rf} = 28,7$ MPa
I	$\sigma^{5} \cdot N = 10^{13}$	$\sigma_{L,Rf} = 32,3$ MPa

In Figure 15 all experimental test results and their regression lines are shown.

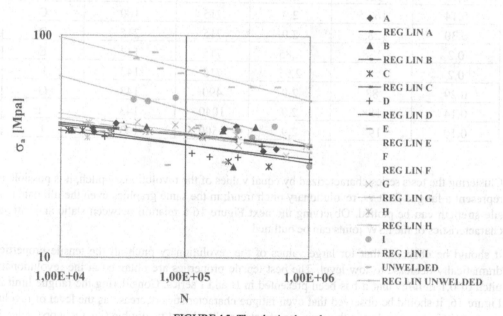

FIGURE 15. The obtained results

Analyzing Figure 15 and Table 3 some significant observation can be carried out: for large load levels all series – except ID I – have shown the same number of cycles at fracture, this result suggests that welding parameter did not strongly influence this fracture parameter; furthermore, for low load levels (high cycles number) all series converge to the same resistance value.

This two observations suggest that process parameters less influence fatigue life than the tensile strength. In fact, if a unwelded/welded fatigue limit ratio is considered, all test specimens show a resistance in the range 63-52% with respect to the base material, instead of 60-27% shown for tensile strength ratio [4].

The process parameters that strongly influence the mechanical properties of the FSW joint are rotation speed R and the tool feed rate V_f; this two parameters define the specific heat supplied to the joint. On the other hand, tool penetration has a slight influence on mechanical properties. Following the latter observations an unique process parameter characterizing the conferred power was utilized, namely the revolutionary pitch, defined as the ratio between the tool feed rate and the tool speed rotation.

In Table 4 the data in terms of revolutionary pitch, ultimate tensile strength (σ_r) and process parameters for all test specimens are reported.

TABLE 4. Revolutionary pitch, ultimate tensile strength and process parameters for all the test specimens

Revol. pitch.	σr [MPa]	Δh [mm]	R [rpm]	V_f [mm/min]	ID
0,2	172,2	2,9	715	143	A
0,10	125	2,9	715	71	B
0,14	167,5	2,9	715	100	C
0,30	128,3	2,9	715	215	D
0,2	138	2,85	715	143	E
0,2	-	2,95	715	143	F
0,29	88	2,9	490	143	G
0,14	188	2,9	1040	143	H
0,10	194	2,9	1500	143	I

Clustering the tests series characterized by equal values of the revolutionary pitch, it is possible to represent a fatigue limit vs. revolutionary pitch trend; in the same graphics even the ultimate tensile strength can be plotted. Observing the next Figure 16 a relation between static and fatigue characteristics of the FSW joints can be outlined.

It should be observed that for large values of the revolutionary pitch all the tensile properties dramatically decrease to low level. The best tensile properties are obtained at the revolutionary pitch of 0,1 mm/rev that it has been presented in B and I series. Considering the fatigue limit in Figure 16, it should be observed that even fatigue characteristics decrease as the level of revolutionary pitch increases but in this case the effect is really almost negligible (i.e. the worst value is 15% smaller than the best value).

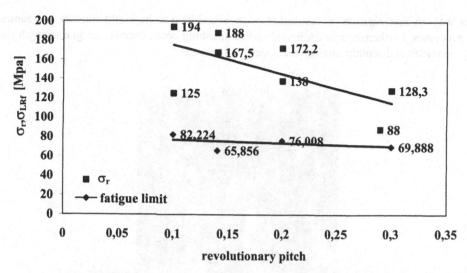

FIGURE 16. Fatigue and tensile limit curve vs. revolutionary pitch

In Figure 17 the fatigue curves obtained by clustering the experimental data for classes of revolutionary pitch are reported. It should be observed that linear regression lines characterized by similar slopes are obtained.

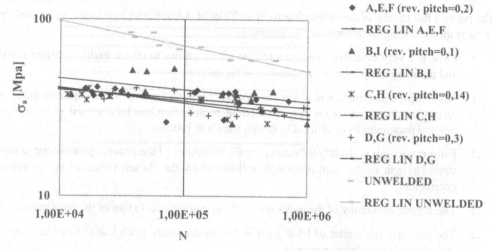

FIGURE 17. Fatigue curve for different revolutionary pitch values

The fracture location of any joint is a direct reflection of the weakest part of the joint; as shown in Figure 18 the fracture start from a point in HAZ in the retreating side of joint and continue through cross section. Crack path is influenced by the geometrical discontinuities due to the FSW process and present on joint surface: actually fracture starting point is always on the upper side of joint where are even present residual stresses and geometry discontinuities that represent stress concentration. From such starting point the crack growths through the joint and changes direction in its evolution during the fatigue test.

In this way an investigation on residual stresses would help to highlight fracture mechanics in FSW processes. Furthermore an additional surface finishing could improve fatigue strength eliminating geometrical discontinuities on the upper joint surface.

FIGURE 18. Fractured joint

4   CONCLUSIONS

In the paper a full review of the research activity in FSW of AA6082-T6 butt joints is reported. In particular the following statements can be assumed:

- ❑   FSW is a very effective joining technique which allows to obtain highly resistant joints and avoiding the presence of typical welding defects.

- ❑   Several process parameters affect the process mechanics and as a consequence the joint resistance. In particular the specific thermal contribution can be assumed as an overall process parameter to be utilized to design the FSW process.

- ❑   Fatigue strength is slightly influenced by the variations of the process parameters; in turn static strength of the joint is strongly influenced by the chosen values of the operative parameters.

- ❑   The fatigue efficiency of the joints is of 60% in comparison to that of the base material.

- ❑   The principal parameter of FSW joint is the revolutionary pitch that define heat supply and, accordingly, defect presence.

- ❑   Two different positions of fracture have been found. One along the welding line and one in the zone HAZ, function of the effectiveness of the process parameters. Fractures along the welding line indicate the presence of shape defects in the joining, namely the tunnel defects.

ACKNOWLEDGMENTS

This work was made using MIUR (Italian Ministry for University and Scientific Research) funds.

REFERENCES

1. Guerra, M., Schmidt, C., McClure, L.C., Murr, L.E., Nunes, A.C., (2003), Flow patterns during friction stir welding, Materials characterization, Vol. 49, 95-101.
2. Rhodes, C.G., Mahoney, M.W., Bingel, W.H., Spurling, R.A., Bampton, C.C., (1987), Effects of Friction Stir Welding on Microstructure of 7075 Aluminum", Scripta Materialia, Vol. 36, No. 1, 69-75.
3. Liu, G., Murr, L.E., Niou, C.S., McClure, J.C., Vega, F.R., (1997), Micro-structural aspects of the friction-stir welding of 6061-T6 aluminum alloy, Scripta Materialia, Vol. 37, 355-361.
4. Barcellona, A., Buffa, G., Fratini, L., (2004), Process parameters analysis in friction stir welding of AA6082-T6 sheets, keynote paper of the VII ESAFORM Conference, 371-374.
5. Shigematsu, I., Kwon, Y.J., Suzuki, K., Imai, T., Saito, N., (2003), Joining of 5083 and 6061 aluminum alloys by friction stir welding, J. of Mat.s Science Letters, Vol. 22, 343-356.
6. Lee, W.B., Yeon, Y.M., Jung, S.B., (2003), The improvement of mechanical properties of friction-stir-welded A356 Al alloy, Material Science & Engineering, Vol. A355, 154-159.
7. Liu, H.J., Fujii, H., Maeda, M., Nogi, K., (2003), Tensile properties and fracture locations of friction-stir-welded joints of 2017-T351 aluminum alloy, J. of Materials Processing Technology, Vol. 142, 692–696.
8. Pao, P.S., Gill, S.J., Feng, C.R., Sankaran, K.K., (2001), Corrosion–fatigue crack growth in friction stir welded Al 7050, Scripta Materialia, Vol.45, 605-612.
9. Dickerson, T.L., Prydatek, J., (2003), Fatigue of friction stir welds in alluminium alloys that contain roots flaws, International Journal of Fatigue, Vol.25, 1399–1409.
10. Sutton, M., Reynolds, A., Yang, B., Taylor, R., (2003), Mixed mode I/II fracture of 2024-T3 friction stir welds , Engineering Fracture Mechanics, Vol.70, 2215-2234.
11. Ericsson, M., Sandstrom, R., (2003), Influence of welding speed on the fatigue of friction stir welds, and comparison with MIG and TIG, International Journal of Fatigue, Vol.25, 1379–1387.
12. Bassu, G., Irving, P.E., (2003), The role of residual stress and heat affected zone properties on fatigue propagation in friction stir welded 2024-T351 aluminum joints, International Journal of Fatigue, Vol.25, 77–88.
13. John, R., Jata, K.V., Sadanada, K., (2003), Residual stress effects on near-threshold fatigue crack growth in friction stir welds in aerospace alloys, Internal Journal of Fatigue, Vol.25, 939-948.
14. Peel, M., Steuwer, A., Preuss, M., Withers, P.J., (2003), Microstucture, mechanical properties and residual stresses as a function of welding speed in aluminium AA5083 friction stir welds, Acta Materialia, Vol.51, 4791–4801.
15. Reynolds, A.P., Tang, W., Gnaupel-Herold, T., Prask, H., (2003), "Structure, properties and residual stress of 304L stainless steel friction stir welds, Scripta Materialia, 48, 1289–1294.
16. Song M., Kovacevic, R., (2002), A New Heat Transfer Model for Friction Stir-Welding, Transaction of NAMRI/SME, SME, Vol. 30, 565-572.
17. Song, M., Kovacevic, R., (2003), Thermal modeling of friction stir welding in a moving coordinate system and its validation, Int. J. of Machine Tools & Manufacture, Vol. 43, 605-615.
18. Schmidt, H., Hattel, J., Wert, J., (2004), An analytical model for the heat generation in friction stir welding, Modeling and Simulation in Materials Science and Eng., Vol.12, 143-157.
19. Chao, Y.J., Qi, X., Tang, W., (2003), Heat transfer in friction stir welding – Experimental and numerical studies, Transaction of the ASME, Vol.125, 138-145.
20. Chen, C.M., Kovacevic, R., (2003), Finite element modeling of friction stir welding – thermal and thermomechanical analysis, Int. J. of Machine Tools & Manufacture, Vol. 43, 1319-1326.
21. Lockwood, W.D., Reynolds, A.P., (2003), Simulation of the global response of a friction stir weld using local constitutive behaviour, Materials Science and Engineering, Vol. A339, 35-42.

22. Xu, S., Deng, X., (2002), A three-dimensional model for the friction-stir welding process, Proceedings of the 21th Southestern Conference on Theoretical and Applied Mechanics.
23. Xu, S., Deng, X., (2003), Two and Three dimensional finite element models for the friction stir welding process, Proceedings of the 4th International Symposium on Friction Stir Welding.
24. Deng, X., Xu, S., (2001), Solid mechanics simulation of friction stir welding process, Transaction of NAMRI/SME, SME, Vol. 29, 631-638.
25. Buffa, G., Fratini, L., (2004), Friction Stir Welding Of AA6082-T6 Sheets: Numerical Analysis And Experimental Tests, Proceedings of 8th NUMIFORM Conference, 1224-1229.
26. Fratini, L., Buffa, G., (2004), CDRX modelling in friction stir welding of aluminium alloys, accepted for publication on Int. J. Machine Tools & Manufacture.
27. Jata, K.V., Semiatin, S.L., (2000), Continuous dynamic recrystallization during friction stir welding of high strength aluminum alloys, Scripta Materialia, Vol.43, 743-749.
28. Beccari, S., D'Acquisto, L., Fratini, L., Salamone, C., (2005), Thermal Characterization Of Friction Stir Welded Butt Joints, submitted to the Scientific Committee of Shemet '05 Conference.
29. Barcellona, A., Buffa, G., Fratini, L., Palmeri, D., (2005), On microstructural phenomena occurring in friction stir welding of aluminium alloys, submitted for publication on Int. J. Machine Tools & Manufacture.

UNCONVENTIONAL CUTTING TECHNOLOGIES IN ORTHOPAEDIC SURGERY

J. A. McGeough[1], A. Okada[2]

[1] School of Engineering and Electronics, The University of Edinburgh, U.K.
[2] Department of Mechanical Engineering, Okayama University, Japan

KEYWORDS: Orthopaedic surgery, Unconventional cutting, Bone

ABSTRACT. In this paper, unconventional methods of cutting based on ultrasonics, laser, and water jet mechanisms that might be used in orthopaedic surgery are discussed. Their advantages include no or little contact with bone or tissue that is to be cut, the application of much less mechanical force compared to traditional techniques, and reduction in localised heat effect in the region undergoing surgical cutting. A critical analysis is performed to identify their limitations and merits.

1 INTRODUCTION

Many advancements have been made in understanding the anatomy of the body in surgical operations. However surprisingly little development has occurred in the devices and methods needed for surgery. Surgical tools such as scalpels, hacksaws and drills have undoubtedly been improved in terms of the materials that are used, and the mechanisms that are employed [1]. However they still adhere to the basic designs and mechanical methods adopted in previous centuries. They also face the inherent difficulties of accuracy of cutting, sterilisation, and thermal damage to neighbouring tissue, which can delay or prevent healing for example killing cells, necrosis.

Industry has also had to consider the limitation of traditional methods of cutting, for instance to deal with materials that are difficult to cut by conventional techniques, or to achieve acceptable accuracy. In the most recent half-century, unconventional methods of cutting and machining have had to be developed to overcome these difficulties. Their main attraction is that material removal or cutting may be obtained, not by high mechanical force, but by relying on alternative means of energy transfer or chemical reaction [2].

In this keynote paper, the prospects for utilising these unconventional methods in orthopaedic surgery are explored, although these techniques may also be applied to other areas of medicine that require surgical cutting.

2 STRUCTURE OF BONE

Figure 1 shows the bone structure. Bone may be considered to consist of two distinct parts. The outer cortical shell is hard and compact, and the inner, particularly the ends where stresses become more complex, is termed "cancellous" which is an elaborate construction of trabeculae which are orientated in the direction of the applied load. Bone is a also composite material made from two components. Firstly, it contains an inorganic mineral, a calcium phosphate and an

Published in: E. Kuljanic (Ed.) *Advanced Manufacturing Systems and Technology*,
CISM Courses and Lectures No. 486, Springer Wien New York, 2005.

organic matrix composed of 90% collagen and 10% ground substances (glycosaminoglycans and glycoproteins). The inorganic component is organized as small crystals in the matrix. Hydroxyapatite crystals are stiff and brittle, but the collagen fibres are more elastic, and this property gives the composite much better properties.

Bone structure is very complex, its properties can be difficult to measure, and can vary greatly from person to person. The values of bone properties vary with gender, age and loading (according to Wolff's law).

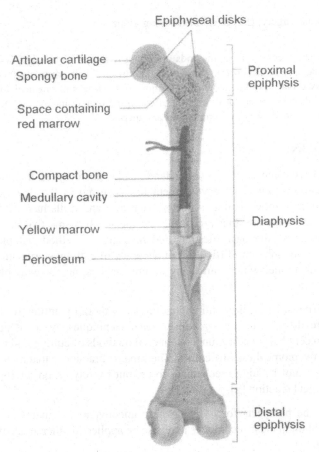

FIGURE 1. Femur structure

3 CONVENTIONAL BONE CUTTING

Mechanical bone and tissue cutting tools have been used by surgeons for centuries. Typical examples include scalpels, gouges, bonefiles, chisels, saws, drills, and reamers. Some examples are shown in Figure 2. Tried and tested, these tools enable the surgeon to perform almost any orthopaedic procedure, with cutting being achieved at a most angles and in different planes. Surgeons become skilled in the use and control of these mechanical tools. Most of these devices are made from surgical grade stainless steel, owing to its strength and durability, its high resistance to corrosion. Other materials used include chromium and titanium.

Disadvantages of mechanical surgical tools are that often surgeon has to exert considerable force on bone in order for a section to be cut; for instance the force require to cut bone with hand saws is more than 30N, during which its temperature can rise to more than 70°C (Bone cell damage usually begins between 42°C and 48°C). Such surgical tools that generate friction have to be lubricated to minimize thermal damage during an operation. Powered saws greatly reduce the forces required. The blades usually vibrate while a force is applied in the perpendicular plane by the surgeon so that only a small force is needed. Indeed□ electrically (by battery) or pneumatically- powered surgical tools are now commonplace in the operating theatre.

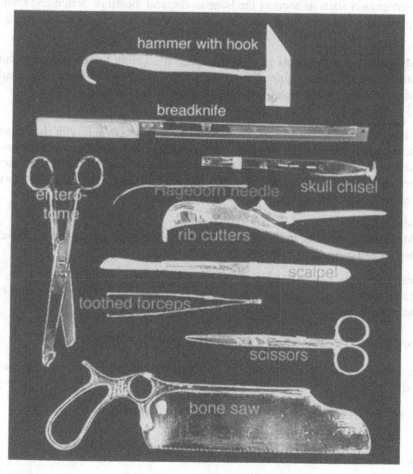

FIGURE 2. Orthopaedic cutting tools

Despite these advances, mechanical surgical tools still have the following drawbacks.

1. Infection
    The instrument relies on physical contact with the patient, to cut through bone and tissue, giving rise to the risk of infection. Sterile tools can only be guaranteed by use of disposable instruments (which is sometimes not possible especially with expensive instruments). The handles and motors of the instrument must be sterilized; and this sterilisation cannot be fully guaranteed (CJD is an example of a disease caused by a protein, which was not destroyed in normal sterilisation procedures). Moreover, as surgical tools are sharp, there is the risk that if the surgeon's skin is cut, with the further prospect of spread of infection.

2. Temperature increase of bone
    As noted above, conventional tools can readily give rise to temperatures of more than 70°C, and therefore cause irreparable damage to bone cells and necrosis. Bone regeneration starts as soon as the bone is cut, and therefore design and choice of the appropriate tool to facilitate bone structure recovery is a major consideration.

3. Accuracy
    Manually-controlled tools rely on the surgeon's ability to achieve the accuracy and precision of cutting required. Bone is a hard material: the tools may break during surgery, owing to the high frictional and applied force needed.

4. Bone debris [3],[4]
    The cutting action can give rise to emission of bone debris. With chisels and gouges, the debris tends to be large and visible, and readily removable. However the geometry of saw teeth and drill channels lead to the formation of smaller particles. These particles are more difficult to extract from the wound, and may even enter the blood stream leading to complications in the heart and lungs.

5. Invasive surgery
    Mechanical methods are invasive. They may lead to lengthy patient recovery times, higher cell damage, and risk of infection.

6. Limited life time
    Even with the advanced modern materials, mechanical tools wear out and eventually need to be replaced.

4   BONE CUTTING BY UNCONVENTIONAL METHODS

The main prospects for unconventional, in comparison with established cutting, methods are considered to be:

(i)   no or minimal contact between tool and bone, depending on the technique.

(ii)   little or no mechanical force applied to the bone

(iii) localised heat affected zone on the bone

Unconventional methods in orthopaedic surgery have hitherto received little attention. This section of the paper is devoted to an evaluation of three methods.

4.1 ULTRASONIC CUTTING [5],[6]

The term "ultrasonic" defines sound waves of frequencies above the threshold of human hearing, 20 kHz. Ultrasound offers a wide range of applications in both the medical and industrial fields. The use of ultrasound in medicine began in the 1940s. Dussik in Austria in 1942 described transmission of ultrasound for investigation of the brain. From the 1960s, the introduction of commercially available systems allowed the further development of ultrasound. Advances in electronics and piezoelectric materials provided further improvements from still to real-time moving images. The development of Doppler ultrasound had been progressing in parallel with the imaging technology. The combination of the two technologies in Duplex scanning and the subsequent development of colour Doppler imaging provided scope for investigating the circulation and blood supply for example to organs and tumours. The advent of the microchip in the 1970s and increases in processing power allowed faster and more powerful systems incorporating digital beam-forming, more enhancements of the signal and new ways of interpreting and displaying data, such as power Doppler and three-dimensional imaging.

In consequence ultrasound is now being used for numerous medical applications. Examples include cancer detection, fetal monitoring, gynaecology, cardio, physiotherapy pain relief, high-intensity focused ultrasound in tumor ablation. Ultrasound has also superseded invasive surgery for procedures such as liver surgery. In the removal of kidney stones, the ultrasound probe uses sound waves to break up the stones in place of a procedure in which a surgical incision is made in the back [7],[8]. In orthopaedic surgery, procedures for using ultrasound in conjunction with traditional mechanical cutting tools such as the scalpel, chisel, saw and drill, are emerging.

With the ultrasonic scalpels shown in Figure 3(left, centre), ultrasound aided cutting of bone was

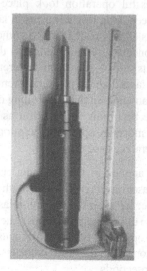

FIGURE 3. 20kHz (left) , 35kHz (centre) scalpels, and experimental apparatus (right)

investigated. The ultrasonic scalpel consisted of an electrically ultrasonic generator, a transducer and a cutting probe. The apparatus assembled is shown in Figure 3(right). The bone was firmly clamped to a wooden base. The latter was then mounted on a measurement platform which recorded vertical and horizontal displacements encountered by the bone during the cut. The thermal camera was placed above the sample in order to monitor the temperature variation. For cutting with the scalpel at two ultrasonic frequencies, the force required, cutting rate and temperature variation were measured.

The experiments showed that the forces required to cut bone with an ultrasonic scalpel under a properly controlled frequency conditions are smaller than that without ultrasonics. The first tests also indicated that a higher frequency, less force is required. Further research is now needed to determine the frequency that yields fast cutting at low force. Temperatures above 40°C were also noted in the vicinity of the cut. These experiments were carried out without coolant. The use of coolant (saline solution) is expected to lower temperature. Test with a saline flow of 80ml/min confirmed a decrease in the cutting temperature.

Further work is now required to attain the high performance needed to make ultrasonically-assisted surgical tools acceptable in practice. The main aims of this work are likely to be

(i)   more precision and control in cutting

(ii) reduction in tissue damage to the bone, thereby more rapid healing

(iii) decreasing time of cutting

(iv) enhance presently-used mechanical methods of cutting, without the need for further training in their use

4.2 LASER CUTTING [5],[9],[10]

The first working laser device was produced about 1960, and shortly thereafter medicine adopted this new technology. In 1965 the first successful operation took place with lasers used for corrective eye surgery. Since then, there have been countless advances in laser technology and its use in surgery to the extent that many operations cannot be performed without a laser. Indeed, the improvement of the laser owes much to the pioneering research done by the medical profession. The laser can treat an enormous range of ailments such as the chronic snorer, relief being achieved by a simple procedure. In dermatology, the laser can be used for removing moles, tattoos, birthmarks, wrinkles, and skin resurfacing. It is anticipated that controlling or even destroying skin cancer is now possible. Dental procedures are becoming less painful as the laser replaces the mechanical drill. The laser is proving to be an indispensable surgeon's tool because of its broad application, efficiency, and improvement in overall ease of use.

At present, many kinds of oscillator are being applied in various industrial fields, such as CO_2, Argon, YAG, Excimar lasers. Generally, each laser has its own wavelength. The laser with shorter wavelength commonly has the smaller spot size: for example the argon laser has a much smaller spot size than that of the CO_2 laser, but they have different applications. The effects of a laser beam can be altered by pulsing the beam, rather than operating with a continuous mode. "Q-switching" describes the quality switching of the optical resonator in the laser. It creates very high peak powers in millions of Watts and delivered in a few nanoseconds.

Laser machining in surgery offers many advantages, such as no contact with bone and aseptic effects. Sterilisation of operative sites, low surrounding cell trauma, reduced blood loss, precision operation, and reduction in postoperative pain. The cumulative effect is a reduction in medical costs with less need for hospitalization and an increase in numbers of out-patients. Although the use of lasers in the operating theatre is fast becoming commonplace, little research has been conducted. A preliminary investigation of cutting bone and surrounding tissue with lasers is described below.

The influences of laser power on the rate and the depth of cut, temperature effects have been analysed. Figure 4 shows the laser apparatus Opus20 used. This is primarily a dental laser and incorporates Er:YAG and CO_2 oscillator in a single device. Pulse repetition rates, duration and energy of respectively 10-20pulses/sec, 200-500µsec, 0.9-1.9J/pulse and, CO_2 (10.6µm) or Er:YAG (2.94µm) were selected. The laser beam is delivered through an optical fiber cable. An air/saline solution mix is directed from the hand piece. This bio-acceptable mix provides extra cooling for conditions where acceptable cell temperature has to be ensured, as well as protecting against infection and parching of the bone during irradiation. The pressure of air and solution can be controlled (0-5bar) by valves on the rear of the device. The bone sample is mounted on a platform and clamped in position on a motorized X-Y-Z table.

Experiments indicated that the depth of cut linearly increased with the power and the pulse frequency, and a tapered shape of cut was produced. The temperature variations of bone during laser cutting were measured with the thermal imaging camera. The maximum temperature was found to increase with laser power.

The maximum temperature at the laser incident spot was 70°C under high power condition; with the aid of water/air spray at 1 bar pressure following the pass of a laser the bone temperature was

FIGURE 4. Apparatus for laser cutting of bone

found to reduce rapidly to less than 40°C. Owing to these cooling effects water/air spray function is considered necessary for laser cutting of bone.

These preliminary results have given rise to the following recommendation for laser cutting in bone surgery, especially healing behavior of the bone is of concern. The laser should be used on low power adequate for the procedure, in order to reduce charring of the bone edges and the heat affected zone (HAZ). In order to achieve least HAZ, a laser with extremely short pulse and wavelength such as femto-second and Cu-vapour types appears to be suitable for surgery. For operations where less precision and larger amount of damaged tissue is allowed, higher speed of cutting can be achieved if the laser power rate is raised. Lasers such as the dye and second harmonics YAG where variable wavelengths are available might facilitate the operation. The advantages of laser cutting in orthopaedic surgery are;

(i)   there is no contact between laser tool and the body part

(ii)   no mechanical force is required to cut

(iii)   it provides a precise accurate targeted focus for dissection

(iv)   there is better freedom of geometry with flexible optic fibres

(v)   operating times are lowered

(vi)   it has aseptic benefits and can provide sterilisation in the region of operation

(vii)   by haemostatic action it can reduce blood loss

(viii) it can reduce effect of trauma in neighbouring cell

4.3 WATER JET CUTTING [10]-[13]

Water jet cutting is a relatively new process. Water pressures of 400MPa are commonly used for industrial purposes, although pressures as high as 1400MPa are possible [14]. The nozzle diameter can vary between 0.05mm and 1mm. The outcome is a fine water jet that can cut a narrow groove. Nowadays, water jets are combined with computer-controlled systems to provide highly accurate machining. Materials for which water jet cutting is used include cardboard, insulating materials, rubber, soft gasket material, fabrics, plastics, and food produce.

Plain water is typically used for cutting softer such as non-metallic materials. The addition of abrasives to the stream allows almost any material to be cut. In comparison with plain water, abrasive water jet cutting provides a better surface finish, deeper depth and faster rate of cut. Abrasive water jets are used to cut metals, glass, ceramics, and stone.

The water jet technique is increasingly popular for medical applications. In 1982, a new water jet method for stone destruction in the common bile duct was reported in Germany. Since then applications in dentistry, surgical cleaning and liver surgery have been found.

Its attraction is based on the water jet ability to cut through any animal or human body parts including bone, provided the appropriate pressure is utilised. A suitably small nozzle diameter and pressure setting allow for a precise, accurately targeted action for dissection. The depth of cut achievable may be calculated from the pressure and nozzle diameter.

Control over pressure then allows the surgeon to cut through layers of tissue leaving underlying layers intact. As the tool does not directly touch the patient during the operation the risk of infection is greatly reduced. In the light of these prospects for water jet cutting, an experimental programme was performed to assess whether water jet cutting can be a suitable in modern orthopaedic surgery.

A schematic diagram of the apparatus is shown in Figure 5. The water is filtered before entering the system to ensure that no scale deposits build up at the orifice due to mineral content in the water. The water is fed to the intensifier where it is pressurized to the high pressures required for cutting. The intensifier is a double-acting piston that pressurizes the water as it is driven back and forth by the hydraulic fluid.

The key feature in the jet nozzle assembly is the jewel made from ruby or diamond shown in Figure 6(right). The high pressure water jet is forced a hole of diameter between 0.050 and 1mm in the jewel which concentrates the flow into a fine jet. Increase in the material removal rate and a deeper cut occur with rise in flow rate. The diameter size directly affects the flow rate of the water in the exiting jet and consequently the depth of cut achieved. The nozzle also ensures a satisfactory flow quality and dynamic performance. A force acts on the orifice due to the pressurized flow through the nozzle. The force experienced by the nozzle increases as the standoff distance is reduced.

The effects of pressures, the nozzle diameter, and standoff distances on cutting were investigated. The water jet was found to be effective in cutting bone. Higher pressure, smaller diameter of nozzle, and shorter standoff distance produced a deeper and narrower width of cut. The temperature of the bone during cutting reached 31°C.

The water jet machine utilized during these experiments was designed for industrial use and

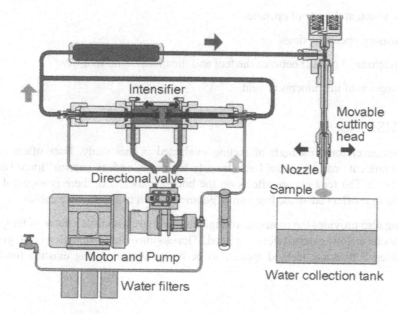

FIGURE 5. Schematic diagram of water jet equipment

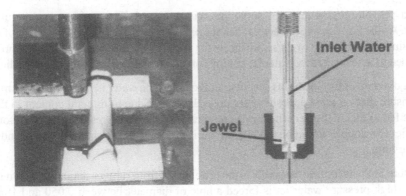

FIGURE 6. Orifice and sample apparatus

therefore is unsatisfactory for use in surgery. An early development has to be improvement in its low freedom of geometry. Utilization of water jet technology in the surgical environment as a hand held instrument is advocated. A form of water jet scalpel would be ideal. The use of water jet technology in surgery is a feasible alternative to traditional methods due to the following advantages;

(i)   low thermal heating, no trauma to surrounding tissue

(ii)   reduced blood loss

(iii) precise, accurately-targeted action for dissection

(iv)   clear vision at the site of operation

(v)   decreasing operating times

(vi)   no mechanical contact between the tool and the material, no tool wear

(vii) elimination of instruments in field

5   CONCLUSIONS

Of the three unconventional methods of cutting evaluated in this study, laser offers the best prospects. Accurate cuts can be attained at relatively high rate; and no external force has to be applied to the bone. The temperature effects on the bone are localised; there is apparently little debris. The technique offers adequate freedom of geometry, with use of flexible cables.

Water jet cutting also provides the prospects of high cutting rates with little effects of temperature on the bone; there is low external force applied. Development of mechanisms for geometry freedom is needed. Ultrasonic method appears to be best suited as aiding existing mechanical methods.

Other unconventional methods of cutting will offer their own advantages and limitations. Further researches are needed to establish the potential for unconventional method of cutting in orthopaedic surgery.

ACKNOWLEDGMENTS

The authors wish to thank I.Edward, R.Hannah, P.McMullen, B.Moore, K.Neish, R.Rankin, and R.Rankine for their assistance with the experiments needed to perform this project, and Mr.D.Ross, N.Rachmanis, F.Borocin with the benefit of their experience in water jet, laser and ultrasonic cutting. The Government of Japan is thanked for the MEXT Overseas Advanced Educational Research Practice Support Program Fellowship awarded to Dr. A.Okada which enabled him to research this topic with Prof. McGeough at the University of Edinburgh.

REFERENCES

1. (1989), Changing Role of Engineering in Orthopaedics, Proceedings of the Institution of Mechanical Engineers International Conference, 14-15.
2. L.Slatineaunu et al, (2004), NonTraditional Manufacturing Processes, Tehnica Info Publishing House, Moldova.
3. C.M.Robinson, (2001), Current Concepts of Respiratory Insufficiency Syndromes after Fracture, The Journal of Bone and Joint Surgery, Vol 83-B/6, 781-791.
4. J.A.McGeough, H.Rasmussen and J.Christie, (2005), Effect of Bone Reaming in Oxygen Transport, (in preparation).
5. Amaral,J.F. and C.Chrotskeic,(1997), Experimental Comparison of the Ultrasonically Activated Scalpel to Electrosurgery and Laser Surgery for Laparoscopic Use, Min. Invas. Ther, Allied Tech.[Vol 6, 324-331.
6. Horlon,J.E., Tapley,T.M. and Jacpway,J.R., (1981), Clinical Applications of Ultrasonic Instrumentation in the Surgical Removal of Bone, Oral Surgery, Vol 51, 236-241.
7. B.S.Khambay and A.D. Walmsley, (2000), Investigation in to the Use of an Ultrasonic Chisel to Cut Bone -Part 1-, Journal of Dentistry, 28.
8. B.S.Khambay and A.D. Walmsley, (2000), Investigation in to the Use of an Ultrasonic Chisel to Cut Bone -Part 2-, Journal of Dentistry, 28.
9. G.T. Absten and S. N. Joffe, (1985), Lasers in Medicine – An Introductory Guide, Chapman and Hall.
10. J.Meiger, (2002), Laser Micromachining, Chapter 8 in Micromachining of Engineering Materials, Marcel Dekker, USA (Editor J.A.McGeough).
11. The Development of the Water Jet Scalpel with Air Pressure, Symposium on Jet Cutting Technology, BHRA, England, 39-52.
12. J.A.McGeough, (1988), Advanced Methods of Machining (Chapter 5 Laser-jet machining and Chapter 7 Water-jet machining), Chapman and Hall, United Kingdom.
13. V.K.Jair, (2002), Advanced Machining Processes, Allied Publishers PVT Ltd, India.
14. Kalpakjian. S. and Schmid.S. R., (2001), Manufacturing Engineering and Technology, 4th Ed., Prentice Hall.

ASSEMBLY OF MICROPRODUCTS: STATE OF THE ART AND NEW SOLUTIONS

M. Santochi[1], G. Fantoni[1], I. Fassi[2]

[1] DIMNP - Department of Mechanical, Nuclear and Production Engineering, University of Pisa, Italy
[2] ITIA - Institute of Industrial Technologies and Automation, National Research Council, Italy

KEYWORDS: Microassembly, Microfactories.

ABSTRACT. Nowadays, miniaturization is playing an important role in product redesign while complex microproducts are leaving their traditional domains. These two aspects concur to the need for increasing their production and reducing their cost. One of the main production problems is represented by microassembly. This keynote paper deals with the microassembly problems, shows the state of the art of the research, highlights the most promising R&D areas in this field and finally presents some new solutions for hybrid microproducts assembly.

1 INTRODUCTION

During the last decade, more and more micro products spread out in the medical and biomedical field (e.g.: pacemakers, analysis equipments, micro tweezers for minimally invasive surgery, micro drug delivery systems[1]), in the automotive field (sensors for safety in cars e.g. electro-static field sensors for controlling airbags [2]), in aeronautics and aerospace (lightweight distributed sensors for micro crack detection [3]) or in the IT field (ink jet printers, reading caps for hard disk but also micropumps for microprocessor cooling) and finally in the telecommunication area (micro optical switches, etc..). Microproducts are now leaving those traditional applications to invade other fields such as entertainment and sport equipment (noise canceller ear plugs, variable stiffness tennis racket, skis equipped with piezoelectric active dampers).

1.1 MOEMS AND HYBRID MICROPRODUCTS

The term "microproducts" includes at least two different classes of devices: MOEMS and hybrid microproducts. MOEMS (Micro Opto Electro Mechanical Systems) are manufactured by silicon technology and their final structure is nearly planar (2D ½); hybrid microproducts have a high 3D structure and consist of several components, made by different materials. The wide development of micro products has been mainly due to the influence of the electronic world and silicon based technologies which allow the production of high quantities of low cost sensors, chips etc.. Be-cause of the well known limits of the silicon technology, at present it seems difficult to imagine a 3D micro product composed by different materials produced through an up-scaling of electronic technologies. For this reason a lot of hybrid devices have been developed in research centres, but only few of them are commercially available and/or produced in small and medium series [4]. Some of these prototypical products are: a micro-touching probe [5], a micro-stepping, an elec-tromagnetic [6] and a linear variable reluctance micro motor, a micro harmonic drive [7], a

Published in: E. Kuljanic (Ed.) *Advanced Manufacturing Systems and Technology*,
CISM Courses and Lectures No. 486, Springer Wien New York, 2005.

micromixer, a microcar [8], a micro electro mechanical pump [9] etc... The main reason of their scarce diffusion does not depend on their performances or on their manufacturing but on the difficulties of their massive, automated and economical assembly. As stated by several authors [1], [3], [10], [11] the exploitation of this huge market depends on the ability to industrially produce and assembly new low cost devices, making them available to a large number of customers. For that reason, the development and manufacturing of hybrid microproducts represent new urgent issues to be solved.

The process chain of hybrid microproducts involves the classical aspects of every product: materials, parts, products and the technologies as planning, manufacturing, assembly, control. Laser microcutting, metal sheet microforming [12], polymers microinjection molding are some of the technologies used for manufacturing microparts. The still open issues in manufacturing technologies are not the actual obstacles to a massively parallel production of thousands of microparts by batch processes, but the real bottleneck for mass production is represented by assembly. It is mainly due to the difficulties of automation, necessary for maintaining high quality, reliability and reproducibility [11]. The development of a complete automation in assembly [3] and/or massively parallel micro-assembly systems [1] are considered key issues to reduce production time and costs. Several attempts of automation have been developed in the last years: some microfactories for production and assembly have been conceived [10], but, as stated in [1], "the main challenges [...] for a successful development of these processing centres are [...] feeding/handling/assembling".

2   PROBLEMS DURING MICROASSEMBLY

In microassembly, new problems, different from the traditional assembly process, emerge: the dominance of surface forces (electrostatic, surface-tension and van der Waals forces) causes adhesion phenomena, particularly important in handling. Fragile components can prevent multiple handling. Magnification lenses cause loss of depth of field and small dimensions produce difficulties in control and measurement. However, the small scale allows to exploit chemical and physical principles (from acoustic handling to laser tweezers, from electrostatic grippers to SAW[1]) not suitable for handling standard macro components.

- Adhesive forces

It has been proved that superficial forces (in particular surface tension, Van der Waals [13], electrostatic and/or triboelectrification forces [14]) can increase friction [15] and sticking effects, which negatively affect both production and handling phases. Several papers have proposed and validated analytical models for the meaningful forces in microassembly ([16], [17], [18], [19], [20]), others suggest alternative innovative handling strategies ([1], [21], [22]). A quantitative comparison of interaction forces between bodies ([1], [23]) showed that capillary forces are relatively strong as compared with the other ones and the electrostatic forces can overcome capillary forces in proper conditions. While Casimir and Van der Waals forces [23] are relatively weak in

---

[1] Surface Acoustic Waves [25], generated by applying radio-frequency electric pulses to the chips, realize a nanopump able to obtain a precise electronic control of chemical reactions on the surface of a biochip.

the range of microns, magnetic forces may become significant for long-range interactions only with ferromagnetic parts.

Between 1 mm and 1 μm, as experiments and theory demonstrate, when humidity increases (HR>50%) a liquid bridge between two solid surfaces appears and capillary forces, increasing the reciprocal attraction, can make the releasing of a grasped object difficult or even impossible. The electrostatic charges on a component, caused by triboelectrification, play a relevant role and furthermore attract the dust that can make some assembly task impossible [24], can damage the component or reduce the performances of the whole product. However some of the described problems can be avoided by using a clean room or at least a protected and humidity controlled environment.

- Depth of field

A low size of parts causes difficulties in observation. In particular an increase of the magnification has several drawbacks [1]: the observed area becomes small, the depth of field is reduced if compared with the width of observed area, the distance between the lenses and the component generally causes problems in the component handling. It is clear that these problems negatively affect the measurement and control phase.

- Measurement and control

Microassembly requires on line control [31] of handling, feeding, etc. necessary to assure the correctness of the whole process. Although product control (measurement of dimensions, masses, etc.) is fundamental, process control [38] and process parameters monitoring (measure of small forces and fields [59], [60], [61]) are necessary to guarantee high standards of quality in microassembly.

- Methods and models

Open problems still remain in process characterisation as in simulations and in FEM tools. Even if large efforts have been spent in the last years [26], force characterization and modelling have to be done in order to evaluate the role of each force and the applicability of physical and chemical principles, not usable in standard assembly.

3 PROPOSED SOLUTIONS

The approaches to the assembly of hybrid systems are numerous. Some of these techniques exclude the others, while others could be used jointly in order to exploit the advantages and reducing the drawbacks. Neglecting the manual assembly approaches ([27], [28]), often used to assembly prototypes and demonstrators in research centres, the most interesting automated solutions are: self and parallel assembly, macrofactories, microfactories (Figure 1).

3.1 PARALLEL ASSEMBLY AND SELF-ASSEMBLY

The concept of simultaneous assembly of a large number of components is the base for self and parallel assembly, where components self locate in stable pre-determined positions and are then assembled at the same time. A prototype of self assembly has been obtained [32] by using a substrate with an array of pockets, with a suitable shape to maintain the components in the right position. These parts, suspended in a fluid, are dispensed over the substrate. If the components enter into the pockets they remain assembled, on the contrary they flow down and are re-circulated. The idea in general (see also [32], [33], [34], [35], [36], [37]) is to create a structured potential (gravitational, electrostatic, etc..) pattern toward which components move randomly until they are captured by the field: in [33] the pattern is based on an electrostatic structured field, while in [34] and [35] hydrophilic and hydrophobic areas guide the components in parallel self positioning and orienting.

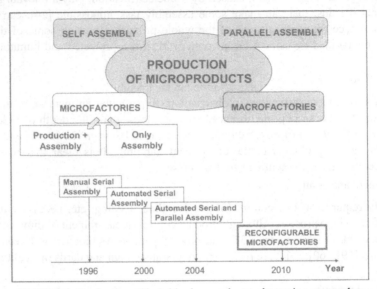

FIGURE 1. Production of hybrid microproducts: alternative approaches

3.2 MACROFACTORIES

The idea of macrofactories is based on the concept that traditional machines can be adapted to the microproduction. In fact, in order to produce thousands and thousands parts, standard big machines may be more economical. Examples of this concept are electronic assembly machines, modified to assembly hybrid microproducts [5], or standard small robots where some actuated extra degrees of freedom are added in the gripper to perform fine positioning [76]. The use of standard robots reduces certainly the costs but increases the dimensions of the clean and controlled environment, necessary to reduce sticking problems, and shows numerous limits in performing assembly operations where more then one device at time is requested: in fact the simultaneous presence of different devices, systems and robots causes a systematic lack of room.

3.3 MICROFACTORIES

In literature some attempts to aggregate a series of machines and devices in a microfactory have been also shown. The microfactory has been defined as a micro fabrication system [29] and two different categories can be found: manufacturing and assembly microfactories, and assembly microfactories.

MANUFACTURING AND ASSEMBLY MICROFACTORIES

They consist of a series of micromachines for microcomponents manufacturing (microlathe, micropress, etc.), of an arm for handling operations and some microgrippers. By using a series of microCCD cameras the operator controls in a monitor the operations performed by each machine tool and manually assembles the microparts. In [8] a desktop microfactory is used to assembly different variants of a microcar (scale 1:87) and includes all production steps and assembly modules.

ASSEMBLY MICROFACTORIES

The automated assembly microfactory ([30], [31]) consists of a series of devices as microgrippers for grasping, microfeeders for transporting, sieves/riddles for selecting, robots for executing a set of consecutive operations. Operations performed in an assembly microfactory include also gluing, soldering, control etc.. Therefore glue dispensers, laser for soldering, micro friction stir welding machines, microscopes, sensors etc.. a very large number of devices must simultaneously work in a very small area. For example a gripper grasps a component, positions and keeps it in its final position on the microproduct while the component is soldered and the process controlled. The few existing assembly microfactories are often product oriented, semiautomatic and rigid. Their miniaturisation and their flexible and reconfigurable automation are the challenges of the future research.

The comparison of the different proposed approaches, based on their advantages and drawbacks (Table 1), shows that the reconfigurable assembly microfactory is one of the most promising concepts. In fact it potentially allows to automatic assembly, in a small clean room, hundreds and hundreds of products by using flexible modules easy to be reconfigured by rapid set up operations when the production changes.

TABLE 1. Strength and weakness of the different approaches

	Advantages	Drawbacks
Manual or Tele-operated assembly	Flexible structure with possibility of assembly different and complex structures	Time consuming activity, problems of visual depth of field, trial and error approach
Self assembly and or Parallel assembly	Very large number of parts assembled at the same time	Rigid structures with ad hoc devices designed for specific components. Low number of components per inch ratio [35]. Often stochastic process.
Macro-factories	Use of standard robots for coarse positioning and devices on hand for fine positioning.	Need for a very large clean area with controlled humidity, temperature and dust; limited accuracy of conventional robots.
Micro-factories	Precision and accuracy of micro devices, micro robots, etc.. Small clean room with a controlled environment.	Unsolved problems of miniaturisation of the devices. Need for flexibility and reconfigurability.

As affirmed in [8], "To guarantee a holistic view [3] it is necessary to map the whole microproduction process chain from planning to final assembly". So, for assembly microfactories (Figure 2), research should focus on assembly methods, systems and planning; microassembly stations and microassembly devices, models (theoretical and numerical) and software for force evaluation and analysis of downscaling effects, control and monitoring within the scope of microproduction.

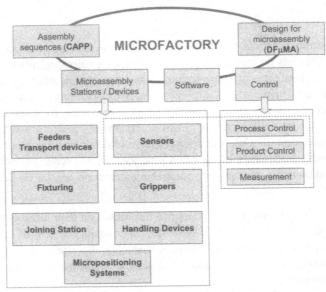

FIGURE 2. Microfactory: conceptual framework

Many papers dealt with the solution of single tasks or microassembly elementary operations, while the integration of so many devices in a flexible and automated microfactory is still at an early stage (even if some very promising attempts exist [10], [8], [7]).

4   MICROASSEMBLY ELEMENTARY OPERATIONS

The main effort of future research will be the automation of single processes and their integration. An approach able to sum the advantages of the other techniques in Table 1 (especially parallel and self assembly) with the flexibility of the microfactory can be very interesting and appreciated. In order to assembly complex hybrid 3D microproducts in a microfactory (Figure 2), a series of single elementary operations as singularising, orienting, aligning, handling, grasping, releasing, joining, precise positioning, measuring etc... have to be performed.

Because of the adhesive forces cause sticking problems, the design of each new gripper has to face both the grasping and the releasing phase. Table 2 has been created to map the gripping principles and the proposed releasing strategies. It describes the state of the art and highlights some interesting areas (coloured cells) for future development of new grasping-releasing techniques.

- Grasping: it is the most investigated problem [22] because a macro to micro approach has been followed bringing to numerous microgrippers obtained by downscaling. But this kind of micro-grippers ([39],[40]) showed immediately their limits in releasing of microparts [1], the most complex problem to be solved as shown in Table 2. Contact less handling strategies overcome the traditional approaches because "releasing problems does not exist without grasp". Some tens [42] of research works in literature exist, but contactless handling still seems a promising field to be explored and exploited for its interesting characteristics.

- Feeding: in order to connect the different microdevices of the microassembly factory ([1], [3], [31]), a necessary activity could be the development of a device able to align, orient and transport microparts. Interesting attempts for transporting powders [67], microcomponents [68] and feeding miniparts [47] allow to perform only some of the previous tasks.

- Joining and connecting methods: systems as mechanical joining [52], microsnaps [58], Velcro systems [72] and gluing [59] have been successfully tested, but methods of stable connection of thin metal sheets as laser microwelding and Micro Friction Stir Welding are also promising.

- Robots: two different approaches have been proposed, standard small robots (e.g.: SCARA) with fine positioning systems integrated in the gripper [69] and microrobots [70] able to perform precise positioning dedicated to microassembly.

- Metrology and control: systems for real time control (vision systems [5], [29]), for process characterisation and process setting up and devices for final control (sensors) of the whole assembled product are also an urgent problem to overcome.

In order to come to a complete automated microfactory the development of grippers with additional characteristics, capabilities, degrees of freedom for dexterous manipulation, the development of compliant grippers, devices and hand effectors (as indicated in [3], [42]), joining systems etc.. are only some of the requirements to be solved.

5 NEW SOLUTIONS

The DIMNP and ITIA have started a joint research activity on microassembly. The research is dealing with the study and development of a contactless handling system [43], of modular feeders [44], of adhesive [45], electrostatic and mechanical grippers.

5.1 CONTACTLESS ELECTROSTATIC FEEDER OF MICROPARTS

At the DIMNP, a contactless electrostatic feeder has been developed for an increase of automation and for the reduction of the assembly time. Moreover the design of the feeder, able to align and transport microparts to an assembly station, was oriented to achieve a high reconfigurable structure [46] to answer the need for flexibility of the assembly microfactory.

TABLE 2. Gripping principles and the proposed releasing strategies

Gripping Principles \ Releasing Principles	Gluing the part to the substrate on the right place	By positive mechanical engagement	Micro snaps	Micro Velcro	Injection of gas: a small puff of gas (blow away)	Mechanical release with needle	Destruction of the gripper mechanism	Vibration of the gripper	Stripping off the component against a sharp edge	Auxiliary tool	Rolling the gripper, change its shape to decrease the contact surface	Electrostatic repulsion	Reduction of surface tension	Pressure generation by thermal variation into microholes	Roughness variation	Micro pyramids	Thermal control of Surface tension forces due to air humidity	Spread and retract the liquid (alcohol)	Contactless handling
Friction	[52] [59]	[52]																	
Jaw / Mechanical		[52]				[52], [54]						[21]			[21]		[21]		
Magnetic		[41]														[60]			
Ice							[56]	[57]											
Van der Waals									[54]	[54]	[54]								[59]
Electrostatic											[51]¹	[53]							[47]
Adhesive (liquid)					[52] [54]						[19] [45]					2		3	
Suction					[30]														
based on Bernoulli principle																			
Pressure generated by an acoustic source								4											[25] [48]
Pressure generated by a laser beam																			[49] [50]

Micro snaps: Need for a structured surface [64]
Micro Velcro: Need for a structured surface [65], [66]

---

¹ In [51] the authors roll the discharged gripper. The modification of the liquid meniscus by electro-wetting [71] may be interesting to be developed.

² Spread the liquid trough the micro-pyramids and after retract the liquid.

³ Variation of the surface characteristics (hydrophobic vs. hydrophilic conditions) - concepts based on [62], [63].

⁴ Releasing by acoustic pressure (a sonar wave move the component and detach it from the gripper).

The contactless feeder is based on the property of materials to be attracted, by charge induction, towards regions with a higher electric field in the same way as a dielectric/conductive plate is attracted inside the two charged plates of a capacitor. The feeder (Figure 3) consists of a high voltage supply (1) that, by a PLC based switching system (5), supplies a series of V-shaped parallel wires (the electrodes (4)) mounted over a conducting and vibrating platform (2). The vibration has been used to make the microparts free to move on the working surface following the existing potential fields (in this case electrostatic). Therefore, in order to minimize the energy of the whole system, the component is attracted and oriented under the charged electrode.

The electrodes are charged one after the other according to a proper sequence that creates a stepping electrostatic field (that is a travelling capacitor) able to attract and transport the microparts when aligned.

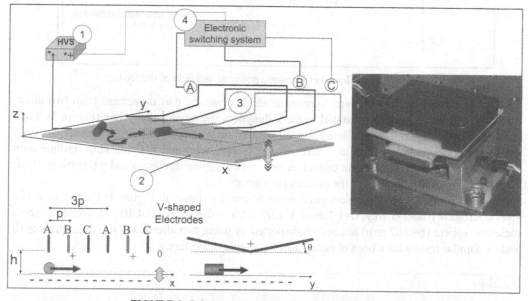

FIGURE 3. Scheme and photo of the feeder (3 phases)

Three distinct steps can be observed even if they could simultaneously happen: during the first step the component is attracted and oriented under the electrode, then it moves towards the V-tip, after that, it is transported along linear or angular trajectories following the supplied field.

Even if the presence of the component locally modifies the electric field, however general assessments continue to be valid. In fact, for example, the direction of the force acting on the component can be evaluated, neglecting the field perturbations introduced by the part, by analysing the unmodified electric potential and field. Actually the force is proportional to the electrostatic field or to the gradient of the electrostatic potential (Figure 4): the electrostatic field and the resulting force increase where the potential varies rapidly.

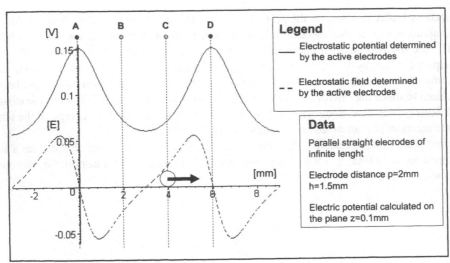

FIGURE 4. Evaluated electrostatic potential and field of the feeder

Two CCD cameras have been used to monitor and analyse the part movement from two directions: the first one has its axis parallel to the x direction, the other to the y one (Figure 3). Each
frame has been captured by a tailored software, based on a commercial vision system, has been
the component recognised and its centre of gravity measured and plotted. This configuration
allowed to observe the component behaviour on both x (feeding direction) and y (centring effect)
corresponding to the variation of the process parameters.

The typical results of feeding micro parts along a line are shown in Figure 5. The feeding of a
steel microgear (mass=3.1mg, d=1.20mm, l=1.05 mm) is shown in FIGURE 5a, while the same
dielectric sphere (φ=0.62 mm) has been transported by using two alternative supply sequences (b
and c). Similar results have been obtained also for angular trajectories.

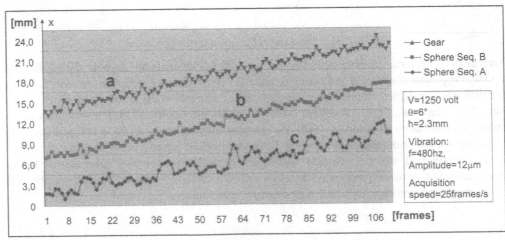

FIGURE 5. Typical results of feeding different microparts

Several electrode layouts have been tested (3 distances for p; 3 and 4 phases devices) but also different supply sequences and switching periods T (single active electrode or superposition of two consecutive active electrodes).

The components used in the experiments were cylinders, small glass fibres, spheres, SMDs (Surface Mounted Devices), micro gears, pins, screws etc.. with at least one dimension less than one millimetre. Both conductive and dielectric materials have been successfully tested.

In order to answer the requirements of flexibility and reconfigurability of an assembly microfactory, the concept of the feeder has been transferred in a pre-engineered version based on modular elements which can be reconfigured when required by production. By rapid set up changes it is possible to obtain the desired path by connecting straight and angular modules.

Not only layout variations can be satisfied, but also parts different in dimensions, materials and shapes can be fed through the adjustment of mechanical elements and/or electrical parameters. In particular: vibration amplitude and frequency can be chosen among those sustaining a random Brownian motion of the component, voltage can be varied according with the component mass and geometry, the horizontal position of each platform can be adjusted by regulating three grains. Also the switching period and sequences can be easily modified as the pitch of the electrodes that can be changed by replacing the PCB structure where the electrodes are patterned.

The future developments of this research work are oriented to increase the feed rate: the idea is to design and test a parallel feeder able to sort, align and transport many components simultaneously. Moreover, even if the feeder shows a good reliability, it will be improved creating a confined controlled environment.

5.2 ADHESIVE VARIABLE CURVATURE MICROGRIPPER

Among the numerous developed grippers, shown in Table 2, an example one is represented by an adhesive variable curvature microgripper developed by ITIA [45].

The gripper has been theoretically studied and a first prototype developed. Firstly the physical properties of the liquid that is responsible of the capillary force (e.g. the density of the liquid and its surface energy) and its compatibility with the task to be performed (i.e. objects may be contaminated by the chosen liquid) were considered. Then the physical approach to control the magnitude of the capillary force was defined to realise a reliable gripper. The capillary force can be influenced in several ways: the influence of the surface characteristic parameters on the capillary force is well documented in the literature [16]. These parameters are related to the geometry [73, 74] (i.e. roughness, curvature radius, dimensions) of the objects in contact with the liquid and the electrochemical interactions (i.e. electrostatic charge distribution, contact angle, line tension) between them and the liquid. The working principle of the gripper is based on the control of the radius of the gripper: by decreasing the curvature radius, the force of adhesion due to the liquid bridge between the component and the gripper is reduced, until the releasing of the microparts is ensured.

Therefore the capillary force between a spherical gripper and a flat object was investigated. Starting from [19], a more complex analytical model was developed and it was compared with the results obtained from a finite element model of the gripper.

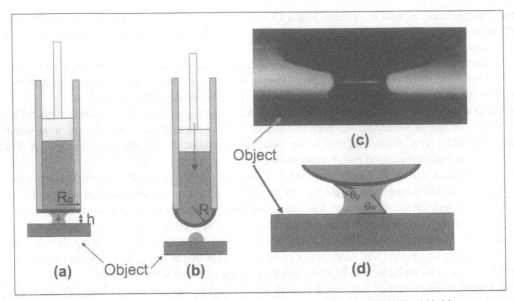

FIGURE 6. The working principle of the gripper; (a) when the gripper is flat the liquid bridge exerts a force able to lift the object; (b) the curvature radius is decreased until the liquid bridge is broken and the object released; (c) an image of the membrane and the capillary bridge; (d) the contact angles θ_p and θ_w

Considering the influence on the capillary force of the distance between the gripper tip and the object (h), the curvature radius of the gripper (R) and the amount of liquid deposited (V), it can be observed that, as the volume (V) of the drop increases:

1. if the distance between the membrane and the plane (h) is zero, the force decreases;
2. if h is greater than zero but negligible with respect to the radius of the wet area, the force decreases likewise the behaviour obtained for a null distance
3. if h is not negligible with respect to the radius of the wet area, the force increases.

Moreover as the curvature radius increases the capillary force increases (see also Table 3), in agreement with the theoretical predictions.

The theoretical results confirmed the possibility of using capillary forces as a mean for the manipulation through a gripper with variable curvature. The gripper consists of a circular membrane of radius R_g (Figure 6) able to decrease its curvature radius from a flat configuration (R = ∞) to a hemispherical configuration (R=R_g) as a result of stresses induced by an external force (for example a pressure) or by potential energy (electro-active materials). The maximum capillary force is exerted by the flat configuration which can consequently lift up the maximum handable mass. The minimum force is produced by hemispherical configuration and it can release the minimum handable mass. In Table 3 the maximum and the minimum handable mass and the corresponding R_g for a certain amount of deposited liquid (V), in analogy with [75], have been reported.

The surface tension (γ) of the liquid can be exploited in order to shift the handlable range of weights: the lower the surface tension of the liquid, the lighter the maximum and minimum liftable masses. Liquids with too low values of surface tension and low molecular weight, which easily evaporate, can lead to unstable handling. The values of the contact angles (θ_p, θ_w

in Figure 6d) can be controlled using appropriate coatings on the surfaces in direct contact with the liquid.

TABLE 3 The theoretical maximum and the minimum handable mass and the corresponding Rg and V with h = 0 mm, γ = 0.073 mN/mm (water), θp = θw = 30°

R_g [mm]	V [mm³]	Mass min [mg]	Mass max [mg]
0.05	1.00E-04	2.67	37.7
0.1	8.00E-04	5.34	75.4
0.2	6.40E-03	10.7	150
0.5	1.00E-01	26.7	377
0.8	4.10E-01	42.8	603
1	8.00E-01	53.5	754

The experimental gripper has of 2 d.o.f. (vertical and horizontal translation) and an elastic membrane fixed on its bottom tip. A hydraulic actuation system controls the shape of the membrane and, in particular, realizes the transition between the flat and the hemispherical configuration (Figure 6). An external vision system was used to monitor h and R.

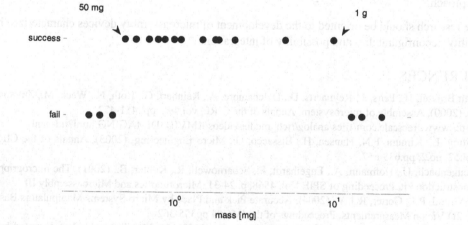

FIGURE 7. Summary of the handling tests performed with parts of different mass (each circle represents a different component). Gripping was considered a successful if both gripping and releasing succeeded

The experiments consisted in a pick and place operation of components of different mass. The experiments were conducted in an environment without restrictions on cleanness and humidity; temperature was around 25°C and variable conditions of relative humidity (20% up to 60%) were encountered during the experimental sessions. In order to exclude the influence of other parameters, the components handled were all rectangular flat components of variable size and weight, made of silicon with a mirror polished face.

The results of the experiments (Figure 7) were compared to the theoretical predictions and the handable weight range determined (50mg – 1000mg) was in good agreement with the analytical evaluations (considering that the prototype has a membrane radius Rg equal to 1.3 mm). Liquids with different surface energy (water, oil, soapy water, alcohol) were examined and the results confirmed the theoretical prediction: the gripping principle is valid for any liquid and the magnitude of the capillary force changes following the variation of the surface tension of the liquid.

The main advantages of the gripper are the use of a force of high magnitude in the micro-domain such as the capillary force; moreover, the axial compliance due to the low stiffness of liquid meniscus can be exploited for assembly operations. Furthermore the choice of an adhesive gripping system decreases the risk of damaging fragile components. In addition, since the gripping principle is valid for any interposed liquid, functional fluid such as lubricant or bonding agent can be adopted.

6 CONCLUSIONS

At present many microdevices remain as prototypes in research institutes and their engineered versions do not reach their potential market. In the future the market of hybrid microproducts will certainly grow up and micromanufacturing, and specially microassembly, will play a strategic role for their high quality and low cost.

This paper has discussed the open problems in microassembly, the state of the technique, the most relevant and interesting research areas. Two examples of the research on microassembly complete the paper showing how research in this strategic field requires a fundamentally new multidisciplinar approach.

Future research should be oriented to the development of microassembly devices characterized by reliability, reconfigurability and possibility of integration.

REFERENCES

1. Van Brussel, H., Peirs, J., Reynaerts, D., Delchambre, A., Reinhart, G., Roth, N., Weck, M., Zussman, E., (2000), Assembly of microsystem, Annals of the CIRP, vol 49/2 pp. 451-472.
2. http://www.freescale.com/files/analog/rich_media/videos/RMVID3DIMAGING.html?tid=tanl
3. Alting, L., Kimura, F.N, Hansen, H., Bissacco, G., Micro Engineering, (2003), Annals of the CIRP, vol.52, no.2, pp.635-657.
4. Gengenbach, U., Hofmann, A., Engelhardt, F., Scharnowell, R., Köhler, B., (2001), The microgripper Construction kit, Proceeding of SPIE Vol. 4568, p. 24-31, Microrobotics and Microassembly III
5. de Grood, P.J., Gorter, R.J.A., (2004), Accurate Pick and Place by Micro Systems Manipulators Based on 2D Vision Measurements, Proceedings of the IMG04 pp355-362
6. Stefanini C., Dario P., Carrozza M.C., D'Attanasio S., (1998), "A Mobile Microrobot Actuated by a New Electromagnetic Wobble Micromotor". IEEE/ASME Transactions on Mechatronics, 3 (1) 9-16.
7. www.nanomotor.de
8. www.microcar-karlsruhe.de
9. Croquet, V. Delchambre, A., (2004), Innovative Implantable Drug Delivery System: Design Process, Proceedings of the International Precision Assembly Seminar, Bad Hofgastein, Austria
10. Fleischer, J., Volkmann, T., Weule, H., (2003), Factory Microplanning Methodology for the Production of Micro Mechatronical System, CIRP Seminar on Micro and Nano Technology 2003, Copenhagen, November 13-14, pp. 17-20.

11. Reinhart, G., Höhn, M., (2001), Cost Efficient Assembly of Microsystems Using Positioning Strategies of Endpoint Sensing and Actuating, Production Engineering, Vol. VII/2.
12. Geiger, M.; Kleiner, M.; Eckstein, R.; Tiesler, N.; Engel, U., (2001), Microforming. Keynote Paper. Annals of the CIRP, 50/2, 445-462.
13. Autumn, K., Liang, Y. A., Hsieh, S. T., Zesch, W., Pang Chan, W., Kenny, T. W., Fearing, R., Full, R. J., (2000), Adhesive force of a single gecko foot-hair, Letter to Nature, NATURE, Vol 405.
14. Allen, R. C., (2000), Triboelectric Generation: Getting Charged, EE-Evaluation Engineering Desco Industries, pp.8 S-4-S-10.
15. Hitoshi Suda, (2001), Origin of Friction Derived from Rupture Dynamics, Langmuir, Vol.17, No.20, pp. 6045-6047.
16. Fearing, R.S., (1995), Survey of Sticking Effects for Micro Parts Handling, Proc. IROS '95, IEEE/RSJ Int. Conf on Intelligent Robots and System, 2:236-241
17. Menciassi, A., Eisinberg, A., Izzo, I., Dario, P., (2004), From "Macro" to "Micro" Manipulation: Models and Experiments, Transactions on Mechatronics, IEEE, June 2004.
18. Feddema J.T., Xavier P. and Brown R., (1998), Assembly Planning at the Micro Scale, Proceeding of the Workshop on Precision Manipulation at the Micro and Nano Scales, Proceedings of IEEE International Conference on Robotics and Automation, Leuven, Belgium, May 16-20.
19. Pagano, C., Ferraris, E., Malosio, M., Fassi, I., (2003), Micro-handling of parts in presence of adhesive forces, CIRP Seminar on Micro and Nano Technology 2003, Copenhagen, pp.81-84
20. Lambert, P., Delchambre, A., (2003), Forces acting on Microparts: Towards a numerical approach for gripper design and manipulation strategies in Microassembly, Proceedings of the International Precision Assembly Semininar, Bad Hofgastein, Austria
21. Arai, F., Ando, D., Fukuda, T., Nonoda, Y., Oota, T., (1995), Micro manipulation based on micro physics-strategy based on attractive force reduction and stress measurement, Proceedings of the International Conference on Intelligent Robots and Systems, pp.236-241.
22. Tichem, M., Lang, D., Karpuschewski, B, (2003), A classification scheme for quantitative analysis of micro-grip principles, Proc. of the Int. Precision Assembly Semininar, Bad Hofgastein, Austria.
23. Shu-Ang Zhou, (2003), On forces in microelectromechanical Systems, International Journal of Engineering Sci. 41, 313-335.
24. Fujita H., Omodaka A., (1988), The Fabrication of an Electrostatic Linear Actuator by Silicon Micromachining, IEEE Transactions on Electron Devices, Vol 35, NO. 6, pp. 731-734
25. Hélin, P., Druon, C., Sadaune, V., (1996), A Microconveyor Using Surface Acoustic Waves in the HF Band, Proc. Mecatronics '96, 580-582.
26. Grutzeck, H., Kiesewetter, L., (2002), Downscaling of gripper for micro assembly, microsystem Twehcnologies (8), pp.27-31 ©Springer-Verlag
27. http://www.aist.go.jp/MEL/
28. Sato, K., Koyano, K., (1993), Novel Manipulator for Micro Object Handling as Interface between Micro and Human Worlds, IEEE/RSJ International Conference on Intelligent Robots and Systems
29. Ooyama Naotake et al., (2000), Desktop Machining Microfactory, Proc.2nd Int.workshop on microfactories, Switzerland, p 14-17
30. Geiger, M., Egerer, E., Engel, U., (2002), Cross Transport in a Multi-Station Former for Microparts, Production Engineering, Vol. IX, No.1, pp. 101-104.
31. Onori, M., Barata, J., Lastra, J., Tichem M., (2002), European Precision Assembly - Roadmap 2010, Assembly-Net, ISBN 91-7283-637-7.
32. H. J. Yeh and J. S. Smith, (1994), Fluidic assembly for the integration of GaAs light-emitting diodes on Si substrates, IEEE Photon. Technol. Lett., vol. 46, pp. 706–709.
33. K. Böhringer, K. Goldberg, M. Cohn, R. Howe, and A. Pisano, (1998), Parallel microassembly with electrostatic force fields, in Proc. IEEE Int. Robot. Automat. Conf., pp. 1204–1211.

34. Xiong, X., Hanein, Y., Fang, J., Wang, Y., Wang, W., Schwartz, D. T., Böhringer, K. F., (2003), Controlled Multi-Batch Self-Assembly of Micro Devices. ASME/IEEE Journal of Microelectromechanical Systems 12(2):117-127.

35. Del Corral, C., Zhou, Q., Albut, A., Chang, B, Franssila, S., Tuomikoski, S., Koivo, H.N., (2003), Droplet Based Self-Assembly of SU-8 Microparts, Proceedings of 2nd VDE World Microtechnologies Congress, MICRO.tec 2003, Germany, pp. 293-298.

36. Böhringer, K. F. Srinivasan, U., Howe, R. T., (2001), Modeling of Fluidic Forces and Binding Sites for Fluidic Self-Assembly. IEEE Conference on Micro Electro Mechanical Systems (MEMS), pp. 369-374, Interlaken, Switzerland, January 21-25,.

37. Terfort A., Whitesides, G. M., (1998), Self-assembly of an operating electrical circuit based on shape complementarity and the hydrophibic effects, Adv. Mater., vol. 10, no. 6, pp. 470–473.

38. Reinhart,G., Hohn, M., Growth into Miniaturization - Flexible Microassembly Automation, Annals of CIRP Vol. 46/1/97 p.7-10

39. Weck, M., Hümmler, J., Petersan, B., (1997), Assembly of Hybrid micro systems in a large-chamber electron microscope by use of mechanical grippers, Proc. of SPIE, Micromachining and microfabrication Process Technology III, 3223:223-229.

40. Carrozza, M.C., Dario, P., Menciassi, A., Fenu, A., (1998), Manipulating biological and mechanical microobjects using LIGA-microfabricated End-effectors, Proc. ICRA, 1811-1816.

41. Ahn, C.H., Allen, M. G., (1994), A Fully Integrated Micromachined Magnetic Particle Manipulator and separator, Proc. IEEE MEMS, pp. 91-96.

42. Lambert, P., Vandaele, V., Delchambre, A., (2004), Non contact handling in micro-assembly: state of the art, IPAS 2004, February 12-14.

43. Biganzoli, F., Fantoni, G, (2004), Contactless Electrostatic Handling of Microcomponents, Proc. Instn. Mech. Engrs. Vol 218 Part B: Journal of Engineering Manufacture, pp1795-1806.

44. Fantoni, G., Santochi, M., (2004), A contactless electrostatic device for linear movement of mini and microparts, Proceedings of the IMG04, Genova, Italy, pp. 343-348.

45. Biganzoli, F., Pagano, C., Fassi, I., (2005), A micro-manipulation system based on capillary force, to be published.

46. Fantoni, G., Santochi, M., (2005), Modular contactless feeders for mini and microparts, Annals of the CIRP, vol.54/1.

47. Gengenbach, U., Boole, J., (2000), Electrostatic feeder for contactless transport of miniature and Microparts, Microrobotics and Micro-manipulation, Proceeding of SPIE, pp. 75-81.

48. Reinhart, G., Hoeppner, J., (2000), Non-Contact Handling Using High-Intensity Ultrasonics, Annals of the CIRP Vol. 49/1/2000

49. Bancel, P.A., Cajipe, V.B., Rodier, F., Witz, J., (1998), Laser seeding for biomolecular crystallization, Journal of Crystal Growth, Vol. 191, pp. 537-544.

50. Rambin, C.L., Warrington, R.O., (1994), Micro-assembly with a focused laser beam, IEEE MEMS, pp.285-290.

51. Nakao, M., Tsuchiya, K., Matsumoto, K., Hatamura, Y., (2001), Micro Handling with Rotational Needle-type Tools Under Real Time Observation, Annals of the CIRP, vol. 50/1.

52. Bark, C., Binneboese, T., (1998), Gripping with low viscosity fluid, IEEE Int. workshop on MEMS, pp.301-305.

53. Hesselbach, J.; Büttgenbach, S.; Wrege, J.; Bütefisch, S.; Graf, C., (2001), Centering electrostatic microgripper and magazines for microassembly tasks. Microrobotics and Microassembly 3, Proc. of SPIE, vol.4568, Newton, USA.

54. Zesch, W., Brunner, M., Weber, A., (1997), Vacuum tool for handling microobjects with a NanoRobot, Proc. ICRA, 1761-1776.

55. Arai, F., Fukuda, T., (1997), A new pick up and release method by heating for micromanipulation, IEEE MEMS, 383-388.

56. El-Khouy, M., (1998), Ice gripper handles microsized component, Design News, September, 8.
57. Saitou, K., Wang D., Wou S.J., (2000) Externally Resonated Linear Vibromotor for Microassembly, Journal of Microelectromechanical systems, vol. 9, no. 3, 336-345
58. Prasad, R., Böhringer, K.-F., MacDonald, N. C., (1995), Design, fabrication, and characterization of single crystal silicon latching snap fasteners for micro assembly, in ASME Int. Mech. Eng. Congr. Expo., 917–923.
59. Shimada, E., Thompson, J.A., Yan, J., Wood, R., Fearing, R.S., (2000), Prototyping Millirobot using dexterous microassembly ad folding, Proc. ASME IMECE/DSCD November 5-10. More details http://robotics.eecs.berkeley.edu/~eshimada/micro/index.html
60. Arai, F. Andou, D. Nonoda, Y. Fukuda, T. Iwata, H. Itoigawa K., (1996), Micro endeffector with micro pyramids and integrated piezoresistive force sensor, Proceedings of the 1996 IEEE/RSJ International Conference on Intelligent Robots and Systems.
61. Thompson, J.A., Fearing, R.S., (2001), Automating Microassembly with Ortho-tweezers and Force Sensing, IROS 2001, Maui, HI.
62. Karl F. Böhringer, Srinivasan, U., Roger T. Howe, (2001), Modeling of Fluidic Forces and Binding Sites for Fluidic Self-Assembly. IEEE Conference on Micro Electro Mechanical Systems (MEMS), Interlaken, Switzerland, 369-374.
63. http://www.ee.washington.edu/research/mems/selfassembly/
64. Dechev, N., Cleghorn, W. L., Mills, J. K., (2003), Construction of a 3D MEMS Microcoil Using Sequential Robotic Microassembly Operations, Proceedings ASME International Mechanical Engineering Congress and R&D Expo 2003, Washington, D.C, Nov 15-21.
65. Han, H., Weiss, L. E., Reed, M.L., (1991), Design and Modeling of a Micromechanical Surface Bonding System, Transducers, p. 974-977.
66. Cohn, M. B., Böhringer, K. F., Noworolski, J. M., Singh, A., Keller, C. G., Goldberg, K. Y. Howe, R.T., Microassembly technologies for MEMS, in Proc. SPIE Microfluid. Devices Conf., vol. 3515, Santa Clara, CA, Sept. 1998, pp. 2–16.
67. Moesner, F.M., Higuchi T., Electrostatic Devices for Particle Microhandling, (1999), 530 IEEE Transactions on Industry Applications, Vol. 35, No. 3.
68. Konishi, S., Fujita, H., (1994), A conveyance system using air flow based on the concept of distributed micro motion systems, journal of micromechanical Systems; Vol.3, No.2, pp.54-58.
69. http://www.ind.tno.nl
70. Fahlbusch, S.; Fatikow, S.; Seyfried, J.; Buerkle, A., (1999), Flexible microrobotic system MINIMAN: design, actuation principle and control 1999. International Conference on Advanced Intelligent Mechatronics, Proceedings, IEEE/ASME, 156-161.
71. Lee, J., Moon, H., Fowler, J., Chang-Jin K., Schoellhammer, T., (2001), Addressable micro liquid handling by electric control of surface tension, The 14th IEEE International Conference on Micro Electro Mechanical Systems, 499-502.
72. Han, H., Weiss, L. E., Reed, M.L., (1991), Design and Modeling of a Micromechanical Surface Bonding System, Transducers, 974-977
73. Ata, A., Y. I. Rabinovich, R. K. Singh, (2002), Role of surface roughness in capillary adhesion, Journal of Adhesion Science and Technology, vol. 16, no. 4, 337-346
74. Marmur, A., (1993), Tip-Surface Capillary Interactions, Langmuir, vol. 9, 1922-1926.
75. Grutzeck, H., Kiesewetter, L., (2002), Downscaling of grippers for micro-assembly, Microsystem Technologies,n° 8, pp 27-31
76. Höhn, M., Robl, C., (1999), Qualification of standard industrial robots for micro-assembly, in Proc. of the IEEE International Conference on Robotics & Automation (ICRA'99), Detroit, USA.

IMPROVING PRODUCTIVITY IN INTERRUPTED FINISH TURNING OF Ti6Al4V

K. Sørby[1], K. Tønnessen[2]

[1] Department of Production and Quality Engineering, Norwegian University of Science and Technology, Trondheim, Norway
[2] SINTEF Technology and Society, Production Engineering, Trondheim, Norway

KEYWORDS: Interrupted Turning, Titanium, High-Pressure Cooling.

ABSTRACT. The paper presents results from machining tests that are carried out to investigate the type of tool wear that occurs in interrupted cutting of Ti6Al4V under different process conditions. It has been found that the type of wear is similar to the wear in continuous cutting, but the wear develops much faster. The tests show that by using high-pressure coolant (30 MPa), the cutting speed can be increased by 40 to 50 percent. Edge preparation of cutting tools is of great importance. The best results are achieved with a cutting edge that is slightly rounded. Carbide tools are recommended for the turning operations. Tests show that PCD, CVDD, and CBN have limited usefulness in interrupted cutting.

1 INTRODUCTION

The paper presents results from a co-operative project involving the Norwegian University of Science and Technology, SINTEF, and the company Volvo Aero Norge AS that produces components for the aircraft engine industry. In the manufacture of aircraft engines, turning accounts for up to 85 % of the machining time [1]. Turning operations in titanium are characterized by long machining time and large tool consumption, especially in interrupted cutting.

Ti6Al4V is the most common titanium alloy, comprising about 50 % of the total titanium production. The alloy is widely used in aerospace engine components, due to its high strength-to-weight ratio, good corrosion resistance and heat resistance. While the efficiency of most machining operations has increased dramatically during the last years, this is not true for titanium turning.

In turning of jet engine components typical finishing cuts are undertaken with depth-of-cut $a_p = 0.25 - 0.30$ mm, and feed $f_n = 0.15$ - 0.20 mm/rev. It is often necessary to use tools with an entering angle (tool cutting-edge angle, κ_r) larger than 90° and nose radius $r_\varepsilon = 0.8 - 1.2$ mm due to the component geometry. This insert geometry is advantageous with respect to low cutting forces, but it is less desirable than for example round inserts with respect to tool life due to higher temperature at the tool nose.

In turning operations, interrupted cutting is undesired due to mechanical shock and thermal fatigue. Interrupted turning of titanium must be performed with low cutting speed in order to achieve acceptable tool life. This paper presents results from a study on methods for enhancing the productivity of this type of turning operations. The workpiece geometry and the cutting data

Published in: E. Kuljanic (Ed.) *Advanced Manufacturing Systems and Technology*, CISM Courses and Lectures No. 486, Springer Wien New York, 2005.

used in the machining tests are typical for a turning operation in the manufacturing of jet engine components.

2   CUTTING TEST DATA

The purpose of the machining tests presented in this paper is to investigate the tool wear mechanisms in interrupted titanium turning, and to test methods to enhance the productivity. Tests are performed with micrograin carbide tools, diamond tools, and cubic boron nitride tools. The tests are carried out with conventional flood flushing and with the use of high-pressure coolant supply.

Flank wear and tool chipping were measured and registered during the tests. Details on the experimental data are shown in Table 1.

TABLE 1. Experimental data

Machine tool	Vertical turning lathe, Hessapp DV 80
Cutting unit	Coromant Capto C6 PCLNR-45065-12 Entering angle: 95°
High-pressure nozzles	2 nozzles, $d = 0.6$ mm Distance from cutting edge: 60 mm

Inserts	Geometry: Rhombic insert 80°, Insert clearance angle = 0°, Nose radius = 0.8 mm
	
	Uncoated carbide: CNMG 120408, SECO grade 883 (K20-K30)
	                               CNMG 120408, SECO grade 890 (K10-K20)
	                               CNGG 120408, SECO grade 890 (K10-K20)
	Polycrystalline diamond, PCD: CNMS 120408F, Kennametal grade KD100
	CVD diamond, CVDD: CNMS 120408E, Kennametal grade KD1450
	Cubic boron nitride, CBN: CNGA 120408S, Kennametal grade KD050

Test piece	Rigid workpiece, $d = 580$ mm with 6 radial slots (10 mm wide) Ti6Al4V, HRC 36
Operation	Face turning, Depth-of-cut: $a_p = 0.25$ mm Feed: $f_n = 0.20$ mm/rev Cutting speed: $v_c = 35 - 70$ m/min Comparative tool life tests are performed on equal diameter of the test piece.
Cutting fluid	Cimstar 560, Emulsion 4 %
Coolant pressure	Flood flushing: 2 nozzles, 0.7 MPa (flow rate: 26 l/min) High-pressure coolant: 2 nozzles, 15 - 30 MPa

3 TEST RESULTS, CARBIDE TOOLS

3.1 INITIAL TESTS: CARBIDE TOOLS WITH CONVENTIONAL FLOOD FLUSHING

The purpose of the initial tests was to study the tool wear characteristics in interrupted titanium turning, and to compare the tool wear rate of continuous and interrupted turning. As expected, the tool wear is large in interrupted cutting. No catastrophic tool failure occurred with the carbide tools, but small chipping was observed. At the start the flank wear is slow and equal along the tool edge. When flank wear reaches $VB_{max} \approx 0.12$ mm the wear rate increases very fast. Figure 1 shows the flank wear for a carbide tool immediately before and after the end of tool life. Except for the rapid wear rate, the tool wear characteristics of interrupted cutting are quite similar to characteristics found in continuous cutting for the chosen cutting data.

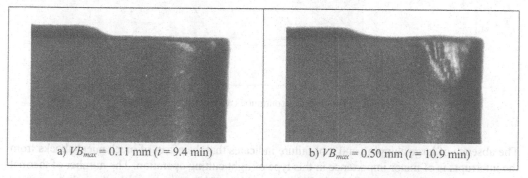

a) $VB_{max} = 0.11$ mm ($t = 9.4$ min) b) $VB_{max} = 0.50$ mm ($t = 10.9$ min)

FIGURE 1. Flank wear of a carbide insert a) before and b) after the end of tool life

The relation between cutting speed and tool life is commonly expressed by the Taylor-equation, $v_c T^n = C$. In tests carried out by the authors [2] the exponent n is found to be 0.17 in continuous turning of Ti6Al4V, which is a low value, compared to turning of steels and other common materials. A low value of the exponent n expresses that the tool life is very sensitive to changes in the cutting speed. The results presented in Figure 2 shows that n is small ($n = 0.19$) also in interrupted turning of Ti6Al4V.

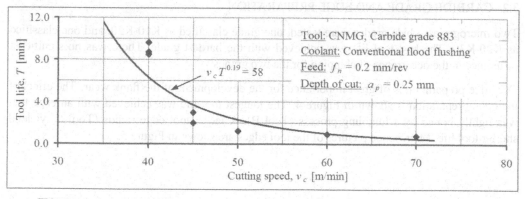

FIGURE 2. Tool life of carbide tools in interrupted turning. Tool life criterion: $VB_{max} = 0.2$ mm

Additional tests with continuous turning were carried out and compared with interrupted turning. The results are shown in Figure 3. The difference in tool life for the two cases is considerable. The tool wear tests at continuous turning at 70 m/min were stopped after 20 minutes to conserve test material.

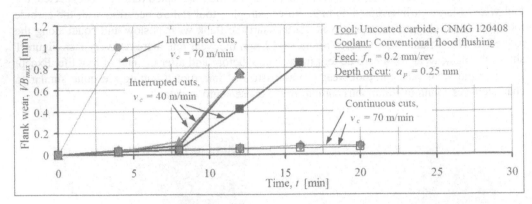

FIGURE 3. Tool wear in continuous and interrupted turning

The absence of large chipping and tool failure indicates that the effect of mechanical shocks from the interrupts is of minor importance in this type of light cutting operation. The number of thermal cycles is relatively low, approximately 2700 for tool life of 10 min at $v_c = 40$ m/min. It is assumed that thermal fatigue has a small effect on the tool wear.

Most probably the large tool wear is caused by the cyclic formation and tearing off of build-up material on the tool. By this mechanism particles are removed from the tool material. It can be assumed that the mechanism is particularly strong in interrupted cutting, due to the cyclic welding and cooling. Adhesion tests carried out by Nabhani [3] showed that the welding joint is stronger than the tool material, and that the rupture therefore will occur inside the tool material.

3.2 CARBIDE GRADE AND EDGE PREPARATION

Two micrograin carbide grades were tested, one grade classified as K10-K20, and one classified as K20-K30. The best tool life was achieved with the hardest grade. There was no significant difference in the occurrence of chipping for the two grades.

The edge preparation is of great importance for the development of the flank wear. The effect of the edge preparation is shown in Figure 4. The longest tool life was achieved with an edge that was lightly honed by a tumbling process (Tool B). A larger tool edge radius (Tool C) yielded shorter tool life. Microscope pictures of the tool edges are shown in Figure 5.

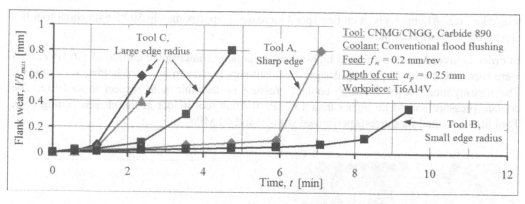

FIGURE 4. Tool wear for inserts with different edge preparations

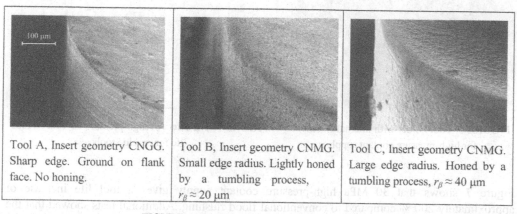

| Tool A, Insert geometry CNGG. Sharp edge. Ground on flank face. No honing. | Tool B, Insert geometry CNMG. Small edge radius. Lightly honed by a tumbling process, $r_\beta \approx 20\ \mu m$ | Tool C, Insert geometry CNMG. Large edge radius. Honed by a tumbling process, $r_\beta \approx 40\ \mu m$ |

FIGURE 5. Edge preparation of the tested inserts

As shown by Bouzakis et al. [4] tool-chip contact length is large for large edge radius. Similarly, large edge radius yields large contact length between tool flank and workpiece. Therefore, larger tool edge radius results in increased cutting forces and increased heat generation near the tool edge, which yields faster tool wear. It could be mentioned that the demand for relatively small edge radius is one of the factors that explains why coated tools in general are not better than uncoated tools in titanium turning.

3.3 THE EFFECT OF HIGH PRESSURE COOLANT ON TOOL LIFE, CARBIDE TOOLS

The technique of using high-pressure coolant to enhance tool life in titanium turning is described in several publications [1, 2, 5, 6], and to a certain extent the technique is used in industrial applications. In continuous turning of Ti6Al4V a tool life increase of typically 300 - 400 % can be observed when applying high-pressure cutting fluid (10 - 30 MPa) between the chip and the tool rake face. Alternatively, a cutting speed increase of approximately 50 % with equal tool life is possible in these examples.

In order to investigate the effect of high-pressure coolant in interrupted cutting of Ti6Al4V, tests were carried out with a specially designed coolant supply system and tool holder, see Figure 6. The investigation shows that high coolant pressure is desirable with respect to tool life. The coolant pressure should be higher than 15 MPa for this tool holder at the chosen cutting data. Tool life was equal for the tests performed with 30 and 40 MPa.

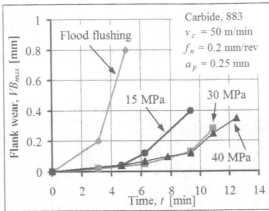

FIGURE 6. Tool holder with internal channels for high-pressure coolant supply

FIGURE 7. The effect of high-pressure coolant on tool life for carbide tools

Figure 7 shows that 30 MPa high-pressure coolant supply gives a tool life increase of approximately 200 % compared to conventional flood flushing. Additional tests showed that the tool life at cutting speed $v_c = 35$ m/min with conventional fluid flushing is shorter than tool life at $v_c = 50$ m/min with 30 MPa high-pressure coolant. From these results it can be concluded that high-pressure cooling make possible a cutting speed increase of more than 43 % at equal tool life.

4 TEST RESULTS WITH DIAMOND AND CUBIC BORON NITRIDE TOOLS

Additional tests were carried out to investigate the possibility of using new superhard tool materials in interrupted cutting of Ti6Al4V. The tests included polycrystalline diamond (PCD), chemical vapour deposition diamond (CVDD), and cubic boron nitride (CBN). Tests were carried out with conventional flood flushing and high-pressure coolant in the range 15 - 30 MPa. The tool wear as a function of travelled distance (spiral cutting length) is plotted in Figure 8.

The best results were achieved with polycrystalline diamond, PCD. However, the tool life was not considerably better than the tool life observed with micrograin carbide, grade 890. The tool wear of PCD was characterized by a fast initial wear, followed by a relatively long steady-state stage.

This tool wear characteristic indicates that the initial edge preparation geometry is not optimal for this type of turning operation.

The CBN inserts showed a disappointing performance in the tests. Catastrophic failure occurred almost immediately. Possible explanations for this result are the undesirable negative rake angle and nonoptimal edge preparation. The observed tool life results for PCD, CVDD, and CBN corresponds well to results from continuous cutting tests performed by Hoffmeister and Wessels [7]. The results also agree with Hartung and Kramer [8] who found that PCD was more wear resistant than carbide, and that CBN was less wear resistant.

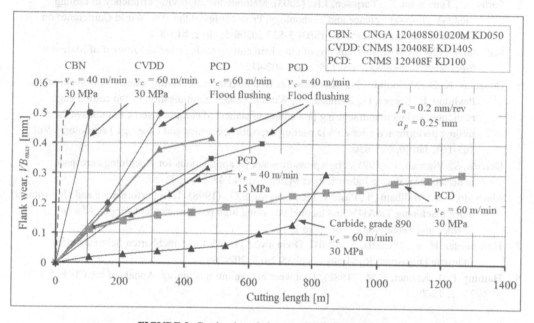

FIGURE 8. Cutting length for superhard tool materials

5 CONCLUSIONS

In finishing turning with interrupted cuts, low cutting speeds must be used in order to achieve acceptable tool life and cutting length.

The test results presented in this paper show that optimal edge preparation and efficient cooling are important factors for high productivity in interrupted titanium turning. Good results were achieved with a carbide tool that was lightly honed by a tumbling process, to edge radius $r_\beta \approx 20$ µm. The tests show that use of high-pressure coolant makes possible a cutting speed increase of 40 - 50 % compared to conventional flood flushing. PCD, CVDD, and CBN have limited usefulness in interrupted cutting.

At the moment, it seems that the most important research topic for increased productivity is the high-pressure cooling technique. For example, enhanced cooling by high-pressure coolant on both tool rake face and tool flank face is one possible method that should be investigated further.

ACKNOWLEDGEMENTS

The authors thank The Research Council of Norway for supporting this work.

REFERENCES

1. López de Lacalle, L.N., Pérez-Bilbatua, J., Sánchez, J.A., Llorente, J.I., Gutiérrez, A., Albóniga, J., (2000), Using high pressure coolant in the drilling and turning of low machinability alloys, International Journal of Advanced Manufacturing Technology, Vol 16, No 2, 85-91.
2. Sørby, K., Tønnessen, K., Torjusen, J.E., (2003), Methods for improving efficiency in turning Ti6Al4V, Ti-2003 Science and Technology, Proceedings of the 10th World Conference on Titanium, Hamburg, Germany, ISBN 3-527-30306-5, No 2, 815-822.
3. Nabhani, F., (2001), Wear mechanisms of ultra-hard cutting tools materials, Journal of Materials Processing Technology, Vol 115, No 3, 402-412.
4. Bouzakis, K.-D., Michailidis, N., Skordaris, G., Kombogiannis, S., Hadjiyiannis, S., Efstathiou, K., Pavlidou, E., Erkens, G., Rambadt, S., Wirth I., (2003), Optimisation of the cutting edge roundness and its manufacturing procedures of cemented carbide inserts, to improve their milling performance after a PVD coating deposition, Surface and Coatings Technology, Vol 163-164, Jan 30, 625-630.
5. Derrien, S., Vigneau, J., (1995), High pressure water jet applications for machining aerospace materials, Mécanique Industrielle et Matériaux, Vol 48, No 1, 31-34.
6. Machado, A.R., Wallbank, J., Pashby, I.R., Ezugwu, E.O., (1998), Tool performance and chip control when machining Ti6Al4V and Inconel 901 using high pressure coolant supply, Machining Science and Technology, Vol 2, No 1, 1-12 .
7. Hoffmeister, H.-W., Wessels, T., (2001), Drehen von TiAl6V4 mit hochharten Schneidstoffen, Industrie Diamanten Rundschau, Vol 35, No 3, 202-208.
8. Hartung, P.D., Kramer, B.M., (1982), Tool wear in titanium machining, Annals of the CIRP, Vol 31, No 1, 75-79.

CUTTING PROCESS OPTIMIZATION PRACTICAL PROCEDURE

N. L. Coppini, E. A. Baptista

Production Engineering Post-Graduation Program, Methodist University of Piracicaba, Brazil

KEYWORDS: Machining, Optimization, Cutting Process.

ABSTRACT. Cutting process optimization could be considered in different levels and approaches. How to select tools and how to take care of evident waste of time and costs can not be considered as a competitiveness market factor any more. Everyone responsible for cutting process planning must be up to date with these procedures and the tool makers are ready to help their customers. Otherwise, after the best selected tool and cutting conditions are settled and the process is running in shop floor many things could be done to go deeper in the cutting process optimization. One of them is how to consider the cutting process system and scenario influence on cutting condition, more specifically, on the cutting speed, which is usually to select from data published in catalogues and normally based on the pair tool/workpiece. The purpose of this paper is to present and discuss, step by step, a mathematical method and practical procedure to search the best cutting condition in different industrial scenarios.

1 INTRODUCTION

With With a market more and more competitive, the need to practice a more efficient manufacture, with right decisions based on more precise information, becomes crucial for the existence of any manufacturing industry.

Specifically, in the cutting processes, the optimization of the cutting parameters has already been accomplished for a long time, based mainly on the reduction in time of production, that is gotten with the application of the correct tool, with advances and depth of cutting applicable at the machine-tool-workpiece system [1, 2].

The optimization of the cutting speed also allows to obtain a larger rate of chip removal and consequently, it reduces the time of machining. However, the increase of the cutting speed has a negative effect on the tool life, increasing the machining time, due to the need of more frequent tool changes.

The cutting speed optimization can be based on the Maximum Efficiency Interval, MEI, in order to select the optimized cutting speed [3]. However, the use of the (x) and (K) values, exponent and constant of the Taylor's tool life equation, obtained from the literature or other sources, as a laboratory. The test, done in ideal conditions, can make the cutting speed optimization a harder task, because they are different from the real machine-tool-workpiece system, causing mistakes in all cutting speed optimization.

Published in: E. Kuljanic (Ed.) *Advanced Manufacturing Systems and Technology*,
CISM Courses and Lectures No. 486, Springer Wien New York, 2005.

Another option is the use of these constants obtained from real tests done in shop floor. In this case, the MEI determination will be more trustful, because the optimization will be accomplished in real time with the process, obtained data straight the machine-tool-workpiece system.

Usually, the cutting parameters optimization in machining is limited to the correct choice of the tool with advances and depth of cutting, and cutting speed indicated by sources that publish the experience accumulated in the subject (generally manufacturing catalogs) [2]. The main concern, in optimization terms, is machining the workpiece respecting the design specifications. The practical reports that show a concern of allying to the care above mentioned are not frequent. This way, operational conditions of machining are used by means of inaccurate data application, in spite of being notorious the fact that this production process depends greatly on a number of influence factors such as the practice is. For this reason, generator of possible great mistakes which are perpetuated in shop floor, producing and adding eventual great damages that stay as if they did not exist. Thus, the optimization more frequently found in the machining practice, is based on " adopting the machining conditions", managing the passive time, as the machine operator would have to change the tool or recovery the tiredness.

Among the cutting parameters, the depth of cutting and the feed are the most easily optimized: it can work with values close to the ideal considering the power of the machine, the dimension of the material to be removed, the form of the chip and the final quality of the workpiece demanded for the specific operation goal of the optimization.

However, the cutting speed besides exercising more significant influence than the feed and the depth of cutting about the tool life, also has more influence in the cutting time and in the machining cost considering the same operation. For the fact of being possible to cut relatively with a large range of cutting speed, different wear mechanisms can exist, resulting different behavior of the tool life. During the machining, different values of the cutting speed can imply in different times and total machining costs for the workpiece. These are consequences of the combined influence of the corresponding cutting time and the tool change time.

As consequence of the above exposed, it is evidenced that the knowledge of the optimized cutting speed for a machine-tool-workpiece system can only be acquired after the accomplishment of optimizations that consider data obtained in process.

The present work presents a practical procedure, that can be used in shop floor, that allows to optimize the cutting process adapting the cutting speed based on the MEI, which is defined with the real values of (x) and (K) obtained from the machine-tool-workpiece system.

For this the practical procedure to optimize the cutting process is presented in this paper, together with the analysis of the influence of some variables, of the productive environment, that take action directly in its results. The adaptations of the practical method to the different productive scenery are also presented.

2   PRATICAL PROCEDURE TO OPTIMIZE CUTTING PROCESS

To cutting parameters optimization, it was used of the Maximum Efficiency interval, MEI, already consecrated by the specialized literature [1]. That is the interval among the minimum cost

(V_{cmc}) and maximum production (V_{cmxp}) machining condition, as described below in the equations 1 e 2.

$$V_{cmxp} = \left\{ K / \left[(x-1) * T_{ft} \right] \right\}^{(1/x)} \tag{1}$$

$$V_{cmc} = \left\{ K * (S_h + S_m) / 60 * (x-1) * \left[K_{ft} + ((S_h + S_m) * T_{ft}) \right] \right\}^{(1/x)} \tag{2}$$

Where (K) and (x) are constants and coefficient of the Taylor's tool life equation, T_{ft} is the tool change time, S_h and S_m they are respectively, operator's wage and machine cost both per hour and K_{ft} are the tool cost.

The optimized cutting speed, should have as reference the MEI references as V_{cmxp}, V_{cmc} or V_{cmclim}, which is the cutting speed of minimum cost limit, as showed in the equation 3. This definition should respect the productive scenery, that depending on its configuration, it indicates which cutting speed should be used, therefore the use of the V_{cmxp} as reference to optimization, results in a high production with the smallest possible time, but with a relatively high cost, being above the minimum cost. The use of the cutting speeds tending to V_{cmc}, originate a production relatively smaller with the larger machining time, when compared with the V_{cmxp}, even so the production cost is the smallest as possible.

$$V_{cmclim} = \left\{ K * (S_h + S_m) / 60 * (x-1) * K_{ft} \right\}^{(1/x)} \tag{3}$$

Other factors that act directly in the calculation of V_{cmxp}, V_{cmc} and V_{cmclim}, influencing the optimized cutting speed, is the tool change time and costing system [3,4].

When the tool change time tends to a small value, even to zero, causes cutting speed on high values, and some problems can occur by the technical limits as the machine power and cutting speed availability, for example.

In this case V_{cmclim} can be used [5], as reference, because the tool change time is so small that its product for the sum of the operator's wage plus machine cost is worthless in relation to the initial tool cost (see equations 2 and 3). This value will always be an appropriate reference, because, it will never be smaller than V_{cmc} and it will always be smaller than V_{cmxp}, for the small tool change time, even so, different from zero. Therefore V_{cmclim} belongs to MEI, being between V_{cmc} and V_{cmxp}.

The cost system also has direct influence in the determination of MEI, because of the definition of V_{cmc} and V_{cmclim} values. It uses, among others, the operator's wage value and machine cost. The cost system should determine these values accurately, because, to the opposite, the results will be affected. Therefore the use of V_{cmc} or V_{cmclim} should be in concomitance with the Activity Based Cost System, ABC [4, 5].

The machining optimizing methodology consists on the determination in shop floor of (x) and (K) values from Taylor's tool life equation, described in the equations 4 and 5. The machining is executed for a peculiar number of workpieces, with cutting speed selected in catalogs. Soon after the tool life as well the cutting time for each speed are measured during the production, until that the system of two equations for two unknown quantities (4 and 5) allowing calculate the values of the (x) and (K) values above mentioned.

$$x = \left[\log(Z_1 / Z_2) / \log(V_{c2} / V_{c1}) \right] + 1 \tag{4}$$

$$K = Z_1 * V_{c1} * T_{c1} \tag{5}$$

Where, Z_1 is the number of workpieces, V_{c1} is the cutting speed value, and T_{c1} is the cutting time for the cutting speed V_{c1}. The index (2) represents a different situation from the situation (1).

The tool life can be measured in time (minutes), effective cutting length and number of workpieces. In all the cases and for each test accomplished to obtain the tool life, the significant size of the sample can be determined by the application from Student's distribution, conferring more reliability to the obtained results.

In the scenery where there are quality problems or variation of material properties, both the workpiece and the tool, the procedure here described when applied, cannot produce good results.

2.1 INFLUENCE OF VARIABLES OF PROCESS IN MEI

For the minimum cost cutting speed (V_{cmc}) analysis, the optimization results accomplished by Malaquias will be used [6]. Variations will be implemented in the costs and in the tool change time, T_{ft}. The data and results of the accomplished optimization were: tool cost, K_{ft} = R\$ 3,28, T_{ft} = 0,58 min; operator's wage, S_h = R\$13,60; machine cost, S_m = R\$15,00; first cutting speed used in the test, V_{c1} = 175 m/min; tool life in number of workpieces for V_{c1}, Z_1 = 16; second cutting speed used in the test, V_{c2} = 210 m/min; tool life for V_{c2}, Z_2 = 9. With the mentioned data it was calculated the MEI, which values are: V_{cmc} = 175 m/min; and cutting speed maximum production, V_{cmxp} = 325 m/min.

Based on those results, different situations were analyzed. In the first, only K_{ft} value was changed in ±10% in relation to the real situation. Table 1 presents the results obtained with this procedure. It must be observed that the line in prominence refers to the real condition. Figure 1 refers to the graph elaborated from the Table 1 data.

In the second situation operator's wage and machine cost values were considered together (S_h + S_m). Again it was made a ±10% variation in relation to the real situation. The other parameters were kept unaffected. Table 2 presents the obtained results. It must be observed that the line in prominence refers to the real condition. Figure 2 refers to the graph elaborated from Table 2 data.

Being analyzed the Table 1, with its respective graph (Figure1), it is possible to observe, in the case used as example, that V_{cmc} is not very sensitive (variation corresponding ~2%) due to the alterations of K_{ft}, value at least for a ±10% variation of this parameter. For a ±10% variation on the sum of the (S_h + S_m) values (Table 2 and Figure 2), it is observed that V_{cmc} also presents little around 2% variation.

After analyzing the behavior of V_{cmc} in function to the alteration of the values of the involved costs, the influence of T_{ft} was verified. For this, it was also varied the initial value of T_{ft} in ±10% and the other values were maintained unaffected. The results are described in the Table 3, whose graph is represented by the Illustration 3.

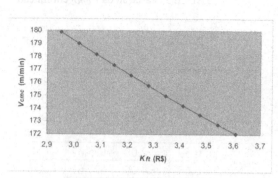

FIGURE 1. Table 1 Chart

TABLE 1. K_{ft} Variation

K_{ft}	%	V_{cmc}	%
2,95	- 10,00	179,88	+ 2,36
3,02	- 8,00	179,01	+ 1,86
3,08	- 6,00	178,16	+ 1,38
3,15	- 4,00	177,33	+ 0,91
3,21	- 2,00	176,53	+ 0,45
3,28	0,00	175,74	0,00
3,35	+ 2,00	174,97	- 0,44
3,41	+ 4,00	174,21	- 0,87
3,48	+ 6,00	173,47	- 1,29
3,54	+ 8,00	172,75	- 1,70
3,61	+ 10,00	172,05	- 2,10

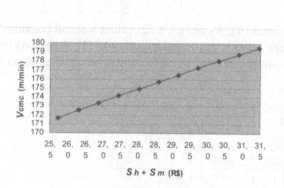

FIGURE 2. Table 2 Chart

TABLE 2. $(S_h + S_m)$ Variation

$(S_h + S_m)$	%	V_{cmc}	%
25,74	- 10,0	171,66	+ 2,36
26,31	- 8,0	172,51	+ 1,86
26,88	- 6,0	173,34	+ 1,38
27,46	- 4,0	174,15	+ 0,91
28,03	- 2,0	174,95	+ 0,45
28,60	0,0	175,74	0,00
29,17	+ 2,0	176,51	- 0,44
29,74	+ 4,0	177,27	- 0,87
30,32	+ 6,0	178,02	- 1,29
30,89	+ 8,0	178,76	- 1,70
31,46	+ 10,0	179,48	- 2,10

Based on the results described in Table 3, it is possible to verify that V_{cmc} also presents a small 2% variation, for ±10% variation in T_{ft}.

After the three analyses accomplished until the moment, it still fits a question: What can happen if the K_{ft}, $(S_h + S_m)$ and T_{ft} values are changed at the same time? Tables 4 and 5 were elaborated to give the answer.

In Table 4 the T_{ft}, K_{ft} and $(S_h + S_m)$ values were changed simultaneously and gradually in ±2% up to the difference among the ends reached ±10%. Table 5 presents an randomized combination among the T_{ft}, K_{ft} and $(S_h + S_m)$ real values and the extreme values presented in Table 4.

When analyzing Table 4 it is possible to notice that V_{cmc} has practically stable behavior (±0,19%).

TABLE 3. Variação da V_{cmc} em função do T_{ft}

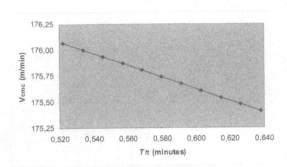

t_{ft}	%	V_{cmc}	%
0,52	- 10,00	176,07	+ 2,36
0,53	- 8,00	176,00	+ 1,86
0,55	- 6,00	175,94	+ 1,38
0,56	- 4,00	175,87	+ 0,91
0,57	- 2,00	175,80	+ 0,45
0,58	0,00	175,74	0,00
0,59	+ 2,00	175,67	- 0,44
0,60	+ 4,00	175,61	- 0,87
0,61	+ 6,00	175,54	- 1,29
0,63	+ 8,00	175,48	- 1,70
0,64	+ 10,00	175,41	- 2,10

In Table 5, however, the maximum and minimum T_{ft}, K_{ft} and $(S_h + S_m)$ values were all combined each other. It is possible to verify that V_{cmc} can suffer larger variations, up to 4,75%, depending on the considered combination. (this behavior can be explained by mathematical procedures, but this will not be showed in this work).

TABLE 4. Behavior of V_{cmc} due to alterations in values do T_{ft}, k_{ft} e $(S_h + S_m)$

%	T_{ft}	k_{ft}	(s_h+s_m)	V_{cmc}	%
- 10,00	0,52	2,95	25,74	176,07	+ 0,19
- 8,00	0,53	3,02	26,31	176,00	+ 0,15
- 6,00	0,55	3,08	26,88	175,94	+ 0,11
- 4,00	0,56	3,15	27,46	175,87	+ 0,07
- 2,00	0,57	3,21	28,03	175,80	+ 0,04
0,00	0,58	3,28	28,60	175,74	0,00
+ 2,00	0,59	3,35	29,17	175,67	- 0,04
+ 4,00	0,60	3,41	29,74	175,61	- 0,07
+ 6,00	0,61	3,48	30,32	175,54	- 0,11
+ 8,00	0,63	3,54	30,89	175,48	- 0,15
+ 10,00	0,64	3,61	31,46	175,41	- 0,19

2.2 T_{ft} INFLUENCE IN V_{cmxp}

The V_{cmxp} does not depend on the production costs, for this reason, only the T_{ft} influence was analyzed. Table 6 and its graph (Figure 4) show the results.

TABLE 5. V_{cmc} behaviour in function of t_{ft}, k_{ft} e $(S_h + S_m)$ values

T_{ft}	$(s_h + s_m)$	k_{ft}	V_{cmc}	%	T_{ft}	$(s_h + s_m)$	k_{ft}	V_{cmc}	%
0,52	25,74	2,95	176,07	+ 0,19	0,58	31,46	3,28	179,48	+ 2,13
0,52	25,74	3,28	171,95	- 2,15	0,58	31,46	3,61	175,74	0,00
0,52	28,60	2,95	180,25	+ 2,57	0,58	25,74	3,61	168,03	- 4,39
0,52	28,60	3,28	176,07	+ 0,19	0,58	31,46	2,95	183,67	+ 4,52
0,52	25,74	3,61	168,29	- 4,24	0,64	31,46	3,61	175,41	- 0,19
0,52	31,46	2,95	184,09	+ 4,75	0,64	25,74	3,61	167,77	- 4,53
0,52	31,46	3,61	176,07	+ 0,19	0,64	31,46	2,95	183,26	+ 4,28
0,52	28,60	3,61	172,34	- 1,93	0,64	25,74	2,95	175,41	- 0,19
0,52	31,46	3,28	179,85	+ 2,34	0,64	31,46	3,28	179,12	+ 1,92
0,58	28,60	3,28	175,74	0,00	0,64	28,60	3,61	171,75	- 2,27
0,58	25,74	3,28	171,66	- 2,32	0,64	28,60	3,28	175,41	- 0,19
0,58	28,60	2,95	179,88	+ 2,36	0,64	25,74	3,28	171,37	- 2,49
0,58	25,74	2,95	175,74	0,00	0,64	28,60	3,61	171,75	- 2,27
0,58	28,60	3,61	172,05	- 2,10					

FIGURE 4. Table 6 Chart

TABLE 6. V_{cmxp} behaviour in function of t_{ft}

T_{ft}	%	V_{cmxp}	%
0,522	- 10,00	333,29	+ 1,03
0,534	- 8,00	331,53	+ 1,02
0,545	- 6,00	329,82	+ 1,02
0,557	- 4,00	328,16	+ 1,01
0,568	- 2,00	326,53	+ 0,49
0,580	0,00	324,95	0,00
0,592	+ 2,00	323,40	- 0,48
0,603	+ 4,00	321,90	- 0,94
0,615	+ 6,00	320,42	- 1,39
0,626	+ 8,00	318,99	- 1,83
0,638	+ 10,00	317,58	- 2,27

For the analyzed situation it can be verified that V_{cmxp} suffers a small change, around 2%, when Tft varies in ±10%. Obviously, if the Tft value suffers larger reductions, where its value can be practically or even zero, the Vcmxp value will tend to increase significantly, surpassing the limit of the machine-tool-workpiece system. In this case, this limit determines the maximum production condition. This fact is frequently unknown in the productive industrial environment, because it is very common to find operational conditions where cutting speed is practiced with value below the limit of the system and sometimes besides smaller than the minimum cost speed. The optimization methodology in analysis allows that such misunderstandings are avoided and that at least the production can happen with cutting speeds larger than the minimum cost condition.

2.3 APLICABILITY OF METHODOLOGY IN PRODUCTIVE SCENARIES

The scenery was classified by the batch size, the workpieces diversity, the material type and geometric forms. Thus, the following scenery is identified as:

a) mass production – this scenery has a big batch size, which allow that the machining tests, to (x) and (K) determination, be accomplished a lot of times, as much as necessary. This is the ideal scenery to apply the optimization methodology proposed in this work. In it, the MEI can be defined more accurate and the statistical validation can be done easily;

b) flexible flexible production where workpieces have the same material and similar geometry – though flexible this scenery allows that all workpieces be treated like just one batch. In this case, the total workpieces quantity permits the tests to be accomplished like the mass productio;

c) flexible production where workpieces have the same material, but different geometry – this scenery will be treated in the same way of the last scenery. However, in function of the difference in geometry, the results can be influenced by the tool wear down variation. Although this distortion possibilities on the result this scenery will be treated like the last scenery. But, in this case, the tool life must be expressed in time units, due the difference in the geometry. The statistical validation can be followed by a bigger dispersion when the cutting speed optimized will be used. This dispersion occurs in function of the sequence of the different geometry in the batch, although of the pair tool/material is keep constant;

d) flexible production where the material and geometry of the workpieces are different, but that use the same tool to machining the same operation – this scenery have one more complicated factor when compared with others, described before, because the sequence of the workpieces can vary during the tests execution. However, it allows that the number of the tool spend be managed, calculating the tool life share consumed for each workpiece using the x and k values, which can be obtained from literature. In this case their reliable is delicate. But also, they can by obtained in the machining process development, using the optimization methodology proposed in this work, which is more trustful. In function of the tool wear down variation, caused by the different material in process, the results can show some variation. Thus, this approach gives us conditions to specify the number of the tool that will be necessary to machine all batch of workpieces, and the operator needs to report the real number of the tool spent. The statistical validation, in this case, is damaged, so, the results must be used like an orientation to the operator, who will know the tool replacement time. The advantage of this procedure is to void the responsibility about the tool replacement of the operator.

3 ACTUAL WORKS

Actually the authors are working in the web based system development, which will be used to aid the cutting process optimization accomplishment in shop floor. The procedure described in this work was used as reference in the system development.

The referred system can be used by Internet and can allow that the cutting process optimization is made easily.

4 CONCLUSIONS

Based on the described in this work it can be concluded that:

- the optimization methodology is reliable and can be applied in several scenery of industrial environments, allowing, thus, its use to turn more efficient both the machine-tool and the tool;

- when the tool change time is very small the V_{mxp} is not recommended as reference speed to optimization;

- ABC cost system results more reliability when the V_{cmc} or V_{cmclim}, is used as reference speed to optimization;

- the optimization procedure described in this work can be adapted to be used in the several types of scenery found in shop floor;

- based on the scenery where the application of the described methodology is recommended, the authors are developing and improving computacional systems that involve different types of scenery, being possible its application in different manufacturing environment.

REFERENCES

1. RODRIGUES, A, C. S.; DINIZ, A, E.; COPPINI, N. L. - Análise das condições operacionais visando a obtenção das condições de usinagem otimizadas. 7O. Seminário de comando numérico no Brasil, São Paulo, SP, 1987.
2. DINIZ, A, E.; COPPINI, N. L.; VILELLA, R. C.; RODRIGUES, A, C. S. - Otimização das condições de usinagem em células. Máquinas e Metais, junho,1989, pags. 48 à 54.
3. MALAQUIAS, J.; COPPINI, N. L. - Velocidade de mínimo custo como condição suficiente para seleção da velocidade de corte otimizada. IV Congresso de engenharia mecânica norte-nordeste, Recife, PE, 1996.
4. MALAQUIAS, J. C.; COPPINI, N. L. - Seleção da velocidade de corte em usinagem com base na velocidade de mínimo custo, EME'97 I - encontro de mestrandos em engenharia da UNIMEP, Santa Bárbara d'Oeste, SP, 1997.
5. COPPINI, N. L.; ARAÚJO, G. A - Sistema de apoio na escolha de condições operacionais para processos de usinagem. Anais em CD do COBEM'97, Baurú, SP, 1997.
6. COPPINI, N. L.; MALAQUIAS, J. C.; MARCONDES, F. C. - Otimização em usinagem, uma visão gerencial. Trabalho submetido ao CEM-MNE'98 sob número 0341. UFPB - João Pessoa, PB, 1998.

ACKNOWLEDGMENTS

The authors would like to thank FAPESP and Sandvik Coromant do Brasil Ind. e Com.

OPTIMIZATION OF MACHINING PROCESS USING EVOLUTIONARY ALGORITHMS

G. Cukor[1], E. Kuljanic[2], B. Barisic[1]

[1] Department of Production Engineering, Faculty of Engineering of the University of Rijeka, Croatia
[2] Department of Electrical, Management and Mechanical Engineering, University of Udine, Italy

KEYWORDS: Cutting, Optimization, Evolutionary Algorithms.

ABSTRACT. Advanced manufacturing requires a powerful tool for reliable modeling and solving the complex machining optimization problems. This paper proposes a non-conventional approach using evolutionary algorithms inspired by Darwinian findings about the evolution of the biological species and the survival of the fittest organisms (i.e. natural selection). It is illustrated with an experiment of longitudinal hard turning. Genetic programming (GP) is used to develop the models of both the surface roughness and the tool life considering the cutting speed, the feed and the depth of cut as predetermined cutting parameters. Finally, genetic algorithm (GA) is applied for their optimization.

1 INTRODUCTION

Manufacturing firms typically determine cutting parameters, namely cutting speed v_c, feed f and depth of cut a_p, based on the machinist's experience or handbook recommendations. With the increasing use of CNC machine tools, which involve large capital expenditures, the use of an economic analysis combined with technological considerations becomes an imperative. The dissolving of national and regional boundaries also drives the need to manufacture a product of high quality, at the lowest possible cost and rapidly. Hence, optimum cutting parameters play a key role in competitiveness in the market.

Procedures reported so far to determine the optimum cutting parameters are graphical techniques [1], linear programming [2], sequential quadratic programming [3], geometric programming [4], dynamic programming [5], numerical and direct search techniques [6], neural networks [7], and approaches based on traditional mathematical optimization (gradient method, etc.) [8]. This task has proven to be surprisingly difficult. It requires intricate mathematical analysis and computer assistance, and depends on quantitatively reliable mathematical models for the machinability performance measures (such as tool life T, surface roughness R_a, etc.) and detailed specifications of the machine tools, cutting tools and components, which act as constraints on the feasible cutting parameters. Furthermore, conventional approach is to use an ordinary least squares (OLS) regression analysis for developing the machinability models.

In the last decade, the use of evolutionary computation methods, or also called the genetic methods, based on imitation of Darwinian natural selection has become widespread. This is due to fact that many systems are too complex to be successfully optimized by the use of conventional deterministic algorithms. On the contrary, the evolutionary algorithms (EA) involve probabilistic operations. Genetic algorithm (GA) is the most successfully implemented evolutionary

Published in: E. Kuljanic (Ed.) *Advanced Manufacturing Systems and Technology*,
CISM Courses and Lectures No. 486, Springer Wien New York, 2005.

computation method for optimizing of various machining processes [9, 10]. On the other hand, out of the evolutionary computation methods, genetic programming (GP) is probably the most general evolutionary approach [11]. However, its implementation for developing the machinability models has not been reported yet.

In this paper, an integrated GP-GA concept for modeling and optimization of various machining processes is proposed, Figure 1. GP modeler replacing the standard OLS regression analysis is used if the machinability model is unknown. Modeler input consists of experimental values of the independent variables (cutting parameters) and associated values of the dependent variable (machinability performance). As evolution takes place in virtual computer space, it is possible to analyze a great number of different solutions within a relatively short time. Modeler output consists of one (or more if desired) randomly generated machinability model(s) corresponding as much as possible to the experimental data. GA optimizer is used if the optimization model of respective machining process is already known. Optimizer output consists of the randomly generated set of optimized cutting parameters. The proposed concept is illustrated on a longitudinal hard turning example.

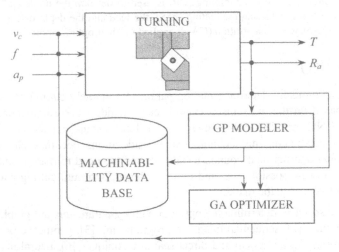

FIGURE 1. Integrated GP-GA concept for turning

2 MODELING BY GP

In this section we are going to explore GP for a specific problem: finding a good regression function for a set of data. Suppose we have a set of data pairs (x_i, y_i) for $i = 1, ..., N$. If the data "look" linear, we might try OLS linear regression: find a and b such that $y = ax + b$ is the best-fit line (in the sense of least squares) to the data.

But what if the relationship is not at all linear (usually a case when dealing with machining processes)? Then we must guess the form of the regression equation (e.g., quadratic, log-linear, etc.), then figure a way to find its coefficients. This may be hard, or worse, arbitrary. There may also be several independent variables, x_1, x_2, x_3, etc.

What if we don't think least squares is the best criterion to use? It should be pointed out, OLS regression has some nice properties, but it is commonly used just as much for the fact that it is easy to compute.

It's not hard to think of a regression equation as a "computer program" that produces some output (a y value) given some input (a set of x_i values). The regression equation can be simple or horribly complicated. We don't care so long as the predicted values of y for all the x's in the domain of interest are pretty accurate.

So here is the pitch. We'd like a program that can figure out both the functional form and the appropriate coefficients. We'll use GP to do this.

The processes that make up a complete run of GP can be divided into a number of sequential steps:

1. Create a random population of the programs (organisms or chromosomes) of various forms and lengths using the genes provided. These genes are independent variable symbols (x_1, x_2, x_3, ...), numerical constants and mathematical operators (+, -, *, /, etc.).

2. Evaluate each program assigning a fitness value according to a pre-specified fitness function that measures the ability of the program to solve the problem.

3. Assure survival of more fit programs and their advance in unchanged form into the next generation using the reproduction (i.e. selection) operator.

4. Genetically recombine the new population with the crossover operator from a randomly chosen set of parents, Figure 2.

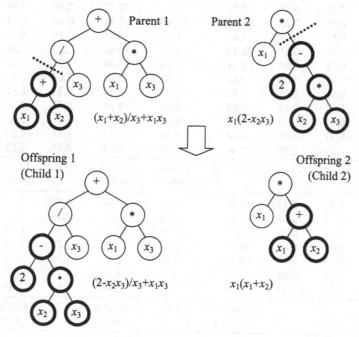

FIGURE 2. Crossover in GP

5. Introduce the minor random changes in the gene structure of developed organisms using the mutation operator.

6. Repeat steps 2 onwards for the new population until a pre-specified termination criterion has been satisfied or a fixed number of generations has been completed.

7. The solution to the problem is the genetic program with the best fitness within all the generations.

3 EXPERIMENTAL DETAILS

Longitudinal hard turning experiments were carried out on a conventional Oerlikon DM-3-1-635 lathe using triple coated (TiCN + Al_2O_3 + TiN) carbide (WC) round inserts of the ISO shape RNMM 1606 and tool holder PRGCR 2525 M16.

The workpiece material was low-alloyed steel 42CrMo4 (DIN) over-welded with submerged arc welding procedure in 2 passes. Cromecore 430 (0,043% C, 0,83% Mn, 0,84% Si, 0,005% P, 0,005% S and 17,97% Cr) wires with diameter of 2,4 mm were used. Hence, built-up welded coat was high-alloyed martensite material with measured hardness of 48 - 49 HRC. Row bar dimensions were Ø233 × 780 mm.

TABLE 1. Comparison between experimental data and developed predictive models

No.	v_c m/min	f mm/rev	a_p mm	$T^{(E)}$ min	$T^{(GP)}$ min	$T^{(LR)}$ min	$R_a^{(E)}$ μm	$R_a^{(GP)}$ μm	$R_a^{(LR)}$ μm
1	39	0,21	0,25	139,63	139,63	101,32	5,46	5,46	5,54
2	81,5	0,21	0,25	7,23	7,20	10,23	5,05	4,96	4,83
3	39	0,53	0,25	55,44	55,46	50,85	6,62	6,64	6,40
4	81,5	0,53	0,25	3,61	3,62	5,14	6,16	6,15	5,58
5	39	0,21	0,65	22,43	22,44	30,06	2,39	2,24	2,16
6	81,5	0,21	0,65	3,11	3,14	3,04	1,95	2,02	1,89
7	39	0,53	0,65	9,81	9,51	15,09	2,64	2,51	2,50
8	81,5	0,53	0,65	1,92	1,95	1,52	2,33	2,31	2,18
9	60	0,37	0,45	9,84	10,33	8,24	2,78	2,85	3,13
10	60	0,37	0,45	10,16	10,33	8,24	2,87	2,85	3,13
11	60	0,37	0,45	10,44	10,33	8,24	2,81	2,85	3,13
12	60	0,37	0,45	9,46	10,33	8,24	2,83	2,85	3,13
			$\bar{S}$		1,94%	24,95%		1,97%	7,26%
			R		0,99997	0,95173		0,99898	0,9847
13	49,5	0,37	0,35	28,64	25,93	20,64	4,49	3,87	4,16
14	49,5	0,37	0,55	21,42	11,56	11,62	2,28	2,47	2,66
15	60	0,53	0,25	18,67	20,78	13,32	6,36	6,32	5,91
16	60	0,21	0,45	18,46	17,33	12,56	2,29	2,57	2,87
17	60	0,53	0,45	8,23	7,56	6,31	3,22	3,13	3,31
18	81,5	0,37	0,35	3,57	3,31	4,38	4,15	3,65	3,79
19	81,5	0,37	0,55	2,56	2,46	2,46	2,03	2,33	2,43
			Total $\bar{S}$		6,08%	25,45%		4,65%	9,20%
			Total R		0,99687	0,94765		0,9898	0,97746

The feasible space of the cutting parameters was selected by varying v_c in the range 39 - 81,5 m/min, f in the range 0,21 - 0,53 mm/rev, and a_p in the range 0,25 - 0,65 mm. A simple two level fractional factorial design of experiments with included centre points runs ($2^k + n_0$) was adopted for experimentation. Two sets of experimental data were used: the first set of $2^3 + 4 = 12$ experimental data to determine prediction models and the second set of 7 experimental measurements to verify the models, as listed in Table 1.

The machined surface roughness was measured by a profile meter TAYLOR-HOBSON Surtronic 3+. The average surface roughness R_a, which is the most widely used surface finish parameter in industry, is selected in this study. It is the arithmetic average of the absolute value of the heights of roughness irregularities from the mean value measured within the sampling length of 8 mm (DIN 4762). Also, herein the tool life T is defined as the period of cutting time until the average flank wear land V_B of the tool is equal to 0,3 mm. In the experiments, the flank wear land was measured by using an optical tool microscope. All measurements are listed in Table 1.

4 ANALYZIS OF RESULTS

It is very important to know that modeling of both the tool life and the surface roughness using GP involves finding both the functional form and the numerical coefficients of the mathematical model. It should be noted that no assumption is made here in advance about the size, shape and complexity of the eventual satisfactory solutions. These facts make this study completely different from the conventional approach of using OLS linear regression where the goal is to discover merely a set of numerical coefficients for a function whose form has been pre-specified.

Table 1 shows a comparison between experimental data (E) and prediction models developed genetically (GP), and by OLS linear regression (LR). First, for the learning set of 12 cases, the following models were generated:

$$T^{(GP)} = \frac{1}{6,92249}\left[\frac{2}{f} + \frac{1}{a_p} + \frac{1}{f\,a_p^2}\left(\frac{915,615}{v_c + f + a_p} - 10,7855\right)\right] + 1, \tag{1}$$

$$T^{(LR)} = e^{13,08931}v_c^{-3,11061}f^{-0,7446}a_p^{-1,27162}, \tag{2}$$

$$R_a^{(GP)} = \frac{3}{v_c + f + a_p}\left[\frac{3v_c}{a_p(v_c + f + a_p)} + 1\right] +$$

$$+ \frac{1}{a_p}(f + 1,47057) + a_p(3 - f + a_p) - 3, \tag{3}$$

$$R_a^{(LR)} = e^{1,27179}v_c^{-0,18615}f^{0,15548}a_p^{-0,98511}. \tag{4}$$

It was decided that the termination criterion for GP modeller in learning phase is reached when the average percentage deviation

$$\overline{S} = \frac{1}{N} \sum_{i=1}^{N} \frac{\left| y_i^{(E)} - y_i^{(model)} \right|}{y_i^{(E)}} \cdot 100\% \tag{5}$$

is less than or equal 2%, where N is the size of sample data. Also, the regression coefficient was used for comparison:

$$R = \sqrt{1 - \frac{\sum_{i=1}^{N} \left[y_i^{(E)} - y_i^{(model)} \right]^2}{\sum_{i=1}^{N} \left[y_i^{(E)} - \overline{y}^{(E)} \right]^2}} \cdot \tag{6}$$

Obviously, GP gave better solution since $\overline{S}^{(GP)} >> \overline{S}^{(LR)}$ and $R^{(GP)} > R^{(LR)}$ in Table 1. After that, the predictive capabilities of developed models were tested with additional set of 7 cases. For the total of 19 cases the respective average percentage deviations and the regression coefficients were calculated. Again, GP gave better solutions as can be seen from results in Table 1. Especially much better solution is recorded for the tool life model, which is very important for the reliability of machining process optimization.

5   GA OPTIMIZATION

When considering a single pass turning operation, the weighted objective function to be minimized is [12]:

$$\varphi(v_c, f, a_p) = w_1 \frac{cost}{cost^*} + w_2 \frac{time}{time^*} + w_3 \frac{R_a^{(GP)}}{R_a^*}, \tag{7}$$

$$cost = c_1 \times time + c_2 \left[\frac{t_2}{T^{(GP)}} \right], \tag{8}$$

$$time = t_2 + t_3 \left[\frac{t_2}{T^{(GP)}} \right], \tag{9}$$

$$t_2 = \frac{\pi D \ell}{10^3 v_c f}, \tag{10}$$

$$c_2 = \frac{cost\ of\ holder}{400} + \frac{cost\ of\ insert}{0,75 \times number\ of\ cutting\ edges}, \tag{11}$$

subjected to $v_{cmin} < v_c < v_{cmax}$, $f_{min} < f < f_{max}$, $a_{pmin} < a_p < a_{pmax}$ and $R_a < R_a^*$, where w_i are the weighting factors ($0 < w_i < 1$, $w_1 + w_2 + w_3 = 1$), $cost^*$ is the expected machining cost limitation per piece (EUR), *time*[*] is the expected machining time limitation per piece (min), R_a[*] is the surface roughness limitation (µm), c_1 is the operating cost of lathe (EUR/min), c_2 is the cost of cutting edge (EUR), t_2 is the machining time (min), t_3 is the tool change time (min), D is the cutting diameter (mm) and ℓ is the length of pass (mm).

An important problem in the implementation of GA for machining process optimisation is the construction of a fitness function. In this study penalty terms corresponding to the constraint violation are added to the objective function and fitness is obtained. Penalty terms are added only if the constraints are violated. Hence, the fitness function is:

$$\phi(v_c, f, a_p) = \varphi(v_c, f, a_p) + \left[\min\left\{0, g_i(v_c, f, a_p, R_a)\right\}\right]^2, \tag{12}$$

where g_i are the inequality constraints to the optimisation problem. Note that in the feasible space of solutions, the contributions from the penalty terms are 0 (zero).

The following input data for GA optimizer were set: $D = 233$ mm, $\ell = 780$ mm, $t_3 = 1$ min, $c_1 = 1,67$ EUR/min, $c_2 = 3,2$ EUR, $w_1 = 0,3$, $w_2 = 0$, $w_3 = 0,7$, $cost$[*] $= 65$ EUR and R_a[*] $= 2,5$ µm. After genetic search (Figure 3) with number of generations $= 60$, crossover probability $= 0,85$ and mutation probability $= 0,05$, the founded optimal cutting parameters were: $v_c = 71,6$ m/min, $f = 0,46$ mm/rev and $a_p = 0,65$ mm. Also, the genetic evolution history is described in Figure 4.

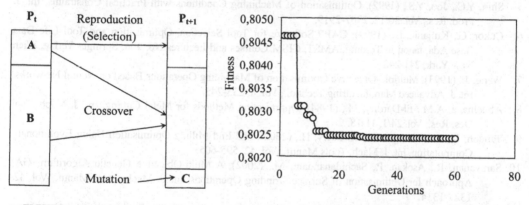

FIGURE 3. Evolutionary strategy   FIGURE 4. Fitness evolution in turning

6   CONCLUSIONS

In this study, the evolutionary algorithms were applied for modeling and optimization of the machining process using an integrated GP-GA concept. It imitates the natural evolution of living organisms, where in the struggle for natural resources the successful individuals gradually become more and more dominant, and adaptable to the environment in which they live, whereas the less successful ones are present in the next generation rarely.

In the proposed GP approach the programs, i.e. mathematical models, undergo adaptation. The evolution spontaneously finds the secret of hidden information within the experimental data and develops a mathematical model. The presented research shows that the average percentage deviation in the case of the tool life model obtained by standard OLS linear regression is

25,45% and only 6,08% in the case of simulated evolution. Therefore, the model developed without the influence of human intelligence is about 4,19 times more precise than one whose form has been pre-specified.

Also, the application of GA approach to obtain optimal cutting parameters has proved to be efficient and robust. It will be quite useful at the computer-aided process planning (CAPP) stages in the manufacturing of high quality goods with tight tolerances by a variety of machining operations, and in adaptive control of intelligent machine tools.

REFERENCES

1. Kiliç, S.E., Cogun, C., Sen, D.T., (1992), A Computer-Aided Graphical Technique for the Optimisation of Machining Conditions, Comp. Ind., Vol. 20, 319-326.
2. Ermer, D.S., Patel, D.C., (1974), Maximization of the Production Rate with Constraints by Linear Programming and Sensitivity Analysis, Proc. 2nd NAMRC, Madison, 436-443.
3. Chua, M.S., Loh, H.T., Wong, Y.S., Rahman, M., (1991), Optimisation of Cutting Conditions for Multi-Pass Turning Operations Using Sequential Quadratic Programming, J. Mater. Proc. Technol., Vol. 28, 253-262.
4. Eskicioglu, H., Nisli, M.S., Kiliç, S.E., (1985), An Application of Geometric Programming to Single-Pass Turning Operations, Proc. 25th Int. Mach. Tool Des. Res. Conf., Birmingham, 149-157.
5. Shin, Y.C., Joo, Y.S., (1992), Optimisation of Machining Conditions with Practical Constraints, Int. J. Prod. Res., Vol. 30/12, 2907-2919.
6. Cukor, G., Kuljanic, E., (1999), CAPP Software for Tool Selection, Optimisation and Tool Life Data Base Adaptation in Turning, AMST, CISM Courses and Lectures No. 406, Springer Verlag, Wien New York, 241-248.
7. Wang, J., (1993), Multiple-Objective Optimisation of Machining Operations Based on Neural Networks, Int. J. Advanced Manufacturing Technol., Vol. 8, 235-243.
8. Abuelnaga, A.M., El-Dardiry, M., (1984), Optimisation Methods for Metal Cutting, Int. J. Mach. Tool Des. Res., Vol. 24/1, 11-18.
9. Tandon, V., El-Mounayri, H., Kishawy, H., (2002), NC End Milling Optimisation Using Evolutionary Computation, Int. J. Mach. Tools Manuf., Vol. 42, 595-605.
10. Saravanan, R., Asokan, P., Sachidanandam, M., (2002), A Multi-Objective Genetic Algorithm (GA) Approach for Optimisation of Surface Grinding Operations, Int. J. Mach. Tools Manuf., Vol. 42, 1327-1334.
11. Brezocnik, M., Kovacic, M., (2002), Integrated Evolutionary Computation Environment for Optimising and Modelling of Manufacturing Processes, Proc. 6th TMT, Neum, 107-110.
12. Cukor, G., (1999), Optimisation of Machining Process for Advanced Machining Systems, PhD Thesis, Faculty of Engineering of the University of Rijeka.

OPTIMIZATION OF TURNING OPERATION: NIMONIC VALVES CASE

C. T. Watanabe[1], M. R. V. Moreira[1], M. V. Ribeiro[1], J. L. Nogueira[2]

[1] Machining Study Laboratory, Department of Materials Engineering, FAENQUIL, Lorena/SP, Brazil
[2] National Institute for Space Research, INPE, Cachoeira Paulista/SP, Brazil

KEYWORDS: Machining, Optimization, Turning, Manufacture Cell.

ABSTRACT. The choice of the tool for one determined operation and the correct determination of the machining conditions, represent an important factor in the metals working, being this determinant for the evolution of tool-machine and cutting tools. Such fact accents in the serial manufacturing, where divergences in the choice of the cutting speed and the tool may cause notables variations in manufacture costs. The decisions for tools selection, machining parameters determination and tool exchange times are carry by process planners, programmers and machine operators in different periods of manufacture. This share of responsibility and the lack of interaction with the process have difficult very the taking of decisions of a form optimized in relation to the tool rack and the process in itself. The considered work aims at to promote the optimization of the machining processes in turning in a manufacture cell, of automobile parts factory, using for this optimization methodology of the process based on the maximum production condition, through the determination of the coefficients of the Taylor's equation of tool life in shop floor environment. The objective is to provide to real increases of productivity and quality in industries, without the requirement of investments in new means of production, and valuing the use of the information inside it productive system.

1 INTRODUCTION

The introduction of new products and the modern demands with relationship to the competitiveness are sources of disturbances for the production, being reflected in the methods, process and production means. The structure of the production systems and your physical disposition (layout) are placed continually in check, having need to do frequent evaluations for the productivity are maintained inside of the corporate target. Many other layout problems grow in elapsing of the time for countless reasons, to which they are evidenced by production bottleneck, arrears and inexplicable idle times, stock excessive and high set-up (machine preparation time) [1]. The use of the cellular manufacture, that had beginning starting from the of 90 years, it appeared with objective of reaching the benefits of the linear arrangement in manufacture systems, with a varied production mix and small batches [2]. Like this, each manufacture cell, composed by a group of different machines, it is responsible for the production of a group of similar items or a family of items. It allows although the machines are to each other close and that a smaller number of employees operates the same number of machines. In what concerns the arrangement of the machines in the cell, the most usual formation is that in " U ", where the items enter on one side

Published in: E. Kuljanic (Ed.) *Advanced Manufacturing Systems and Technology*,
CISM Courses and Lectures No. 486, Springer Wien New York, 2005.

and they leave for the other [3]. The cellular layout allows the reduction of the stock in process, of the movement costs and of the lead times (total time of the workpiece production) [4].

For the formation of the cells several algorithms exist based on the group technology, however none of these methods can be adopted in any case. The cells of group technology are formed by equipments non similar, but that together can be considered as an only process in that the parts can be processed through the total cycle of manufacture in an only place.

2 OPTIMIZATION METHODOLOGY

When one want to work with optimization models for the machining conditions, is need first to express the tool life in function of the cutting speed. Such procedure implicates in the definition of what it is life and that sorts things out it can be quantified. Is called tool life the time that the same works indeed, until losing its cutting capacity, inside of a criterion previously established. The loss of the cutting capacity is usually evaluated through a certain wear degree.

The definition of the life criterion is very important due influences of the experimental coefficients x and K of the Taylor's equation, for instance. The tool life time (T), in minutes, it can be calculated by the Equation 1, where v_c is the cutting speed, in m/min. The change of the criterion or the use of an inadequate criterion for a specific situation can alter significantly values obtained by any model.

$$T = \frac{K}{v_c^x} \tag{1}$$

The methodology of optimization applied is based on the determination of the maximum production speed (v_{mxp}) starting from the definition of the Taylor's equation coefficients of the tool life in industrial environment. For so much, the following procedures should be taken [5]:

- Statistic measures of the cutting time for workpiece (t_c) and of the time for edge change or the tool change time (t_{Tf});
- To the adopted conditions, to make the workpiece in subject determining the medium number of workpieces for cutting edge (z') and,
- To adopt value of speed (v_c'') 20% different that (v_c') and, for this new cutting speed, to determine the medium number of workpieces for cutting edge (z'').

The maximum production speed (v_{mxp}) is based on the fact that, with the increase of the cutting speed, reduce the cutting time and the relative costs to the machine and the operator. However, it reduces the tool life simultaneously, causing an increase of the relative time to change and a relative increase to the part of cost of the tool. In that way, conditions exist in which the total time of production is minimum, what corresponds to the maximum production [6]. The cost/benefit relation can be determined by calculating the piece manufacturing cost for the selected cutting speed, based on the speed where maximum speed production (v_{mxp}) for manufacturing each specific piece is achieved in the industrial environment [7]. Like this being, the v_{mxp} is the cutting speed for the one which the sum of the tool change times and of machining is minimum, in other words, it is the speed in the which happens a balance among the positive and negative effects associated to the use of cutting high-speeds, which are the largest rate of material removal and the largest frequency of tool changes, respectively.

To obtain this value, through the Equation 2, it should be to calculate the coefficients of the Taylor's equation (x and K) and measure the tool change time (t_{Tf}), in minutes [6].

$$v_{mxp} = x\sqrt{\frac{K}{(x-1)\cdot t_{Tf}}} \tag{2}$$

When a workpiece is machined by a turning process, can occur that cutting speeds in the operation are fixed, but in the others situation the values of rotation during the machining process can be constant. However, when both are variables, the utilization of Equation 4 is needed to establish the medium speed (v_m) for this operation. After through the Equation 3 and Equation 5 the calculation of the Taylor's coefficient (x and K) can be done.

$$x = 1 + \left[\frac{\log(z'/z'')}{\log(v_m''/v_m')}\right] \tag{3}$$

where:

z' = life of the edge for the first condition, in piece.
z'' = life of the edge for the second condition, in piece.
v_m' = medium cutting speed of the first condition, in m/min.
v_m'' = medium cutting speed of the second condition, in m/min.

The medium speed in, m/min can be calculated by the relationship among the cutting length (l_c), in meters and the cutting time, in minutes.

$$v_m = \frac{l_c}{t_c} \tag{4}$$

For to calculate the K coefficient the Equation 5 is applied:

$$K = z \cdot v_m^x \cdot t_c \tag{5}$$

If the v_{mxp} is between v_c' and v_c'' this become denominated of selected cutting speed (v_{CSEL}), but it doesn't happen a new rehearsal it should be proceeded and to follow the calculation of the new v_{mxp}, until that the found value of the speed is in the interval. To proceed of ownership of the data already picked until then, it is possible to make the calculation of the machining cost for workpiece (C_{tp}), being used the Equation 6, second Ribeiro (1999), expression this developed starting from

Ferraresi's cost equations (1977), where C_{um} is cost of use of the machine, in US\$/min; P_{sup} is the price of the holder, US\$; T_{sup} is your life in terms of the number of edges; P_{ins} is the price of the inserted, US\$, N_{ins} is the number of edges that inserted it possesses; Z_T is the life of an edge, in number of workpieces.

$$C_{tp} = C_{um} \cdot t_{um} + \left(\frac{\dfrac{P_{sup}}{T_{sup}} + \dfrac{P_{ins}}{N_{ins}} + C_{um} \cdot t_{Tf}}{Z_T} \right) \tag{6}$$

The utilization time of the machine (t_{um}), in min/workpiece, it is given by the Equation 7, where t_i, in minutes, it is handling and unproductive time and Z, in pieces, refers to the batch produced in one month.

$$t_{um} = t_c + t_i + \left(\frac{t_c}{T} - \frac{1}{Z} \right) \cdot t_{Tf} \tag{7}$$

The value of the cost/benefit of the employment of a certain tool can be calculated by the Equation 8. In this the tool cost is used (C_{tp})ref, dictates of reference, that actually is the current condition of the process, and the optimized cost of the other tested conditions (C_{tp}).

$$R_{CB} = \frac{(C_{tp})_{ref} - C_{tp}}{(C_{tp})_{ref}} \times 100\% \tag{8}$$

The tool edge life (T_A), in minutes, it is given by the Equation 9

$$T_A = Z_T \cdot t_c \tag{9}$$

The total tool life (T_T), in minutes it is given by the Equation 10.

$$T_T = Z_T \cdot t_{um} \tag{10}$$

The amount of consumed edges by shift (A_t), where (t_t) it is the duration of the work shift, in minutes it is given by the Equation 11.

$$A_t = \frac{t_t}{T_T + t_{Tf}}$$ (11)

The size of the batch produced by shift (Z_{shift}), in workpieces it is given by the Equation 12.

$$Z_{shift} = Z_T \cdot A_t$$ (12)

The machining cost for the lot of a shift (C_l), US$ it is given by the Equation 13.

$$C_b = Z_{shift} \cdot C_{TP}$$ (13)

Using this methodology is possible to supply the manufacturing cell optimized conditions, and select tools, besides always doing forecast of costs with the objective of its use in shop floor, where it will directly be able to receive the information from the generating source, in other words, of the machines of the production line.

3 EXPERIMENTAL PROCEDURE

The test consisted of the accompaniment of the turning operation of semi-finish in a numeric control lathe (CNC) of an admission valve for a motor of internal combustion Figure 1. The valve was forged previously, and it is composed of two different materials: the stem is made of steel and the base with a nickel alloy, both parts are welded. The machining in subject was only limited to the area of the nickel alloy, being accomplished the following operations: head diameter (in two passes), chamfer, and facing (in two passes). The used tool was a cemented carbide uncoating SEGN 120308, BG31 CDC class.

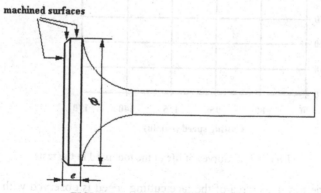

machined surfaces

FIGURE 1. Admission Valve V5293

With relationship to the end of tool life criterion, the superficial roughness was used, however your determination depended on the subjective evaluation of the operator, because there was not an available surface roughness tester to take place the control of the roughness of the machined workepieces. The dimensional quality of the workpieces was also used as end life criterion and it was frequently verified in relation to the thickness (e) of the valve base and the head's diameter (Ø) according to the Figure 1. This was accomplished through a control device with dial indicator, with resolution of 0.02 mm and it allowed although if made the corrections in the process, in the sense of compensating the suffered wear for the cutting tool.

4 RESULTS

A battery of tests was made, being used 4 initial cutting speeds in the facing (80, 95, 115 and 140 m/min), and three cutting speeds for the diameter and chamfer (120, 140 and 160 m/min), for a cutting length of 15,14 m and maximum rotation of 1,600 rpm, staying constant the feed rate(0.1 mm/rev) and the machining depth (0.4 mm). The measurement change time was 0.75 min. The work shift was of 8 hours, but as there was the time for the meals and other permissions, was considered that were worked 7 hours. For indeed the calculation of costs was considered a batch of 4,000 workpiece/month. With the tests was verified that the cutting operation of the diameter and of the chamfer didn't determine the end of tool life. Therefore chose to pick the data using the four speeds of cut of the facing and the speed of 160 m/min for the diameter and chamfer. To follow the curve of the tool life used in the tests is presented Figure 2.

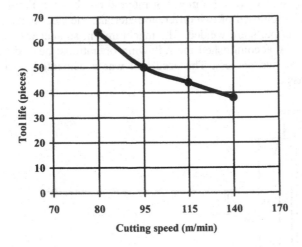

FIGURE 2. Curves of life of the tool used in the tests

In the Figure 3 the performance area of the face cutting speed is observed with the variation of the diameter of the valve head, for a maximum rotation of 1,600 rpm. It is noticed that as increases the cutting speed, the space that she stays constant is smaller. In the cutting speed of 140 m/min, the speed stays constant just in the space of diameter of 32.5 up to 27.8mm, that is, only 4.7mm in the diameter. Therefore, to increase the cutting speed above 140m/min doesn't become advantageous, because the space of performance becomes very small.

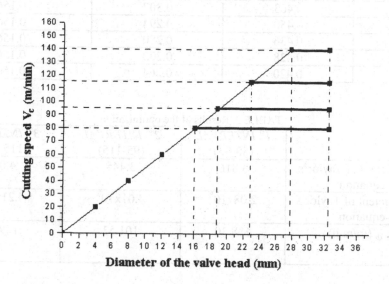

FIGURE 3. Area of performance of the cutting speed with the diameter variation

In the Table 1 consist the measured values for the machining cycle times which are: the cutting time, the handling and the unproductive times. All quantified in minutes. The first line of this table the values obtained with the current conditions of the production process are presents. As the facing operation was shown more critic, chose in centering the focus of the optimization in this operation. The Table 2 shows the results of this optimization.

In the Table 3 the values of tests costs, obtained by the adopted methodology are presented. And in the Table 4 more results are showed about the machining tests in the shop-floor.

To illustrate the advantages for the use of the new cutting conditions in relation to the reference condition, in other words, usually used (face cutting speed same to 80 m/min, and diameter cutting speed of 120 m/min), maximum rotation of 1,600rpm, was set up the table 5, in the which contain the values obtained with the substitution of the reference condition by the others, being taken into account a lot with 4,000 workpieces/month.

TABLE 1. Values of time of the tests

Initial cutting speed (m/min)	Cycle time (min)	Cutting time (min)	Handling and unproductive times (min)
80	0.463	0.307	0.156
80	0.450	0.294	0.156
95	0.436	0.280	0.156
115	0.428	0.272	0.156
140	0.420	0.264	0.156

TABLE 2. Results of the optimization

	1st INTERVAL (80-95)	2nd INTERVAL (95-115)	3th INTERVAL (115-140)
x coefficient of Taylor's equation	3.816	3.445	4.000
K coefficient of Taylor's equation	2.08×10^8	4.01×10^7	4.21×10^8
v_{mxp} (m/min)	138.30	101.53	129.44

TABLE 3. Values of costs of the tests

Initial cutting speed – V_{CI} (m/min)	Machine cost (US$/part)	Tool cost (US$/part)	Total cost (US$/part)	R_{CB} (%)
80	0.702	0.0698	0.772	-
80	0.681	0.0698	0.751	2.72
95	0.663	0.0894	0.752	2.59
115	0.651	0.1016	0.753	2.46
140	0.641	0.1176	0.759	1.68

TABLE 4. Values of production of the tests

V_{CI} (m/min)	Z (#parts)	T (min)	T_T (min)	Edges for shift (A_t)	Workpieces for shift (Z_{shift})	Cost of the batch (US$)
80	64	19.65	29.95	13.68	875	675.50
80	64	18.82	29.06	14.09	901	676.65
95	50	14.00	22.10	18.38	919	691.09
115	44	11.97	19.10	21.16	931	701.04
140	38	10.03	16.23	24.73	940	713.46

TABLE 5. Values with the substitution of the cutting speed

Substitution	Bonus (US$)	Bonus in inserts	Tools for 4,000 workpieces
80-80	84	12	32
80-95	80	12	40
80-115	76	11	46
80-140	52	7	53

With aim to ilustrated the turning operation in the Figure 4 and Figure 5 are showed the tool wear and formed chip respectivily for reference condition (80 m/min in the facing and 120 m/min for diameter). Can be observed the difference among the wear size for different tool regions, such as the region when facing are made (4a and 4b) and the region when the diameter are turning (4c). With related the formed chip in the Figure 5 it showed a predominant form of the obtained chip ins this operation, the mainly characteristic is the segmented form in its all lenght.

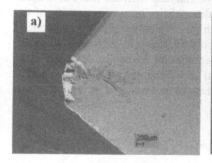 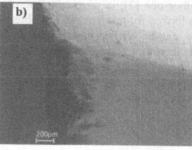 

FIGURE 4. Micrography of tool wear for 80 m/min
a) Facing b) Its flank and c) Diameter [8]

FIGURE 5. Micrography of obtained chip for 80 m/min [8]

5 ANALYSIS AND COMMENTS

The adopted methodology if it showed plenty of practice, because the results were obtained without if it caused the smallest upset to the productive process, in other words, the performance of the system was the least intrusive possible, arriving to the point of to the opposite, to increase the production of the machine in that shift, because the cutting speed was varied. The data of the test were picked inside of a work shift (8 hours). In the tests it was obtained a decrease in the costs and in the production time of the lot. For a lot of 4,000 workpieces it was obtained a decrease of almost 3 hours in the time, and a reduction of US$52.00 in the costs for the lot production. As it can be observed in the Figure 3, the maximum rotation of 1,600 rpm was a limiter in the optimization. A next step would be to increase the maximum rotation until a compatible value, and to pick new data. As this was an initial work, in other words, in the first results of the optimization are only meditated, is waited for the next period that the new resulted to converge for more favorable situations in relation to the economy obtained in the process. We can also increase that this work is a complementation of another one [7], in which the optimization of the turning operation of the valve "filler" area, this was looked for that was not test object in this evaluation. For the projections done for the selected speed, can be verified that is possible to obtain a certain economy, when the selected condition is used, therefore same consuming a larger amount of edges, stopping more times to accomplish the change of edges, diminishes the time of use of the machine, that has a very great influence in the total cost in this case. In the times of great competitiveness in that we lived, any reduction of costs salutary become.

ACKNOWLEDGMENTS

To FAPESP (Process 01/00759-0) and CNPq by financial support and scholarship.

REFERENCES

1. Hassan, M.M.D.-Machine layout problem in modern manufacturing facilities. *International Journal Production Research*, v. 32, n.11, p. 2559-2584, 1994.

2. Burbidge, L. J. Change to group technology: Process organization is absolete. *International Journal of Production Research*, v. 30, n. 5, p. 1209-1219, 1992.

3. Black, J. T.; Schroer, Bernard J. Simulation of an apparel assembly cell with walking workers and decouplers. *Journal of Manufacturing Systems*. v. 12, n. 2, 1994.

4. Giffi, C.; Roth, A.; Seal, G. Competing in world-class manufacturing: American's 21st Century Challenge. National Center for Manufacturing Sciences, 1990, 410 p.

5. Ribeiro, M.V Optimisation of the Cutting Conditions Aided by Computer During the Process Development, College of Mechanical Engineering, State University of Campinas, Ph.D. Thesis (in portuguese), 1999, 138.

6. Ferraresi, D. *Fundaments of metals machining*. 1st ed; São Paulo: Ed. Edgard Blücher, 1977, 751 p. (in portuguese).

7. Ribeiro, M.V and Coppini, N.L. "An Applied Database System for the Optimization of Cutting Conditions and Tool Selection", *Journal of Materials and Process Tecnology*, v.92/93, p. 371-374, 1999.

8. Watanabe, C.T; Ribeiro.; "Stud of cellular manufacture ambient turning optimization", MSC. Department of Materials Engineering, FAENQUIL-USP/DEMAR Brazil, 2004, 80p. (In portuguese).

MODELING OF CUTTING FORCES IN PLUNGE MILLING

M. Al-Ahmad, A. D'Acunto, C. Lescalier, O. Bomont

Laboratoire de Génie Industriel et Production Mécanique (LGIPM - EA 3096)
CER ENSAM – Metz Technopôle – 4 rue Augustin Fresnel – 57 078 Metz cedex 3

KEYWORDS: Plunge Milling, Cutting Forces, Cutting Geometry.

ABSTRACT. This paper deals with the modeling of cutting forces in plunge or vertical milling. The operation of rough and semi rough machining allows realizing the walls of moulds and matrices. This new technique of milling greatly contributes to shortening the entire production process, and therefore, saving significant production time and costs. In this paper, we propose a cutting model that includes the determination of tool geometry, trajectories of cutting edge, chip geometry (radial engagement, chip thickness) and the evaluation of the cutting forces. This model is validated experimentally through tests run in the machining centre. In this study, plunge cutting was carried out on the 40CrMnMo8 material using end milling.

NOMENCLATURES

R: Tool diameter (mm)

f_z : Feed rate per tooth (mm/tooth)

N: Spindle speed (rpm)

V_c : Cutting speed (mm/min)

α : Lead angle (deg.)

γ : Rake angle (deg.)

λ_s : Helix angle (deg.)

κr: Angle of edge direction (deg.)

ae : Radial engagement (deg.)

P : Pick feed (mm)

AB: Radial engagement in position θ (mm)

A: Cutting section (mm2)

θ : Cutter rotation angle (deg.)

θ_e : Entry angle (deg.)

θ_s : Exit angle (deg.)

h: Thickness of cut (mm)

Q : Material removal rate (cm3/min)

$F_x(\theta)$, $F_y(\theta)$, $F_z(\theta)$: Force components in X, Y and Z directions (N)

$F_t(\theta)$, $F_r(\theta)$, $F_p(\theta)$: Tangential, radial and axial forces (N)

K_t, K_r, K_p : Cutting force coefficients in tangential, radial and axial directions (N/mm2).

2   INTRODUCTION

The mechanistic approach of end milling was developed by Tlusty [1] and improved later on by DeVor [2] and Altintas [3]. It was expanded to model the ball-end milling process [4]. Yucesan and Altintas [5] presented a semi-mechanistic model which predicts the shear and friction load distribution on the rake and flank faces of the helical ball-end mill. Many studies

Published in: E. Kuljanic (Ed.) *Advanced Manufacturing Systems and Technology*,
CISM Courses and Lectures No. 486, Springer Wien New York, 2005.

were conducted in the area of mechanistic modelling for the conventional milling process. The qualities and limitations of this process are essentially dependent on the determination of the coefficients relating to chip geometry and cutting forces. As for plunge milling, little work was done and presented in literature. In the following, we propose the modelling of the plunge milling procedure by the mechanistic approach.

3   MECHANISTIC CUTTING FORCE MODEL IN PLUNGE MILLING

3.1   CUTTER GEOMETRY

The cutter geometry (Figure 1a) and the tool angles (Figure 1b) are defined in the cutter coordinate system (X, Y, Z). This tool geometry is mainly marketed by Mitsubishi [6].

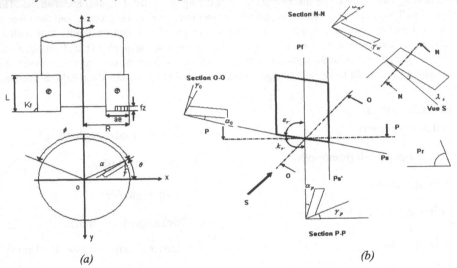

(a)   *(b)*

FIGURE 1. Cutter geometry

3.2   TOOTH TRAJECTORY

The relative movement of the cutter with respect to the workpiece consists of a rotation around the tool axis and a feed movement along the Z-axis.

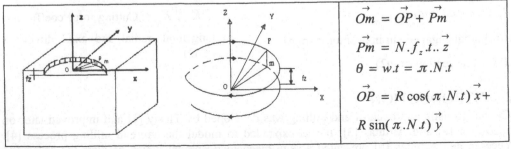

$$\vec{Om} = \vec{OP} + \vec{Pm}$$

$$\vec{Pm} = N.f_z.t..\vec{z}$$

$$\theta = w.t = \pi.N.t$$

$$\vec{OP} = R\cos(\pi.N.t)\,\vec{x} + R\sin(\pi.N.t)\,\vec{y}$$

FIGURE 2. Tooth trajectory

The trajectory of a tooth in the matter follows a cycloid curve determined by the movement of a point fixed on the cutting edge (Figure 2). The equation defining the trajectory of a tooth in the matter is given by: $\vec{Om} = R\cos(\pi.N.t)\,\vec{x} + R\sin(\pi.N.t)\,\vec{y} + N.f_z.t.\vec{z}$ (1)

3.3 CHIP GEOMETRY

The chip geometry during plunge milling is affected by different factors, main of which are the ratio of feed movement to cutting speed, the cutting geometry and the tool angles. In our study, the determination of cutting forces requires a good specification of the chip cross section. In general, we determine this cross section by multiplying the thickness of cut (h) by the radial engagement (AB).

RADIAL ENGAGEMENT

In our study, we distinguish two cases during plunge milling: (i) a simple case corresponding to the first passage of the cutter through the matter and (ii) a case corresponding to the radial engagement by taking into account the preceding passage of the cutter in the matter.

RADIAL ENGAGEMENT DURING THE FIRST PASSAGE (Figure 3)

The cutting geometry makes it possible to determine the radial engagement (AB) in a position $\theta \in [\theta_e, \theta_s]$ of the cutter. We consider the cutter completely engaged in the workpiece. The radial engagement is given in function of the rotation angle of the cutter by the following relation: $AB = R - (\dfrac{R - a_e}{\sin\theta}) = \dfrac{R\sin\theta - R + a_e}{\sin\theta} = \dfrac{R(\sin\theta - 1) + a_e}{\sin\theta}$ (2)

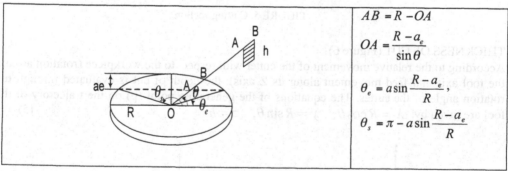

$$AB = R - OA$$
$$OA = \frac{R - a_e}{\sin\theta}$$
$$\theta_e = a\sin\frac{R - a_e}{R},$$
$$\theta_s = \pi - a\sin\frac{R - a_e}{R}$$

FIGURE 3. Cutting section

RADIAL ENGAGEMENT IN CASE OF MULTIPLE PASSAGES

In this case, the chip geometry is evaluated according to the rotation angle $\theta \in [\theta_e, \theta_s]$ during a complete rotation of the cutter (Figure 4). The entry and exit angles of the tooth in the matter are defined by: $\theta_e = a\cos\dfrac{p}{2R}$ and $\theta_s = \pi - a\sin\dfrac{R - a_e}{R}$ (3)

The radial engagement (AB) is given according to the rotation angle of the cutter θ (Figure 5). The relation is as follows: $AB = R - (\sqrt{R^2 - (P.\sin\theta)^2} + P.\cos\theta)$ (4)

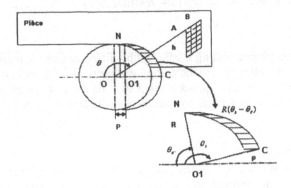

FIGURE 4. Chip geometry

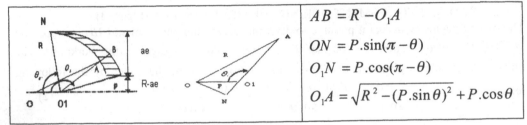

$$AB = R - O_1A$$
$$ON = P.\sin(\pi - \theta)$$
$$O_1N = P.\cos(\pi - \theta)$$
$$O_1A = \sqrt{R^2 - (P.\sin\theta)^2} + P.\cos\theta$$

FIGURE 5. Cutting section

THICKNESS OF CUT (Figure 6)

According to the relative movement of the cutter with respect to the workpiece (rotation around the tool axis and feed movement along its Z axis), the depth of cut is evaluated through the rotation angle of the cutter. The equations of the generating point (p) on the trajectory of the tool are given by: $X = R\cos\theta, \quad y = R\sin\theta, \quad z = h$ (5)

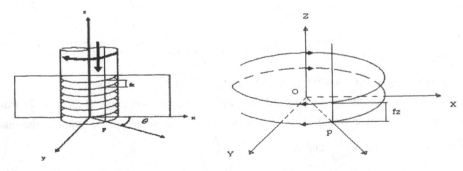

FIGURE 6. Cycloid trajectory of a tooth in the metal

To determine the thickness of cut, we considered three configurations of the tooth (Table 1) (at the entry, in the medium and at the exit).

TABLE 1.

Configurations	Engagement	Equations
1 (Figure 7a)	the cutter engages in the workpiece : $\theta \le 2\pi$	$\tan \alpha = \dfrac{fz}{2\pi} = \dfrac{h}{\theta} \Rightarrow h = \dfrac{\theta f_z}{2\pi}$
2 (Figure 7b):	the cutter completely engaged in the workpiece $2\pi \prec \theta \prec (n-1)\pi$	$h = f_z$
3 (Figure 7c):	the cutter leaving the workpiece: $\theta \succ (n-1)\pi$	$\tan \alpha = \dfrac{fz}{2\pi} = \dfrac{h}{2\pi - \theta} \Rightarrow h = f_z - \dfrac{\theta f_z}{2\pi}$

We distinguish two cases: one stationary (configuration 2) and the other non-stationary (configurations 1 and 3). For the determination of the cutting forces, we will place ourselves in configuration 1. The effects of the entry and exit into the workpiece are not taken into account. We assume that the stationary configuration is similar to that for drilling.

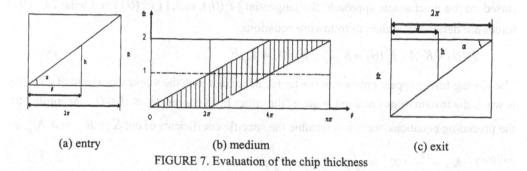

(a) entry   (b) medium   (c) exit

FIGURE 7. Evaluation of the chip thickness

DETERMINATION OF THE CUTTING SECTION

For the two configurations, the evolution of the chip section (Figure 8) is a function of the instantaneous rotation angle of the cutter.

We suppose that the maximum thickness of cut is ($hmax = fz$).

(i)   Full engagement in the metal (Figure 8.a): $A = f_z \cdot \dfrac{R(\sin \theta - 1) + ae}{\sin \theta}$ (6)

(ii)   Radial engagement by taking into account the preceding passage of the cutter Figure 8.b: $A = f_z \cdot (R - (\sqrt{R^2 - (P.\sin \theta)^2} + P.\cos \theta))$ (7)

We are using as well the second configuration to determine the cutting forces.

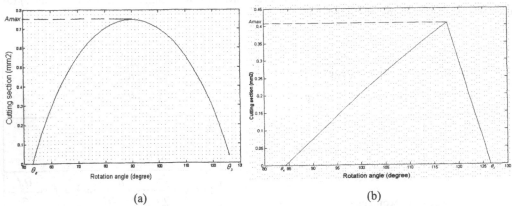

(a)   (b)

FIGURE 8. The evaluation of the cutting section with the rotation angle, ($R=25mm$,
$fz=0.15mm/rev.$, $ae=5mm$)

3.4 CUTTING FORCES FOR FLYCUTTING

The determination of the cutting forces enables us not only to make sound choices for the machine tools (power and torque) but also to pre-dimension the machine base where the piece is to be fixed.

Based on the mechanistic approach, the tangential ($F_t(\theta)$), radial ($F_r(\theta)$) and axial ($F_p(\theta)$) forces are defined according to following equations:

$$F_t(\theta) = K_t A \; ; \; F_r(\theta) = K_r F_t \; ; \; F_P(\theta) = K_P F_t \tag{8}$$

The cutting forces appear only when the tool is in contact with the workpiece (zone of cut), that is when the instantaneous rotation angle of the cutter lies between: $\theta_e \le \theta \le \theta_s$. According to the preceding equations, we can determine the specific coefficients of cut K_t, K_r and K_p as

follows : $\quad K_t = \dfrac{F_t}{A}, \; K_r = \dfrac{F_r}{A K_t}, \; K_p = \dfrac{F_p}{A K_t}$ \hfill (9)

We project the cutting forces $F_t(\theta)$, $F_r(\theta)$, $F_p(\theta)$ representing respectively the tangential, radial, and axial forces in the Cartesian coordinate system (Figure 9). The following equations determine $F_t(\theta), F_r(\theta), F_p(\theta)$ in function of F_x, F_y, F_z and the angular position of the cutter:

$$\left.\begin{cases} F_r = F_x.\cos\theta + F_y.\sin\theta \\ F_t = -F_x.\sin\theta + F_y.\cos\theta \\ F_p = F_z \end{cases}\right\} \Rightarrow \left\{ \begin{bmatrix} F_r(\theta) \\ F_t(\theta) \\ F_p(\theta)_z \end{bmatrix} = \begin{bmatrix} \cos\theta & \sin\theta & 0 \\ -\sin\theta & \cos\theta & 0 \\ 0 & 0 & 1 \end{bmatrix} . \begin{bmatrix} F_x(\theta) \\ F_y(\theta) \\ F_z(\theta) \end{bmatrix} \right\} \tag{10}$$

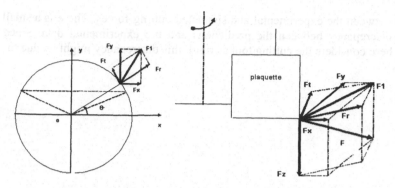

FIGURE 9. Analysis of the forces applied on the piece by tooth

4 VALIDATION OF THE MODEL

4.1 EXPERRIMENTAL SYSTEM

A set of experiments was performed on the Univer 700 vertical milling machine for the model validation. The workpiece material used was 40CrMnMo8. The cutter was (Mitsubishi, PMR 405003A22R) with diameter of 63 mm, on which an insert was mounted throughout the experiment. The conditions set for this experiment were selected as the common conditions for plunge milling. The generated tool path is transferred to the data server of the CNC system via an ethernet communication. A tool dynamometer (Kistler) is mounted on a tilting vice under the workpiecc to measure the cutting force signal (Figure 10). This signal is amplified through a charge amplifier and recorded on a data recorder. The sampling rate for data acquisition is 5 kHz. All cutting conditions are the same as those used in the simulation.

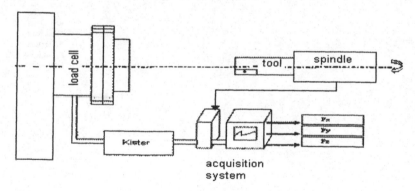

FIGURE 10. Testing configuration

4.2 CALIBRATION RESULTS AND VALIDATION

Typical experimental cutting forces and the corresponding model predictions are shown in (Figures 11-12). These figures represent the simulated forces for one revolution of the cutter. The comparison between experimental and simulated forces is chosen for cutting conditions similar to those for rough machining in plunge milling. In general, one can notice a good agreement between the experimental and simulated cutting forces. There is a small amount of amplitude discrepancy between the predictions and the experimental data. Since the model developed here considers the cutting tool as rigid, this discrepancy might be due to cutting tool deflections.

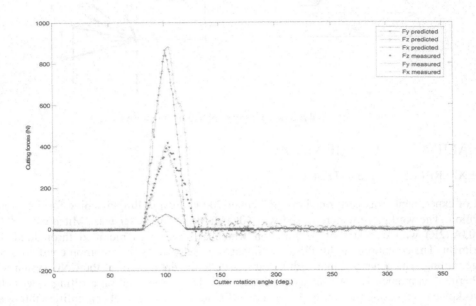

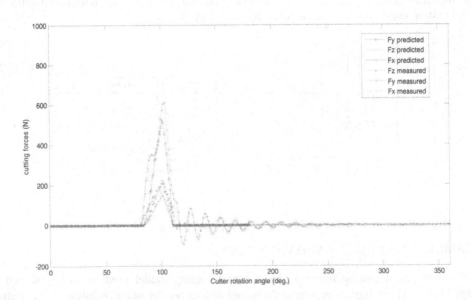

FIGURE 11. Measured and predicted cutting forces. Cutting conditions: ae=2 mm, fz=0.15mm/tooth, VC=150mm/min, (a) P=10mm, (b) P=5mm

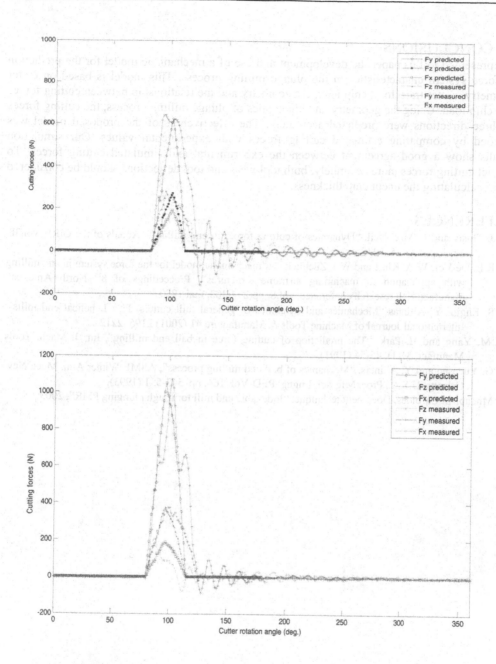

FIGURE 12. Measured and predicted cutting forces. Cutting conditions: P=10 mm,
fz=0.15mm/tooth, VC=150mm/min, (a) ae=2mm, (b) ae=3mm

5 CONCLUSIONS

We presented in this paper the development and use of a mechanistic model for the prediction of force system characteristics in the plunge milling process. This model is based on cutter geometry, cutter trajectory, chip load, cut geometry and the relationship between cutting forces and chip load. Using the geometry and kinematics of plunge milling process, the cutting forces in three directions were predicted accurately. The effectiveness of the proposed model was verified by comparing estimated cutting forces with experimental values. Our simulation results show a good agreement between the experimental and simulated cutting forces. To predict cutting forces more accurately, both tool wear and tool deflections should be considered when calculating the uncut chip thickness.

REFERENCES

1. J. Tlusty and P. Mac. Neil, "Dynamics of cutting forces in end milling". Annals of the CIRP, vol.30, No.1, pp.21 -25, 1975.
2. R.E. De Vor, W.A. Kline and W.J. Zdeblick. "A mechanistic model for the force system in end milling with application to machining airframe structures". Proceedings of 8th North American manufacturing research conference, vol.8, pp.297-303, mai 1980.
3. S. Engin, Y. Altintas "Mechanics and dynamics of general mill cutters. Part I: helical end mills". International Journal of Machine Tools & Manufacture 41 (2001) 2195–2212.
4. M. Yang and H. Park. "The prediction of cutting force in ball end milling", Int. J. Mach. Tools Manufact. 31(1), 45-54 (1991).
5. G. Yucesan and Y. Altintas, "Mechanics of ball end milling process", ASME Winter Ann. Meet, New Orleans, U.S.A., Proc. Mfg Sci. Engng. PED-Vol. 164, pp. 543-551 (1993).
6. Mitsubishi Carbides, Document technique, "Indexable end mill for rough plunging PMR", 2003.

INVESTIGATION OF TOOLPATH INTERPOLATION METHODS IN THE HIGH-SPEED MANUFACTURE OF MOLDS AND DIES

K. Schützer, A. L. Helleno

Laboratory for Computer Application in Design and Manufacturing (SCPM),
Methodist University of Piracicaba (UNIMEP), Santa Bárbara d'Oeste, Brazil

KEYWORDS: Milling, Spline, Toolpath Interpolation.

ABSTRACT. The application of high speed machining of molds and dies with their high complex geometry have made the traditional methods of toolpath interpolation normally used by CAM systems, i.e., linear and circular interpolations, the bottle neck of the whole process. These traditional interpolation methods increase the machining time, have a negative influence in the final quality of the surfaces and are a technological limitation for the full application of HSC. As a result, the study of new methodologies on toolpath interpolation is becoming one of the main areas of research in manufacturing of molds by HSM. Among the different methodologies of toolpath interpolation there are the linear, circular and polynomial. The milling experiments were made using work pieces of AISI SAE P20 Steel and the geometry from the NC Gesellschaft test part.

1 INTRODUCTION

Within the CAD/CAM/CNC chain applied to the development of molds and dies, the toolpath is determined based on a tolerance defined by the user of the CAM system, which is applied to the geometric model of the product to be manufactured. In Figure 1 one can observe that this tolerance is directly related to the accuracy of the toolpath, to the size of the NC program and, consequently, to the calculation time.

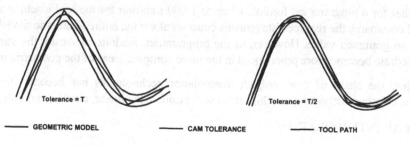

FIGURE 1. Toolpath in function of CAM tolerance

The commercially available CAM systems offer diverse interpolation methods in order to obtain a toolpath in the manufacturing of complex surfaces that better adapts to the tolerance range stipulated by CAM.

Published in: E. Kuljanic (Ed.) *Advanced Manufacturing Systems and Technology*,
CISM Courses and Lectures No. 486, Springer Wien New York, 2005.

In traditional manufacturing, in which the demands on the feedrate are significantly less, the tool-path interpolation methods were disregarded as a resource for the CAM systems. This turned linear interpolation into the standard due to its mathematical simplicity and ease of use for the programmer.

However, with the application of HSM Technology where the demands on the feedrate are higher, the linear interpolation method begins to create various limitations, primarily in relation to the rate of machining and the accuracy of the geometric model. This has resulted in the reconsideration of the utilization of toolpath interpolation in the generation of the NC program [1 - 5].

Figure 2 illustrates the values of the real feedrate in function of the increase in programmed feedrate obtained in an experiment at the Laboratory for Computer Application in Design and Manufacturing (SCPM/UNIMEP). In these tests the NC program was generated using linear interpolation.

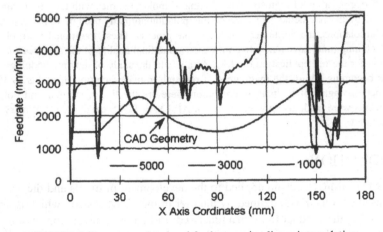

FIGURE 2. Programmed and real feedrates using linear interpolation

Observe that for a programmed feedrate of up to 1,000 mm/min for the tool machine and tested numerical command, the real feedrate remains constant along the entire geometric model and with the same programmed value. However, as the programmed feedrate increases, the variations of the real feedrate become more pronounced in the more complex areas of the geometric model.

Due to this, the study of new toolpath interpolation technologies has become fundamental. Among the existing methodologies, this article will evaluate the linear, circular and polynomial.

1.1 LINEAR INTERPOLATION

In linear interpolation, the CAM system determines the toolpath through the interpolation of straight line segments that best adapt to the tolerance range of the CAM system (see Figure 1). These straight line segments are represented by the G01 command of the ISO 6983 programming language. Because linear interpolation uses straight line segments to represent the toolpath, it has a simpler mathematical representation than other methods. For this reason it has become the most popular method for representing toolpaths.

However, as can be observed in Figure 2, the increased feedrate from the new HSC technology have resulted in making the use of linear interpolation for representing the toolpaths technologically limiting.

This fact occurs primarily because of one technical characteristic of the numerical command called block processing time (BPT). This is the time that the CNC takes to read a block of information, process and transmit this information for the machine to execute the movement [6].

Therefore, despite the difficulty in determining the processing time of the machine tool, its study is of extreme importance, for if the block processing time is longer than the block execution time, the machine will reach the end point of the segment before the information for the next movement is available.

In this case the more modern CNCs automatically reduce the programmed feedrate to be compatible with the processing time, resulting in a lower real feedrate and, consequently, a longer machining time.

1.2 CIRCULAR INTERPOLATION

Through this method, the CAM system determines the toolpath through an association of straight line segments and arcs that best adapt to the tolerance limits of the CAM system. These segments are represented by the G01, G02 and G3 commands of the ISO 6983 language (see Figure 3).

While with linear interpolation the complex surfaces are represented only by straight line segments, in the circular interpolation these small segments, when possible, are substituted by arcs, resulting in a smoother toolpath, smaller NC programs and, consequently, better real feedrate performance.

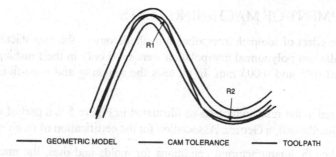

FIGURE 3. Circular interpolation representation

1.3 POLYNOMIAL INTERPOLATION

The application of polynomial interpolation was the beginning of a new phase in the methodologies of toolpath interpolation. The toolpath would no longer be represented through the use of elementary geometric elements (straight lines and arcs), as occurs in linear and circular interpolation, but by segments of curves (C0, C1, ..., Cn) based on the mathematical models used by the CAD system in the representation of complex surfaces. With this, CAM systems can determine a smoother and more precise toolpath that can be adapted to the tolerance limits of the CAM system (see Figure 4).

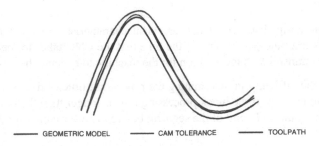

FIGURE 4. Polynomial interpolation representation

In the case of polynomial interpolation, these curves are defined by a polynomial function of degree 3 or 5, where the variables are the coefficients of the polynomial function and the p variable is the parameter of the generation of the curve, as illustrated in the following equations [8]:

$$f(p) = a_0 + a_1 \cdot p + a_2 \cdot p^2 + a_3 \cdot p^3 \tag{1}$$

$$f(p) = a_0 + a_1 \cdot p + a_2 \cdot p^2 + a_3 \cdot p^3 + a_4 \cdot p^4 + a_5 \cdot p^5 \tag{2}$$

The generated NC program no longer contains the G01, G02 and G03 commands of the ISO 6983 language, but rather a specific language for each numeric command that represents these polynomials.

2 DEVELOPMENT OF MACHINING TESTS

To investigate the effect of toolpath interpolation methodology in the manufacture of molds and dies, linear, circular and polynomial interpolation were used only in the finishing operation with CAM tolerance of 0.05 and 0.005 mm. In all tests the roughing and semi-finishing operations were identical.

The test model used in the machining tests as illustrated in Figure 5 is a part of a test model created by the NC-Gesellschaft, a German Association for the certification of machining centers [7].

In order to simulate the manufacturing conditions for molds and dies, the manufacturing tests were done in AISI-SAE P20 steel provided by Aços Villares S.A..

FIGURE 5. Test model used in experiments [7]

The machining was done at the Romi Machining Center, Discovery 760 model, equipped with Siemens 810D command configured especially for the SCPM to develop research tests using complex methodologies of toolpath interpolation (Polynomial, Spline and NURBS).

The finishing operation being studied was done with a ball-end mill cutter with a diameter of 16 mm with two cutting edges for interchangeable inserts and class P10A inserts, provided by Sandvik do Brasil S.A.. The following parameters were used in the finishing operation:

- rotation of tree axis (n): 10,000 min^{-1};
- cutting speed (V_c): 502 m/min;
- feedrate (V_f): 5,000 mm/min.

A comparative analysis of toolpath interpolation methodologies was carried out according to the following characteristics:

- machining time of test model;
- behavior of the real machining feedrate along the geometry of the test model;
- accuracy of the test model geometry in comparison with the geometric model developed by the CAD system;
- surface quality.

3   ANALYSIS OF THE RESULTS

3.1   MANUFACTURING TIME

As can be observed in Figure 6, the polynomial interpolation presented better results with a reduction of 6% and 32% in relation to linear interpolation with a CAM tolerance of 0.05 mm and 0.005 mm, respectively.

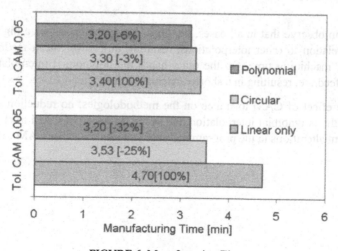

FIGURE 6. Manufacturing Time

In comparing the interpolation methodologies and the effect of the CAM tolerance, one can observe that:

- the polynomial interpolation, in which the toolpath is represented by curve segments, does not show alterations in machining time in function of CAM tolerance;

- the reduction of machining time in function of tolerance could only be verified for linear and circular interpolations.

3.2 PERFORMANCE OF THE REAL FEEDRATE

Figure 7 and 8 show that despite the programmed feedrate of 5,000 mm/min, the real feedrate fluctuated along the geometry of the test model. Also, different characteristics were found among the toolpath interpolation methodologies.

This reduction in real machining feedrate occurred primarily in function of the programmed segment size and the block processing time of the machine tool.

For linear, circular and polynomial interpolation the programmed segment size for each of the NC programs was determined and compared with the real rate of machining.

Equation 3 supplies the estimate of the maximum feedrate in function of the processing time and of the size of the movement segment [6].

$$V_f = \frac{\Delta x}{BPT/60} \tag{3}$$

Through this equation, it is possible to obtain the size of the minimum segment for a certain feedrate and to compare it with the size of the segments contained in the toolpath of the NC program.

Considering a feedrate of 5,000 mm/min (V_f) and a block processing time of 12 ms (BTP), calculated through preliminary tests, it is possible to determine the size of the minimum segment equal to 1 mm (Δx).

From this, one can observe that in all cases, the linear interpolation shows a higher reduction in real feedrate in relation to other interpolations, resulting in a lower mean feedrate and, consequently, a longer machining time. On the other hand, the polynomial interpolation shows less reduction in real feedrate, resulting in a shorter machining time.

In relation to the effect of CAM tolerance on the methodologies, no reduction on real feedrate was observed in the polynomial interpolation due to the influence of CAM tolerance since there were no significant alterations in the programmed segment size for either CAM tolerance.

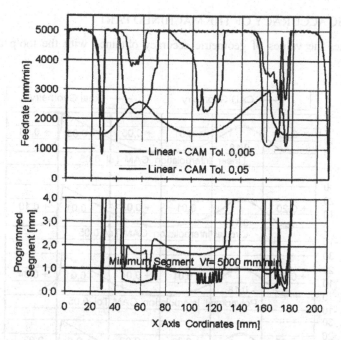

FIGURE 7. Performance of the real feedrate - Linear Only (Black line: CAD geometry)

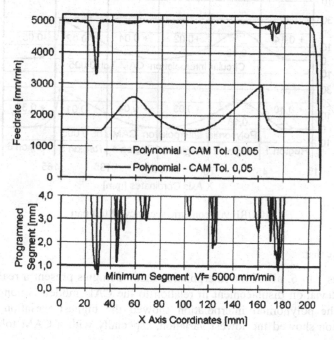

FIGURE 8. Performance of the real feedrate - Polynomial (Black line: CAD geometry)

3.3 GEOMETRIC ACCURACY OF THE MACHINED PART

Figure 9 illustrates the values of geometric accuracy obtained with the toolpath interpolation methodologies.

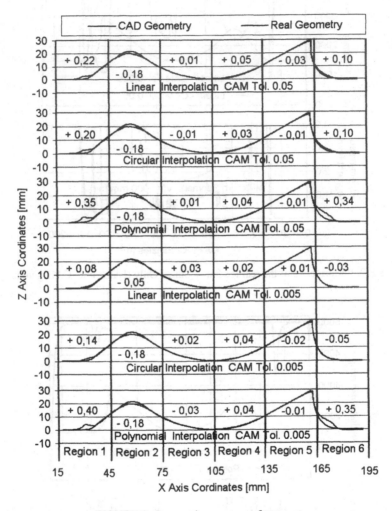

FIGURE 9. Geometric accuracy of test part

These results show that:

- for regions 1 to 6, all of the interpolation methodologies present a real geometry with outward deviation displacement in relation to the CAD model. Among these methodologies, the polynomial interpolation showed the highest variation and the linear interpolation showed the lowest variation, especially with a CAM tolerance of 0.005 mm;

- in region 2 all of the interpolation methodologies show a real geometry with inward deviation displacement in relation to the CAD model. Similar geometric variations of around 0.18 mm appeared in all of the methodologies with the exception of linear interpolation with a CAM tolerance of 0.005 mm, which showed a variation of around 0.05mm;

- regions 3, 4 and 5 presented a real geometry similar to the CAD model for all of the interpolation methodologies. The variation of the real geometry at the end of region 5 was disregarded due to lack of a good contact region for carrying out the measurement.

3.4 SURFACE ASPECT

By measuring the test model roughness, it was possible to observe that:

- all of the toolpath interpolation methodologies presented roughness values inferior to 0.50 μm Ra;

- variations of CAM tolerance in the interpolations did not affect surface roughness.

In addition to the roughness measure, a comparison of superficial aspects was carried out where it was possible to observe a "faceted" aspect on the complex machined surfaces with Linear and Circular interpolation.

4 CONCLUSIONS

Based on the results obtained, the following conclusions can be drawn:

- the CAM tolerance used in the generation of the toolpath influences the performance of the toolpath interpolation methodologies in different ways. While for Linear Only and Circular Interpolations a reduction in CAM tolerance results in a significant increase in machining time and in a reduction in deviance from the real geometric model, in Polynomial Interpolation the variation in tolerance significantly influenced only the surface aspect. This occurred because Linear and Circular Interpolation use a large quantity of small straight line segments to represent complex surfaces with small CAM tolerance ranges;

- Linear Interpolation was shown to be a more precise methodology primarily when there was a reduction in CAM tolerance. However, the generation of straight line segments smaller than the minimum established for the processing time of the machine tool and the faceted surface aspect in complex surfaces makes this methodology a limitation in the high-speed manufacture of molds and dies;

- Polynomial Interpolation has become one of the solutions for the manufacture of molds and dies, due to its performance in relation to machining time. However, high deviance in accuracy related to a faster machining feedrate and, consequently, a higher demand for dynamic behavior of the machine tool, can compromise the subsequent steps of adjustment and polishing of the mold, causing an increase in the lead time of the product;

- Circular Interpolation by showing a better performance in machining time and in accuracy of the real geometric model when compared to Linear and Polynomial Interpolations, as well as a better surface aspect when a greater CAM tolerance is applied, can be considered a solution in the search for optimizing the manufacturing process of molds and dies. However, because it also uses straight line segments to represent the toolpath, it presents the same problems of Linear Interpolation in the manufacture of molds and dies. This makes it also a limitation in the high-speed manufacture of molds and dies.

The objective of studies on toolpath interpolation methodologies is the manufacture of molds and dies with the HSC technology. Results demonstrate that even when machining at feedrates within the transition range for HSC Technology there are limitations in the application of Linear Interpolation as well as benefits that can be found with the use of other interpolation methodologies.

ACKNOWLEDGMENTS

This study was done with the support of CAPES, a Brazilian federal agency that promotes scientific research.

REFERENCES

1. Lartigue, C., Tournier, C., Ritou, M., Dumur, D., 2004, High-performance NC for HSM by means of polynomial trajectories, Annals of the CIRP, 53/1:317-320.
2. Altintas, Y., Erkorkmaz, K., 2003, Feedrate optimization for spline interpolation in high speed machine tools, Annals of the CIRP, 52/1:297-302.
3. Koninckx, B., Van Brussel, H., 2002, Real Time NURBS interpolation for distributed motion control, Annals of the CIRP, 51/1:315-318.
4. Yau, H.T., Kuo M.J., 2001, NURBS machining and feed rate adjustment for high-speed cutting of complex sculptured surfaces, International Journal of Production Research, 39/1:21-41.
5. Langeron, J.M., Duc, E., Lartigue, C., Bourdet P., 2004, A new format for 5-axis toolpath computation using Bspline curves, Computer-Aided Design, 36:1219-1229.
6. Arnone, M., 1998, High performance machining, Hanser Gardner Publications, Cincinnati, USA
7. NC Gesellschaft Recommendation, 2000, Milling Cutters and Machining Centers, NCG2004/Part1
8. Sinumerik 840D/840Di/810D, 2000, Programming Guide Advanced
9. Stoeckhert, K., Menning, G., 1998, Mold making Handbook, Hanser/Gardner Publications, Ohio, USA

WIPER TOOLS IN TURNING FINISHING
OF QUENCHED STEEL

C. Borsellino[1], E. Lo Valvo[2], V. F. Ruisi[2]

[1] Department of Industrial Chemistry and Materials Engineering, University of Messina, Italy
[2] Department of Technology and Mechanical Production, University of Palermo, Italy

KEYWORDS: Turning, Wiper insert, Cutting Parameters.

ABSTRACT. In the last years, the research on metal machining has been focused on the development of tools that can support an increment of the cutting parameters instead of the increment of tool life; this strategy is focused to obtain a relevant productivity increment without getting worse the final product quality. The new design of the tool corner radius has allowed preserving the same product quality due to the "wiper effect" generated during the cut.

The employ of such re-designed tools can allow substantial productivity improvements and, by employing the same feedrate, it is possible to achieve excellent surface finish and eliminate the grinding operation. With the aim to verify the possibility to realize good finishing on hardened materials with the above mentioned new tool-manufacturing technology the Authors conducted a set of turning tests on a quenched steel.

The quality of the manufactured products are compared with the one obtained by grinding. The final product qualities are compared in terms of micro geometrical characteristics and tolerances of the obtained workpiece surfaces are investigated to completely determine its mechanical performances.

1 INTRODUCTION

Highly-stressed steel components, e.g. gear and bearing parts, require hardening, to increase wear resistance. Even today, mechanical finishing still relies on energy and cost-intensive grinding processes. It was only with the development of modern superhard cutting materials with geometrically defined cutting edges (PCBN, ceramics) that an alternative to grinding emerged. Substantial advantages in terms of more flexible, lower cost, more environmentally-friendly production can be achieved by replacing grinding with this innovative technology.

Despite these evident advantages, industrial realization of hard machining has remained slight in comparison with the potential spectrum of applications. The clearly unsatisfactory industrial acceptance of hard machining technology can be attributed partly to insufficient knowledge of the component behavior of hard machined technical surfaces and partly to the uncertainty about the attainable accuracies-to-size-and-shape [1].

In addition to uncertainties with regard to the properties of hard turned surfaces, very high shape and accuracy requirements and surface quality requirements have frequently represented a stumbling block to industrial application of hard turning technology. Simple grinding quality, i.e. geometric tolerances corresponding to IT6 and surface qualities of Ra 2-3 μm can be regarded as the maximum level of quality which can be obtained using the currently available precision lathes. According to Taniguchi's diagram, the development of achievable machining accuracy, grinding as a typical "precision" and "high precision" machining process can be placed in terms

Published in: E. Kuljanic (Ed.) *Advanced Manufacturing Systems and Technology*,
CISM Courses and Lectures No. 486, Springer Wien New York, 2005.

of the levels of accuracy achievable between conventional hard-turning and ultra-precision machining of non ferrous metals with mono crystalline diamonds ("ultra-precision machining"). These levels of accuracy are insufficient to permit hard machining to be used as a substitute for many precision grinding operations [2].

In recent times the tools that can exploit the wiper effect have been produced [3-4]. The new design of the tool corner radius has allowed preserving the same product quality due to the "wiper effect" generated during the cut. They allow substantial productivity improvements or, which is the same, to reduce the time of cut and then to reduce production costs. In particular, if the feed can be increased to double, the cutting time is halved and almost twice as many components are made in the same time maintaining the same surface finish [5]. Otherwise, by employing the same feedrate it is possible to achieve excellent surface finish and eliminate, in such cases, the grinding operation. This last objective can be reached if also high rigidity CNC lathe are employed.

With the aim to verify the possibility to realize good finishing on hardened materials with the above mentioned new tool-manufacturing technology the Authors conducted a set of turning tests on a quenched steel.

The quality of the manufactured products are compared with the one obtained by turning with grinding. The final product qualities are compared in terms of micro geometrical characteristics of the obtained surfaces (roughness).

2 EXPERIMENTAL SET-UP

2.1 MATERIAL

The cutting tests were conducted on a "Graziano SAG 12" CNC lathe. The workpiece material and geometry characteristics are reported in the following Table 1:

TABLE 1. Workpiece material characteristics and geometry

Chemical composition %	C = 0.13-0.18; Mn = 0.8-1.00; Si = 0.15-0.4; P_{max} = 0.035; S_{max} = 0.035; Cr = 0.8-1.1; Ni = 0.8-1.1
Hardness	HRc 20 - HV = 500 [Kg/mm²]
Tensile strength	R = 680 N/mm²
Geometry	Diameter= 30 [mm]; Length=140 [mm]

The workpieces undergone to a carburizing treatment and a quenching to obtain a final external hardness of 58 HRc (1050 HV) for a depth of about 1 mm.

2.2 TOOLS

Two commercially available wiper inserts were selected for the tests, according to the insert number WNGA0804 04/08; in particular two different tools materials were employed, their characteristics are reported in Table 2.

TABLE 2. Wiper inserts

Tool	Material	Radius
A	CBN 7020	0.4 and 0.8
B	Ceramic 6050	0.4 and 0.8

The tested inserts were mounted on a commercial tool holder (MWLR 2525M 08) with the following geometry:

- o rake angle $\gamma = 6°$
- o clearance angle $\alpha = 6°$
- o minor clearance angle $\alpha' = 6°$
- o side cutting edge angle $\psi = -5°$
- o working approach angle $\psi' = 5°$
- o inclination angle $\lambda = -6°$

2.3 CUTTING PARAMETERS

Each kind of insert has been tested three times with the following cutting parameters:

- o depth of cut: 0.05 mm
- o feeds: $f_1 = 0.1$ mm/rev; $f_2 = 0.2$ mm/rev
- o speeds: $V_{1C} = 2.5$ m/s; $V_{2C} = 4.2$ m/s

2.4 MEASUREMENTS

At the end of each cut, the image of the secondary edge was acquired employing a processing image system as shown in Figure 1. It is composed by a CCD camera (Sony AVC-D50CE monochrome, high resolution) connected with a PC, properly equipped to store and process the acquired images.

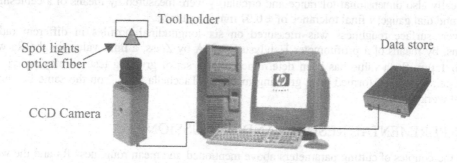

FIGURE 1. The image processing system employed for the tool wear evaluation

On the acquired image, the value of the secondary flank wear was measured employing the commercial software "Inspector 5.1" by Matrox. This measure was performed ad fixed time interval until the value of VB'$_B$ = 0.2 mm was reached, for each test.

Among the wear criteria, only the value of the secondary flank wear has been taken into account, because the crater and primary edge wear was absolutely absent as it is possible to notice in the following Figure 2.

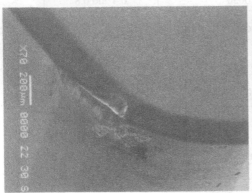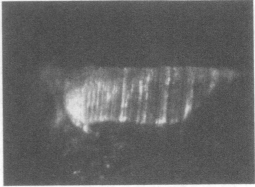

FIGURE 2. The wear of a secondary edge - Tool A

On the left a SEM photograph is shown where it is possible to notice that the wear is only localized on the secondary edge into a flank wear and a small crater on the chamfer between the secondary plane and the rake face; on the righ side the image of the secondary flank acquired by the image processing system shown in Figure 1.

In this last the presence of grooves is evident. During the cut the grooves number and their depth augment and the measured Ra value is very low; when the wear phenomena increase the groove number decreases but also their size grow until they begin to interfere each other leaving a zone of the minor cutting edge heavily worn with chipping, causing a massive decay of the surface finishing conditions [6-7].

Periodically also dimensional tolerance and circularity were measured by means of a centesimal caliper and dial gauge; a final tolerance of ± 0.01 mm is assumed.

Moreover, surface roughness was measured on six longitudinal profiles in different radial positions, by means of a profilometer Handysurf E-35A by Zeiss, a limit value for the Ra was fixed to 1 μm. Such value has been determined after a set of grinding test, employing typical cutting parameters, performed on a grinding machine "Tacchella 612U" on the same hardened material workpiece.

3   EXPERIMENTAL RESULTS AND DISCUSSION

For all the couples of cutting parameters above mentioned, the mean roughness Ra and the wear trends are shown in Figure 3a, b for the two kind of tools employed, called A and B.

The lifetime is evaluated on the base of a flank wear limit VB'$_B$ = 0.2 mm or on the roughness limit Ra= 1μm.

It is evident from the results shown in Figure 3 that the end of the tool life is determined by different condition at varying tool radius: all tools with radius 0.4 mm reach the Ra limit when the value of the flank wear is still much lower than its limit value.

The tools with radius 0.8 instead allow to obtain very low roughness values even if the secondary flank wear reaches the limit.

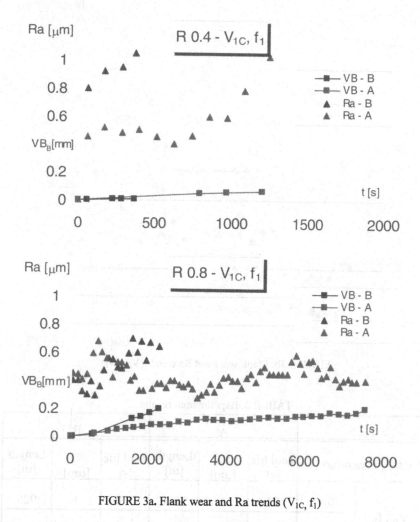

FIGURE 3a. Flank wear and Ra trends (V_{1C}, f_1)

Both kind of tools exploit the wiper effect producing always very good finishing, for the examined cutting parameters; the different behavior can be explained considering that the tools with higher radius offer a much extended area of the tool tip involved in the cut that can still perform a good wiper effect also if the flank wear is much consistent.

In Table 3 the obtained tool lives, mean roughness and the correspondent cutting lengths are summarized for all the above-mentioned tools and cutting parameters V_{1c} and V_{2c} at feed f_1.

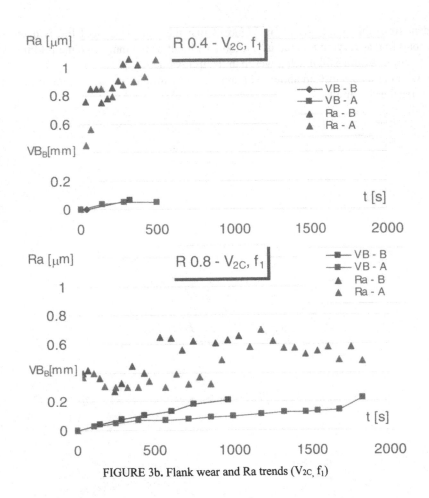

FIGURE 3b. Flank wear and Ra trends (V$_{2C}$, f$_1$)

TABLE 3. Experimental results

Tools		A			B		
Cutting Parameters		Tool life [s]	Ra [μm]	Length [m]	Tool life [s]	Ra [μm]	Length [m]
V$_{1C}$, f$_1$	0.4	1230	1	3080	370	1	925
	0.8	7580	0.4	18950	2280	0.65	5700
V$_{2C}$, f$_1$	0.4	500	1	2080	285	1	1190
	0.8	1820	0.6	7590	960	0.6	4000

The tool-life values measured from the experimental trends are reported in the next Table 4.

TABLE 4. Tool life [s]

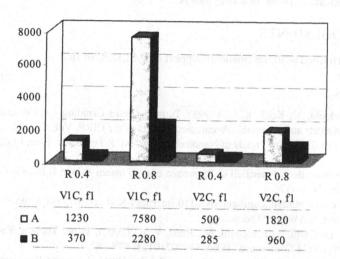

	R 0.4 V1C, f1	R 0.8 V1C, f1	R 0.4 V2C, f1	R 0.8 V2C, f1
☐ A	1230	7580	500	1820
■ B	370	2280	285	960

The tool lives of A tools are always greater than the B ones, from two to more than three times, moreover it is evident that tools with radius 0.8 mm always show longer lives than the 0.4 ones. The increase in speed as usual reduces the tool life of all tools of about two-three times.

With the aim to verify the possibility of increase productivity, also the value of f_2 with a speed V_{2C}, was tested (even if the theoretical value of Ra, ottenibile with a conventional tool radius, would be surely greater than the imposed limit). With this high value of feed both A and B tool lives are reached because the VB'$_B$ limit is overcame; the tools with radius 0.4 do not allow a satisfactory finishing, while the 0.8 ones reach the imposed limit after very low cutting times (about 150 s the B and 450 the A - being the corresponding cutting length 625 and 2080 m).

To be able to perform a deepened comparison of the mechanical characteristics of the finished parts, also the residual stress status will be investigated from the Authors.

4 CONCLUSIONS

On the basis of the obtained results we can draw that:

- The employ of wiper tools allows the effective cost reduction of finishing operation on hardened steel mechanical parts; by turning with high precision and rigidity lathes is possible to obtain very low roughness and geometric tolerances.

- The high tool lives of the inserts employed and the corresponding cutting length show that a mechanical part can be finished or super-finished in few seconds allowing a great increasing in productivity.

- Tool's quality and geometry and cutting parameters have to be correctly chosen to perform good finishing quality. In particular, for the material selected for the tests, the tool B

(Ceramic Wiper insert) has a cost of about 1/10 of tool A (CBN Wiper insert). On the converse it shows a lifetime of about 1/3 of tool A, then the cost for each part finished with a tool B is about 1/3 of the one with tool A.

ACKNOWLEDGEMENTS

This work was performed with the financial support of M.I.U.R. of Italy.

REFERENCES

1. Konig, W., Berktold, A., Koch, K.-F., (1993), Turning versus Grinding - A Comparison of Surface Integrity Aspects and Attainable Accuracies, Annals of the CIRP, Vol. 42/1/93, 61-64.
2. Taniguchi, N., (1992), Future Trends of Nanotechnology, Int. J. Japan Soc. Prec. Eng., Vol. 26, No. 1, 1-7.
3. N.N., (2001), Sorprendenti Potenziali di Risparmio con gli inserti Wiper, Il Mondo delle Lavorazioni Meccaniche, 2, 34-37.
4. http://www.coromant.sandvik.com/sandvik/0110/Internet/I-Kit1/se02673.nsf/Alldocs/ Wiper*Inserts*2AWiper*Technology
5. Borsellino, C., Lo Casto, S., Piacentini, M., Ruisi, V. F., (2003), Wiper Tools in Turning of Carbon Steels, 6th AITeM Conference, ISBN 88890210121, 578-582.
6. Borsellino, C., Lo Valvo, E., Piacentini, M., Ruisi, V. F., (1996). A new on-line roughness control in finish turning operation, Advanced Manufacturing Systems and Technology, CISM Courses and Lectures No. 372, Kuljanic, E. (Ed.), Springer Wien New York, 661-668.
7. Borsellino, C., Piacentini, M., Ruisi, V. F., (1999), Minor Cutting Edge Wear Influence in Finish Turning Operations, Advanced Manufacturing Systems and Technology, CISM Courses and Lectures No. 406, Kuljanic, E. (Ed.), Springer Wien New York, 163-170.

A STUDY OF FACTORS THAT AFFECT THE BUILD-UP MATERIAL FORMATION

N. Tomac[1], K. Tønnessen[2], F. O. Rasch[3] and T. Mikac[4]

[1] Department of Polytechnic, Faculty of Philosophy, University of Rijeka, Croatia
[2] SINTEF Production Engineering, Trondheim, Norway
[3] NTNU Norwegian University of Science and Technology, Trondheim, Norway
[4] Faculty of Engineering, University of Rijeka, Croatia

KEYWORDS: Machining, BUE, Formation Factors.

ABSTRACT. A feature when cutting many alloys is that workpiece material adheres to the cutting tool at the sliding contact surfaces, between the work material and the tool. This built-up material formed during cutting is of fundamental importance in machining operations, because it may significantly affect the surface roughness, tool wear, workpiece dimensions and tolerances, tool forces, and chip form. The agglomeration of the work material to the tool appears to be analogous to cold welding, metal transfer in tribology and dead zone in extrusion. In machining terminology this phenomenon is often called "built-up edge" (BUE). Several important factors affect the built-up material formation, e.g. cutting temperature, cutting speed, strain hardening, adhesion between the work material and the tool, micro-crack formation, plastic flow of the work material in the vicinity of the cutting edge, etc.

1 INTRODUCTION

The built-up material formed during cutting is of fundamental importance in machining operations, because it may significantly affect the surface roughness, tool wear, workpiece dimensions and tolerances, tool forces, chip form, etc. Figure 1 schematically depicts a model of cutting process with the three main forms of material build up and various factors which may affect on their formation.

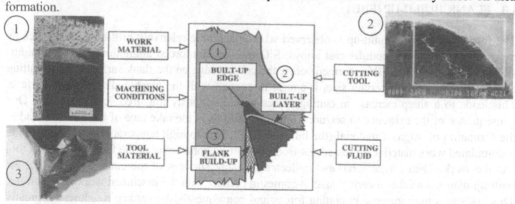

FIGURE 1. Model of cutting zone with three main forms of built-up material and factors which affect their formation

Published in: E. Kuljanic (Ed.) *Advanced Manufacturing Systems and Technology*,
CISM Courses and Lectures No. 486, Springer Wien New York, 2005.

1.1 BUILT-UP EDGE (BUE)

When machining a number of alloys a built-up edge may form on the rake face of the cutting tool and this can have considerable influence on surface quality, dimensional tolerances, cutting forces, tool wear and chip shape. The built-up edge is a semi stable body of workpiece material, which usually overhangs the cutting edge and periodically fractures away adhering to the machined surface and/or chip. The introduction of new tool materials and machine tools, which enable high cutting speeds at which the large BUE is not formed, has diverted attention from the BUE problems. Regardless of this, the accuracy of machined parts is not always as could be expected, especially in finish operations, despite the high accuracy of the machine tools, and carefully chosen tooling. This may be explained by the fact that a small BUE is always formed on the cutting tool and causes a deterioration of the accuracy of the machined operation.

1.2 BUILT-UP LAYER (BUL)

During the machining of free-cutting alloys inter-metallic and/or non-metallic build-up layers are formed on the tool face causing an improvement in surface roughness of the machined part and tool service life by decreasing crater and/or flank wear and lowering tool forces. The BUL provides additional lubrication between the contact surfaces of the tool and workpiece, reducing their contact areas and causing a reduction in tool forces and temperature, thus enabling an increase in metal removal rate for a given tool life. A BUL can be detected on the PCD tool inserts. Energy disperative X-ray (EDX) was used to estimate the micro content of the BUL on a PCD tool. The manganese content in the BUL was found to increase reaching a value of about 20%, in comparison to the workpiece material, where the manganese content is 0.17 % [1]. Microscopic examinations of the magnesium alloys revealed the presence of hard MnAl particles in their microstructures. Analyzed chemical composition of the BUL area proves that the MnAl compounds have a high affinity to the tool materials. Micro hardness tests values as high as 150 HV were observed in the BUL of an AZ 91 magnesium alloy compared with 70 HV in the workpiece itself.

1.3 FLANK BUILD-UP (FBU)

The formation of flank build-up is observed when machining relatively brittle materials such as magnesium cast alloys, nodular cast alloys, SiC whisker reinforced aluminium and aluminium-silicon cast alloys. The FBU is formed below the cutting edge on the flank surfaces of the cutting tool. When it reaches a certain size, it comes in contact with the machined workpiece surface. This leads to a sharp increase in cutting forces with a seriously impaired surface quality. One consequence of the existence of seizure friction conditions at the rake face of the tool can lead to the formation of stagnant material (the built-up edge). Due to high temperature and pressure the accumulated work material is pressed out from the cutting zone between the flank surfaces of tool and the work. Then it is adhered and collected on flank surfaces of the cutting tool. When the built-up material reaches a certain size, it comes in contact with the machined workpiece surface. This leads to a high increase in cutting forces and consequently the surface roughness is greatly impaired.

2 FACTORS THAT AFFECT THE BUILD-UP MATERIAL FORMATION

Many researchers have pointed out several important factors, which affect the built-up material formation, e.g. cutting temperature, cutting speed, strain hardening, the adhesion property between the work material and the tool, microcrack formation, plastic flow of the work material in the vicinity of the cutting edge, etc. The factors that are influential on the built-up material formation are complex, varied, and interactive.

2.1 EFFECT OF ADHESION

The central question of the fundamental study of the built-up material is why and how it is bonded to the tool. Three facts are undeniable:

1. The temperatures in the cutting zone and the tool are lower than the melting point of the work material.

2. The "welded" work material is severely deformed.

3. Contaminant layers are not present at the interface between the work material and the tool.

The adhesion between the tool and the workpiece material was proposed as one of the important factors, which affect the formation of built-up material [2]. The mechanism of adhesion is very complex and not completely understood. A number of hypotheses have been proposed to explain its fundamental nature. More recently, some researchers have investigated atomic or molecular interactions at the interface of two solids based on the electron configuration of atoms. The electronic theory of adhesion approaches an understanding of the phenomenon. The adhesion properties of materials are important and efforts to understand them have led to a deep insight into the fundamental structure of many solids. In machining operations the conditions at the chip/tool interfaces are especially convenient to adhesion or bonding. Two aspects of adhesion are commonly considered in assessing the material transfer in tribology and metal cutting, namely, mechanical interlocking, and physicochemical adhesion. The mechanical aspect of adhesion is related to the development of large areas of the intimate contact between the work material and the tool. The work material in the cutting zone, particularly near the cutting edge, is subjected to a high temperature and pressure. Measurements of the tool forces indicate that the compressive stress on the face surface of the tool is very high over the contact area. It has been shown that a typical value of the mean compressive stress for a carbon steel is around 770 MPa (7700 bar). In the presence of a built-up edge when cutting steel temperature at the tool/chip interface may be 600 $^{\circ}$C [2]. Due to this extreme pressure and relatively high temperature, the work material will plastically deform and create a complete contact in both "hills and valleys" of the tool surface. This mechanical interlocking of asperities can certainly be a significant factor in adhesion. It has been demonstrated that higher roughness of the contact surfaces can promote the material transfer or adhesion. Beside the built-up material phenomenon, adhesion is associated in metal cutting with mechanisms of tool wear, discontinuous and serrated chip formation and tool vibrations. In general, adhesion wear is explained as a consequence of the formation and destruction of the adhesive junctions. In the regions of seizure and sliding, the work material bonds to the cutting tool. The adherent material may be periodically torn away, usually taking small particles of the tool material. Palmer [3] has shown that the formation of discontinuous chips is closely related to adhesion of the work material (chip) to the tool. Results of examination indicated that a considerable part of the chip body was static relative to the tool and appeared to form a built-up edge on the tool.

2.2 EFFECT OF TEMPERATURE

No matter how we consider the process of built-up material, the cutting temperature is a major factor [2]. The appearance and disappearance of the built-up material are caused by the variations of the temperatures of cutting zone and chip/tool interface. In practice we detect the formation and disappearance as a result of variations of the cutting conditions, such as cutting speed and feed. As the cutting speed of the cutting process changes, the temperature of the cutting zone, tool and workpiece changes. Experimental evidence indicates that the BUE does not form at very low and high cutting speeds. The BUE was detected to form when cutting steels at temperatures between about 350 $^{\circ}$C and 500 $^{\circ}$C. Iwata [4], who studied the effect of microcrack generation on the built-up material formation, indicated that above this temperature the work material recovers its ductility, which inhibit crack formation. It has been also shown that the sudden disappearance of the BUE coincides with the incipient austenite formation of the work material that generates more heat and accelerates the softening of the BUE [5]. Our cutting experiments also confirm that the cutting temperature is a very important factor in the formation of the flank build-up. In the first paper is shown that the FBU in machining magnesium alloys is formed when the cutting speed (temperature) exceeds a critical limit, e.g. 650 m/min for AZ 91 magnesium alloy. When magnesium alloys were machined at relatively low cutting speeds the FBU was not formed. During machining of magnesium alloys at cutting speeds over 600 m/min temperatures between 400 $^{\circ}$C to 480 $^{\circ}$C have been measured. We have also demonstrated that the FBU was completely eliminated at any examined cutting speed with the use of a water based cutting fluid [6] or polycrystalline diamond (PCD) tooling [7]. It was presented that machining with the use a PCD tool and a water based cutting fluid result in lower cutting temperatures. In particular, it was shown that PCD tools generate less heat during cutting than similar cemented carbide tools due to PCD's low friction coefficient, high thermal conductivity and very sharp cutting edge. A water based cutting fluid is used mostly to cool the workpiece, tool and chips, lowering the cutting temperature.

2.3 EFFECT OF WORK MATERIAL PROPERTIES

A number of researchers have indicated the importance of work hardening on the built-up material formation. Takayama and Ono [8] have supposed that materials with higher work-hardening exponent have a greater tendency for the BUE formation. Williams and co-operators [9] have shown that work-hardening is not a sufficient explanation for the generation and growth of the built-up material. From a large number of cutting tests they observed that pure metals and single-phase alloys (such as Cu30Zn brass) do not normally form built-up edges, however high their work-hardening rates. The authors showed that built-up material is formed only when cutting alloys with more than one phase, e. g. steel, cast iron, some nickel-based, copper based and aluminium based alloys. The absence of the built-up material when cutting pure metals and single-phase alloys has been explained by the postulation that such materials under all cutting conditions behave as a homogeneous mass in the cutting zone, causing a separation between the chip and tool. This boundary condition in metal cutting is usually interpreted as sliding. Experimental findings [10] also confirm theory of the multi-phase dependent built-up material formation. Namely, it is demonstrated that all the magnesium alloys, which form the FBU have a heterogeneous microstructure (multi-phase alloys) while all the magnesium alloys, which do not, have a homogeneous microstructure (single-phase alloys). It was observed that the presence of the low melting intermetallic (3-phase in magnesium matrix is responsible for the difference in the tendency for the FBU formation. Brittle materials such as magnesium, nodular cast iron, particle reinforced and silicon cast aluminium have shown a tendency to the flank build-up formation. In contrast, the built-up edge is usually formed during machining ductile materials. It is shown that the built-up edge forms at relatively low cutting speeds, but the flank build-up is formed at very high cutting speeds, e.g. magnesium. A viable explanation of this difference may lie in the adhesion properties of work materials, which in turn depend on the characteristics of the work material. The results of experimental investigation [1] clearly show that the FBU is not formed at relatively low cutting speeds when cutting magnesium alloys. Visual examination of tool inserts revealed that only traces of the work material are present on the contact area of the tool face. This observation suggests that the ductility of magnesium in the cutting zone was low, causing sliding at the tool/chip interface. As mentioned previously, a FBU is formed when certain magnesium alloys are machined at very high cutting speeds and at higher cutting temperatures. Measurements of the cutting temperatures during machining of the AZ 91 magnesium alloy at cutting speed over 600 m/min have shown to be in the 350 $^{\circ}$C to 480 $^{\circ}$C temperature range. These somewhat approximate mean temperatures are around the equilibrium temperature (437 $^{\circ}$C) in the magnesium aluminium diagram, where the 3-phase melts. In the absence of real comparative hardness and plasticity data of magnesium phases at the high temperatures encountered during machining, the equilibrium magnesium aluminium diagram was used to indicate conditions of different phases at elevated temperatures. At such high cutting speed (temperatures) the micro-smelting of the R-phase will cause in the macro scale, a softening of the work material that increases its plasticity. In turn this will result in an increased area of the contact between the work material (chip) and the tool. The intimate contact between the work material and tool can result in a very strong adhesion (i.e. solid weld is formed). When the bonds between the tool and chip are "stronger" then chip, the shearing will occur within "softer" chip. This boundary condition of friction at the chip/tool interface in metal cutting is interpreted as the seizure. It is obvious that seizure friction conditions at the tool/chip contact areas lead to the formation of a built-up material. The findings that the FBU is not formed at low but at very high cutting speeds also confirm that material properties in the cutting zone and especially in the secondary deformation zone vary dramatically with the cutting speeds. Because of the adhesion (seizure) formed primarily at the rake face and the cutting edge a small volume of the material remains stationary in the front of the tool. This is a well known phenomenon in all metal forming processes, and termed as "dead zone". The experimental results indicate that the dead material zone increases with friction along the face surface. Several investigators have suggested that the built-up material is a consequence of the existence of the stationary material (dead zone), which is always present in cutting processes. When cutting magnesium alloys the stationary material cannot withstand the compressive stress imposed by the cutting action. This stationary material is pressed out from the cutting zone between the workpiece and the flank surface of the tool.

2.4 EFFECT OF PLASTIC FLOW

The formation of the built-up material during cutting is closely related to the generation of plastic flow of the work material in contact with the tool in vicinity of the cutting edge [11]. The nature of flow pattern or geometry of the system depends on the work material, cutting conditions and tool but also on the strength of the bond between chip and tool. The examination of the metallurgical sections normal to the cutting edge of the "quick stop" specimens revealed that undeformed grains of the work material are deformed in the primary zone to such an extent that grain boundaries appear as single lines. These lines form a pattern that suggests the direction and extent of metal movement during deformation. At the rake face of the tool such lines exhibit a point of inflection clearly caused by a chip-tool adhesion. Figure 2 shows a metallurgical section through quick-stop for 60/40 brass at 120 m/min.

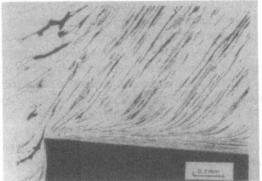

FIGURE 2. Metallurgical section through quick-stop for 60/40 brass at 120 m/min

2.5 EFFECT OF SHARPNESS OF THE CUTTING EDGE

The sharpness of the cutting edge has an important effect on the formation of the built-up material. A series of experimental investigation of the effect of the edge sharpness on the flank build-up formation were performed at the Research Laboratories of the NTH-SINTEF. These investigations have shown that a tool with a larger edge radius has a higher tendency to the FBU formation. Because, increasing the radius of the cutting edge will increase the volume of the stationary work material, the deformation zone and friction at the cutting edge. It can be seen that the cutting length decreases as the edge radius of the tool increases. It is worth to note that diamond tools, which do not form the FBU at any examined cutting speed, have an average value of the cutting edge radius 40 to 50 nm [12].

2.6 EFFECT OF MICROCRACKS

Several authors have suggested that formation of microcracks in the deformation zone of two-phase alloys plays a significant role in the mechanism of built-up material. Using machining inside a scanning electron microscope Iwata and Ueda [13] have detected two types of microcracks associated with the formation of the built-up material: First microcracks form under the cutting edge on the flank surface of the tool and then another subsequently forms at the back of the cutting edge on the tool rake face. It has been suggested that microcrack formation in the cutting zone is mainly due to the difference in the elastic and plastic properties in the second phase, inclusions and the matrix that can lead to the separation at the interfaces so that cavity forms around the second phase and/or inclusions. It has been also shown that the presence of microcracks in the deformation zone of a material leads to an increase in its ductility when it is sheared in the presence of a high compressive stress, as in machining. Luong [14] with a help of a quick-stop device studied machining of two steels and pure copper. He observed the absence of microcracks in the deformation zone when pure copper was cut in contrast to the steels. From these results he concluded that microcracks are restricted to the two- or multiphase materials.

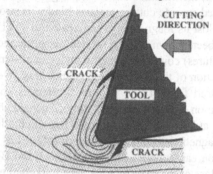

FIGURE 3. Model of cutting zone with stagnant zone and deformation lines

This observation is in agreement with the Williams and Rollason [15] theory that built-up formation seems to be limited to two- or multi phase materials. It is interesting to note that the porosity of the workpiece has a significant effect on the FBU formation. Visual examination of the workpiece (an AM 20 magnesium alloy) revealed surface areas with a numerous pores and cracks. No FBU was formed on the non porous area of the workpiece surface at any examined cutting speeds. But it was detected on the small porous area at cutting speeds > 800 m/min. Results of machining test [10] indicate that pores in the cutting zone behave in a similar way as microcracks.

2.7 EFFECT OF MAGNETIC FIELD

The existence of an electric current in the cutting process has been recognized for many years, but it is not clear what the exact role it has on the cutting process. The existence of the thermoelectric-currents in the cutting zone generates a magnetic field. The existence of the magnetic field might contribute to explanation of several unresolved phenomena observed in metal cutting processes. Many workers [16], [17] have indicated the beneficial effect of MnS inclusions, which are incorporated in free machining alloys, on machinability. These inclusions have a high affinity to the cutting tool and deposit as layers on the wear tool surfaces, although the reason for it has not been understood. The MnS layers may act to reduce the tool-chip friction resulting in lower cutting temperatures, tool forces and wear. In addition, these layers may also act as diffusion barriers to reduce tool wear. A similar behavior of the MnAl particles, which are incorporated in magnesium matrix, has been observed at the Research Laboratories of the NTH and SINTEF. Metallographic examination of the tool inserts used in cutting magnesium revealed a high concentration of manganese on contacting areas between the work material and the tool. The MnAl compounds are present in the magnesium alloys as very hard particles [1]. Although manganese, sulphur and aluminium are not ferromagnetic by themselves, they form intermetallic compounds MnAl and

MnS that are ferromagnetic. The MnAl specimens that were produced at SINTEF Metallurgy Division showed magnetic properties, namely, they were attracted by a permanent magnet. Naerheim and Trent [18] have suggested that the crater worn in the rake face when cutting steel with cemented carbide tools at high speeds is primarily the result of atomic diffusion of cobalt, which was contained in the tool, into the work material flowing over the surface. Ono and Lee [19] have observed in diffusion tests that cobalt becomes highly concentrated at the interface of a carbon steel specimen and a tungsten carbide tool. The reason why cobalt penetrate by diffusion into steel chip and is concentrated at the interface has not been given are not given. Cobalt, which is commonly used in the cemented carbide cutting tools as binder, is strong magnetic material. It is interesting to note that cobalt lost its magnetism when heated to $1121\ ^{\circ}C$. This critical temperature for iron is $770\ ^{\circ}C$. It has been shown that the sudden disappearance of the steel BUE at high cutting speed (cutting temperatures) coincides with the incipient austenite formation on the under surface of chip. As an explanation of the BUE disappearance was given that the austenite generates more heat due to increased friction that in turn accelerates softening of the BUE. It is remarkable that the transformation of (α-Fe (ferrite), body centered cubic (bcc), into γ-Fe (austenite), face centered cubic (fcc), at $770\ ^{\circ}C$ is accompanied by/with the loss of magnetism. In other words austenite is not ferromagnetic substance. Pekelharing [20] has indicated that vibration of the tool in the cutting direction can suppress the formation of the BUE. Beside a reduction of friction, vibrations can cause loss of magnetism. One very important example of the influence of an external magnetic field on wear and friction of metals was given by Hiratsuka and co-workers [21]. The authors investigated the friction and wear of different pure metals combinations using a pin-on-disc test in air and argon. The experiments performed in air have shown a significant wear reduction of Fe/Fe and Ni/Ni combinations, and a wear increase of Cu/Fe and Zn/Fe combinations by applying an external magnetic field in comparison with ordinary friction tests (without the magnetic field.

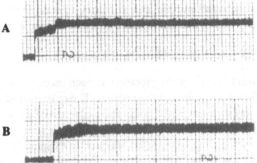

FIGURE 4. Variation of the cutting force (A magnetized insert, B ordinary insert)

A simple cutting test has been performed at the Research Laboratories of the NTH-SINTEF in order to demonstrate the influence of magnetism on the cutting process. An AZ91 magnesium alloys was turned at cutting speed 550 m/min with a magnetized and an ordinary tool insert. At this cutting speed range the protective built-up layer is formed in the ordinary cutting that decreases tool forces and vibrations. Figure 4 shows the cutting force as a function of cutting length for (A) a magnetized and (B) an ordinary tool insert. As it can be seen from this Figure when cut with the magnetized insert the cutting force signal from the dynamometer was smoother than obtained with the ordinary insert. This can be largely attributed to a greater built-up layer formed on the magnetized insert that reduces vibration. Such big layer simultaneously indicates a higher attraction of the work material, especially MnAl particles, to the magnetized insert. We believe that some of these previously mentioned aspects may be related to the cutting process and the built-up material. The investigation of the effect of the magnetic field on the cutting process was beyond the scope of this work and it will be the aim of the further investigations.

2.8 EFFECT OF OTHER CUTTING PARAMETERS

There are a number of other parameters that can effect the formation of built-up material, such as: tool geometry, tool material, machine tool, cutting fluid etc. The tool geometry has a strong influence on the frictional conditions and hence on the build-up of material. For example, an increase of rake angle will reduce the cutting force and the compressive stress on the rake face. It was also demonstrated that an increase of clearance angle increases the critical cutting speed or completely eliminate the FBU formation. It was found that the thermal conductivity of the tool material has an influence on the build-up of material. Decreasing the conductivity increases the tool/chip interface temperature and hence conditions of the built-up material formation. It was reported that machine-tool vibrations in some way change the proportions of seizure and sliding at the tool/chip interface [12]. For the FBU formation it is worth to indicate that a change of the cutting conditions in such way that lowers the temperature of a cutting operation will reduce the possibility of its formation.

3   CONCLUSIONS

The formation mechanisms of the three main forms of built-up material are principally of the same type. In order for the built-up material to be understood the adhesion of the metal surface has been considered in detail. The adhesion forces of the tool face surface and the lower surface of chip may be so high that then cause seizure or cold welding.

The results of the present investigation support the hypothesis that the built-up materials forms when cutting two or multi-phase alloys. It was noted that the presence of the low melting intermetallic β-phase in magnesium matrix is responsible for the difference in the tendency for the FBU formation.

The findings that the FBU is not formed at low but at very high cutting speeds also confirm that material properties in the cutting zone and especially in the secondary deformation zone vary dramatically with the cutting speeds. At very high cutting speeds (cutting temperatures) the work material in the cutting zone may become soften and ductile.

A larger edge radius has a higher tendency to FBU formation because; increasing the radius of the cutting edge will increase the volume of the stationary work material, the deformation zone and friction at the cutting edge.

Attempts have been made to explain the effect of the magnetic properties of the work and tool materials on the built-up material formation. It is noted that MnS and MnAl particles, which are incorporated in the matrix of the work material, are ferromagnetic and have a high affinity to the cutting tool.

REFERENCES

1. Tomac, N., Tonnessen, K., / submitted by Rasch, F. O., (1991), Formation of Flank Build-Up in Cutting Magnesium Alloys, Annals of the CIRP, Vol. 41/l: 98-42.
2. Kuznetsov, V. D., (1966), Metal Transfer and Build-up in Friction and Cutting, Pergamon Press, Oxford, London.
3. Palmer, W. B., (1967), Plastic Deformation when Cutting into an Inclined Plane, Journal Mechanical Engineering Science, Vol. 9/No 1: 1-10.
4. Rowe, G. W., (1964), Friction and Metal-Transfer, Wear, Vol. 7: 204-216.
5. Nakayama, K., Shaw, M. C., Brewer, R. C., (1966), Relationship Between Cutting Forces, Temperatures, Built-up Edge and Surface Finish, Annals of the CIRP, Vol. 24: 211-223.
6. Tomac, N., Tønnessen, K., Rasch, F. O., (1991) The Use of Water Based Cutting Fluids in Machining of Magnesium Alloys, Proceedings of International Conference on Innovate Metal Cutting Processes and Materials ICIM'91, Turin, Italy, 105-113.
7. Tomac, N., Tonnessen, K., Rasch, F. O., (1992), PCD Tools in Machining of Magnesium Alloys, European Machining, May/June: 12-16.
8. Takeyama, H., Ono, T., (1968), Basic Investigation of Built-up Edge, Journal of Engineering for Industry, May: 35-342.
9. Williams, J. E., Rollason, C. E., (1970), Metallurgical and Practical Machining Parameters Affecting Built-up Edge Formation in Metal cutting, Journal of the Institute of Metals, Vol. 95: 144-153 [15] Bailey, A. J., 1977, Surface Damage During Machining of Annealed Nickel Marring Steel, Wear, Vol. 42: 277-296.
10. Videm, M., Hansen, R. S., Tomac, N., Tønnessen, K., (1994), Metallurgical Considerations for Machining of Magnesium Alloys, Proceedings of SAE International Congress & Exposition, Detroit, Michigan, USA, 23-30.
11. Wright, P. K., Horne, J. G., Tabor, D., (1979), Boundary Conditions, at the Chip-Tool Interface in Machining: Comparison Between Seizure and Sliding, Wear, Vol. 54: 371-390.
12. Form, G. W., Beglinger, H., (1970), Fundamental Consideration in Mechanical Chip Formation, Annals of the CIRP, Vol. 18: 153-168.
13. Rowe, G. W., (1964), Friction and Metal-Transfer of Heavily-Deformed Sliders, Wear, Vol. 7: 204-216.
14. Luong, L. H. S., (1980), Influence of Microcracks on Machinability of Metals, Metals Technology, November: 465-471.
15. Williams, J. E., Rollason, C. E., (1970), Metallurgical and Practical Machining Parameters Affecting Built-up Edge Formation, Journal of the Institute of Metals, Vol. 95: 144-153.
16. Joseph, R. A., Tipping, V. A., (1975), The Influence of Non-metallic Inclusions on the Machinability of Free-machining Steels, Proceedings Conference: Influence of Metallurgy on Machinability, American Society for Metals, USA, 12-14 October 1975, 55-88.
17. Ramalingam, S., Thomann, B., (1975), The Role of Sulphide Type and of Refractory Inclusions in the Machinability of Free Cutting Steels, Proceedings Conference: Influence of Metallurgy on Machinability, American Society for Metals, USA, 12-14 October, 111-129.
18. Naerheim, Y., Trent, E. M., (1977), Diffusion Wear of Cemented Carbide Tools when Cutting Steel at High Speeds, Metals Technology, December: 548-555.
19. Lee, M., Richman, M. H., (1974), Some Properties Of TiC-coated Cemented Tungsten Carbides, Metals Technology, December: 538-546.
20. Pekelharing, A. J., (1974), Built-Up Edge [BUE]: Is the Mechanism Understand?, Annals of the CIRP, 23/2: 207-212.
21. Hiratsuka, K., Sasada, T., Norose, S., (1993), Wear of Metals, in a Magnetic Field, Wear, Vol. 160: 119-123.

PATTERN FORMATION AND WAVINESS IN SURFACE GRINDING

T. Jansen, O. Webber

Department of Machining Technology, University of Dortmund, Germany

KEYWORDS: Surface Grinding, Pattern Formation, Waviness.

ABSTRACT. Surface ground workpieces can exhibit quality impairments in the shape of unwanted pattern formation. Experimental investigations show, that pattern formation in surface grinding is independent of self excited vibration and that the pattern wave length coincides with the tangential feed. The observed patterns can therefore be attributed to e. g. either residual grinding wheel imbalance, roundness error or non-uniform grinding wheel surface topography. The visibility of the patterns mainly depends on the magnitude of the tangential feed. Increasing the feed reduces pattern visibility. Due to the underlying formation mechanism, pattern formation goes hand in hand with a waviness profile, which is dominated by waves with a wavelength equal to the pattern wave length. By measuring and adjusting the phase shift between the grinding wheel rotation during consecutive grinding passes, both the workpiece waviness and the visibility of the surface patterns can be reduced during sparking out.

1 INTRODUCTION

Grinding processes are fine machining processes generally employed for machining workpieces that often have to fulfil high quality standards with respect to geometric accuracy and surfaces quality. In case of components that will be visible during their later application, surface quality also includes appearance. The formation of visible patterns on the workpiece surface therefore may lead to the workpiece being rejected.

The quality of ground workpieces is influenced by many factors. The main influencing factors include the structure and composition of the grinding wheel, the workpiece material, type an application of the cooling lubricant and the cutting parameters. An unfavourable selection of these factors will lead to defective workpieces and therefore these parameters are under controlled during production. The formation of visible patterns on ground surfaces is generally attributed to either self excited or forced vibration or imperfections of the grinding wheel in the shape of geometrical errors of an uneven surface topography around its circumference. Pattern formation, unfortunately, is known to occur unexpectedly during series production, which prompted the investigations presented in this paper.

This paper will at first introduce the experimental setup used during the investigations. Following onto this, results regarding the dependency of pattern formation after rough grinding and the employed process parameters will be presented. Periodic surface patterns and the corresponding waviness profiles exhibit the same dominating wavelength. Based on this, a method for reducing

Published in: E. Kuljanic (Ed.) *Advanced Manufacturing Systems and Technology*, CISM Courses and Lectures No. 486, Springer Wien New York, 2005.

waviness was adapted to surface grinding and its effectiveness for reducing waviness as well as the formation of visible pattern formation was studied.

2   EXPERIMENTAL SETUP

The experiments were carried out on a surface grinding machine type Geibel & Hotz FS 635-Z CNC with a maximum spindle speed of 9000 min^{-1}, a maximum spindle power of 9 kW and a maximum feed speed of 35 m/min. The machine features a belt driven, CNC controlled table axis. For investigating the process dynamics, the machine was equipped with additional sensors for recording time series of the process forces as well as momentary table position and grinding wheel rotation angle. The experimental setup is shown in Figure 1.

FIGURE 1. Experimental setup

The workpiece material used was hardened X210CrW12 at approx. 62 HRC. The material was machined using an SG-grinding wheel Type 3A 54 G/H 162 V635. This combination of workpiece material and grinding wheel was chosen in order to be able to observe self excited vibration under laboratory conditions, i. e. without having to remove large material volumes. A rotating diamond dressing tool was applied to ensure constant dressing conditions over the duration of the project. The process parameters were varied within the following ranges:

Cutting speed: $v_c = 20 - 40$ m/s

Depth of cut: $a_e = 5 - 20$ μm

Related material removal rate: $Q'_w = 1 - 4$ mm^3/mm s

In order to facilitate the analysis of surface pattern, the surface grinding process was carried out with zero transverse feed, i. e. a plunge cutting operation was chosen.

The visibility of surface patterns is difficult to measure and in practice the visible appearance decides as to whether a workpiece becomes a reject or not. To evaluate the degree of pattern formation therefore four independent assessors classified the workpieces in six categories, from 1 = no visible pattern to 6 = strong pattern formation. Surface waviness was determined by tactile scanning of the workpiece surface in grinding direction.

3   EFFECT OF PROCESS PARAMETERS ON PATTERN FORMATION

To determine the influence of the cutting parameters on pattern formation, experiments were carried out according to a central composite design plan within the above mentioned parameter range. The probability of the degree of pattern formation was statistically modelled using a proportional odds model based on the classification results determined by the four independent assessors [1]. The results are exemplified in Figure 2.

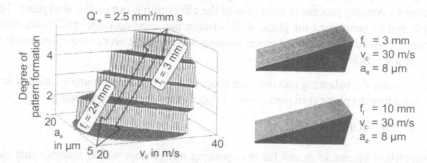

FIGURE 2. Degree of pattern formation after rough grinding as a function of process parameters. Workpiece material: X210CrW12 (ca. 62 HRC), SG grinding wheel type 3A G/H 162 V635, emulsion 5%, down grinding

For a fixed value of $Q'_w = 2.5$ mm^3/mm s, the graph in Figure 2 shows the distribution of the most likely degree of pattern formation to result. The visibility of the periodic patterns evidently shows the maximum gradient along the diagonal indicated. For a constant related material removal rate Q'_w, reducing the cutting speed v_c as well as depth of cut a_e leads to less pattern formation. Reducing both parameters is equivalent to a reduction of the tangential feed f_t. This implies that the tangential feed is the major influencing factor regarding the formation of visible patterns. Through cutting experiments with constant tangential feed values of $f_t = 3$ mm and $f_t = 10$ mm, under variation of the cutting speed v_c and depth of cut a_e, this result could be verified. Two example workpieces surfaces from this test series are shown in Figure 2, on the right.

From the time series of the process forces recorded during machining, the process could be furthermore classified with respect to the occurrence of self excited vibration. In accordance with the fact, that in surface grinding self excited vibrations mainly occur in the shape of grinding wheel regenerative chatter [2], no correlation between the occurrence of chatter and the formation of visible patterns could be detected.

A comparison of the periodic surface patterns with the measured waviness profiles showed, the both are dominated by the same wavelength which is identical to the tangential feed f_t. A single pattern segment is therefore produced during one grinding wheel revolution which may be attributed, as mentioned above, to imbalance, run-out or an uneven surface topography along the grinding wheel circumference.

4   PATTERN FORMATION AND WAVINESS

As increasing the tangential feed leads to an increased roughness depth, it is not in every case a feasible method for reducing pattern formation. The observation, that in surface grinding pattern wavelength and domination waviness profile wavelength both coincide with the tangential feed, suggests that the visibility of patterns may also be reduced by reducing surface waviness.

The options for reducing waviness in surface grinding are either reducing run-out by improving the balancing an dressing process or reduction of the effect of run-out on the workpiece. This can be accomplished by controlling the phase shift between the grinding wheel oscillation during the current grinding pass and an assumed corresponding waviness left on the workpiece surface during the preceding grinding pass [3][4].

Through improving the balancing and dressing process, no reduction of pattern visibility could be achieved. Therefore the phase shift method was implemented for surface grinding.

4.1   KINEMATIC MODEL

Using a simplified kinematic model for the sparking out process which assumes that machine deflection is constant and proportional to the mean normal force during each overrun and grinding wheel run-out f_R leads to a sinus shaped waviness profile of the workpiece, an approximation of the resulting final waviness profile can be constructed as the lower envelope curve of these partially overlapping theoretical surface profiles. From this model the dependency of the surface waviness on the phase shift between the grinding wheel rotation during consecutive grinding passes can be calculated. Figure 3 illustrates the construction of the theoretical workpiece surface for up or down grinding. For calculating the assumed constant deflection during each spark out pass, the recorded time series of the normal force were used.

Model assumptions:

a) Machine deflection occuring during sparking out is proportional to mean normal force furing each grinding pass.

b) Influence of dynamic normal force components is being neglected.

Process parameters of example:

Depth of cut: a_e = 14 μm
Tangential feed: f_t = 3 mm
Wheel run-out: f_R = 2.5 μm
Phase shift: $\Delta\varphi$ = 72°
Machine stiffness: k_{yy} = 0.0514 μm/N

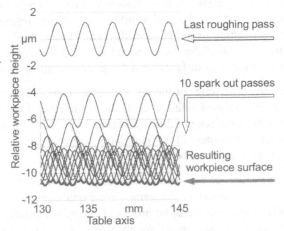

FIGURE 3. Construction of the theoretical workpiece surface using a simplified kinematic model

From the theoretical workpiece surface resulting from this model, the mean waviness depth W_a can be calculated. Figure 4 shows the influence of the phase shift on the workpiece waviness und variation of the grinding wheel run-out f_R and number of spark out passes n.

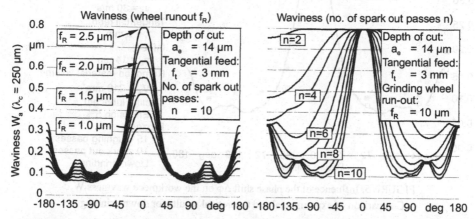

FIGURE 4. Influence of grinding the wheel run-out f_R and the number of spark out passes n on the theoretically resulting workpiece waviness W_a

The development of waviness is symmetric to 0° phase shift and shows two prominent maxima at 0° and ±180° in case of a sufficiently high number of spark out passes. Otherwise minima occur at ±180°.

4.2 EXPERIMENTAL RESULTS

In order to verify the simplified kinematic model, grinding tests were carried out. As the basic model is only applicable to either up or down grinding, a modified oscillation process design was adopted. Material was only removed during every second grinding pass resulting in an oscillating down grinding process. The kinematic simulation was performed after finishing the machining test. Therefore the recorded force data could be used to calculate the deflection during sparking out. Furthermore, as the actual grinding wheel run-out at cutting speed could not be measured, the necessary model parameter f_R was determined by fitting the model to the experimental data.

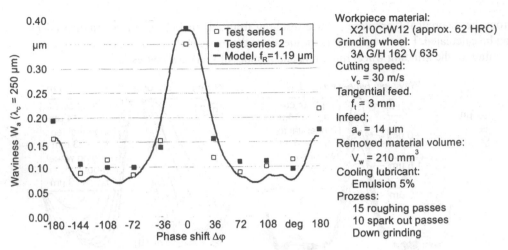

FIGURE 5. Influence of the phase shift $\Delta\varphi$ on the workpiece waviness W_a.
Comparison of experimental and theoretical results for down grinding

Figure 5 shows a comparison of the experimental results with computational results from the kinematic model. The computational results correlate well with the experimental data. Figure 6 shows exemplary workpiece surfaces produced during the test series. The workpiece surfaces shown on the left and on the right hand side of Figure 6 stem from processes carried out with a phase shift of $\Delta\varphi = 0°$ and $\Delta\varphi = \pm180°$, respectively. These show a degree of pattern formation corresponding to the relatively high W_a values which can be obtained from Figure 5. The surface depicted in the centre stems from a process carried out with phase shift of $\Delta\varphi = 108°$. It shows a very low level of pattern formation in correspondence with a low waviness.

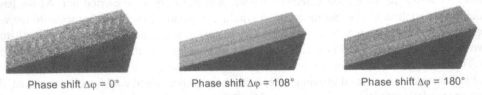

FIGURE 6. Influence of phase shift $\Delta\varphi$ on pattern formation.
Cutting parameters, see Figure 5

As mentioned above, the basic kinematic model is only applicable to either up or down grinding processes during which the tangential feed motion has the same orientation relative to the workpiece. This is a prerequisite for calculating a phase shift between consecutive grinding passes. During oscillating grinding though, feed direction changes with every grinding pass.

In order to adapt the phase shift method to oscillating grinding, the phase shift must be calculated not between consecutive grinding passes but between every other grinding pass. Furthermore, one has to differentiate between the table oscillation period T_t being $T_t = 2u\,T_s + (\Delta\varphi/360°)\,T_s$, with grinding wheel oscillation period T_s and $u \in \{1, 2, ...\}$, or the table oscillation period being $T_t = (2u+1)\,T_s + (\Delta\varphi/360°)\,T_s$.

In the first case, the time it takes the table to travel between both turning points, assuming a phase shift of $\Delta\phi = 0°$, is $T_t/2 = uT$. During consecutive grinding passes, the minima of assumed sinus shaped surface profile produced during these overruns, would occur in the same location on the workpiece. In the second case, again assuming a phase shift of $\Delta\phi = 0°$, the time the table takes between turning points is $T_t/2 = uT_s + \frac{1}{2}T_s$. In this case, the minima during consecutive grinding passes would occur with an offset of $f_t/2$. In order to account for this effect, the effect of a phase shift on the workpiece waviness has considered as being periodic with 720°. A phase shift of $\Delta\phi = 0°$ hereby corresponds to $T_t = 2uT_s$. A phase shift of $\Delta\phi = \pm360°$ corresponds to $T_t = (2u+1)T_s$.

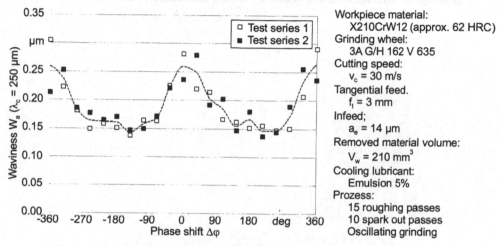

FIGURE 7. Influence of the phase shift $\Delta\phi$ on the workpiece waviness W_a.
Experimental results for oscillating grinding

Figure 7 shows experimental results for oscillating plunge cut surface grinding with controlled phase shift. Again, a distinct effect on the workpiece waviness can be observed. The effect on pattern formation is similar to the one found for down grinding as illustrated in Figure 6.

5 SUMMARY

The formation of visible patterns on ground workpiece surfaces may lead to these being rejected. The effect of the cutting parameters on pattern formation during rough grinding was investigated. The tangential feed f_t was found to be of main influence. Increasing f_t leads to a reduction of pattern visibility. The occurrence of pattern formation was observed not to correlate with the occurrence of chatter vibration. Pattern wavelength as well as the dominating wavelength of the waviness profiles are identical with the tangential feed. By adjusting the phase shift of the grinding wheel rotation between consecutive grinding passes, workpiece waviness could be reduced for oscillating plunge cut down grinding as well as true oscillating plunge cut grinding. In both cases a reduction of the waviness correlated with reduced pattern visibility.

REFERENCES

1. Weinert, K., Mehnen, M., Webber, O., Henkenjohann, N., (2004), Optimization of the Surface Grinding Process by Means of Modern Methods of Statistical Design of Experiments, Annals of the German Academic Society for Production Engineering, Vol. XI/1, 49-54
2. Inasaki, I., Karpuschewski, B., Lee, H. S. (2001), Grinding Chatter – Origin and Suppression, Annals of the CIRP, Vol. 50/2, 515-534
3. Foth, M., (1989), Erkennen und Mindern von Werkstückwelligkeiten während des Aussenrundschleifens, Dissertation RWTH Aachen, Germany
4. Trmal, G. J., Holesovsky, F., (2001), Wave-shift and its effect on surface quality in super-abrasive grinding, International Journal of Machine Tools and Manufacture, Vol. 41, 979-989

EMPIRICAL MODELLING AND OPTIMISATION OF PRECISION GRINDING

P. Krajnik, J. Kopac

Faculty of Mechanical Engineering, University of Ljubljana, Slovenia

KEYWORDS: Grinding, Modelling, Optimisation.

ABSTRACT. Assessment of the grinding process quality includes the micro geometrical quantities. The efficient surface texture modelling and optimisation is therefore essential for the competitive grinding processes. The objectives of this paper refer to development of empirical surface texture models and optimisation of precision centreless grinding (CG) with recently developed, highly-efficient, sintered aluminium-oxide-nitride (AlON) abrasive. Investigation has been carried out to study specific effects of the CG gap set-up, the dressing condition and the process kinematical factors on component surface texture parameters. Central composite design (CCD) of experiments and linear regression analysis (LRA) has been employed to develop surface texture models. The optimisation goals are combined in an objective function called desirability and solved by numerical nonlinear programming.

1 INTRODUCTION

Efficient utilization of the new abrasive tools is dependant on accurate predictions of the ground surface texture. Development of such predictive capability is founded on empirical surface texture models, which are used for CG system optimisation. Fixed abrasive grinding represents a machining process, whose complex characteristics determine the technological output and quality. There are many different methods for development of particular grinding process models and solving optimisation problems [1].

AlON abrasive has been essentially employed to enhance the efficiency of the precision CG process. It is able to preserve its cutting efficiency and self-sharpening at decreased wheel porosity, which indirectly enables better form holding and overall grinding quality. Given its improved hot hardness and brittleness, it has much less tendency to dull than other alumina based grits in low force regimes [2].

The experimental investigation is based on CCD of experiments. The development of empirical surface texture models is founded on LRA, which is an efficient statistical technique for establishing the relationship between the controllable CG system factors and specific metrological detectable surface texture responses. The major goal of LRA is determination of the regression coefficients that provide the best fit of the predictive polynomials to a measured set of surface texture data. All relevant data analysis is computer aided and includes design matrix evaluation, models analysis and numerical optimisation. A constrained optimisation problem of the CG system relates to minimization of surface roughness. These goals are combined into an overall desirability function, which has to be maximized. For optimal selection of controlled CG system factors, nonlinear programming has been employed.

Published in: E. Kuljanic (Ed.) *Advanced Manufacturing Systems and Technology*,
CISM Courses and Lectures No. 486, Springer Wien New York, 2005.

2   CENTRELESS GRINDING

Plunge centreless grinding has been widely employed as a high-rate precision process in industries that require large batches of complex, rotationally symmetrical components.

Surface texture, has been investigated according to the following CG system factors:

- CG gap set-up factor: the component centre height, h,
- Dressing factor: the longitudinal dressing feed-rate, f_d,
- Kinematical factor: the control wheel speed, n_r,
- Kinematical factor: the cutting speed, v_c.

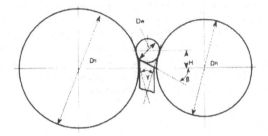

FIGURE 1. CG gap

CG is a complex process with a large number of influencing factors, which are nonlinear, interdependent and difficult to quantify. The primal nonlinear plunge CG effect relates to the contact loss between the component and the grinding wheel.

3   SURFACE TEXTURE ASSESSMENT AND PREDICTION

The concept of surface texture relates to manufacturing procedure, measurement and component performance; see Figure 2.

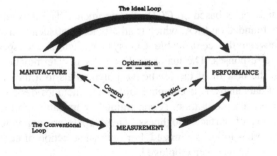

FIGURE 2. Surface texture concept [3]

Inspection of the surface texture on a ground component will often reveal unsuitable wheel topography, incorrect grinding gap set-up or wrong kinematical engagements.

Within the presented research, the discussion of surface texture is limited on two internationally recognised (ISO 4287:1997 standard) and most widely used amplitude parameters R_a (arithmetic average height) and R_z (ten-point height). R_a is the average absolute deviation of the roughness profile from the mean line over one sampling length as shown in Figure 3.

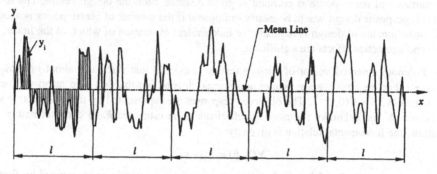

FIGURE 3. Arithmetic average height R_a [4]

R_z is the difference in height between the average of the five highest peaks and the five lowest valleys along the assessment length of the roughness profile. Figure 4 shows the definition of ten-point height parameter.

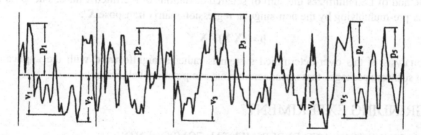

FIGURE 4. Ten-point height parameter R_z [4]

The objective of empirical surface texture modelling is to develop a predictive capability for grinding performance with respect to discussed amplitude parameters, which are closely allied to the quality and functionality of a ground component. In this way, surface texture can be predicted in a term of empirical model founded on an appropriately designed experiment. Each empirical model is limited and is nontransferable. This means that is only valid for the considered grinding wheel-component-grinding machine combination.

Our research employs a CCD of experiments with five factor levels, namely two-level full facto-rial with n_f factorial points, augmented with additional n_0 centre and $2p$ two-level axial points. Axial points are located at a specific distance α from the design centre in each direction in each axis. The factorial points represent a first-order model, while centre points, set to the midpoint of each factor range and axial points provide information about curvilinear existence.

Desirable properties of such designs involve rotatability and orthogonality. A design is rotatable if the variance of the response is constant at given distance from the design centre. The rotatable central composite design would be nearly orthogonal if the number of centre points is about five [5]. Orthogonality of design is requisite for independent evaluation of which of the linear, quad-ratic and interaction effects are significant..

Let Y denote observed vector of surface texture responses that fluctuates about unknown true response. For the error we are assuming a normal distribution that has zero mean and variance homogeneity, $\varepsilon \approx N(0, \sigma^2)$. The observed responses are represented by the controllable system regressors X and unknown regression coefficients to be estimated β. In compact matrix-vector notation, the fundamental relation is given by:

$$Y(X, \beta) = X\beta + \varepsilon \tag{1}$$

In the context of applied LRA surface texture responses are usually approximated by first-order or second-order polynomials:

$$\hat{Y}(X, \hat{\beta}) = X\hat{\beta} \tag{1}$$

The prediction of surface texture responses $\hat{Y}$ is dependent on the input regressors X and esti-mated regression coefficients $\hat{\beta}$. The regression matrix X contains the corresponding polynomial terms of the p system factors. $\hat{\beta}$ is the least square (LS) estimator of the regression coefficients β. The method of LS minimizes the sum of squared deviations of the fitted values. The $\hat{\beta}$ is calcu-lated by pre-multiplying by the non-singular regression matrix transpose X':

$$\hat{\beta} = (X'X)^{-1}X'Y \tag{3}$$

Data analysis of this over-determined system is rather straightforward with the application of analyst software packages that include numerous computational algorithms.

4   GRINDING EXPERIMENT

4.1 MACHINE-TOOL AND EXPERIMENTAL COMPONENTS

Grinding experiments were conducted on a Schaudt Mikrosa BWF - Kronos M centreless grind-ing machine, shown in Figure 5. The machine is equipped with a Sinumerik 840D CNC controller and a dynamic grinding wheel balancing unit. It is important to conduct the experiments under chatter-free conditions and to keep the in-feed speed (60 µm/s), specific material removal rate (1.05 mm³/mm s), the grinding depth (0.2 mm), the depth of dressing (0.02 mm), the spark-out time (0.1 s) and the coolant flow constant.

The component material was 9SMn28 (DIN standard), free-cutting unalloyed steel. The work rest blade ($\beta = 30°$) was specially made to support experimental component, shown in Figure 6.

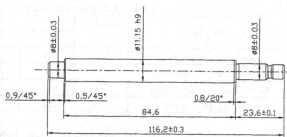

FIGURE 5. Grinding machine-tool

FIGURE 6. Experimental component

4.2 GRINDING AND CONTROL WHEEL

A vitrified grinding wheel, 3LA60L6V63L, with an AlON abrasive was used. Wheel dimensions were 500 x 88 x 304.8 mm. Moreover, a standard rubber bonded control wheel of 300 x 103 x 304.8 mm dimensions was employed.

4.3 MEASUREMENTS

The measurements were carried out with a Mitutoyo Surftest SJ-301 stylus type surface texture measuring instrument. Nonperiodic roughness profile evaluation has been conducted according to ISO standard, which employs a high-pass Gaussian filter, a sampling length of 0.8 mm and evaluation length of 4 mm. Each ground component was measured three times. The surface texture responses, summarized in Table 2 are the average readings of three consecutive measurements.

4.4 DESIGN OF EXPERIMENT

Design of experiment is required if we wish to extract meaningful conclusions from the measured responses. The proper design point determination, selection of the responses the experimental interest require competent process knowledge. In this way, the expert approach to design matrix appointment has been used. Experimental investigations require precise CG system set-up for each run to ensure that the trials are done according to plan. Errors and inaccuracies at this stage could nullify experimental validity.

The experiment includes four controllable CG system factors, ($p = 4$), whose levels are presented in Table 1. Here we follow the convention of coding the factor levels so the design points have coded levels for each factor. The region of interest, coded $\{-1, 1\}$, is a region determined by lower and upper limits on factor level setting combinations that are of major interest. The CCD extends the region of interest to the region of operability, coded $\{-2, 2\}$, which is determined by lower and upper factor level setting combinations that can be operationally achieved with acceptable safety and that will output a testable component.

In this research, 29 sets of experiments are sorted using the standard ordering and are carried out according to experimental design matrix, shown in Table 2.

TABLE 1. Levels of grinding system factors

Grinding factors	Symbol	Code				
		-2	-1	0	1	2
Component centre height [mm]	A: h	10	11.5	13	14.5	16
Dressing feed-rate [mm/min]	B: f_d	100	200	300	400	500
Control wheel speed [rpm]	C: n_r	36	41	46	51	56
Cutting speed [m/s]	D: v_c	40	45	50	55	60

TABLE 2. Experimental design matrix and responses

Run	A: h [mm]	B: f_d [mm/min]	C: n_r [rpm]	D: v_c [m/s]	R_a [μm]	R_z [μm]
1	13	300	56	50	0.45	3.52
2	16	300	46	50	0.37	2.72
3	14.5	200	41	55	0.38	2.67
4	13	100	46	50	0.34	2.61
5	11.5	200	51	55	0.38	2.71
6	14.5	400	41	45	0.39	2.87
7	11.5	400	51	55	0.37	2.60
8	14.5	400	51	45	0.38	2.81
9	11.5	200	41	55	0.39	2.84
10	11.5	200	51	45	0.42	2.86
11	14.5	200	51	55	0.38	2.73
12	11.5	400	41	55	0.37	2.77
13	13	300	46	50	0.43	3.18
14	14.5	200	51	45	0.32	2.41
15	11.5	200	41	45	0.44	3.10
16	14.5	400	51	55	0.37	2.67
17	13	500	46	50	0.46	3.17
18	10	300	46	50	0.36	2.60
19	13	300	46	60	0.46	3.32
20	14.5	200	41	45	0.39	2.95
21	13	300	46	50	0.45	3.41
22	11.5	400	51	45	0.39	2.77
23	13	300	46	50	0.44	3.13
24	13	300	46	40	0.43	3.22
25	13	300	46	50	0.45	3.21
26	13	300	36	50	0.50	3.66
27	14.5	400	41	55	0.39	2.84
28	11.5	400	41	45	0.38	2.82
29	13	300	46	50	0.44	3.13

5 EXPERIMENTAL RESULTS

5.1 DESIGN EVALUATION

The CCD evaluation is founded on advanced regression matrix analysis of presumed second-order model. First, we have to ascertain that the selected CCD enables the estimation of regression coefficients. The evaluation algorithm found no aliases for the chosen second-order model. This means that there are enough unique design points for sufficient estimation of all regression coefficients.

In some CCD, there can be one or more nearly linear dependencies, among the model coefficients, which can seriously affect the regression coefficients estimates. Multicollinearity is indicated by the variance inflation factor (VIF), which measures how much the variance of regression coefficient is inflated by the lack of orthogonality in the design. VIF exceeding ten indicate problems due to multicollinearity. The VIF values of employed CCD are 1 for linear and interaction terms and 1.08 for quadratic terms. From this we can conclude that the design is nearly orthogonal.

Another simple design evaluation criterion refers to the G efficiency, which is the average prediction variance as a percentage of the maximum prediction variance. The aim of adequate design is the G efficiency at the least 50%. The calculated G efficiency from the design points is 88.7%, which is sufficient.

In addition to numerical design evaluation it is useful to plot the standard error of the design over the investigated design space. The plot shows how the error in the predicted response varies over the region of operability. The shape of the plot depends only on the design points and the selected model degree. The plot, shown in Figure 7, exhibits circular contours and a symmetrical 3D shape, which indicates design rotatability. Another noticed feature is the relatively low error around the centre points.

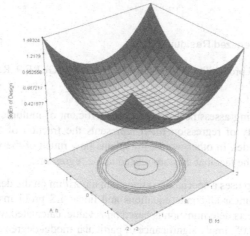

FIGURE 7. Standard error of CCD

5.2   MODEL FITTING AND ADEQUACY CHECKING

In the first phase of model adequacy checking it is necessary to verify that none of the LS regression assumptions are violated. The residuals from the LS fit, therefore play an important role. The useful graphical diagnostics involves the normal probability plot of the residuals, shown in Figure 8, which is used for checking the normality assumptions. Normal probability plots have been checked for both responses. Their residuals all plot approximately along a straight line; hence the normality assumptions are satisfied. Another graphical diagnostics tool is a plot of residuals versus predicted response, shown in Figure 9. The residuals of all responses scatter randomly, suggesting that the variance of observations is approximately constant.

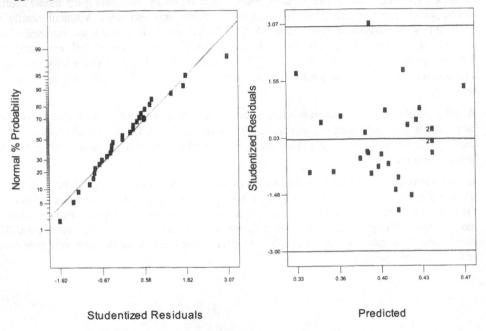

FIGURE 8. Normal probability of R_a residuals   FIGURE 9. Residuals vs. predicted R_a

Numerical model fitting assessment is based on coefficient of multiple determination, R^2, which determines the quality of regression fit. It represents the fraction of total variation in the data accounted by the model. In other words, it explains how much of the variability in the response can be explained by the fact that they are related to the regressors.

The model fitting step uses a special decomposition algorithm on the design matrix, which is used for solving various linear algebraic equations and linear LS problems. Model significance and significant factor effects determination is based on P value, calculated with analysis of variance. P values smaller than 0.05, imply significance of particular model degree or CG system factor.

TABLE 3. Evaluation of non-reduced surface texture models

Response	Mean	St. dev.	Model degree	Model P value	Significant model terms	R^2
R_a	0.40	0.035	quadratic	0.1419	h^2, f_d^2	0.6427
R_z	2.94	0.250	quadratic	0.1166	h^2, f_d^2	0.6580

5.3 SURFACE TEXTURE MODELS AND RESPONSE SURFACE PLOTS

Surface texture models have been developed in a form of non-reduced polynomials in terms of dimensionless coded factors. The models can be used to predict specific surface texture parameter at particular design points.

$$R_a = 0.44 - 0.005h + 0.0075f_d - 0.0092n_r - 0.0008v_c - 0.026h^2 - 0.017f_d^2 + 0.002n_r^2 -$$
$$0.0055v_c^2 + 0.011hf_d - 0.005h\,n_r + 0.01h\,v_c + 0.005f_d\,n_r + 0.0038n_r\,v_c \quad (4)$$

$$R_z = 3.21 - 0.012h + 0.042f_d - 0.066n_r - 0.023v_c - 0.190h^2 - 0.130f_d^2 + 0.040n_r^2 -$$
$$0.040v_c^2 + 0.061hf_d - 0.0075h\,n_r + 0.031h\,v_c + 0.025f_d\,n_r - 0.0013f_d\,v_c + 0.030n_r\,v_c \quad (5)$$

The nature of the models response surfaces depends on the signs and magnitudes of the model coefficients, comprised in upper equations. The second-order coefficients play a vital role. Figure 10 shows the 3D response surfaces of the investigated surface texture parameters plotted against the two most influential factors, while the remaining two factors are set to their central level.

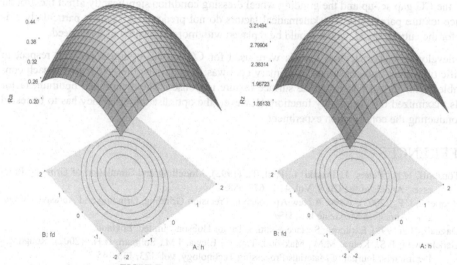

FIGURE 10. R_a and R_z response surface plots

6   CG SYSTEM OPTIMISATION

The goal of the CG system optimisation is to simultaneously minimize R_a and R_z within the region of interest. This problem is solved by a numerical nonlinear programming that searches for a combination of factor levels that simultaneously satisfy the requirements placed on the response and factors. This technique is based on a desirability function, which has to be maximized [5].

The maximization of desirability function is adjusted by the assignment of 0.1 goals weights and selection of 100 optimisation cycles. The minimization goal importance was set to a highest value of 5, while the factor importance was equal and set to midvalue of 3.

TABLE 4. CG system optimisation solution

h	f_d	n_r	v_c	Desirability	R_a	R_z
14.5 mm	200 mm/min	51 rpm	45 m/s	0.984	0.34 μm	2.64 μm

7   CONCLUSIONS

The development of empirical surface texture models has been founded on CCD of experiment and computer aided LRA, employed to fit appropriate models from experimental data. The numerical nonlinear optimisation algorithm produced a system of constraints that were solved via a penalty function approach in a downhill simplex search.

The insufficient experimental replicates caused a relatively low R^2, significant model lack of fit and relatively large standard error of the design. The real run replicates at the same system settings have to be included in all experimentations. The tentative stepwise elimination of insignificant model terms did not drastically improve their quality; hence it was omitted and not discussed.

Only the CG gap set-up and the grinding wheel dressing condition significantly affect the ground surface texture parameters. The kinematical factors do not predominate. This is particularly evident for the cutting speed, which should be replaced with more influential in-feed speed.

The developed surface texture models were used for CG system optimisation with respect to specific objectives and constrains. A primary goal was to determine the best level for each controllable factor in order to minimize surface texture responses. The determined optimum factor levels maximized the desirability function. However, the optimisation efficiency has to be tested by conducting the confirmation experiment.

REFERENCES

1. Tönshoff, H.K., Peters, J., Inasaki, I., Paul, T., (1992), Modelling and Simulation of Grinding Processes, Annals of the CIRP, Vol. 41/2, 677-688.
2. Roquefeuil, F., (2003), Abral: A New Approach to Precision Grinding, Grinding and Abrasives Magazine, Abrasives Magazine Inc., USA.
3. Dagnall, H., (1998), Exploring Surface Texture, Taylor Hobson Limited, England.
4. Gadelmawla, E.S., Koura, M.M., Maksoud, T.M.A., Elewa, I.M., Soliman, H.H., (2002), Roughness Parameters, Journal of Materials Processing Technology, vol. 123, 133-145.
5. Myers, H., Montgomery, D.C., (2002), Response Surface Methodology, Process and Product Optimization Using Designed Experiments, John Wiley & Sons, Inc., USA.

AN INVESTIGATION ON GRINDING PROCESS OF NATURAL STONES USING ARTIFICIAL NEURAL NETWORKS

A. Di Ilio[1], A. Paoletti[1]

[1] Department of Management, Energetics and Mechanical Engineering,
University of L'Aquila, Italy

KEYWORDS: Grinding, Natural Stones, Neural Networks.

ABSTRACT. Predictions on the tool condition and the surface finish of workpiece in grinding process of metal materials have been studied in the past years using physical and empirical models. In this paper, the feasibility of using neural networks, based on signals detected by multi-sensorial system to monitor tool and workpiece surface conditions in grinding operation of natural stones, has been investigated. Grinding wheel wear evaluation has been carried out measuring flat area percentage on the active surface of the tool by means of a vision system. Workpiece surface roughness has been assessed by means of a mechanical profilometer. Neural network models have allowed to predict grinding wheel cutting ability and workpiece surface finish by measuring on-line the grinding forces and the surface wheel temperature variation.

1 INTRODUCTION

In the processing of natural stones, the production of modular elements for the building sector has been facilitated by the machine and tool developments of last years. Segmented diamond tools have revolutionized the whole stones sector allowing increased material removal rates compared to conventional gang sawing techniques [1]. In natural stones cutting, it must be stated that the material removal process is a grinding process, although the term sawing is commonly used. Stone sawing by means of diamond impregnated segments consists in wearing away its mineral constituents by passing rigid grits over the machined surface. The diamond crystals are firmly held in a matrix, which erodes progressively exposing fresh particles while those, protruding sufficiently to cut, are subjected to mechanical degradation to be finally dislodged [2]. Stone processing is a multidimensional problem of great complexity, where the sawing method affects to a great extent the conditions of debris removal and retention of diamond grits [3, 4]. The supervision of tool wear and workpiece surface finish are the most difficult task in the context of tool condition and workpiece surface quality monitoring for machining processes. This fact can be ascribed to the large number of interrelated parameters that influence the cutting process and make it extremely difficult to develop a proper model [5]. Based on a continuous acquisition of signals with multi-sensor systems, it is possible to estimate or to classify certain wear parameters by means of neural networks. Neural networks have shown to be the highly flexible modelling tools with capabilities on learning the mathematical mapping between input variables and output features. In the field of machining operations multi-layer perceptrons, based on back-propagation technique, have been employed for monitoring and modeling the processes [6]. Mathematical models for tool wear and life

Published in: E. Kuljanic (Ed.) *Advanced Manufacturing Systems and Technology*,
CISM Courses and Lectures No. 486, Springer Wien New York, 2005.

have been obtained from the experimental data using multiple regression analysis [7]. The results have shown that using the force-wear equation derived from regression analysis is a fairly accurate way of predicting the attainment of prescribed tool wear. However, the use of a neural network analysis can further improve the accuracy of the tool wear prediction, particularly when the functional dependency is non linear [8]. In addition to artificial neural network, fuzzy systems or combinations of both can be used to model the non linear dependencies between features extracted from the measured signals of cutting force, temperature and vibration on the one hand and tool wear or workpiece surface finish on the other hand [9, 10].

This article aims to investigate the feasibility of using neural networks, based on signals detected by multi-sensorial system, to monitor tool and workpiece surface conditions in machining by abrasion of natural stones. Grinding wheel wear assessment has been carried out measuring grinding forces by means of a piezoelectric dynamometer, flat area percentage and temperature variation on the active surface of the tool by means of a vision system and an infrared pyrometer, respectively. Workpiece surface roughness has been evaluated by means of a mechanical profilometer.

2 EXPERIMENTAL TESTS

The experimental apparatus consists of a horizontal spindle surface grinder, equipped with a two component piezoelectric dynamometer, clamped on the grinding table, two charge amplifiers and two analog filters [11]. A vision system and an infrared pyrometer have been placed on the head of the grinding machine. Signals of grinding forces and grinding wheel temperature have been acquired by means of an A/D card, adopting a sampling rate equal to 3 kHz. The monitoring of grinding forces, flat area percentage and temperature of grinding wheel surface allows to obtain an on-line assessment of progressive grinding wheel wear. Experimental tests have been carried out on two different natural stones, named Bianco Carrara Marble, whose compressive strength, σ_R, is equal to 41.6 ± 3.5 N/mm^2 and Perlato Royal Coreno, whose compressive strength, σ_R, is equal to 85.2 ± 4.6 N/mm^2. Specimens with dimensions 18 mm width, 57 mm length and 42 mm height, have been cut from plates and ground in direction perpendicular to that of sedimentation. The tool employed for the tests has been a diamond grinding wheel, type ASD126R75B99 (external diameter, d = 200 mm; width, b = 15 mm, maximum speed = 35 m/s), which has been dressed using a soft Al$_2$O$_3$ dressing stick. Dry conditions have been adopted and a two-levels factorial design has been chosen for the cutting parameters, as shown in Table 1.

TABLE 1. Grinding parameters values

Grinding parameters	Low Level	High level
Depth of cut, a [mm]	0.02	0.04
Wheel peripheral speed, Vs [m/s]	10	20
Workpiece speed, Vw [mm/s]	150	300

In order to find the relationship between the flat area percentage of the grinding wheel, the two components of force, the grinding wheel temperature variation and the surface roughness of workpiece, firstly, a linear regression analysis has been carried out on the experimental data. Processed values have been obtained during the first up-grinding pass: in this condition, the specific material removal, V'_w is equal to 1.14 mm³/mm. The results concerning the tool wear are shown in Figures 1 and 2, while the ones with regard to workpiece surface roughness are reported in Figures 3 and 4.

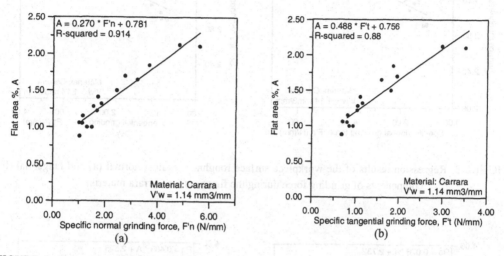

FIGURE 1. Regression results of the flat area against normal (a) and tangential (b) components of grinding force during the first pass for Carrara material

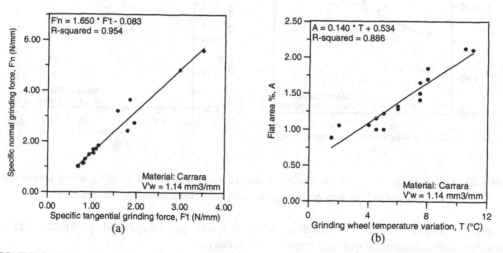

FIGURE 2. Regression results of normal component against tangential component (a) and flat area percentage against grinding wheel temperature variation (b) during the first pass for Carrara material

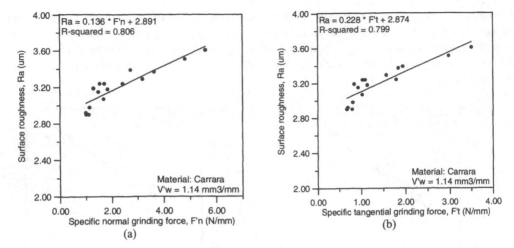

FIGURE 3. Regression results of the workpiece surface roughness against normal (a) and tangential (b) components of grinding force during the first pass for Carrara material

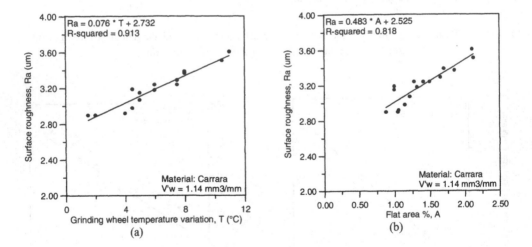

FIGURE 4. Regression results of the workpiece surface roughness against grinding wheel temperature variation (a) and flat area percentage (b) during the first pass for Carrara material

Figure 5 depicts the trend of grinding wheel flat area percentage and workpiece surface roughness as a function of material compressive strength.

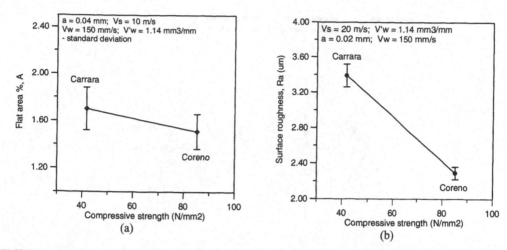

FIGURE 5. Grinding wheel flat area percentage (a) and workpiece surface roughness (b) as a function of workpiece material mechanical properties during the first pass

As can be seen, lower flattening of tool diamond grains and better surface finish of workpiece have been obtained for Coreno. This result is more evident for the workpiece surface roughness and is due to Coreno higher compressive stress. The cutting mechanism of marble can be described as the plastic deformation and the brittle fracture of stone, which is affected by the cutting conditions and the properties of material. Carrara surface is characterised by higher plastic deformation, which produces a worst finish.

3 IMPLEMENTATION OF NEURAL NETWORK MODEL

Data obtained by experimental work have been used to train the neural network, whose structure is a three layers (input, hidden and output), feed-forward, back-propagation type (Figure 6). This is one of the most popular models in the area of in-process monitoring [12].

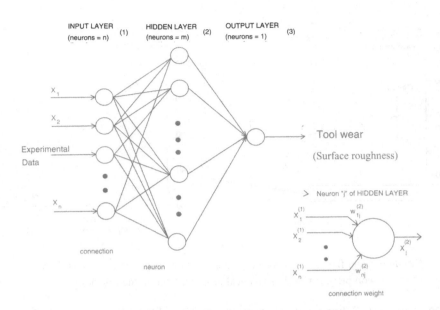

FIGURE 6. Structure of neural network

The input layer receives data of workpiece material properties, force and temperature features, the hidden layer performs features extraction from the input data, and the output layer generates the correspondent tool wear and surface roughness values. Each layer contains a number of interconnected processing elements, namely nodes or neurons. Each node has several input paths which are: experimental data $(x_1, x_2, \dots x_n)$, for neurons in input layer, and outputs from previous neurons, modified by the connection weights (w_{ij}), for nodes in hidden and output layers $(x_1^{(1)},$ $x_2^{(1)}, \dots x_n^{(1)},$ and $x_1^{(2)}, x_2^{(2)}, \dots x_m^{(2)}$, respectively). The sum of the signals is then modified by a sigmoid type activation function. Saturation of the activation function has been prevented normalizing data employed to train the neural network. The back-propagation algorithm minimizes the error between the calculated output of the network and desired output. The learning rate and the weights are adaptively varied during training [13]. A limit error equal to 0.01 and a maximum number of iterations equal to 2000 have been chosen. Neural networks with two different structures have been tested; in particular, the following ones have been taken into consideration: a) four inputs: the features of workpiece compressive strength, mean values of tangential and normal components of grinding force and maximum value of grinding wheel surface temperature variation, and considering the flat area percentage on the grinding wheel, as output.

b) Five inputs: the features of workpiece compressive strength, mean values of tangential and normal components of grinding force, maximum value of grinding wheel surface temperature variation and flat area percentage on the grinding wheel, and considering the workpiece surface roughness, as output.

Training of the neural network has been performed on experimental data obtained according to the plan shown in Table 1 and acquired during the first up-grinding pass, just after grinding wheel dressing operation. Verification has been carried out on a second set of data referred to the same experimental plan but acquired during the hundredth pass, in correspondence of which, the specific material removal, V'$_w$ is equal to 114 mm^3/mm.

4 RESULTS AND DISCUSSION

The comparisons between the predicted grinding wheel flat area percentage and workpiece surface roughness against the measured flat area and surface roughness are shown in Figures 7 and 8 for Carrara and Coreno material specimens, respectively. A perfect prediction would see all the plotted points sitting on the 45° line.

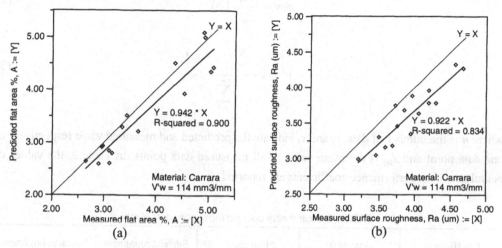

FIGURE 7. Diagrams of predicted grinding wheel flat area (a) and workpiece surface roughness (b) against measured flat area percentage and surface roughness during the hundredth pass for Carrara material

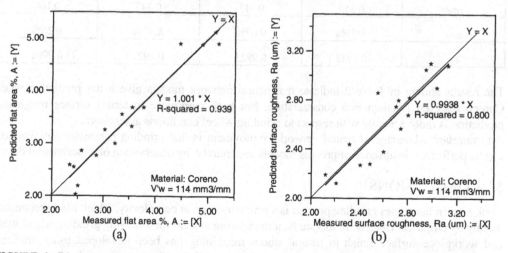

FIGURE 8. Diagrams of predicted grinding wheel flat area (a) and workpiece surface roughness (b) against measured flat area and surface roughness during the hundredth pass for Coreno material

In order to monitor the performance of neural networks analyses, the root mean square error (*rms*), the coefficient of variation in per cent (*cov*) and the absolute fraction of variance (r^2) have been used. Value of r^2 equal to 1 denotes perfect prediction. The terms above cited are defined as follows:

$$rms = \sqrt{\frac{\sum_{i=1}^{n}(y_i - x_i)^2}{n}} \tag{1}$$

$$cov = \frac{rms}{x_m} \cdot 100 \tag{2}$$

$$r^2 = 1 - \frac{\sum_{i=1}^{n}(y_i - x_i)^2}{\sum_{i=1}^{n}x_i^2} \tag{3}$$

where n is the number of data, y_i and x_i indicate the predicted and measured value respectively of one data point and x_m is the mean value of all measured data points. In Table 2, the values of neural networks performance coefficients are reported.

TABLE 2. Neural networks performance coefficients

Coefficients	Flat area (Carrara)	Flat area (Coreno)	Surface roughness (Carrara)	Surface roughness (Coreno)
rms	0.359	0.313	0.347	0.166
cov	9.57%	9.07%	8.87%	6.26%
r^2	0.9913	0.9923	0.9922	0.9961

The results shown in Table 2 indicate that neural network models give better prediction when Coreno material is taken into consideration. For both workpiece materials surface roughness prediction is more accurate with respect to grinding wheel conditions assessment.

An important advantage of neural network employment is that grinding parameter optimization can be performed in absence of process models and, purely, by observation of experimental data.

5   CONCLUSIONS

Tool wear in the stones grinding process is a problem of great complexity, which has been treated mainly by empirical models. In this study, a monitoring system for assessing grinding wheel wear and workpiece surface finish in natural stones machining has been developed using artificial neural networks. By measuring the grinding forces and the surface grinding wheel temperature variation, the progressive loss of tool cutting ability can be monitored without interrupting the machining operation. Hence the decision making for grinding wheel dressing operation becomes less time consuming and maximizes the total effective tool life. The surface finish of workpiece can be also predicted by using the processed data of grinding forces, surface temperature variation and flat area percentage on the active surface of the tool. Simulation results confirm the feasibility of this approach and show a good agreement with experimental results for the range of machining conditions under test. Neural networks performances have been evaluated by taking into consideration some statistical coefficients.

ACKNOWLEDGMENTS

The authors would like to thank MIUR (Italian Ministry of Education, University and Research) for financial support of this work through grant PRIN 2002 No. 2002092122-002.

REFERENCES

1. Tonshoff, H.K., Apmann, H., Asche, J., (2002), Diamond Tools in Stone and Civil Engineering Industry: Cutting Principles, Wear and Applications, Diamond and Related Materials, Vol. 11, 736-741.
2. Konstanty, J., (2002), Theoretical Analysis of Stone Sawing with Diamonds, Journal of Materials Processing Technology, Vol. 123, 146-154.
3. Di Ilio, A., Togna, A., (2003), A Theoretical Wear Model for Diamond Tools in Stone Cutting, International Journal of Machine Tools & Manufacture, Vol. 43/11, 1171-1177.
4. Di Ilio, A., Paoletti, A., Togna, A., Turchetta, S., (2004), Tool Wear in Stone Cutting: Theoretical Model and Experimental Validation, Roc Maquina Magazine–Dimension Stone Industry, Vol. 52, 10-15.
5. Szecsi, T., (1999), Cutting Force Modelling using Artificial Neural Networks, Journal of Materials Processing Technology, Vol. 92, 344-349.
6. Tsai, K.M., Wang, P. J., (2001), Prediction on surface Finish in electrical Discharge Machining based upon Neural Network Models, International Journal of Machine Tools and Manufacture, Vol. 41, 1385-1403.
7. Tosun, N., Ozler, L., (2002), A Study of Tool Life in Hot Machining using Artificial Neural Networks and Regression Analysis Method, Journal of Materials Processing Technology, Vol. 124, 99-104.
8. Lin, J.T., Bhattacharyya, D., Kecman, V., (2003), Multiple Regression and Neural Networks Analyses in Composites Machining, Composites Science and Technology, Vol. 63, 539-548.
9. Sick, B., (2002), On-Line and Indirect Tool Wear Monitoring in Turning with Artificial Neural Networks: a Review of More Than a Decade of Research, Mechanical Systems and Signal Processing, Vol 16/4, 487-546.
10. Korosec, M., Balic, J., Kopac, J., (2005), Neural Network based Manufacturability Evaluation of Free Form Machining, International Journal of Machine Tools & Manufacture, Vol. 45, 13-20.
11. Di Ilio, A. Paoletti, A., (2004), Monitoring of the Grinding Process of Natural Stones, Proceedings of ICME-4, Sorrento, 431-436.
12. Lin, S.C., Ting, C.J., (1996), Drill Wear Monitoring using Neural Networks, International Journal of Machine Tools and Manufacture, Vol. 36/4, 465-475.
13. Simpson, P.K., (1991), Artificial Neural Systems, Foundation, Paradigms, Applications and Implementations, General Dynamics, Electronic Division, San Diego.

TOOL WEAR AND SURFACE ASPECTS WHEN TURNING TITATIUM AND ALUMINUM ALLOYS

M. V. Ribeiro, M. R. V. Moreira, E. A. Cunha

Machining Study Laboratory, Department of Materials Engineering, FAENQUIL, Lorena/SP, Brazil

KEYWORDS: Machining, Wear, Roughness, Titanium, Aluminum.

ABSTRACT. This paper studies the effects of cutting parameters in the tool wear and surface quality (roughness). The tests were carried out on a CNC lathe, using uncoating and coating cemented carbide tool under finish cutting conditions. Microestructural characterization of tool wear and surface was made using scanning electron microscopy (SEM). The turning of the Ti (6Al-4V) alloy is very difficult due the rapid tool wear. Such behavior result of its low thermal conductivity in addition the high reactivity with the cutting tool. After the machining, the aluminum samples were analyzed with the aim to investigate the mechanisms, which influence the finish of machining surfaces. In the all analyzed samples were observed typical characteristic of machining surfaces, as feed marks, typical deformations of built-up edge (BUE) beyond deposited particles in the machining surface. A typical formation of the side flow mechanism was also observed.

1 INTRODUCTION

Titanium alloys is very used in the aeronautical industry, mainly, due high strength-to-weight ratio, lower density of titanium when compared with stell, creep and fatigue resistance, corrosion resistance in high temperatures, etc. Advanced engineering materials, such as structural ceramics, titanium alloys, inconel alloys, offer unique combination of properties like high strength at elevated temperaures, resistance to chemical degreadation and wear resistance. The titanium structure alpha phase is hexagonal close-packed (hcp), at temperature of 980 °C (Beta transus), the alpha phase forms into beta phase, structure to the body centred cubic (bcc).

The alfa/beta alloys, in private have, in most part, alpha phase at room temperature, but they do have more of the beta phase, the high temperature allotrope, than the former class of alloys. These alloys, include Ti-6Al-4V (Ti-6-4), Ti -6Al-6V-2Sn (Ti -6-6-2) and Ti -6Al-2Sn-4Zr-6Mo (Ti -6-2-4-6). The titanium is used thoroughly in the production advanced industrial equipments, in the generation of energy and in the transport. Ti-6-4 is the workhorse of the titanium industry, does it accounts go about 60% of the total titanium production [1]. Considering the two-phase alloy (Ti-6Al-4V), can identified two evolution types be: the first improved the creep resistance and the second improved the mechanical resistance and the resistance to low cycle fatigue [2]. The machining of the titanium alloy is difficulted basically due to its high chemical reaction with the materials of the tool and its low thermal conductivity (about 7,3 W/m K) generating high temperature in chip/tool/workpiece interface (Bhaumik, 1995), wich favors the wear mechanisms. About 80% this generated heat, it is retained in the tool and 20% in the chips [3].

Published in: E. Kuljanic (Ed.) *Advanced Manufacturing Systems and Technology*,
CISM Courses and Lectures No. 486, Springer Wien New York, 2005.

During the machining of titanium alloy, tool wear progress rapidly because of the high cutting temperature and strong adhesion between the tool and the material work, owing to their high chemical reactivity [4]. Some specific studies in tool failure modes and wear mechanisms when machining titanium alloy have been conducted [5,6,7]. Due to the low elasticity module of the titanium the tool is subject to a pulsation load (spring back) during the machining causing attrition and vibration [3]. The Ti-6Al-4V alloy it possess a low thermal conductivity doing with most of the heat generated during the machining it is retained in the tool. The combination of pulsant efforts and high temperature accelerates the wear mechanisms of the tool, these factors contribute to its low acting.

In machining, three different types of chips are know to forms: continuous chips, segmented chips and completely separated segmented if extreme cutting conditions are used. Ti-6Al-4V forms segmented chips for a wide range of cutting parameters [8,9]. For turning of titanium alloy, the cemented carbide (WC/Co) represent the best cutting tool for machining of Ti-6Al-4V [10]. The high chemical reactivity titanium alloy results in diffusion and excessive carter wear [11]

The aluminum stands out for possessing a significant importance due to its great production potential and its intrinsic physical and mechanics characteristics, as for instance, its high relationship between resistance / weight. The actual need of products production, every time smaller, lighter, with larger precision, reliability and, if possible, with low cost, it has constantly been throwing challenges to the production engineering [12].

Inside of the production processes, the machining belongs it a group of processes that, it takes more in consideration the dimensional precision and the surface finish of the produced workpieces. In this case, larger control easiness is had on the variables with influence in the quality of the produced surfaces, mainly in what it refers to the surface texture.

They are several the machining parameters that influence the finish surface, among them can be mention: workpiece and cutting tool geometries; machine tool stiffness; workpiece and cutting tool materials; cutting conditions. With relation to the cutting conditions, the more influential parameters are the feed rate (f), nose ratio of cutting tool (r_ε), the cutting depth (ap), and finally the cutting speed (vc). The finish surface machining depends a lot on the relationship among feed (f) and nose ratio (r_ε of cutting tool, which can be given by the following Equation 1:

$$R_y = \frac{f^2}{8 \cdot r_\varepsilon} \tag{1}$$

Therefore if the feed be varied, staying the constant nose ratio of the cutting tool, there will be a variation of the theoretical maximum roughness ($R_{y,theor.}$) that will be directly proportional to this new feed value. The theoretical roughness is the highest possible roughness of being obtained in a turning process. The obtained real roughness is usually larger (and sometimes very larger) than obtained using it the Equation 1 due to factors such as vibration, chip deformation, side flow, etc. [13]. The roughness of the final surface, obtained during the turning operation, can be considered as the sum of two independent effects [14].

- The roughness of the theoretical surface (standard), which is the result of the geometry of the cutting tool and feed rate;

- The roughness of the real surface, which is the result of the cutting operation irregularities.

For this work specifically, the influence of the feed rate was evaluated taken account of roughness in aluminum alloy AA7050 turning. For this tests were accomplished varying the feed of the tool and staying constant the other cutting parameters. Starting from the tests analysis, it was possible to evaluate which the behavior of the cutting tool, as well as of the machine surface with relation to damages caused during the process evolution.

2 MATERIALS AND METHODS

2.1 TOOLS

Titanium: the machining tests were accomplished through external cylindrical turning using a titanium alloy (Ti-6Al-4V) with tools VBMT 110304 PF ISO P10 class, titanium based cermet (Sandvik CT 5015), VBMT 110304 MF ISO M15 class, cemented carbide (Sandvik GC 1025) coating by TiAlN (PVD) with thickness of 4 µm, which provide toughness and wear resistence to cutting tool and VBMT 110304 UF ISO S15 class, uncoating cermented carbide (Sandvik H13A). The tests were conducted without coolant.

Cutting tools used in aluminum machining testes were: VCGX 110204 – Al ISO N15 class, uncoating cermented carbide (Sandvik H10); VCGT 110304F – Al KX ISO N15 class, uncoating cermented carbide, but with polish surface; finally VCGX 110204 – Al ISO N15 class, cemented carbide coating by diamond (Sandvik 1810), which provide stability and wear resistance to cutting tool. The tests were carry out with coolant.

2.2 MICROSTRUCTURE

Figure 1 show tests alloys microstructures. In dark region there is beta phase and clear region there is alpha phase. The initial microstructure consists of equiaxed alpha grains and intergranular beta phase, Figure 1a. The heterogeneity microstructure with various grains size are show in Figure 1b, deformed to the direction rolling.

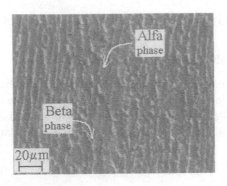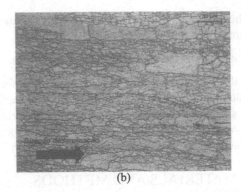

FIGURE 1. a) Titanium alloy microstructure (Ti-6Al-4V) b) Aluminum alloy microstructure (AA 7050 - T7451)

During the tests, the temperature in the contact area between workpiece and the cutting edge was measured using a infrared radiation pyrometer, model Cyclops-52, manufactured by Minolta-Land. The characterization of the tool wear and chip was accomplished with the aid of an electronic microscope of scanning LEO, model 1450 - VP, in mode of secondary electrons and an optic microscope, Leica model DM coupled IRM an image analyzer Leica QWIN for microstructure and morphologic analyses. All of the cutting tests were conducted in a CNC lathe (ROMI), with maximum rotation of 4000 rpm and power of 10 KW. The roughness was mensured using a portable roughness meter coupled to machine tool, the mensurement was made in 60° intervals with relation to workpiece axis. The used cutting parameters are shown in Table 1.

TABLE 1. Turning test conditions

	Titanium	Aluminum
Cutting speed v_c (m/min)	85; 100; 120	800
Feed rate f (mm/rev)	0,1	0,02; 0,05; 0,1; 0,2; 0,3; 0,4; 0,5; 0,9
Depth cutting a_p (mm)	0,5	1,5
Cutting fluid	no	yes
Tools tested (ISO Class)	H13A (ISO S15) CT5015 (ISO P10) GC1025 (ISO M15)	H10 (ISO N15) Polish KX (ISO N15) 1810 (ISO N15)

3 RESULTS AND DISCUSSION

3.1 TITANIUM

There are usually three criterion used for discussing machinability: tool life, surface finish (roughness) and power required to cut.

Since it was a finishing turning, the tool life criterion was the surface roughness of the machined surface. The criterion of roughness was 0,9 μm.

Three series of tests for turning Ti-6Al-4V were conducted to measure the progress of roughness. Figure 2a, Figure 2b and Figure 2c shows the effect of speede of cutting in surface roughness. Two distincts regions are observed Figure 2a. Firstly the tendency to roughness is similar (v_c= 85 m/min and 100 m/min) with length of cutting of 700m ($R_{a,máx}$=0,9 μm). The second aspect is related with speed of 120 m/min where the workpiece present is more several roughness (3 μm). The life of cutting tool is very small, around 120 m.

The excessive increase of the roughness its observed using ISO P10 tool Figure 2b up to 120 m/min. Only using speed of 85 m/min (l_c=170 m) the roughness was down of 0,9 microns.

Can be observed in Figure 2b for ISO M15 tool a down tendency of obtained roughness for all tested speeds (85 and 100 m/min). In the case of 100 m/min, in particular, was a fall until a minimum value (0,6 μm) and next an abrupt grown in the roughness value (1,8 μm) due a cutting edge deformation Figure 5c and Figure 5d.

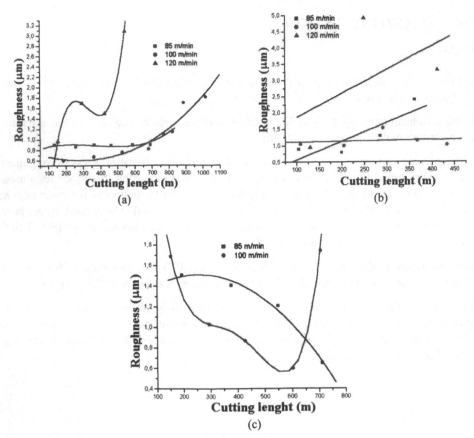

FIGURE 2. The surface roughness in turning Ti-6Al-4V: a) Using ISO S15, b) Using ISO P10, c) Using ISO M15

Figure 3a, Figure 3c and Figure 3e shows the comparison of the cutting edge using cermented carbide with cutting speed v_c=85, 100 and 120 m/min. Crater wear was observed in all conditions in function of the adhesion and to remove of the chip in tool surface. With the formation of notch and chips deposited on the tool surface, Figure 3b, Figure 3d and Figure 3f, the flank wear occour mainly due the grains agglomerates removal of tool material by the adherent chip or workpiece, which was the major contributor to the flank wear.

Through the tool micrographies above can be possible to observ the minor deterioration of cutting edge occurs to ISO S15 tools, Figure 3, the same was present the better results in relation to its behaviour roughness, in the other hand, the others cutting tools, ISO P10 and ISO M15, present a great degree deterioration of the cutting edge, in special the ISO M15 tools, Figure 5, shows a great flank wear, together chipping and plastic deformation.

Severe wear of the cutting edge was observed, in Figure 4b, Figure 4d and Figure 4f, with chip adhesion in cutting edge Figure 4a and Figure 4c and surface of the tool Figure 4e. The excessive flank wear can be observed in Figure 4b, Figure 4d and Figure 4f. In the ISO P10 tool, didn´t observed plastic deformation, but its cutting edge was deformed by flank wear and chip adhesion to making worse the chip formation Figure 6b. A better behaviour of roughness obtained can be explained by the maintenance of cutting edge form for more time than ISO M15 tool.

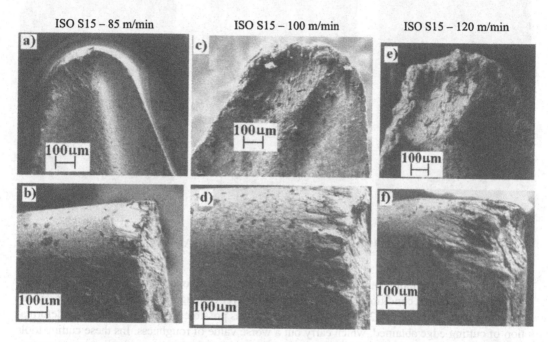

FIGURE 3. Tool surface and flank wear in the machining of Ti-Al-4V for different cutting speeds using ISO S15

ISO P10 – 85 m/min   ISO P10 – 100 m/min   ISO P10 – 120 m/min

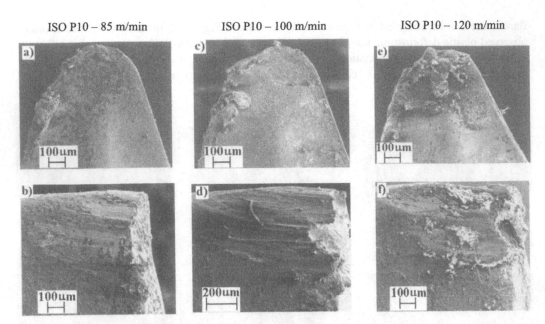

FIGURE 4. Tool surface and flank wear in the machining of Ti-Al-4V for different cutting speeds using ISO P10

About the presented tool wear for the ISO M15 Figure 5 can be observed the great chipping wear in 85 and 100 m/min conditions, after the removal of layer coating the tool wear increase provides these results, in the case of 120 m/min only one step was possible to make, due the great deformation of cutting edge obtained, which carry out a worse value of roughness. Ins these cutting tools can be noted a certain degree of plastic deformation in its cutting edges.

ISO M15 – 85 m/min   ISO M15 – 100 m/min   ISO M15 – 120 m/min

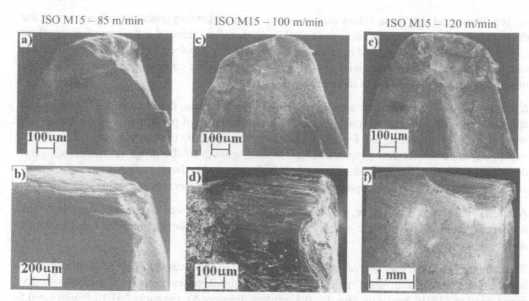

FIGURE 5. Tool surface and flank wear in the machining of Ti-Al-4V for different cutting speeds using ISO M15

In the Figure 6 can be observed the generated chip in the better cutting condition (S15 ISO in 85 m/min), which shows a very uniform morphology in relation to other test conditions. In this condition the chip presents very distinct zones of primary and secondary deformations, which characterize a major easily in the cutting. Can be noted the others conditions, M15 ISO and P10 ISO both in 85 m/min, which produce very deformed chips, thus the major severity in cutting operation was demonstrated.

ISO S15 – 85 m/min   ISO M15 – 85 m/min   ISO P10 – 85 m/min

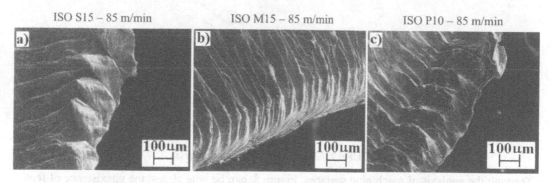

FIGURE 6. Chip formed in the machining of Ti-Al-4V for different cutting tools

3.2 ALUMINUM

In the analysis, Figure 7, it was possible to note that, for smaller feed than f = 0,1 mm/rev., the obtained roughness presented larger values than foreseen them by the calculations of the theoretical roughness and that these values came more or less in a landing constant, independent of the decrease of the feed rate. This fact can be explained, therefore below this feed rate landing, there is not significant contribution of the roughness generated by the feed of the cutting tool, but the roughness created by the superficial defects generated during the cutting operation, defects these, inherent to the geometric factors like as nose ratio mainly.

It was possible to note also that with the increase of the feed rate, the measured roughness tends approaching of the theoretical roughness and that the best strip of performance of the three tools was among f = 0,1 mm/rev. and f = 0,3 mm/rev. (recommended by the manufacturer). This can be explained, therefore for low progress values, a small growth of the progress generates a great decrease of the specific pressure of cutting (Ks), therefore the shavings formation is facilitated and the workpiece roughness approaches of the ideal, that is the theoretical roughness.

Above f = 0,3 mm/rev., however, the roughness has a gradual growth in relation to the theoretical roughness, mainly due to larger severity of the process, what a larger number of defects in the surface generates.

The roughness tests in function of the progress were led to the breaking of the cutting tool and it was possible to notice different resistance degrees for the three tested tools. The tools with diamond coating "1810" broke up after f = 0,3 mm/rev. (region A); the tools " H 10 " with f = 0,5 mm/rev. (region B) and, finally, the "Polished" tools, with f = 0,9 mm/rev. (region C), demonstrating that the "Polished" tools had a very larger resistance to the generated impact when the cutting process begins.

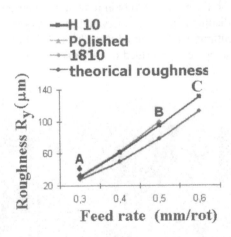

FIGURE 7. Effect of feed rate on roughness

Through the analysis of machining surfaces, Figure 8, can be note almost the unexistence of feed marks left by cutting tool, but occur only some superficial defects, such as welding material in surface (a, b, c). Next, it is possible to note a very growth phenomena relation to operation stiffness, like as side flow (f) and surface material of workpiece pulling out (g), which can be contributed for roughness growth.

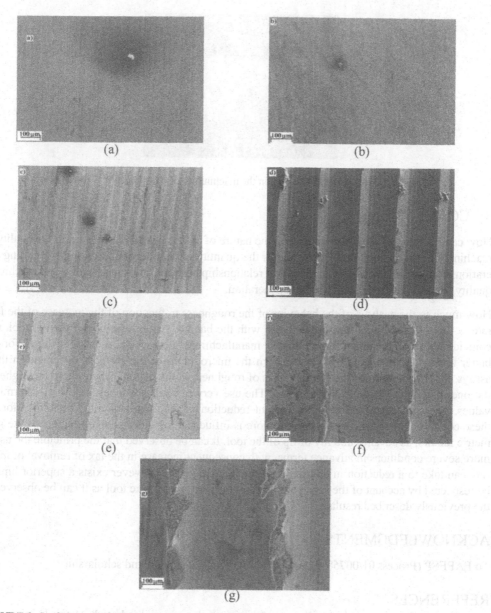

FIGURE 8. Surfaces generated with increase of the feed in the machining operations with cut fluid and "H 10" tools, (a) f = 0,02 mm/volta; (b) f = 0,05 mm/volta; (c) f = 0,1 mm/volta; (d) f = 0,2 mm/rev.; (e) f = 0,3 mm/rev.; (f) f = 0,4 mm/rev. (g) f = 0,5 mm/rev

The microstructure of the chip in both can be observed the cut conditions, Figure 9.

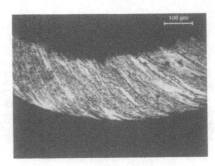

FIGURE 9. Micrograph of the chip in the machining with cut fluid, Vc = 800 m/min

4   CONCLUSIONS

How can be noted by the obtained results, the nature of chip formation influences in the quality of machining surfaces, in the tool wear and in the quantity of heat generated during the cutting operations. Thus, can be possible to identify a relationship between deformation chip and machining quality, such as low roughness and heat generation.

How much to the analysis of the behavior of the roughness in function of the increase of the feed rate, a fact of raised importance has to see with the band of better behavior of the cut tool, that coincide with the recommended one for the manufacturer. Another factor that contributed for one better understanding of the results had been the micrographs of the surfaces generated in these assays, that had confirmed the found values of roughness in the regions where these are higher of the one than of the theoretical roughness. The use very reduced advances, below of determined values, does not contribute for a significant reduction of the values of roughness, therefore in these conditions the superficial finishing more is influenced by other people's factors to the geometric factors, as advance and ray of tip of the tool. It can be observed that the principle the use of more severe conditions in advance terms, with consequent increase in the tax of removal of material, can take to a reduction in the running time reducing costs, however exists a superior limit to be respected by account of the finish surface and the resistance of the tool as it can be observed in the previously described results.

ACKNOWLEDGMENTS

To FAPESP (Process 01/00759-0) and CNPq by financial support and scholarship.

REFERENCES

1.   Boyer, R.R.; "An overview on the use of titanium in the aerospace industry". Materials Science and Engineering A213 (1996) 103-114.
2.   Vigneau, J., " Obtaining high productivity in the usinagem of titanium alloys "., you Conspire and you Put, n. 380, p. 16-32, September 1997.
3.   Ezugwo, E.O, Wang, Z. M., " Titanium alloys and their machinability", Journal of materials processing technology, V.68, P.262-274, 1997.
4.   Zoya, Z.A.; Krishnamurthy, R.; "The performance of CBN tools in the machining of titanium alloys". Journal of Materials Processing Technology 100 (2000) 80-86.

5. Dearnley, P.A., Greason, A.N. Mater. Sci. Technol. 2 (1986).

6. Hartung, P.D.; Kramer, B.M. Ann. CIRP 31 (1) (1982) 75-80.

7. Ezugwo, E.O. Technical Report, University of Warwick, 1988.

8. Siemers, C.; Mukherji, D.; Bäker, M.; Rösler, J.;"Deformation and microstructure of titanium chip and workpiece". Z. Metallkd march, 2001.

9. Komanduri, R.; Reed, W., "Evaluation of carbide grades and a new geometry for machining titanium alloys", WEAR, 92, 1983.

10. Lopez of Lacalle, L N, Llorente, J.I., Sánchez, J.A., 1998, " Improving the cutting Parameters the machining of Nickel and Titanium alloys ", Annals of the CIRP vol. 47 CD-ROM.

11. Ribeiro, M.V.; Moreira, M.R.V.; Ferreira, J.R.;" *Optimization of titanium alloy (6Al-4V) machining*". *Jornal of materials processing technology,*; 143-144 (2003) 458 – 463.

12. Heinz, A., Haszler, A., Keidel, C., Moldenhauer, S., Benedictus, R. e Miller, W. S. (2000). "Recent development in aluminum alloys for aerospace applications". Materials Science & Engineering, vol. A, No. 280, pp. 102-107.

13. Diniz, A. E., Marcondes, F. C., Coppini N. L (1999). " Technology of materials machining". São Paulo: MM Editora. (In portuguese).

14. Cunha , E.A., Ribeiro, M.V.; *Stud of alloy ASTM 7050T7451 machining*"; MSc, Department of Materials Engineering - FAENQUIL-USP/DEMAR- Lorena (Brazil), 2004, 86p.. (In portuguese).

MODELLING AND SIMULATION CREEP FEED GRINDING PROCESS

M. Dieye, A. D'Acunto, P. Martin

Laboratoire de Génie Industriel et Production Mécanique (LGIPM - EA 3096)
CER ENSAM – Metz Technopôle – 4 rue Augustin Fresnel – 57 078 Metz cedex 3
[mamadou.dieye, alain.dacunto, patrick.martin]@metz.ensam.fr

KEYWORDS: Creep Feed Grinding Process, Modelling, Computer Simulation, Surface Waviness, Workpiece/Tool/Machine System (WTM).

ABSTRACT. The aim of our article is to propose a model of simulation of the dynamic behaviour of the creep feed grinding process. Specifically, we seek to predict the waviness of the bonded surface (depth of cut). We propose, initially, a mechanical modelling taking account of the operational parameters and the geometrical and mechanical characteristics of the workpiece, the tool and the machine-tools. The models are founded on the Newton's second law and on the Work Energy principles with various assumptions on the contact wheel workpiece. In a second time, computer simulations (under Simulink) are proposed in order to evaluate the waviness of the profile and the stability and the precision of these models.

1 INTRODUCTION

The creep feed grinding process is characterized by a great depth of cut (ap = 1 - 6 mm), a high dimensional accuracy (quality 5) and a low table speed (v_w = 20 - 80 mm/min). It allows the high mechanical characteristics materials machining (materials tools, ceramics...) with a high geometrical degree of accuracy [1]. The high cost of its operations is justified by the fact that the grinding process comes in last to process planning. The need for controlling this process leads to model whole or part of the Workpiece/Tool/Machine (WTM) system [2], in order to study the principal influence parameters on the quality of the workpiece.

The control of the geometrical specifications (form, position, waviness and roughness) and of surface integrity of the material (residual stress, burns...) of rectified surfaces requires a good comprehension of the influence of the parameters:

- externals to the process that are the mechanical characteristics of the workpiece and the grinding wheel, the performances of the machine tool (geometrical and kinematics quality, vibrations);

- internals that are the operating conditions of piloting of the process (speeds of the grinding wheel and the workpiece, real depth of cut, conditions of dressing, characteristic of the lubricant),

- connection variables of the process (equivalent depth of cut, cutting force, power, chatter, applications conditions...).

We do not treat the aspects relating to the residual state of the skin of the workpiece. Our work consists in studying the influence of the vibrations on real depth of cut in order to predict the

Published in: E. Kuljanic (Ed.) *Advanced Manufacturing Systems and Technology*,
CISM Courses and Lectures No. 486, Springer Wien New York, 2005.

bonded surface waviness starting from computer simulations resulting from the analytical and phenomenological models. Many works introduce models which bind the grinding wheel topography to the roughness [3] [4] [5] [6], of the vibrations of the spindle or the table of the machine tool [7] [8].

There exists today little work concerning the prediction of the surface waviness in grinding. We propose starting from existing models (formulation, equivalent depth of cut, characteristic of the grinding wheel and the spindle like those of the workpiece) a simulation of the relative positions of the grinding wheel compared to the workpiece during a grinding operation.

2   NOMENCLATURE

TABLE 1. Simulation Parameters of surface grinding process

Symbol (unit)	Description
A (mm²)	Grinding wheel/workpiece contact area
δ_0, δ (mm)	Initial, Instantaneous depth of cut
k_w (N/m)	Stiffness of Workpiece
k_s (N/m)	Stiffness of Wheel
k_b (N/m)	Stiffness of Spindle
c_w (N.s/m)	Damping Factor of Workpiece
c_s (N.s/m)	Damping Factor of Wheel
c_b (N.s/m)	Damping Factor of Spindle
G (m/s²)	Gravity
m (kg)	Mass of Wheel and Spindle
E (GPa)	Young's Modulus (Steel)
t_s (m)	Width of Wheel
t_w (m)	Thickness of Workpiece
R (mm)	Radius of Wheel
v_w (mm/min)	Feed Rate of Workpiece
v_s (m/s)	Velocity of Wheel Surface
ρ (kg/m3)	Workpiece Density (Steel)
K (-)	Empirically Determined Coefficient
η (-)	Constant Material Property
μ (-)	Force Proportionality Constant
$F_{c,o}$ (N)	Initial Cutting Force

3 STEP OF MODELLING

3.1. GEOMETRICAL AND KINEMATICAL MODEL

During the surface grinding process, the relative position grinding wheel/workpiece is represented by the diagram hereafter, the grinding wheel assembled on the animated spindle of a rotational movement (v_s) and the workpiece immobilised on the horizontal table raving a translator movement (v_w). Like working hypothesis, we consider that the grinding wheel geometry and the flatness of initial workpiece surface are perfect. The value of the instantaneous depth of cut (d), is equivalent to the nominal depth of cut added the vertical difference between the points A and B (Figure 1):

$$\delta(t) = \delta_o + y_B(t) - y_A(t)$$

(1)

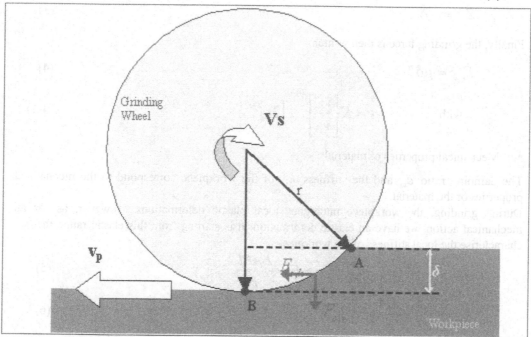

FIGURE 1. Basic Components of the Creep Feed Grinding System

3.2. DATA CONFIGURATION

The data configurations result from models or experiences relative to various researches in grinding process. In our study, we are interested on the nature of the cutting force characterising the grinding wheel / workpiece contact [7]. In addition, we take account of the behaviour of the spindle in grinding [8]. The mechanical properties of material to be rectified are those of the 40 Cr Mn Mo 8 treaty (hardened and tempered) with 50 HRc.

• Grinding wheel/workpiece contact

The grinding force wheel/workpiece corresponds to the mechanical action of the grains of grinding wheel (ploughing, friction, shearing) along the contact surface. We are interested in the normal component of the grinding force which acts on the surface of the contact wheel/workpiece, and use the expression established by Werner [9].

$$F_{c,v} = K\left[\frac{Z'}{v_s}\right]^{2\eta-1} [\delta]^{1-\eta}[2r]^{1-\eta} \tag{2}$$

The metal removal rate per unit width (Z') is equivalent to the product of the table speed (v_p) and the depth of cut (δ) :

$$Z' = v_w \delta \tag{3}$$

Finally, the grinding force is then written:

$$F_{c,v} = \mu \delta^\eta \tag{4}$$

with $$\mu = K\left[\frac{v_w}{v_s}\right]^{2\eta-1} [2r]^{1-\eta} \tag{4-1}$$

• Mechanical properties of material

The damping ratio c_w and the stiffness k_w of the workpiece correspond to the mechanical properties of the material.

During grinding, the workpiece undergoes local plastic deformations. However, as for all mechanical action, we have an elastic deformation. It is starting from this elastic range that we characterise the local stiffness of the workpiece.

$$\sigma = E\varepsilon \rightarrow F/_A = E\,\Delta l/_l \rightarrow F = \frac{EA\Delta l}{l} \tag{5}$$

$$k_w = \frac{EA}{l} = \frac{EA}{t_w} \tag{6}$$

$$k_w = \frac{Et_s\sqrt{2r\delta - \delta^2}}{t_w} \tag{7}$$

Concerning the damping ratio of the workpiece, we take into account the results of the experiences of [10].

• Data relating to the dynamic behaviour of the spindle carries grinding wheel / grinding wheel

The stiffness k_b and the ratio damping c_b of the spindle carries grinding wheel / grinding wheel are determined by deflection tests of the spindle. The values result from the works of [11].

3.3. MODEL OF CONFIGURATION

The dynamic behaviour of grinding process can be assimilated and discredited with sets of masses, springs and dampers in series and/or parallel (Figure 2) [10].

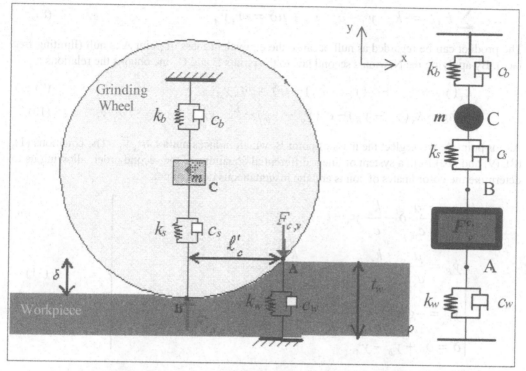

FIGURE 2. Creep Feed Grinding components and their dynamic model equivalents

From a geometrical point of view, our unidirectional model is defined in normal plan (X,Y) on the bonded surface passing by the point C, belonging to the grinding wheel axis. The points A and B are the extreme points of the contact surface between the grinding wheel and the workpiece. The point C is assimilated to the centers of the grinding wheel and of the spindle.

3.4. MECHANICAL MODEL

We propose to model the mechanical behaviour founded on the Newton's second law applied to the point and on the work energy principle. These principles are applied to the various points A, B, and C. The equations of the relative movement in each point are $y_A(t)$, $y_B(t)$ and $y_C(t)$.

We determine the depth of cut (δ) which defines the vertical distance during the displacement of the grinding wheel between the co-ordinates of points A and B: The formulation is as follows

$$\delta(t) = \delta_o + y_B(t) - y_A(t)$$

(1)

where δ_o is the nominal depth of cut.

NEWTON'S SECOND LAW

As for us, we exclusively apply the Newton's second law to the considered point. By applying the Newton's second law to point A (figure 2), the relation is:

$$\sum F_{A,y} = -k_w \cdot y_A - c_w \cdot \dot{y}_A + \mu\delta = m_A \ddot{y}_A \tag{8}$$

The product can be regarded as null because the equivalent mass of point A is null (limiting free face). By applying the Newton's second law to the points B and C, one obtains the relations :

$$k_s(y_C - y_B) + c_s(\dot{y}_C - \dot{y}_B) + \mu\delta = m_B \ddot{y}_B \tag{9}$$

$$-mg - k_s(y_C - y_B) - c_s(\dot{y}_C - \dot{y}_B) - k_b \cdot y_C - c_b \cdot \dot{y}_C = m \ddot{y}_C \tag{10}$$

As for point A, we neglect the mass of point B, which induces nullity $m_B \ddot{y}_B$. The equations (1), (8), (9) and (10) form a system of linear differential equations of the second order, allowing us to determine the co-ordinates of points and the instantaneous depth of cut.

$$\begin{cases} \dot{y}_A = \dfrac{\mu}{c_w}\delta - \dfrac{k_w}{c_w} y_A \\[2ex] \dot{y}_B = \dfrac{\mu}{c_s}\delta - \dfrac{k_s}{c_s} y_B + \dot{y}_C + \dfrac{k_s}{c_s} y_C \\[2ex] \ddot{y}_C = -g + \dfrac{c_s}{m}\dot{y}_B + \dfrac{k_s}{m} y_B + \dfrac{-c_s - c_b}{m}\dot{y}_C + \dfrac{-k_s - k_b}{m} y_C \\[2ex] \delta = \delta_0 + y_B - y_A \end{cases} \tag{11}$$

WORK ENERGY

A formulation of work-energy principles is given by :

$$\frac{d}{dt}(E_{sys}) = \frac{d}{dt}(W_{ent}) - \frac{d}{dt}(W_{sor}) \tag{12}$$

Total energy of the system E_{sys} (workpiece and grinding wheel) is made up of the kinetic, potential and dissipative energies.

Contrary to the assumptions of preceding modelling, we introduce a linear density (ξ) with the grinding wheel / workpiece contact. At point A, the kinetic energy (EC.) and potential energy (EP) are thus obtained by

$$EC = \frac{1}{2}mv^2 = \frac{1}{6}\xi\dot{y}^2 = \frac{1}{6}\rho A\dot{y}_A^2 \tag{13}$$

$$EP = \frac{1}{2}ky^2 = \frac{1}{6}k_w \dot{y}_A^2 \tag{14}$$

where ρ is density of the workpiece, and A a point of the wheel/workpiece contact surface.

In substituting the equation (12) one obtains:

$$\frac{d}{dt}\left(E_{sys}\right) = \frac{d}{dt}\left(EC+EP\right) = \frac{d}{dt}\left(\frac{1}{6}\rho A\dot{y}_A^2 + \frac{1}{2}k_W y_A^2\right) = \dot{y}_A\left(k_W y_A + \frac{1}{3}\rho A\ddot{y}_A\right) = \frac{dW_{ent}}{dt} - \frac{dW_{sor}}{dt}$$

(15)

Work in entry results from the efforts of grinding for an elementary displacement along the Y axis. It is defined by :

$$dW = Fdy$$

(16)

$$\frac{dW_{ent}}{dt} = F_c\frac{dy}{dt} = \mu\delta\ddot{y}_A$$

(17)

Work exits the system comes from the dissipative aspects. Consequently we can write

$$\frac{dW_{sor}}{dt} = F_d\frac{dy}{dt} = \left(c_w\dot{y}_A\right)\dot{y}_A = c_w\dot{y}_A^2$$

(18)

In substituting the equations (17) and (18) in the equation (15) and while dividing by $\dot{y}_A$ one obtains:

$$\frac{1}{3}\rho A\ddot{y}_A = -c_w\dot{y}_A - k_w y_A + \mu\delta$$

(19)

A second system of linear differential equations can be formed by combining the equations (19), (9) and (10), and while replacing d according to the equation:

$$\begin{cases} \ddot{y}_A = -\dfrac{3c_w}{\rho A}\dot{y}_A - \dfrac{3k_w}{\rho A}y_A + \dfrac{3\mu}{\rho A}\delta \\[2mm] \dot{y}_B = \dfrac{\mu}{c_s}\delta - \dfrac{k_s}{c_s}y_B + \dot{y}_C + \dfrac{k_s}{c_s}y_C \\[2mm] \ddot{y}_C = -g + \dfrac{c_s}{m}\dot{y}_B + \dfrac{k_s}{m}y_B + \dfrac{-c_s - c_b}{m}\dot{y}_C + \dfrac{-k_s - k_b}{m}y_C \\[2mm] \delta = \delta_0 + y_B - y_A \end{cases}$$

(20)

The equations obtained are different from the first modelling. Only the component defining the behaviour of point A is modified. It is due to the consideration of the mass of the workpiece and the contact area.

MODEL OF COMPUTER SIMULATION

A model of computer simulation of the grinding process, based on the results of the mechanical model describes higher, is created in the Matlab programming environment [13].

The created model of computer simulation uses the interface Simulink of Matlab. Simulink is a package which provides a graphic interface and has devices specifically intended for simulations in real time.

Models were created for the two systems of equations of the movement: equations (11) based exclusively on the Newton's second law and the equations (20) based on a combination of the Newton's second law and the work energy principles. A graphic example representative of the model (Simulink) is shown in the figure 3.

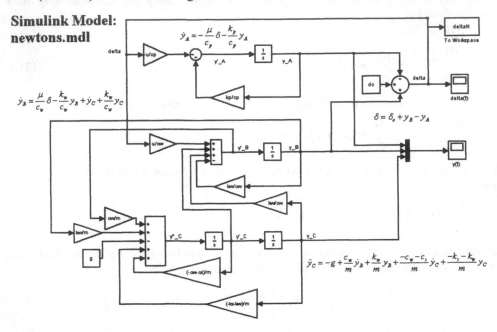

FIGURE 3. Representative graphic example of Simulink model interface (Second Law system)

The choice of the solver methodology of the systems of differential equations in Matlab is made for the computer method. Consequently we use ode15s Matlab Solver.

4 RESULTS OF SIMULATIONS

The initials simulation parameters of surface grinding process are: v_s=30m.s^{-1}, v_w=50mm.min^{-1}, a=5mm and E=210*10^{9}Pa.

4.1 COMPARISON BETWEEN THE TWO MECHANICALS MODELS

The solutions resulting from δ are presented (Figure 4), with the solutions for the system of the second law (broken line) and the solutions for the work energy (straight line).

The frequencies of vibrations of δ are the same ones for the solutions of the two systems. There is a notorious difference between the forms envelopes of the two systems. As long as the solution of the system of the work energy keeps a constant amplitude around 5.002 mm, the solution of the Newton's second law have an amplitude of 1.4 μm which also varies with time around 5.002mm.

In spite of that, the amplitudes of the vibrations are narrowly limited between 5.0013 and 5.0027 mm (with an average thus of 5.002 mm), and the two systems do not provide radically different solutions.

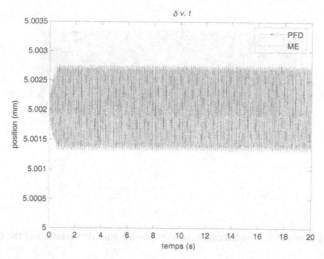

FIGURE 4. Plots of delta for the Second Law and the Work-Energy systems

The two systems provide appreciably equal solutions for the behaviour of point A. The difference between the two forecasts is slightly attenuated by the scale of clearances : 5×10^{-8} mm consequently, the behaviour of affects hardly the depth of the cut. The intuitive computer solution of the two systems provides that point A will be guided to the top take into account thus the deflection of the spindle and the deflection of the grinding wheel at the time of the entry in contact of the grinding wheel with the workpiece. The force of grinding should always act as bottom on point A.

Moreover, the frequency of the vibrations of point A envisaged by the work energy system is considerably larger than for the Newton's Second law system. The work energy system was developed in order to reduce the frequency vibration of point.

4.2 STABILITY AND PRECISION

Simulations were led for the two systems. The period of the two systems (Newton's second law and work energy) was 70 seconds. The results, represented on figures 5 and 6, were employed to evaluate the stability of the system. The Newton's second law system does not seem to converge towards a single value finished nor diverging with infinity when time tends towards infinity. This implies that the system is stable. It is interesting to note that the magnitude variation of the Newton's second law system does not seem to have any periodicity of the second degree. The work energy system does not show any change of magnitude vibratory, which suggests that this system is also stable.

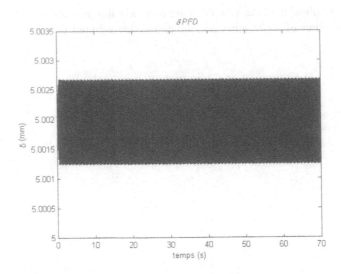

FIGURE 5. Depth of cut for the Second Law system-(Magnitude variation of the depth of cut)

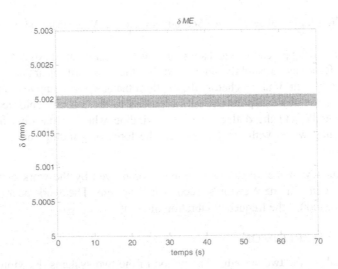

FIGURE 6. Depth of cut for the Work-Energy system-(Magnitude variation of the depth of cut)

It is important to consider the predicted values depth of cut in the context of nominal depth of cut to evaluate the percentage error. Consequently, the resolutions of $\left(\delta - \delta_o\right)/\delta_o$ were created for simulations of both system for 70 seconds time fields. The results, represented on figures 7 and 8, indicate the maximum percentage error of the two systems from 0.04 to 0.05%.

This high precision illustrates well the fact that the creep feed grinding process is supposed being a cutting process of very good dimensional accuracy.

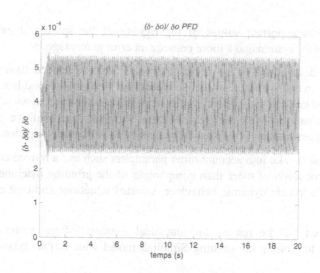

FIGURE 7. Plots of $\left(\delta - \delta_o\right)/\delta_o$ for the Second Law system

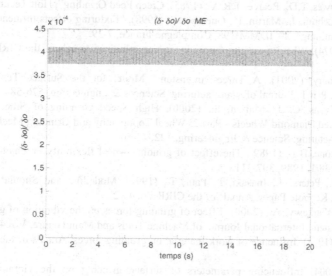

FIGURE 8. Plots of $\left(\delta - \delta_o\right)/\delta_o$ for the Work-Energy system

5 DISCUSSION

We have showed that the model was stable for our particular parameters of simulation. The relative error of the nominal depth of cut about 0.05% is relatively correct for our original parameters of simulation. We suggested that our estimated value for the rigidity of the spindle was near to the values of rigidity of grinding machines usually used [11].

Simulations with less important values of the rigidity of the spindle showed a magnitude vibratory increase of the system and a more consequent error percentage.

Two mathematical descriptions of the grinding process were developed having for result two systems different from the equations of the movement: the Newton's second law system based on the Newton's second law and the work energy system also based on the Newton's second law but also taking into account the work energy principles. Simulink is an intuitive graphic interface, easily extensible for new models which would be applied for future investigations.

Thus, we will be able to take into account other parameters such as : a movement in two or three dimensions, an introduction of more than components of the grinding machine; a more precise representation of the spindle dynamic behaviour; external vibrations and chatter, material states of burn...

Moreover an interest will be put on the analytical solutions of the systems of differential equations, in order to prolong the existing Simulink model with various types of behaviour of grinding process.

REFERENCES

1   Andrew, C., Howes, T.D., Pearce, T.R.A., (1985), Creep Feed Grinding", Holt Technology.
2   D'Acunto, A., Lebrun, J., Martin, P., Gueury, M., (1998), "Fixturing effects influence on workpiece quality in milling". 2nd IDMME'98. Compiègne-France, 7-29.
3   Verkerk, J., (1974), Wheel wear and forces in surface grinding, Annals of the CIRP Vol. 23/1, 81-86.
4   Erik, J., Salisbury, (2001), A Three-Dimensional Model for the Surface Texture in Surface Grinding", Part 1, Journal of Manufacturing. Sciences & Engineering, 576-584.
5   Hwang, W., Evans, C. J., Malkin, S., (2000), High Speed Grinding of Silicon Nitride With Electroplated Diamond Wheels - Part 2: Wheel Topography and Grinding Mechanisms, Journal of. Manufacturing. Science & Engineering, 122, 42-50.
6   Baylis, R.J., Stone, B.J., (1989), The effect of grinding wheel flexibility on chatter Annals of the CIRP Vol 38/1, 1989, 307-311.
7   Tonshoff, H.K., Peters, J., Inasaki, I., Paul, T., (1992), Modelling and Simulation of Grinding Processes, Keynote Paper, Annals of the CIRP Vol 41/2.
8   Alfares, M., Elsharkawy, A., (2000), Effect of grinding forces on the vibration of grinding machine spindle system, International Journal of Machine Tools and Manufacture, Vol 40, 2003-2030.
9   Werner, P.G., (1978), Influence of work material on grinding forces, Annals of the CIRP Vol 27/1, 243-248.
10 Dièye, M., (2004) Influencing parameters of surface grinding on the vibratory level of the Workpiece/Tool/Machine system (WTM), 5nd IDMME'04. Bath, United Kingdom.
11 Orynski, F., Pawlowski; W., (2002), The mathematical description of dynamics of the cylindrical grinder, International Journal of Machine Tools and Manufacture, Vol 42/7, 773-780.
12 Ginsberg, J.H., (1998), Advanced Engineering Dynamics", Cambridge Press.
13 Simulink, Numerical computer with Matlab, (http://www.mathsworks.com).

MODEL-TO-PART: A ROAD MAP FOR THE CNC MACHINE OF THE FUTURE

P. Gray[1], G. Poon[1], G. Israeli[2], S. Bedi[1], S. Mann[2], D. Miller[3]

[1] Department of Mechanical Engineering, University of Waterloo, Canada
[2] School of Computer Science, University of Waterloo, Canada
[3] Department of Electrical Engineering, University of Waterloo, Canada

KEYWORDS: CNC Machining, Simulation, Intelligent Control.

ABSTRACT. In the paradigm for CNC machining, tool path programming and online control of the machine are separate tasks. Though this paradigm has proved successful to date, it is now showing its limitations particularly in 5-axis machining, high-speed machining, direct model-to-part manufacturing and automated error detection and prevention. This paper describes work to implement a new paradigm for CNC machining described by Gray et al. [1] for direct model-to-part manufacture by providing the part definition to the NC controller. The intention here is to integrate a simulation module into the controller to detect and prevent errors by merging the physical workspace of the machine with the virtual workspace of the simulator. This paper describes the architecture for this concept and the simulation module that has been developed. Also, a machining test was conducted on a special-built machine to analyze the machine and simulation components.

1 INTRODUCTION

Today's CNC machining process paradigm (Figure 1) begins with the CAD model of the part. A plan is then made to determine how to machine the part according to the stock material form, the machining operations and the availability of machines and tools. Tool paths are generated accordingly and frequently the paths are simulated to detect potential interference problems between the machine tool and the workpiece as well as to check feed rates and tool dimensions. This paradigm has changed little from its inception except perhaps for the addition of the simulation process. Limited computing power at that time necessitated the division between CNC control and tool path generation, however, the cost and performance of today's computers no longer warrants this division of labor. The paradigm is starting to show its limitations with motion control issues in 5-axis machining and high-speed machining, which cannot be properly addressed with the restrictive ISO G-code programming standard. Also, the sensory capabilities and intelligence of the controller are no longer adequate to permit the development of direct model-to-part manufacturing automation.

Another major drawback with the division of labor is that the process is generally open-loop. If there are problems at any stage, it is up to the operators or programmers to close the loop by identifying and fixing the problems. Fixing problems often requires reprogramming some of the tool paths or changing tooling or fixtures. Not only is this time consuming and expensive, it is error prone because intense manual intervention and judgment are required.

Published in: E. Kuljanic (Ed.) *Advanced Manufacturing Systems and Technology*,
CISM Courses and Lectures No. 486, Springer Wien New York, 2005.

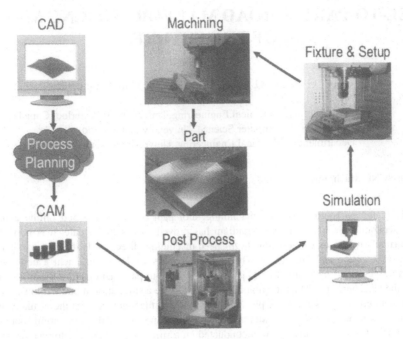

FIGURE 1. CNC machining paradigm

1.1 MACHINING ERRORS

Errors in CNC machining can be catastrophic in nature due to the large forces, high velocities and the high cost of the machines, tools and workpiece stock material. Though the simulation process is designed to assist in error prevention, it can only detect tool path errors. Also, it assumes that the process parameters in simulation are identical to those of the actual machine, tool and workpiece setups [1]. Touch probes and laser tool measuring systems add to the security of the system but they too cannot detect in-process problems that may occur such as tool wear and breakage, incorrect clamping positions, workpiece movement vibrations and manual operation errors. It is clear that today's CNC machine is blind in that the only object it knows exists in its workspace is the cutting tool. It has no information about the stock material, fixturing and clamping, or of the part it is supposed to make except for the relative coordinate frame of the part with respect to the machine coordinate frame. It blindly follows the commands given to it even if it is commanded to destroy itself.

1.2 OPEN ARCHITECT CONTROL

Open architect control (OAC) is the most promising solution to improve machine intelligence. It would also improve the rate of development and open new markets by allowing end-users to integrate new sensor and control technologies and new manufacturing and inspection strategies without significant re-investment in a completely new machine or controller. It would also permit greater customization of the CNC machine for the end-users' specific applications. Some initial work in the use of OAC has been presented but much of the focus has been on optimizing cutting parameters during machining [2][3]. The extent of interaction with the OAC is mainly shared memory access with the machining process variables to adjust feed rates. Our goal for the development of an intelligent machine to achieve direct model-to-part manufacture with minimal human intervention requires greater demands on the controller.

Direct model-to-part manufacture has been the focus of several of the competing OAC standards groups such as OSACA, JOP, OMAC [4]. However, although the development of a world-wide standard is an important task, it will take time to settle on an architecture [5] and demonstrate its usefulness to industry. STEP-NC [6] has similar objectives for automated CNC manufacturing but the focus is more on the development of data transfer standards rather than on OAC for manufacturing. Industrial acceptance will be difficult to win largely due to high initial startup capital costs, learning curves and skepticism of system robustness.

Regardless of the standards, work must continue on the concepts required to achieve direct model-to-part manufacture. Improvements in the intelligence and sensory capabilities of CNC machines for automation of tool path generation and error prevention are necessary components that can be studied as general concepts without having to agree on an architecture standard. One may argue that a working prototype should be developed before trying to set a standard to ensure that all communication protocols and components will be present and sufficient for an automated system. Thus, our approach is to construct a working system to develop and demonstrate the concepts and to expose the communication protocols, sensory requirements and intelligence that will be required. The system components we develop will consist of general concepts and solutions that can be drafted into different system standards providing they are sufficiently open.

1.3 PROJECT OBJECTIVES

Our goal is to map a virtual workspace of the machine to the machine's actual workspace to predict and prevent errors for direct model-to-part manufacturing. To achieve this goal, models of all objects residing inside the machine's workspace must be modeled in the virtual workspace. A simulation system must be integrated into the controller to predict errors before they are made so that a tool path generation system residing in the controller can then be used to alter the tool path to prevent errors. Additional sensory capabilities will eventually be necessary to make the system robust such as 3D cameras and tool wear and breakage detection. Naturally, the scope of this project is large and will require many years of development to completely address all the complex issues associated with industrial CNC manufacturing. Thus, we have developed a simplified CNC machine to begin this project, which allows us to develop individual components of the project in an incremental fashion. The focus of this paper is to demonstrate the mapping of the simulator's virtual space to the machine workspace, to test of our simulator and to present the work to date for the integration of the simulator into the machine controller.

This paper builds upon the concepts presented by Gray et al. [1] who declared that parameter mismatch between simulation and the machine's workspace along with events that are not considered in the simulation can lead to these catastrophic errors. The objective here is to introduce the CAD model into the CNC controller. A simulation module can then be applied to maintain a live model of the current remaining stock material. With such a model, the control could then predict problems like excessive cutting forces, tool deflections and interference.

This paper is arranged in the following order: We begin by describing a simple machine that these concepts will initially be implemented and tested on. The following section describes the current features of the simulation module and the features we intend to develop for a complete implementation. A machining test is then presented along with an evaluation of accuracy by comparing a 3D laser scan model of the machined part to the simulation. The paper concludes with remarks on how these components will be integrated into a single intelligent machine controller in future work.

2   THE SINGLE-AXIS LATHE

To begin developing the concepts of the new paradigm, we constructed a machine with reduced complexity compared to that of a commercial CNC machine to reduce the initial magnitude of the problem. The Single-Axis Lathe is a simple machine originally constructed with the intention of providing woodworking hobbyists an inexpensive approach to CNC machining of complex ornamental wood pieces by Kaplan et al. [7]. Figure 2 shows the Single-Axis Lathe. A router tool (E) serves as the machine spindle whose motion is controlled by a stepper motor, perpendicular to the cylindrical workpiece's longitudinal axis according to the depth of cut for a given tool position. The router axis is mounted on a platform (C) that moves longitudinally along the workpiece's rotary axis. This axis is mechanically linked to the motor shown (B) which is linked to the rotation of the workpiece. Thus, the tool path simplifies to a helical path around the part. Figure 2 shows some example parts made with this machine.

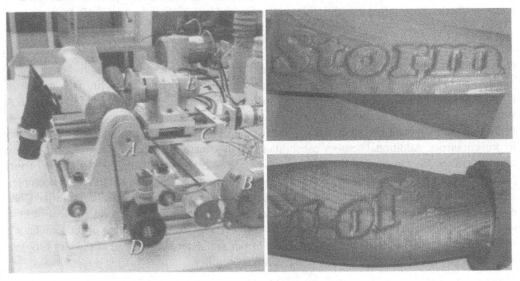

FIGURE 2. Single-axis lathe machining examples

3   TOOL POSITIONING

The focus of Kaplan et al.'s work was to provide a set of design tools to the wood worker and to map a tool path onto the designed part. The tool path is then downloaded to the machine control and the part is cut directly. Though this is a valid method for direct model-to-part paradigm, it is limited to parts that have ornamental patterns according to their design tools. A fundamentally different approach is proposed in this paper that is applicable to a more generalized control architecture designed to address many of the issues related to today's CNC manufacturing paradigm.

Instead of using the simplified design tool developed by Kaplan et al. [7], in the work for this paper, the model of the part was generated with an external CAD software package. The input to our tool path generator is an STL file containing triangle vertices. The tool path is then generated directly from the triangle data set, which is downloaded to the machine. Since the tool path trajectory of the Single-Axis Lathe is a simple helix, the generation of a tool path directly from the part model is a straight forward task. Generalizing automatic tool path generation is an enormous task that will require the development of intelligence algorithms. This will be the next stage of our research project where we redesign our test machine for more degrees of freedom to warrant decision-making in the tool path planning.

4 ERROR PREDICTION, DETECTION AND CORRECTION

To correct an error, it must first be predicted or detected. The key component in the new paradigm as proposed by Gray et al. [1] is the development of a simulator. The simulator has two roles: To maintain a current model of the remaining stock material and to predict problems with future tool motions as dictated by the tool path. Our simulator uses a simple geometric algorithm first developed for toroidal tools by Roth et al. [8] and later generalized for any surface of revolution by Bedi and Mann [9] to compute the discrete points known as grazing points on the swept surface that a cutting tool will leave as it moves through space (Figure 3). It is a generalized algorithm for any surface of revolution so that any cutting tool can be modeled for any spatial motion. By connecting the grazing points, the swept surface left by the cutting tool can be represented as a set of triangles.

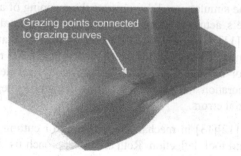

FIGURE 3. Grazing curves

To model the remaining stock material at any point in time, the swept volume of the tool path must be subtracted from the stock material. The approach we use in our simulator for Boolean subtraction of two volumes is the Z-map (Figure 4), which was pioneered by Van Hook [10]. The Z-map consists of vectors that represent the stock material all pointing in the same direction, preferably along the tool axis direction. The vectors are clipped at their intersections with the swept surface. The tips of the vectors are connected into a set of triangles for display purposes to model the remaining stock material. Our simulation module can model 3-, 4-, and 5-axis machine with the Z-map. However, for the Single-Axis Lathe the simulator uses a radial ray casting direction from the center of the stock material (Figure 5) as opposed to the single common direction for 3-axis machining. Figure 6 shows an example of the simulation for the Single-Axis Lathe.

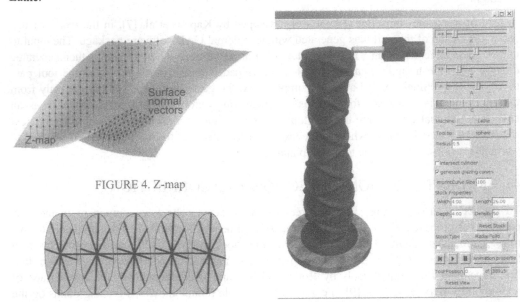

FIGURE 4. Z-map

FIGURE 5. Lathe ray casting vectors   FIGURE 6. Lathe material removal simulation

At the moment our simulator resides outside of the CNC controller of our machine. It accepts the tool path as input and can generate an STL file of the simulated machined part for output. Our work here is to evaluate the simulator and demonstrate the mapping of a virtual workspace of the machine and the machine's actual workspace with the eventual goal of error prediction and prevention. Austin et al. [11] provide a discussion on the use of triangulated models and the accuracies required in triangulated models for gouge-detection in CNC machining. Ultimately we will integrate the simulator into the CNC controller and add error detection algorithms to the simulator. A tool path generation and planning system will also be integrated into the machine's controller to correct potential errors.

The work by Roth et al. [12][13] in mechanistic modeling of cutting forces will be useful for predicting cutting loads and tool deflection. Roth et al.'s approach used the computer's graphics hardware to perform the ray casting and intersections to model the remaining stock material. The depth buffer of the computer's graphics hardware is typically used for hidden surface removal; as a pixel is rendered to the graphics card, its corresponding depth in view space is compared to the point that was last rendered to that pixel coordinate. If the new pixel is closer to the eye position in the view space than the previous pixel, the new pixel color replaces the previous color in the frame buffer. The depth buffer value for that pixel is then updated to the new value. For machining material removal simulation, the viewing volume is setup as shown in Figure 7 and the depth buffer is set to render objects furthest from the eye position. That way, the tool paths that cut the deepest will determine the stock current material at any point. This graphics-assisted approach to simulating the machined part can be applied to our 3- and 5-axis machining simulations to accelerate some of the calculations and transfer some of the computational loads from the CPU to the graphics hardware's GPU.

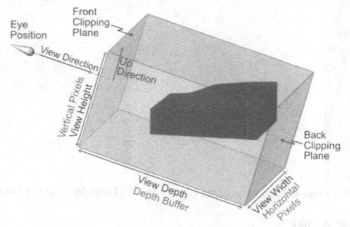

FIGURE 7. Viewing volume setup

5 MACHINING EXPERIMENT

A machining experiment was conducted to demonstrate the Single-Axis Lathe and to demonstrate the mapping between the simulation's virtual workspace the machine's actual workspace. The part was designed in SolidWorks and a triangulated STL model file was generated. The STL file was used by the tool path generator, which automatically generated a tool path that was downloaded to the machine controller. The cutting tool was a ball nose with a diameter of 6.35mm and the workpiece material was Cedar. The tool path used for machining was cut short to accommodate the stock material, which was shorter than the designed part. The simulation was performed and an STL file of the simulated workpiece was generated. The part was machined and a 3D laser scan of the part was made using a Minolta Vivid 900 3D camera. The machined data was registered to the simulated tool path to show the differences between the two spaces.

To test the machine accuracy, the scanned data was registered to the STL CAD model data using Raindrop Geomagic Studio software. Raindrop Geomagic Qualify software was used to compare the aligned models. The results are shown in Figure 8. The results indicate some discrepancy between the simulation and the machined part, which was not unexpected. The part was finish machined directly, which means that the smaller diameter section near the end of the tool path resulted in an extremely large depth of cut, particularly for our machine. At this point in the tool path, visual tool deflections were observed. Since there is no encoder on the controlled tool axis, with the high loads experienced at the end zone of the part, it is possible that some of the motor pulse counts were lost and a cumulative error would have resulted. Also, we checked the alignment between the rotary axis of the workpiece with respect to the cross-travel axis of the spindle platform and found that the two axes are slightly skewed in such a way as to bias the machining results in a fashion similar to that observed in the comparison.

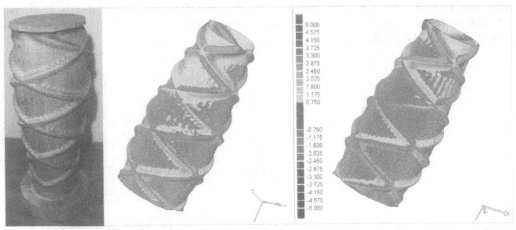

FIGURE 8. Photo and two views of the machined part 3D scan data registered to simulator data

6   FUTURE WORK

The discrepancies measured between the simulated and machined part clearly indicate that the machine must be precisely measured and any errors in construction must be mapped in the simulator and tool path generator. The limitations of this machine due to its original purpose to provide an inexpensive and simple machine for woodworking hobbyists have now become quite apparent in its new application as a platform for developing the concepts of the new paradigm for CNC manufacturing.

We are currently constructing a second generation lathe machine in a more precise fashion with a more rigid structure to improve on the current model (Figure 9). The new machine will eliminate the fixed helical tool path since it has three independent axes with encoders, which will permit the testing of tool path planning intelligence algorithms.

Our next goal is to fully integrate the simulator and tool path generator into the controller. Error detection and prevention algorithms will then be developed for the simulator and tested with the new machine.

Finally, the observed tool deflections can be addressed in several ways: The structural rigidity of the machine can be improved; the cutting loads can be predicted to dynamically adjust the tool position to machine the part more accurately; and the tool path planning can be used to adjust the axial depth of cut and number of cuts used to finish the part.

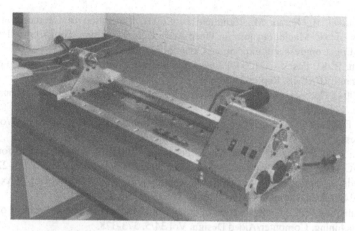

FIGURE 9. New 3-axis lathe (fabrication not completed)

7 CONCLUSIONS

In this paper we have presented a methodology to realize the new paradigm for CNC manufacturing as proposed by Gray et al. [1]. The paradigm calls for the mapping between the simulator's virtual workspace with the machine's actual workspace to detect and prevent errors and to automate the machining process for direct model-to-part manufacturing.

The registration of data sets from the simulation module and the 3D scanned part showed some discrepancies in the high depth of cut area of the part. The discrepancies can be attributed to the inaccurate build of the machine and excessive cutting forces causing severe tool deflections. These problems stem from the adaptation of this machine from its original intent to provide an inexpensive and simple CNC machine for woodworking hobbyists to a platform for developing and testing concepts for direct model-to-part manufacturing. Nevertheless, the results indicate that the machine geometry must be accurately measured and the cutting forces should be predicted and accounted for in the tool path generation and simulation modules.

ACKNOWLEDGEMENTS

The authors would like to thank Kevin Moule for assisting in the 3D part scans and data registration and for other various programming assistance throughout the project. We would also like to thank the funding agencies that have generously supported this project: Natural Sciences and Engineering Research Council of Canada, the Ontario Innovation Trust and the Canada foundation for Innovation.

REFERENCES

1. Gray, P., Bedi, S., Mann, S., (2003), A New Paradigm for Numerically Controlled Machines, Proc. of Virtual Concept 2003, Biarritz, France, 128-134.
2. Mori, M., Yamazaki, K., Fujishima, M., Liu, J., Furukawa, N., (2001), A Study on Development of an Open Servo System for Intelligent Control of a CNC Machine Tool, Annals of the CIRP, Vol 50/1, 247-250.

3. Jerard, R., Fussel, B., Ecran, M., Hemmett, J., (2000), Integration of Geometric and Mechanistic Models of NC Machining into an Open-Architecture Machine Tool Controller, Proc. of Intl. Mech. Eng. Congress and Exposition, Orlando, Florida.

4. Katz, R., Min, B.-K., Pasek, Z., (2000), Open Architecture Control Technology Trends, University of Michigan ERC/RMS, Report 35.

5. Nacsa, J., (2001), Comparison of Three Different Open Architecture Controllers, Proc. of IFAC MIM, Prague, 134-138.

6. Feeney, A., Frechette, S., (2002), Testing STEP-NC Implementations, Proc. of World Automation Congress.

7. Kaplan, C.S., Bedi, S., Mann, S., Israeli, G., Poon, G., (2004), A New Paradigm for Woodworking with NC Machines, J. of Computer-Aided Design and Applications, Vol 1, 217-222.

8. Roth, D., Bedi, S., Ismail, F., (2001), Surface Swept by a Toroidal Cutter during 5-Axis Machining of Curved Surfaces, Computer-Aided Design, Vol 33/1, 57-63.

9. Bedi, S., Mann, S., (2002), Generalization of the Imprint Method to General Surfaces of Revolution for NC Machining, Computer-Aided Design, Vol 34/5, 373-378.

10. Van Hook, T., (1986), Real-Time Shaded NC Milling Display, Computer Graphics Proceedings ACM SIGGRAPH, Vol 20/4, 15-20.

11. Austin, S., Jerard, R., Drysdale, R., (1997), Comparison of Discretization Algorithms for NURBS Surfaces with Application to Numerically Controlled Machining, Computer-Aided Design, Vol 29/1, 71-83.

12. Roth, D., Ismail, F., Bedi, S., (2003), Mechanistic Modelling of the Milling Process using an Adaptive Depth Buffer, Computer-Aided Design, Vol 35, 1287-1303.

13. Roth, D. Ismail, F., Bedi, S., (2005), Mechanistic Modelling of the Milling Process using Complex Tool Geometry, Int. J. Adv. Manuf. Technol., Vol 25, 140-144.

DESIGN OF A ROBOTIC VISION SYSTEM

A. Biason[1], G. Boschetti[2], A. Gasparetto[1], A. Puppatti[1], V. Zanotto[1]

[1] Department of Electrical, Management and Mechanical Engineering, University of Udine, Italy
[2] Department of Mechanical and Management Innovation, University of Padova, Italy

KEYWORDS: Robotics, Stereo Vision, Localization.

ABSTRACT. The paper presents the design and the implementation of a vision system for a Cartesian robot located in the Mechatronics laboratory of the Dept. of Electrical, Management and Mechanical Engineering of the University of Udine. The system hardware is made of two monochromatic analogue cameras with charge coupled sensor JAI CV-A50 and of an image acquisition board NI PCI-1409. The software code has been implemented in MATLAB and in the NI MAX environment. The vision system can control the two cameras by making a direct image acquisition and calibrating the cameras using the Faugeras method. Moreover, it is possible to obtain the three-dimensional localization of a selected object through stereoscopic vision techniques. The vision system described in this paper will be soon integrated with the Cartesian robot located in the laboratory, so as to have a powerful tool which turns out to be very useful for several application of conventional and applied robotics.

1 INTRODUCTION

Vision systems are very important in the modern industry because they represent one of the more powerful instruments for the observation of the surrounding environment. An industrial robot can interact with the external environment by means of an appropriate vision system, like a man can do with his sight. Vision systems are very attractive for industries since they could mean very high automation. There are many types of vision systems; the most common are the two-dimensional ones for their simple programming and computational time consuming while the three-dimensional ones represent the last borderline. Three-dimensional vision systems are used to acquire the shape of a scene to obtain geometric information of it; they are non-contact methods that utilize the light reflective propriety of the surface of bodies to reconstruct their structure. In this paper is designed a three-dimensional vision system in order to locate a point of interest and a known object positioned in the work-space of a portal Cartesian robot. The purpose is to carry out a simple three-dimensional vision system to be insert in the robot control system to create automated applications in the future.

2 THE STRUCTURE OF THE VISION SYSTEM

The purpose of the work is to create e simple three-dimensional vision system that allows the robot to recognize the position of points of interest in the robot workspace. This system is principally formed by two sub-systems: one is used to set cameras proprieties, acquire and save images and calibrate cameras; another is used to locate, by stereoscopic vision, the three-dimensional position of a point or, as extension of this last function, a known object. The supervision of an

Published in: E. Kuljanic (Ed.) *Advanced Manufacturing Systems and Technology*,
CISM Courses and Lectures No. 486, Springer Wien New York, 2005.

operator is an integral part of the system. The procedure is designed so that the computing time is not too long and the possibility of failure is not too high.

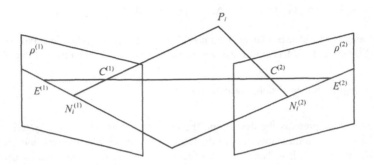

FIGURE 1. Stereo configuration and epipolar geometry

The vision system is composed by the illumination device, two cameras with their relative lens systems, a frame grabber and the host computer. The illumination device is a simple halogen reading lamp of 20W (very cheap in comparison with other devices). Illumination is an important aspect for a vision application and it is often undervalued: a well-devised algorithm can fail in case the illumination is insufficient in the observed scene. The model adopted to schematize the functioning of the camera is the pinhole one: the light coming from the scene passes through the lenses system and hits the charge coupled device (CCD) giving rise to the image.

The vision system has a sub-system consenting the setting of the camera characteristics like focus, iris, bit-depth and so on (in this case focus and iris are manually regulated) and consenting the acquisition and data management of the image data. This sub-system is the National Instrument's Measurement and Automation Explorer. The cameras controlled by NI MAX are two: the right and the left camera of the stereo apparatus. Cameras are positioned as illustrated in Figure 1. Each of them can frame the image of the work-space of the robot and the acquired images are very similar.

After the calibration of the cameras, the three-dimensional localization of a point can be obtained through the three basic steps of the stereo vision theory: rectification of cameras and images, calculation of the corresponding points and reconstruction of the three-dimensional shape. The basic function of this system is the localization of one point of interest selected by the operator. The location of a known object consists in the localization of three particular points.

3   CAMERA CALIBRATION

Calibration is a fundamental step in a vision system. Its effectiveness considerably affects the results of following operations. For this reason the calibration has to be well done. In this work cameras are calibrated with the Faugeras method which allows to obtain the intrinsic and extrinsic parameters of the cameras. Camera is schematized with the pinhole model (see Figure 2).

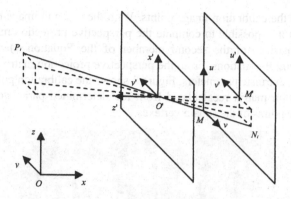

FIGURE 2. The pinhole camera model

The intrinsic parameters are: the skew angle θ, the focal length in horizontal pixels α_u, the focal length in vertical pixels α_v, the horizontal position of the principal point $u_{M'}$, the vertical position of the principal point $v_{M'}$; these values allow to compute the first matrix in the second member of Equation 1. The extrinsic parameters are: Cardan's angles α, β, γ, which describe the rotation of the real frame of reference as regards the camera standard frame of reference, and the three coordinates x'_O, y'_O, z'_O of the origin of the standard frame of reference; these values are synthesized in the second matrix of the second member of Equation 1; this rotation-translation matrix allows to calculate the point coordinates referred to the standard frame of reference, from the coordinates referred to the real frame of reference. By Equation 1 it is so possible to know the real coordinates of a point from the coordinates of its image, known the scale factor z' corresponding to the distance of the point from the camera focal plane. The lenses radial distortion has been neglected in this calibration procedure because of the good quality of the used lenses. The result shows that this hypothesis is acceptable (see Figure 3).

$$
z'_{P_i}\begin{bmatrix} u_{N_i} \\ v_{N_i} \\ 1 \end{bmatrix} = \begin{bmatrix} \alpha_u & -\alpha_u \cot \vartheta & u_{M'} & 0 \\ 0 & -\alpha_v / \sin \vartheta & v_{M'} & 0 \\ 0 & 0 & 1 & 0 \end{bmatrix} \begin{bmatrix} R_{11} & R_{12} & R_{13} & x'_O \\ R_{21} & R_{22} & R_{23} & y'_O \\ R_{31} & R_{32} & R_{33} & z'_O \\ 0 & 0 & 0 & 1 \end{bmatrix} \begin{bmatrix} x_{P_i} \\ y_{P_i} \\ z_{P_i} \\ 1 \end{bmatrix}
\tag{1}
$$

The calibration process requires the acquisition of the image of the calibration grid. This grid is composed by two panels on which are painted several squares. The positions of the vertices of these squares are known by construction. By processing the image it is possible to obtain the image coordinates of the same vertices. In this last step the points of interest are determined with the following procedure: find the square contours with an edge detection, determine the lines containing them with the MATLAB Radon transform, compute the analytical intersections of these lines in order to obtain the calibration image points. With the real and image coordinates of several no-coplanar points it is possible to compute the perspective projection matrix (given by the product of the two matrixes of the second member of the Equation 1) resolving an over-determined linear system of equations. From the perspective projection matrix it is so possible to extract the intrinsic and extrinsic parameters. Figure 3 shows the calibration points computed with the perspective projection matrix recalculated from the determined parameters. The points are quite close to the image points of the square vertexes.

FIGURE 3. Calibration results for the two cameras

4   RECTIFICATION OF THE STEREO IMAGES

Corresponding points are those points, in a pair of stereo images, that are the image of the same real point. For example in Figure 4 are illustrated a pair of stereo images and are marked two corresponding points which are the images of a vertex of the carton. From the epipolar geometry these points are constrained to belong to the epipolar lines. To make easy and quick the calculation of the corresponding points is suitable to transform the images so that these lines are horizontal, i.e. the corresponding points are searched along the raster lines of the images as it is shown in Figure 5. Unfortunately this transformation applied to the pair of images is rather time-spending for the bilinear interpolation between adjacent pixels, so it is suitable to select the regions of the images that contain the information of interest in order to restrict the amount of pixels to process. This is done by superimposing the two images with two different colors so that the operator can see the information of the two images synthesized in one image only; then he can select in one time the portions of the images with the point under investigation. This procedure also allows to reduce the error in the evaluation of the corresponding points. The rectification operation could be avoided by setting cameras as illustrated in Figure 6: in this case, it is necessary an accurate construction of the stereo apparatus that reduces the flexibility of the system. Rectification is done with the hypothesis that cameras are identical (with the same intrinsic parameters). After this operation two new perspective projection matrixes are obtained for the two cameras.

FIGURE 4. A pair of stereo images with a pair of corresponding points marked

FIGURE 5. A pair of stereo images after rectification: the epipolar lines are horizontal

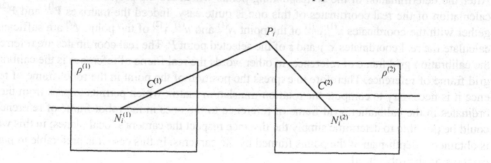

FIGURE 6. Rectified configuration

5   CALCULATION OF THE CORRESPONDING POINTS

The calculation is done with an area based method which allows to obtain the corresponding points with the precision of a pixel. This method compares pairs of $2m+1$ x $2n+1$ windows of the two images and then determines which of them are more similar. More in detail when a point is selected by the operator in the first image, the $2m+1$ x $2n+1$ window around this point is compared with the same size windows around the points along the same raster line in the second image. The comparison can be easily done with the Equation 2. D is the adopted similarity function which is the sum of the squared light intensity differences between two windows corresponding to the sum-squared error between the windows. The windows more similar are those which have D minimum. The presence of identical minimums can be possible; in this case the algorithm fails. Fixed $u^{(1)}$ and $v^{(1)}$ and the window size (i.e. m and n), D is a function depending by $u^{(2)}$ and $v^{(2)}$ or, after rectification, only by $u^{(2)}$, being $v^{(2)}$ equal $v^{(1)}$.

$$D = \sum_{k=-m}^{m} \sum_{l=-n}^{n} \left[I^{(1)}\left(u^{(1)} + k, v^{(1)} + l\right) - I^{(2)}\left(u^{(2)} + k, v^{(2)} + l\right) \right]^2 \tag{2}$$

Equation 2 consents to detect corresponding points in regions of the image where there is a strong discontinuity in pixel light intensity along the horizontal direction. The existence of the corresponding points is guaranteed by the correct operator's selection of the areas of interest in the two images. The size of the window can be changed in the program: it must be neither too little nor too large, in fact little windows introduce an error correlate to signal noise, while large windows introduce an error correlated to depth variation.

6   CALCULATION OF THE REAL COORDINATES OF A POINT

After the determination of the corresponding points which are the projection of a real point, the calculation of the real coordinates of this one is quite easy. Indeed the matrixes $\mathbf{P}^{(1)}$ and $\mathbf{P}^{(2)}$ together with the coordinates $u^{(1)}$, $v^{(1)}$ of the point $N^{(1)}$ and $u^{(2)}$, $v^{(2)}$ of the point $N^{(2)}$ are sufficient to calculate the real coordinates x, y and z of the selected point P. The real coordinates are referred to the calibration grid frame of reference, in other words the real frame of reference is the calibration grid frame of reference. Therefore to express the position of the point in the robot frame of reference it is necessary to compute the rotation-translation matrix which permits to pass from the coordinates in the calibration grid frame of reference to the ones in the robot frame of reference. It could be possible to determine simply the distance respect the camera's focal planes; in this way it is obtained a depth-map of the points framed by the cameras. In this case it is preferable to put the cameras on the robot hand.

7   CALCULATION OF THE LOCATION OF A KNOWN OBJECT

The location of a known object consists in the identification of the position of the object frame of reference as regards the robot frame of reference, i.e. in the computation of the rotation-translation matrix which permits to pass from the coordinates of the object frame of reference to the ones of the robot frame of reference. Since the coordinates of the object points, as regards the object frame of reference, are known, this last matrix allows to compute the coordinates of the objects points as regards the robot frame of reference. This matrix is computed by the individuation of three points: the origin of the object frame of reference, a point belonging to its x axis and a point belonging to its y axis. These points are easily recognizable thanks to a white label stuck on the object as it is shown in Figure 7. The individuation of these points only is enough to allow the system to recognize the object with a partial view of it, i.e. the one which frames the label.

FIGURE 7. Localization of three points to individuate the object frame of reference

8   EXPERIMENTAL TESTS

In Figure 8 are shown the results of the planned system. In the first case is detected a point, in the second is detected another point belonging to an object. In both cases the error is about some centimeters. The object is a dark carton on which is stuck a white label: this colors are chosen to make the point detection unfailing. In the output graphics illustrated in Figure 8 black points represent the robot frame of reference, while the red point represents the detected point. In the output graphic on the right, blue points represent the object frame of reference recognized by the detection of three points, while green ones represent the carton edges.

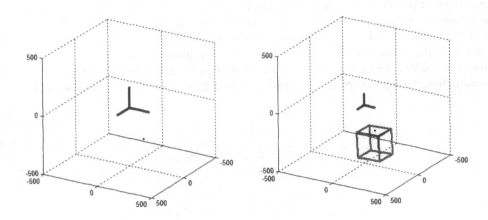

FIGURE 8. Localization of a point and of a known object

9 CONCLUSIONS

The vision system described in this article allows the Cartesian robot to locate a point or a known object. It can be used in some industrial applications like handling or assembling. The implemented system low accuracy limits its use to rough works. The accuracy can be improved by perfecting the steps illustrated above, e.g. using a better illumination device, removing the MAX dependence, using a nimbler calibration points acquisition procedure, computing exactly the cameras parameters, considering the lenses radial distortion, improving the rectification and the search of the corresponding points.

It would be desirable to automate the procedures removing the operator intervention and integrating the vision system in the control one.

Cameras are fixed outside the workspace of the robot for reasons of safety. In the future it would be interesting to mount cameras on the robot hand, before the end effector and after the wrist; in this way it is possible to inspect the object with more accuracy, from several points of view and at a short distance.

This work is prevalently based on [1], [2], [3], [4] and [5]. More details about camera calibration are in [6], [7], [8], [9]. [10], [11], [12], [13], [14], [15], [16], [17], [18] dealt with the rectification of the stereo images. [19] regards the calculation of the corresponding points.

REFERENCES

1.   Fusiello, A., (2002), Visione artificiale: Appunti delle lezioni.
2.   Faugeras, O., (1993), Three-Dimensional Computer Vision: A Geometric Viewpoint, Cambridge (USA), The MIT Press.
3.   Forsyth, D.A., Ponce, J., (2003), Computer Vision: A Modern Approach, USA, Prentice Hall.
4.   Shirai, Y., (1987), Three-Dimensional Computer Vision, Germany, Springer-Verlag.

5. Lanzetta, M., (1998), Visione Tridimensionale nei Processi di Produzione - Parte I "Stato dell'Arte", Automazione e Strumentazione Elettronica Industriale, Vol XLVI/2, 155-165.

6. Carpile, B., Torre, V., (1990), Using Vanishing Points for Camera Calibration, International Journal of Computer Vision, Vol 4, 127-140.

7. Robert, L., (1996), Camera Calibration without Feature Extraction, Computer Vision, Graphics and Image Processing, Vol 63/2, 314-325.

8. Tsai, R., (1987), A Versatile Camera Calibration Technique for High-Accuracy 3D Machine Vision Metrology Using Off-the-Self TV Camera and Lenses, IEEE Journal of Robotics and Automation, Vol 3/4, 323-344.

9. Trucco, E., Verri, A., (1998), Introductory Techniques for 3-D Computer Vision, Prentice-Hall.

10. Dhond, U.R., Aggarwal, J.K., (1989), Structure from Stereo - A Review, IEEE Transactions on Systems, Man and Cybernetics, Vol 19/6, 1489-1510.

11. Fusiello, A., Trucco, E., Verri, A., (2000), A Compact Algorithm for Rectification of Stereo Pair, Machine Vision and Applications, Vol 12/1, 16-22.

12. Hartley, R., Gupta, R., (1993), Computing Matched-Epipolar Projections, Proceedings of the IEEE Conference on Computer Vision and Pattern Recognition, 549-555.

13. Hartley, R.I., (1997), Kruppa's Equations Derived from the Fundamental Matrix, IEEE Transactions on Pattern Analysis and Machine Intelligence, Vol 19/2, 133-135.

14. Hartley, R.I., (1997), Theory and Practice of Projective Rectification, International Journal of Computer Vision, Vol 35/2, 1-16.

15. Zhang, Z., (1998), Determining the Epipolar Geometry and its Uncertainty: A Review, International Journal of Computer Vision, Vol 27/2, 161-195.

16. Isgro, F., Trucco, E., (1999), Projective Rectification without Epipolar Geometry, Proceedings of the IEEE Conference on Computer Vision and Pattern Recognition, Vol I, 94-99.

17. Loop, C., Zhang, Z., (1999), Computing Rectifying Homographies for Stereo Vision, Proceedings of the IEEE Conference on Computer Vision and Pattern Recognition, Vol I, 125-131.

18. Pollefeys, M., Koch, R., VanGool, L., (1999), A Simple and Efficient Rectification Method for General Motion, Proceedings of the IEEE International Conference on Computer Vision, 496-401.

19. Fusiello, A., Roberto, V., Trucco, E., (1997), Efficient Stereo with Multiple Windowing, Proceedings of the IEEE Conference on Computer Vision and Pattern Recognition, 858-863.

3D LOCATION OF CIRCULAR FEATURES FOR ROBOTIC TASKS

M. Sonego[1], P. Gallina[2], M. Dalla Valle[1], A. Rossi[1]

[1] Dipartimento di Innovazione Meccanica e Gestionale, University of Padova, ITALY
[2] Dipartimento di Ingegneria Meccanica, University of Trieste, ITALY

KEYWORDS: 3D location estimation, vision systems, robotics,

ABSTRACT. In many robotic applications, 3D location estimation of an object with respect to a reference frame is required. Circular markers located on the object are often employed for this purpose. This paper addresses the problem of 3D location estimation of circular features from a single camera image. A mathematical closed-form solution of the problem allows to determine the normal vector to the circle feature with respect to the camera reference frame. High calculation accuracy is guaranteed by a Kalman filter which performs a sub-pixel parametric estimation. A set of experimental results shows the validity and the accuracy of the process involved in 3D angular estimation. The theory is applied to a real industrial problem, namely the 3D location of a car rim. In fact, this information is necessary to perform a pick-and-place robotic task.

1   INTRODUCTION

Estimation of 3D information from 2D images is a fundamental task both in machine vision and robotic vision. This problem exists in two forms: the direct and the inverse. The firmer concerns with the case in which the camera parameters (the extrinsic ones: 3D position and orientation of the frame, and the intrinsic ones: focal length, lens distortion factor, etc.) are known. 2D images allow to calculate 3D location of objects, landmarks or features. The inverse problem involves the calculation of camera parameters (extrinsic and intrinsic ones for a fixed camera, or just the extrinsic ones for a moving camera) from one or more camera frames. In the applied literature this problem is referred to as camera-calibration.

The problem of 3D location estimation of an object in a scene has been extensively studied. Many mathematical models, based on point features analysis, have been developed for location estimation. The line features location estimation problem has been studied also, but the most interesting case is related the quadratic-curved features, particularly the circular ones. Indeed, a circle feature is a very common shape both in theoretical studies and real applications (many manufactured objects have circular holes or surface circular contours). A circle shows the interesting property that its perspective projection in any arbitrary orientation is always an exact ellipse.

Several methods have been proposed to solve circular-feature-based 3D angular-location estimation of an object: a interactive method has been proposed by Malgaonkar [1], a closed-form solution based on linear algebra was developed [2,3], and a closed-form mathematical

Published in: E. Kuljanic (Ed.) *Advanced Manufacturing Systems and Technology*,
CISM Courses and Lectures No. 486, Springer Wien New York, 2005.

solution based on 3D analytical geometry of circular features was proposed by Y.C. Shin et al. [4] and R. Safaee-Rad [5]. The drawback of all these methods consists in the fact that they do not take into consideration quantization errors.

In fact, since the image of a circular feature is an ellipse, the accuracy of the 3D angular-estimation of the feature is strictly related to a good parametric identification of the 2D ellipse on the image (or camera) frame. The finite camera resolution causes the image frame to be affected by a quantization error. It becomes necessary to identify an appropriate algorithm which may be able to take into account for reconstruction errors. In literature, several solutions are proposed: mainly they make use of least-square fitting algorithms [6], border dependent cost functions [7] or extended Kalman filters [8,9,10]. The extended Kalman filter gives a high accuracy parametric fitting of the elliptical feature, but at each step, the filter requires a linearization of the quadratic equation at the current processed point. As a consequence it leads to a high processing time.

This paper brings together the above-quoted works by presenting a complete solution of circular feature 3D location estimation. As the matter of fact, 3D location estimation involves the calculation of the centre of the circular feature as well as the angular position of the normal to the plane the circular feature belongs to. These two problems can be split into two separate procedures. Moreover, once the angular estimation is accurately performed, the estimation of the circular feature's centre becomes trivial. For this reason and for sake of clarity in this work only the problem of the angular estimation is taken into account.

Given a single image frame of a circular feature, the proposed method provides a simple estimation algorithm of the ellipse. In a second step, by using the equation of the estimated quadratic curve, the angular position of the circle is calculated.

This paper is organized as follows. Section 2 gives a review of the geometrical solution of the circular feature angular-estimation problem by using the image ellipse equation [11]. The third one describes a method to extract the equation of an elliptical feature form an image frame by means of a Kalman filter. The last Section of the paper reports experimental results. A car rim is used as the circular feature and a gray-level camera is employed in a vision system. By directly measuring the angular position of the rim, it has been possible to evaluate the accuracy of the whole process involved in the 3D angular estimation. We stress the fact that this vision module is part of a complete robotic system employed for pick-and-place operations of car rims in an industrial environment.

2   ANGULAR LOCATION ESTIMATION OF A CIRCULAR FEATURE: ANALYTICAL FORMULATION

In order to estimate the 3D location of the circular feature, the camera is modelled by means of the pinhole model. According to this model, when the camera grabs an image, the circular feature is projected on the CCD screen through the cone surface. The vertex of such a cone is the centre of the camera's lens. Its position is known and it is given by the coordinates c_x, c_y and c_z with respect to the CCD frame $<X,Y,Z>$ (See Figure 1). The projection of a circular feature is a ellipse on the CCD plane (screen). The actual focal length of the camera is assumed to be known.

This Section is devoted to the calculation of the normal to the circular feature once the five basic parameters (a',h',b',g',f',d') that define the ellipse in the image-coordinate frame (i.e. the estimated ellipse) are given. In fact, the general equation of an ellipse is $a'x^2 + 2h'x\,y + b'y^2 + 2g'x + 2f'y + d' = 0$.

In the next Section it will be shown how to estimate such parameters.

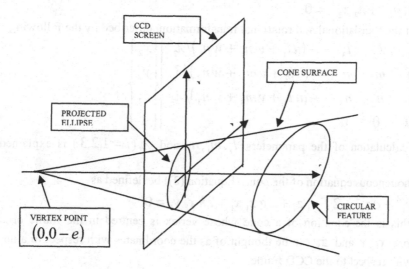

Figure 1. Cone construction.

Therefore, the purpose of the present Section is to estimate the orientation of the circular feature with respect to the camera frame (also called CCD frame). This is equivalent to solve the following 3D geometrical problem: given a 3D cone surface defined by a base (the perspective projection of circular feature on the image plane, i.e. the ellipse) and a vertex (the centre of the camera's lens) with respect to reference frame, determine the orientation of a plane (with respect to the same reference frame, i.e. the CCD frame) such that it intersects the cone and generates a circular curve.

The coordinates of the centre of the camera's lens with respect to the camera CCD frame is $\{0,0,e\}^T$ where e is the focal length.

The equation of a cone, whose vertex is the point $\{0,0,e\}^T$ and whose base is defined by the ellipse

$$F(x,y) \overset{\Delta}{=} a'x^2 + 2h'x\,y + b'y^2 + 2g'x + 2f'y + d' = 0$$
$$z = 0$$

$$(1)$$

can be expressed as [11]

$$ax^2 + by^2 + cz^2 + 2f\,yz + 2g\,zx + 2h\,xy + 2u\,x + 2v\,y + 2w\,z + d = 0 \qquad (2)$$

where

$$a = e^2\,a', \quad b = e^2\,b', \quad c = d'$$

$$d = e^2\,d', \quad f = -e\,f', \quad g = -e\,g' \qquad (3)$$

$$h = e^2\,h', \quad u = e^2\,g', \quad v = e^2\,f', \quad w = -e\,d'$$

Equation (2) can be reduced to the following compact form

$$\lambda_1 x_2^{\,2} + \lambda_2 y_2^{\,2} + \lambda_3 z_2^{\,2} = 0 \qquad (4)$$

by means of the translational and rotational transformation, described by the following

$$\begin{Bmatrix} x \\ y \\ z \\ 1 \end{Bmatrix} = \begin{bmatrix} l_1 & l_2 & l_3 & -(u\,l_1 + v\,m_1 + w\,n_1)/\lambda_1 \\ m_1 & m_2 & m_3 & -(u\,l_2 + v\,m_2 + w\,n_2)/\lambda_2 \\ n_1 & n_2 & n_3 & -(u\,l_3 + v\,m_3 + w\,n_3)/\lambda_3 \\ 0 & 0 & 0 & 1 \end{bmatrix} \begin{Bmatrix} x_2 \\ y_2 \\ z_2 \\ 1 \end{Bmatrix} \qquad (5)$$

where the calculation of the parameters l_i, m_i, n_i and $\lambda_i\,(i = 1,2,3)$ is explained in the following.

Let the homogeneous equation of the cone (Equation (2)) be defined as

$$ax_1^{\,2} + by_1^{\,2} + cz_1^{\,2} + 2f\,y_1 z_1 + 2g\,z_1 x_1 + 2h\,x_1 y_1 = 0 \qquad (6)$$

Note that this is the equation of a cone whose vertex is centred in the frame origin. It is remarked that x_1, y_1 and z_1 can be thought of as the coordinates with respect to a new frame translated with respect to the CCD frame.

Let us define now a rotational transformation T_1

$$\begin{Bmatrix} x_1 \\ y_1 \\ z_1 \end{Bmatrix} = \begin{bmatrix} l_1 & l_2 & l_3 \\ m_1 & m_2 & m_3 \\ n_1 & n_2 & n_3 \end{bmatrix} \begin{Bmatrix} x_2 \\ y_2 \\ z_2 \end{Bmatrix} = T_1 \begin{Bmatrix} x_2 \\ y_2 \\ z_2 \end{Bmatrix} \qquad (7)$$

such that Equation (6) is reduced to the following form

$$\lambda_1 x_2^{\,2} + \lambda_2 y_2^{\,2} + \lambda_3 z_2^{\,2} = 0 \qquad (8)$$

The easier way to perform this simplification is to express the cone Equation (6) in the following form

$$X^T A X = 0 \qquad (9)$$

where

$$X \triangleq \begin{Bmatrix} x_1 \\ y_1 \\ z_1 \end{Bmatrix}, \quad A \triangleq \begin{bmatrix} a & h & g \\ h & b & f \\ g & f & c \end{bmatrix} \qquad (10)$$

A is a square matrix that can be diagonalized. This operation allows us to find the eigenvectors $\mathbf{v}_1, \mathbf{v}_2$ and $\mathbf{v}_3$ and the eigenvalues λ_1, λ_2 and λ_3 [12].

By defining V (such that the columns of V are the vectors $\mathbf{v}_1, \mathbf{v}_2$ and $\mathbf{v}_3$):

$$V \overset{\Delta}{=} \left[\mathbf{v}_1 \middle| \mathbf{v}_2 \middle| \mathbf{v}_3\right] \tag{11}$$

the matrix A can be expressed in the following form

$$A = V \begin{bmatrix} \lambda_1 & 0 & 0 \\ 0 & \lambda_2 & 0 \\ 0 & 0 & \lambda_3 \end{bmatrix} V^{-1} = V D V^{-1} \tag{12}$$

By replacing Equation (12) into (9), it yields

$$\{x_1 \quad y_1 \quad z_1\} V D V^{-1} \begin{Bmatrix} x_1 \\ y_1 \\ z_1 \end{Bmatrix} = \{x_2 \quad y_2 \quad z_2\} D \begin{Bmatrix} x_2 \\ y_2 \\ z_2 \end{Bmatrix} \tag{13}$$

where

$$\begin{Bmatrix} x_2 \\ y_2 \\ z_2 \end{Bmatrix} = V^{-1} \begin{Bmatrix} x_1 \\ y_1 \\ z_1 \end{Bmatrix} \tag{14}$$

so the needed rotational transformation T_1 is $T_1 = V$. In the $< x_2, y_2, z_2 >$ reference frame the equation of the cone has the form expressed by Equation (8). It represents a cone whose vertex is centred in the origin frame and whose principal axis is parallel to a cartesian axis. It has been proven that two of the three parameters λ_1, λ_2 and λ_3 of Equation (8) are positive with one always negative. If λ_3 is assumed to have a negative value, then the principal axis of the central cone would be the z_2 axis of the $< x_2, y_2, z_2 >$ frame.

By considering Equation (8), we must now calculate the equation of a plane

$$l x_2 + m y_2 + n z_2 = p \tag{15}$$

whose intersection with the cone originates a circle. In fact the vector $\{l, m, n\}^T$ represents the normal to the circular feature. It is recalled that the estimation of $\{l, m, n\}^T$ from the parameters (a', h', b', g', f', d') is the goal of present Section. In the next Section it will be shown how to estimate the aforesaid parameters.

Let us defines the matrix T_2 [11]

$$
T_2 \stackrel{\Delta}{=} \begin{bmatrix} \dfrac{-m}{\sqrt{l^2+m^2}} & \dfrac{-lm}{\sqrt{l^2+m^2}} & l \\[2ex] \dfrac{l}{\sqrt{l^2+m^2}} & \dfrac{-mn}{\sqrt{l^2+m^2}} & m \\[2ex] 0 & \sqrt{l^2+m^2} & m \end{bmatrix} \tag{16}
$$

By applying the rotational transformation

$$
\begin{Bmatrix} x_2 \\ y_2 \\ z_2 \end{Bmatrix} = T_2 \begin{Bmatrix} x_3 \\ y_3 \\ z_3 \end{Bmatrix} \tag{17}
$$

to Equation (8), the new Z axis will be normal to plane defined by Equation (14). We are now looking for the values of l, m and n such that the intersection between the cone and the plane $z_3 = p$ will be a circle.

By operating the above substitution and by defining $z_3 = p$, Equation (8) becomes

$$
A x_3^2 + B y_3^2 + C x_3 y_3 + D x_3 + E y_3 + F = 0 \tag{18}
$$

The necessary and sufficient conditions for the quadratic Equation (18) to be a circle, are

$$
A = B, \quad C = 0 \tag{19}
$$

By adding a normalization condition

$$
l^2 + m^2 + n^2 = 1 \tag{20}
$$

we obtain a complete three equations system.

The vector $\mathbf{u} = \{l \quad m \quad n\}^T$ is the required normal vector to the circle with respect to the central cone reference frame $< x_2, y_2, z_2 >$. To express the vector with respect to the translated camera reference frame $< x_1, y_1, z_1 >$, it is necessary to apply the rotational transformation T_1

$$
\begin{Bmatrix} l' \\ m' \\ n' \end{Bmatrix} = T_1 \begin{Bmatrix} l \\ m \\ n \end{Bmatrix} \tag{21}
$$

The whole computational procedure for the proposed analytical solution consists of the following steps:

a) starting from the ellipse equation (Equation (1)) and the focal length, calculate the cone equation (Equation (2));

b) calculate his matrix form equation (Equation (9)) by considering the homogenous equation of the cone equation (Equation (6));

c) find the eigenvalues $\lambda_1, \lambda_2, \lambda_3$ and the eigenvector $\mathbf{v}_1, \mathbf{v}_2, \mathbf{v}_3$;

d) calculate the coefficients l, m and n;

e) express l, m and n with respect to the translated camera frame by using Equation (21).

3 PARAMETRIC IDENTIFICATION OF A ELLIPTICAL FEATURE

In the previous Section, the 3-D location estimation problem has been solved under the hypothesis that the parametric equation of the ellipse image is already known. The main purpose of this Section, as well as the main and original contribution given by this paper, consists in developing a suitable Kalman filter, which extracts the equation of the elliptical feature from a single camera frame. In other words, this Section is devoted to explain the estimation of the parameters (a', h', b', g', f', d') from the grabbed image.

First of all, we assume to be able to extract the set of border pixels form the image. Since a real camera has a finite resolution, an error (up to pixel dimension) occurs in determining border features. The Kalman filter must both calculate the parametric equation of the ellipse image and reduce the reconstruction errors.

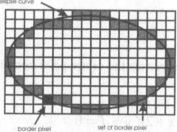

Figure 2. Pixels switched on the CCD screen.

Before starting to explain the filter implementation, we must give this simplified definition: a digital image is a matrix of pixels that have a binary configuration (on or off). The image of a circular feature is affected by the quantization error due to camera CCD. Estimating the equation of the quadratic feature from a digital image is equivalent to find the equation of an ellipse (there are infinite), whose feature is contained in the set of border pixels.

We define *set of border pixels* the contour pixels that are "switched-on" in the image frame.

Each pixel of set of border pixels contains a segment of *ellipse curve*. The set of points that belong to the segment contained in the pixel j (the pixel j is assumed to belong to set of border pixels) can be defined as

$$I_j \overset{\Delta}{=} \left\{ (x_p, y_p) \mid x_{cj} - dx \leq x_p \leq x_{cj} + dx, y_{cj} - dy \leq y_p \leq y_{cj} + dy \right\} \tag{22}$$

where dx and dy are half of pixel dimension and (x_{cj}, y_{cj}) are the centre pixel coordinates.

Let us define the following *point's border matrix* W

$$W \overset{\Delta}{=} \begin{bmatrix} x_{p1} & y_{p1} \\ \vdots & \vdots \\ x_{pj} & y_{pj} \\ \vdots & \vdots \\ x_{pN} & y_{pN} \end{bmatrix} = \begin{bmatrix} x_{c1} & y_{p1} \\ \vdots & \vdots \\ x_{cj} & y_{pj} \\ \vdots & \vdots \\ x_{cN} & y_{pN} \end{bmatrix} = \begin{bmatrix} x_{c1} & y_{c1}+e_1 \\ \vdots & \vdots \\ x_{cj} & y_{cj}+e_j \\ \vdots & \vdots \\ x_{cN} & y_{cN}+e_N \end{bmatrix} \qquad -dy \le e_j \le dy \qquad (23)$$

where N is the number of border pixels and $e_j = y_{pj} - y_{cj}$. Each row of W represents the point coordinates belonging to ellipse to be reconstructed. At the same time, let us define

$$W_e \overset{\Delta}{=} \begin{bmatrix} x_{c1} & y_{c1} \\ \vdots & \vdots \\ x_{cj} & y_{cj} \\ \vdots & \vdots \\ x_{cN} & y_{cN} \end{bmatrix} \qquad (24)$$

In this case each row of W_e represents the centre point coordinates of a pixel the ellipse "is crossing".

Let us introduce the equation of the ellipse as

$$a'x^2 + b'y^2 + 2h'xy + 2u'x + 2v'y + d' = 0 \qquad (25)$$

It can be rewritten as

$$y = -\frac{a'}{2v'}x^2 - \frac{b'}{2v'}y^2 - \frac{h'}{v'}xy - \frac{u'}{v'}x - \frac{d'}{v'} \overset{\Delta}{=} a''x^2 + b''y^2 + 2h''xy + 2u''x + d''$$

(26)

By replacing the coordinates of a single point (23) into Equation (25), the following equation is obtained

$$y_{cj} + e_j = a''x_{cj}^2 + b''(y_{cj} + e_j)^2 + 2h''x_{cj}(y_{cj} + e_j) + 2u''x_{cj} + d'' =$$
$$= a''x_{cj}^2 + b''y_{cj}^2 + b''e_j^2 + b''y_{cj}e_j + 2h''x_{cj}y_{cj} + 2h''x_{cj}e_j + 2u''x_{cj} + d'' \qquad (27)$$

where $j = 1, \ldots, N$.

At this point the following hypothesis is introduced: the term in e_j^2 in the second member of the Equation (27) is assumed to be neglected, so the equation becomes

$$y_{cj} = \begin{Bmatrix} x_{cj}^2 & y_{cj}^2 & 2x_{cj}y_{cj} & 2x_{cj} & 1 \end{Bmatrix} \begin{Bmatrix} a'' \\ b'' \\ h'' \\ u'' \\ d'' \end{Bmatrix} + \tilde{e}_j, \quad j = 1, \ldots, N \qquad (28)$$

where $\widetilde{e}_j = \left(2b''y_{cj} + 2h''x_{cj}\right)e_j$

Experimental results will show the validity of this hypothesis.

The previous Equation can be expressed in a more compact form as

$$y_{cj} \stackrel{\Delta}{=} H_j X_j + \widetilde{e}_j, \quad j = 1, \ldots, N \tag{29}$$

where $H_j = \left\{x_{cj}^2 \quad y_{cj}^2 \quad 2x_{cj}y_{cj} \quad 2x_{cj} \quad 1\right\}$ and $X_j = \left\{a'' \quad b'' \quad h'' \quad u'' \quad d''\right\}^T$.

From now on, we will refer to X term as *vector state*, to $\widetilde{e}_j$ term as *error* and to y_{cj} term as *data*.

By adding the state updating equation $X_{j+1} = X_j$, the following mathematical description of system is obtained

$$
\begin{aligned}
X_{j+1} &= X_j \\
y_{cj} &= H_j X_j + \widetilde{e}_j, \quad j = 1, \ldots, N
\end{aligned}
\tag{30}
$$

which allows an easy Kalman filter implementation. The problem of ellipse parameter estimation has been transformed in a general problem of estimation of the state of a discrete process governed by the linear stochastic difference equation (30). The next step in filter implementation consists in calculating the *covariance matrix of the error* and to define the *initial conditions*.

The filter system assumes the following form

$$
\begin{cases}
\hat{X}_{j+1} = \hat{X}_j \\
y_j = H_j \hat{X}_j + e_j
\end{cases}
\tag{31}
$$

with initial condition $\hat{X}_0$ and P_0.

Starting form the initial condition, the filter processes the set of border points (one at step) by applying a recursive algorithm.

At each step, the vector $\hat{X}_j$ contains the last updated estimation parameters of ellipse.

From the previous analysis results that the Kalman filter implementation involves two different main step :

a) Estimation of the initial condition of the filter: five points from the border vector are chosen and the $\hat{X}_0$ vector and the P_0 matrix are calculated;

b) Filtering of whole border points set: at each step the filter process a single border point. The values y_j, H_j, λ_j, L_j are calculated, then the state vector X_j and the error covariance vector P_j are updated. The algorithm is repeated until the estimation is sufficiently accurate and/or the whole border is processed.

4   EXPERIMENTAL RESULTS

The theory of the 3D location of a circular feature is implemented in a real vision system module. This module is part of a robotic system which is employed for the spatial manipulation of car rims. Namely, the robotic system has to locate the orientation and the position of the rim, then grasps the rim and finally places the rim in a new location. The most critical phase of this pick-and-place operation is the 3D location of the rim. Since the rim has a circular shape, at least the outer border, the method of a 3D location of a circular feature presented in the previous sections represents an effective tool for accomplishing the task.

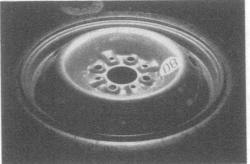

Figure 3. Picture of the rim car.

Therefore a rim of a car wheel was addressed as a target. By applying an appropriate illumination it is possible to emphasize the rim border on the black tyre background (Figure 3). In a such a way it becomes possible to extract the border pixels positions from a camera image frame. The experimental apparatus is made up of the following major components:
- a frame (which is referred to as *ground frame*) which allows an accurate angular position (angular error position = 0.01degrees) in a defined angular range (±4degrees). The car rim is mounted over the frame;
- a gray-level camera: Sony XC-73CE with 2/3-in CCD and effective pixel 752Hx582V, CCD cell size 6.5x6.25μm.;
- lens Computar M8513 8.5 mm;
- Euresys Piccolo acquisition Card (frame grabber) and Pentium II 350MHz system.
A calibration procedure is applied to the camera in order to find a rotational transformation from the camera reference frame to the *ground frame*. Since the position of the camera does not appear in any of the relationships presented in this paper, the calibration process involves only the angular camera-position estimation.
Each measurement is performed as follows. First the ground frame is accurately set in such a way that the unit vector perpendicular to the car rim assumes the orientation defined by the actual values (α_a, β_a). Then the camera acquires an image of the rim and the optimized algorithm estimates the measured values (α_m, β_m). This procedure is repeated several times, for different couples (α_a, β_a).
Experimental results are reported in Table 1. As it can inferred, the maximum error is about 0.15 [deg]. Therefore the accuracy obtained by simulation results is confirmed by experimental ones.

It has been observed through experimentation that the method developed in this paper works reliably. However there exist two factors which affect the process accuracy: the size of the feature in the image plane and the relative orientation of the rim with respect to the camera image plane (let us define Θ as the angle between camera direction and a normal vector to circle feature plane).

The former one is affected by the size of the feature, the distance between the camera and the feature and the camera focal length. The second one is affected by the amount of deformation of the feature in camera image. For a small angle Θ, the amount of deformation is quite small, and it will be very difficult to detect this change from a gray-level image (the accurate estimation of the ellipse's parameters depends on how accurate the deformation of the ellipse can be detected).

TABLE 1. Experimental results

Actual rim car normal vector (α_a, β_a) [deg]	Estimated rim car normal vector (α_m, β_m) [deg]
(0,0)	(0.04,-0.04)
(0,0.4)	(-0.02,0.46)
(0,0.8)	(-0.02,0.91)
(0,1)	(0,1.04)
(0,1.4)	(0.01,1.37)
(0,1.8)	(0,1.88)
(0,2)	(0.01,2.12)
(0,2.5)	(0.04,2.59)
(0,3)	(0.04,3.15)
(-2.8,0)	(-2.71,-0.05)
(-2.4,0)	(-2.44,0.03)
(-2,0)	(-2.03,-0.03)
(-1,0)	(-1.04,-0.09)
(1,0)	(1.03,-0.09)
(1.7,0)	(1.79,-0.05)

5 CONCLUSION

In this paper the 3D angular-estimation problem of circular feature is addressed. The proposed analytical method is based on a geometrical analysis of the problem. A closed-form analytical solution is described. Since the accuracy of the whole process is strongly related to a good parametric estimation of the ellipse image, a suitable optimized fitting algorithm is developed. The required accuracy is guaranteed by a Kalman filter and a speed improvement is performed by applying a algorithm optimization step.

Both simulation and the experimental result show the validity of the whole process involved in the 3D angular estimation. The applicability of the process in a industrial environment is shown by means of a car rim as target.

REFERENCES

1. P.G. Malgaonkar, Analysis of perspective line drawings using hypothesis based reasoning, Ph.D. dissertion, Virginia Polytech. Inst. and Sate Univ., Blacksburg, VA, 1984.
2. H. S. Sawheney, J.Oliensis, and A. R. Hanson, Description from image trajectores of rotational motion, in Proc. 3rd IEEE Int. Conf. Comput. Vision (Osaka Japan), Dec. (1990), pp. 494-498.
3. D. H. Marimount, Inferring statial structutr from feature correspondence, Ph.D. dissertion, Stanford Univ., Stanford, CA, Mar. 1986.
4. Y. C. shin and S. Ahmad, 3-D location of circular and spherical features by monocular model-based vision, in Proc. IEEE Int. Conf. Syst. Man Cybern. (Los Angeles), Nov. (1990), pp. 215-220.
5. R. Safaee-Rad and I.Tchoukanov, Three-Dimensional location estimation of circular feature for machine vision, in Proc. IEEE Int. conf. Syst. Man. Cybern. (Boston), Nov. (1989), pp. 576-581.
6. W.Gander., GH.Golub., R.Strebel., Least-squares fitting of circles and ellipses,BIT. vol.34, no.4; Dec. (1994), p.558-78
7. G.Montilla, C.Roux, V.Barrios, V.Torrealba, N. Range, V.Subacius, L.Miliani, C.Vasquez, Model-based, knowledge-based epicardial boundary detector, roceedings. Computers in Cardiology 1993 (Cat. No.93CH3384-5). IEEE Comput. Soc. Press, Los Alamitos, CA, USA; 199, pp.205-8
8. J.Porrill, Fitting ellipses and predicting confidence envelopes using a bias corrected Kalman filter, Image-and-Vision-Computing.vol.8, no.1; Feb. (1990), p.37-41.
9. MS.Nixon, Circle extraction via least squares and the Kalman filter, Computer Analysis of Images and Patterns.5th International Conference, CAIP '93 Proceedings. Springer-Verlag, Berlin, Germany; 1993; xvi+857 pp. p.199-207.
10. N. Werghi-N, C. Doignon, Contour feature extraction with wavelet transform and parametrization of elliptic curves with an unbiased extended Kalman filter, ACCV '95.Second Asian Conference on Computer Vision. Proceedings. Nanyang Technol. Univ, Singapore; (1995); 3 pp.186-90 vol.3.
11. R-Safaee-Rad, B. Benhabib, and K. C. smith, Perspective projection and perspective distortion of circular-landmark features, in Proc. CSECE Canadian Conf. Electrical Comput. Eng (Ottowa, ON, Canada), Sept. (1990), pp. 45.3.1-45.3.5.
12. E. Mosnat, Problemes de Géométrie Analytique, vol. 3, 3rd. ed., Paris: Vuibert, 1921.
13. Roberto Moresco, Elementi di algebra e di geometria,3rd ed. Padua, Progetto 1992.

A MOBILE ROBOT PLATFORM FOR FERROMAGNETIC BASE PLATES

N. Tomac[1], B. Solvang[2], T. Mikac[3]

[1] Department of Polytechnic, Faculty of Philosophy, University of Rijeka, Croatia
[2] Department of Industrial Engineering, Narvik University College, Norway
[3] Faculty of Engineering, University of Rijeka, Croatia

KEYWORDS: Mobile robot, Construction, Linear Rotational Movement.

ABSTRACT. In this paper a prototype of a robot platform designed for movement on large ferromagnetic plates is presented. The intention was to build a platform with a high positioning accuracy, which have an unrestricted working area and carry a universal robot on its top plate. Throughout this paper the main elements of the mobile system is presented. The robot platform has a hydraulic drive motor, 7 cylinders and six electromagnetic legs. The movement of robot can be linear or rotational in both directions, which enable it to reach any position on an extended surface.

1 INTRODUCTION

There are many types of industrial areas in which mobile robots can be used to increase productivity and to spare human for drudgery of repetitive or dangerous work. Large distances, often encountered in an industrial environment, require mobile robots that can operate at such ranges [1]. For example, in the shipbuilding industry, the need of using mobile robots is very considerable. Currently, many workers such as arc cutters, welders, grinders and painters work on the scaffolding around the hull of the ship [2]. Manual labour under these conditions, often high above ground, is both dangerous and expensive due to the large amount of infrastructure needed (walkways, necessary tool-support etc.). In addition, an advanced movable robot also be very useful on steel bridges, buildings, and constructions [3]. The question is how to fully automate operations such as: welding, grinding, painting and others on big ferromagnetic plates. Especially, it is a very large interest in the shipyards industry to develop robot system for grit or water blasting in hull cleaning [4]. Even from the productivity point of view, operations like for example arc welding, on well defined extended surfaces, are more efficient carried out by a robot system compared with a human work under such difficult conditions [5].

The well known industrial robot, in its original setup, due to its stationary base has a limited usage under such working conditions. It simply cannot cover a big enough working area. The mobility of entire robot can take different forms and common implementations are tracked, gantry and whiled mobile robots [6]. All three types of robots cannot be used on big ferromagnetic plates because of several drawbacks. Robot with linear base or tracked robot is an ancillary device often used to provide an additional position axis, a linear movement of the entire robot along the floor. Gantry robots are very expensive and take a lot space [7]. Both, before mentioned robots have

Published in: E. Kuljanic (Ed.) *Advanced Manufacturing Systems and Technology*,
CISM Courses and Lectures No. 486, Springer Wien New York, 2005.

limiting working areas. On the other hand wheeled mobile robots cannot deal with big forces and is normally incapable of climbing onto inclined surfaces.

This work deals with the autonomous climbing and walking robot, which uses a special concept of motion on metallic based structures, see Figure 1. Our idea was to design a sophisticated platform, which is small, strong and stable, such that working range can be enlarged. In addition, the idea was that this platform can be moved on horizontal, vertical and inclined ferromagnetic plates with the help of electromagnetic legs. To maintain stability of the platform it is designed with six hydraulic legs. Three legs always are in contact with the ground.

FIGURE 1. Mobile platform

2   LINEAR MOVEMENT

Motion pattern is shown in Figure 2. The platform linear motion, one step forward, is illustrated in six positions:

Point 1: Starting point. All magnetic feet are in contacts with the ground and the outside frame
          is in the left outermost position.
Point 2: Outside feet is drawn up.
Point 3: Outside frame is moved to right outermost position.
Point 4: Outside feet are put down and inside feet are drawn up.
Point 5: Inside carriage is moved to the right outermost position.
Point 6: Inside feet are put down. Platform has moved forward.

With this procedure the platform moves one step to the right. The platform can go both forward and backward.

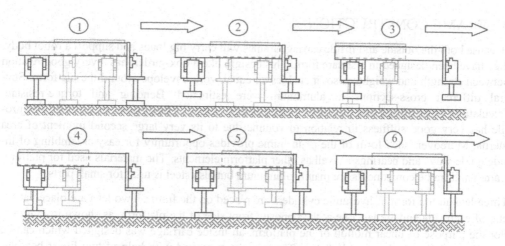

FIGURE 2. Linear movement

3 ANGULAR MOVEMENT

The platform can sweep round in both directions for 45^0, when inside legs are in contact with the base plate and outside legs are elevated. Figure 3 illustrates a rotation of the platform from $+45^0$ to -45^0. Rotational movement is carried out by a motor/gear system placed in the inner frame.

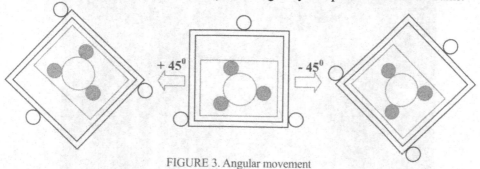

FIGURE 3. Angular movement

The characteristic moving features of the platform can be summarised:

- Can perform linear and angular movement in every direction of a plan.

- Inside and outside legs can not come in a conflict with each other.

- Can be used on rough base plane.

- Can always hold a surface in the same plane with a constant distance to the base plane.

4   FRAME CONSTRUCTION

Because both the outside and inside carriage frames will carry big loads and support a robot body, they have been designed in a square form. This design, like a box-girder, has a very good relation between strength and weight. Also, it enables an equal work envelope around the platform. Several different cross-sections in aluminium were estimated. Bending and torque-resistant calculations have reviled that the open-U cross section gives the best result. In addition, this profile has very good stiffness in relation to volume due to its very large second moment of area inertia. Moreover, such form of the main frame provides opportunity for easy assembling of inside guide-ways and bearings as well as other platform elements. The materials used for platform frames and frame components are mainly aluminium but also steel is used for small parts.

Three legs in the form of hydraulic cylinders are placed on the frame. Two legs are placed at the end of each side and the last one at the opposite front side of the platform, as shown in Figure 4. For the purpose of linear motion of the platform, an inside carriage was designed which can be moved in relation to the outside frame. The carriage is moved with help of two linear bearings and a hydraulic cylinder. In addition, to enable the platform to rotate, an inside pipe equipped with a gear and hydraulic motor was implemented. Three inside legs are placed on this pipe. To achieve the best possible compact construction the hydraulic revolving motor is placed inside of the pipe and the three inside legs are placed on the outside of the pipe.

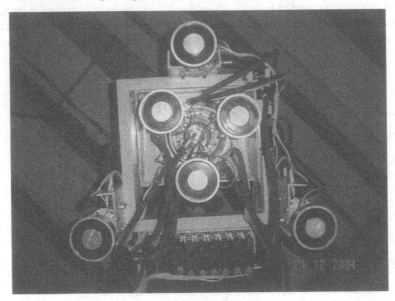

FIGURE 4. Platform view from below

5 LEG AND ROTARY EQUIPMENT

The platform has six legs. Every leg is composed of a hydraulic cylinder with a ball joint and an electromagnet at its end (see Figure 5). The used hydraulic cylinders are specially designed to withstand rough working conditions. Due to movement of the platform on vertical and inclined surfaces; shear forces appear across all piston rods of the hydraulic cylinders. Normally, piston roads are not designed for direct impact of shear forces. Such diagonal forces will apparently cause large wear and leakage after very short time. To avoid these problems all rods are strengthened with an extra bearing. Electromagnets are used for a strong attach of robot feet onto the ferromagnetic base plate. The ball joint between the piston rod of the cylinder and the electromagnets is used for avoiding of extensive "rigidity". With that, the platform is not dependent of complete flatness of the base plate.

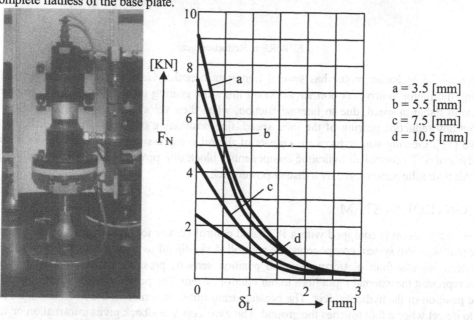

a = 3.5 [mm]
b = 5.5 [mm]
c = 7.5 [mm]
d = 10.5 [mm]

FIGURE 5. Hydraulic cylinder, holding force versus thickness of plate

Numerous calculations have been done to identify the necessary strength of the hydraulic leg system. Different positions of outside and inside legs, on both horizontal and vertical base plates, have been evaluated. Based on such evaluations the selected hydraulic cylinder is very short and can reach a maximal piston rod speed of 1.5 m/s. To achieve good position accuracy the hydraulic cylinders are equipped with a feed-back transducer (linear potentiometer, which is wire-wound with a conductive layer of plastic). The potentiometer is built into the piston rod and has the necessary accuracy. The platform must, at all times, hold on to the ferromagnetic base plate. Therefore, on the end of each leg strong electromagnets are used. Figure 5 shows the relationship between the thickness of a base plate and the interspaces between the electromagnet and the base plate, in relation to the electromagnetic force. A hydraulic motor and a reduction gear enable angular motion (rotation) of the platform. Figure 6 shows the top end of the reduction gear.

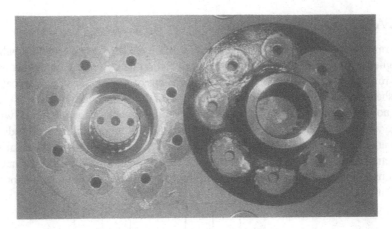

FIGURE 6. Reduction gear

Gearing of the hydraulic motor has several important benefits. Gearing increase the strength, enabling the motor to drive its normal workload times the gearing ratio. Also, hydraulic motors become unstable in speed, due to internal friction, when their velocity is below approx. 5 rpm. Without any reduction gearing of the joint speed, fine positioning of the platform is difficult to achieve [11]. Gearing also reduces the impact of the oil compressibility (bulk modulus) in the joint dynamics. To control all hydraulic components a block of 8 proportional hydraulic valves is used. All hydraulic parts are fed by a distant power pack.

6 CONTROL SYSTEM

The mobile platform is equipped with a PC-based control system for supervising the movement of the platform. All system components are controlled via digital to analog and analog to digital converters. Signals from platform mounted position sensors, pressure transmitters, and gyroscopes represent the external input lines to our control system. The position sensors give feedback on the position of the hydraulic feet. The pressure transmitters mounted in the hydraulic lines are used to detect when a foot touches the ground. The gyroscopes feedback gives information on the levelling of the platform in relation to its foundation. User communication with the control system is conducted via a PC-screen and a keyboard. The control system software was designed to operate in two different modes, either manual mode or automatic mode.

6.1 MANUAL MODE

The control system allows the user to operate the platform manually by using selected keys on a keyboard. Control signals are calculated in the platform software and fed out to the hydraulic proportional valve block. This functionality allows an operator to drive the platform to a starting point of a working operation. After the operation was completed the user guided the platform back to the starting point on easy and secure way. Manual mode can be used to override the automatic mode process in such way that the operator can reposition the platform. This action can be performed for avoiding of obstacles, without stopping the ongoing working process.

6.2 AUTOMATIC MODE

The general idea here is to integrate the platform control system with the control system of a suitable industrial robot. To enlarge the robot-workspace, most industrial robots today have the possibility to incorporate external axes in their control architecture. This feature enables us to configure the platform rotary and linear axes as such two external axes. These external axes represent extra two degree of freedom (DOF) of an industrial robot. The angular rotation of the frame represents the first DOF and the second DOF is the linear movement of the frame.

Usually, industrial robots also have a good set of functionality within various areas of applications. For example in the area of welding a wide range of process support tools are available like welding torches ready for tool canter point mounting, calibration tools for identification of new tool centre point, cleaning-, and wire cutter equipment. These are all together items that ensure a high productivity in the welding process. Reducing down-time for maintenance is especially important in applications where it takes time to bring in the equipment for a work over, or it is difficult to bring the man to the machine. The tracking system is another important feature which enables the robot to locate the welding groove dynamically.

In automatic mode the platform is taking full advantage of the capability of modern industrial robots enabling a coordinated movement of the platform joints and the robot joints. This is especially important to enable a continuous speed of, for example, the welding torch in relation to the welding groove.

7 CONCLUSIONS AND FURTHER WORK

In this paper the construction details of a mobile robot platform designed for motion on ferromagnetic plates is given.

The platform outside and inside carriage frames are designed to carry big loads and support robot body.

It has six hydraulic legs, with electromagnets, making it capable to climb on vertical and inclined surfaces.

The control system software was designed to operate in either manual mode or automatic mode.

It can perform linear and angular movement in every direction of a plan.

The ranges of applications can vary from shipbuilding to bridges and other larger ferromagnetic structures.

Generally, the usage of the platform is only limited by the capability of the selected equipment (robot) mounted on the platform top.

The movable platform has been built and tested successfully at Narvik Institute of Technology.

Further work is related to an installation and system integration with an industrial robot.

ACKNOWLEDGEMENTS

Funding for this work was provided by Narvik University College. This project would not have been possible without the devoted hard work of M. B. Krøtø, J. H. Luther, W. Silden, T. J. Stavang, and Ø. Øyen.

REFERENCES

1. M. Vincze, M. Ayromlou, C. Beltran, "A System to Navigate a Robot into a Ship Structure" Machine Vision and Applications, vol. 14, pp. 15–25, 2003.
2. Pi-Cheng Tung , Ming-Chang Wu, Yean-Ren Hwang, "An image-guided mobile robotic welding system for SMAW repair processes," International Journal of Machine Tools & Manufacture, vol. 44, pp. 1223–1233, 2004.
3. S.B. Nickerson, P. Jasiobedzki, D. Wilkes, "The ARK project: Autonomous mobile robots for known industrial environments," Robotics and Autonomous Systems, vol. 25, pp. 83-104, 1998.
4. Iwan Ulrich, Johann Borenstei, "The Guide Cane — Applying Mobile Robot Technologies to Assist the Visually Impaired," IEEE Transactions on Systems, Man, and Cybernetics, —Part A: Systems and Humans, Vol. 31, No. 2, pp. 131-136, 2001.
5. R. Ceres, J.L. Pons, "Design and Implementation of an Aided Fruit-harvesting robot (Agribot)," Industrial Robot, Vol. 25, No. 5, pp. 337-346, 1998.
6. W.G. Rippey, J.A. Falco, "The Nist Automated Arc Welding Testbed" Proceedings of 7th International Conference on Computer Technology in Welding San Francisco, CA, July 8-11, 1997.
7. B. Pekkari, "Environmental Concerns are driving the Development of the Welding Processes and Applications," ESAB AB, Box 8004, 402 77 Götebog, 2003.
8. K. Okamura, "Ultra High-speed Arc Welding (4 m/min)," Industrial Robot, Vol. 25, No. 3, pp. 185-192, 1998.
9. C. Balaguer, A. Gimenez, J.M. Pastor, "A Climbing Autonomous Robot for Inspection Applications in 3D Comlex Environments," Departamento de Ingenieria Electrica, Electronica y Automatica (DIEEA) University Carlos III of Madrid.
10. B. Rooks "Robots Score at Grinding and Polishing," Industrial Robot, Vol. 25, No. 4, pp. 252-255, 1998.
11. Solvang, B., Lien, T.K., Thomessen, T. (1999). "A high precision underwater manipulator". 30th International Symposium on Robotics. Tokyo, Japan.

COMPUTER AIDED DESIGN OF MAIN SPINDLE AND FEED DRIVES FOR NUMERICALLY CONTROLLED MACHINE TOOLS

Z. Pandilov[1], V. Dukovski[1]

[1] Department of Production Engineering, Faculty of Mechanical Engineering, University "Sv. Kiril i Metodij", Skopje, Republic of Macedonia

KEYWORDS: Computer Aided Design, NC Machine Tool, Feed Drives, Main Spindle Drives.

ABSTRACT. Characteristics of main spindle and feed drives for the NC machine tool highly depend upon skillfulness of composing variable speed motors (AC or DC) and mechanical transmission elements. In order to enable interactive computer aided design and analysis of different design variants of NC machine tool drives, original computer programs have been developed.

1 INTRODUCTION

A special feature of NC machine tool drives is use of variable speed (AC or DC) motors which provide continuous changing of cutting speed and feed rate.

The use of variable speed motors creates a question of their appropriate composing with mechanical transmission elements in order to get better output characteristics and to satisfy some conditions as a system.

2 THEORETICAL CONSIDERATIONS AND COMPUTER PROGRAM FOR DESIGNING MAIN SPINDLE DRIVES FOR NC MACHINE TOOLS

Main spindle drives for NC machine tools must provide constant power at wide range of speeds on the output of the main spindle. They consist of three parts: 1. variable speed motor, 2. mechanical transmission elements which provide appropriate output characteristics of the main spindle and 3. main spindle.

Usually mechanical transmission elements consist of: belt transmission or combination of belt transmission with gearbox (with two, three or four speeds).

Intensive development of quality tool materials enable using of very high cutting speeds and power.

Necessary output power on the main spindle can be calculated as:

$$P = \frac{Ft \cdot v}{60 \cdot 10^3} \quad [kW] \tag{1}$$

where: Ft-tangential cutting force component [N]; v-cutting speed [m/min].

Published in: E. Kuljanic (Ed.) *Advanced Manufacturing Systems and Technology*, CISM Courses and Lectures No. 486, Springer Wien New York, 2005.

NC machine tools are used for production of workpieces with different shapes, dimensions and materials, with wide range of cutting data.

For ensuring these requirements the speeds on the main spindle must be regulated in very wide range,

$$Rms = \frac{n_{max}}{n_{min}} = \frac{v_{max}}{v_{min}} \cdot \frac{D_{max}}{D_{min}} = Rv \cdot Rd \tag{2}$$

where: Rms-range of regulation of output main spindle speeds; Rv -range of regulation of cutting speeds; Rd -range of diameters of the parts, or of the cutting tools; n_{max}, n_{min} -maximal and minimal main spindle speed; v_{max}, v_{min} -maximal and minimal cutting speed; D_{max}, D_{min} - maximal and minimal diameters of the parts or the cutting tools.

According to our empirical investigation the range of regulation of main spindle speeds for NC machine tools usually is within Rms=20-350 (exclusively rare to 600). Such kind of wide regulation of main spindle speeds needs particular attention in selection of variable speed motors and mechanical transmission elements.

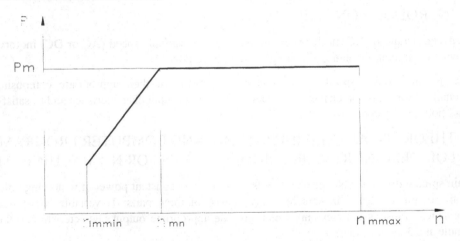

FIGURE 1. Power-speed diagram of variable speed AC motor

Fig.1 presents power-speed diagram of variable speed motor, where nmmin, nmn and nmmax are minimal, nominal and maximal speed of the motor, and Pm is nominal power of the variable speed motor.

Usually range of regulation of speed at constant power of variable speed motors is (2-8) (sometimes reaches values 12-16), which is far bellow required range Rms=20-350.

The overall range of regulation of output main spindle speeds can be calculated as:

$$Rms=Rmsm \cdot Rmsp \tag{3}$$

where: Rmsm=2-50 -range of regulation of main spindle speeds at constant torque; Rmsp=2-45 (exclusively rare 70) -range of regulation of main spindle speeds at constant power.

There are two alternative methods of obtaining wide range of main spindle speeds at constant power: overrating of the AC or DC motor or combining the motor with gearbox with two, three or four speeds.

The second solution with two, three or four speed gearbox is widely used at the NC machine tools.

Selecting the number of steps Z of the gearbox is in the range of regulation of the variable speed motor with constant power Rmp, while with using the range of variable speed motor with constant torque Rmm=Rmsm, the whole range of regulation of output speeds of the main spindle is obtained [10, 1, 5].

Because of that, we can write:

$$Rmsp = Rmp \cdot Rz \qquad (4)$$

where: Rmp -range of regulation of variable speed motor with constant power; Rz -range of regulation of the gearbox.

Variable speed motor can be treated as a particular group of gearbox with continuous changing speeds, which is first in the kinematic chain, with infinitely large number of transmissions, with transmission ratios which obtain geometrical progression with progression ratio $\varphi \rightarrow 1$ and range Rmp.

Gearbox can be treated as a transmission group which extend the speed range of the motor at constant power. Because of that characteristic of a transmission group φ_z is:

$$\varphi_z = Rmp \cdot \varphi \qquad (5)$$

Because $\varphi \rightarrow 1$, we obtain

$$\varphi_z = Rmp \qquad (6)$$

As,

$$Rz = \varphi_z^{(z-1)} \qquad (7)$$

we can write

$$Rz = Rmp^{(z-1)} \qquad (8)$$

With the substitution equation (8) in (4), we get

$$Rmsp = Rmp \cdot Rmp^{(z-1)} = Rmp^z \qquad (9)$$

where: Z-number of speeds of the gearbox.

With known Rmsp and Rmp, using the equation (9), we can calculate the necessary number of speeds of the gearbox Z_E:

$$Z_E = \frac{\log Rmsp}{\log Rmp} \qquad (10)$$

The equation (10) is recommended also in the literature [10,1, 5].

Because Z_E is usually a decimal number, it is round to the nearest full number.

If $Z > Z_E$ we get characteristic with overlapping speeds (fig.2a).

In case $Z < Z_E$ we get characteristic with step decrease of the power ΔP (fig.2b).

For example, if from equation (10) we get $Z_E=2.5$, than Z can be 2 or 3.

In case of Z=3 we get P-n characteristic as in the fig. 2a, and if is accepted Z=2 we obtain characteristic as in the fig. 2b.

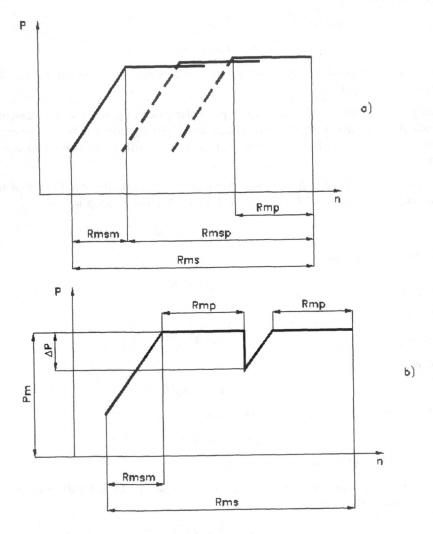

FIGURE 2. Diagram P-n of the main spindle
a) with Z=3 and b) with Z=2

Percentual decrease of the power ΔP in relation with the nominal power Pm of the motor, when $Z < Z_E$ can be calculated with the equation (11),

$$\frac{\Delta P}{Pm} = \left(1 - Rmp \cdot z - \sqrt[z]{Rmp / Rmsp}\right)100 \ [\%] \tag{11}$$

Usually $\Delta P/Pm$ should not be greater than 30% [10,1,5].

The above algorithm is implemented in an original computer program written for PC in C language which enables interactive design and analysis of main spindle drives with different design characteristics.

3 THEORETICAL CONSIDERATIONS AND COMPUTER PROGRAM FOR DESIGNING FEED DRIVES FOR NC MACHINE TOOLS

The feed drive consists of an electromotor and mechanical transmission elements. The mechanical transmission elements comprise all the machine parts which lie in the torque (power) transmission flow between the servo motor and the tool or workpiece. In different design variants the following mechanical transmission elements are most frequently used: clutches, ball lead screw and nut units, rack and pinion units, bearings, gears, gearboxes (planetary, cycloidal, harmonic), toothed belt gears, guideways etc.

The main task in the feed drive design is a selection of a servo motor and mechanical transmission components. During this process the drive angular nominal frequency ω_{od} and nominal angular frequency of the mechanical transmission elements ω_{omech} are calculated.

In order not to affect the properties of the highly dynamic AC or DC servo motor, the nominal angular frequency of the mechanical transmission ω_{omech} elements must be higher than the drive nominal angular frequency ω_{od}.

According to [11,12,5]

$$\omega_{omech} / \omega_{od} \geq 2 \tag{12}$$

is recommended.

To satisfy the requirements and to enable a long exploitation period particular attention has to be paid to the selection of feed drive servomotors. An improper servo motor selection results in a less efficient operation of machine tool and a short exploitation period.

Total load torque Mtot can be calculated as:

$$Mtot = Mmf + \sum Mfl \quad [Nm] \tag{13}$$

where: Mmf is a torque caused by the machining force [Nm]; $\sum Mfl$ is a sum of torques caused by friction and losses [Nm].

The next step is a calculation of the necessary motor speed ne for a rapid feed rate.

The selection of a variable speed motor can be from a catalogue, or from an appropriate data base, developed during the investigation [5].

The total moment of inertia Jtot can be calculated as:

$$Jtot = Jm + Jext \quad [\,kgm^2\,] \tag{14}$$

where: Jm is a motor moment of inertia $[\,kgm^2\,]$; Jext is an external moment of inertia reflected on motor shaft $[\,kgm^2\,]$.

Equations necessary for calculation of Mtot, n_e and Itot for different design variants are given in details in [11,5].

After calculation (Mtot and n_e), for the selected servo motor an analysis of dynamic behavior must be performed.

With a dynamic behavior analysis, we calculate the acceleration time to rapid traverse feed rate for loaded motor t_a, nominal angular frequency of the drive ω_{0d} and position loop gain Kv.

The acceleration time to maximal speed for loaded motor t_a can be calculated as:

$$t_a = \frac{Jtot \cdot n_m}{9.55 \cdot Ma} \cdot 10^3 = \frac{(Jm + Jext) \cdot n_m}{9.55 \cdot Ma} \cdot 10^3 \quad [ms] \tag{15}$$

where: n_m is maximal motor speed $[\,min^{-1}]$; Ma is acceleration torque [Nm].

The acceleration time to maximal speed of unloaded motor t_b is:

$$t_b = \frac{Jm \cdot n_m}{9.55 \cdot Ma} \cdot 10^3 \quad [ms] \tag{16}$$

The acceleration time to the maximal speed of unloaded motor t_b is given in a motor catalogue. If t_b is not given directly, it can be calculated indirectly by the maximal angular acceleration of the motor shaft α [rad/ s^2].

Because

$$Ma = Jm \cdot \alpha \quad [Nm] \tag{17}$$

equation (16) becomes

$$t_b = \frac{n_m}{9.55 \cdot \alpha} \cdot 10^3 \quad [ms] \tag{18}$$

With the substitution of equation (16) in (15)

$$t_a = t_b \cdot \frac{(Jm + Jext)}{Jm} \quad [ms] \tag{19}$$

If t_a is greater than a permitted value, corrections are made in mechanical transmission components (transmission ratio, feed screw lead etc.), in order to reduce t_a and to satisfy the necessary value.

For an approximate mathematical calculation of nominal angular frequency of the drive ω_{od}, model shown in fig.3 [11], can be used.

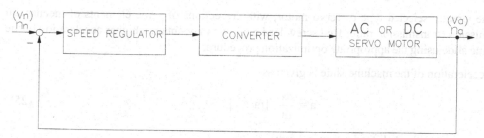

FIGURE 3. A block diagram of speed controlled AC or DC servo drive [11]

According to the model given in [11], and shown in fig.3

$$\omega_{od} \approx \frac{1}{Teld} \cdot \left(1 + \frac{1}{2 \cdot \dfrac{Tmech}{Teld}}\right) \quad [s^{-1}] \tag{20}$$

where: Teld is a drive electrical time constant [s]; Tmechd is a drive mechanical time constant [s].

Another important element which can be approximately calculated is the position loop gain Kv.

The position loop gain Kv is a ratio of nominal speed v_n [m/min] and difference between nominal and actual position Δx [mm].

$$Kv = \frac{v_n}{\Delta x} \quad \left[\frac{m/min}{mm}\right] \tag{21}$$

$$Kv = \frac{1000}{60} \cdot \frac{v_n}{\Delta x} \quad [s^{-1}] \tag{22}$$

The analysis in [11] shows that in ideal condition the optimal value of Kv must lie in the range of:

$$0.2 \cdot \omega_{od} \leq Kv \leq 0.3 \cdot \omega_{od} \tag{23}$$

For real conditions it is recommended:

$$Kv < (0.2\text{-}0.3) \cdot \omega_{od} \tag{24}$$

The calculated values for Kv from equations (23) and (24) are approximate. The exact value can be obtained experimentally by the numerical control of machine tool [5,6,7,8,9,2,3].

One of the most important requirements for good dynamic behavior of the feed drive is high acceleration of the NC machine tool slide due to the demand for minimal mechanical time constant [11,4,5]. Magnitude of inertial forces which directly affect the accuracy depends on the magnitude of slide acceleration.

Acceleration limits are recommended [11,4,5]:

-for machine tools with normal accuracy (a_{per} = 0.8-1.5 m/s^2),

-for machine tools with greater accuracy (a_{per} = 0.2-0.4 m/s^2).

For the already selected type of servo motor, with corrections of some elements of mechanical transmission (transmission ratio, feed screw lead etc.), we may obtain a higher acceleration of the machine slide using the appropriate optimization procedure.

The acceleration of the machine slide is given as:

$$a = \frac{dv}{dt} \ [m/s^2] \tag{25}$$

For the variant with a ball feed screw and nut:

$$a = \alpha_1 \cdot \frac{h \cdot i}{2\pi} \ [m/s^2] \tag{26}$$

and for the rack and pinion variant:

$$a = \alpha_1 \cdot r_p \cdot i \ [m/s^2] \tag{27}$$

where: v is a rapid traverse feed rate [m/min]; h is a feed screw lead [m]; r_p is a radius of the pinion [m]; i is a transmission ratio; α_1 is an angular acceleration of the loaded motor shaft [rad/s^2].

The angular acceleration of loaded motor shaft α_1 is

$$\alpha_1 = \frac{Ma}{Jtot} \ [rad/s^2] \tag{28}$$

where: Ma-is an acceleration torque of the selected motor [Nm].

In that case equations (26) and (27) are transformed into:

$$a = \frac{Ma}{Jtot} \cdot \frac{h \cdot i}{2\pi} \ [m/s^2] \tag{29}$$

$$a = \frac{Ma}{Jtot} \cdot r_p \cdot i \ [m/s^2] \tag{30}$$

The optimization of a transmission ratio i, feed screw lead h or radius of the rack pinion r_p will be done by using the following procedure:

1. For every standard value of the feed screw lead h or radius of the pinion r_p we can calculate the transmission ratio range $i_1 \leq i < i_2$, to satisfy the following conditions:

-the calculated necessary motor speed n_e for the desired rapid traverse feed rate must be smaller or equal to the maximum motor speed n_m ($n_e \leq n_m$),

-the total load torque Mtot must be smaller or equal to the nominal motor torque Mn (Mtot≤Mn).

2. In the range [i1,i2] we will find the maximum of the function of acceleration a=f(i) at constant h or r_p :

$$\max\{a(i)\} = \max\{a(i1),a(i2),a(e1),.....,a(ej)\} \tag{31}$$

where e1,......,ej are extremes in the range [i1,i2].

The extremes can be found by the equation

$$\frac{da(i)}{di} = 0 \tag{32}$$

The function of the acceleration a=f(i) in the range [i1,i2] may have one, more or no extremes (fig.4).

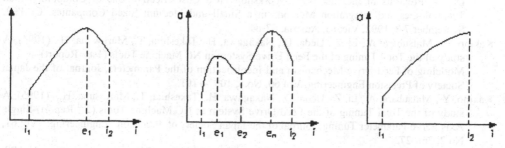

FIGURE 4. Possible forms of the function of acceleration

When the function of acceleration a=f(i) gets a maximal value for the constant feed screw lead h or radius of the pinion r_p the transmission ratio obtains the relative optimal value iop.

3. To get the absolute optimum of the transmission ratio iopt and optimal values of the feed screw lead hopt or of the pinion radius ropt we should repeat n times procedures described above in 1 and 2 for all standard values for h and r_p. In that way we will get n relative optimal transmission ratios iopi for the appropriate n different standard values for hi or rpi, where i=1,...,n.

The pair (iopi,hi) or (iopi,rpi) that gives the maximal value for the acceleration function, will provide the absolute optimum for the transmission ratio iopt, and, the optimal value for the feed screw lead hopt or for the radius of the pinion ropt.

It means

$$\max\{a(iop,h)\} = \max\{a(iop1,h1),...,a(iopn,hn)\} \tag{33}$$

or

$$\max\{a(iop,rp)\} = \max\{a(iop1,rp1),...,a(iopn,rpn)\} \tag{34}$$

Using equations (33) and (34) we obtain the pair (iopt,hopt) or (iopt,ropt) which provides a maximal value for the function a=f(i).

This optimization procedure is different from procedures shown in [11,4], where the relative optimal transmission ratio iop is calculated using equation (32) without taking in consideration that ne≤nm and Mtot≤Mn.

4 CONCLUSIONS

The created computer programs enable an efficient interactive and optimal design and analysis of different variants of NC machine main spindle and feed drives. The presented software also reduces the design time and modernizes the design process.

REFERENCES

1. Dukovski, V., Pandilov, Z., (1992), Computer Aided Design of NC machine Tool Main Spindle Drives, Preprints of the first IFAC-Workshop on a Cost effective Use of Computer Aided Technologies and Integration Methods in a Small and Medium Sized Companies, CIM'92, September 7-8, 1992, Vienna, Austria, 57-64.
2. Kakino, Y., Matsubara, A., Li ,Z., Ueda, D., Nakagawa, H., Takeshita, T., Maruyama, H., (1994), A study of the Total Tuning of the Feed Drive System in NC Machine Tools (1st Report)-Modeling of Feed Drive Mechanism and Identification of the Parameters, Journal of the Japan Society of Precision Engineering, Vol.60, No.8, 1097-1101.
3. Kakino, Y., Matsubara, A., Li, Z., Ueda, D., Nakagawa, H., Takeshita, T., Maruyama, H., (1995), A study of the Total Tuning of the Feed Drive System in NC Machine Tools (2nd Report)-Single Axis Servo Parameter Tuning, Journal of the Japan Society of Precision Engineering, Vol.61, No.2, 268-272.
4. Motika, R., Ciglar, D., (1986), Optimization of feed drives for NC machine tools, Strojarstvo, Vol.28, No.4 , 229-233, (In Croatian).
5. Pandilov, Z., (1993), Computer Aided Design of regulated drives for NC machine tools, M.Sc.Thesis, Faculty of Mechanical Engineering, Skopje. (In Macedonian).
6. Pandilov, Z., Dukovski, V., Dudeski, Lj., (1993), Computer Aided Design of Main Spindle and Feed Drives for NC Machine Tools, Preprints of Synopsis Papers of the 1st Balcan IFAC Conference on Applied Automatic Systems, 27-29 September, 1993, Ohrid, Republic of Macedonia, 73-74.
7. Pandilov, Z., Dukovski, V., (1993), Computer Aided Optimal Feed Drives Design for NC Machine Tools, Proceedings of the 2nd Conference on Production Engineering CIM'93, November, 1993, Zagreb, Croatia, H19-H26.
8. Pandilov, Z., Dukovski, V., Dudeski, Lj., (1995), Computer aided design of drives for NC machine tools, Proceedings of 10th International Conference of Engineering Design, Vol.4, Praha, Czeh Republic, August 22-24, 1995, 1587-88.
9. Pandilov, Z., Dukovski, V., (1995), One approach towards analytical calculation of the position loop gain for NC machine tools, Proceedings of the 3rd Conference on Production Engineering, CIM'95, 23/24 November 1995, Zagreb, Croatia, G77-G83.
10. Push, E.,V., (1986), Machine tools, Machinostroenie, Moscow. (In Russian).
11. Stute, G., Bobel, K., Hesselbach, J., Hodel, U., Stof ,P., (1983), Electrical feed drives for machine tools, Edited by Hans Gross, Chichester, John Wiley & Sons.
12. Weck, M., (1984), Handbook of Machine Tools, Volume 3, Automation and Controls, Chichester, John Wiley & Sons.

CONSTRAINED H∞ DESIGN OF PID CONTROLLERS

F. Blanchini [1], A. Lepschy [2], S. Miani [3], U. Viaro [3]

[1] Dept. of Mathematics and Informatics, University of Udine, Italy
[2] Dept.. of Information Engineering, University of Padova, Italy
[3] Dept. of Electrical, Managerial and Mechanical Engineering, University of Udine, Italy

KEYWORDS: Process control, PID controllers, H_∞ norm, Stability margins, Crossover frequency

ABSTRACT. The paper presents an efficient procedure for synthesizing PID controllers so as to maximize Smith's vector stability margin, i.e., to minimize the H_∞ norm of the sensitivity function, while ensuring specified values of the phase margin and gain crossover frequency. In this way, a trade-off between dynamic performance and robustness is achieved. The considered problem would be hard to solve with existing techniques; the suggested method, instead, leads to the desired result with remarkable computational simplicity, as shown by meaningful examples.

1 INTRODUCTION AND PROBLEM STATEMENT

According to surveys on the state of the art of industrial practice, the large majority of control loops resorts to simple PI or PID controllers [1]. It has authoritatively been stated [2] that, even if the subject dates back to the 1940's, there is still "a need for substantial theory in PID control" and, in fact, the last five years have witnessed a revival of interest in this field [3], [4]. In particular, a characterization of all stabilizing PID controllers has been derived and numerical search techniques have been employed to single out, among the admissible candidates, a controller that optimizes a selected performance index [2].

In this paper, reference is made to the H_∞ norm of the sensitivity function, which is related to the inverse of Smith's vector stability margin m_v (minimum distance of the open-loop Nyquist diagram from $-1 + j0$). As is well known, this index well accounts for system robustness against plant parameter variations and disturbances.

It has been shown [5], [6] that, on each plane of the three-dimensional controller parameter space characterized by a given proportional gain k_P, the region where the H_∞ norm does not exceed an upper bound U_H, consists of the union of convex sets; obviously, the sets corresponding to bound U_H are subsets of those corresponding to a larger bound. Also, as U_H tends to infinity, the convex sets become (convex) polygons (possibly, with one side at infinity), inside each of which the numbers of left half-plane (LHP) and right half-plane (RHP) closed-loop poles remain the same [6]. Clearly, for design purposes, only *stability* regions are of interest. A simple alternative procedure to determine the regions with the same closed-loop pole distribution is outlined in Section 2.

Published in: E. Kuljanic (Ed.) *Advanced Manufacturing Systems and Technology*,
CISM Courses and Lectures No. 486, Springer Wien New York, 2005.

Besides robustness, other requirements concerning response promptness and transient accuracy are usually taken into account. In most industrial plants, particularly those for which PID controllers are suited, response promptness is strictly related to the so-called gain crossover frequency ω_A, and transient accuracy to the phase margin m_φ. In fact, as ω_A increases, the closed-loop passband also increases (and, thus, the rise time decreases) and, as m_φ increases, the step response overshoot decreases. That is why many classic design procedures involving PID controllers or lag-lead networks are based on ω_A and m_φ. On the other hand, the integral action exerted by a PID controller ensures that the loop type is (at least) one, so that the steady-state error for constant set-points is zero and step disturbances entering the process are asymptotically rejected, independently of the open-loop Bode gain.

This paper presents a computationally efficient method to determine the parameters of a PID controller in such a way that the specifications regarding ω_A amd m_φ are satisfied and vector margin m_v is maximized. The paper is organized as follows: Section 2 formulates the specifications in analytic and geometric terms; Section 3 describes the main steps of the suggested design procedure; Section 4 applies the method to a pair of meaningful examples; Section 5 briefly compares the present technique with alternative design procedures.

2 BASIC EQUATIONS

Let us assume that the plant to be controlled is satisfactorily described by a strictly-proper *rational* transfer function $P(s)$ without poles and zeros on the imaginary axis, as is the case in most industrial processes, and denote by

$$C(s) = \frac{k_I + k_P s + k_D s^2}{s} \tag{1}$$

the PID controller transfer function and by $G(s) = C(s)P(s)$ the open-loop transfer function.

The specifications on ω_A amd m_φ can be reformulated in terms of the following *interpolation* condition:

$$G(j\omega_A) = e^{j(m_\varphi - \pi)} , \tag{2}$$

which, however, does not guarantee that the Nyquist diagram of $G(j\omega)$ does not intersect the unit circle at other points closer to $-1 + j0$ than (2) (or even that the system is stable). Therefore, it will be necessary to check whether the resulting controller parameters actually ensure the desired system behaviour.

By indicating with $R(\omega)$, $jI(\omega)$ and $M^2(\omega)$, respectively, the real part, the imaginary part and the square magnitude of the process harmonic response $P(j\omega)$, for $\omega_A \neq 0$ equation (2) can be separated into the two equations:

$$R(\omega_A)(k_I - \omega_A^2 k_D) - k_P \omega_A I(\omega_A) = \omega_A \sin m_\varphi , \tag{3}$$

$$I(\omega_A)(k_I - \omega_A^2 k_D) + k_P \omega_A R(\omega_A) = -\omega_A \cos m_\varphi , \tag{4}$$

from which:

$$k_P = \hat{k}_P := -\frac{R(\omega_A)\cos m_\varphi + I(\omega_A)\sin m_\varphi}{M^2(\omega_A)} \tag{5}$$

(note that $M^2(\omega_A) \neq 0$ under the considered assumptions) and

$$k_I = \omega_A^2 k_D + \hat{q} \tag{6}$$

where

$$\hat{q} := \omega_A \frac{R(\omega_A) \sin m_\varphi - I(\omega_A) \cos m_\varphi}{M^2(\omega_A)}. \tag{7}$$

Clearly, $\hat{q}$ corresponds to the integral gain of the pure PI controller satisfying the specifications on ω_A and m_φ; in other words, these specifications completely determine such a controller. In the case of a PID controller, equation (6) defines a straight line on the plane $k_P = \hat{k}_P$ of the three-dimensional controller parameter space. Therefore the design problem entails finding a point on this straight line and, in particular, on its segments belonging to the stability regions (if any).

The closed-loop pole distribution changes when poles cross the imaginary axis. First of all, this occurs when parameter k_I passes from positive to negative values, or vice versa, because then a pole crosses the imaginary axis *through the origin*; consequently, straight line $k_I = 0$ separates regions of the relevant (k_D, k_I)-plane characterized by different numbers of LHP poles. Also, when the relative degree δ_P of $P(s)$ is equal to 1, a pole passes from the RHP to the LHP, or vice versa, *through the point at infinity* for $k_D = -1/K_E$, where K_E is the Evans gain of $P(s)$, i.e., the ratio of the leading coefficients of its numerator and denominator polynomials; therefore, if $\delta_P = 1$, straight line $k_D = -1/K_E$ separates regions with different pole distributions.

In most cases, however, the pole distribution changes when *a pair* of complex conjugate poles crosses the imaginary axis for finite non-zero values of ω. At these angular frequencies $G(j\omega) = C(j\omega)P(j\omega) = -1$, i.e.,

$$Im[G(j\omega)] = 0, \; Re[G(j\omega)] = -1. \tag{8}$$

The first of conditions (8) implies that

$$k_I - \omega^2 k_D = k_P \omega \frac{I(\omega)}{R(\omega)}. \tag{9}$$

By substituting (9) into the expression of $Re[G(j\omega)]$, the second of conditions (8) is equivalent to:

$$k_P M^2(\omega) + R(\omega) = 0. \tag{10}$$

For $k_P = \hat{k}_P$, equation (10) supplies the (real) values, if any, of ω, say ω_k, $k = 1, 2, \ldots, N_k$, that correspond to the above-mentioned crossings of the imaginary axis when the proportional gain is equal to $\hat{k}_P$. By substituting ω_k for ω and $\hat{k}_P$ for k_P in (9), the following straight-line equations are obtained:

$$k_I = \omega_k^2 k_D + \hat{k}_P \omega_k \frac{I(\omega_k)}{R(\omega_k)}, \; k = 1, 2, \ldots, N_k. \tag{11}$$

These lines too separate (k_D, k_I)-regions with different closed-loop pole distributions. In this way, plane $k_P = \hat{k}_P$ is partitioned into *convex* polygons (possibly, with one side at infinity) whose maximal number can easily be determined according to the pancake-cutting formula [6]. To ascertain whether they correspond to stable behaviour, it is enough to check one point only inside each polygon. Clearly, polygonal regions having one side in common with a stability region may not correspond to stable behaviour and can be neglected. By way of example, Fig. 1 shows the 6 distinct regions with the same pole distribution in the upper half ($k_I > 0$) of plane $k_P = 2$ for the process transfer function [6] :

$$P(s) = \frac{10s^3 + 9s^2 + 362.4s + 36.16}{2s^5 + 2.7255s^4 + 138.4293s^3 + 156.471s^2 + 637.6472s + 360.1779}. \tag{12}$$

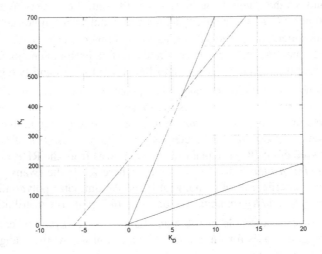

FIGURE 1. Polygonal regions with the same pole distribution

Any point of a stability region corresponds to a specific open-loop transfer function $G(s)$ and, therefore, to a particular value of vector margin m_v. According to the results in [6], the points characterized by the same value $\hat{m}_v$ of m_v form the boundary of a convex region inside which $m_v \geq \hat{m}_v$. The boundary degenerates to a point or a line when $\hat{m}_v$ is the maximal value achievable within the region. From a geometrical point of view, the considered design problem consists in finding where line (6) on plane $k_P = \hat{k}_P$ touches, without intersecting, a constant-m_v contour; the coordinates of the point of contact supply the desired values of k_D and k_I. From the operative point of view, the solution can simply be obtained through a one-dimensional search limited to the intersection between (6) and the stability regions.

3   DESIGN PROBLEM SOLUTION

The square distance $\rho^2(\omega; k_P, k_I, k_D)$ of the generic point on the open-loop Nyquist diagram of $G(j\omega)$ from $-1 + j0$ turns out to be:

$$\rho^2(\omega; k_P, k_I, k_D) := |1 + G(j\omega)|^2 = \frac{1}{\omega^2} \Big\{ \big[(k_I - \omega^2 k_D)^2 + \omega^2 k_P \big] M^2(\omega)$$
$$+ \omega^2 + 2\omega(k_I - \omega^2 k_D)I(\omega) + 2\omega^2 k_P R(\omega) \Big\} \tag{13}$$

which for $k_P = \hat{k}_P$ and k_I as in (6) becomes:

$$\rho_A^2(\omega; k_D) = \frac{1}{\omega^2} \left\{ \left[(\omega_A^2 - \omega^2) k_D + \hat{q} \right]^2 M^2(\omega) + \omega^2 \hat{k}_P^2 M^2(\omega) + \omega^2 + \right.$$
$$\left. 2\omega \left[(\omega_A^2 - \omega^2) k_D + \hat{q} \right] I(\omega) + 2\omega^2 \hat{k}_P R(\omega) \right\}. \tag{14}$$

Rational function (14) exhibits one or more minima for any value of k_D. If the closed-loop system is stable, the square root of the overall minimum gives the vector margin m_v (inverse of the H_∞ norm of the sensitivity function). The problem is then to maximize m_v with respect to k_D inside the stability regions, if any. Since the m_v-contour for a given value of the vector margin is contained into the one for a smaller value and the encircled area is convex, function (14) admits a unique maximum inside every stability region. Once the best value $\hat{k}_D$ of k_D has been found, the corresponding value $\hat{k}_I$ of k_I is supplied by (6); as already pointed out, the value $\hat{k}_P$ of k_P is directly provided by the interpolation condition.

The solution of the design problem could be obtained by determining the analytic expressions of the m_v-contours in the regions of interest of plane $k_P = \hat{k}_P$. To this purpose, by setting $k_I - k_D \omega^2 = q(\omega)$ and $\rho = m_v$, equation (13) with $k_P = \hat{k}_P$ can be rewritten as the second-degree polynomial equation in q:

$$M^2(\omega) q^2 + 2\omega I(\omega) q + \omega^2 \left[M^2(\omega) \hat{k}_P^2 + 1 - m_v^2 + 2R(\omega) \hat{k}_P \right] = 0 \tag{15}$$

whose roots are real if and only if its discriminant is nonnegative, i.e., for $\omega > 0$:

$$\Delta(\omega) := M^2(\omega) m_v^2 - \left[R(\omega) + \hat{k}_P M^2(\omega) \right]^2 \geq 0. \tag{16}$$

For any value of m_v, (16) supplies the ω-intervals $J_k, k = 1, 2, \ldots, N_J$, (called *active intervals* in [6]) where $\Delta(\omega)$ is nonnegative. As shown in the quoted reference, the m_v-contours on plane $k_P = \hat{k}_P$ lie on the curves described by the parametric equations:

$$k_D = -\frac{1}{2\omega} \frac{\partial}{\partial \omega} q(\omega), \; \omega \in J_k \tag{17}$$

$$k_I = -\frac{\omega}{2} \frac{\partial}{\partial \omega} q(\omega) + q(\omega), \; \omega \in J_k \tag{18}$$

where $q(\omega)$ denotes the (real) solutions of (15), which depend on ω. Equations (17) and (18) represent the envelope of the family of straight lines:

$$k_I = \omega^2 k_D + q(\omega), \; \omega \in J_k. \tag{19}$$

On the basis of the previous equations, a Matlab program has been developed. The results obtained in some meaningful cases are illustrated in Section 4.

4 EXAMPLES

4.1 EXAMPLE 1

Let us refer to the plant tansfer function:

$$P(s) = \frac{1}{(1+s)\left(1+\frac{s}{5}\right)\left(1+\frac{s}{10}\right)\left(1+\frac{s}{20}\right)} \tag{20}$$

and to the specifications: $m_\varphi = \pi/4$, $\omega_A = 3$, which are compatible with the use of a PID controller whose phase is included between $-\pi/2$ and $\pi/2$ for $k_P, k_I, k_D > 0$; in fact, $\arg[P(j\omega_A)] = -2.2298$ so that $\arg[C(j\omega_A)] = \arg[G(j\omega_A)] - \arg[P(j\omega_A)] = m_\varphi - \pi + 2.2298 = -0.1264$. According to (5) and (7), the interpolation condition (2) leads to $\hat{k}_P = 3.8622$ and $\hat{q} = 1.4722$; therefore (6) particularizes to:

$$k_I = 9k_D + 1.4722 . \tag{21}$$

The roots of (9) with $k_P = 3.8622$ turn out to be: $\omega_1 = 3.6157$ and $\omega_2 = 19.2854$. On plane $k_P =$

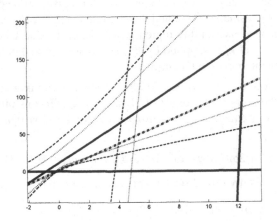

FIGURE 2. Stability region (border in bold), straight line (21) (bold, dot-dashed) and constant m_v contours for $m_v = 2/3$ (thin dashed) and $m_v = 1/2$ (thin dotted)

$\hat{k}_P$, the straight lines (11) separating regions with the same closed-loop pole distribution are: $k_I = \omega_1^2 k_D + 11.4858$ and $k_I = \omega_2^2 k_D - 4477.6$, besides straight line $k_I = 0$. They are depicted in Fig. 2 together with straight line (21). The unique stability region in the considered parameter plane is the triangle whose sides lie on these lines; the intersection of this triangle with (21) is given by the segment S from point $(-0.1636, 0)$ to point $(12.3415, 112.5460)$. The constant-m_v contours inside the stability region are displayed in the same figure. A simple search along S shows that the vector margin is maximum at point $(0.8769, 9.3643)$ where $m_v = 0.7547$ Therefore, the transfer function of the resulting optimal controller is:

$$C(s) = \frac{9.3643 + 3.8622s + 0.8769s^2}{s} . \tag{22}$$

The open-loop Nyquist diagram together with the $m_v = 0.7547$ locus is drawn on the left of Fig. 3 and the closed-loop step response on the right of Fig. 3.

4.2   EXAMPLE 2

Let us now assume that the process transfer function is given by (12) and that the desired phase margin and crossover frequency are $m_\varphi = 1.5759$ and $\omega_A = 44.1766$ (these rather queer specifi-

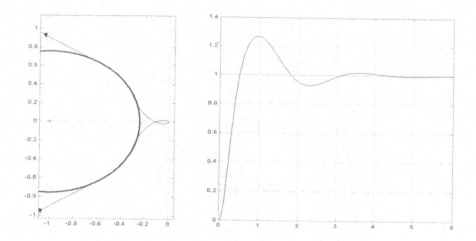

FIGURE 3. Nyquist plot and circle with radius $m_v = 0.7547$ (left) and closed loop step response (right) for the open loop transfer function $C(s)P(s)$ of example 4.1

cations have been adopted to allow comparison with the controller derived in [6]). The related interpolation condition supplies $k_P = 2$ and $q = -16951.67$ so that:

$$k_I = \omega_A^2 k_D - 16951.67 . \qquad (23)$$

There are two stability regions on plane $k_P = 2$, as shown in Fig. 4. Line (23), also depicted in Fig. 4, intersects them both.

Inside the *upper* stability region, m_v achieves its maximum value $\hat{m}_v = 1/3$ at $(9, 612.47)$ where the envelope of one family of straight lines (19) corresponding to $\hat{m}_v$ crosses the envelope of another family corresponding to the same value of m_v (but to a different active interval). Precisely, at this intersection, the value of the current parameter along the first envelope is $\omega = 5.764$ and along the second envelope is $\omega = 8.489$. As a consequence [6], the open-loop Nyquist diagram for the resulting controller transfer function:

$$C(s) = \frac{612.47 + 2s + 9s^2}{s} \qquad (24)$$

is tangent to the circle with centre $-1 + j0$ and radius $\hat{m}_v$ at these two angular frequencies, as shown in Fig. 5 (left). Note that he same controller was chosen in [6] without reference to the above-mentioned interpolation condition (which, as already said in Section 1, well accounts for response promptness).

Along the relevant segment of line (23) inside the *lower* stability region, m_v increases as $k_I \to 0$. However, to achieve a reasonably short settling time, it is convenient to choose a not too small value of the integral gain. At point $(k_I = 69.936, k_D = 8.722)$ the vector margin is

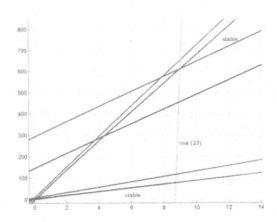

FIGURE 4. Constant m_V contours for $m_V = 1/3$ for example 4.2

$m_v = 0.6642$. The open-loop Nyquist diagram corresponding to the resulting controller transfer function:

$$C(s) = \frac{20 + 2s + 8.9s^2}{s} \tag{25}$$

is represented in Fig. 5 (right). The closed-loop step response obtained using controller (24) is compared with the one obtained using controller (25) in Fig. 6.

Let us only observe that the two step responses are characterized by almost the same rise and settling times, but the values of the integral and derivative gains of controller (25) are (much) smaller.

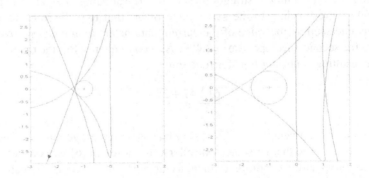

FIGURE 5. Open loop Nyquist plot correponding to controlller (24) and circle with radius $m_V = 1/3$ (left); open loop Nyquist plot correponding to controlller (25) and circle with radius $m_V = 0.6642$ (right)

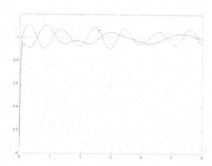

FIGURE 6. Closed loop step responses obtained using controller (24) (dashed) and (25) (solid)

5 CONCLUSIONS

A simple procedure for synthesizing PID controllers so as to maximize Smith's vector stability margin while ensuring specified values of the phase margin and gain crossover frequency, has been presented. It draws on a previous paper by the present authors [6], where a characterization of the parameter regions inside which the H_∞ norm of the sensitivity function does not exceed given upper bounds, was provided. The suggested method has been implemented using Matlab. Its operation has been illustrated by means of a pair of representative examples in Section 4: the results obtained show that it can successfully be applied even when the process transfer function is rather complicated.

REFERENCES

1. K. J. Åström, and T. Hägglund, *PID controllers: Theory, Design, and Tuning*. Research Triangle Park, NC: Instrument Society of America, 1995.

2. A. Datta, M. T. Ho, and S. P. Bhattacharyya, *Structure and Synthesis of PID Controllers*. London, U.K.: Springer–Verlag, 2000.

3. Special Section on PID control, *Proc. IEE, Control Theory and Applications*, vol. 149, no. 1, Jan. 2002.

4. W. K. Ho, T. H. Lee, H. P. Han, and Y. Hong, "Self tuning IMC-PID control with interval gain and phase margins assignment", *IEEE Trans. Contr. Syst. Technol.*, vol. 9, no. 3, pp. 535-541, May 2001.

5. M. T. Ho, "Synthesis of $\mathcal{H}_\infty$ PID controllers: a parametric approach", Automatica, Vol. 39, no.6, pp. 1027–1036, July 2003.

6. F. Blanchini, A. Lepschy, S. Miani, and U. Viaro, "Characterization of PID and lead/lag compensators satisfying given $\mathcal{H}_\infty$ specifications" IEEE Trans. Aut. Contr. , vol. 49, no. 5, pp. 736–740, May 2004.

CONCLUSIONS

REFERENCES

MULTI-AGENT MANUFACTURING SYSTEM DESIGN BASED ON EMERGENT SYNTHESIS APPROACH

Z. Car, T. Mikac

Department of Mechanical Engineering, Faculty of Engineering, University of Rijeka, Croatia

KEYWORDS: Emergent Synthesis, Multi-agent, Layered BMS, Multi-Objective Design and Reconfiguration.

ABSTRACT. This paper presents a approach for multi-objective design and reconfiguration of the current manufacturing system. A proposed model for multi-objective design and adaptable reconfiguration is based on the multi-agent principles and Emergent Synthesis concept. Emergent Synthesis is introduced in the adaptable reconfiguration within implementation of the layered based architecture and utilizing the multi-agent concept. Based on the proposed approach a simulation model has been built and evaluated. For the purpose of analysis of this approach it was simulated production in the shop-floor of the manufacturing system with the two different initial configurations. For evaluation of here presented approach the case study was performed using realistic data; it was simulated production of the cellular phone devices. From obtained results of the simulation it can be conclude that a proposed model has a high level of multi-objective adaptation of the manufacturing system shop-floor.

1 INTRODUCTION

The manufacturing enterprise is the subject of new pressures demanding changes in the way that companies compete, how they are organised and structured and the technologies that they adopt. Studies [1,2] have identified three clusters of the pressures acting on the manufacturing based enterprises; these are: globalization, benign products / processes, business and organizational structure. Those new environmental characteristics and requirements have challenged manufacturing enterprises to improve the efficiency and productivity of their production activities. Today's manufacturing systems have to be able to produce products with low production cost and high quality on time. In addition to that manufacturing system should be able to adjust and respond rapidly to changes in product design and demand with none or low cost investment. Such kind of market behaviour requires shortening the manufacturing system leading time, which means the design and development/reconfiguration time of manufacturing systems [3,4]. It can be said that the lead time for reconfiguration and development of manufacturing systems today has become a bottleneck.

Until now the concept of computer integrated manufacturing (CIM) has been promoted as the concept that will somehow deal with request for high operational flexibility and the ability of a system to transform its internal structure and technology at will. However, Warnecke [5] pointed out that with introduction of the CIM it was often wrongly assumed (most of the implementations of the CIM) that organizational structure and procedures can be automatically improved by CIM implementation. It was also recognized, during process of the implementation of the CIM, the aim of total integration (automation) and optimization of the system, and systematizes the activities of the company [6]. Because of that there is a high initial cost of the system, long lead time, and generation of rigid systems due to large size of the centralization. The consequences are that the rigidity and emphases on automation may lead to loss of the flexibility and adaptability to the changing environment [7]. As correction to these situations conceptual solutions with notation which address today's fundamental

Published in: E. Kuljanic (Ed.) *Advanced Manufacturing Systems and Technology*,
CISM Courses and Lectures No. 486, Springer Wien New York, 2005.

problems of manufacturing enterprise (the lack of the flexibility in an organizational structure and operating norms to easily be adapted to changing environments) were proposed: fractal, holonic and biological concept [4,5]. Detail discussion about conceptual, design and operational features can be found in Ref. [7].

However, numerous of researches have noted that adaptation of the manufacturing enterprises on only information and control level is not enough for the manufacturing system to survive in chaotic and dynamic environment [8]. There is increase need for manufacturing systems to have ability to dynamically adjust their shop-floor layout in order to respond rapidly to changes in product design and demand with none or low cost investment. Some of the solutions (approaches) for dynamical adjustment of the manufacturing system shop-floor which does not uses only classical approaches for production planning and control are: Dynamic Plant Layouts, Dynamic Cellular Manufacturing, Reconfigurable Manufacturing Systems (RMS), Lineless manufacturing systems, etc [9,10,3,11].

The aim of this paper is to present approach for multi-objective design and adaptable reconfiguration of the manufacturing system shop-floor, which is based on Emergent Synthesis Approach [12]. Emergent Synthesis; is introduced in the adaptable reconfiguration within implementation of the layered based architecture with applying concept of the Biological Manufacturing Systems (BMS) – lineless manufacturing, Layered BMS concept. Lineless manufacturing is based on implementation of Biological Manufacturing System (BMS) concept, and it is one of the implementation of the BMS concept in the manufacturing environment.

2 EMERGENT SYNTHESIS APPROACH AND ITS IMPLEMENTATION IN PROPOSED MODEL

Emergent Synthesis approaches can be literally defined by its name. Literal definition of the synthesis is "combined elements into a whole". Term synthesis here is closely related to human activities for creation of the artificial things. However, in this case synthesis is also closely related to the analysis. This is because all elements of the system to be able to utilise pragmatic structure, "new whole", before that need to be analysed. This is needed to define their functional interrelationship, nature, inside proposed pragmatic structure [12].

Term emergence can be defined from linguistically, evolutionary, physical, mathematical, etc. point of view. Oxford dictionary defined emergence as something to become known, be revealed. The term emergence has, in recent years, been used in many fields to describe qualitatively different observations. According to Cariani [13], the various definitions of emergence can be broken down into three broad categories: computational (mathematically based concept), thermodynamic (physically based materialism) and relative to a model (functionally based hylomorphism). New direction in Artificial Life was introduced by Langton [14]. He defines emergence in terms of a feedback relation between the levels in dynamical system; local micro-dynamics cause global macro-dynamics while global macro-dynamics constrain local micro-dynamics. This definition implies that implicit global complexity emerges from explicit local simplicity.

According to above given definitions and explanations it can be said that Emergent Synthesis related approaches are being using both bottom-up and top-down features, unlike traditional analytical, deterministic approaches based on top-down problem decomposing; such as operational research, symbolic artificial intelligence, etc.

Emergent Synthesis was introduced in proposed model for design and adaptable reconfiguration within two separate levels. In first level it was applied concept of the BMS (lineless manufacturing), and on the second level it was proposed characteristically structure of the model, within implementation of the layered based architecture. With this approach it is possible to achieve synthesis between the simulation of lineless manufacturing systems and the real ones through layered architecture with introduction of the constraints, behaviour and flow of information.

2.1 BIOLOGICAL MANUFACTURING SYSTEM (BMS)

Biological Manufacturing System (BMS) concept is based on essential biological organisms' characteristics and behaviour. Biological organisms insure their existence and there are capable of adapting themselves in dynamic living environment by showing functions such as self-recognition, self-development, self-recovery, adaptability and evolution. Fusion of biological information with individuals makes living systems complex but adaptive [11]. Lineless manufacturing systems are one of the implementations of BMS idea in the manufacturing environment. By straightforward definition it can be said that main feature of the lineless manufacturing systems is that all production entities, such as machines and AGVs, can move freely in the manufacturing shop-floor.

2.2 LAYERED ARCHITECTURE

With introduction of the multi-objective design and adaptable reconfiguration based on the layered architecture and multi-gent concept there is a trade between high adaptability of lineless production and reduction of technical difficulties in implementation of the lineless production in real manufacturing environment with introduction of constraints for movement of elements of manufacturing system and number of different products on shop-floor. With implementation of this approach it is obtained emergence of the candidate structure. In order to keep paper to an acceptable length, here it will be given short description of the Layered architecture. Detail discussion about this proposed architecture can be found in Ref. [15]. The proposed model has following characteristics. The model consists of three interdependent layers where multi-agent architecture is introduced. Firstly three colonies of multi-agents have been introduced: virtual Biological Manufacturing Systems, Control and Reconfiguration, and the Real Manufacturing System. Through layered architecture with multi-agent concept it is achieved a synthesis between the simulation of lineless manufacturing systems and the real ones by introducing the constraints. Virtual Biological Manufacturing System is an extension of real manufacturing system based on the concept of BMS. In this layer it is used the simulation of the lineless manufacturing system for simulation of the production of the new order. Here it is used highly adaptable features of the BMS in the dynamic manufacturing environment, floor level. The control of the production in the lineless manufacturing system is too complex to be controlled with traditional methods, because of the large degree of freedom of the system. In here presented model it is proposed the two phases's segregated self-organisation [15] for control of the lineless manufacturing system in production. Layer of the control and reconfiguration is used for introduction and analysis of the new proposed structure, configuration of the manufacturing system, in the reconfiguration mode, and control of the real manufacturing system in work mode. In this layer analysis of the proposed structure is performed in order to recognise function within manufacturing environment. This layer is the key layer in here presented approach for adaptable reconfiguration. With introduction of the layered architecture and in particular this layer gives ability to the model to breach gap between simulation and real manufacturing system. This layer has two very important characteristics. Firstly, through this layer it is indirectly introduced reality in the simulation by introducing positions constraints for the movements of the elements of the manufacturing system. Secondly, through this layer it is introduced noise from surrounding manufacturing system environment. Level of reality, complexity and uncertainty which it will be introduced in the simulation it is filtrated through layer of the real manufacturing system.We can say that with introduction of moderately autonomous layers and autonomous agents which exist in the layers we have produce cooperative behaviour in the system.

3 SIMULATION MODEL

The behaviour was analysed, within corresponding results, of the proposed model for multi-objective design and adaptable reconfiguration of the manufacturing system shop-floor. In this phase evaluation of the proposed model is based on the simulation performed with realistic data. For this analysis of the model case study was performed; production of the cellular phone devices.

3.1 SETUP OF THE SIMULATION MODELS FOR CASE STUDIES

It was simulated production of the cellular phone devices. For the initial configuration we have been using circle initial configuration (FIGURE 1). This example is taken because of the following reasons. Manufacturing system for production of the cellular phones needs to survive in extremely dynamic environment. Market trends dictate fast changes of the models for production, considering for example Japanese market; companies are pressured to change models few times per year. In the same moment manufacturing based companies they are under increasing pressure from product distributors for mass-customization in production. This market behaviour produces very much uncertain, complex and dynamic manufacturing environment. Manufacturing systems for production of the cellular phone devices need to produce high volume of the same and/or similar type of the products in short time span, and product types are undergoing very short life-time. The setup of the simulation mode was defined as follow: production process consists from high number of the operations (up to 25, depends from type of order); with high number of the jobs in the one order between 3000 to 5000 jobs. On the shop-floor in the same time can coexist up to 15 different orders. The number of orders which coexist on the shop-floor is depended of there type. The production times for different operations are defined in the Table 1.

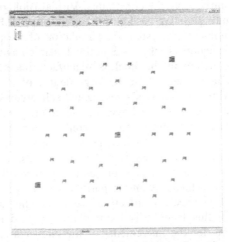

FIGURE 1. Outlook of the initial configurations

evaluation is based on the differential between total production time and due date, and also priority of each order is taken in consideration.

3.2 METHODS AND HEURISTICS

The emerging behaviour, two phases's self-organisation [15], in the Virtual Biological Manufacturing System is based on the scheduling rule and topological position of machines on the shop-floor. The presented model in this paper it will evaluate the behaviour based on the two different objectives for design and adaptable reconfiguration: reduction of the transportation distance, and minimizing deviation from the due date. In the case of the first objective the distance between input and output of the material is taken as reference for optimisation. In the layer of the Virtual BMS are presented two types of scheduling problems: scheduling for dispatching of the new order for reconfiguration and production scheduling, production in the lineless simulation. In the layer of the control and reconfiguration (and real manufacturing system) there are also two scheduling problems scheduling for dispatching of the new order for production and production scheduling. However, the production scheduling is obtained from predefined machine sequence which emerges in the Virtual BMS.

$$I_j = \frac{w_j}{p_j} \exp\left(-\frac{\max(d_j - p_j - t, 0)}{k\,\overline{p}} \right) \tag{1}$$

where are:
w_j weight of job j, priority factor,
p_j processing time on the of job j,
d_j due date of job j,
t time,
k scaling factor, and,
$\overline{p}$ average of the processing time of the remaining jobs.
The scheduling for dispatching product on the manufacturing system shop-floor (for reconfiguration and production process) is purely of the production nature, and it is very important problem. This is because if there is high number of orders which coexist on the same shop-floor influence of the dispatch rule to the overall production time of the order can be crucial. Due to this reason ATC heuristic was used as a technical solution. The Apparent Tardiness Cost (ATC) heuristic is presented in the equation [16] (1).
ATC heuristic gives us ability of the dynamic adaptation through time scale. ATC combines Weighted Shortest Processing Time first (WSTP) and Minimum Slack first (MS). These characteristics are given as a possibility for the adaptation of scheduling rules. WSTP is a static rule, a function of the job and/or machine. MS is a dynamic rule; a function of job j and time t and job with minimum slack is scheduled [17].
If parameter k is very large, the ATC rule reduces to the WSTP rule. If k is very small, the ATC rule reduces to the MS rule when there are no overdue jobs and to the WSTP rule for the overdue jobs otherwise. In our model it was used a scaling factor k for tuning the schedule rules in a dynamic environment. To utilise information which later it will be used in the process of the global interaction a production-topology factor (P_t) was introduced. The P_t factor (2) it was used to define scheduling rule in the lineless simulation. Through P_t factor it is expressed the relationship between production factors and topology of the shop-floor. Production-topology factor is adapted for the use in scheduling

evaluation is based on the differential between total production time and due date, and also priority of each order is taken in consideration.

3.2 METHODS AND HEURISTICS

The emerging behaviour, two phases's self-organisation [15], in the Virtual Biological Manufacturing System is based on the scheduling rule and topological position of machines on the shop-floor. The presented model in this paper it will evaluate the behaviour based on the two different objectives for design and adaptable reconfiguration: reduction of the transportation distance, and minimizing deviation from the due date. In the case of the first objective the distance between input and output of the material is taken as reference for optimisation. In the layer of the Virtual BMS are presented two types of scheduling problems: scheduling for dispatching of the new order for reconfiguration and production scheduling, production in the lineless simulation. In the layer of the control and reconfiguration (and real manufacturing system) there are also two scheduling problems scheduling for dispatching of the new order for production and production scheduling. However, the production scheduling is obtained from predefined machine sequence which emerges in the Virtual BMS.

$$I_j = \frac{w_j}{p_j} \exp\left(-\frac{\max(d_j - p_j - t, 0)}{k \bar{p}}\right) \tag{1}$$

where are:
w_j weight of job j, priority factor,
p_j processing time on the of job j,
d_j due date of job j,
t time,
k scaling factor, and,
$\bar{p}$ average of the processing time of the remaining jobs.

The scheduling for dispatching product on the manufacturing system shop-floor (for reconfiguration and production process) is purely of the production nature, and it is very important problem. This is because if there is high number of orders which coexist on the same shop-floor influence of the dispatch rule to the overall production time of the order can be crucial. Due to this reason ATC heuristic was used as a technical solution. The Apparent Tardiness Cost (ATC) heuristic is presented in the equation [16] (1).

ATC heuristic gives us ability of the dynamic adaptation through time scale. ATC combines Weighted Shortest Processing Time first (WSTP) and Minimum Slack first (MS). These characteristics are given as a possibility for the adaptation of scheduling rules. WSTP is a static rule, a function of the job and/or machine. MS is a dynamic rule; a function of job j and time t and job with minimum slack is scheduled [17].

If parameter k is very large, the ATC rule reduces to the WSTP rule. If k is very small, the ATC rule reduces to the MS rule when there are no overdue jobs and to the WSTP rule for the overdue jobs otherwise. In our model it was used a scaling factor k for tuning the schedule rules in a dynamic environment. To utilise information which later it will be used in the process of the global interaction a production-topology factor (P_t) was introduced. The P_t factor (2) it was used to define scheduling rule in the lineless simulation. Through P_t factor it is expressed the relationship between production factors and topology of the shop-floor. Production-topology factor is adapted for the use in scheduling for different objectives. Factor c which defines the influence of the topology of the manufacturing shop-floor was defined iteratively.

$$P_t = \begin{cases} cI_j + (1-c)D_j & \text{- for min. dev. from due date} \\ cD_j + (1-c)I_j & \text{- for red. of trans. distance} \end{cases} \tag{2}$$

where are:
c a distance influence coefficient,
z_j number of processes which need to be done on the one job,
I_j expression of the production factors, ATC heuristic,
D_j a distance influence factor.
where D_j is define as:

$$D_j = \frac{d_{s\,max} - d_s}{d_{s\,max}} \cdot 100 \tag{3}$$

d_{smax} a maximum distance from the based object to the other production entities (machine, AGVs), and
d_s a distance from the based object and a possible production entity.

As it have been noted earlier in text a distance influence coefficient (c) is defined iteratively. Here it will be given definition of the distance influence coefficients for analyses of the proposed concept for design and adaptable reconfiguration,

$$c = \begin{cases} 0.6 & \text{if} \quad np_j \le 0.7(d_j - t) \quad \text{and} \quad 1 \le z_j \le 18 \\ 0.85 & \text{if} \quad np_j \le 0.7(d_j - t) \quad \text{and} \quad 18 < z_j \le 25 \\ 0.9 & \text{if} \quad 0.7(d_j - t) < np_j \end{cases} \tag{5}$$

where are:
p_j processing time on the of job j,
d_j due date of job j,
t time,
z_j number of processes which need to be done on the one job,
n_j number of jobs in the one order.

4 RESULTS AND DISCUSSION

To analyse behaviour of the proposed model for multi-objective design and adaptation two types of the simulations analyses have been performed. To be able to present detail behaviour of the model under dynamic environment it will be presented results of the numbers of the simulation runs and specific adaptation indexes were introduced. For this analysis it has been used results of the production for 100 orders for the each case of the defined objective. The adaptation indexes are defined as follow. Adaptation index for reduction of the transportation distance of material through shop-floor:

$$I_a = \frac{D_{br}}{D_{ar}} \tag{7}$$

D_{ar} transportation distance after multi-objective design and adaptable reconfiguration, and

D_{br} transportation distance before multi-objective design and adaptable reconfiguration.

Adaptation index for minimization of the deviation of the due date:

$$I_d = \begin{cases} |p_{bri} - p_{ari}| & \text{for} \quad c_i \leq d_i \\ |p_{bri} - p_{ari}| - 2.5 & \text{for} \quad c_i > d_i \end{cases} \tag{8}$$

p_{bri} deviation from due date of the order i, before reconfiguration,

p_{ari} deviation from due date of the order i, after reconfiguration,

c_i completion time for order i, and,

d_i due date for order i.

The I_d does not show total information of the behaviour of the simulated production in manufacturing system, due to that reason exact deviation from the due date is supplemented on the figures.

4.1 CASE STUDY

For the case studies only circle initial configuration was used as initial condition. This is due to reason that this initial configuration has presented most adaptable ability in compare with others.

In the FIGURE 3 are presented adaptation indexes in a case of the cellular phone production. The standard deviation of the adaptation index for reduction of the transportation distance of material through shop-floor (FIGURE 3a) is between 1.1 and 1.5, where 91% of the introduce orders finishes inside this window. Also can be seen that 72% of the all order have more then 25% reduction of the reduction of the transportation distance of material through shop-floor. However, the number of the jobs with low or without positive adaptation index is 4%. These results suggest that proposed model have high adaptation ability for the objective of the reduction of the transportation distance through manufacturing system shop-floor. In a case of the adaptation index for min. deviation from the due date (FIGURE 3b), it can be seen that adaptation index is high, standard deviation of the index is between 7.06 and 15.03. This shows high adaptation of the system for the objective of the minimization of the due date deviation. The deviation of the due date obtained in simulation is supplemented, and can be seen that 83% of the introduced orders finishes inside ±7% deviation from the due date. In the same moment it can be observed from the presented results that only 6% of the introduced orders have deviation from due date higher then 12%. This confirms conclusion that proposed model for adaptable reconfiguration and with circle initial configuration is highly adaptable in dynamic manufacturing environment.

From the presented results can be concluded that for the second case study it is obtained high adaptation ability in the case of the orders which have been assigned objective for minimization of the deviation from the due date. For orders which was assigned objective for reduction of the transportation distance adaptation ability is varied and model cannot provide satisfied adaptation.

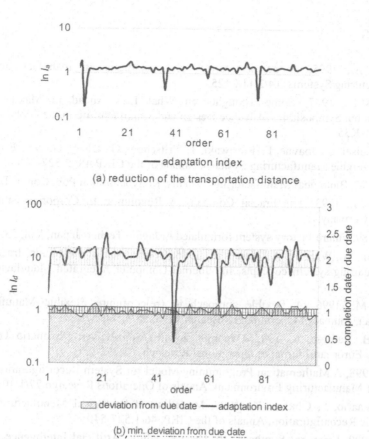

FIGURE 3. The adaptation indexes in a case of the cellular phone production

5 CONCLUSIONS

In this paper it was presented approach for multi-objective design and adaptable reconfiguration of the manufacturing system shop-floor. Proposed approach is based on utilisation of the Emergent Synthesis Approach through multi-agent organisation, with introduction of the Biological Manufacturing System and Layered based architecture. According to presented approach simulation model have been built in eM-Plant programming environment, and analysed with usage of the realistic (case studies) input data. In a case of the simulate cellular phone production results suggest that proposed approach have high adaptation abilities for the both introduced objectives. From those results can be concluded that here presented approach for multi-objective design and adaptable reconfiguration has high adaptation ability for both objectives the adaptation of the production where proportion of the single processes times in not high in overall production time. The main contribution of this approach in is that simulation model has capability in cooperation between production entities and surrounding manufacturing environment to emerge new configuration of the manufacturing system shop-floor and production scheduling.

REFERENCES

1. Buzacott, J.-A., 1995, A perspective on new paradigms in manufacturing. Journal of Manufacturing Systems, 14/2:118-125.
2. Merchant, M.E., 1997, Some Thoughts on What Lies Ahead in Manufacturing, CIRP International Symposium – Advance Design and Manufacturing in the Global Manufacturing Era, K31-K35.
3. Koren, Y., Heisel, U., Joavne, F., Moriwaki, T., Pritschow, G., Ulsoy, G., Van Brussel, H., 1999, Reconfigurable Manufacturing Systems, Annals of the CIRP 48/2: 527-540.
4. Ueda, K., 1994, Biological Manufacturing Systems, Kogyochosakai Pub. Comp. Tokyo, Tokyo.
5. Warnecke, H.J., 1993, The Fractal Company A Revolution In Corporate Culture, Springer-Verlag, Germany.
6. Suda, H., 1989, Future factory system formulated in Japan, Techno Japan, Vol. 23:15-25.
7. Tharumarajah, A., Wells, J.-A., Nemes L., 1996, Comparison of the bionic, fractal and holonic manufacturing system concepts, Int. Journal Computer Integrated Manufacturing, 9/3:217-226.
8. Zubair, M.-M., 1996, A flexible approach to (re)configure Flexible Manufacturing Cells. European, Journal of Operational Research 1996; 95/3:566-576.
9. Montreuil, B., Laforge, A., 1992, Dynamic Layout Design Given a Scenario Tree of Probable Futures, European Journal of Operational Research, 63:271-286.
10. Chen, M., 1998, A Mathematical Programming Model for System Reconfiguration in a Dynamic Cellular Manufacturing Environment. Annals of Operations Research 77/1:109-128.
11. Ueda, K., Vaario, J., Ohkura, K., 1997, Modelling of Biological Manufacturing Systems for Dynamic Reconfiguration, Annals of the CIRP, 46/1:527-540.
12. Ueda, K., 2000, Emergent Synthesis, In: Ueda K, editor. Artificial Intelligence in Engineering, 15/4:319-320.
13. Cariani, P., 1991, Emergence and Artificial Life. In: Langton CG, Taylor C, Farmer JD, Rasmussen S, editors, Artificial Life II, Addison-Wesley:767-775.
14. Lengton, C., 1989, Artificial Life, In: Lengton CG, editor. Artificial Life. Addison-Wesley, 1989. p. 1-48.
15. Car, Z., Hatono, I., Ueda, K., 2002, Reconfiguration of Manufacturing Systems based on Virtual BMS. The 35th CIRP International Seminar on Manufacturing Systems, 15-24.
16. Pinedo, M., 1995, Scheduling: Theory, Algorithms and Systems. Prentice Hall, New Jersey.
17. Brucker, P., 1998, Scheduling algorithms. Springer, Berlin, Germany.
18. Hwang, Y., Ahuja, N., 1992, A Potential Field Approach to Path Planning, IEEE Transactions on Robotics and Automation, Vol. 8:23-32.
19. Bazargan-Lari, M., 1999, Case Study: layout design in cellular manufacturing, European Journal of Operational Research, Vol. 112:258-2.
20. Tam, K.-Y., Li, S.-G., 1991, A hierarchical approach to the facility layout problem, International Journal of Production Research, Vol. 29/1, pp. 165-184.

DESIGN OF A PRODUCT CONFIGURATION SYSTEM BY THE OBJECT-ORIENTED APPROACH

D. Antonelli[1], N. Pasquino[2], A. Villa[1]

[1] Department of Production System and Economics, Politecnico di Torino, Italy
[2] University of Naples, Italy

KEYWORDS: Mass Customization, Product Configuration, Object-Oriented.

ABSTRACT. Mass customization can be approached from many points of view. When the focus is centered on the order acquisition and on the fulfillment process, the recourse to a computer support system, called Product Configurator, is widely adopted. To use it as a support to the design and the development of new products is a challenging application, seldom implemented in current industrial systems. Here, concepts taken from object programming are applied: namely the hereditariness of the class attributes by the children classes. The aim is to implement a hierarchic approach to the customization of the product with the advantage of a thorough applicability of postponed operations along the entire supply chain. In the work the conceptual framework and the application issues of the hierarchic Product Configurator are detailed. In order to evaluate the advantages (simplicity, flexibility and quickness) and the drawbacks (explosion of varieties and difficult stock minimization) of this methodology, it has been applied to case studies taken from production.

1 INTRODUCTION

In the case of order acquisition and fulfillment process, the best way to allow the customer to customize the product in presence of an high variety of product features is to have recourse to a software support system called "product configuration system" (or Product Configurator for sake of brevity). Presently, Product Configurator is largely applied in the computer, automotive and clothes companies to help the customer to configure a personalized version of the product and conversely to ensure a fast, complete and reliable transmission of the order to the production ([1]).

As a matter of fact, the growth of customizable features of the products forced a far larger growth in the number and kind of product feature data which have to be transmitted from the customer to the company to execute the order. Therefore it is helpful to use the support of an automatic system dealing with the number and the variety of product features and their related data. Making reference to Figure 1, borrowed by [1] it is possible to make a distinction between the application of the Product Configurator to help the customer in selecting among the numerous product alternatives already existing in the company catalogue or the intervention as a support to help the technicians to design and develop a new product on the basis of the customer specifications. The second alternative is far more challenging and not yet implemented in current industrial systems. In the second case, the Product Configurator is used twice: first to allow the customer to know if the desired product variant is feasible and which is its estimated cost, second to fulfill the order by adding the new designed parts to the existing ones in order to generate the Bill of Materials (BOM).

Published in: E. Kuljanic (Ed.) *Advanced Manufacturing Systems and Technology*,
CISM Courses and Lectures No. 486, Springer Wien New York, 2005.

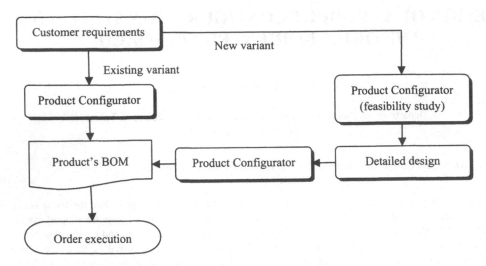

FIGURE 1. Product Configurator applications

Authors believe that there is a large number of industrial sectors whose manufacture could benefit from a mass customization strategy but many of them are somewhat constrained by the requirement of set of alternatives far too large to be organized in a product family. As an example we mention the furniture companies which make a large use of modularity principles, especially for the making of libraries or kitchens, but which cannot predict all the possible combinations of modules, therefore requiring a skilled sale support to help the customer in the process of customization.

Therefore, present study focuses on production cases where either the supply chain is severely branched or the product tree is very complex: the car manufacture, the building industry, the furniture industry. In these cases classic strategies, like modularity and postponement are insufficient to implement a true mass customization approach. The benefits of modularity are not completely deployed and the postponement conflicts with the complexity of production programming and scheduling, being applied only on parts that are processed at the final assembly level (e.g., in the houses: sanitary brands, pavement types; in the car: color, wheels, interior trim).

The study aims to produce the guidelines for the design of a new product in the framework of an existing product family, designing also the consequent multi-stage production process so that the number of product alternatives be maximum and the number of devoted branches and of differentiation points be minimum. The result is both a first attempt of process plan (to be detailed by technical personnel) and an evaluation of the economic effectiveness of adding the new product to the existing family. The underlying idea is to substitute the classic rigid product family concept or the more recent platform approach ([2-4]) with a dynamic product family class, where the family elements are not enumerated but only the rules to generate a new family element are defined.

The methodology proposed to accomplish the proposed task is a derivation of the object oriented approach, applied to the design of a product. By using this approach the automation of the product tree design can be greatly simplified. The aim is to implement a hierarchic approach to the customization of the product that, respect to the modular design should have the advantage to allow a thorough applicability of postponed operations along the entire supply chain.

In the work the conceptual framework and the application issues of a hierarchic Design for Mass Customization (DFMC) are detailed together with a mathematical model of the supply chain management oriented to MC. In order to evaluate the advantages (simplicity, flexibility and quickness) and the drawbacks (risk of explosion of unnecessary varieties and difficult stock minimization) of this methodology, it has been applied to case studies taken from production.

The concepts which are applied in present DFMC proposed methodology are: modularity, commonality, postponement and object oriented approach ([5-8]). All of these concept are already been proposed in order to obtain DFMC ([9-12]), but they have been never employed together.

The modular conception aims at designing product elements with functional over-capabilities (they are able to satisfy a redundant set of requirements), in such a way to be interchangeable: *modules*. The *commonality* is the generalized use of the same module inside many different products. Modularity depends on two product characteristics: similarity among physic and functional architectures, minimization of the interaction among the components. Some researches ([13-18]) have been already presented which propose methods aimed at building modular product families. In particular, Huang and Kusiak [13] propose a matrix decomposition of the product tree in order to highlights the interchangeable or independent elements. Jiao et Tseng [14] propose a method to develop an architecture of product families by applying three different points of view, functional, technical and physical, to the classification of product variant.

In the *object oriented* (OO) approach, an object is a bundle of related variables and methods. *Objects* are a model of the real-world objects, with a state and a behavior. A *class* is a blueprint or prototype that defines the variables and the methods common to all objects of a certain kind. Every object is obtained by *instancing* a class. Classes can be defined in terms of other classes. Each lower level class (subclass) inherits the state from the upper level class (superclass).

2 OUTLINE OF THE DFMC PROBLEM

At the starting point is the demand for a new product, whose degree of innovation is defined by means of one or both of the following cases:

1. the product requires the activation of new functions with respect to the set of function satisfied by present production;

2. the product requires the insertion / substitution of new components

The Product Configurator recognizes the request by translating it in terms of function or component trees (see chapter 3), receiving the input data from the customer and integrating from the product database of the company. It is assumed that the company already has its own production organized by families and that a component tree is linked to every product.

It is possible to outline three main issues, representing consecutive steps during the product defi-

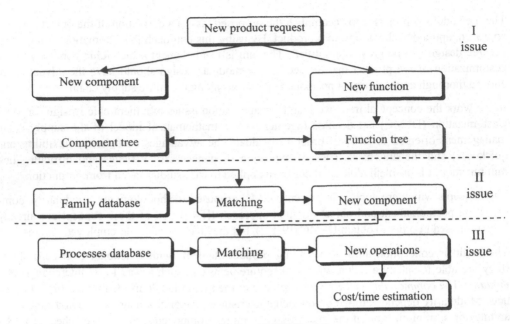

FIGURE 2. Scheme of DFMC Problem

nition and the feasibility study. They are summarized in Figure 2.

I issue: Selecting the component tree of the new product. The new functions and the new components are outlined and inserted in a component tree.

II issue: From the logical tree of the components, paying attention to the corresponding new functions, selecting the components that constitute the product tree (the normal tree and not the meta-tree).

III issue: Defining the new process planning. The product tree must be converted into a process planning and the production time and the related costs are estimated.

3   MODELLING AND USING THE PRODUCT FAMILY TREE

In order to apply the object oriented approach to our methodology, we consider a family of products with a still undefined number of components. Each product configuration is represented by the BOM, i.e. the product tree. Every item inside the BOM is an object defined by a number of geometric and functional attributes (states) and by the techniques used to produce/acquire/assembly it (behaviors). Firstly, the company is required to design the super-class that enclose all of the product family. This superclass is represented by a product meta-tree, that is a product tree with only the branch structure and the levels but without instancing the items corresponding to one specific product. The connections among different items on different levels are described by means of logic operators: AND, OR. Other operators are applicable in general, here the operators are reduced to the main two for sake of simplicity. The family tree is a meta-tree because it represent all the possible product trees of a product family by giving the connection rules. In other terms, the meta-tree is a graphic representation of the platform approach, a widely applied design techniques for the development of new product families ([12-14]). In Figure 3 a sample family tree is represented. On the same time the functional tree must be produced with a direct correspondence between function and component.

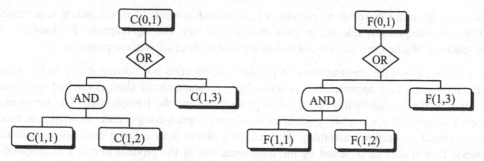

FIGURE 3. Example of the component tree and of the function tree

As far as now we did not define a single product but a set of functional requisites matched with corresponding physical components. The pairs of meta-trees give a thorough description of a family of products in terms of the functionalities which can be performed by different product instances and the contribution of every component to the product operations, as well as in terms of the assembly configuration of each product. A product family is described in terms of a function vector, i.e. the list of functions related respectively to the product family $F(0,1)$ and to each instance j on a given detail level I, i.e. $F(i,j)$. Therefore the function tree can be modeled by a connection matrix, i.e. an incidence matrix which specifies both the existence of a connection between two nodes in the tree and the type of connection. With reference to the example of Figure 3, a connection in the function tree can represent:

> The union of two complementary functions to give rise to a more complete one;
> The alternative between two functions, each one being used to generate more complex functions on an upper level.

TABLE 1. Connection matrix of function tree

From \ To	F(0,1)	F(1,1)	F(1,2)	F(1,3)
F(0,1)				
F(1,1)	1		AND	OR
F(1,2)	1			OR
F(1,3)	1			

The connection matrix MF is generated by the application of the following rules:

> MF{F(i,j),F(n,m)} = 1 if F(i,j) is a sub-item of F(n,m);
> MF{F(i,j),F(n,m)} = AND if F(i,j) has to be joined with F(n,m);
> MF{F(i,j),F(n,m)} = AND if F(i,j) has to be joined with F(n,m);

The connection matrix is non symmetrical and it is made of a lower triangular matrix with the product structure and a upper triangular matrix with the connections, as can be seen in Table 1.

Now, let assume the customer to propose a new variant in terms of new function requirements and/or new components, selected in the component library. Having reference to chapter 2, the configuration of a new product variant is decomposed in three subsequent problems.

The first problem is the construction of a product and function tree corresponding to the desired product variant. In our approach this is equivalent to instancing the class of product trees corresponding to the product family, i.e. the component tree and the function tree. The components selected should follow some MC guidelines: re-use of common components, minimal cost, modular, postponed production / assembly. The second problem is to define a draft process planning method. This should be obtained by the transformation of the product tree in a corresponding manufacturing sequence. The method is used to determine the lead time required to produce / assembly every component and the corresponding manufacturing costs. The third problem is to define a schedule of the production process of the new product variant in order to estimate its production time and the production cost.

The solution of the first problem is obtained by a thorough application of the OO approach:
> A mapping of the functions of the new project onto the functions of the product family individuating the functions not anticipated in the existing tree.
> The masking of existing rows and columns in the connection matrix, corresponding to function non required by the customer.
> The mapping of the new product matrix onto the reduced function matrix, introducing the new connections corresponding to the new functions and the mirroring of the function matrix onto the corresponding component matrix.

The first step is made possible by the rule of inheritance inside a object class. As a matter of fact, if the customer ask for a new function which substitute an existing one, all the functions which are in a lower hierarchical level will be inherited by the new function. Its position in the hierarchy will determine the connections with the other functions on the same level and the relationships of subordination with the functions on different levels. In the second problem, the structure of the product tree is directly converted in a manufacturing tree by applying concepts of "pattern recognition". The third problem can be solved by applying a planning algorithm to the process plan embedded in the product tree, by following the OO paradigm.

For every operation the production lead times and costs are given, therefore it is possible to build a Gantt diagram of the production sequence in order to recognize the release time of every operation and to define the completion time of the entire product. Similar approach will lead to the formation of the productive cost.

4   APPLICATION OF THE METHOD

The case study considered is an electric mixer to be employed in the preparation of different kinds of foods with different way of mixing together the ingredients of the recipe. The high level component tree is provided in Figure 4, where a circle substitutes the AND a rhombus the OR.

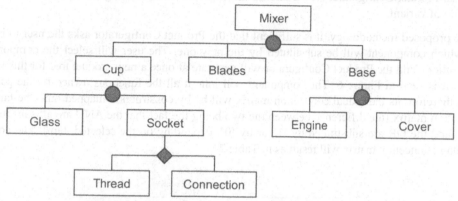

FIGURE 4. Example of the component tree of an 'electric mixer'

The mixer is obtained by assembling together all of the individual items with the choice of the socket type that is produced in two variants: thread for slow but secure connection, plug for a fast connect / disconnect operations at the sake of a less secure connection. The tree is limited to the higher levels but can be further detailed at will.

Similarly the function tree is generated with a strict correspondence between component and function (Figure 5). Often a component has different functions, let say that an object owns a whole set of functions. As an example, the cover shares the esthetical function, the electric insulation of the user, the containing of the engine, the handling for the whole device, etc. For sake of shortness, every set of functions has been condensed in one. Furthermore, it must be remembered that functions are inherited among objects along the relationship chain. For example, the cover inherit the supporting function from its parent, the base of the mixer.

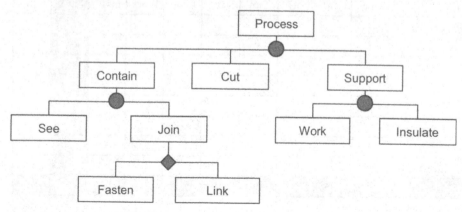

FIGURE 5. The function tree for the product 'mixer'

Now let's suppose that our customer has a unsatisfied need of a universal adapter able to allow using different types of cups on the same mixer. A universal adapter is a common component, disposable in the company database of items, because it is used in a number of other electric devices, but this customization variant has not been provided for the present product family. Usually a normal Product Configurator would discard the request by not allowing the customer to ask for this kind of variant.

By the proposed methodology, it is sufficient that the Product Configurator asks the user to indicate which component will be substituted by the new one. The user will select the component "connection" and the Product Configurator will generate at once a new product tree for this variant, as it is seen in Figure 6. The component will inherit all the functions owned by the parent items, therefore its individual connection matrix will be by construction mapped onto the family connection matrix (the difference between the two being the fact that the AND are all substituted by '1' and the OR are substituted by '1' or by '0', depending on the selected item). The corresponding connection matrix will result as in Table 2.

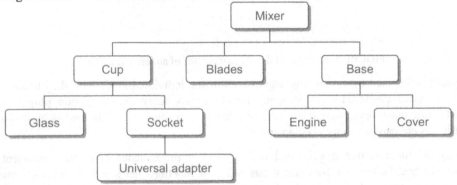

FIGURE 6. Product tree of the new variant, obtained by instancing the component tree

TABLE 2. Variant connection matrix mapped onto family matrix

From \ To	Proc	Cont	Cut	Sup	See	Join	Work	Ins	Fast	Link
Process	░									
Contain	1	░	A	A						
Cut	1		░	A						
Support	1			░						
See		1			░	A				
Join		1				░				
Work				1			░	A		
Insulate				1				░		
Fasten										
Std. link						1				░

Now a significant use is made of the OO property which allows the combination of data and methods. In the present application, methods are represented by the productive functions associated to each component. Therefore the passage from the product tree to the process plan is straightforward.

The product tree become a process tree, i.e. a tree constituted by production processes in the different nodes. Every process is represented in terms of lead time. A procedure of operational research will allow to recognize the precedence in the execution of the processes. The procedure results in the scheduling of the processes on a Gantt diagram, as depicted in Figure 7. The diagram has been generated by using the bottom-up sequence in the building phase. As an example, the assembly phase of engine and cover can span all the interval until the end of the V period, because the only precedence is the final assembly of the mixer, but it has been executed at the very start of the production process. Different strategies would require information about factory capacities, surely unavailable in the design stage. By assigning the processing program to appropriate cost drivers it is possible to have an estimate of the costs. This is a common application of aggregate production planning procedures.

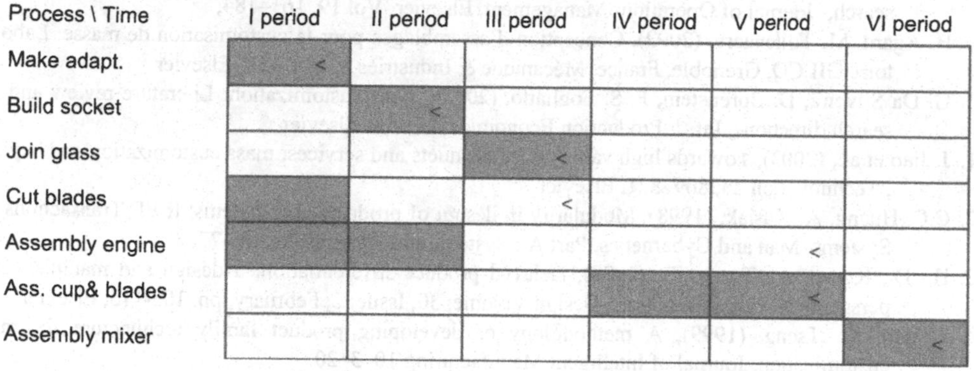

Process \ Time	I period	II period	III period	IV period	V period	VI period
Make adapt.	<					
Build socket		<				
Join glass			<			
Cut blades			<			
Assembly engine					<	
Ass. cup& blades				<		
Assembly mixer						<

FIGURE 7. Gantt diagram for the production programming

5 CONCLUSIONS

The term "mass customization" is interpreted usually as a flexible way of programming the production and assembly phase, postponing some activities in order to produce the product on the customer demand. The importance of the initial phase of order acquisition is often neglected. In many industrial sectors, the difficulty in acquiring customized orders is due to the necessity of a strong and expensive interaction with the selling personnel. That is why many companies, when selling a product on internet, are obliged to stiffen the order acquisition process by reducing the number of eligible alternatives in the product specifics. The Product Configurator presented here extends the platform concept to the order acquisition and integrates it with OO methods. The result is the possibility of acquisition of customized orders without the need of predicting all the possible variants in the product family.

REFERENCES

1. Forza C., Salvador F., (2002), Managing for variety in the order acquisition and fulfilment process: The contribution of product configuration systems, Int. J. Production Economics 76, 87–98, Elsevier
2. Meyer, Marc H.; Dalal, Dhaval, (2002), Managing platform architectures and manufacturing processes for nonassembled products, The Journal of Product Innovation Management Volume: 19, Issue: 4, July, pp. 277-293, Elsevier
3. Farrell, Ronald S.; Simpson, Timothy W., (2003), Product platform design to improve commonality in custom products, Journal of Intelligent Manufacturing 14, Issue: 6, December, pp. 541-556
4. Muffatto, Moreno, (1999), Introducing a platform strategy in product development, International Journal of Production Economics Volume: 60-61, April 20, pp. 145-153, Elsevier
5. Juliana Hsuan, (1999), Impacts of supplier-buyer relationships on modularization in new product development, European Journal of Purchasing & Supply Management 5, 197-209, Pergamon Press
6. Markham T., Frohlich A., Westbrook R., (2001), Arcs of integration: an international study of supply chain strategies, Journal of Operations Management 19, 185–200, Elsevier
7. Steve Brown, (2001), Managing process technology — further empirical evidence from manufacturing plants, Technovation 21, 467–478, Elsevier
8. Van Hoeka, R.I., (2001), The rediscovery of postponement a literature review and directions for research, Journal of Operations Management, Elsevier, Vol 19, 161–184,
9. B. Agard, M. Tollenaere, (2002), Conception d'assemblages pour la customisation de masse, Laboratoire GILCO, Grenoble, France, Mécanique & Industries 3, 113–119, Elsevier
10. G. Da Silveira, D. Borenstein, F. S. Fogliatto, (2001), Mass customization: Literature review and research directions, Int. J. Production Economics 72 1-13, Elsevier
11. J. Jiao et al., (2003), Towards high value-added products and services: mass customization and beyond , Technovation 23, 809–821, Elsevier
12. C.C. Huang, A. Kusiak, (1998), Modularity in design of products and systems, IEEE Transactions on Systems, Man and Cybernetics, Part A : Systems and Humans 28, 66–77.
13. He D.; Kusiak A., Tseng, T., (1998), Delayed product differentiation: a design and manufacturing perspective , Computer-Aided Design Volume: 30, Issue: 2, February, pp. 105-113, Elsevier
14. J. Jiao, M. Tseng, (1999), A methodology of developing product family architecture for mass customization, Journal of Intelligent Manufacturing 10, 3–20.
15. Salvador, F.; Forza, C.; Rungtusanatham, M., (2001), Modularity, product variety, production volume, and component sourcing: theorizing beyond generic prescriptions, Journal of Operations Management, pp. 549-575, Elsevier
16. Hsu, Hsi-Mei; Wang, Wen-Pai, (2004), Dynamic programming for delayed product differentiation, European Journal of Operational Research, pp. 183-193, Elsevier
17. Dobrescu, Gabriel; Reich, Yoram, (2003), Progressive sharing of modules among product variants, Computer-Aided Design Volume: 35, Issue: 9, August, pp. 791-806
18. Ricardo Ernst, Bardia Kamrad, (2000), Evaluation of supply chain structures through modularization and postponement, European Journal of Operational Research 124, 495-510, Elsevier
19. Christiansen, Kåre; Vesterager, Johan, (1999), Engineering Bereitschaft as an enabler for concurrent engineering, Robotics and Computer-Integrated Manufacturing Volume: 15, Issue: 6, December, pp. 453-461
20. Jianxin Jiao, Qinhai Ma, Mitchell M. Tseng, (2003), Towards high value-added products and services: mass customization and beyond, Technovation 23, 809–821

OPERATIONAL SOLUTIONS FOR SHORT-TERM PRODUCTION PLANNING AND CONTROL IN A MAKE-TO-ORDER COMPANY

M. De Monte[1], E. Padoano[1], D. Pozzetto[1]

[1] Department of Mechanical Engineering, University of Trieste, Italy

KEYWORDS: Production Control, Load-Oriented Manufacturing Control, Simulation.

ABSTRACT. The purpose of this paper is to propose a method for short-term production planning, focused on control and regulation of work-in-process in a make-to-order production system. The possible improvements in terms of performance indices (production lead time and reliability of delivery dates) of the job shop are pointed out. The solution implied a revision of the production process and the application of input-output control methods for operations management at job shop level. The production stages that were independent from customers' orders were decoupled from activities that could be performed before an order entry. For workload control the load-oriented manufacturing control (LOMC) method was applied to a manufacturing area that produces order-specified items. The experimental phase, on the one hand, gave insights on improvements that can be obtained with the proposed method and, on the other hand, made it possible to define dispatching rules that were effective for work order management in the case study.

1 INTRODUCTION. PRODUCTION REALITY UNDER STUDY

In make-to-order (MTO) production systems a company produces a product only after having received the order from the customer. Generally associated with this type of set-up are production systems of limited repetitiveness such as the job shop, where each order is characterized by a technological cycle that requires the use of several different machines, and with respect to which the processing sequence may be different from order to order. In these systems external dynamics are present which often imply variations to the work in process (WIP). This may cause significant processing problems as it is directly or indirectly linked to almost all of the main features of the system: average flow time, machine utilization and delivery date.

In this paper these problems are analyzed with reference to the Special department of a production plant which produces food preparation and distribution equipment and ventilation equipment for foodservice systems. All of the equipment is produced according to the MTO system. An important part in the personalization of the solution and job planning is played by the product design activities which always precede production.

Once the order has completed the planning phase, production involves the following work phases: storing of the raw materials, shearing, punching, bending, assembly and packaging. Within the system manufacturing occurs for individual pieces or at best for small batches; when all of the components of the equipment have completed the processing phase, they are sent by trolley to the next phase. Between one phase and the next interoperational buffers are present to

Published in: E. Kuljanic (Ed.) *Advanced Manufacturing Systems and Technology*,
CISM Courses and Lectures No. 486, Springer Wien New York, 2005.

receive the queued orders waiting to be processed. Semi-manufactured products, components and accessories are acquired at certified external operators.

2   THE MANUFACTURING OF ORDERS IN THE SPECIAL DEPARTMENT

The manufacturing of orders within the special department is carried out with the help of an ad hoc computer program. With the exclusion of some of the secondary modules the program can be said to mainly involve production planning and the advancement of the orders in the department.

After an initial phase of receiving confirmation of the order from the sales department, the second phase involves planning the orders. Thus it can be decided which orders should be produced internally and which externally, and delivery times requested by the sales department can be confirmed or modified. After product design planning, which makes use of the software's control system, confirmation that the design phase has been carried out is given and the order is advanced to the state of design completion.

External dynamics can lead to changes in the planning of the department. A frequent case is the work practices of the sales representatives, who generally place orders and assign planning priorities while keeping waiting times under control, with the aim of improving customer service. In cases where waiting times are considered to be excessive, the sales reps also insert potential orders. In this manner some incomplete orders manage to enter into the planning system of special orders – the current manufacturing system in fact includes all orders in the planning process. This creates confusion in the system in that urgent orders are slowed down. After the planning phase, the planned order is ready to be produced and the software indicates the date on which the production cycle is scheduled to begin.

During the production phase many orders are delayed. In fact, very few orders with top priority meet the deadlines set during the initial planning phase. The aim of production planning therefore is reduced to making up the delays accumulated in the design phase, thus limiting as much as possible the delays in delivery times of the orders already present in the production cycle. In order to do so, the foreman of the special department reprograms order production using subjective criteria.

All of the variations made during the planning, product design and production phases are destined to continually modify the order delivery dates until the execution of the order. This leads, on the one hand, to difficulties in determining waiting times, and on the other, a push towards an urgency oriented method of production [1], aimed at finalizing the most frequently requested order. In this scenario production is not carried out on the basis of the real needs dictated by appropriate planning, but rather by following the urgencies which from time to time crop up.

So as to provide an overall assessment of the situation, the performance of the system was determined through the following criteria [2]:

- average flow time of an order in the production system (Flow - days);
- total number of orders put out by the system late with respect to the delivery date agreed upon with the customer (Nlate);
- average lateness (Alat), defined as the relationship between the sum of delays (in days) and the total number of orders;

- delivery date reliability (Ddr), defined as the relationship between the number of orders put out by the system after the delivery date and the total number of orders met.

The results obtained from a significant and representative data sample enabled the typical weeks in the context analyzed to be identified. The following average values of the performance indices for the internal production system were obtained: Flow = 31,6 (working days), Nlate = 11 (orders), Alat = 8,7 (working days), Ddr = 14%.

The manufacturing system developed in the special department for production planning does not allow general performance in line with management expectations to be obtained. As a result, a study was undertaken to develop a production planning system capable of providing improved general performance and above all delivery date reliability values higher than the current ones.

3 THE PROPOSED SOLUTION

On the basis of the problems and the aims of the special department, a system of production planning based on load control, and in particular on the Bechte model (load-oriented manufacturing control – LOMC) was adopted [3] [4]. The choice of a load control methodology is linked to the fact that this approach, when accurately planned and regulated, enables a significant improvement in the flow time of the production orders and greater compliance with delivery dates to be obtained [5] [6]. The choice of using the Bechte method was influenced by the results of earlier studies [7] which showed that Bechte's "probabilistic" model, although inferior to other models in deterministic environments, is the most appropriate in highly dynamic environments (such as MTO systems: unavailability of machines, variability in processing times, etc.), owing to its greater robustness.

It should be noted that the procedure for implementing LOMC requires at least one year for the implementation of each step. In order to offer management some useful data in support of this course of action, a pilot study of the method was carried out. For the pilot study to be adequate and measure the possible margins for improving the general performance of the department (and in particular delivery date reliability), a simulation model was adopted, thus overcoming the difficulties of directly assessing those effects in the field. A number of simplifications which are summarized below were introduced into the model, which nonetheless maintains all of the main characteristics of the actual system.

3.1 NUMBER OF WORK CENTERS

Although seven work centers are involved in the production of the special orders, the order passes through three main phases: product design, processing (punching, shearing and bending programs) and assembly (assembly and packaging). In the production model introduced, these three main phases were made to coincide with three "macro" work centers, each of which seen as individual machines, thus reducing the number of work centers from seven to three. In the production system being studied two critical stages were identified (coinciding with the bending and assembly work centers) and the schematization "individual machine" is valid in all the situations in which one of the stages is more critical (with regard to production capacity): the other stages cause no variation in the performance parameters of the shop floor taken as a whole.

An analysis of the special items most frequently processed at each work center revealed that the most critical stage of the processing phase is bending. Indeed, with only a couple of exceptions, bending accounts for 60% of the time required for retooling in the special department. Similarly, the assembly phase is critical with respect to the packaging phase. In most cases, given is bulky dimensions the equipment tends to be packaged in the assembly islands where they take up assembly resources.

3.2 ALLOCATION OF THE JOBS TO THE MACHINES

In some work centers several machines work in parallel (design, bending and assembly); in this case the buffer function is entrusted to queues dedicated to individual machines. In this study, however, a different structure was adopted (one already used in previous works [8]) to facilitate the successive phase of implementing the LOMC system: the machine queues were suppressed and for each work center a "phase" queue was assumed which functioned as a collection point for the jobs awaiting processing. This solution was considered to be in line with the real-life situation (after numerous observations carried out in the field) in that it allows the machines to "choose", on the basis of an appropriate rule, the type of job among those awaiting processing, without performing pre-defined allocations.

The adoption of the "phase" queues, rather than queues dedicated to individual machines, involves two distinct cases:

- "job chooses machine";
- "machine chooses job".

In the first case when processing of an order is terminated on one machine, it enters the queue of the following work center downstream. If no machine at the new center is free, the job chosen by dispatching priority remains in the queue, whereas if one or more machines are in waiting, the job approaches the machine identified by the priority rule adopted. This rule, which is defined following observations of real-life production, associates a weight to each machine and, more precisely, a greater weight and therefore a higher priority to the most efficient machine.

In the second case when a machine has an insufficient workload available to cover the daily work period, it becomes available for accepting the jobs chosen by the dispatching priority until it reaches full capacity, whereas if there are no jobs in the queue, after having processed the current job it places itself on standby. In order to make these criteria operative, two algorithms were developed which are added to the general algorithm created for reproducing the mechanism of job advancement on the shop floor, and the algorithm handling the jobs at the individual machines (Figures 1 and 2).

The advancement of the simulation occurs by events and the cyclically analyzed phases are assembly, processing, design, with respect to which dispatching is performed, and the overall queued jobs with respect to which the loading is performed. During each workday – the reference time for describing the real progress of the work in the department – the simulator performs these phases one at a time and in the order indicated.

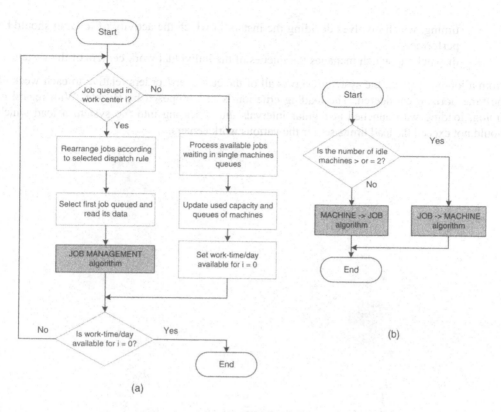

FIGURE 1. General (a) and machine management (b) algorithms

A series of observations performed in the department identified the resources effectively available. On that basis, in the model a number of machines operating in parallel at each work center were considered:

- 2 machines in the product design work center (3 designers);
- 2 machines in the processing work center (2 press-brake operators);
- 4 machines in the assembly work center (4 assembly workers).

In order to find the daily work hours to be associated with each machine a precise study was performed to determine the productivity of each special production worker for a sufficiently significant period of time which enabled the average daily design times per unit, average daily assembly times per unit and the use of machines present in the processing phase of special orders.

The planning and control phase of the analyzed production system can be broken down in the following way:

- loading, which consists in determining which of all the jobs to be produced should enter into the system in a given moment and in what order;

- timing, which involves deciding the instant in which the activity of loading should be performed;
- dispatching, which manages the queues of the individual work centers of the system.

From a job-entry phase the system receives all of the customers' orders relative to each week of the time horizon considered. The loading criterion is that proposed by Bechte. With regard to timing, loading was launched at regular intervals thus releasing into the system a load which would not exceed the load limits set for the various work centers.

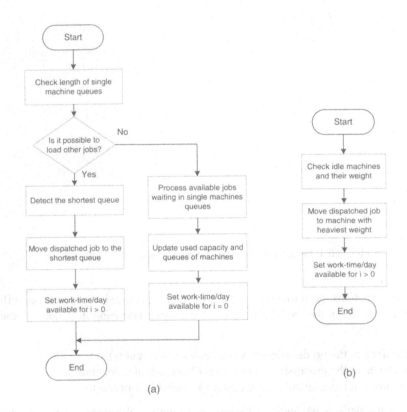

FIGURE 2. Job chooses machine (a) and machine chooses job (b) algorithms

The Bechte model proposes dynamically maintaining the real lead times at the planned levels, limiting the WIP and at the same time obtaining satisfactory machine utilization coefficients. The model is characterized by the presence of load limits for the phases, which constitute the load threshold value assignable to each phase. At fixed time intervals the loads already present in the system are calculated and the possibility of entering new jobs into the system itself is assessed.

The first job in the job pool waiting to enter the system is considered and the additions to the load that it causes in the individual phases is calculated: if for none of the phases the sum of the load due to the job and of the loads already in the system is greater than the load limits chosen, the order enters the system and the state of the loads is updated. In a similar fashion all the jobs in the job pool are analyzed in sequence. The loading procedure is completed when no other job can enter without exceeding one or more of the phase load limits.

The calculation of the additions to the load in each phase, expressed in terms of total hours admissible per work center, is determined by the relationship [8]:

$$C(m,i) = T(m,i) \cdot \prod_{k \in S} \left(\frac{100}{LL(k)} \right)$$

dove:
$C(m,i)$ = addition to load of the m job in the i phase;
$T(m,i)$ = processing time of the m job in the i phase;
$LL(k)$ = load limit of the k phase, expressed as a percentage of the production capacity of the phase in the period considered;
S = total number of phases which the m job needs to pass through before arriving at the i phase.

The criterion devised takes into account, therefore, the position of the job within the system, penalizing the workloads in relation to the distance of the job from the phase considered (as the distance increases the probability of the job reaching that phase before the end of the current period decreases). With regard to dispatching two rules were implemented:

1. FCFS (First Come First Served), according to which the job which has been the longest time in the queue is chosen;
2. SPT (Shortest Processing Time), according to which the job with the shortest processing time in the current phase is chosen. This rule is capable of minimizing the average flow time and the average lateness. It has the advantage of suffering little from the randomness of processing times, although it has the disadvantage of excessively delaying the jobs with longer processing times.

4   THE PILOT STUDY

After having set the planning period to a week, all of the real data were gathered relative to the incoming orders of the special department in relation to the sample week, identifying therefore for each confirmed order the product design, processing and assembly times. After that, with the help of the workers the initial state of the system at the beginning of the week was identified, which corresponds to the state of the various "phase" queues before each work center, as well as those before each individual machine.

In carrying out the simulations the following hypotheses were introduced:

- a safety capacity of 5% was taken into consideration. This reduction in capacity was introduced to compensate for possible errors in the assessment of the model introduced for describing the department;

- the same time criterion was used, so that the state of the system was identified at equal and constant intervals for the purposes of activating the loading procedure;
- the specific parameters of the loading criterion were set following calibration tests.

With the current production planning and control criteria the following values for the performance indices were obtained (compare Tab. 1): Flow = 30,4 (workdays), Nlate = 9 (orders), Alat = 9,4 (workdays), Ddr = 10%.

The Nlate index in this case was calculated on the basis of the confirmed delivery time of 21 workdays in the month.

Table 1 summarizes the results obtained from the simulation of the LOMC system with FCFS and SPT dispatching priorities and the following system parameters obtained from calibration:
- planning period of 1 week;
- control frequency of releases equivalent to one workday;
- load limit of the product design work center equivalent to 200%;
- load limit of the processing work center equivalent to 200%;
- load limit of the assembly work center equivalent to 250%.

The performance indices were:
- for FCFS
- Flow = 25,7 (workdays), Nlate = 3 (orders), Alat = 1,7 (workdays), Ddr = 70%
- per SPT
- Flow = 14,4 (workdays), Nlate = 0 (orders), Alat = 0 (workdays), Ddr = 100%

TABLE 1. Summary of the results of the simulation

Order confirmed	Actual		LOMC and FCFS		LOMC and SPT	
	Lead time (days)	Lateness (days)	Lead time (days)	Lateness (days)	Lead time (days)	Lateness (days)
1	27	6	25	-1	16	-10
2	37	16	24	-2	13	-13
3	11	-10	23	-3	12	-14
4	26	5	28	2	20	- 6
5	35	14	25	-1	13	-13
6	40	19	25	-1	12	-14
7	30	9	26	0	14	-12
8	30	9	26	0	14	-12
9	24	3	27	1	14	-12
10	25	4	28	2	16	-10

A comparative analysis of the results obtained in the various simulations and under actual operations demonstrates the LOMC system is superior to the system currently in use with regard to all of the indicators examined. In particular, the LOMC system with the FCFS dispatching priority enables a reduction in the average flow time of the system of 15.5% (from 30.4 to 25.7

workdays); with regard to the LOMC system with the SPT dispatching priority a 52.6% reduction in average flow time of the system (from 30.4 to 14.4 workdays) is achieved.

The LOMC system also proved superior in terms of delivery date reliability, registering a significant increase in this value: the LOMC system (FCFS priority) achieved a delivery date reliability of 70%, compared with the 10% figure of the current system. It should be borne in mind, however, that in calculating the delivery date reliability value, given the impossibility of agreeing on delivery dates with the customer, the total average lead time of the system at the beginning of the planning period was used as a comparison term in the calculation of the Nlate index, as this was the only significant and measurable value available. In fact this value should provide an indication of the state of the current load of the production system. However, in the case of the LOMC system with SPT priority the figure indicating the total average lead times of the system is not very significant: the average flow time of the system (14 workdays) diverges significantly from the total average lead times in the observation period (from 26 to 31 workdays). In contrast, in the case of the LOMC system with FCFS priority there was a significant correlation between average flow time of the system (26 workdays) and the total average lead times of the system in the observation period (from 25 to 31 workdays).

5 CONCLUSIONS

The results obtained demonstrate experimentally that the performance of the LOMC system is superior to the manufacturing system currently in use. Also noted in the simulations, however, was the presence of the undesired phenomenon of increased idle time at work centers owing to a shortage of jobs to be processed at a work center. However, this was expected, since a previous study [9] demonstrated that in the job shop systems, where there is a dominant flow in operations (as is the case in the flow shop), the performance of the system diminishes with instances of a backlog of jobs caused by certain work centers forming a bottleneck.

The study demonstrates how the FCFS dispatching priority achieves an overall performance inferior to that obtained with SPT. This was to be expected, as was the fact that the SPT system would delay the delivery of orders with greater processing times.

In summary, the results of the pilot study show that:
1. the LOMC system is superior to the system in current use with regard to all of the performance indicators considered;
2. SPT dispatching priority guarantees improved performance in terms of the average flow time of the system, although it significantly delays the delivery of orders with higher processing times;
3. FCFS dispatching priority guarantees a positive balance between the workloads of the various work centers of the system, with the variations in average lead times for each center in the order of three workdays;
4. with FCFS priority significant results can be obtained in terms of planning (and therefore delivery date reliability); this is because the FCFS priority is the best at applying the Bechte concept to the case study.

REFERENCES

1. Stalk G., Hout T. M., (1990), Competing against time: how time-based competition is reshaping global markets, Free Press, New York.
2. Kume H., (1987), Statistical methods for quality improvement, Gilmour Drummond Publishing, UK.
3. Bechte W., (1988), Theory and Practice of Load-Oriented Manufacturing Control, International Journal of Production Research, Vol. 26/3, 375-395.
4. Bechte W., (1994), Load-Oriented Manufacturing Control, Just-in-Time production for Job Shop, Production Planning & Control, Vol. 5/3, 292-307.
5. Wiendahl H.P., (1995), Load-Oriented Manufacturing Control, Spring Verlag, Berlin Heidelberg.
6. Land M.J., Gaalman G.J.C., (1996), Workload Control Concepts in Job Shop: A Critical Assessment, International Journal of Production Economics, Vol. 46-47, 535-548.
7. Bergamaschi D., Cigolini R., Perona M., Portioli A., (1997), Order review and release strategies in a job shop environment: a review and a classification, International Journal of Production Research, Vol. 35/2, 339-420.
8. Perona M., Pizzoli M., Pozzetti A., (1993), Scheduling basato sul controllo del carico: una proposta per sistemi di assemblaggio di schede elettroniche, XX Convegno Nazionali di Impiantistica industriale, Napoli, 163-187.
9. Oosterman B., Land M.J., Gaalman G.J.C., (2000), The Influence of Shop Characteristics on Workload Control, International Journal of Production Economics, Vol. 68/1, 107-119.

IMPROVING PROCESS PLANNING
THROUGH SEQUENCING THE OPERATIONS

N. Volarevic[1], P. Cosic[1]

[1] Department of Industrial Engineering, Faculty of Mechanical Engineering and Naval Architecture,
University of Zagreb, Croatia

KEYWORDS: Process Planning, Sequencing the Operations, Applied Matrix Procedure.

ABSTRACT. The intention of this paper is to give some methodical approach in process planning. Within this it covers the problem of defining the sequence of operations. Quality of the product, production time and production cost determine sequence of operations. The purpose is to analyse their influences and how to apply them in making decisions. Matrix of anteriorities sometimes generates multiple solutions. Understanding these influences could lead to solution of these problems.

1   INTRODUCTION

One of the purposes of our research is to promote such systematic thinking among students in the field of process planning. The solution has to systematize process planning making it easier to analyze machining processes needed to produce the part considering costs, time of production and functional demands. It is also a step forward to define criteria that could be integrated in intelligent process planning.

From a study on process planners, it appears that they rely on experience and intuition. As different process planners have different experience, it is no wonder that for the same part, different process planners will design different processes. The experienced process planner usually makes decisions based on comprehensive data without breaking it down to individual parameters. There is no time to analyse the problem. Understanding and a methodical thinking flow will improve the performance of the process planner. Good interpretation of the part drawing includes mainly dimensions and tolerances, geometric tolerances, surface roughness, material type, blank size, number of parts in a batch, etc. Logical approach in a process planning, as very complicated, multilevel and comprehensive approach of generating alternative process plans could be divided to:

a) selection of primary processes,

b) selection of machine tools and tools,

c) sequencing the operations according to precedence relationships.

Intention is to discuss part of process planning regarding sequencing the operations according to precedence relationships.

Process planning can be defined by a sequence of activities. Managing of a company [1] calls for many economic decisions such as the economics of manufacturing, a certain product,

Published in: E. Kuljanic (Ed.) *Advanced Manufacturing Systems and Technology*,
CISM Courses and Lectures No. 486, Springer Wien New York, 2005.

capital investment and cash flow needs, type and number of machines needed, number of employees, due date of delivery, layout etc. A decision implementation has to be based on intuition, on partially estimated data or accurate data. Process planning has to provide the background for economic evaluation. For example, when a new product is introduced in the company, the finance department wants to know its manufacturing costs. Today, two methods are used to try to overcome this dilemma and to shorten the process generation time. One is to use a computer, i.e. CAPP. Hovewer, although research and development efforts in the field of CAPP over the past two decades have resulted in numerous experimental CAPP systems, they have had no significant effect in manufacturing planning practice. Another method is to constantly improve the process planner's intuition, knowledge and expertise. [2]

2   QUALITY, PRODUCTION TIME AND COST

Process planning could be presented like a balance between producing a part meets functional requirements, minimal production time and minimal production cost. Relation between part manufacturing, production time and cost certainly exists but is not always very clear. All three have same inputs and in most cases they are machine type, selection of tool, how a part is fixed during process, machining parameters and few others. All this inputs affect this criterion. It was presumed that achieving a functional part should be the most important criterion. If the process is not set to be able to produce a part that meets all geometric and dimensional constrains, surface roughness, surface hardness and many other requirements then there is no sense to talk about cost and production time. Defining the sequence of operations is led primarily by this criterion, to make the process plan that is able to produce a part of required quality.

Often there is a need for additional criterion. A part feature or a number of features could be made according to different process plans. All of this process plans give a part of sufficient quality. To be able to evaluate this process plans it is necessary to take into consideration production time and production costs. In the paper mentioned principal was shown in making decisions related to sequence of operations. It was presumed that loss of time caused by machine change or tool change or fixture change affects price of part production. It also affects quality but within allowed limits.

3   IMPORTANCE OF SEQUENCING THE OPERATIONS

The operations defined in process planning have to be put in certain order according to precedence relationships based on technical or economical constraints. Operations sequencing depends on many influences like:

  a)   nature of the material,

  b)   general shape of the part,

  c)   required level of accuracy,

  d)   size of the raw material,

  e)   number of parts in the batch,

f) possible choice of machine tools, etc.

One of the possible approaches is to classify different categories in the following way:

a) dimensional precedence – dimensions with a datum as anteriority

b) geometric precedence – geometric tolerances with data references as anteriorities,

c) datum precedence – case to the choice of a datum,

d) technological precedence – case of a technological constraint,

e) economic precedence – economic constraints that reduce production costs and wear or breakage of costly tools.

To achieve the nominated goal for definition of sequencing the operations is very complicated, multi-level, particular problem. Therefore, the expected difficulties in the process of solving this problem can be: pattern recognition, selection of datum, connection between machining surfaces and type of operations, machining tools, tools and positioning and work holding, etc. So, as the first step in process sequencing is selection of the simplified approach. It includes definition of:

a) codes for machining surfaces,

b) number of passes,

c) type of fine (F) / rough (R) machining,

d) definition the relevant anteriorities different types (dimensional, geometric, technological, economic).

As it is obviously, this approach expects the experienced process planner. One of the well-known methods of finding the order of precedence of the operations is based on the use of a matrix.

Having defined all the anteriorities, it is now possible to find the right sequence of operations for machining. The consistency of the anteriorities depends heavily on the experience of the process planner. Solution is result of weighted category of anteriorities, minimal number of precedence operations and finishing of precedence operations. The chosen order of anteriorities implementation is result of higher priority associated to dimensional and geometrical features then economical aspects. The difficulty can come from the assessment of the anteriorities, which can result in contradictory conditions. In this case the process planners have to introduce additional criterion in order to solve this contradictions. At the same time process planer defines anteriorities needed to establish a matrix, he makes a table that contains possible machining processes, machines, fixture devices and tools for every feature. To solve contradictory situation the feature that precedes according to matrix is compared with the momentarily possible features in the matrix. "Values" in the table that belong to features are compared. The feature whose "values" from table are the most similar to "values" of preceding feature has advantage. The logic in this approach is that as much as possible number operations in a sequence should be done by same process on the same machine in the same fixture and using same tool.

4 MATHRIX METHOD – BOLT EXAMPLE

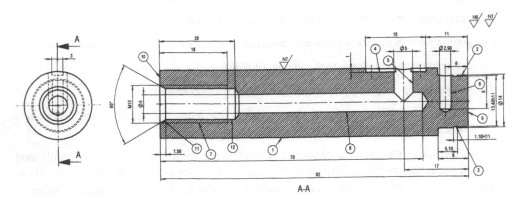

Quantity: 14 000 pcs.
Material: St60-2
Taking into account geometry of the product the primary shape would be a bar Φ20.

FIGURE 1. Bolt drawing

The first step is to analyze the part drawing and "divide" the part into features. Features are made by different machining operations. According to geometric shape, tolerances, surface quality and other information a drawing contains we can select possible machines and tools by which a specific feature could be produced. For the example in Figure 1 a selection was made and is presented below in Table 1.

TABLE 1. Surface analysis

Surface[1] (Feature)	Description	Surface quality	Machine	Tool
1	$\Phi18k6 \left(\begin{smallmatrix} +15\,\mu M \\ +2\,\mu M \end{smallmatrix} \right)$	Ra 1.6μm	Lathe	
2	$\Phi14$	Ra 6.3μm	Lathe	
3	$\Phi13,40h11$	Ra 6.3μm	Lathe	
4	Counter bore 2mm	Ra 6.3μm	Mill	
5	$\Phi5$	Ra 6.3μm	Drilling machine Mill	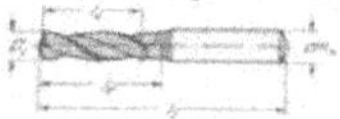
6	$\Phi5$ dpth.70	Ra 6.3μm	Drilling machine Mill	
7	M10 dpth.18	Ra 6.3μm	Drilling machine Mill	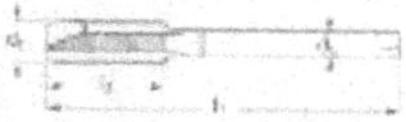
8	$\Phi2.90$ dpth.9	Ra 6.3μm	Drilling machine Mill	
9	82 (right side)	Ra 6.3μm	Lathe	
10	82 (left side)	Ra 6.3μm	Lathe	
11	1.5x60°	Ra 6.3μm	Mill Drilling machine	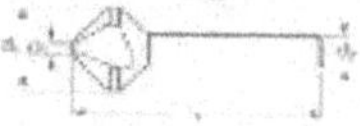
12	$\Phi8.4$ dpth.20	Ra 6.3μm	Drilling machine Mill	

The problem that appears next is which feature should be machined first and *more important in which order should features be done*. Certainly there are restrictions regarding technology, geometric and dimensional tolerances, datum, economy (reduce production costs and wear or breakage of costly tools). Taking into account all this restrictions another table (Table 2) is made in which it is clear which features must precede before other features.

---

[1] See Figure 1

TABLE 2. Table of anteriorities

Surface (Feature)		Anteriorities
1R[2]	$\Phi18k6 \left(\begin{array}{c} +15\,\mu M \\ +2\,\mu M \end{array} \right)$	10R
2R	$\Phi14$	1R,10R
3R	$\Phi13,40h11$	2R
4R	Counter bore 2mm	1R,10R
5R	$\Phi5$	1R,4R,6R
6R	$\Phi5$ dpth.70	10R,1R,3R,2R
7R	M10 dpth.18	6R,12R,10R,11R
8R	$\Phi2.90$ dpth.9	2R
9R	82 (right side)	1R,2R,3R
10R	82 (left side)	
11R	1.5x60°	6R,10R,12R
12R	$\Phi8.4$ dpth.20	6R,10R,1R

If Table 2 is presented in matrix (Table 3), advantages from this approach are now clear. It is easy to see that the first feature to be machined is 10R. When 10R is removed from table it sets free other features that were "blocked" by it. [3]

---

[2] R - roughing

TABLE 3. Matrix of anteriorities

execute this operations

before this operations

	1R	2R	3R	4R	5R	6R	7R	8R	9R	10R	11R	12R	
1R										X			1
2R	X									X			2
3R		X											1
4R	X												1
5R	X			X		X							3
6R	X	X	X							X			4
7R						X				X	X	X	4
8R		X											1
9R	X	X	X										3
10R													0
11R						X				X		X	3
12R	X					X				X			3

We are not going to analyze all the steps in the determination of operation sequence. It would be interesting to look at the situation when two or more features are not preceded by any other feature that needs to be done before. This means that all of them can be done at the same time. But this is not possible because only one feature can be machined in time. One of them must go first and then the other. In this example this situation occurs in the third step. This situation is shown in Table 4. The feature that was done before is 1R (Table 4a). In this step we have to decide which feature is going to be machined first 2R or 4R (Table 4b). To make this decision we need more data. Therefore another table was made, shown in Table 5.

TABLE 4. Matrixes of anteriorities for second step and third step

execute this operations

before this operations

	1R	2R	3R	4R	5R	6R	7R	8R	9R	11R	12R
1R											
2R	X										
3R		X									
4R	X										
5R	X			X		X					
6R	X	X	X								
7R						X				X	X
8R		X									
9R	X	X	X								
11R						X					X
12R	X					X					

a)

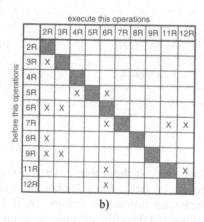

execute this operations

before this operations

	2R	3R	4R	5R	6R	7R	8R	9R	11R	12R
2R										
3R	X									
4R										
5R			X		X					
6R	X	X								
7R					X				X	X
8R	X									
9R	X	X								
11R					X					X
12R					X					

b)

In the Table 5. a few additional criteria were brought out. In order of significance they are:
- same machine (if we change the machine we change all other factors: process, fixture and tool)
- same process (if we change the process we change fixture type, tool and sometimes machine).
- same fixture (changing fixture needs more time than changing tool and it is recommended to do as much operations as possible in one fixture because it is more precise)
- same tool (the least significant factor in this list)

TABLE 5. Additional criterions for solving conflict situations

	Surface (Feature)	Process	Machine	Fixture	Tool
1R	$\Phi 18k6 \left(\begin{array}{c} +15\,\mu M \\ +2\,\mu M \end{array} \right)$	Turning	Lathe	▼10 ▼9	
2R	$\Phi 14$	Turning	Lathe	▼10 ▼9	
3R	$\Phi 13,40 h 11$	Turning	Lathe	▼10 ▼9	
4R	Counter bore 2mm	Milling	Mill	▼1	
5R	$\Phi 5$	Drilling	Drilling machine Mill	▼1	
6R	$\Phi 5$ dpth.70	Drilling	Drilling machine Lathe	▼1	
7R	M10 dpth.18	Threading	Drilling machine Mill	▼1	
8R	$\Phi 2.90$ dpth.9	Drilling	Drilling machine Mill	▼1	
9R	82 (right side)	Turning	Lathe	▼1 ▽10	
10R	82 (left side)	Turning	Lathe	▼1 ▽9	
11R	1.5x60°	Countersinking	Mill Drilling machine	▼1	
12R	$\Phi 8.4$ dpth.20	Drilling	Drilling machine Lathe	▼1	

If we look at the Table 5 we can see that feature 1R that proceeded was done by turning process on lathe. Since feature 2R is also done by turning on lathe which means by the same machining process as feature 1R it has advantage before feature 4R. Feature 4R requires milling and therefore different tool and fixture.

5 ANALYSIS OF RESULTS

Experience and knowledge of process planer has lot of influence on decision which features precede other features. Trying to put this in matrix of anteriority does not always give unique solution. The shape of part is usually very complex so process planer can miss or not see some relations. Knowledge and experience are limited. That is the reason why table of anteriorities is not always set up to give unique answer. This example shows logical approach that can be used to solve conflict situations in decision making regarding sequencing of operations. This approach reduces influence of intuition and gives more methodical approach suitable for intelligent process planning. [4]

6 CONCLUSIONS

One of problems process planning solves is setting up optimal sequence of operations. Sequence is set up following certain rules. Some operations have to precede other operations regarding requirements for dimensional and geometric tolerances, restrictions that come from technology, economical aspects and some other. For instance before thread cutting operation a hole should be cut and that is technology restriction. Systematizing these rules is a difficult task. The other big task is how to consider all the information we get from a technical drawing in making a process plan. This information include geometric shape, dimension and their tolerances, geometric tolerances, surface roughness, material type and its hardness, size of raw material, number of parts that need to be produced. The aim is to produce functional part at lowest price. One of the methods that can help process planers in making their decision is using a matrix of precedence. But it was shown that there are situations when matrix does not give exact answer. Two or more operations appear with the same level of precedence regarding matrix. This might be caused by process planer's lack of experience and knowledge or incomplete drawing. Anyway the matrix was not set up to give unique answer. To solve these situations other criteria was set up. It says that as much as possible number operations in a sequence should be done on the same machine by same process in the same fixture and using same tool.

ACKNOWLEDGEMENTS

This project is a part of the scientific project titled Intelligent Process Planning and Reengineering 0120-029 financed by the Ministry of Science and Technology of the Republic of Croatia. We express gratitude for the financial support for the project.

REFERENCES

1.   Halevi, G., (2003), Proocess of Operation Planning, Kluwer Academic Publishers, Dordrecht.
2.   Halevi, G., Weill, D., R.., (1995), Principles of Process Planning, Chapman & Hall, London.
3.   Weill, R., Spur, G., Eversheim, W., (1982), Survey of Computer-Aided Process Planning System, Analls of the CIRP Vol. 31/2/1982, pp 539-551.
4.   Chang, Tien-Chien, (1990), Expert Process Planning for Manufacturing, Addison-Wesley Publishing Company, New York.

IMPLEMENTATION OF AN ON-LINE SUPERVISION SYSTEM FOR SCHEDULING AND CONTROLLING A PRODUCTION PLANT

C. Giardini[1], E. Ceretti[2]

[1] Department of Design and Technology, University of Bergamo, Italy
[2] Department of Mechanical Engineering, University of Brescia, Italy

KEYWORDS: Supervision System, Production Control, Industrial Application.

ABSTRACT. This paper focuses on the definition of a supervision and data acquisition system in a manufacturing plant. In particular, the work here presented refers to the implementation of a SCADA system in a company which produces components by injection molding. The supervision and data acquisition system is based on an electronic structure connecting the production machines with the supervision computer where all the production and management data are collected and treated by a suitable software (MES). Different solutions have been analyzed; the one implemented is based on a PLC logic and it is also the cheapest one able to guarantee the future expansion to additional production machines.

1 INTRODUCTION

Nowadays, the market is demanding good products at low cost so competition between companies is getting stronger. Within this working scenario company efficiency is a fundamental requirement. The company must assure good quality of its products, satisfy the customer delivery time and decrease the production costs in order to remain competitive. To reach this goal a good knowledge of the productive system, in terms of machine occupancy or availability or in general workshop efficiency, is fundamental. The knowledge of daily process information is essential in performing a correct production scheduling and to guarantee the overall production control.

The advantages of a data acquisition system are both technical and economical. In fact, the knowledge of the production data allows the identification of the plant *bottle necks*, the improvement of the production processes so determining more reliable working machines, an improvement of the workshops internal management and, finally, a production cost reduction for the changes adopted. Traditionally, where performed, these data were collected manually once a day by the workshop chief, who was able to select the important information from all the production data, and then given to the costs office. In detail, the data necessary for a production control are: machine working time, machine idle time, machine availability time, set up time, loading and unloading time, machine alarms, number of produced parts, cycle time, ... These is daily workshop information and are firstly related to the machine efficiency, secondly with the operator ability and finally with the machine set up. The advantages of using of a data collection system are the possibility for the cost analyst office to determine precisely the production costs and of the workshop chiefs to identify the machine efficiency and to improve it.

Published in: E. Kuljanic (Ed.) *Advanced Manufacturing Systems and Technology*, CISM Courses and Lectures No. 486, Springer Wien New York, 2005.

A step toward the company innovation and efficiency is the automatic detection of the production data by using the new capabilities of computer and telecommunication sciences. In fact, an electronic data acquisition system is fast, sure, improves the data collection and management, reduces the time necessary to insert the data manually with a saving in cost and in the possibility of errors. In addition, this system ensures a real time monitoring of the workshop activities by identifying immediately the critical events (machine or part failure, inefficiency and so on), and speeds up the distribution and the knowledge of the production information within the company structure and in particular between the users (to reschedule the production or to update the production due date).

2 SYSTEM FOR PRODUCTION MANAGEMENT, PLANNING AND CONTROL

A typical system for production management (see Figure 1), planning and control is mainly characterized by three software (each one is then supported by other software):

1. a management software (ERP – Enterprise Resource Planning) which builds and elaborates the production orders on the basis of "unlimited capacity", that is, the real capacity of the company to satisfy the market is neglected and the company resources are considered unlimited;

2. a scheduling software, which elaborates the production orders of the ERP program and plans the production on the basis of the actual plant working capacity by allocating the jobs to the available resources (machines, workers, fixtures, …);

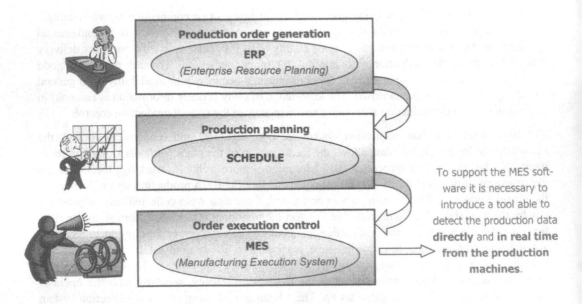

FIGURE 1. Architecture of production management, planning and control software

3. a supervision software, which controls the production execution (MES – Manufacturing Execution System) and verifies the schedule defined by the two above mentioned software types.

Once the production orders are generated, scheduled and executed, it is necessary to obtain a feed back from the production. The system which furnishes information on the machines activity to the MES software is called SCADA (Supervisory Control And Data Acquisition). SCADA automatically and in real time detects the production data from the working machines [1] [2] [3] [4] [5] [6] [7] [8].

It is obvious that if the MES system is not supported by a SCADA it is impossible to control the production plan because a manual data implementation will not allow a real time production control. As a consequence the MES will lose its efficiency and scheduling software could only be used to allocate activities but not to follow them during the production cycle. Without SCADA all the production management, planning and control computerised system will lose most of its efficacy.

For the above reasons, this work refers to the realization (hardware and software) of a SCADA system in a company which produces components by injection molding (Figure 2). The supervision and data acquisition system is based on an electronic structure connecting the production machines with the supervision computer where all the production and management data are collected and treated by suitable software (MES).

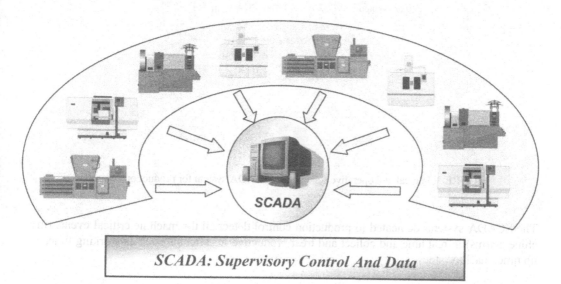

FIGURE 2. The SCADA control system

2.1 THE SCADA SYSTEM

The name SCADA, *Supervisory Control And Data Acquisition,* is used for electronic/computer systems dedicated to on line data acquisition. The main activities of a SCADA are the creation of an interface between several users (hardware and software) and the machine in question, and the storage of the collected information [4] [5] [6] [7] [9] [10].

Usually, a SCADA operates on two levels (Figure 3):
- Electric/electronic level, through different devices it detects electric signals from the machine;
- Information level, by means of software it treats the electronic signals it detects and produces data, available for all the users (an example is the MES system).

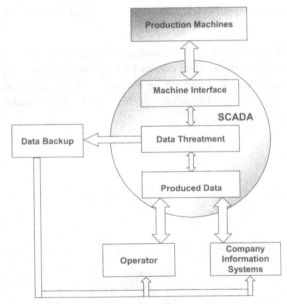

FIGURE 3. Logical and operative scheme of a SCADA system for production control

The SCADA systems dedicated to production control detect all the machine critical events (machine alarms) in real time and collect and treat productive information such as working time, set up time, machine stops.

3 PROJECT DESCRIPTION

3.1 THE WORKSHOP AND THE MACHINES

The area of the company in which we developed the project is the injection molding workshop where all the plastic components of Metalwork SpA are produced. Metalwork is an Italian company working in the field of pneumatic valves and actuators. The number of different items is around 600, and the number of the working machines is 18 (produced by two press builders).

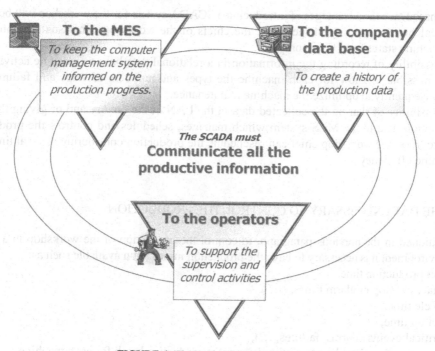

FIGURE 4. The goals of SCADA system

The injection moulding presses operate in two different ways:

- Automatic, that is they produce parts in series and automatically unload the part;
- Semi-automatic, that is they produce parts in series and wait for the operator for the manual unloading of the part.

Every machine can manufacture different items according to the raw material, the die and the software The machine set up is done by a skilled operator, who controls the quality of the first produced pieces before starting the production, and this lasts from 20 to 40 minutes. The number of machine set ups depends on the batch dimension. Each machine has a control panel which shows the cycle time, the parts produced, the clock, the alarm list, the injection material temperature, the lubrication conditions. In the case of failure, the machine automatically stops the production and produces a visual and acoustic alarm. The possible machine alarms are: wrong part unloading, end of raw material, bad die lubrication, too high working temperature, machine mechanic or electronic failure, bad die design or failure of the mechanical and electrical devices needed for production (thermocouples, humidifier, cooling system, etc.). The number of alarms for each machine during the 24 working hours ranges from 10 to 100 and, in most cases, the machine operator is able to solve the problem and to restart the production. The Kanban technique is used to schedule the production. Since this methodology aims to produce *Just In Time* it is fundamental to have the information about the plant efficiency and the machine availability.

As a consequence the advantages of introducing a SCADA system for the workshop will be:
1. Real time monitoring of the plant; the chiefs of the workshop can visually control the machine status from their computer.
2. Possibility of recording the information in a relational data base of machine activities; it is possible to know for each machine the types and numbers of stops and failures and consequently to optimize the machine maintenance.
3. Possibility of sharing the collected data in the LAN of the factory and of giving the production data to the MES system which manages, schedules and controls the production progress; the workshop chief can reschedule the production considering the real time machine efficiency.

3.2 THE DATA NECESSARY TO CONTROL THE PRODUCTION

As mentioned in the previous paragraph, to control the production of the workshop in a just in time environment it is necessary to have the production information available such as:
- net production time,
- machine stop or alarm times,
- cycle time,
- set up time,
- critical events (alarms, failures, …),
- number of produced parts within the current productive batch for each machine.

This information can be manually detected and then manually loaded into the MES system, or, to be really efficient, automatically collected. In the following all the steps necessary to implement a SCADA system in the workshop will be analysed and described.

3.3 HOW TO DETECT DATA FROM THE PRODUCTION MACHINES

The first problem that arises during the implementation of a SCADA system is related to the information exchange between operating machines and computers. This communication problem is not related to the exchange of information between machine and external environment (all the machines are equipped with HMI, that is with a *Human Machine Interface* to furnish productive parameters and data to the operator) but between the machine electronic devices and a Personal Computer (it is necessary to have physical devices and drivers able to translate the machine language into a standard PC based language). This is fundamental if the productive information is to be reachable by several technicians within the company via PC.

Within the workshop of Metalwork SpA three different machine status had been found:
- Machines that can not communicate with the external world,
- Machines that can communicate using ad hoc protocols,

– Machines that communicate using standard protocols.

The solutions are:

1. To install on the serial ports of the machine ad hoc drivers,
2.a. To detect electric signals via a PLC (Programmable Logic Controller),
2.b. To detect electric signals via an industrial PC.

The advantage of the first solution is the possibility of exchanging all types of information between machines and SCADA. The disadvantages are the high costs and realization times together with an uncertainty about the final result due to the lack of information related to the communication protocols.

The advantages of the second solution are the lower costs and realization times and the confidence of obtaining good results since the machine communication protocols are bypassed. The disadvantages are related to the possibility of obtaining information only from the detected signals and to the need for increasing the number of cables to obtain more data.

4 THE PROJECT AS REALISED

Considering the above mentioned advantages and disadvantages for each different technical solution, the relative cost analysis was also carried out in order to identify the best technical and economical solution. Finally, the solution is based on a PLC logic which is also the cheapest that allows the future expansion to additional production machines [11] [12].

Each machine is equipped with a PLC. Eight electric signals are collected by the machine PLC. Customized software is active on each PLC to transform the electric signals into digital ones. All the PLC are connected with a hub via the Ethernet as illustrated in Figure 5. The hub controls the flow of the data coming from each PLC and gives these data to the SCADA server where all the software needed to translate the PLC signals and to elaborate and publish the SCADA information is present. This system is developed with the OPC technology which enables the data exchange between PLC and the software applications.

The SCADA detects for each machine:

– Machine status,
– Number of performed cycles within the production batch,
– Cycle time for each injection cycle,
– Machine alarms,
– Cycle time alarm (the imposed cycle time is overridden).

Starting from the acquired information the SCADA system elaborates additional production data. All the information is displayed with a user friendly interface on the SCADA PC client located in the workshop or to other authorized users (Figure 5). At the same time these data are stored in a Server data base and then given to the MES system.

The information exchange between MES and SCADA is fundamental in order to avoid collecting data from machines that are under set-up or not in production. In fact, these machines would be considered under continuous alarm for the SCADA.

Figure 6 shows the output of the SCADA; all the production machines are displayed and the active alarms are immediately visible by the chiefs of the workshop (through the client SCADA) or by the authorized users. The productive data that the SCADA furnishes to the MES system are fundamental in running a new production scheduling and in avoiding delay in the batches due date.

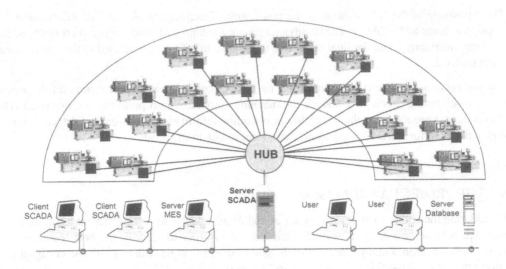

FIGURE 5. The system as realized

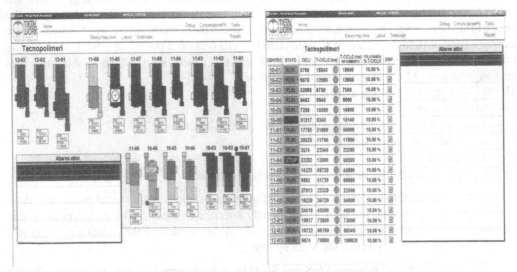

FIGURE 6. Graphical output of the SCADA

5   CONCLUSIONS & FUTURE WORKS

The implementation of a SCADA system in the injection moulding workshop of Metalwork SpA completely changed the procedure of collecting and elaborating production data, furnishing many advantages related to production control and supervision (*Supervisory Control*) and data acquisition (*Data Acquisition*).

With regard to production supervision and control the system is independent of all the other company computer systems. The SCADA collects in real time and continuously the productive data and immediately updates the information for the workshop chiefs and the users. In this way the system is a support tool for the workshop operators giving them a continuous control over the machine status and efficiency. The final result is an increase in the productive time.

With regard to data acquisition, the SCADA works together with other computerised systems and furnishes information to the company database. This production database is updated automatically and furnishes information to the MES system by saving time and work loads of the operators responsible for the manual data acquisition. The MES uses the SCADA data to produce reports on the production status; in this way it is possible to identify the production history of each batch and to calculate the production costs. In addition the MES can calculate the performance indexes of the machines, of the operators and of the workshop. Another advantage is the historical database which gives the possibility of identifying the critical moulds and machines and to improve them. This increases the productive time too.

Thanks to the SCADA implementation the information exchange within the company is faster and, since it elaborated automatically, is free from human error.

A possible future activity will be the extension of the SCADA system to the metal cutting workshop: this will imply a duplication of the choices made for the injection moulding workshop.

Secondly, it is possible to improve this SCADA by installing new SCADA clients running on pocket PC and wireless net, so eliminating the need for a fixed PC.

In addition, it could be interesting to improve the efficiency of the PLC alarms by connecting them to acoustic devices so as to advise the operators in a faster and more efficient way.

Finally, it would also be interesting to connect the assembly workshop with the SCADA system. In this area the operations are semiautomatic being performed manually by operators supported by electronic machines.

In this way it could be possible to have a real time control of all the company.

REFERENCES

1. Clarke G., Reynders D., (2004), Practical Modern SCADA Protocols: DNP3, 60870.5 and Related Systems, Newnes.
2. Bailey D., Wright E., (2004), Practical SCADA for Industry, Newnes.
3. Park J., Mackay S., (2003), Practical Data Acquisition for Instrumentation and Control Systems, Newnes.

4. Park J., (2003), Practical Data Communications for Instrumentation and Control, Newnes.
5. Wiebe M., (2003), A Guide to Utility Automation: AMR, SCADA, and It Systems for Electric Power, Pennwell.
6. Boyer S.A., (2003), SCADA: Supervisory Control and Data Acquisition, ISA.
7. Williams R.I., (2002), Handbook of SCADA systems: for the oil & gas industry, Elsevier Advanced Technology
8. Lopez O., (2003), 21 CFR Special Edition Including SCADA Systems, Sue Horwood Publishing Limited.
9. Park J., Mackay S., Wright E., Reynders D., (2003), Practical Industrial Data Networks: Design, Installation and Troubleshooting, Newnes.
10. Strauss C., (2003), Practical Electrical Network Automation and Communication Systems, Newnes.
11. Weigant J., (2003), Creating HMI/SCADA Industrial Applications Using Microsoft Access, Industrialvb.
12. Bolton W., (2003) Programmable Logic Controllers (Third Edition), Newnes.

LOW-COST TRANSFORMATION OF A CONVENTIONAL MILLING MACHINE INTO A SIMPLE FSW WORK STATION

M. Ponte, J. Adamowski, C. Gambaro, E. Lertora

Department of Production Engineering, Thermo-energetic and Mathematical Modelling, University of Genoa, Italy.

KEYWORDS: Friction Stir Welding, Milling, Aluminium Alloys.

ABSTRACT. Many of the typical problems related to fusion welding of aluminium alloys can be avoided by using non conventional joining techniques. One of particular interest is the Friction Stir Welding (FSW) in which the joined material remains at solid state, thus allowing to avoid solidification phenomena such as formation of dendritic structures or liquation cracking. Since FSW is governed by two completely mechanical processes, i.e. mixing and forging, it is possible to join heterogeneous aluminium alloys without filler metal. Using specially designed tools and machines up to 75mm of aluminium can be welded in a single pass. This paper describes a low-cost method of transforming a conventional milling machine into a simple FSW work station. The machine adaptation process includes the implementation of a temperature measurement and a downward and horizontal forces monitoring systems, as well as designing of a tool fixing system and the tool itself. Finally are presented the results obtained using the modified FSW work station along with the cost related considerations.

1 INTRODUCTION

Welding of aluminium based alloys has been an industrial practice for some decades now. Two welding methods have particularly proved to be capable of producing sound welds in aluminium alloys: Metal Inert Gas (MIG) and Tungsten Inert Gas (TIG) techniques, respectively known in English nomenclature as Gas Metal Arc Welding (GMAW) and Gas Tungsten Arc Welding (GTAW). Both processes being based on fusion of welded material and possible filler metal are characterized by the presence of metallurgical features influencing the quality of welds, such as formation of dendritic, solidification structure, liquation cracking or cold bonding; besides the results are highly dependent on the manual skills of the welder. A solution can be to use alternative joining methods in which the material remains in solid state. One such method, particularly applicable for joining aluminium alloys is Friction Stir Welding (FSW), presented schematically in figure 1.

Published in: E. Kuljanic (Ed.) *Advanced Manufacturing Systems and Technology*,
CISM Courses and Lectures No. 486, Springer Wien New York, 2005.

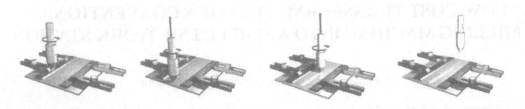

FIGURE 1. Schematic representation of FSW process

This process invented, developed and patented in the early 90's by the TWI, UK [1], uses a special tool, composed typically of two components: a shoulder and a probe (fig. 2). The welded material is heated up thanks to the friction heat generated between the surface of welded material and a rotating tool. That heat is used to soften the edges of the material to be joined which facilitates the action of the probe, that mixes together the plasticized material, creating a joint.

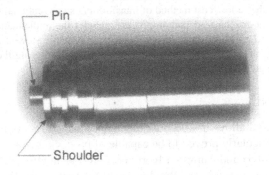

FIGURE 2. FSW tool

There are two phenomena that govern the FSW process: mixing and forging, and it is possible to join heterogeneous aluminium alloys without filler metal. Using custom made tools and machines up to 75mm of aluminium can be welded [2].

The mechanics of the FSW is similar to that of a conventional milling process, hence the equipment used in FSW has general characteristics of a milling machine [3,4]. There are FSW machines available commercially, for both research and production applications (fig. 3). They differ by dimensions and possible applications, in terms of maximum weldable material thickness, material type (aluminium alloys, magnesium alloys, ferrous alloys, etc.) or the possibility of performing welds in 3D. Their common characteristic is however a relatively high cost, which especially in an initial phase can be a significant obstacle for whoever wants to study the new process.

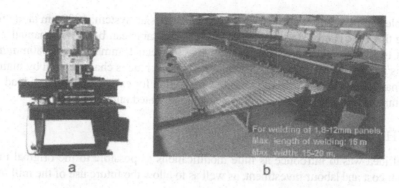

FIGURE 3. FSW machines: a – ESAB [5], b – Hydro Marine Aluminium [6]

Alternatively a conventional milling machine can be adapted to perform FSW joints. Such mill must typically be characterized by:

- spindle drive power sufficient to provide constant rotational velocities,
- transverse drive power sufficient to provide constant travel velocities,
- low vibration,
- good stiffness.

2   ORIGINAL MACHINE

The machine chosen to be converted into FSW workstation, presented in figure 4 is Arno Nomo F3 vertical milling machine, equipped with two replaceable spindle heads powered by a 10kW electric motor: one using a gear transmission, the other a belt transmission. The main spindle head can be rotated in the XZ plane (where X is the longitudinal axis of the table and Z is the vertical axis). The gear transmission offers twelve distinct rotational velocities, in the range between 40 and 1700 rpm.

FIGURE 4. Arno Nomo milling machine before transformation

The part table can move in all three directions of the Cartesian system: 1000mm along the X axis, 300m along the Y axis and 420mm along the Z axis. The table can be moved manually and automatically at twelve distinct travel rates in the range between 10mm/min and 880mm/min. Table movement is guaranteed by a 3kW electric motor. The machine is characterized by high rigidity – in fact the maximum deformations of the frame are 0.1mm for each 10t of vertical load. There are no force control or rotation speed control systems incorporated into the machine.

3   MACHINE TRANSFORMATION

The general idea was to introduce as little modifications as possible to the original machine, in order to limit cost and labour investment, as well as to allow the future use of the mill for machining operations.

In order to monitor the process, two or more load cells would be necessary. One of those was necessary to investigate the load on the Z axis, also called "vertical load", and the other one was necessary to measure the load on the X axis.

From many possible solutions one was chosen, that could generally be divided into the following steps:

- designing of tool, tool holder and downward force measurement system.

- designing of part clamping and axial force measurement system.

- designing of temperature measurement system and connection to a PC.

The above mentioned steps are hereafter described.

3.1   TOOL HOLDER AND DOWNWARD FORCEMEASUREMENT SYSTEM

One of the ideas behind the machine modification process was the will of investigating on the characteristics of a FSW process, among which one of the most importance is the downward force applied to the tool. Two possible solutions of force monitoring were considered:

- force sensor (toroidal load cell) incorporated into a rotating tool holder,

- downward loads on the spindle read by one or more load cells integrated in a parts' fixing system located on the workbench.

The second solution was abandoned because although being valid, it would require higher cost investment.

Consequently the other option was chosen designing a custom made tool holder that would incorporate a force measurement system. It was preferred not to introduce force sensors into the preexisting spindle, because it would have required modification of the machine head.

The analysis of the FSW process allowed to identify the characteristics that the rotating tool holder should have:

- A possibility to transfer the rotational movement from the tool shoulder grip to the pin.

- A good stiffness of the all structure with the aim to have no vibrations.

- A modular structure to allow modification of individual parts in case of necessity.

The location of the load cell in the rotating spindle implied a choice of a toroidal load cell, with external diameter of 80 mm, internal diameter of 50 mm and the maximum load of 50 kN.

The load cell was fixed between the spindle and the tool shoulder grip. It was supported axially by two thrust ball bearings and radially by a needle roller bearing. In this way the rotational movement of the tool holder wasn't transferred to the load cell. That solution was imposed by the presence of the signal cable of the load cell.

Exploded and assembled views of the tool holder are represented in Fig. 5. In the image it's possible to observe four principal parts: spindle, tool shoulder grip, shoulder and pin.

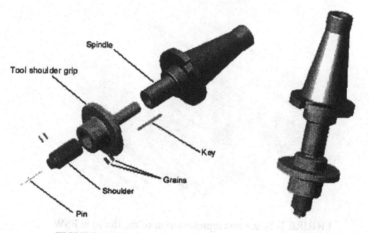

FIGURE 5. 3D view of the spindle and the tool holder (AutoCAD)

An advantage of this choice is the possibility of using the spindle system on other milling machines equipped with a ISO 50 conical tool holder.

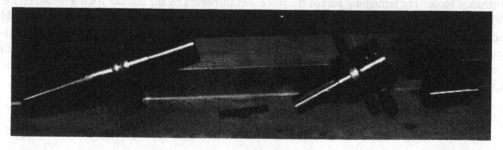

FIGURE 6. View of the spindle and the tool holder

3.2 CLAMPING AND AXIAL FORCE MEASUREMENT SYSTEM

In Friction Stir Welding the correct position of the parts to be joined and also the clamping system used to fix them is of a great importance. During the process high thermo-mechanical forces are generated [6]. For that reason the parts must be locked in all the directions. The three principal reactions are the following (Fig.7):

- Reaction to the vertical force

- Reaction to the travel movement of the tool

- Reaction at the clamps dues to the thermical expansion

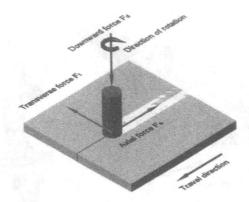

FIGURE 7. Schematic representation of the forces in FSW

The rotation of the pin penetrated between the faying edges is the main reason for the lateral expansion of the plates and must be contained by the clamps. The two parts to be joined must be free to move axially (towards the load cell) thus allowing the acquisition of the major part of the horizontal force. The remaining part of the force was dispersed by the contact friction between the plates and the fixture. A flexible clamping system was realized because of the necessity to weld parts of various dimension. In this way it's possible to weld different thickness, with the maximum envelope of 400x1200 mm (W x L). Fig. 8 shows the 3D model of the clamping system used to fix the plates to the workbench. The system is composed of a steel plate (1500x500x15mm), the load cell, three angular and three vertical constraints.

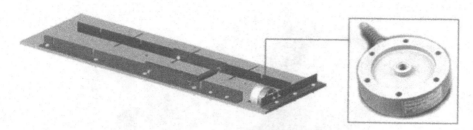

FIGURE 8. Clamping system and the load cell

The steel plate was anchored to the bench of the milling machine by two M16 bolts positioned on the central axis of the plate.

Two of the three angular constraints (green parts in Fig. 8) were fixed to the bench using M10 plugholes in order to guarantee the stiffness of the system. Only one of the three angular parts (blue, Fig. 8) was fixed to the bench using three elongated plugholes in order to have the possibility to be adjusted to the dimensions of the plates to be welded. Two vertical constraints (red, Fig. 8), and the cylinder (black, Fig. 8) limit the vertical movements of the welded parts at the same time guaranteeing their longitudinal movements. The position of the axis of the load cell could be modified to be aligned with the central axis of the weld.

Another issue was related to the temperatures that the load cell can support. During welding the aluminium plates can reach temperatures around 400°C [7] while the maximum temperature that the load cell can withstand is 100°C. In order to protect the cell, a disk of stainless steel was positioned between the cell and the plates to be welded. Stainless steel was chosen as material for the disk due to its low thermal conductivity.

3.3 TEMPERATURE MEASUREMENT SYSTEM

The system able to monitor the temperature variation during the process was composed of a set of "K" type thermocouples located near the welding area. The signals from the thermocouples and the load cells were collected by a connector block (Fig.9) able to cut out noise, filter all the electrical signals in input and reduce its intensity to micro impulses.

In order to minimize the influence of the power supply line (electromagnetic field) the PC during acquisition was supplied from the battery. Moreover all the cables were shielded and auxiliary resistances were implemented to reduce signal noise. The two load cells required an external power supply of 12.5V DC. This value had to be stabilized due to the fact that the output signal was measured in mV and even a slight variation of input voltage could compromise the accuracy of the measurement.

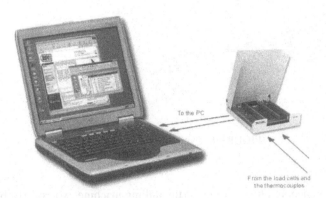

FIGURE 9. Schematics of connections of the acquisition system

There are 68 channels for input signals and 68 pin connectors for the output signals in the connector block. This output signals are in a range between −10V and +10V and up to 800mA. An interesting characteristics of the connector block is the possibility to use it in various climatic conditions (temperature range 0÷70°C and humidity range 0÷90%). Fig. 10 shows an example of the block diagram, the program part permitting the set up of the input signals.

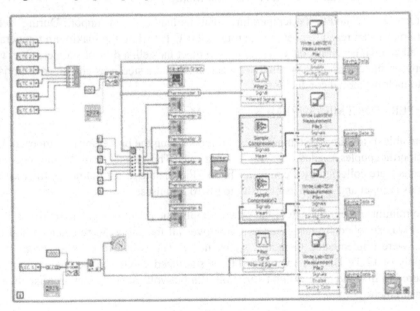

FIGURE 10 Block diagram of Labview 7 Express

The acquired data were transferred to a PC and elaborated using Labview 7 Express software. One of the main characteristics of the program is its ability to simulate even very complex measurement systems – it can be set to monitor several physical measures at the same time, i.e. temperature, force, resistance, etc. Setting the software for different applications is relatively easy. The measured values are automatically stored in electronic form as text files. The software can also be connected in a loop-back system to be used as a remote control device.

4 CONCLUSIONS

The project to transform a conventional milling machine into a simple Friction Stir Welding work station was successfully concluded.

It was demonstrated that it's possible to produce sound FS welds using the adapted milling machine.

Cost and labour investment were contained and the conversion process didn't show particular problems.

The instrumentation produced is extremely simple and thanks to its modular structure it can easily be modified and additional control components can be added.

The adapted milling machine has successfully been used to carry out a test campaign of the weldability of AA6082-T6 aluminium alloy by FSW [8].

REFERENCES

1. Thomas W.M., Nicholas E.D., Needham J.C., Murch M.G., Templesmith P., Dawes C.J., (1993), International Patent Application N° PCT/GB92/02203.
2. http://www.twi.co.uk/j32k/unprotected/band_1/fswmat.html.
3. Diane L. Hallum, (2000), Stirring Up Welding, Forming and Fabricating Vol. 7 No. 10.
4. Zelinski P., (2002), Rapid traverse-Welding On A Machining centre, MMS online.
5. http://www.esab.com.
6. http://www.hma.hydro.com.
7. Thomas J. Lienert, William L. Stellwag, Jr., (2001), Determination of load, torque and tool temperature during friction stir welding of aluminum alloys, AWS Convention 2001, Paper 11B, pp. 152-155.
8. Adamowski J., Gambaro C., Ponte M., Lertora E. (2005), Investigation on Friction Stir Welding weldability of aluminium alloy AA6082-T6, Presented for acceptance at AITEM 2005.

CAD/CAM IN A COMPLEX INDUSTRIAL ENVIRONMENT

D. David[1], M. Ermacora[1], G. Totis[2]

[1] Manufacturing Department of Danieli & C., Buttrio, Italy
[2] Department of Electrical, Management and Mechanical Engineering, University of Udine, Italy

KEYWORDS: CAD, CAM, CAPP.

ABSTRACT. Manufacturers' main targets (high productivity, high quality and low costs) are found in a complex economic environment demanding flexibility and spirit of adaptability. The achievement of these goals is strongly dependent on the logistic technological architecture making up the planning stage, which regulates the production process. First, the internal framework of Danieli & C. Manufacturing Department is described, and the main functions and tasks of the various divisions are pointed out. A perspective based on CAM is adopted, since the CAM division plays a central role in this structure. The present situation is then compared to the CAPP approach. Finally a CAD/CAM 3D integrated approach is proposed for a marginal field.

1 INTRODUCTION

Today, the worldwide market requires production to dynamically adapt to the requirements of market demand in the shortest time possible. In other words, manufacturers have to produce a high quality product as quickly as possible, at the lowest possible cost. In such a competitive environment only those who show an open-minded attitude will be able to survive, develop and grow. Therefore, we must pay attention to world market trends, not opposing them but adapting to them, even if this requires significant effort and investment. In the last decades, some large companies, such as Danieli & C., have taken this challenge. In fact, Danieli & C. has adopted an aggressive behavior to avoid being overcome by market trends, but rather to encourage and lead them. Some meaningful examples of this enterprising conduct are the investments made to build new installations and industrial plants in foreign countries such as India and China. In fact, these countries are now growing and they will probably influence or even dominate the market in the near future. Another promising activity is training new generations of skilled workers, mostly Italian but also from all over the world, who will head the foreign subsidiaries. The initial European advantage, consisting of high level know-how, has to be considered as a basis to build the future economic and industrial configuration of the company.

There are several aspects which can be interpreted as an expression of know-how: one is the technological-logistic architecture of the manufacturing department which is supported by a huge amount of data and specialized knowledge accumulated throughout the years. This data constitutes the technical-historical heritage of the production department. This department is made up of various divisions accomplishing different tasks and run through the application of strict internal regulations. Respecting these rules and using instruments such as a powerful C.A.M. language is fundamental to efficient programming.

Published in: E. Kuljanic (Ed.) *Advanced Manufacturing Systems and Technology*,
CISM Courses and Lectures No. 486, Springer Wien New York, 2005.

The purpose of this paper is to explore and analyze the structure of the Danieli & C. Manufacturing Department, considering the key role played by the CAM division. First the production/planning stage will be analyzed, paying particular attention to the input/output data exchanged between CAM and other divisions. This architecture will be briefly compared to a different configuration, in order to highlight the advantages and disadvantages of the current structure, in comparison with the alternative one. Finally a CAD/CAM 3D integrated approach will be proposed for marginal production, the sculptured figures. This idea will be examined from a theoretical point of view. This speculation will serve as a starting point to plan an experiment, which is necessary to assess the efficacy and applicability of the method.

2 PRODUCTION PLANNING ARCHITECTURE

The Danieli & C. Manufacturing Department can be defined as a huge organized system made up of several internal divisions; it takes advantage of synergetic interaction between them to give a precise response to external inputs, that is to say to produce high quality outputs with minimum cost and time.

The planning stage will be considered in this chapter, according to a pattern that leads to the production of a generic mechanical piece. Figure 1 illustrates the general architecture made up of various divisions and layers.

The first step is carried out in the CAD division, where the CAD operators deal with project and design problems; their task consists in generating a model of the mechanical piece, fulfilling the customer's specifications and structural engineering requirements. This model has to collect and represent geometry, form and dimensional tolerances, surface roughness indications, in short all the geometrical and technological aspects that characterize the piece. Two different CAD instruments are mainly used to develop such a model: AUTOCAD (Autodesk inc.) and ProEngineer (Parametric Technology Corporation). Although the company is starting to use a 3D CAD system, AUTOCAD is still the most widely used: indeed it covers 90% of production. This can be justified by the fact that the usual target of the company can be represented by the common orthogonal projection views and sections, and it can be obtained by 2.5 axes machining phases, i.e. 3 axes not simultaneously interpolated. This means that all the surfaces and desired features can be achieved by a sequence of machining phases, characterized by tool feed motions having 2 degrees of freedom, and one positioning movement. On the contrary, dealing with solid and sculptured figures, which are formed by very complex surfaces that can require up to 5-axis machining, a 3D model is necessary and unavoidable [1]. However, this particular issue will be faced in the last paragraph. The focus is now on the most common components represented in 2D.

The job order is then passed to the UPP division (production planning office), whose main task consists in working out production planning. It also deals with make or buy decisions, it feeds a separate small division dedicated to the rough component engineering, it creates an empty work order which is an open document, ready to be filled during the following steps.

The Methods division, receiving mechanical drawings and inputs from UPP, works out the component cycle. The component cycle consists of all those thermal and machining processes, dimensional validations and checks that must be applied to the rough component in order to achieve the final product. Each operation making up the component cycle refers to a specific machine or treatment. Moreover, a mechanical operation could be further split into phases, corresponding to different workpiece positioning, carried out on the same center.

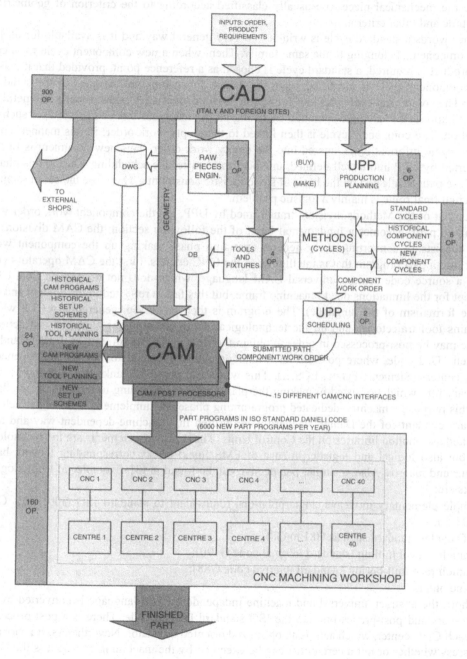

FIGURE 1. Present architecture

It is important to note that the selection of this list is aided and guided by a set of unsubjected standard Danieli cycles, simply called standard cycles, associated to various product families, where the mechanical pieces are usually classified according to the criterion of geometrical similitude and other criteria.

In other words a standard cycle is written in a very general way and it is available for all the new components belonging to the same family. Then, when a new component cycle for a certain product is required, a standard cycle is chosen as a reference point, provided that it exists. The component cycle is identified when the actual dimensions are assigned to the standard cycle. The component cycle provides for all the possible alternative paths, equally competitive. This is expressed in a single operation by listing all the machines that can be used for such an operation. The component cycle is then linked to the empty work order: in this manner, component cycle information is entered into the empty work order. This new document is finally converted by UPP into a well-defined, univocal route, after the scheduling. During this phase, a precise path is selected on the basis of some logistic constraints. Therefore the submission of the actual final route is mainly a logistic problem.

The output of the Methods division, transformed by UPP, i.e. the component work order with submitted path, becomes a fundamental input of the following section: the CAM division. Its goal consists in converting every generic operating phase making up the component work order into a part program that is intelligible to the CNC centers. First the CAM operators generate a source code using a universal CAM language, which does not depend on final CNC (except for the limitations due to machine frame, but this fact is reflected in the content and not in the formalism of the language). The program is then processed, generating a CLF which contains tool trajectories and all the technological data corresponding to CAM instructions. These may be post-processed in order to obtain the final part program written in ISO standard Danieli. This code, where possible, is similar or almost identical for all the heterogeneous CNC centers (Siemens, Fanuc, ECS...). This was made possible thanks to a demanding preliminary job, which consisted in making the plant uniform, according to the Danieli standard. For this purpose a machine-dedicated programming phase was implemented, during which the primary elements of the Danieli standard were written in a machine-dependent way and then installed as function libraries on the Control Units. These primary elements are the technological (but also logical and logistic, in case of FMS lines) bricks corresponding to both basic actions and macros. Also, a similar perspective can be found in [3]. Examples of technological bricks are:

- simple elementary movements/interpolations (equivalent to standard ISO G00, G01, G02, G03...);
- ISO cycles standard such as G81 to G86...;
- Danieli helical milling (standard internal code G190);
- Danieli face mill-boring (standard internal code G88);
- some others.

In short, the abstract, universal and machine independent CAM language is converted by the processors and post-processors into the ISO standard Danieli code. There is a post-processor for each CNC center, which acts as an optimized oriented translator. Nevertheless, it is not able to assess whether or not a certain list can be executed by the machining center; it is the CAM operator who has to establish the CAM list content, working out an executable program by taking into account machine features and capabilities.

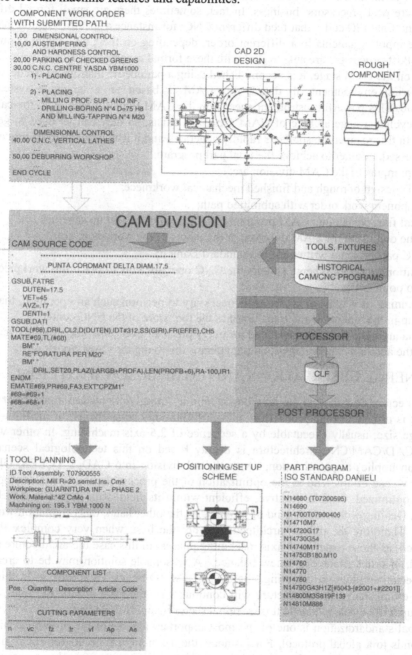

FIGURE 2. INPUT/OUTPUT streams of CAM division

Therefore, if the submitted path has to be modified and another machine is chosen to carry out a specific operation, the CAM source code should be slightly adjusted. Formal aspects, on the contrary, are post processors' business. In fact, sometimes there are small formal differences between the final ISO codes that feed different CNC: for instance a specific Danieli macro may accept the input arguments in a different order, depending on the CNC mounted on a given machine. Post-processors are able to deal with these formal aspects.

Focusing on the CAM stage, it is accomplished using a parametric instrument inside a codified manufacturing environment. In detail, this CAM is based on a language system called TOOL2000. This powerful instrument speeds up the CAM stage. In fact, as in the case of the standard cycles, a huge archive of CAM programs exists, so they can be used to generate a new one. In this manner the source is rapidly generated and it is now ready to be processed and post-processed, in order to achieve the final part program.

To sum up, inputs of the CAM division are:

- the CAD design of rough and finished mechanical workpiece;
- the component work order with submitted path;
- historical files containing: CAM programs, tool planning and set up schemes.

Finally, the outputs of the CAM division, or the inputs of the workshop, are:

- the CNC part program written in ISO standard Danieli;
- the positioning/set up scheme (it tells the CNC operator how to place the workpiece, where the zero point is…);
- tool planning (it is a list of all the tools necessary to perform such an operation, that have to be set up in the tool room and transferred to the tool store of the CNC center);

These data are developed and produced for every phase of every operation of the cycle. Figure 2 shows the above-mentioned input/output streams involving the CAM division.

3   GENERAL OBSERVATIONS AND COMPARISON WITH CAPP

The architecture described in the previous section is strongly oriented to a particular segment. In fact it is oriented to the production of high technology components, of medium-large and even huge size, usually executable by a sequence of 2,5 axis machining. In other words, the present CAD/CAM/CNC architecture is deeply based on this technological scenario. This orientation implies a specialization, hence the subdivision into CAD, CYCLES, UPP, CAM sections, whose efforts tend to the optimization of the process. In short, this configuration is strongly optimized, flexible, reactive, efficient within its field, with regard to a well defined segment. Dealing with another kind of product, on the other hand, this orientation would loose efficacy. Therefore the present architecture is not the best, when very complex figures are considered, as those requiring 5 axis machining phases. In this case the architecture should be modified, in order to achieve a new balance. A reasonable solution may be to create a new small specialized division, adopting the CAD/CAM 3D integrated approach. In this manner, the well established present structure will continue with the main production lines, while the new group will be able to handle this marginal sector in an effective way.

The global standardization is one of the most important aspects of the present architecture; it corresponds to a global protocol. For instance, the technological bricks represent an essential part of this protocol. Basically, we can distinguish two different approaches to program CNC machines: the first is the so called American approach, which is based on a powerful CAM on the top of the structure. It generates a long part program, by using only elementary instructions. The European approach, on the contrary, tends to misuse the specific macros mounted on the machine. The approach adopted by Danieli is balanced between the afore said solutions: it uses both elementary and macro instructions (the bricks). This approach does not refuse the advantages of the macros, and contemporary it establishes a well-defined and homogeneous CNC interface layer. In short, this is an efficient architecture based on standardization, organized into layers which stratify the divisions and regulate the input/output streams between them, whose aim is the optimization of the whole process.

This allows to use a powerful CAM language on the top of the planning chain, setting aside the matters arising from the heterogeneous CNC environment. However, to generate such an effective configuration a great amount of energy was invested, while establishing the Danieli standards on the CNC machines. Besides, a minimum energy is continually needed to maintain its internal order and coherence, and to upgrade the system, as in the case of installing a new machine. In this context, there are also some sociological aspects to be taken into account, which can influence the efficacy of the whole system.

Besides, it has to be pointed out that this architecture is capable to capitalize the working hours spent to release the historical cycles and CAM/CNC programs, since they are easily worked out to generate new ones.

It is interesting to compare the present situation with the CAPP (Computer Aided Process Planning) approach described in literature [1,5-7], where the CAD-CAM functions are all integrated into only one software package, and they are executed almost automatically. That is to say the CAPP operator has to generate the 3D complete model of the component, equipped with all the necessary geometrical technological features. This represents a new commission, in comparison to the traditional task accomplished by a CAD designer. The CAPP software transforms the model into an executable part program, written in the machine language (ISO code), while the operator only has to carry out supervision, control and correction functions. It could seem, in theory, that the CAPP approach is absolutely the best one in terms of productivity and efficacy. Neverthless, it has not yet been actually realized in the considered segment. It probably depends on the difficulty of embracing, comprehending and working out all the aspects of this complex field, assuring the same level of competitiveness associated to the present architecture. Moreover, the preliminary phase characterizing the implementation and development of a CAPP method is very heavy and onerous, especially when the variant approach [1], based on parametrical families, is adopted. The latter is not very flexible but quite static, difficult to regenerate and upgrade.

It can now be observed that the existing architecture is made up of about 900 CAD operators, 24 CAM operators and 160 CNC operators. CAD designers deal with project and design problems, whereas the CAM operators are involved in the process planning stage. The latter are very skilled, since they have gained considerable experience; thus they have developed a systematic vision of these interconnected fields. They represent the core of this architecture and their job is essential in this scenario. From a CAPP perspective, on the contrary, all the previous skills and capabilities should be included in the CAPP system, and partly condensed into the CAPP operator, because the way in which the model is generated usually influences and determines the manufacturing process. The CAPP operator should therefore be more sensitive to the manufacturing problems, in comparison to the CAD designer, who usually deals only

with project and design problems. Therefore, in this respect, at least 900 skilled CAPP operators are needed, and likely more, to carry out the new mission. Hence, it seems the CAM
division only to be moved and added to the CAPP division. Thus an effective advantage, especially in terms of human resources optimization, has not been actually achieved.

Finally, some notes about the role played by post processors. Their quality greatly influences
productivity since it determines the degree of optimization of the ISO code. Post processors of
a CAPP system are supposed to be less adaptable and therefore less performing than those
developed in a specific CAM system.

However, it can be considered as a feasible approach, which is up to now the closest to CAPP,
the so called CAD/CAM 3D integrated approach (with both CAD and CAM operator). In fact,
some CAD/CAM 3D interesting powerful softwares already exist. Undoubtedly such packages
are the only way to handle complex 3D geometry, and they could be surely adopted in the
2.5D segment. Nevertheless, in the case of Danieli, the existing architecture is still the best
one, in terms of global input/output balance, in light of the subjects discussed above.

However, the aim of this section was simply to highlight some of the most common misunderstandings and contrasts, arising between theoretical principles and industrial reality, when
speaking about CAD/CAM/CAPP. In particular, it has been pointed out that the applicability
of the CAPP to a system has to be carefully evaluated, comparing it to the present situation
with a critical and realistic attitude. Indeed, the effective complexity of the real industrial
background can't be neglected.

4 A 3D APPROACH APPLIED TO SCULPTURED FIGURES

The CAD/CAM 3D integrated approach could be successfully applied in the field of sculptured figures, which are characterized by very complex geometry, requiring machining phases
at 3-5 axes simultaneously interpolated [2,8]. This is a very small segment for Danieli & C. (in
all its history less than 10 sculptured figures were committed and produced inside), but it could
play a major role, especially in the near future, with the prospect of job orders on behalf of
third parties from the energy field. Up to now these jobs have been completed in the following
way: for each sculptured object a group of skilled CAM operators is formed to create a single
complete parametric CAM macro. This macro generates an optimized CNC part program when
the input arguments are assigned. This macro applies only to the objects of this particular family. Several intelligent working hours are necessary to work out such a complex macro (for
instance the case of Figure 3 requires about 90 hours). On the contrary, by adopting a
CAD/CAM 3D integrated approach it is likely that the generation of the CNC part program
would be strongly aided by the more oriented and effective CAD/CAM 3D instruments. This
method does not replace the CAM manual phase carried out by skilled CAM operators, but it
will be considerably simplified and sped up. First a 3D model is generated using solid modeling software (as ProEngineer, introduced in [4]). Next, the model geometry is directly
imported from the native file into the CAM integrated environment, where the CAM operator
plans the machining phase, using the CAM 3D instruments. Finally, the generated CAM
source is processed and post-processed in order to obtain the CNC part program.

There are two disadvantages:

- a less optimized and performing final ISO code;
- the tendency to restart from the beginning of the 3D process, even for similar components; that is, the working hours are not capitalized as in the previous method.

However these disadvantages are offset by the need to handle with simpler and more adequate instruments, when dealing with such complex geometries. Anyway, these conjectures have to be further developed, in order to plan a design of experiments, whose task is to assess and confirm their validity.

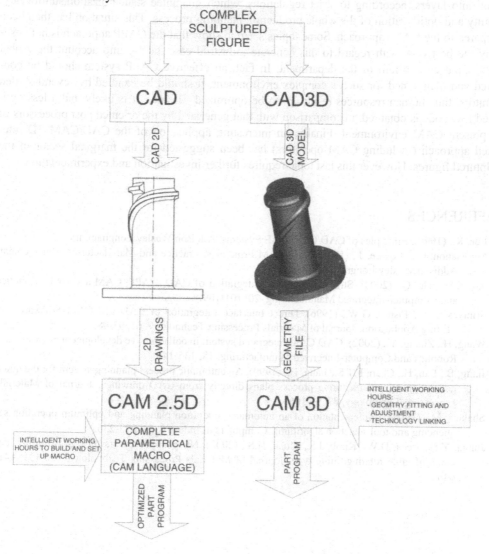

FIGURE 3. Present and alternative method to deal with sculptured figures

5   CONCLUSIONS

The present architecture of the Danieli & C. Manufacturing Department has been described and examined in terms of internal structure. The connection and interdependence between the various divisions have been shown, an input-output analysis has been accomplished. A perspective based on CAM has been adopted, since the CAM division plays a central role in this structure. It has been pointed out that the present architecture assures a satisfactory level of productivity, quality and low costs, since it is based on a strong specialization of the different divisions, a fine stratification into layers, according to strict regulations, which guarantee standardization, uniformity, stability and optimization of the whole production planning process. This situation has then been compared to the CAPP approach. Some topics seem to show that the CAPP approach is not absolutely the best one, with regard to this concrete industrial case, taking into account the global input/output equilibrium of the department. In fact, an effective CAPP system should be conceived and maintained for such a complex environment. It should be handled by several skilled operators, thus human resources result not to be optimized. Besides, it is likely that a less optimized ISO code is obtained, in comparison with that generated by the oriented post processors of the present CAM environment. Finally, an interesting application of the CAD/CAM 3D integrated approach (including CAM operators) has been suggested for the marginal sector of the sculptured figures. However this last topic requires further investigation and experimentation.

REFERENCES

1. Lee, K., (1999), Principles of CAD/CAM/CAE Systems, Addison Wesley Longman, Inc.
2. Mc Mahon, C., Browne, J., (1998) CAD CAM Principles, Practice and Manufacturing Management, Addison Wesley Longman Limited.
3. Xu, X.W., He, Q., (2004), Striving for a total integration of CAD, CAPP, CAM and CNC, Robotics and Computer-Integrated Manufacturing, 20, 101-109.
4. Srinivasan, V., Fisher, G.W., (1996), Direct Interface Integration of CAD and CAM Software - A Milling Application, Journal of Materials Processing Technology, 61, 93-98.
5. Wang, H., Zhang, Y., (2002), CAD/CAM integrated system in collaborative development environment, Robotics and Computer-Integrated Manufacturing, 18, 135-145.
6. Jiang, B., Lau, H., Chan, F.T.S., Jiang, H., (1999), An automatic process planning system for the quick generation of manufacturing process plans directly from CAD drawings, Journal of Materials Processing Technology, 87, 97-106
7. Shakeri, M., (2004), Implementation of an automated operation planning and optimum operation sequencing and tool selection algorithms, Computers in Industry, 54, 223-236.
8. Junga, Y.H., Leea, D.W., Kimb, J.S., Mokc, H.S., (2002), NC post-processors for 5-axis milling machine of table rotating/tilting type, Journal of Materials Processing Technology, 130-131, 641-646.

OPTIMIZED HOTWIRE CUTTING ROBOTIC SYSTEM FOR EXPANDABLE POLYSTYRENE FOAM

P. Gallina[1], R. Mosca[1], P. Pascutto[1]

[1] Dipartimento di Ingegneria Meccanica, University of Trieste, Italy

KEYWORDS: Force Control, Delayed Reference Control, Hotwire Cutting.

ABSTRACT.This paper presents the ideation and implementation of a 2-axes robotic system for hotwire cutting of polystyrene plates. In particular, since the quality of the cutting process is strongly affected by, among others, the interaction force between the hotwire and the workpiece, an accurate force control is required. The force control module, which is referred to as *delayed reference control* (DRC) belongs to the category of non-time based controllers. It is recalled that in a time-based control, the desired input reference is described as a function of time only, which is referred to as reference time: $x_d(t)$. What is relevant is that such a reference can never be modified by any external event. Conversely, according to the DRC theory, the desired input reference x_d is a function of the time and a variable, which plays the role of a time delay: $x_d(t-T)$. The time delay T is properly calculated on-line according to the measured force signal in such a way to improve the quality cutting process during the interaction phase. DRC theory and its practical implementation on a 2-axes robot are presented as well as an accurate description of the cutting process. In fact, experimental results validate theoretical predictions.

1. INTRODUCTION

This paper deals with the optimization of hotwire cutting process of EPS (expandable polystyrene) foam. Although such a technology is simple to be implemented and had become widely spread in industry, only a few research works tried to gain an insight into the physical phenomena of the cutting process [1, 2]. So far, hot wire cutting robots for EPS are mainly employed to produce sheets, plates, containers and other simple geometric shape workpieces.

In most cases it is not required high accuracy during cutting process, but also exist some promising rapid prototyping (RP) techniques which employ EPS foam [3], where optimizing the cutting process is of primary importance. The accuracy and the quality of the cut surface depend on a number of parameters such as wire geometry and length, wire material, temperature of the wire, EPS thickness, mechanical tension on the wire and cutting tool speed. The latter is the most important parameter that rules the process, and so it has to be defined properly in order to achieve good results[4]. In fact, if the tool speed is too high, the wire heat source could not be sufficient to melt the front foam causing the wire to bend. As a consequence the cut surface results irregular showing a typical circular pattern. On the other hand, if the tool speed is too low, the melting zone around the wire tends to increase and to change its shape from a long and narrow area to circular-like one. This is because the EPS can not dissipate the excessive heat.

Published in: E. Kuljanic (Ed.) *Advanced Manufacturing Systems and Technology*, CISM Courses and Lectures No. 486, Springer Wien New York, 2005.

Moreover, low tool speed decreases productivity. Therefore there exists an optimum maximum speed that depends on all the aforesaid parameters.

One way to solve the problem of the optimum speed consists in measuring the force exerted by the EPS on the wire, and then employing this parameter to correct the cutting process parameters. Different strategies have been implemented: adaptive controllers [5], which estimate the parameter involved in the force process model on-line and adjust the force controller gains accordingly; sometimes they exhibit an undesirable dynamic behaviour, mainly because of sensitivity to unmodeled dynamics [6]. Elbestawi et al. [5] developed a deadbeat controller. To this regard the non linear force process has been approximated with a linear term coupled with a third order static nonlinear feed term. In order to regulate the machining force, Harder [7] linearized the force process about a nominal feed. This way, standard control techniques can be applied. Because of the high number of parameters involved in the process also neural network [8], and fuzzy logic [9] controllers have also been applied.

A common feature of all the aforesaid controllers is that they are time-based. Although tracking control is generally aimed at ensuring that the output of a system follows a desired path, or reference input, defined as a function of time (*trajectory following*), when dealing with some advanced applications (e.g. contouring), following a desired path defined only through space (*path tracking*) becomes the primary goal. Tracking controllers can be basically classified into two categories: time based, and non-time based controllers.

The formers are required to guarantee the convergence of the state of the system to a desired state (reference input) which is a pre-planned function of time, commonly calculated off-line by means of a process called pathplanning. During the task, the planned function x_d can never be modified by any event or circumstance: as underlined in Tarn et al. [11], time-based approaches do not represent the best solution for path tracking problems.

Successful examples of non-time based controllers can be found in different fields. The velocity field control approach has been used by Li [12] in coordinated contour control for machine operation. In the velocity field control paradigm, instead of requiring the system to be at a specific location at each instant of time, the system is required to track a velocity reference which is a function of the position of the end-effector. An interesting event-based control strategy has been recently introduced and applied to several different control problems: robot motion control [11], multi-robot coordination [13], force and impact control [14], robotic teleoperation [15] and manufacturing automation [16]. The basic idea behind such a method is considering the reference input a function of an "action reference parameter" instead of a function of time; the action reference parameter is relevant to the sensory measurement and the task. In the event-based approach the role the planner changes: in traditional time-based planning and control, the planner is responsible for preplanning the input reference as a function of the time, while in the event-based control, an "action reference block" is introduced to compute the action reference parameter on-line, based on sensory measurement. The planner generates the desired reference value according to the value taken by the action reference parameter.

In the following a non-time based control scheme for force control in hotwire cutting robotic systems is presented. This new control is based on the modification of the desired input reference x_d: it is a function of time and a variable which plays the role of a time delay: $x_d(t-T)$. The time delay is properly calculated on-line according to the measured force signals in such a way to improve the interaction with the environment. In fact, the DRC consists in an outer force feedback loop around an inner position feedback loop. The effectiveness of the controller has been proven by means of a 2 dof robotic system. Eventually experimental results are presented.

2. THEORY

Basically, the 2 dof robotic system analyzed in this work is made up of two axis linear ball bearing slides, which move in two perpendicular directions. Each axis is operated by a DC motor. One axis moves the cutting tool, while the other one moves the EPS workpiece. A sketch of the system is presented in Figure 1. In order to gain an insight into the problem of cutting process optimization, it is assumed in this section that the workpiece is fixed while the cutting tool moves horizontally along the feed direction. As the matter of fact the workpiece can move along a perpendicular direction for contouring tasks. The position of the cutting tool is referred to as x.

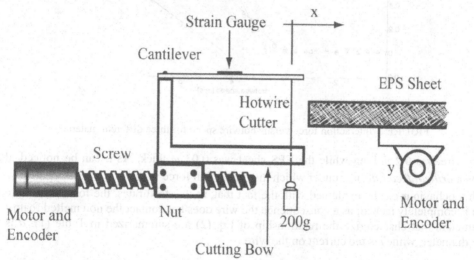

FIGURE 1. Sketch of the controlled axis which moves the cutting tool along the feed direction

As the cutting tool progresses along the feed direction X, the workpiece exerts an interaction force on the tool. In accordance with experimental results, it is assumed that the cutting force acting on the tool is directly related to the feed rate $\dot{x}$.

A good approximation of the function f is:

$$f_p(\dot{x}) = \begin{cases} 0 & if \quad \dot{x} \le v_m \\ q_m + c_m \dot{x} & if \quad \dot{x} > v_m \end{cases} \tag{1}$$

where v_m and c_m are constants and depend on the cutting process and the material, and $q_m = -c_m v_m$: figure 2 shows results obtained employing three different wires: a steel wire, a copper wire and a constantan wire.

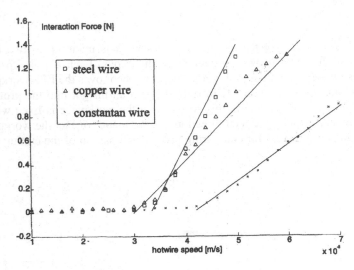

FIGURE 2. Interaction force versus hotwire speed for three different materials

Each wire was 0.2 m long while the EPS sheet was 0.03 m thick. As it can be noticed, there exists a *critical feed rate* v_m, under which the interaction force is null.

Such a behaviour can be explained with the fact that, at low feed rates, the foam in front to the wire is completely melted; as a consequence the wire does not contact the non melted foam.

Parameters that characterize the relationship of Eq. (2) are summarized in Table (1): d is the wire diameter, while i is the current on the wire.

TABLE 1. Parameters that define the relationship between interaction force and feed rate

Wire material	c_m [Ns/m]	v_m [m/s]	q_m [N]	d [m]	i [A]
Steel	835	$3.3 \cdot 10^{-3}$	-2.78	$0.25 \cdot 10^{-3}$	1.31
Copper	476	$3 \cdot 10^{-3}$	-1.43	$0.25 \cdot 10^{-3}$	3.82
Constantan	310	$4.1 \cdot 10^{-3}$	-1.28	$0.2 \cdot 10^{-3}$	0.41

Defining m the moving mass along the X direction, c the damping coefficient, F_c the control force and F_p the interaction force, the equation of motion along the X direction is considered:

$$m\ddot{x} + c\dot{x} = F_c - F_p \qquad (2)$$

Let us assume that the position of the X-axis is controlled by means of a PD controller: defining x_d and x the desired and the actual position of the cutting tool, k_P and k_D the proportional and the derivative gains it follows:

$$F_c = k_P\left(x_d - x\right) + k_D\left(\dot{x}_d - \dot{x}\right) \qquad (3)$$

By replacing Eq. (4) into (2), and considering that F_p is neglectable with respect to the control action force, it yields

$$m\ddot{x} + \left(c + k_D\right)\dot{x} + k_P x = k_P x_d + k_D \dot{x}_d \qquad (4)$$

During cutting phase, high contact forces could be generated. In order to avoid this problem, in the following it will be shown how to improve cutting process by the DRC module.

While usually the desired trajectory of tool is planned off-line and the desired displacement is a function of time $x_d = g(t)$, DRC strategy defines the desired displacement as a function of $l = t - T$. The DRC basically introduces a delay T on the time that defines the desired displacement; assuming that there is a sensor capable of measuring the force F_p, time delay is calculated online by the DRC module accordingly to the following expression:

$$T = \int_0^t \alpha\, F_p\, dt \qquad (5)$$

with $\alpha \in R$. The desired displacement becomes:

$$x_d(t - T) = g(t - \alpha \int_0^t F_p\, dt\,) \qquad (6)$$

The whole control scheme of the system is shown in Figure 3.

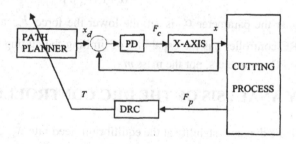

FIGURE 3. DRC control scheme

By replacing Equations (6) into (4), the closed-loop non linear dynamic equation of the system is carried out

$$m\ddot{x}+(c+k_D)\dot{x}+k_Px =k_Pg(t-\alpha\int_0^t F_p dt)+k_D\frac{d}{dt}\left(g(t-\alpha\int_0^t F_p dt)\right) \tag{7}$$

In order to establish the dynamic conditions at the steady state equilibrium, Equation (7) is derived:

$$m\ddot{x}+(c+k_D)\ddot{x}+k_P\dot{x} =k_P\frac{dg}{dl}(1-\alpha F_p)+k_D\left(\frac{d^2g}{dl^2}(1-\alpha F_p)^2-\frac{dg}{dl}\alpha\dot{F}_p\right) \tag{8}$$

If the system was stable, at the equilibrium point, both the third derivative of x and acceleration would be null. Considering the fact that at the steady state equilibrium $\dddot{x}=\ddot{x}=\dot{F}_p=0$ and that, if path trajectory is smooth enough, $\frac{d^2g}{dl^2}\ll 1$, from Eq. (8) the equilibrium feed rate is

$$\dot{x}_{eq}=\frac{dg}{dl}(1-\alpha F_p)=\beta(1-\alpha F_p) \tag{9}$$

where $\beta=dg\ dl$. If the theoretical relationship between F_p and the feed rate $\dot{x}$ is given by equation (1), the equilibrium feed rate is

$$\dot{x}_{eq}=\begin{cases}\beta & if \quad \beta\leq v_m \\ \dfrac{\beta(1-\alpha\ q_m)}{1+\beta\ \alpha\ c_m} & if \quad \beta>v_m\end{cases} \tag{10}$$

Therefore $\dot{x}_{eq}$ represents the steady state equilibrium point of the system. In other words, at the equilibrium, the feed rate is constant and related to the parameter α,β and c_m. The main improvement introduced by the DRC controller is that, at the equilibrium point, the force is

$$F_{eq}=f_p(\dot{x}_{eq})=\frac{1+q_m\ (c_m\beta)}{\alpha+1\ (c_m\beta)} \tag{11}$$

Therefore, the higher the parameter α is set, the lower the force F_{eq} at the equilibrium is. In conclusion the DRC controller has the capability of limiting the cutting force. Note that F_{eq} is not affected by the PD parameters, nor the mass m.

3. STABILITY ANALYSIS OF THE DRC CONTROLLER

In order to analyze the dynamic stability at the equilibrium feed rate $\dot{x}_{eq}$, the desired displacement function $g(\cdot)$ will be linearized around the equilibrium condition:

$$x_d = g(l) = g(l_{eq}) + \frac{dg}{dl}(l - l_{eq}) = g(l_{eq}) + \beta(l - l_{eq})$$

$$\dot{x}_d = \frac{dg}{dl}\frac{dl}{dt} = \beta\left(1 - \alpha F_p\right) \tag{12}$$

$$\ddot{x}_d = -\beta\alpha\dot{F}_p$$

Substituting these values into eq. (8), we obtain

$$m\,\ddot{v} + \left(c + k_D + k_D\beta\alpha\right)\dot{v} + k_P\left(1 + \beta\alpha\,c_m\right)v = k_P\,\beta\left(1 + \alpha\,c_m\,v_m\right) \tag{13}$$

where $v = \dot{x}$. From this equation, it can be easily inferred that the system is always stable, since the Eq. (18) is equivalent to a stable single-degree-of-freedom system, where the equivalent mass is m, equivalent viscous damping is $\left(c + k_D + k_D\beta\alpha\right)$ and equivalent spring constant is $k_P\left(1 + \beta\alpha\,c_m\right)$.

4. EXPERIMENTAL SET-UP

The experimental apparatus is composed of two precision servo controlled slides, a hotwire cutter, a force sensor, and a control software (see Fig. 4).

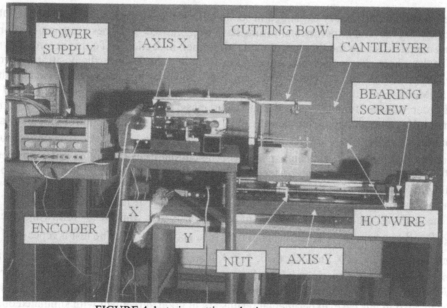

FIGURE 4. hotwire cutting robotic system prototype

The picture shows the front view of the X-slide which moves the hotwire cutter along the X direction. The table where the hotwire is attached to, is moved by means of a high precision ball bearing screw whose pitch is 0.004 m. The screw is directly operated by means of a DC motor connected to one side of the screw by means of an oldham coupling. The motor is operated by a 25 A PWM servo amplifier. A 2000 cpr encoder is clamped to the other side of the screw. The picture shows also the side view of the Y-slide which moves the EPS workpiece; mechanical and electrical components of the Y-slide are the same of the X one. As it can be noticed, X-slide and Y-slide are located at two different heights for a better exploitation of the workspace. The hotwire employed in all the tests is a 200 mm long steel wire; the diameter is 0.2 mm; the wire is heated by a 1.3 A current; the wire is initially pretensioned with a 2 N load. The wire is clamped to a cutter bow whose lower arm is assumed to be rigid while the upper one is realized by means of a 1 mm thick flexible cantilever. The cantilever is 50 mm large and 130 mm long. The interaction force F_p between the EPS foam and the wire is measured by means of two strain-gauges glued to the two sides of the cantilever and connected in a half-bridge configuration conditioner. In fact, when the foam exerts a force to the wire, the cantilever tends to bend.
The two axes are controlled by means of a two axes PCI NI 7344 servo control boards programmed in c language using the NI CVI compiler.

5. EXPERIMENTAL RESULTS

In order to validate the theory explained in section 2 and the benefits introduced by the DRC in terms of quality, several experimental results have been carried out. These tests aim to compare experimental results with the theoretical model described in section 2. To this regard, a dynamic model of the whole electromechanical apparatus, including the X-slide, the position control, the cutting process, the force sensor and the DRC module has been implemented in the Simulink/Matlab environment. The dynamic behaviour of the X-slide is schematized by a second-order transfer function whose equivalent mass and viscous damping coefficient are estimated by a common parameter identification method. The position control block is a simple PD controller whose gains are set to $k_P = 30$N/m and $k_D = 15$Ns/m both in the simulation environment and in the experimental apparatus. The cutting process is represented by means of Eq. (7) taking into account that a steel wire is employed, while the force sensor operates according to Eq. (1). During the first test a 0.03 m thick EPS sheet is fixed to the Y-slide which remains at rest. At the beginning, the wire is located at 0.05 m far away from the EPS sheet. Than the cutting tool is required to move toward the workpiece following the path $x_d = g(l) = \beta\, l$, where β is set constant at 0.01 m/s, and $\alpha = 0.75$. The test is repeated 5 times with increasing values of α, namely 0.75, 1.5, 3, 6 and 10, while β remains constant in all the 5 trials. Simulation and experimental results are shown in Fig 5, which represent the interaction force F_p versus time. The contact between wire and foam occurs approximately at 8 s. As it can be noticed, also in this case, experimental results are in good agreement with simulated ones.

Note that, according with Eq. (11), the higher the value of α is and the lower the steady-state contact force F_p is. This is an important result, since, in order to increase the cutting process quality, it is sufficient to increase the value of α .

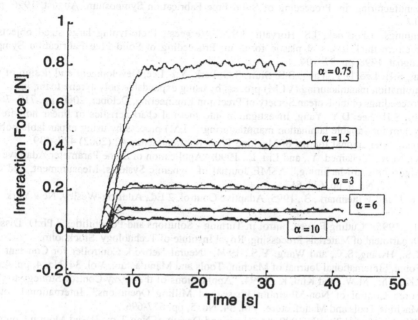

FIGURE 5. Interaction force versus time along a linear path on the X direction, for different values of α ; β =0.01 m/s is constant. Solid lines represent experimental results; dot lines represent simulated results

6. CONCLUSIONS

This paper presented the ideation and implementation of a non-time based force controller called delayed reference control (DRC) applied to a 2-axes robotic system for hotwire cutting of polystyrene plates. Whit respect to other force controllers, the DRC is easy to be implemented since it consists in an outer force control loop around an inner position control loop. Moreover it avoids to replan the path on-line. Simulation and experimental results are carried out in order to validate both theoretical assumptions and benefits. In our opinion, for its simplicity and its effectiveness, this force control strategy could represent a good alternative to traditional force control in many other applications especially for CNC machines.

REFERENCES

1. A. Novc, S. Kaza, Z. Wang, C. Thomas, Techniques for improved speed accuracy in layered manufacturing, in: Proceeding of Solid Free Fabrication Symposium, August, 1996, pp. 609–617

2. A.F. Lennings, J.J. Broek, I.S. Horvath, J.S.M. Vergeest, Proto-typing large-sized objects using freeform thick layers of plastic form, in: Proceeding of Solid Free Fabrication Symposium, August, 1998, pp. 97–104.

3. D.G. Ahn, S.H. Lee, D.Y. Yang, B.S. Shin, S.G. Park, Y.I. Lee, Development and design of variable lamination manufacturing (VLM) process by using expandable polystyrene foam, in: Proceedings of the Korean Society of Precision Engineering, October, 2000, pp. 719–762.

4. D.G. Ahn, S.H. Lee, D.Y. Yang, Investigation into thermal characteristics of linear hotwire cutting system for variable lamination manufacturing (VLM) process by using expandable polystyrene foam, International Journal of Machine Tools & Manufacture 42 (2002) 427–439.

5. Elbestawi, M.A., Mohamed, Y., and Liu, L., 1990, "Application of Some Parameter Adaptive Control Algorithms in Machining," ASME Journal of Dynamic Systems, Measurement, and Control, Vol. 112, No. 4, pp. 611-617.

6. Åström, K.J. and Wittenmark, B., 1995, Adaptive Control, 2 Ed., Addison-Wesley, New York.

7. Harder, L., 1995, "Cutting Force Control in Turning - Solutions and Possibilities," Ph.D. Dissertation, Department of Materials Processing, Royal Institute of Technology, Stockholm.

8. Tang, Y.S., Hwang, S.T., and Wang, Y.S., 1994, "Neural Network Controller for Constant Turning Force," International Journal of Machine Tools and Manufacture, Vol. 34, No. 4, pp. 453-460.

9. Kim, M.K., Cho, M.W., and Kim, K., 1994, "Applications of the Fuzzy Control Strategy to Adaptive Force Control of Non-Minimum Phase and Milling Operations," International Journal of Machine Tools and Manufacture, Vol. 34, No. 5, pp. 677-696.

10. Kang, W., Xi, N., and Tan, J., 1999, "Analysis and Design of Non-Time Based Motion Controller for Mobile Robots", in Proceedings of the 1999 IEEE International Conference on Robotics and Automation, May, Detroit, USA.

11. Tarn, T. J., Bejczy, A. K., and Xi, N., 1993 "Intelligent Planning and Control for Robot Arms", in Proceedings of the IFAC 1993 World Congress, July, Sydney, Australia.

12. Li, P. Y.: Coordinate Contour Following Control for Machining Operation – A Survey, in Proc. of the American Control Conference, San Diego, 1999, pp. 4543-4547.

13. Xi, N, Tarn T. J., and Bejczy, A. K., 1993, "Event-Based Planning and Control for Multi-Robot Coordination", in Proceedings of the 1993 IEEE/ICRA, May, Atlanta, USA.

14. Wu, Y., Tarn, T. J., and Xi, N., 1995, "Force and Transition Control with Environmental Uncertainties", in Proceedings of the 1995 IEEE/ICRA , May, Nagoya, Japan.

15. Tarn, T. J., Xi, N., Guo, C., and Bejczy, A. K., 1995, "Fusion of Human and Machine Intelligence for Telerobotic Systems", in Proceedings of the 1995 IEEE /ICRA, May, Nagoya, Japan.

16. Xi, N., and Tarn, T. J., 1997, "Integrated Task Scheduling and Action Planning/Control for Robotic Systems Based on a Max-Plus Algebra Model", in Proceedings of 1997 IEEE/RSJ International Conference on Intelligent Robotics and Systems, September, Grenoble, France.

REVIEW ON MICROMACHINING TECHNIQUES

E. Gentili[1], L. Tabaglio[1], F. Aggogeri[1]

[1] Department of Mechanical Engineering, University of Brescia, Italy

KEYWORDS: Micromachining, Technologies, Review.

ABSTRACT. Micromachining is related to specific techniques applied to micro and meso scale elements, in order to produce components with high precision and very restrictive dimensional and geometrical tolerances (micron or sub-micron). In the industrial world the interest in microscopic scale manufacturing is exponentially increasing in relation to the rapid growth of Micro Electro Mechanical Systems (MEMS) research. Thus a greater attention is given to improve traditional techniques and developing non-conventional machining methods, in order to obtain more precision. This paper intends to present a review of the actual state of art in micromachining, showing possible different ways to create high aspect ratio patterns or 3D sculptured workpieces. Therefore the most important techniques will be presented, focusing on photolithography (bulk & surface machining), LIGA, laser, micro-EDM, micro-USM and micromechanical machining (microcutting and micromilling). Machinable materials, obtainable working tolerances, limits and problems will be also highlighted for every method.

1 INTRODUCTION

Micromachining technologies are strictly connected to the recent evolution of MEMS. The growth of Micro Electro Mechanical Systems and the related research in different industrial sectors started from microchip technology on silicon wafer, to involve the ones in fields such as the automotive, aerospace, microrobotics, optical and biomedical. A greater interest is focused on the achievement of 3D sculptured surfaces and high aspect ratios with complex fine shapes. Therefore the research is active in characterising new technologies or improve known processes to guarantee fine requested precisions and low costs of manufacturing, so as to facilitate a real growth and diffusion in the industrial world. Obtainable precisions, accuracy, geometrical and dimensional tolerances, thermal and mechanical deformations, surface quality and roughness need to be controlled in order to realise effective processes and to achieve a real improvement in micro-manufacturing technologies. Therefore it is necessary to characterise all the process parameter effects in machining results for different fabricating methods. Hence a review of most representative micromachining technologies will be given, starting from standard lithographic methods to non-traditional technologies and mechanical micromachining techniques.

2 SILICON MICROMACHINING TECHNOLOGIES

Traditional silicon micromachining technologies are based on the lithographic approach, directly from etching and deposing processes used in microelectronics: silicon wafers are machined with chemical or physical etch and elements are realised layer by layer from a silicon substrate. This non-contact method is based on masking and light exposure, without any mechanical interaction

Published in: E. Kuljanic (Ed.) *Advanced Manufacturing Systems and Technology*,
CISM Courses and Lectures No. 486, Springer Wien New York, 2005.

with a tool. The final component is obtained patterning the surface (sub-micron precision) of the workpiece, with a simple bi-dimensional approach [1-2-3]. It's possible to realise thick structures step by step, with a sequence of adding thin films, etching unmasked parts and bonding elements. A pattern of a mask is transferred to a photosensitive layer on the workpiece surface, which is commonly used as a mask itself for pattern transfer on the substrate. The resist layer is exposed to a selective incident light radiation that changes its physical properties and chemical resistance to the developer solution. After developing the resist is used as a mask to protect the underlying layer from etching (structured thin films are used as well as masks), or to realise a selective deposition of new films [1].

2.1 BULK & SURFACE MACHINING

Bulk machining is a subtractive method used for fabrication of silicon devices, based on wet or dry etching techniques. In wet etching the material aggression is achieved by a chemical liquid solution, while the dry etch is obtained by physical or chemical plasma etching. In wet etching processes isotropic (uniform etch rate in all directions) or anisotropic (crystal orientation dependent) etchants can be adopted and doping or etch stop methods are used to control depth machining in silicon wafers. Undercutting effects must be considered in isotropic etch as well as achievable geometries in wet anisotropic etching are limited by crystallographic structure of the silicon substrate. Dry etching involves different physical methods like sputtering and ion milling, chemical approaches and combinations of physical and chemical etch processes like reactive ion etching (RIE), deep RIE and reactive ion beam etching [1-2-3]. Wet etching is mainly used in silicon and glass machining, while dry etching is especially adopted for metals, ceramics and plastic materials [2-3]. While wet etching is a batch method based on immersion of workpieces in an aggressive liquid solution, dry etching can be applied on a single element or a limited number of components in a controlled chamber. Therefore dry etching gives a more expansive technique, primarily adopted for thin layers machining or to obtain particular high quality cross sections [3]. In bulk machining it is difficult to obtain high thickness (up to 100 microns), even if bonding methods (anodic or fusion bonding) can be used to accomplish complex thick structures [1-2-3]. Surface machining is an addictive technique to obtain moveable structures from the starting substrate, based on subsequent depositions and selective etches of thin (from sub micron to several microns dimensions) sacrificial and structural layers. Sacrificial films are used as spacers while structural layers are deposed and patterned to obtain the final element. After the developing of the sacrificial layer the moveable component is then released [2]. Most common deposing methods are chemical vapour deposition (CVD), electrodeposition, epitaxy, thermal oxidation (for sacrificial layers) and physical vapour deposition (PVD - evaporation or sputtering technologies). The main limits of traditional lithographic techniques are related with 2D-based patterning (Figure 1) and process costs. Other problems are also related to machinable materials and masks alignment in layer by layer processes [1-2-3].

 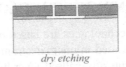 

anisotropic etching in KOH   *dry etching*   *surface machining*

FIGURE 1. Silicon technologies [1]

While these techniques can be used for mass production in microchip manufacturing, industrial applications for complex three-dimensional systems in micro scale are still limited by materials, geometrical limits (design freedom [1]) and fabrication process costs and times.

2.2 LIGA & HIGH ASPECT RATIO MACHINING

High aspect ratio processes are used to obtain high surface micro elements precision with high thickness and straight sidewalls. LIGA technique is the most known method and is primarily used for the production of single moulds for plastics components replication. LIGA (Litographie Galvanoformung Abformung) process is based on lithography with X-ray source. A PMMA layer is coated on a conductive substrate and then exposed to a hard X-ray synchrotron radiation to transfer a pattern from a mask. Using small wavelength (0.2 nm) and low-divergence intense radiation is possible to achieve deep penetrations in the resist. After developing, the polymeric obtained microstructure can be used for a subsequent electroplating process (Figure 2). A high aspect ratio metal element is then produced, which is typically used for hot-embossing or as a mould for micro injection moulding (plastic microstructures) or to replicate metallic (mainly Nickel based) or ceramic components [4]. Obtainable components are characterised by a high aspect ratio (more then 100:1), thickness of thousands micrometers, sub-micron pattern precisions and vertical smooth sidewalls [4-5]. The main limits for this technology are the availability of a synchrotron accelerator and production costs (process time, masks) of the single mould or master element.

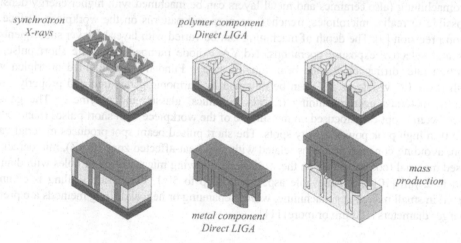

FIGURE 2. LIGA process [5]

Some different techniques, similar to the LIGA process, are now being developed using an UV radiation as exposition source. The S-LIGA method uses different photoresists (SU-8) and allows good precisions (in micron scale) and aspect ratios (up to 60:1) with a much cheaper process then LIGA. Subsequent operations can also be executed obtaining complex multilayered structures. The method can be used both for the production of moulds and for direct electroplating. [6-7] Limits of the process are minor precisions and aspect ratios in regard to LIGA and in obtainable metal components by electroplating (mainly Nickel-based structures).

3   NON-TRADITIONAL MICROMACHINING TECHNOLOGIES

Non-traditional machining techniques are being developed also in microscopic scale, in order to achieve real 3D sculptured complex components. Interesting methods are in particular laser micromachining, micro electro discharge machining and micro ultrasonic machining.

3.1   LASER MICROMACHINING

Laser technology uses light radiation with high energy density as a machine tool and appears as a possible efficient system for micromachining a wide range of materials without any mechanical or chemical interaction with the workpiece [8]. Excimer and Nd:YAG Diode pumped Lasers are mainly adopted for a micromachining operation due to their short wavelength (Figure 3) [9]. Excimer lasers are particular gas-based lasers able to produce short pulses with UV wavelength. The beam profile of an excimer laser is rectangular, with a top hat shape in the long axis and a gaussian profile in the short one. Due to its low coherence characteristics and short wavelength the beam can be exposed on the workpiece surface with a mask projection method (no interference). Complex planar shapes can be obtained by moving the mask as well as the workpiece, allowing large areas machining. High precision (in micron scale) can be achieved and some optical demagnifications are normally used to increase the energy density on the surface, hence the material removal is obtained by ablation. With this technology, typically adopted in polymers micromachining (also ceramics and metal layers can be machined with higher energy densities), is possible to realise microholes, trenches, grooves and patterns on the workpiece surface with micron precision [9]. The depth of machining can be varied with layer by layer subsequent ablations and selective exposure operations. Nd:YAG diode pumped lasers are short pulse, high repetition rate, diffraction limited beam quality lasers. Fundamental, doubled or tripled wavelength (near IR, visible or UV) can be obtained by harmonic generation and properly used in different materials micromachining (metals, ceramics, glasses and polymers). The gaussian-shaped beam is optically focused on the surface of the workpiece with short pulses (nano or pico-second) in high peak power density spots. The short pulsed beam spot produces material vaporisation, avoiding common problems related with high heat-affected zone (HAZ), microcracks and refused material (corona effect) on the workpiece, allowing microdrilling of holes with diameters down to 2 µm [10] and achievable aspect ratios up to 50:1. Percussion drilling is commonly adopted in small microholes machining, while trepanning or helical drilling methods are preferred for larger diameters (100 µm or more) [11].

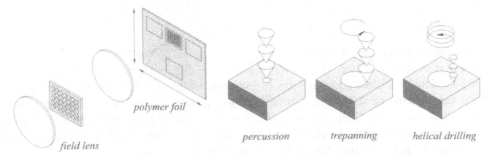

polymer foil

field lens   *percussion*   *trepanning*   *helical drilling*

FIGURE 3. Imaging and focused laser drilling [9]

Femtosecond (i.e. Ti:Sapphire) lasers finally represents the last step in focused laser micro-machining research. With ultra-short pulses (100 fs or less) higher precision can be achieved and micron or sub-micron features can be realised (the pulse terminates without any interaction with the plasma plume generated during ablation) [12-13]. Even if laser micromachining technology is not yet as mature as other methods presented here (see i.e. photolithographic technologies and micro-EDM machining) it appears as a real hopeful way (small manufacturing times) to machine microscopic components (focused beam could allow real 3D shaping by correct motion control) and is already used as well as micro-EDM to drill microholes for different industrial application (medical, optics, electronics and automotive).

3.2 MICRO ELECTRO DISCHARGE MACHINING

In micro electro discharge machining the erosive action of an electrical discharge between con-ductive tool and workpiece is used to remove material. Electro-thermal erosion (material melting and evaporation for every spark) creates small craters both in the piece and in the tool during machining progress, then tool shape is copied in the workpiece with a no contact system [14]. With this technique it is possible to machine not only difficult to cut metals (like hardened steels) or carbides but also semiconductors and conductive ceramics. The discharge results from the application of an electrical voltage between electrodes (tool and workpiece) with a dielectric flushed fluid interposed. The machining process is driven by assigned and controlled gap, volt-age, energy and frequency of discharge. High frequencies (>200Hz) and small energies ($10^{-6} \div 10^{-7}$ J) for every discharge (~100V) are requested to obtain high accuracy and good surface quali-ties (roughness of about 0.1 μm). Sparks are traditionally generated by a relaxation type of power supply (high frequencies achievable and low costs). Micro electro discharge machining is com-monly used to microdrill or to pattern trenches in workpieces, with a micromilling approach and simple shaped tools. Circular rotating micro tools are mainly adopted and on-the-machine made, in order to avoid off-centering and tilting errors and re-chucking imprecisions, improving at the same time roundness quality during machining. Tools are shaped from a rod (i.e. copper or tung-sten carbide) directly by a reverse EDM machining method called wire electro-discharge grinding (WEDG). The process is based on tool direct machining by feeding against a sacrificial wire travelling in a circular guide (avoiding wire bending effects). After the preparation process the tool can be actively used to machine the proper workpiece (Figure 4).

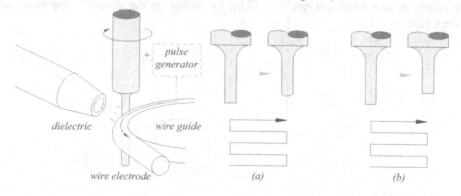

FIGURE 4. WEDG tool shaping [28], tool wear without (a) and with (b) uniform wear method [16]

Tools of 20 μm diameters and smaller have been produced as microdrilled holes with aspect ratio more then 5:1, but higher aspect ratios are achievable with greater diameters [14-15]. Curved features are obtainable by controlling tool-trajectory during the process and correct path generation. Main process difficulties are related with the tool wear control during machining (some problems are also recast layer and microcracks). Particular approaches (i.e. uniform wear method, see Figure 4) could guarantee a continuous shape-recover of the tool during micromilling, due to tool rotation (~2000 rpm) and wear compensation in layer by layer machining with assigned trajectories. Thus tool wear rate and current length of the electrode must be considered to correct the machining process [16]. A particular interest is also focused on the development of a dedicated CAD-CAM tool to improve process capabilities in 3D effective micromachining [17]. Machining time and materials removal rates are mainly limited by small discharge energies applied to guarantee workpiece requested precision. Particular precision motion controls are obviously requested, with machine motors resolution below the micron scale. Some improvements in machining efficiency are also obtainable with ultrasonic tool or workpiece vibration superimposing during drilling operations, in relation to the better removal of debris [18-19]. Despite of the problems related to process optimisation, this technique appears to have good chances both for prototyping and massive production of moulds (real three-dimensional components micro manufacturing seems to be possible [20]) and is yet effectively adopted in the industrial world, i.e. in machining critical fluidic components like inkjet nozzles and diesel injectors.

3.3 MICRO ULTRASONIC MACHINING

Microultrasonic machining is a technology that uses micro tool ultrasonic vibration (up to 30÷40 kHz) to create accurate holes in hard and brittle materials such as silicon, borosilicate glass, quartz and ceramics [21-22-23]. An abrasive slurry is interposed between tool and workpiece and the tool is used as a micromill to obtain drills, trenches or pattern on the workpiece surface. The vibrating tool impacts abrasive grains into the workpiece producing a mechanical removal of the material. Typically the micro ultrasonic machining results as a combined technique, which uses WEDG process to fashion carbide tungsten microtools with on-machine-making method. Circular or stepped rotating tools are used in microdrilling (Figure 5). Microholes as small as 15 micrometer have been obtained (depth of 32 μm) and drills with aspect ratios of 10:1 are achievable with greater tool diameters. Undesired bending effects of the tool during non-machining time (poor quality results or tool breakage) can be avoided by setting up the ultrasonic apparatus on the workpiece table.

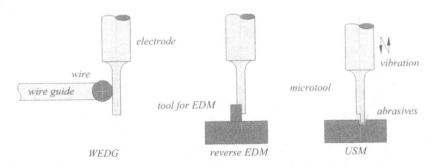

FIGURE 5. MicroUSM procedure[21]

Conditioning process parameters are ultrasonic vibration amplitudes (1÷3.5 μm) slurry concentration (10÷30 % in kerosene or oil medium), abrasive grain materials and sizes (mainly 0.5÷3 μm), rotational speed of the tool (0÷300 rpm) and assigned feed rate (5÷10 μm/min)[21-22]. Good roundness and surface quality (roughness of 0.2 μm) with limited hole diameter variation values are achievable by drilling with small feed rates and correct rotational speed. Microultrasonic machining can be used stand alone as well as a finishing process for microdrilling by μ-EDM [23]. Furthermore the integration of these methods can be adopted to machine a larger set of materials (not only conductive ones) with the same machining structure, reducing working times and improving production quality.

4 MECHANICAL MICROMACHINING TECHNOLOGY

A further field of active research in micromachining technologies involves the optimisation of cutting processes for micromilling, turning and microgrooving operations for a wide range of materials. Due to mechanical interaction between tool and workpiece and dynamic of cutting operations, tool and machine tool characteristics are the fundamental keys to be analysed to guarantee the effectiveness of this technique. An effective CNC support is also obviously required in order to machine curvilinear features and sculptured three-dimensional surfaces. Tool dimensions and cutting edge sharpness represent the first problem in developing this technology in micrometer scale. The microtools must show high robust qualities and be able to carry out correct cutting machining without bend or fail during the process. Grinding operations, normally adopted in producing tools with dimension down to 100 μm, are difficult to be extended to smaller ones (<50 μm) due to relatively high machining forces and obtainable precision [24]. Thus other techniques (Laser, EDM) have been studied and tested to fashion effective micro tools. Recent research developed tools with dimensions smaller then 50 μm (22μm) using Focused Ion Beam Machining (Figure 6). In this way steel, carbide and single diamond microtools with high cutting edge sharpness values (<100 nm) have been successfully manufactured. In micromilling operations ultraprecision machines are commonly adopted (granite and air bearings for all axis), with motions controlled by laser interferometry or linear encoder to achieve submicron resolutions; micromilling tool are chucked in a v-block diamond bearing, in order to minimise radial errors (1 μm). With this set-up trenches and curved features were machined in PMMA, Al and brass workpieces with an aspect ratio from 1 to 2.

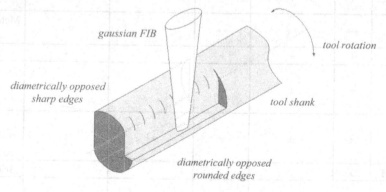

FIGURE 6. FIB micro-end mill shaping [25]

Typical process parameters are high rotational speed (20000 rpm), low feed rates (2÷3 mm/min) and depth per pass (0.5÷1 μm), in order to assure low roughness values (<100 nm on trench bottom) and high machining quality [25]. A similar approach can be adopted in lathe processes, for circular grooves micromachining on planar surfaces and for cylindrical workpieces microgrooving and microthreating [26-27]. Burr (and chip) formation dynamics, as well as tool wear, are relevant problems in micromechanical machining, governed by cutting forces, process parameters and tool geometry (sharpness of cutting edges). Thus main problems in developing this technology give correct tool-shaping, machine tool structure characterisation, process control and right definition of all machining parameters in micron scale. Not only ultraprecision machine tools but also "microfactories" are under active research in order to develop optimised structures for micromechanical machining and improve manufacturing quality. Some interesting experiments were also made in integrating μ-EDM and micromilling processes, using micromilling tools by WEDG and the micro-EDM machine as a milling machine tool [28]. Scaling laws in mechanical micromachining require theoretical high tool rotating speeds and accelerations, however it is still difficult to correctly govern all process parameters, trajectories and mechanical effects during machining. A development of this technique is therefore still required to avoid actual mechanical and electronic technical limits, in order to reach a complete three-dimensional approach to milling operations.

5 DISCUSSION

The development of different techniques instead of lithographic methods for silicon micromachining could open a wide range of possibilities. Microdrilling and planar surface machining operations had achieved a good level of development and different technologies (mainly laser and micro-EDM machining techniques) are nowadays used in effective industrial applications. Nevertheless, despite of great evolutions made in a few years, it is still difficult to gain real three-dimensional micromachining operations in microscopic scale manufacturing.

TABLE 1. Comparison of micromachining techniques

TECNIQUE	MIN. DIM.	ACCURANCY	ASPECT RATIO	GEOM. FREEDOM	ROUGHNESS	MATERIALS
Bulk-Surface	+	+	-	-	+	Semiconductors metals
LIGA	++	++	++	±	++	Metals polymers ceramics
S-LIGA	+	+	+	±	+	Metals
Laser	+	±	±	+	±	Metals polymers ceramics
Micro-EDM	±	±	±	++	±	Metals semiconductors cond. ceramics
Micro-USM	±	±	±	±	±	Brittle materials
Micromilling	±	±	±	±	±	Metals polymers

Lithographic techniques are limited by machinable materials, bi-dimensional approach and real high aspect ratios are difficult to achieve without expensive subsequent masking operations (alignment problems) and bonding methods. The LIGA method appears still too expensive (due to synchrotron apparatus) and single mould production costs are justified only for large scale productions, with a considerable economic consequence and a good yield rate. The S-LIGA technique appears as a possible effective alternative for industrial moulds and massive components production, even if all the limits related with mask-based process still stands. Thus the fast development of alternative presented methods could represent a future solution for the micromanufacturing world, identifying the right technology to adopt in relationship to specific application requests. Nevertheless, problems are still related to the complete comprehension of all different process parameter effects on workpiece output. At the same time actual effective motion control and CNC support will be also necessary for future real complex features and 3D shapes manufacturing. Interesting flexibility improvement and cost savings finally seems to be achievable with the integration of different techniques in a single machine.

6 CONCLUSIONS

In this paper a brief review of micromaching technologies was given. Some of the most common methods recently developed were examined, qualitatively highlighting process capabilities for each of them. All reported results, process limits and related information, came from different studies and university research developed over the last few years. However the rapid growth of new micro manufacturing technologies and fast development of more traditional ones could soon invalidate punctual values as reported and some of the considerations exposed. This work is intended to show the lastest evolutions in the micromanufacturing field, prompted by industrial needs. In conclusion, even if shown technologies are not yet mature enough for a complete surface sculpturing, the near future will probably see an effective approach to real three-dimensional micromanufacturing.

ACKNOWLEDGMENTS

Special thanks to Mrs. Mary Flynn who checked the manuscript.

REFERENCES

1. Lang W., (1996), Silicon microstructuring technology, Materials Science and Engineering, R I7, 1-55.
2. Mehregany M., Dewa A.S., (1993), MCNC Short Course Handbook, Case Western Reserve University, Cleveland, OH, http://mems.cwru.edu/shortcourse/
3. Mems beginner's guide, www.memsnet.org
4. Ehrfeld W., Lehr H., (1995), Deep x-ray lithography for the production of three-dimensional microstructures from metals, polymers and ceramics, Radiat. Phys. Chem. 45 (3) 349-365.
5. Malek C.K., Saile V., (2004), Applications of LIGA technology to precision manufacturing of high-aspect-ratio micro-components and systems: a review, Microelectronics Journal 35, 131-143.
6. Bischofberger R., Zimmermann H., Staufert G., (1997), Low-cost HARM process, Sensors and Actuators A 61, 392-399.
7. Dentinger P.M., Krafcik K.L., Simison K.L., Janek R.P., Hachman J., (2002), High aspect ratio patterning with a proximity ultraviolet source, Microelectronic Engineering 61-62, 1001-1007.

8. Meijer J., Du K., Gillner A., Hoffmann D., Kovalenko V.S., Masuzawa T., Ostendorf A., Poprawe R., Schulz W., (2002), Laser Machining by Short and Ultrashort Pulses, State of the Art and New Opportunities in the Age of the Photons, Annals of the CIRP V.51, 531-550.
9. Paetzel R., (2002), Comparison Excimer Laser – Solid State Laser, Proceedings of ICALEO 2002.
10. Choi S.S., Jung M.Y., Kim D.W., Yakshin M.A., Park J.Y., Kulk Y., (1998), Fabrication of microelectron gun array using laser micromachining, Microelectronic Engineering 41/42, 167-170.
11. Herbst L., Quitter J.P., Ray G.M., Kuntze T., Wiessener A.O., Govorkov S.V., Heglin M., (2003), High peak power solid state laser for micromachining of hard materials, Proc. SPIE V. 4968, 134-142.
12. Semerok A., Chaléard C., Detalle V., Lacour J.-L., Mauchien P., Meynadier P., Nouvellon C., Sallé B., Palianov P., Perdrix M., Petite G., (1999), Experimental investigations of laser ablation efficiency of pure metals with femto, pico and nanosecond pulses, Applied Surface Science 138-139, 311-314.
13. Rizvi N.H., (2003), Femtosecond laser micromachining: Current status and applications, RIKEN Review 50, 107-112
14. Reynaerts D., Heeren P-H., Van Brussel H., (1997), Microstructuring of silicon by electro-discharge machining (EDM) – part I: theory, Sensors and Actuators A 60, 212-218.
15. Rajukar K.P., Narasimhan J., Chandrasekaran V., Yu Z., (2004), Study of EDM generated micro and meso holes, Proc. of JUSFA Japan-USA Symposium on Flexible Automation, US_013.
16. Narasimhan J., Yu Z., Rajurkar K.P., (2004), Tool wear compensation and path generation in micro and macro EDM, Transactions of NAMRI/SME, Vol. 32.
17. Zhao W., Yang Y., Wang Z., Zhang Y., (2004), A CAD/CAM system for micro-ED-milling of small 3D freeform cavity, Journal of Materials Processing Technology 149, 573-578.
18. Murali M., Yeo S.H., (2004), Rapid Biocompatible Micro Device Fabrication by Micro Electro-Discharge Machining, Biomedical Microdevices 6:1, 41-45
19. Huang H., Zang H., Zhou L., Zheng H.Y., (2003), Ultrasonic vibration assisted electro-discharge machining of microholes in Nitinol, J. Micromechanics and Microengineering 13, 693-700.
20. Meeusen W., Reynaerts D., Peirs J., Van Brussel H., Dierickx V., Driesen W., (2001), The Machining of Freeform Micro Moulds by Micro EDM; Work in Progress, Proc. MME 2001, 46-49.
21. Sun X., Masuzawa T., Fujino M., (1996), Micro ultrasonic machining and its applications in MEMS, Sensors and Actuators A 57, 159-164
22. Yan B.H., Wang A.C., Huang C.Y., Huang F.Y., (2002), Study of precision micro-holes in borosilicate glass using micro EDM combined with micro ultrasonic vibration machining, International Journal of Machine Tools & Manufacture 42, 1105–1112
23. Wang A.C., Yan B.H., Li X.T., Huang, (2002), Use of micro ultrasonic vibration lapping to enhance the precision of microholes drilled by micro electro-discharge machining, International Journal of Machine Tools & Manufacture 42, 915-923.
24. C.Friedrich, (1998), Precision Micromanufacturing Processes Applied to Miniaturization Technologies, http://www.me.mtu.edu/~microweb/
25. Adams D.P., Vasile M.J., Benavides G., Campbell A.N., (2001), Micromilling of metal alloys with focused ion beam–fabricated tools, Precision Engineering, Journal of the International Societies for Precision Engineering and Nanotechnology 25, 107–113.
26. Adams D.P., Vasile M.J., Krishnan A.S.M., (2000), Microgrooving and microthreading tools for fabricating curvilinear features, Precision Engineering, Journal of the International Societies for Precision Engineering and Nanotechnology 24, 347–356.
27. Picard Y.N, Adams D.P., Vasile M.J., Ritchey M.B., (2003), Focused ion beam-shaped microtools for ultra-precision machining of cylindrical components, Precision Engineering 27, 59–69.
28. Fleischer J., Masuzawa T., Schmidt J., Knoll M., (2004), New applications for micro-EDM, Journal of Materials Processing Technology 149, 246-249.

ANALYSIS AND MODELING OF LASER BEAM MELT ABLATION

A. Cser, A. Otto

Chair of Manufacturing Technology, University of Erlangen-Nuremberg, Germany

KEYWORDS: Laser ablation, Modeling, Control.

ABSTRACT. Laser beam melt ablation as a non-contact process offers several advantages compared to conventional processing mechanisms. During ablation the surface of the workpiece is molten by the energy of a CO_2-laser beam. This melt is then driven out using the momentum transfer from an additional process gas flow. Although the idea behind laser beam melt ablation is rather simple, the process itself has a major limitation in practical applications: with increasing ablation rate, the surface quality of the processed workpieces declines rapidly. To find an explanation, and to be eventually be able to control the process thorough analyses are necessary. For this, in addition to the measurements of the processed surfaces also the optical process emissions have been recorded and analyzed. These data show a pronounced 1/f-noise-like behavior, which suggests the existence of intermittency during processing.

1 INTRODUCTION

The major advantages of laser beam melt ablation (LBMA) compared to conventional machining are consequences of the method of processing: since there is no mechanical contact no tool wear occurs and even extremely hard or brittle materials, such as tempered steel, glass or ceramics can be processed [1, 2]. There is no need of electrodes as in the case of erosive methods, and no vacuum is necessary as when using electron beam machining [3].

During the process of laser beam melt ablation the surface of the workpiece is molten by the energy of the laser beam and the momentum of an additional process gas is used to drive out the molten material from the laser-material interaction zone. The idea behind the process of melt ablation is rather simple: high laser power enables the removal of a high amount of material.

2 MOTIVATION

Laser beam melt ablation consists of several different part-processes including thermo- and hydrodynamic processes, chemical reactions at the melt-air interface and the influence of the spatio-temporally turbulent gas flow on the melt surface the combination of which results in complex patterns on the processed surface. Depending on the process parameters (laser power, feed rate, focus position, process gas, gas pressure, gas jet geometry, etc.) different types of these patterns can be distinguished on the ablated surface.

With increasing ablation rates, at about ablation rates relevant for industrial applications, surface quality of the workpiece processed declines rapidly which sets a major limitation to practical applications by causing enormous difficulties in process control. If it were possible to improve surface quality and parallel maintain high ablation rates, this technology would repre-

Published in: E. Kuljanic (Ed.) *Advanced Manufacturing Systems and Technology*,
CISM Courses and Lectures No. 486, Springer Wien New York, 2005.

sent a competitive alternative in industrial applications. To achieve the above task it is necessary to acquire a deeper insight into the processes taking place.

3   EXPERIMENTAL SETUP

All experiments have been carried out on a 3-axis, high frequency excited Trumpf TLF700
CO_2-laser, with a maximum output power of 700 W. The laser works in duty cycle mode (with a
cycle frequency of 90 kHz) which means that the pulse frequency as well as the ratio of pulse
and pause can be varied. The material processed is commercial low carbon steel (AISI 1008).
The ablation process itself is restricted to single layer ablation carried out track by track with a
horizontal offset chosen appropriately with a deliberate overlapping of the tracks to enable
smoother surfaces and also a transfer of information. Process gas used in all experiments is
compressed air in order to prevent burning (as in case of pure oxygen) but still enable the oxidation of the melt providing additional energy for the process. During the experiments performed
the average laser power $<P>$ has been varied between 500 - 700 W, the feed rate v between
2 - 9 m/min.

4   ANALYSIS OF DIAGNOSTIC DATA

In order to understand the process it is necessary to gain detailed knowledge about the dynamic behaviour of the observables that can be recorded during processing. Due to its
advantages compared to other (e.g. acoustic) process signals the optical process emission $F(t)$
and in addition the laser power $P(t)$ as input signal have been chosen for this purpose (described in detail in earlier papers (e.g. [4, 5])).

4.1   LINEAR DATA ANALYSIS

Fourier analysis is a standard technique for uncovering periodic structures in data: it splits any
signal into sine and cosine functions, i.e. it classifies and quantifies the periodic signal portions,
which enables the identification of possible system frequencies. For this reason, all measurement signals were submitted to Fourier-Transformation. However it has to be considered that
any frequency detected does not necessarily belong to the process, since also the laser machine itself produces some characteristic vibrations. Also these had to be measured: the system
vibrations have been recorded with the help of perpendicular piezo-sensors. These signals were
then likewise analyzed in the frequency domain.
Figure 1 shows the obtained frequency spectrum of an exemplary emission signal, and below
the frequencies belonging to the laser-machine. The spectrum of the emission signal
$(FFT(F(t)))$ shows a relatively high noise level, but no pronounced characteristic frequencies,
which could not be associated with the laser machine. This may have several causes: on the
one hand it is possible that the signal is simply overlaid with so much noise that no relevant
information can be extracted, but also chaotic processes result in similar spectra. A further possibility would be that the process signal consists of a stochastic and a deterministic part; also
this would result in a continuous frequency spectrum. Nevertheless, the spectrum still carries
certain information: it shows a pronounced 1/f-noise structure.

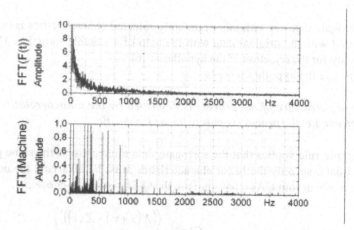

FIGURE 1. Fourier transformed signals. Top: process emission, bottom: machine vibration

Also the power spectra of the measurement signals have been calculated for two different cases: for the same process with and without process gas, in order to determine which part process (melt pool and/or interaction with process gas) results in the above 1/f-noise spectrum. The results are shown in Figure 2.

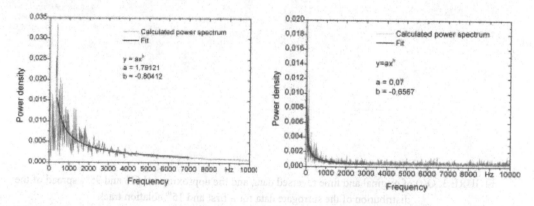

FIGURE 2. Power spectra of measurement signals. Left: with process gas, right: without process gas

It can be clearly seen that 1/f-noise evolves even without process gas, i.e. without the interaction of the gas flow, and this spectrum is even clearer than in the case of a process with process gas. This means that the flicker noise originates in the melt pool, and the gas flow only introduces additional noise.

4.2   TESTING AGAINST LINEARITY

The next step is to test the data for any nonlinearities. For this the method of surrogate data has been chosen. This test assumes the data set to posses certain linear characteristics, based on which a surrogate date set with other randomly selected characteristics is being generated and then compared with the original data with the help of a suitable statistics. This results in a certain probability for the rejection of the hypothesis [6].

In the given case the hypothesis reads:

H1: The entire structure of the process is described by the autocorrelation function and the power spectrum, i.e. the process is purely linear stochastic.

Thus the simple rule applies that the surrogate data must possess the same power spectrum as the original data, so only the linear characteristics must be invariant, the nonlinear properties are not taken into account. As a test statistic the Q-Statistic was chosen:

$$Q(\tau) = \frac{\left\langle (X(s+\tau) - X(s))^3 \right\rangle}{\left\langle (X(s+\tau) - X(s))^2 \right\rangle} \tag{1}$$

where t represents the lag.

The results are shown in Figure 3.

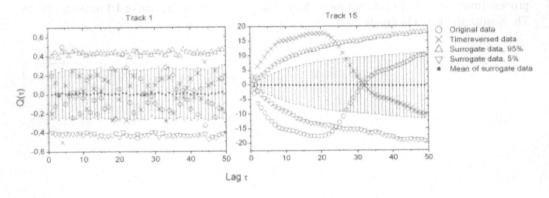

FIGURE 3. Q(t) of original and time reversed data, and the approximated 5% and 95% spread of the distribution of the surrogate data for a first and 15th ablation track

The results show that already in the case of the original data *Q(t)* clearly differs from zero, meaning that the Gauss-distributed original data also posses increments with nongaussian distribution, because of the nonlinear transformation. In addition *Q(t)* is not invariant against a change of the time direction, which clearly excludes a linear-stochastic process. In case of the first ablation track however, the null hypothesis cannot be clearly rejected, since the values are here within the significance barriers. For all other time series however, the value of the Q-statistics is clearly outside of these barriers, so that here the null hypothesis must be rejected.

So it is to be concluded that for a better understanding of the process also nonlinear analyses must be carried out.

4.3 NONLINEAR DATA ANALYSIS

Usually in experiments not all observables of a dynamic system can be measured simultaneously. Most of the time only one physical property is measured as a function of time, and even this is overlaid with noise and other disturbances. Nonetheless the experimenter may gain information about the entire system and reconstruct the dynamics from the scalar time series out of the measurement of even one suitably chosen observable. While recovering the dynamics it is assumed that the special temporal development of the selected observable is caused by the interaction of all influences determining the process. During reconstruction the phase space is used: here the recovery of all phase states results in a geometrical construct called the attractor. In case of models the attractor can be simply recovered by numeric integration and recording of all system variables. However in experimental situations this is not possible. So how is one to abstract the low dimensional manifold from the high dimensional motion equations with the help of only one variable accessible for measurement? The solution of this problem is offered by the embedding of a manifold into a low dimensional subspace.

For this, a singular value decomposition of the trajectory matrix $\underline{X}$ is necessary: $\underline{X} = \underline{S}\,\underline{\Sigma}\,\underline{C}^T$ [7]. The reconstruction takes place in the new orthonormal system spanned by the eigenvectors $\{s_i / I = 1, .., n\}$ of $\underline{S}$ via projection of the original trajectory matrix on the basis $\{c_i\}$ (eigenvectors of $\underline{C}$):

$$\hat{\underline{X}} = \underline{X}\,\underline{C} \qquad (2)$$

Since it is not possible to visualize more than three dimensions, the result of the 3-dimensional reconstruction is presented here (Figure 4).

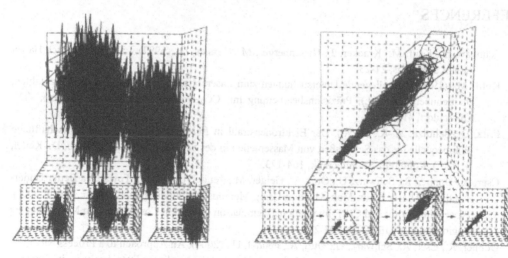

FIGURE 4. Exemplary reconstructed 3-dimensional attractors for a first (left) and a 15th (right) ablation track

There are two different distinct structures which can be recognized: in case of the first tracks the "double clew" (Figure 4 left) and for later tracks the well visible loop structure (Figure 4 right).

The "double clews" evolving for each first ablation track do not show any recognizable structure apart from the distinct splitting into two different parts. So it may either represent a stochastic time series, or a chaotic process which requires a higher dimension in order to be able to unfold. Starting with ablation track number 2 a loop structure can be distinguished referring to a certain periodicity in the process. Here the separate parts are less clearly distinct; the transition from one regime to another is rather a "creeping" one. Still, it can be recognized in both cases that several separate parts of the phase space are used during the process, indicating different kinds of process behavior.

5 CONCLUSIONS

The analysis of the system data shows no distinct frequencies in the linear properties, but a pronounced 1/f-noise-like behaviour, which is often associated with intermittency. Intermittency often occurs, when two or more different types of system behaviour compete with each other. For example in case on fluids: in a wake flow a fluctuating intermittency between laminar and turbulent flow may exist. This may even be considered in the case of laser beam melt ablation: the moving heat source may cause such behaviour.

This assumption is also backed by the results of nonlinear analyses: the separate attractor structures also indicate different competing or coexisting system behaviours.

REFERENCES

1   Schubart, D., Kauf, M., Kracker, J., Hirschberger, M., Gessler, H., (1997), Laser schafft auch Härte-fälle, Laserpraxis, 21-23.

2   Kuhl, M., (1996), Grundlegende Untersuchungen zum Laserabtragen von Werkzeug- und Forstählen, in: Schuberth, S. (ed.), Präzisionsbearbeitung mit CO_2-Hochleistungslasern (Abtragen), VDI, Düsseldorf, 41-49.

3   Fritz, D., Schulze, K.-R., (1996), Der Elektronenstrahl in freier Atmosphäre - eine wirtschaftliche Wärmequelle für das Schweißen von Massenteilen in der Serienfertigung, in: v. Hofe, D., Keitel, S. (eds.), Konferenz Strahltechnik, 164-173.

4   Cser, A., Donner, R., Schwarz, U., Otto, A., Geiger, M., Feudel, U., (2002), Towards a Better Understanding of Laser Beam Melt Ablation Using Methods of Statistical Analysis, in: Teti, R. (ed.), Proceedings of the 3rd CIRP International Seminar on Intelligent Computation in Manufacturing Engineering (ICME 2002), 203-208.

5   Donner, R., Cser, A., Schwarz, U., Otto, A., Feudel, U., (2004), An Approach to a Process Model of Laser Beam Melt Ablation Using Methods of Linear and Nonlinear Data Analysis, in: Radons, G., Neugebauer, R. (eds.), Nonlinear Dynamics of Production Systems, 453-468.

6   Schreiber, T., (1999), Interdisciplinary application of nonlinear time series methods, Phys. Rep., 308, 1-64.

7   Buzug, T., (1994), Analyse chaotischer Systeme. BI-Wissensdhaft-Verlag, Mannheim.

ON THE DEVELOPMENT OF TOOLS FOR MECHANICAL DESIGN OF COLD ROLLING CLUSTER MILLS

E. Brusa[1], L. Lemma[2]

[1] Department of Electrical, Managerial and Mechanical Engineering, University of Udine, Italy
[2] SKF Industrie S.p.A., Airasca (Torino), Italy

KEYWORDS: Cold Rolling Mill, Mechanical Design, Numerical Methods.

ABSTRACT. The paper investigates the main issues of design and modeling of so-called cluster mills for cold rolling of thin and moderate thin steel products, by means of numerical tools. This preliminary discussion summarizes the relevant aspects experimented on few operating cold rolling mills, started in [1], to select models and architecture suitable to build simple tools, to be used for a preliminary prediction of the dynamic response of the whole plant, in presence of known excitation, or even to simulate condition monitoring operation, to support operators in signal processing of sensors equipping bearings and mill frame.

1 INTRODUCTION

The mechanical design of cold rolling mills usually involves several difficulties, due to the amount of phenomena governing the dynamic response of the whole plant, mainly due to: interaction between strip and working rolls, contact between working and back rolls, bearings and supports, frame, damping effects and control systems, where present [1-8]. These aspects motivate the availability of a specialized literature mainly focused on rather simple models, with few rolls, and a large use of a direct regulation of the plant in operating condition, based on vibration monitoring and non destructive examination of the metal strip [5]. Available models demonstrate to be effective in case of 2-high and 4-high mill configurations and allow predicting, at least approximately, the dynamic behavior of rolls and the related effects on the strip. In case of multiple rolls, i.e. cluster mill layout [9], including several supporting and back rolls, coupling effects make more difficult an effective prediction of the dynamic behavior as well as a fast regulation of the mill in operation, to avoid strip faults [4,6]. The latter affect the interpretation of the experimental results concerning the response in frequency domain of several parts of the mill, mainly back rolls, being usually the most monitored because of the accessibility for sensors application. The interaction among multiple rolls as well as the presence of several kinds of bearings and supports, even active suspension systems, joint to the role of localized damping and stiffness effects, made recently higher the industrial demand for process simulation and design tools to analyze the mechanical behavior, especially of cluster mills. It can be noticed that the aim of these tools is, more than an accurate analysis of the whole dynamics, the possibility to provide a qualitative, simple and rather fast prediction of the trend of the process, under effect of known boundary conditions. Actually this need expressed by metallurgical engineering for numerical tools supporting process monitoring, was already experimented by automotive engineering in road test CAE services [10], where to tune vehicle parameters, for handling and comfort, particularly during the recent development of active control systems, the availability of simple models,

Published in: E. Kuljanic (Ed.) *Advanced Manufacturing Systems and Technology*,
CISM Courses and Lectures No. 486, Springer Wien New York, 2005.

even running under Matlab environment, were welcomed. In the case of cluster mill a similar approach is followed: numerical models may be fast, possibly running in few minutes, based on few design parameters, to be tuned by experimental identification on the monitored plant [11-13]. The goal of this research activity is the selection of the numerical models suitable for predicting the relevant phenomena involved in cluster mill operation, to be tuned by direct monitoring of the plant, and to be used to simulate the effect of known excitation, faults or unforeseen actions on the dynamic response of the mill, and to allow operators performing condition monitoring to replicate numerically the generation of symptoms of failure detected on processed signals, to verify their origin, like it was performed by authors in [15].

2   SYSTEM ANALYSIS

To clarify the architecture of the numerical tool suitable for the above mentioned goal a description of a paradigmatic cluster mill is herewith introduced as well as a list of phenomena to be predicted and avoided in operation.

2.1 CLUSTER MILL CONFIGURATION

A particular configuration of rolling mill is the so-called "cluster". Cold rolling occurs at room temperature, by cylindrical and smooth rolls, whose diameter is usually quite small to assure, in absence of benefits due to the temperature in hot rolling process, a suitable pressure on the rolled material. It is well known that rolling load is higher for smaller diameters of the roll, nevertheless slender rolls are prone to bend. Plant layout has to avoid this bending, by supporting working rolls by secondary cylinders. Cluster mill is a typical solution: among several layouts proposed [5,8], Sendzimir mill or "20-high mill" is herewith investigated [Fig.1], together with a more compact version, called "Z-mill" [1]. The number and the layout of rolls may change if another kind of cluster mill is considered, but the main strategy is common: working roll is supported by back rolls, which apply motion and all are supported by a crown of saddles or similar bearings.

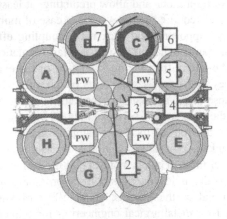
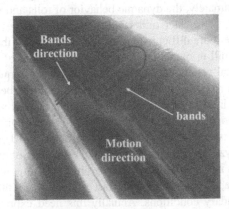

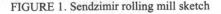

FIGURE 1. Sendzimir rolling mill sketch   FIGURE 2. Examples of surface strip faults

Smaller rolls can roll high resistance metals and steel, to obtain very thin flat and to avoid intermediate thermal process. Layout in fig.1 is suitable to provide the required stiffness to the working rolls (2) because of presence of a crown of additional rolls. The latter are supported by a first set of intermediate rolls (3), which are sustained by a second set (4), still consisting of intermediate rolls, and by the backing bearings (5), whose shafts (6) are supported by saddles (7). The transversal section of the plant looks like in figure 1 for the overall length of the rolls, but for working rolls (2), which have tapered conical ends, allowing a calibration of the plant, performed by moving them along the axial direction of the mill. Saddles (7) rotate instead of the inner rings of the bearings, fit on shafts (6). Rolls referred to as "PW" in figure 1, belonging set called (4), apply motion. Plant layout can be changed by acting on shafts called A, D, E, H, while shafts F and G tune the thickness of the steel flat. To assure the required flat quality, monitoring system measures the steel thickness, before and after cold rolling, and acts on the flat tension. Sometimes, additional hydraulic actuators are used to damp vibration on the intermediate rolls (3), in a closed-loop controlled based on the strip thickness monitoring. The mill may be reversible, if work rolls rotate both clockwise and counterclockwise, and steel strip is rolled in several steps, until the required thickness is achieved. Metal follows "front motion", when it moves from left-side to right-side and is subjected to a front tension, while it "moves back", and is loaded by a back tension, in the opposite direction.

2.2 STRIP FAULTS

Possible strip faults in rolling mill vibration include chattermarks, flatness faults and flat rupture [2-6]. Chattermarks appear often as periodical fault on the strip surface, thickness irregularity or waves [3]. Depending on the frequency interested by vibration problem three typical chatters are usually defined: torsional, appearing at lower values of frequency and depending on applied torque and motors control, so-called "gage or 3rd octave" and "roll or 5th octave". Chatters of 3rd and 5th octave are mainly due to technological and dynamic aspects. Material pre-damage, rotundity and balance errors upon rolls, roll bearings errors and drive irregularities cause usually speed dependent excitation, while a speed-independent evidence on waterfall diagrams acquired by monitoring system is due to self-excited or roll stand vibration, front tension fluctuation, material non-homogeneity, stick-slip, locking and drive control.

2.3 MODELING TASKS

Tools under development are required to include a suitable model of the whole cluster mill, to investigate the interactions occurring among the mechanical parts of the assembly, to predict the dynamic response of the plant and to simulate the monitoring operation of the sensors, whose location may be conveniently defined by running the numerical models to check observability and controllability of the rolls. To proceed straightforward a comprehensive analysis of the whole system to be modeled is sketched in figure 3. The goal of the numerical simulator is to provide a suitable prediction of either: the vibration perceived by the strip due to the motion of rolls, for a given excitation applied anywhere on the assembly, or the dynamic response detected and acquired by sensors, often located at the saddles of back rolls. In practice it consists of the solution of a set of equations of dynamic equilibrium of the assembled parts, with additional control actions, if present. The main transfer function required by condition monitoring operators correlates sensor signals and work roll-strip interaction, namely the force exerted on the work roll, which is usually measured by a load cell. Because of several phenomena occurring in the mill operation, additional excitation may be introduced by every roll composing the whole system. Moreover, if control systems are implemented models have to allow a closed control loop response for a given external action. In fig.3 several control systems may be identified: a rolling motion control, with strip speed feedback, a coiler rotation control with torque feedback, a motor torque control on powered rolls, while in perspective an automatic control may be implemented to correlate the response monitored on sensors (saddles) and motor or even coilers, or, in case of active control of roll vibration (not sketched), an hydraulic control of intermediate rolls, based on rolls position monitoring.

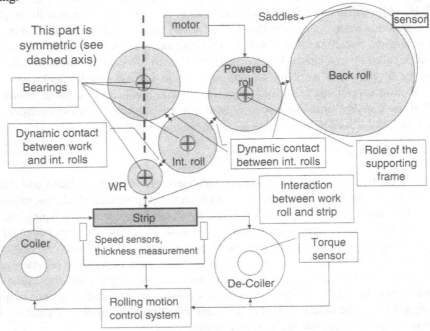

FIGURE 3. Exploded sketch of the Sendzimir layout modeled

Among the faults previously mentioned several may be investigated by the whole simulator. Since chattermarks depend on mass vibration, they may be studied, as well as effects of rotundity and balance errors upon rolls and drive irregularities. In case the model includes mill frame structure, roll stand vibration is predictable. Very poorly predictable are front tension fluctuation, material non-homogeneity, stick-slip and locking. Figure 3 gives an overview of the main aspects to be modeled: vibration and control of rolls in contact; damping of bearings, supports and rolls; plastic behavior of the rolled strip, in contact with work roll, and lubrication; reactions of saddles and back bearings on back rolls; rolling motion and driving torque irregularities. All of these need for a numerical model. Actually two different levels of modeling may be foreseen. The whole system can be interpreted as sum of two subsystems, mutually interacting: the rolling motion with strip and coilers, and the mill with work and supporting rolls. Although in principle they are coupled by strip mechanical response, in practice a complete model, including all controls and parts may be difficult to be run and tuned on the actual parameters of the mill. A preliminary assumption usually applied is that two subsystems are practically uncoupled, at least as first approximation. The latter is acceptable, when the object of the analysis is the dynamic response of mill to an irregularity of the strip action, being required to consider only the mill subsystem, or when a rotundity or balance problem on rolls is simulated, to detect the effect on surface irregularity of the strip and on the rolls vibration. The complete system is strictly required to be analyzed if a MIMO control is applied simultaneously to motor, rolling motion and rolls vibration and effects on strip are studied.

3   MODELING APPROACHES AND CLUSTER MILL PECULIARITIES

According to the previous discussion, which focused relevant issues of the analysis, to proceed further it has to be identified the influence of the whole aspects on the overall dynamic behavior of the mill and the numerical approaches to be implemented into the subroutines. This subject is herewith preliminary developed, by taking into account some experiments performed by authors on a Z-mill, described in [1] and on a Sendzimir mill (figure 4).

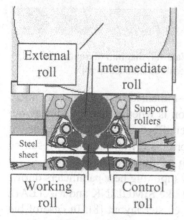 

FIGURE 4. Sketch of the Z-mill and picture of Sendzimir mill used as test cases for experiments

3.1 MILL MODELING BY MULTI-D.O.F.: MULTI-BODY DYNAMICS, ROTOR DYNAMICS AND PLANE LUMPED PARAMETERS MODELS

The available literature provides dynamic models based on the interaction between working roll and strip [5,6,14]. Usually 2-high and 4-high models include equivalent stiffness and inertia for supporting rolls and mill dynamics is numerically evaluated as vibration of work roll on the spring-damper equivalent system, representing the strip behavior. A first important result concerning cluster mill is that the role of all rolls is comparable, therefore model has to be based on multi-d.o.f. system, where inertia of all rolls is considered, to find the critical points of the mill. Moreover, rolls more interested by vibration in chattering phenomenon are often supporting and back rolls in cluster layout, more than or at least as work roll, although this one applies to the strip the action determining surface faults and irregularities. The latter depends on the strip speed and on the frequency of the excitation and on mass and stiffness properties of rolls. Higher values of strip speed are more effective to induce vibration on rolls, by exciting resonances of the assembly. In the case of Z-mill [1] control rolls demonstrated to be the most excited by chatter at the corresponding frequency. As it is shown in figure 5, a wide range of frequencies is excited by chattering phenomenon, with severe amplitude on the control roll sketched in figure 4. It corresponds to the range between 3 kHz – 5 kHz. Amplitude of the dynamics response increases with rolling step, i.e. with strip speed on the mill.

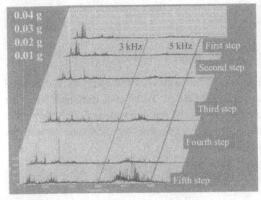

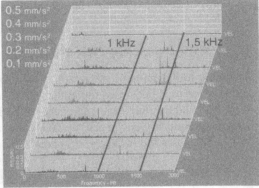

FIGURE 5. Waterfall diagram of Z-mill configuration, measured on intermediate/control rolls

FIGURE 6. Waterfall diagram of Sendzimir mill, measured on backup rolls

A second relevant aspect concerns the use of plane, two-dimensional, models or three-dimensional ones. The question on the corresponding assumption is whether dynamic behaviour of rolls can be considered pure rigid motion or any flexural effect affects the dynamic response of the system, i.e. gyroscopic effect of rolls is relevant for simulation. Plane models, simply resulting from mass, spring and dampers assembly, according to sketch of figure 3, benefit of a lower number of d.o.f., assure fastness and are easy to be modified, by inputting new parameters during tests. They can be implemented in general purpose environment like SIMULINK and MATLAB. These are highly preferred in case of traditional mills, without cluster layout [8]. In case of big back rolls if results of a multibody dynamics code, like ADAMS, and MATLAB models are compared, vibration of rolls show a little influence of gyroscopy, because of the inertia and the layout of the mill, i.e. no conical whirling motion of rolls are apparently excited at the chatter frequency. This is the case of Z-mill [1]. Where inertial properties and stiffness of supports are comparable, like in Sendzimir configuration, the latter result is rather less evident. In particular modal analysis [16] and rotor dynamics [17] may be required to evaluate the presence of vibration modes and whirling motions respectively, dangerous for the frequency excited by the higher values of strip speed in rolling. In practice it can be observed that, because of the contact along the roll between two rolls and the high value of inertia of the whole rotors, usually no critical speed for flexural behaviour falls down the range of frequency explored by rolling operation, but whirling motion can be excited by unbalance, with amplitude sufficient to induce vibration on the adherent rolls, or even to modify friction between surfaces. In addition vibration modes often interest the extremities of the roll, shaped in a such a way that supports and bearings can be fitted. Example of this geometry is reported in figure 7. This effect suggests to discuss early the presence of vibration modes and critical speeds due to rolls, by performing a preliminary analysis of the rotordynamic behaviour of the roll, by FEM as in figure 7. In this sense a three-dimensional model, including gyroscopic and inertial effects of the whole bodies is welcomed, while plane models look unsuitable to predict such aspect. Nevertheless is extremely difficult to include this dynamic effects in the whole mill model: multi-body dynamics codes, usually, do not run rotordynamic analysis subroutines, while compute the dynamic response of the rigid bodies suspended. On the other hand rotordynamics codes [17] basically assume the axis-symmetry of the analysed rotors, i.e. rotors composed by several rotating parts, having a unique rotation axis, and cluster mill configurations exhibit typically multiple rotation axes.

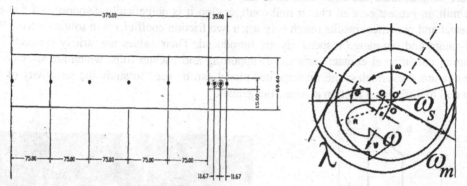

FIGURE 7. FEM model of the end of an intermedi-   FIGURE 8. Unbalanced magnetic pull effect on
ate roll of Z-mill   rotating shaft

3.2 ELECTROMECHANICAL ACTIONS

The above mentioned effect looks critical for powered rolls, where electric motors are connected. It is known that a significant whirl of the rotor induces an electromechanical instability by applying so-called unbalanced magnetic pull, which causes severe damage on motor and rotating parts [15,18,19]. Figure 8 shows unbalanced magnetic pull effect: shaft rotating at angular velocity ω, within the magnetic field rotating with spin speed ω_m may be subjected to a misalignment ε, thus producing a whirling motion with angular velocity λ, and a radial force towards the stator, rotating with speed ω_s . Because of the values very similar of speeds ω, ω_m, ω_s, dynamic effect detected by motor sensors is often a beat response of rotor displacements, whose amplitude depend on the above mentioned spin speeds [18]. This phenomenon explains certain irregularities of rotating shafts in electrical machines, related to the grade of balancing applied to the rotor. It may affects mill waterfall not only at higher speed, but at lower speed too, with irregular peaks of excitation (figure 6). Unbalanced magnetic pull can be simulated in MATLAB subroutines [15].

3.3 MAIN NONLINEAR MODELS

The core of a numerical simulator includes a first model, describing the nonlinear interaction between strip and work roll (Bland and Ford; Johnson and Qi; Lin, Suh, Langari and Noah [14]). It computes stress and rolling pressure on strip and time evolution of vertical displacement of the work roll. The equation of motion of this submodel looks like:

$$m\frac{d^2y}{dt^2} + K(f)y = f = \int_{x1}^{x2} p(x)dx + \int_{x1}^{x2} C_{fr}\tau \tan\phi \, dx \tag{1}$$

where m is the equivalent mass of the whole subsystem, y the vertical displacement of the roll, K the equivalent stiffness, nonlinear function of forcing term f. The main problem of the whole model is that friction coefficient C_{fr} is poorly predictable by numerical methods, always needing for a model tuning on the monitored mill. Moreover while in case of 2-high and 4-high mills literature demonstrated that an average value may be introduced to analyze the dynamic response of the mill, in present case of cluster mill configuration it is numerically demonstrated that experimental and numerical results match only when two friction coefficients in solving integral (1) for in-gauge and out-gauge respectively are introduced. Their values are strictly connected to lubrication condition at contact surfaces of supporting and backup rolls, which are affected by rolls vibration. Nevertheless the implemented model can be used to study the sensitivity of the whole mill to differential friction at in and out gauges.

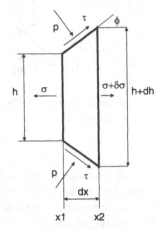

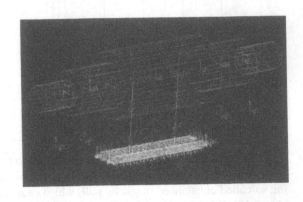

FIGURE 9. Model of roll bite according to [14]   FIGURE 10. Example of ADAMS model of Z-mill rolls, with ends shaped to be supported by bearings

Two additional nonlinear models are introduced to predict the reaction of bearings in terms of elastic and damping forces (Harris [20]; SKF [21]) and the dynamic behavior of contact between rolls (Laursen [22]) into the overall mill model (figure 10). The first one is a typical relation between force and displacement on bearing rings with nonlinear characteristic:

$$\delta = kF^p \rightarrow C = \frac{F}{\delta} = \frac{F^{1-p}}{k} \rightarrow C \cong \frac{dF}{d\delta} = \frac{F^{1-p}}{kp} = \frac{\delta^{\frac{1-p}{p}}}{pk^p} \tag{2}$$

being δ the displacement of ring material, F the applied load, k the stiffness of the whole bearing. Contact dynamics is modeled as impact force, combination of elastic and damping actions:

$$F = \max\left\{ k^*(z_0 - z)^p - c^*\dot{z}\Psi(z_0, d) \right\} \quad z \leq z_0 \quad -1 \leq p \leq 1 \tag{3}$$

where supersigned k^*, c^* are stiffness and damping coefficients of the contact between rolls, z_0 is a reference position (usually the external diameter of roll), z displacement, ψ is a rounded step function whose height is $c \cdot (dz/dt)$ and width d (spanning the range $z_0 - d$ to z_0). Although the two latest models have been introduced in a preliminary implementation of the numerical tool, it can be noticed that coefficients are poorly predictable without a direct identification on the plant [11,12]. Moreover, while in 2-high and 4-high mills the latter stiffness and damping coefficients play a significant role in modeling effectiveness, in case of cluster mills supported by saddles numerical and experimental results show that the highest sensitivity of the model is towards stiffness and damping of saddles, more than bearings and contact among rolls. The most compliant part of the mill assembly, apart from strip, is located at backing rolls. ADAMS model of the test case Z-mill demonstrated that the latter is poorly sensitive to variation of contact (3) and bearing (2) stiffness and damping coefficients, if compared to a variation of values of the equivalent stiffness and damping of backing roll-saddles subsystem. This result is very important from the point of view of monitoring and control operations. About the controllability of the system: since often because of accessibility reasons saddles offer their external surface for sensors location, it can be realized that sensor is non collocated [23] with respect of the input force applied by strip on the work roll. It means that parasitic dynamics is present between sensor and strip due to the dynamic response of all flexible parts included, along the whole path. In fact this cannot be neglected, especially considering that the most active component is the nearest one, including backing rolls and saddles. Any control system operating on the signal processed by these sensors and regulating the force exerted by work roll may suffer significant effects of phase rotation. Moreover the observability of the dynamic behavior of at least intermediate rolls may be significantly distorted by the localized effects occurring at saddles, where sensors operate. This results suggest to operate a preliminary observability and controllability analysis of the multi-body system, by means for instance of a plane model, to locate properly sensors. Currently encouraged solutions already proposed are sensored bearings [SKF,20] or even magnetic suspension technology with self-sensing capabilities [23].

4 CONCLUSIONS

This preliminary analysis was aimed to define some guidelines for performing numerical tools for the mechanical design of cluster mills for cold rolling. It was based on both a preliminary implementation of models available in literature or even developed by authors. Relevant aspects of modelling as well as models have been tentatively listed and a selection among numerical approaches, namely lumped parameters, multi-body dynamics, finite element method, was provided, by performing a screening of the influence of the whole phenomena on the overall mill behaviour. Some key features have been identified from condition monitoring and numerical investigations on test cases Z-mill and Sendzimir mill. The latter include lubrication and friction coefficient determination between rolls and on the strip surface, unbalanced magnetic pull effects at powered rolls, localized deformability of pins at rolls ends and delocalization of the most effective stiffness and damping at backing rolls, on saddles. These preliminary results are currently used to operate a suitable structuring of the numerical tool under development and to select suitable experimental configuration, mainly sensors position, for an effective model updating and identification of the relevant parameters to be tuned for supporting operating condition and monitoring activity.

REFERENCES

1. Brusa, E., Lemma, L., Massetti, D., Miani, F., Sullini, F., (2004) Dynamics of cold rolling mills with multiple rolls, in "Advances in Experimental Mechanics", Proc. of the Int. Conf. Exp. Mechanics, ICEM12, Bari, August 29th – September 2nd, ISBN 88-386-6273-8.
2. Geropp, B., (2003) Vibration problems in Aluminum and Steel mills, ACIDA GmbH, Report.
3. Mackel, J. (2003) Condition monitoring and diagnostic engineering for rolling mills, ACIDA GmbH, Report.
4. Yun, I.S., Wilson, W.R.D., Ehmann, K.F. (1998) Review of chatter studies in cold rolling, Int. J. of Machine Tools & Manufacture, Vol. 38, pp.1499-1530.
5. Ginzburg, V.B., Ballas, R. (2000) Flat rolling fundamentals, Dekker, New York.
6. Yun, I.S., Wilson, W.R.D., Ehmann, K.F. (2002) Chatter in strip rolling process, Mechanical Engineering Lectures, Northwestern University, Evanston, Illinois, USA.
7. Kalpakjian, S. (2001) Manufacturing engineering and technology, 4th Ed., Prentice-Hall, Upper Saddle River.
8. Roberts, W.L. (1978) Cold Rolling of Steel, Dekker, New York.
9. Sendizimir official website: http://www.sendzimir.com
10. Brusa E. et alii (2002) Modelling vehicle dynamics for virtual experimentation, road test supporting and dynamics control, in Computer simulation for automotive applications, SAE Edition.
11. Friswell, M.I., Penny, J.E.T., Garvey, S.D. (1995) Using linear model reduction to investigate the dynamics of structures with local non-linearities, Mechanical Systems and Signal Processing, 9(3), pp.317-328.
12. Friswell, M.I., Penny, J.E.T., Garvey, S.D. (1996) The application of the IRS and balanced realization methods to obtain reduced models of structures with local non-linearities, Journal of Sound and Vibration, 196(4), pp.453-468.
13. Benhafsi, Y., Farley, T.W.D., Wright, D.S. (1999) An approach to on-line monitoring of the 5th octave mode mill chatter to prolong back-up roll life, AMTRI, Machine Tools and Machinery Manufacture, Lectures of the Institute of Materials, London.
14. Lin, Y.J., Suh, C.S., Langari, R., Noah, S.T. (2003) On the characteristics and mechanism of rolling instability and chatter, Trans.ASME, vol.125, pp.778-786.
15. Amati, N., Brusa, E. (2001) Vibration Condition Monitoring of rotors on AMB fed by Induction Motors, Proc. IEEE/ASME Advanced Intelligent Mechatronics AIM'01, pp.750-756.
16. Ewins, D.J. (2000) Modal Testing: Theory, Practice and Application, Taylor and Francis Group; 2nd edition.
17. Genta, G., (2005) Dynamics of rotating systems, Springer Verlag, New York.
18. Holopainen, T., (2004) Electromechanical interaction in rotordynamics of cage induction motors, VTT Publications 543, Otamedia Oy, Espoo, Finland (Replication of PhD thesis dissertation at Helsinki University of Technology).
19. Genta, G., Brusa, E., Amati, N. (2003) Rotordynamics for Electrical Machines, Postgraduate Course on Eletromechanics, Helsinki University of Technology, Dept. El. and Comm. Eng. Espoo, Laboratory of Electromechanics, Lectures Series 11, Otamedia Oy, Espoo, Finland.
20. SKF General Catalogue, 2005.
21. Harris, T.D. (2000) Rolling Bearing Analysis, 4th Edition, Interscience.
22. Laursen, T.A., (2002) Computational Contact and Impact Mechanics, Springer Verlag, New York.
23. Knospe, C., (2002) Control of chatter using active magnetic bearings, Univ. Virginia, USA, Course' Seminar Lecture.

THE EFFECT OF PRESSURE ON THE SURFACES GENERATED BY WATERJET: PRELIMINARY ANALYSIS

M. Monno[1], C. Ravasio[2]

[1] Dipartimento di Meccanica – Sezione Tecnologie e Sistemi di Lavorazione, Politecnico di Milano, Italy
[2] Dipartimento di Progettazione e Tecnologie, Università di Bergamo, Italy

KEYWORDS: Waterjet, Pressure Fluctuation, Cutting Quality.

ABSTRACT. The pressure fluctuation in the waterjet (WJ) process is due to the mechanism of high pressure generation, characterized by a cyclical working, and to physical causes (water compressibility at the exercise pressures). This phenomenon cannot, at present, be removed; however, it can be dampened, depending on the selected constructive system, by installing a pulsation attenuator below the intensifier or by conveniently phasing the pumping cycles of more intensifiers in parallel. In this paper, the effects of the pressure on the WJ process have been investigated. The pressure signals generated by a double-acting reciprocating intensifier pump system have been analysed; the effects of the pressure fluctuation and of the pressure signal form on the cutting quality have been studied through the acquisition of the roughness profiles of the surface of various materials (rubber, polycarbonate, plasticine) generated by waterjet.

1 INTRODUCTION

In WJ/AWJ technology, the high energy of the jet made the cut. In pure WJ cutting, the material removal is due to the high speed of the jet; in AWJ cutting, instead, the jet has only the aim to transfer its momentum to the abrasive particles, whose abrasive and erosive action causes the material removal.

The water pressure is increased up to the operating pressure (at present 380-400 MPa, although there are already proposed and realized systems that reach up to 800 MPa [1]) through a pump system (intensifier).

Intensifier systems operate on the principle of conservation of hydraulic energy. A low pressure, high flow hydraulic region acts on a piston of large area. A large force is generated, which in turn acts on a small plunger area, creating a much higher pressure. An exchange of energy is made where low pressure, high flow hydraulic fluid is converted to high pressure, low flow water. The ratio of the low pressure piston area to the high pressure plunger area is called intensification ratio.

There are two types of intensification systems: the basic design of a standard intensifier consists of a double-acting, hydraulically-actuated piston-plunger ram in which the two opposite-facing cylinders are mechanically coupled and the fluid compression and suction strokes alternate between the two cylinders (Figure 1-a).

As it strokes, creating pressure and output flow in a direction, the opposite side is in its suction stroke. When the piston reaches the end of its stroke, the directional valve is shifted. The output pressure decays due to the response time of the valve and the plunger stroke required to

Published in: E. Kuljanic (Ed.) *Advanced Manufacturing Systems and Technology*,
CISM Courses and Lectures No. 486, Springer Wien New York, 2005.

compress the water to the output pressure. When the water pressure in the high pressure cylinder equals the pump's output pressure, the high pressure check valve opens [2].
Because of the compressibility of water, which can reach up to 15% at 350 MPa [3] [1], every piston covers a part of its working stroke in order to compress the water, without there is water discharge.
In order to reduce the pressure fluctuations, an accumulator of high pressure water is installed. Its volume controls the range of fluctuations. If the accumulator's volume increases, the fluctuation decreases but the cost of the accumulator increases to more than proportional extent. The choice of the volume results from a compromise solution between the benefits from a stable pressure signal and a sensible rise in costs [2, 4].
In order to minimize the pressure fluctuation due to the intensifier, without resorting to the expensive solution of the accumulator, some constructors have opted for the use of *phased pump* (Figure 1-b).

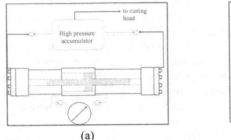

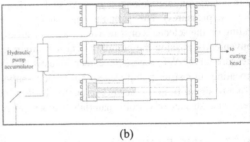

(a)   (b)

FIGURE 1. Double acting intensifier (a) vs. phased pump (b)

---

[1] The ratio $K = \dfrac{\Delta V}{V} = \dfrac{V_w - V_a}{V_a}$ is called water compressibility

$V_w = V_w(P)$ is the volume of water inside the accumulator when the pressure is P
V_a is the volume of accumulator
The empirical relationship between compressibility and pressure of water, in [5], is:

$$K(P) = a_0 + a_1 \cdot P + a_2 \cdot P^2 + a_3 \cdot P^3 + a_4 \cdot P^4 + a_5 \cdot P^5$$

If fluid elasticity coefficient E is the inverse of isothermal compressibility, the relationship, given below, is valid:

$$\frac{1}{E} = K_T = \frac{1}{V}\frac{dV}{dP} = \frac{1}{\rho}\frac{d\rho}{dP}$$

In [6] the authors supposed a linear relationship between E and the pressure P, that is:

$$E(P) = E_0 + aP$$

where $E_0 = E(0) = 20354$ atm at room temperature (25° C). It can be shown that using $a=7$ will yield a good approximation .
From the previous consideration, it comes out that water density is function of pressure with this law:

$$\rho(P) = \rho_0\left(1 + \frac{a \cdot P}{E_0}\right)^{\frac{1}{a}}$$

The constant ρ_0 is the water density when P is atmospheric pressure.

The phased concept is based on at least two (six in [7]) single acting cylinders in parallel in which the plungers' motion can be arbitrarily phased through a combination of hydraulic and electronic circuitry. These systems are timed such that, at the end of the delivery stroke of a piston, there is already another piston in phase of pressurization of the water. That implies the contributory presence of a peak of absorbed power. The concept offers several advantages beyond the lack of need of accumulator, compact size, design flexibility, to name a few [8].

When the high-pressure fluid is created, it is directed into the cutting head, where the pressure energy of the jet is converted into kinetic energy. The energetic conversion is obtained by means of an orifice in sapphire or synthetic ruby; at the exit of the orifice, the jet reaches a rate of up to 900 m/s depending of the selected pressure. The specific energy of the jet depends on the diameter of the orifice, which changes between 0.05 and 0.40 mm (Figure 2-a).

In the AWJ systems, the abrasive is added to the fluid in a mixing chamber (Figure 2-b), placed under the orifice.

(a)　　　　　　　　　　　(b)

FIGURE 2. WJ (a) and AWJ (b) cutting

2 THE EFFECT OF PRESSURE FLUCTUATION ON THE WJ/AWJ CUTTING QUALITY

The water pressure is subjected to cyclical fluctuations mainly due to the intensification mechanism. The control of cyclical pressure fluctuation is very important. In fact, the pressure fluctuation has many effects:

- it causes a periodic stress state in the whole system: every mechanical component of the high pressure circuit is subjected to a fatigue stress and it can be brought about forced vibrations [9];
- the jet is pulsating with possible effects on the cutting quality [9];
- it causes a reduction in the life of many WJ system components [10], for example the nozzle;
- it is often a constraining factor that limits the applications of the WJ technology [10].

The WJ/AWJ (Waterjet/Abrasive Waterjet) removal process is influenced by several technological and fluid dynamics parameters. The surface morphology generated by AWJ has been studied by a few researchers [11-18].

In [11] it had already been developed that the pressure fluctuation is able to emphasize the surface striations, which, in author's opinion, are generated by the characteristic mechanism of metal removal of the water jet technology.

Based on a flow visualization study of the waterjet cutting process, Hashish proposed that a waterjet cut surface consists of two cutting regions [15-16]. The first region (the top cut of the surface) is dominated by the cutting wear mode where penetration occurs in a small impact angle. The second region (the bottom part of the surface) is dominated by the deformation wear mode where penetration occurs in a large impact angle. The surface is smooth in the first region but is marked by striations in the second region.

Afterwards, Hashish proposed that in addition to jet-induced striations, which are due to the deformation wear mode of the waterjet cutting, also traverse-induced striation exists, which are due to the unsteadiness of waterjet pressure or motion of a AWJ traverse system and may appear in both the cutting and deformation wear zones [17].

Therefore, he subdivided the responsible causes of the striation generation in AWJ cutting into:

- process causes: the striation is the result of the material removal mechanism;
- causes involved in the process parameters control: the instability of the process parameters (pressure, abrasive flow rate, feed rate, etc.) is responsible of the striation formation;
- causes involved in the support equipment: the vibrations of the piece and/or the nozzle during the cutting process cause the striation formation.

In [18] the striation formation is believed to be caused by the wavy abrasive particle kinetic energy distribution related to the cut surface. During the cutting in the upper zone of the cut surface most of the particles have higher kinetic energy than the required destruction energy of workpiece material, so a smooth surface can be obtained. As the cutting depth increases, the number of the particles which have enough kinetic energy decreases and the wavy particle kinetic energy distribution profile becomes sharp. This causes the formation of striations. All factors which have an effect on this kinetic energy magnitude and its distribution related to the cut surface must affect the striation patterns and result in striation irregularities.

The interpretation of the mechanism of striation formation in AWJ cutting does not explain the striation formation in the pure WJ process. In fact, along the whole waterjet cutting surfaces, there are irregularities (striations) with dimensions comparable to the jet diameter.

In order to investigate only the effects of pressure fluctuation on the cutting quality, leaving out the effects of the abrasive, only pure water jet (WJ) has been considered. Therefore, the tests have been carried out on workable materials (polycarbonate, rubber, plasticine), with even high thicknesses, without the abrasive injection.

3 EXPERIMENTAL RESEARCH

3.1 THE PRESSURE SIGNAL

The trend of the pressure is a characteristic of the intensification system that generates it.

A double acting intensifier (30 kW) is used. The pressure has been measured through a digital pressure transducer (Inter-Probes HP-48), which is placed in proximity of the cutting head and connected to an acquisition card (Figure 3). The pressure measured at the exit of the intensifier system differs from that acquired for only a constant. This constant represents the pressure losses load inside the pipe lines (Figure 4). The pressure has been measured in 13 levels, from 250 to 370 MPa, with three diameters of nozzle (0.12, 0.2 and 0.3 mm). For every measurement, the acquired pressure signal has been analyzed in time and frequency domain.

FIGURE 3. Arrangement of the pressure transducer

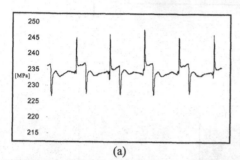
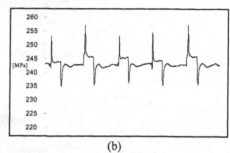

(a)   (b)

FIGURE 4. Water pressure measured with arrangement of the transducer in proximity of
the cutting head (a) and at the exit of the intensifier system (b) (P=250 MPa, dn=0.20 mm)

3.2 THE EFFECT OF THE NOMINAL PRESSURE

Afterwards the measurement of the pressure, the effect of the nominal medium pressure on the
surface quality cutting has been analyzed.

The water pressure is not constant in time but it shows a fluctuation: the effect of this
fluctuation on the surface quality has been verified through the roughness index R_a[2].

Rubber has been cut using the double acting intensifier system, at different level of pressure, at
the following experimental conditions: dn=0.2 mm, u=200 mm/min, sod=2 mm. For every
experimental condition, twelve repetitions have been carried out.

Twenty replications have been realized on polycarbonate test pieces too (P=250 MPa, u=200
mm/min, dn=0.2 mm, sod=2 mm). In fact, rubber is an elastic material and during the profile
acquisition, measurement distortions can compromise the reliability. Polycarbonate has a
higher density and lower elastic characteristic than rubber.

3.3 THE EFFECT OF THE PRESSURE SIGNAL FORM

The effect of the factor "pressure signal form" has been analyzed. In order to do this, the
cutting surfaces obtained using two different pressure signals forms have been compared.
Figure 5 shows the experimental procedure.

---

[2] It is defined R_a the mean of the absolute value of the deviations y(x) of the profile R as regards the

mean line: $$R_a = \frac{1}{l_m} \int_0^{l_m} |y(x)| dx$$ where l_m is the measured length.

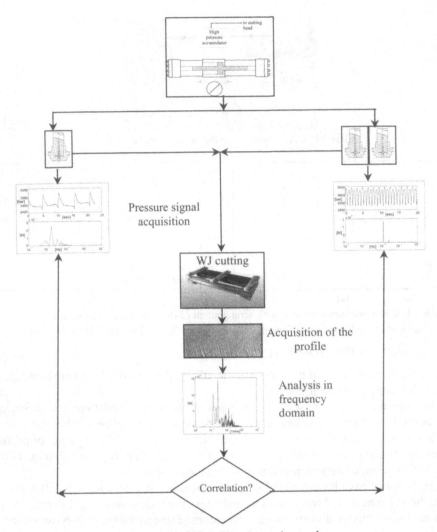

FIGURE 5. Experimental procedure

In order to generate a different pressure signal form, the outflow rate delivered from the intensifier has been varied conveniently by means of the opening of a second jet. This second jet removes a part of the flow rate delivered from the intensifier and increases the pumping frequency. The medium value of the two different pressure signal forms is the same (360 MPa) (Figure 6). In particular, Figure 6-b shows the pressure signal obtained by increasing the outflow rate from the double acting intensifier used.

Rubber has been cut (P=360 MPa, u=200 mm/min). The obtained cutting surface has been analyzed either in terms of synthetic index of roughness (R_{3z}) [3], or by comparing the surface roughness profiles.

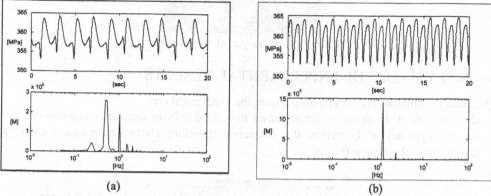

FIGURE 6. Pressure signal obtained with dn=0.2 mm (a) and increasing the delivery outflow (b)

3.4 CHARACTERIZATION OF THE WATERJET IN AIR: THE EFFECT OF THE PRESSURE FLUCTUATION ON DIAMETER OF THE JET

In order to analyse the effect of the level of pressure on the diameter of the jet (dj), some photographic surveys of the jet have been carried out at different values of pressure. Nozzles with diameter of 0.2 and 0.3 mm have been used. The jet has been photographed using a lens that allows to reach a depth of field of 0.05 mm. Then, the images have been acquired through a scanner with 1200x1200 dpi resolution. Every photograph has been elaborated with a software for analysis of images. The diameter of the jet, at 2 mm below the primary nozzle, has been measured, that is at the same distance of the stand off distance used in precision processing (Figure 7).

---

[3] To the aim to make inquiries about the maximum and minimum value on the roughness profile, the parameter R_{3z} is liked better than R_a because it protects from possible local imperfections of the material or exceptional data. It is defined R_{3z} the parameter that considers the average difference between the third maximum and the third minimum of the profile on every base length :

$$R_{3z} = \frac{1}{5}\left(\sum_{i=1}^{5}|y_{3\,MAX(i)}| - \sum_{i=1}^{5}|y_{3\,MIN(i)}|\right)$$

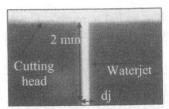

FIGURE 7. Photo analysis of the waterjet

4 ANALYSIS OF THE EXPERIMENTAL RESULTS

The main conclusions that can be drawn from the experiment are:
- the analysis of the pressure signal shows that the dominant frequency corresponds at the switch frequency of the valves: this frequency, therefore, characterizes nearly completely the pressure fluctuation (Figure 8).

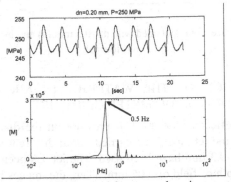

FIGURE 8. Pressure signal in time and frequency domain generated by a double acting intensifier

- When the diameter of the primary nozzle increases, the pumping frequency of the intensifier increases too in order to keep the delivered flow rate continuous (Figure 9).

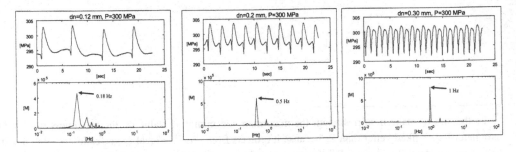

FIGURE 9. Increase of switch frequencies vs. dn

- When the diameter of nozzle increases, the standard deviation of the pressure signal increases too (Figure 10).

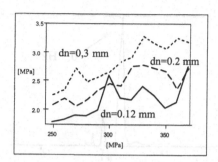

FIGURE 10. Standard deviation of pressure signal vs. dn

- The roughness measurements at the same depth (Table1), carried out on rubber cuts made at different levels of pressure (step of 15 MPa) (section 3.2), do not point out substantial changes in the surface roughness (R_a).

TABLE 1. Mean and standard deviation of R_a on 12 replications vs. pressure (Rubber)

	R_a (P=300 MPa)	R_a (P=315 MPa)	R_a (P=330 MPa)	R_a (P=345 MPa)	R_a (P=360 MPa)
Mean	19.23 μm	18.38 μm	18.35 μm	17.86 μm	17.83 μm
σ	1.36 μm	1.43 μm	1.58 μm	0.91 μm	0.34 μm

- The surface striations have a regular trend and their dimension is comparable with the diameter of the jet. This is pointed out, in particular, in the cuts of very soft materials like plasticine (Figure 11).

FIGURE 11. Plasticine surface generated by waterjet cutting

- The analysis of the surfaces roughness profiles obtained by waterjet cutting of polycarbonate test pieces shows that there is no correlation between the peculiar pulsation of the pressure signal (Figure 12) and the more important frequencies measured on the roughness profile (Figure 13). In fact, the main frequency of the pressure signal is 0.5 Hz (Figure 12); at 200 mm/min, the signal on the surface should have a period of 6.7 mm and therefore a spatial frequency of 0.15 1/mm. This value is not the characteristic frequency of the profiles (Figure 13).

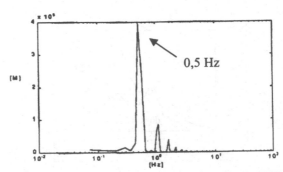

FIGURE 12. FFT of the pressure signal (dn=0.2 mm, P=250 MPa)

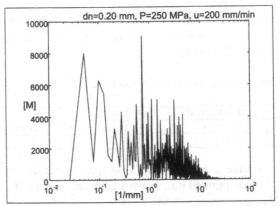

FIGURE 13. FFT of the roughness profile (Polycarbonate)

- The surface morphology obtained with two different pressure signal forms (section 3.3) has
  been compared. Table 2 shows mean and standard deviation of the roughness R_{3z} measured
  on the cutting surface and Figure 14 shows the Fast Fourier Tansform (FFT) of the
  roughness profiles. Case A represents the cut realized with the pressure signal form
  obtained using dn=0.2 mm; Case B, instead, that realized with pressure signal form
  obtained increasing flow rates. The surface quality, in the two cases, has not pointed out
  substantial changes. Moreover, there are no harmonics correlated at the pressure signal
  (Figure 6) that generated the cut.

TABLE 2. Arithmetical mean and standard deviation of R_{3z} at different depth of measure

	Depth: 10 mm		Depth: 20 mm		Depth: 30 mm	
	Mean	σ	Mean	σ	Mean	σ
Case A	75.60 mm	5.39 mm	104.02 mm	8.68 mm	146.02 mm	11.01mm
Case B	74.08 mm	4.55 mm	107.5 mm	4.62 mm	150.42 mm	8.05 mm

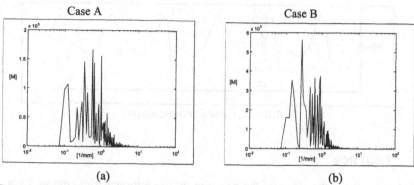

(a)   (b)

FIGURE 14. FFT of the roughness profile obtained by dn=0.2 mm (a) and increasing the delivery flow rate (b)

• The analysis of the photographic surveys of the jet at different level of pressure shows that an increase in the pressure does not cause a substantial fluctuation in the diameter of the jet (Figure 15).

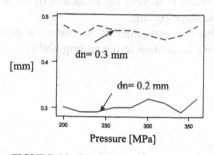

FIGURE 15. Waterjet diameter vs. pressure

Therefore, from the over reported considerations, the pressure fluctuations and the pressure signal form do not have a substantial influence on the surface quality of waterjet cutting.
This can be theoretically explained estimating the penetration time of the waterjet (d_t) inside the material. Through the theoretical waterjet feed valuation V_{th} [19],

$$V_{th} = \sqrt{\frac{2L}{\rho_0(1-C)}\left[\left(\frac{P}{L}+1\right)^{1-C}-1\right]} \qquad C=0.1368 \qquad L=300 \text{ MPa} \qquad (1)$$

the penetration time d_t inside a material with a particular thickness can be estimated. At the process parameters used, the penetration time is some µs while the fluctuation period of the pressure signal is about some seconds; therefore, during the penetration time, the pressure can be considered continuous (Figure 16).

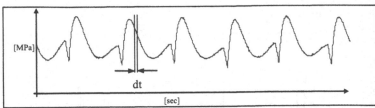

FIGURE 16. Cutting process diagram

5 CONCLUSIONS

This work has analysed the effect of the pressure on the pure waterjet cutting. Cuts of rubber realised by WJ at different value of pressure show negligible changes in the surface roughness. No correlation has been found between the roughness profiles and the pressure signal. Moreover the pressure signal form do not have a substantial influence on the surface quality of waterjet cutting: in fact, at the process parameters used during the penetration time, the pressure can be considered continuous. The striations along all the surface presuppose that there are other factors, for example the vibration of the cutting head and of the work piece [20], that directly influence the quality of the waterjet cutting.

Anyway, most probably, the pressure fluctuation causes a periodic trend of the waterjet specific energy and therefore a fluctuation in the cutting ability.

NOMENCLATURE

P   Pressure [MPa]
u   Feed rate [mm/min]
dn   Nozzle diameter [mm]
sod   Stand off distance [mm]
M   Power Spectrum Density Modulus
ρ_0 Density of the water at atmospheric pressure

REFERENCES

1. Hashish, M., Aspects of abrasive waterjet performance optimization, 8th International Symposium on Jet Cutting Technology, Durham, England, 9-11 Sept., 1996, pp. 297-308
2. Tunkel, R., Double action hydraulic intensifier, 9th American Waterjet Conference, Deaborn, Michigan, 23-26 August, 1997, pp. 373-385
3. Kulekci, M.K., (2002) Processes and apparatus developments in industrial waterjet applications, International Journal of Machine Tools & Manufacture, Volume No 42, pp. 1297-1306
4. Singh, P.J., Computer simulation of intensifiers and intensifier systems, 9th American Waterjet Conference, Deaborn, Michigan, 23-26 August, 1997, pp. 397-414
5. Hu, F., Robertson, J., Simulation and control of discharge pressure fluctuation of ultra-high pressure waterjet pump, 7th American Water Jet Conference, Seattle, Washington, 28-31 August, 1993, pp. 337-349

6. Susan-Resiga, R., Attenuator's volume influence on high pressure's pulsation in a jet cutting unit, 11th International Symposium on Jet Cutting Technology, St. Andrews, Scotland, 8-10 Sept., 1992, pp. 37-45

7. Yie Gene, G., A pulsation-free fluid pressure intensifier, 9th American Waterjet Conference, Deaborn, Michigan, 23-26 August, 1997, pp. 365-372

8. Singh, P.J., Benson, D., Development of phased intensifier for Waterjet cutting, 11th International Symposium on Jet Cutting Technology, St. Andrews, Scotland, 8-10 Sept., 1992, pp.305-318

9. Chalmers, E., Pressure fluctuation and operating efficiency of intensifier pumps, 7th American Water Jet Conference, Seattle, Washington, 28-31 August, 1993, pp. 327-336

10. Fabien, B.C., Ramulu, M., Tremblay, M., (2003) Dynamic Modelling and Identification of a Water Jet Cutting System, Mathematical and Computer Modelling of Dynamical Systems, Volume No:9, Issue: 1, pp. 45-63

11. Hashish, M., Prediction models for AWJ machining operations, 7th American Water Jet Conference, Seattle, Washington, 28-31 August, 1993, pp. 205-216

12. Guo, N.S., Louis, H., Meier, G., Surface structure and kerf geometry in abrasive waterjet cutting: formation and optimization, 7th American Water Jet Conference, Seattle, Washington, 28-31 August, 1993, pp. 1-25

13. Karpinski, A., Louis, H., Monno, M., Peter, D., Ravasio, C., Scheer, C., Südmersen, U., Effects of pressure fluctuations and vibration phenomenon on striation formation in AWJ cutting, 17th International Conference on Water Jetting, Mainz, Germany, 7-9 September, 2004, pp. 123-136.

14. Chao, J., Geskin, E.S., Experimental study of the striation formation and spectral analysis of the abrasive waterjet generated surfaces, 7th American Water Jet Conference, Seattle, Washington, 28-31 August, 1993, pp. 27-41

15. Hashish, M., On the modelling of abrasive-waterjet cutting, 7th International Symposium on Jet Cutting Technology, Ottawa, Canada, 26-28 June, 1984, pp. 249-265

16. Hashish, M., (1989) A model for Abrasive-Waterjet (AWJ) Machining, Transactions of the ASME, Volume No: 111, pp. 154-162

17. Hashish, M., On the modelling of surface waviness produced by abrasive-waterjets, 11th International Symposium on Jet Cutting Technology, St. Andrews, Scotland, 8-10 Sept., 1992, pp. 17-34

18. Chen, F.L., Siores, E., (2003) The effect of cutting jet variation on surface striation formation in abrasive water jet cutting, Journal of Materials Processing Technology, Volume No: 35, pp. 1-5

19. Hashish, M., (1989) Pressure effects in Abrasive-Waterjet (AWJ) Machining, Journal of Engineering Materials and Technology, Volume No: 11, pp. 221-228

20. Monno, M., Ravasio, C., (2005) The effect of cutting head vibrations on the surfaces generated by waterjet cutting, International Journal of Machine Tools and Manufacture, Volume No. 45, Issue 3, pp. 355-363

NEW POSSIBILITIES BY DIRECT LASER MICRO SINTERING FOR MICRO SYSTEM TECHNOLOGY COMPONENTS USING NANOPHASED POWDERS

H. Becker[1], A. Ostendorf[2], P. Stippler[2], P. Matteazzi[3]

[1] Microls, Hannover, Germany
[2] LZH, Hannover, Germany
[3] CSGI and MBN, Italy

KEYWORDS: Microengineering, Nanopowders, Micromanufacturing.

ABSTRACT. Paper describes a new manufacturing method based on the most advanced small scale laser sintering technology able to shape solid metallic objects being composed of nanophased powders. New development in powder feeding will be illustrated being able to feed particles down to three microns medium particle size using combination of different mechanical and physical principles. For hybrid objects possibility with material change on the flight is included. Several peripheral devices for process control are implemented like online temperature control, oxygen control in the process chamber, shape control of processed geometries and integrated observation microscope. Fully atomization is assured by special designed software and industrial standards of PXI technology. Agglomerated nanophased systems were developed to respond to the needs of micro fabrication in terms of phase sets, particle sizes and shapes.

1 INTRODUCTION

Solid freeform fabrication (SFF) [1-3] processes are a group of new technologies which are going to revolution product development and manufacturing. There are common features shared by the various approaches consisting in the ability to produce, without almost any restriction on shape, complex geometry directly from a computer generated model. A typical machine consists of a miniature manufacturing plant representing the convergence of mechanical, chemical, electrical, materials and computer engineering sciences.

In the last ten years such technologies achieved some degree of maturity and some became commercially available to produce three dimensional (3D) parts and tooling using a wide range of materials including paper, polymers, wax, sand, ceramics and metals. The next major step in the development of such technologies is taking place in the area of direct fabrication processes, especially for the low volume production of functional metal, cermets and ceramic components or tooling with the necessary space resolution.

For applications with the need for metallic components with high spatial resolution, economical methods are very limited. Most existing technologies are based on removing material from a bulk like microdrilling with high resolution up to 2 μm but very limited geometry possibilities. Laser assisted manufacturing appear as the most promising candidate for micromanufacturing technology. The main limitation of actually existing technologies for direct fabrication using laser based processes (the others are not competing) rely on the inability to combine: 3D gradient materials; spatial resolution better than 200 μm; careful, on-

Published in: E. Kuljanic (Ed.) *Advanced Manufacturing Systems and Technology*, CISM Courses and Lectures No. 486, Springer Wien New York, 2005.

line control of processing parameters and particularly temperature at the powder/part interaction.

A new advancement in direct micromanufacturing of metallic components has been achieved [4] and is described here. Quite innovative principles have been established concerning high precision Laser Sintering, achieving spatial resolutions better then 50 microns. Innovations were introduced in the areas of optics and Laser source, monitoring and process control, powder feeding and source engineered nanopowders.

2   AGGLOMERATED NANOPOWDERS FOR MICROFABRICATION

The high precision laser based manufacturing required the development of suitable powder grades satisfying some possible criteria:

1)   having particles' sizes selected in the range compatible with desired spatial resolution (in this case about 10 microns);

2)   having phase assemblies suitable for a easy sintering mechanism with limited or none phase transformations, thus conferring stability to the process;

3)   having available fine structures in order to manage the phase assemblies within the 10 microns and in the same time improved mechanical properties.

To satisfy above principles the production process chosen was of mechanochemical synthesis [5], leading naturally to agglomerated structures. The process moreover is capable of delivering industrial, ton range, and quantities of nanomaterials.

Among the phase assemblies chosen there was a very simple system, composed of Fe and Cu having practically no mutual solubility in the solid state. The agglomerated nanopowders were developed with crystal sizes typically in the two phases of 9 nm (Fe) and 16 nm (Cu) as revealed by X-Ray diffraction patterns reported in Figure 1 and confirmed by TEM observation (Figure 2). The agglomerated nanopowders exhibited the expected thermal behavior in showing two melting points at 1090 °C (for copper with some alloying, in Figure 3) and 1540 °C (for Fe). This bimodal behavior and the width in temperature among the two melting points gave the necessary operating window to the process of laser sintering.

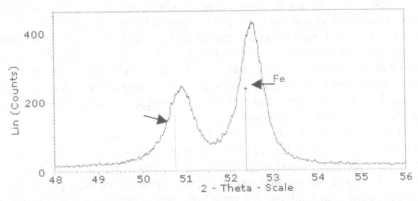

FIGURE 1. X-ray diffraction pattern (CoKα) of nanopowders used for microfabrication.

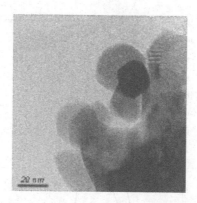

FIGURE 2. TEM image of nanopowders

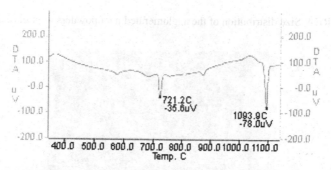

FIGURE 3. DTA traces of the agglomerated Fe-Cu nanopowder

The size distribution of powders for the process of laser sintering had to be accurately adjusted. The wanted size distribution is as narrow as possible within the scope of about 10 microns. To this end various classification methods have been explored. Centrifugal classification was the methodology chosen to classify powders having the broad size distributions coming for the process of synthesis into the desired profile.

Centrifugal classification has been finely tuned to give rise to final sizes' distributions as shown in Figure 4 and confirmed by the direct SEM observations in Figure 5 of powder morphologies.

The width of the distribution is remarkably narrow, within plus or minus 4-5 microns from the average 10 microns.

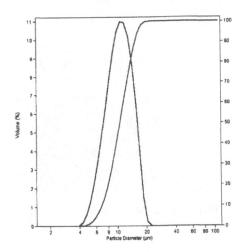

FIGURE 4. Size distribution of the agglomerated nanopowders after classification

FIGURE 5. SEM image of agglomerated nanopowders after classification

3 MICROMANUFACTURING MACHINE: MICROLS MACHINE CONCEPTS

3.1 BASIC MACHINE CONCEPT

Basically the machine concept consists of four main blocks:

- Beam generation, guidance, forming and regulation
- Filler material transport and focusing
- Periphery components
- Automation

The functions of the main blocks are secured by single components and their interactions and are tested constantly on their efficiency. For better illustration the main blocks and their interactions regarding material and information flow are shown in Figure 6.

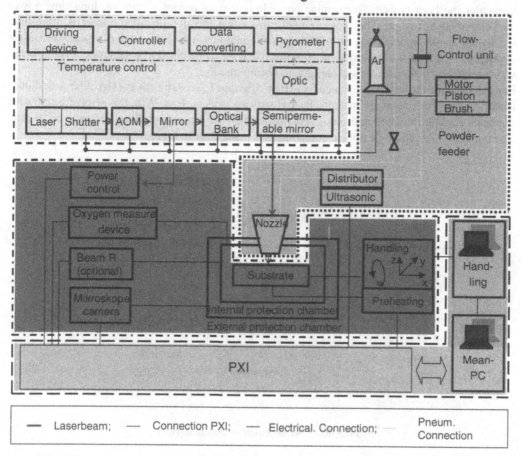

FIGURE 6. Schematic sketch of basic machine concept

In the following the characterization of the single components will be shown.

3.2 BEAM PRODUCTION, GUIDANCE, FORMING AND REGULATION

The generation of radiation is realized by a new concept of a diode pumped thin disc laser which combines the advantages of a low optical degradation of the laser medium and excellent output power scalability.

For characterization beam characteristics of the untreated beam has been measured at three different distances. In the left part of the figure the result of the untreated beam measurements is shown at the named points and is supplemented with the measured intensity profile. In the right part of the picture the intensity distributions have been added for two levels given by the intensity profile under the integration of the Gaussian curve. From the picture it is evident that the beam profile of the disc laser corresponds to a Gauss profile at the first approximation. By repetition measurements after one, four and eight hours the diameter of the untreated beam proved to be stable up to 2.5%.

The conditions for reproducible results are low variations within the laser parameters and in particular the laser power. To examine possible occurring output variations, long time power measurements are made. The measurement proves that the stability of the laser is more than sufficient for the application. From the optical equations it is evident that a possible largest divergence angle or a big numerical aperture (N.A.) is to be chosen to generate a small beam waist diameter by the pre-set wavelength and beam quality.

For extension of the working distance two High-Aperture-Laser-Objectives (HALOs) with diminished spherical aberration are installed behind such an aspheric lens. With this setup a beam waist diameter could be realized and measured smaller than 10 μm in a working distance of 50 mm. The integration of the single optical components is made clear in Figure 7.

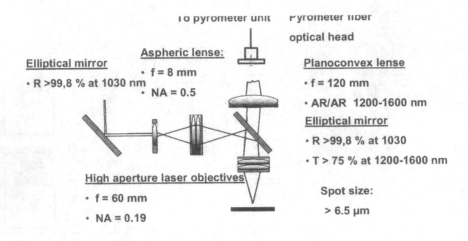

FIGURE 7. Arrangement of optical devices

The measurement of the temperature enables on the one hand side the direct time regulation of phase transition and on the other hand the detection of welding defects in particular and represents in this way an important instrument of the quality assurance of laser beam welding processes. The multi color pyrometric procedure has been chosen as a measuring method because of its flexible application by which the heat radiation emitted from the laser beam interaction zone is detected and transformed into one temperature information.

The processing of the analogous signals delivered by the pyrometer is conducted through a process control software which has been provided by the LZH and which allows control operation at real-time. The program reads the data determined by the quotient pyrometer and conditions them with the help of mathematical functions.

3.3 FILLER MATERIAL TRANSPORT AND FOCUSING

One of the most critical points is the powder feeding of micro particles. Therefore intensive investigations of many possible feeding principles were made. Main problems while feeding were the constant flow of small particles without agglomeration and adhesion at feeding components. Beyond this the demanded very low mass flow eliminated most of the conventional powder feeders. Motivated by [6] describing of possible powder feeding using ultrasonic vibration, investigations started in direction of hybrid-technology feeding. By combination of a modified mechanical feeding principle with stimulation of ultrasonic vibration at certain amplitudes a stable particle flow even of very small particles is possible, proven by several high speed camera recordings. To create gradient components, a change of materials in flight has been realized. Focusing of fed particles has been realized by different implementations of co-axial forming.

3.4 PERIPHERY COMPONENTS

For process control and steadily process observation different periphery components has been integrated which will be explained in the following:

Power control: The power control measures the power while processing and transfers the data for data comparison with deposited tables to the PXI.

Beam R: The Beam R is a new approach to real-time beam profiling. It couples a novel multi-plane real-time slit scanning system with improved slit technology. The possible features are: measuring of the axial misalignment, indicating the current focus direction, estimating the beam waist diameter at the focus, estimating the off-axis distance of the focus, estimating M^2. In case the central measurement plane is positioned in the focal plane, these estimates become measurements of: focus position in z, x-y position, beam waist diameter and irradiance and M^2 beam quality. All measurements are done in real-time whereby beam characterization has been simplified enormously.

Midget microscope: A midget microscope in combination with the software METRIC is used to measure a taken picture of the produced trails. The advantage is that the substrate on which the process takes place must not be taken from the substrate holder.

Preheating of the substrate holder: Optional there is a preheating possibility for the substrate holder heating the substrate. The adjustable and controlled temperature range is between 50 to 400 °C. Preheating could become important in case of avoiding crack effects.

Oxygen measuring device: The oxygen measuring device is to make sure that there is none or less amount of oxygen while processing. It will guarantee reproducible sintering results because of an integrated security check connected to the automation system.

Protection chamber: Former investigations had pointed out that even less oxygen in the processing zone has fatal consequences on processing the exothermic powders. Therefore the protection chamber took centre stage. With the help of the described oxygen measuring device different possibilities were checked. An optimum is found that consists of two combined protection chambers, an internal and an external one. Using both in a certain way the amount of oxygen is below 30 ppm (cp. Figure 8).

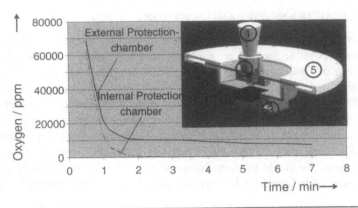

| 1 : Laser beam | 2 : Coaxial Nozzle | 3: Substrate holder |
| 4: Substrat | 5 : Upper sealing | 6 : Lower sealing |

FIGURE 8. Amount of oxygen using combination of internal and external protection gas chamber

3.5 AUTOMATION

Fully automating is assured by specially designed software and industrial standards of PXI technology.

4   MICROMANUFACTURING EXPERIMENTS

Micromanufacturing experiments were preceded by:
1) accurate 3D CAD description of parts to be manufactured;
2) build up, from the CAD files of the slicing procedure and slice by slice description of deposition strategy
3) guess of suitable processing conditions, particularly temperature, in the deposition zone.

Experiments reported here were performed using Fe-Cu based powders previously described, and the processing temperature was fixed at a temperature around 1100 °C, enough to give rise to a partial melting of the powders but still keeping, in each particle, a substantial fraction of material-finely structured.

In Figure 9 and Figure 11 is shown an overview of different components constructed with a spatial resolution better then 25 microns is shown. The sintering mechanisms are highlighted in Figure 11 where it appears to be clear that a liquid phase sintering is operating leaving part of the single particles unmelted and therefore still keeping their shape (detail in Figure 10). X-ray diffraction performed directly on the component shows that (Figure 12) crystal growth occurred (to about 30 nm in both phases) but still conserving part of the original nanostructure. Therefore, the micromanufacturing process resulted in the manufacturing of nanostructured components.

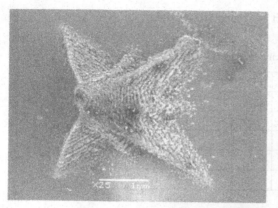

FIGURE 9. Punch: SEM image

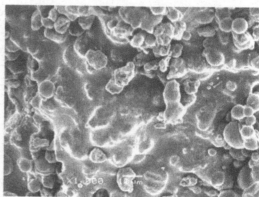

FIGURE 10. Detail of the punch: SEM image

FIGURE 11. Detail of figure 36

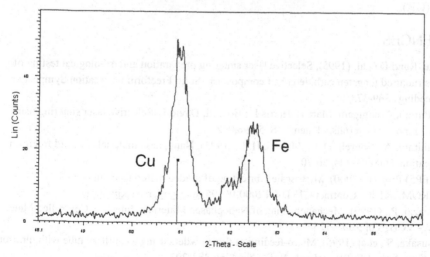

FIGURE 12. X-ray diffraction directly on the part

In the Table 1 some measurements of the object in Figure 11 are listed:

TABLE 1. Measurements on Microparts

Dimension	Standard Deviation
2.603 mm	17 µm
1.633 mm	14 µm
2,611 mm	25 µm
2,886 mm	21 µm
0,244 mm	14 µm

5   CONCLUSIONS AND FUTURE SCENARIO

Direct manufacturing of metallic components with spatial resolution below 20 microns is now reached. Still open are challenges regarding improvements in the machine productivity which is however already beyond the convenience (if not technical capability) of competing technologies.

Quite possible appears the future scenario of combining high resolution direct manufacturing with high productivity low resolution tasks, in an entirely new technological platform for direct manufacturing of materials and components.

ACKNOWLEDGEMENTS

The authors wish to acknowledge the financial contribution given by the EU with the project Micromaking [4], in which several European partners (beyond those listed in the title page) contributed to the innovations introduced: ENISE (F), IMNR (RO), TIL (UK), Siemens (D), INASCO (GR).

REFERENCES

1. Cottle, Rand D et al, (1995), Selective laser sintering preparation and tribological testing of nanostructured tungsten carbide-cobalt composites, Solid Freeform Fabrication Symposium Proceedings 369-373
2. Manthiram, Arumugam; Marcus, Harris L.; Bourell, David L, Selective laser sintering using nanocomposite materials, Patent U.S. 5,296,062.
3. Manthiram, A.; Bourell, D. L.; Marcus, H. L, (1993), Nanophase materials in solid freeform fabrication, JOM 45(11), 66-70.
4. EU (FP5) Project, (2000), Microscale fabrication of graded materials components - MICROMAKING, Contract G1RD-CT2000-00195, www.micromaking.org
5. Matteazzi, P., (1999), Mechanomaking of Nanophased Materials, Interface Controlled Materials, Wiley- VCH, Vol 9 119-125
6. Matsusaka, S., et al (1995), Micro-feeding of fine powders using a capillary tube with ultrasonic vibration, Adv. Powder Technol., Vol 6, No 4, pp 283-293

INCREMENTAL FORMING OF SHEET METAL BY TWO INDUSTRIAL ROBOTS

H. Meier, J. Zhang, O. Dewald

Department of Mechanical Engineering, Ruhr-University Bochum, Germany

KEYWORDS: Sheet Metal Forming, Incremental Forming, Industrial Robot.

ABSTRACT. This paper describes an innovative developed process of incremental sheet metal forming based on two industrial robots. Instead of using expensive dedicated dies simple and reusable tools were designed to form sheet metal into complicated geometrical shapes. These simple tools with flat or hemispherical head are mounted on the end effectors of the robots, while the forming tool on one side is driven by the robot to follow a prescribed path generated from the designed shape, the supporting tool on the other side moves along the path synchronously. by incremental steps, a sheet metal is drawn to its final shape between the forming tool and the supporting tool. As a result, various kinds of sheet metal parts have been formed using this method. The measured profile and strains from the forming process are presented and discussed.

1 INTRODUCTION

The incremental sheet metal forming is an innovative metal-working process for small-lot production and prototyping, sheet metal can be formed into complicated shapes using no conventional dies. This process, along with the introduction of CNC technology has drawn attention to researchers in both academic and industrial fields in recent years. Some new metal forming technologies have been developed. Incremental stretch expanding was studied by Kitazawa[1]. The sheet metal had been hold by a chuck on a modified CNC lathe, while cylindrical tool was mounted on a tool rest. The axisymmetrical shape is stretched from a flat blank by relative motion of the tool around the blank. Matasubara has studied the incremental backward bulge forming with CNC machine [2]. In his study, the sheet metal was formed by the vertical milling machine using two bar tools. The moving tool moves along a designed path on the sheet metal to bend and expand the sheet around the fixed tool post. Jeswiet at Queen's University has investigated further the Matasubara's forming method and, at the same time, also developed a new forming method, the CNC incremental forming[3]. In his new method a support rig, which has a steel plate with a cut-out of the part perimeter, has been used in place of the fixed tool for the CNC forming process. All these processes are characterized by two major advantages that the forming processes are carried out by a tool with simple geometry and the tool path is computer controlled.

This study presents a novel flexible process for incremental sheet metal forming using two industrial robots developed at Ruhr-University Bochum. The sheet metal shaping of regular shapes such as frustum of pyramid, frustum of cone and spherical cup performed by two industrial robots operated by a common PC based controlling program is first introduced. Results of the measured profile and strains from shaping a frustum of pyramid are then presented and discussed. The goal of this study is to demonstrate the feasibility of using two industrial robots instead of conventional dies for incremental forming of sheet metal.

Published in: E. Kuljanic (Ed.) *Advanced Manufacturing Systems and Technology*,
CISM Courses and Lectures No. 486, Springer Wien New York, 2005.

2   THE FORMING PROCESS USING TWO ROBOTS

Figure 1 shows the principle of the incremental forming process by two robots. The flat sheet is held by a clamping device. According to the thickness of sheet metal there are spaces between the forming tool and supporting tool at the beginning of the forming process. The forming tool and supporting tool, which have hemispherical or cylindrical head, are mounted on the end effectors of two robots. The forming tool is driven by the robot on the prescribed path that is generated from the designed shape. On the other side of the sheet metal the supporting tool follows the forming tool synchronously around the part perimeter. The sheet metal is formed between the forming tool and the supporting tool.

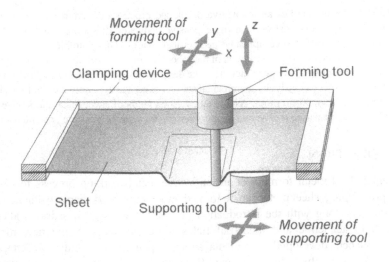

FIGURE 1. Principle of the sheet metal forming process

The forming process itself is mostly realized by the paths on which the tools are moved. To achieve a maximal flexibility, two industrial robots were chosen in the forming process because of the advantage of its six degrees of freedom. The robot enables a performance of complex motion with an independent orientation change of the tools in space. In this way, freeform sheet metal parts will later be achieved by future researches of suitable forming strategy.

The generation of the two tool paths is based on the shape of the selected part and the movement strategy. The paths strategy which is followed by the forming tool and the supporting tool for three standard geometries is shown in figure 2. The tool path direction is indicated by arrows. These movement strategies have been proved out in the experiment work. For example, the tool paths of the frustum of cone are introduced. Before the forming process the forming and the supporting tool move to their starting positions. The supporting tool stands parallel to the forming tool on other side of the sheet. Between the tools there is a calculated offset for the sheet metal. In the first step, the forming tool traces a circular path in x-y plane, on the inner side of the cone's perimeter; the supporting tool is driven along the outer side of the frustum's perimeter synchronously. For the second step, the forming tool moves in radial direction with an incremental feed; the supporting tool stays in the former position. While the third step, the forming tool moves along the new circle in x-y plane, which is calculated from the geometry, and the supporting tool repeats the movement along the perimeter. These steps are repeated analogically until the sheet metal is little by little bent and expanded to the cone's frustum. For the improvement of the shape and the surface further factors must be taken into consideration. To fit the contour to the given shape, tool correction data especially for the hemispherical tool should influence the path generation. At last the calibration of the robot systems according to the clamping situation of the test rig should be included in this calculation.

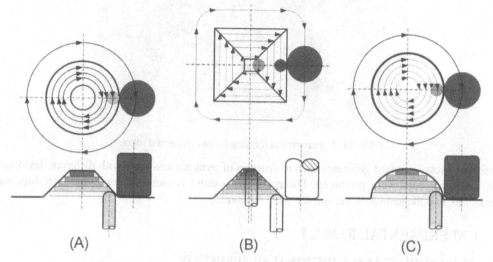

FIGURE 2. Essential tool paths for symmetrical und non-symmetrical shapes

(A) frustum of cone (B) frustum of pyramid (C) spherical cup

3 EXPERIMENTATION

The overview of the incremental sheet metal forming using two industrial robots Manutec R3 is shown in figure 3. The robots are able to transport loads up to 15 kg. Because of the low load of robot, the flat sheet metal used is 0.5-1.0 mm thick. To develop a synchronal motion of the two industrial robots, they were connected by Ethernet card, so the position information of two tools can be transferred for synchronization. The two robots are controlled by a common PC based program which realizes an automatic generation of the robot control programs for the movement in order to simplify the experimental process. An appropriate forming strategy is selected according to the geometry of the work piece, defining the movement path of forming tool and supporting tool. A test rig which stands between the two robots was designed to clamp the sheet metal for the forming process. The working area is 200 mm by 200 mm. The sheet metal used for the experiment is aluminum 99.5 soft. The velocity of the two robots were 10 m/min for the experiment. The forming tool, mounted on the end effector of the robot, is a hemispherical steel shaft of 12 mm diameter; the supporting tool, mounted in portal robot, is a flat steel shaft of 38 mm diameter with ball bearing for reducing the influence of friction between the sheet metal and the tool. The amount of feed of the forming tool in z-direction was 1 mm or 2 mm. The feed plays an important roll in the surface roughness.

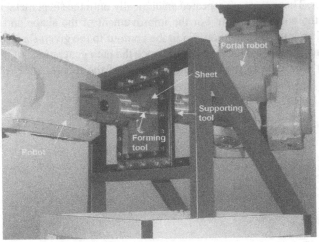

FIGURE 3. Incremental forming by two industrial robots

In the first tests standard geometry such as frustum of pyramid and cone with different deviations and spherical cups were produced. The preliminary cross contour of the formed product was measured, and the strains on the surface were measured.

4 EXPERIMENTAL RESULT

4.1 END PRODUCT AND ADDITIONAL OBSERVATION

A part of the results of kinds of sheet metal parts are shown in figure 4. By using the special way to generate the two corresponding robot programs automatically a great band with of different variants of each of the three geometries were researched.

The photos of the parts show that the surface quality is rather good. However a bulging line can obviously be found on the surface of the spherical cup parts and the frustum of cone. The line results from the feed movement of the forming tool and depends on the feed quantity for each step.

The original flange dimensions of the sheet are maintained. This shows the difference between this forming process and deep drawing. The currently used process is similar to stretch expanding.

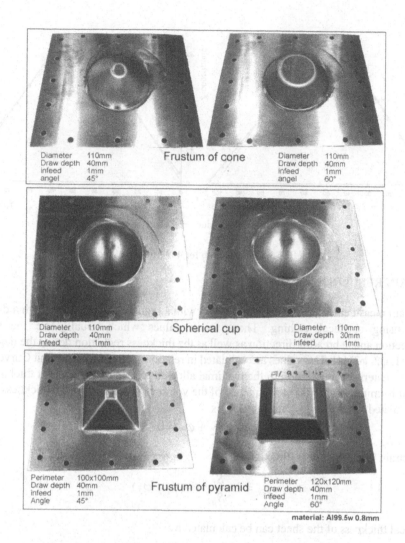

FIGURE 4. Shells of frustum of pyramid and cone, spherical cup

4.2 PROFILE MEASUREMENT RESULTS

For the Evaluation of accuracy of the geometry the two cross section of frustum of pyramid were measured by CMM Zeiss UMM 850. Figure 5 shows the measured contour of the two profiles (x-z plane and y-z plane) for the frustum of 40 mm depth and angle 45°. The upper part of the measured pyramid has a good geometrical accuracy, the maximum error is 0.51 mm. The bottom part of the contour has a bigger error, the error shows several millimeters. The form deviation results from the offset between the forming tool and the supporting tool.

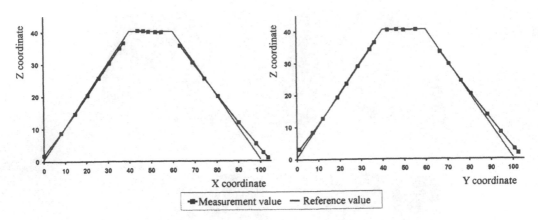

FIGURE 5. Cross section for 45° frustum of pyramid

4.3 STRAIN RESULTS

For the strain measurement, the sheet was marked with a regular grid of circles with a diameter of 2.0 mm, using chemical etching. These strain values which usually are the strains in length-direction and in breadth-direction as well as the thickness reduction define the degree of the forming ($\varphi 1$, $\varphi 2$, $\varphi 3$) [4]. The values are validated in relation to the Forming Limit Curve (FLC) of the selected material or in relation to the maximal allowed reduction of the sheet thickness by the sheet metal forming process. In consequence of the volume constancy law the thickness reduction can be calculated. The volume constancy law is

$$\varphi_1 + \varphi_2 + \varphi_3 = 0 \tag{1}$$

Then the strain in the thickness direction is

$$\varphi_3 = \ln \frac{S_1}{S_0} = -(\varphi_1 + \varphi_2) \tag{2}$$

and the local thickness of the sheet can be calculated by

$$S_1 = S_0 \cdot e^{\varphi_3} \tag{3}$$

Two surface strains, the one is the true strain in length-direction, the another is the true strain in breadth-direction, were measured using the grid on the surface by a traveling microscope. In every faces of the frustum of pyramid the strains of five positions were measured. The measurement positions are illustrated in figure 6. By using the equation (2), the local strains in the thickness direction were calculated, and the thickness reduction is calculated by the equation (3). Examples of the strain measurements and calculations along the sheet surface for the pyramid's frustum of 40 mm depth and an angle of 45° are shown in figure 6.

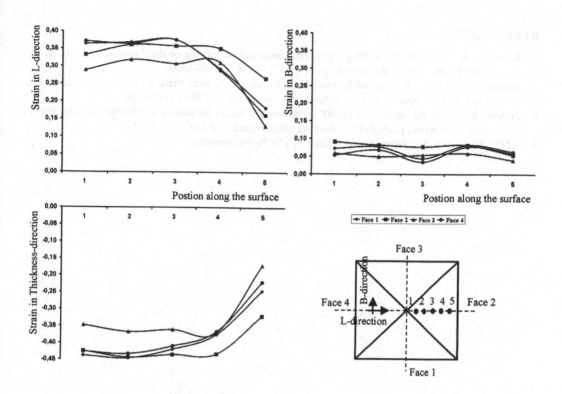

FIGURE 6. Strain for a 45° frustum of pyramid

5 CONCLUSIONS

The experiment confirms that a sheet metal can be successfully formed into simple standard geometrical shapes without using dedicated dies. The corresponding research project studies the fundamentals of the incremental sheet metal forming process using two robots, with the aim to develop a rapid prototyping technique for sheet metal component.

This forming process is similar to stretch expanding forming. Later the clamp device may be modified extending the process to allow a material flow in the flange area.

The geometric accuracy of the sheet part is rather good. The geometric accuracy depends on the paths of two tools.

ACKNOWLEGEMENTS

This work is supported by the German Research Foundation (DFG).

REFERENCES

1. Kitazawa, K., Nakajima, A., (1999), Cylindrical incremental drawing of sheet metals by CNC incremental forming process, Proceedings of 6th ICTP, Vol II, 1495-1500.
2. Matsubara, S., (1994), Incremental backward bulge forming of a sheet metal with a hemispherical tool, Journal of the Japan society for technology of plasticity, Vol 35/1, 1311-1316.
3. Jeswiet, J., Hagan, E., Szekeres, A., (2002), Forming parameter for incremental forming of aluminium alloy sheet metal, Journal of Engineering Manufacture, Vol 216, 1367-1371.
4. König, W.,(1990), Fertigungsverfahren Band 5, VDI Press, Germany.

A FINITE DIFFERENCE MODEL BY A.D.I. METHOD FOR PULTRUSION PROCESS: TEMPERATURE AND DEGREE OF CURE ANALYSIS

P. Carlone, G. S. Palazzo, R. Pasquino

Department of Mechanical Engineering, University of Salerno, Italy

KEYWORDS: Pultrusion, Degree of Cure, Finite Difference Modelling.

ABSTRACT. Pultrusion is one of the most cost-effective processes for composite materials manufacturing. It allows to realize constant section products characterized by remarkable fibres alignment, mechanical properties directionality, good surface quality with high production rate. Thermo-chemical aspects, such as temperature and degree of cure, are very important in order to obtain good mechanical properties of the final product as fast as possible or to realize a post-die shaping of pultruded parts. In this paper a three-dimensional finite difference model based on A.D.I. method and realized in MATLAB language for temperature and degree of cure is proposed. The present model takes in account heat transfer due to heating platens, die-cooler at die entrance and resin exothermic cure reaction. An analysis of the influence of mesh density and Peclet number on model accuracy is presented. Numerical results are compared with experimental data.

1 INTRODUCTION

Pultrusion is a continuous manufacturing process used to form polymeric composite materials into parts of constant cross section. Several reinforcement fibers types, as glass, carbon, and aramid fibers, and thermoset or thermoplastic matrix can be processed by pultrusion. In Figure 1 a pultrusion line is shown. The reinforcement fibers, in the form of continuous strands (rovings) or mats, are placed on creel racks. Fibers are pulled through a guide plate, and then impregnated passing by a resin bath. Uncured material crosses a preform plate system to be correctly shaped before entering the curing die. The die entrance is characterized by a tapered shape to remove the resin excess and make the material inlet easier. A water cooling channel is placed in the first part of the die, to prevent premature material solidification and then some heating platens are placed in top and bottom surfaces of the die to provide the heat for material polymerization. Outside the die, the cured composite material is pulled by a continuous pulling system (caterpillar or reciprocating pullers) and then a traveling cut-off saw cuts the part into desired length.

In recent years, several researchers focussed their attention on different problems related to the pultrusion process. Numerical and experimental analysis were preferred for the complexity of the process. In particular several investigations have been performed on the analysis of the heat transfer and cure, on the pressure rise in the tapered zone of the die, and on problems related to impregnation of reinforcing fibers by thermoplastics.

In [4][5][6][11] several numerical models for microscopic or macroscopic impregnation during the pultrusion process of thermoplastic matrix composites have been proposed. Pressure behavior into the curing die has been investigated by Raper et al. in [8] and by Gadam et al. in [9]. Numerical models, based on Darcy's law for the flow through a porous media, for the pressure and

Published in: E. Kuljanic (Ed.) *Advanced Manufacturing Systems and Technology*,
CISM Courses and Lectures No. 486, Springer Wien New York, 2005.

cal models, based on Darcy's law for the flow through a porous media, for the pressure and velocity fields in a pultrusion die are proposed, evaluating the influence of material properties and
process parameters. The importance of the analysis of heat transfer and cure in pultrusion is
highly recognized to obtain a final product characterized by the desired mechanical properties or
to realize a post-die shaping. A three dimensional model of heat transfer and cure is proposed in
[1] and [2] by Chachad *et al.*. Discretization requirements and the influence of process parameters
are studied in [3] using a bi-dimensional finite element model.

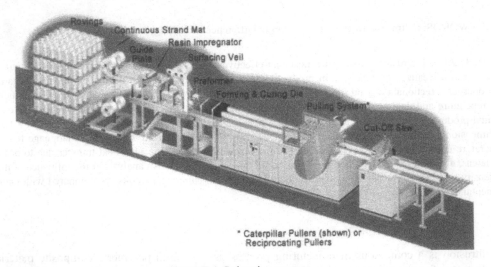

FIGURE 1. Pultrusion process

Thermo-chemical aspects of the process have been investigated by Liu *et al.*, who realized several
numerical models, based on finite difference method, on finite element method, and on nodal
control volume technique, proposed in [7], [10], and [13]. Temperature dependent material properties and resin shrinkage are taken into account in [12] by Joshi *et al.*, who in [14] and [15]
proposed an optimization procedure based on finite element/nodal control technique. In [16] and
[18] finite element simulations of the resin flow in injection pultrusion are proposed. An experimental investigation on bent-pultrusion has been performed by Britnell *et al.* in [17].
This paper deals with a three dimensional finite difference modelling for thermo-chemical aspects
of the process. The model, based on A.D.I. method, takes in account heat transfer due to heating
platens, die-cooler at die entrance and resin exothermic cure reaction. Numerical implementation
is realized using MATLAB software.

2   THERMAL ANALYSIS

2.1   GOVERNING EQUATIONS

A correct modelling of the pultrusion process requires to take into account not only the processing
composite material, but all tools related to heat transfer. Different equations for the energy balance of the heated die and the composite materials are required, taking into account the

exothermic resin cure reaction and convective effect, related to the movement of the processing part. Heat transfer equation for the heated die, in Cartesian coordinate system, can be written as follows:

$$\rho_d c_{p,d} \frac{\partial T}{\partial t} = \frac{\partial}{\partial x}\left(k_{x,d}\frac{\partial T}{\partial x}\right) + \frac{\partial}{\partial y}\left(k_{y,d}\frac{\partial T}{\partial y}\right) + \frac{\partial}{\partial z}\left(k_{z,d}\frac{\partial T}{\partial z}\right),$$

(1)

where T is the temperature, ρ_d is the die material density, $c_{p,d}$ is the specific heat, $k_{x,d}$, $k_{y,d}$, and $k_{z,d}$ are the thermal conductivities in x, y, and z direction respectively.

Taking into account that reinforcing fibers are wetted out and impregnated by the resin before entering the die, it is assumed that the resin does not flow, and then heat transfer equation for the composite part can be written, assuming x axis as the pull direction, as follows:

$$\rho_c c_{p,c}\left(\frac{\partial T}{\partial t} + u\frac{\partial T}{\partial x}\right) = k_{x,c}\frac{\partial}{\partial x}\left(\frac{\partial T}{\partial x}\right) + k_{y,c}\frac{\partial}{\partial y}\left(\frac{\partial T}{\partial y}\right) + k_{z,c}\frac{\partial}{\partial z}\left(\frac{\partial T}{\partial z}\right) + \rho_r V_r Q,$$

(2)

where V_r is the resin volume fraction ($V_r = 1 - V_f$, being V_f the fiber volume fraction), u is the pull velocity, ρ_c is the composite material density, ρ_r is the resin density, $c_{p,c}$ is the specific heat, $k_{x,c}$, $k_{y,c}$, and $k_{z,c}$ are the thermal conductivities in x, y, and z direction respectively, and Q is the specific heat generation rate ([J/Kg]) due to resin exothermic cure reaction.

Composite material density, specific heat, and conductivities can be evaluated using the followings:

$$\rho_c = (1-V_r)\rho_f + V_r\rho_r,$$

(3)

$$c_{p,c} = \frac{(1-V_r)c_{p,f}\rho_f + V_r c_{p,r}\rho_r}{(1-V_r)\rho_f + V_r\rho_r} = \frac{(1-V_r)\rho_f c_{p,f} + \rho_r V_r c_{p,r}}{\rho_c},$$

(4)

$$k_c = \frac{[(1-V_r)\rho_f + V_r\rho_r]k_f k_r}{(1-V_r)\rho_f k_r + V_r\rho_r k_f} = \frac{k_f k_r \rho_c}{(1-V_r)\rho_f k_r + V_r\rho_r k_f},$$

(5)

where subscripts f, r, c, and d are used for fiber, resin, composite, and die material respectively. Several models can be used to describe resin reaction and to evaluate heat generation rate Q, as:

$$\frac{d\alpha}{dt} = K(1-\alpha)^n,$$

(6)

where α is the degree of cure, $d\alpha/dt$ is the rate of the reaction, K is a coefficient related with temperature T by an Arrhenius type equation and n is the kinetic exponent. An other analytic model for resin reaction, due to Kamal and Sourour, can be written as follows:

$$\frac{d\alpha}{dt} = (K_1 + K_2\alpha^m)(1-\alpha)^n.$$

(7)

Exponents m and n are generally assumed as temperature-independent, while K_1 and K_2 are related with temperature T by Arrhenius type equations.

In the present model, indicating as α the degree of cure, defined as the ratio of the amount of heat evolved during the curing process up to time t (indicating as H(t)) to the total heat of reaction, (indicating as H_{tr}), rate of resin reaction can be written as follows:

$$R_r = \frac{d\alpha}{dt} = \frac{dH(t)}{H_{tr}dt}, \tag{8}$$

where:

$$\alpha = \frac{H(t)}{H_{tr}}. \tag{9}$$

Taking into account Equation 8, heat generation can be written as follows:

$$Q = \frac{dH(t)}{dt} = H_{tr}R_r. \tag{10}$$

H_{tr} and R_r can be experimentally evaluated using differential scanning calorimetry (DSC) measurements. Kinetics of the resin can be finally written as follows:

$$R_r = \frac{d\alpha}{dt} = K_0 \, exp\left(-\frac{E}{RT}\right)(1-\alpha)^n, \tag{11}$$

where K_0 is a constant, E is the activation energy, n is the order of the reaction, and R is the gas universal constant (R = 8.363 [J/molK]).
The concentration of the resin species is governed by the following species equation:

$$\frac{\partial \alpha}{\partial t} = R_r - u\frac{\partial \alpha}{\partial x}. \tag{12}$$

2.2 BOUNDARY CONDITIONS

The following assumptions are adopted:
–   the initial degree of cure of the composite is assumed as null and its initial temperature is equal to the resin bath temperature:

$$T_c(x,y,z,t=0) = T_{rb}; \tag{13}$$
$$\alpha(x,y,z,t=0) = 0;$$

–   the temperature and the degree of cure of the processing material in the cross section of the die entrance is imposed always as equal to the above values:

$$T_c(x=0,y,z,t) = T_{rb}; \tag{14}$$
$$\alpha(x=0,y,z,t) = 0;$$

–   the initial temperature of the die is assumed as the room temperature (an initial die temperature closer to expected average temperature makes the convergence easier):

$$T_d(x,y,z,t=0) = T_{amb}; \tag{15}$$

–   the composite cross section, at the die exit, is modeled as adiabatic:

$$\left.\frac{\partial T(x,y,z,t)}{\partial x}\right|_{x=L} = 0; \tag{16}$$

–   constant temperatures are imposed in opportune nodes to simulate the heating platens and the die-cooler:

$$T(y=H/2,x,z,t) = T_{imp}; \tag{17}$$

- convective boundary condition is imposed on external die surfaces:

$$\dot{q}_{cond} = -h_{con}\left(T_{wall} - T_{amb}\right); \tag{18}$$

- taking into account the low pull speed, at the interface between the die and the part, conductive heat flow orthogonal to the die axis is assumed:

$$k_y \frac{\partial T}{\partial y}\bigg|_c = k_y \frac{\partial T}{\partial y}\bigg|_d ;$$

$$\tag{19}$$

$$k_z \frac{\partial T}{\partial z}\bigg|_c = k_z \frac{\partial T}{\partial z}\bigg|_d .$$

3 NUMERICAL MODELLING

The numerical method adopted to evaluate the temperature and the degree of cure, according to the above equations, is the Alternating Direction Implicit (A.D.I.) method. Equations are solved using the Thomas algorithm, which allows to obtain a direct solution by matrix inversion and then it results more efficient than iterative Gauss method for bi-dimensional or three-dimensional problems. An implicit approach has been preferred for problem discretization, for numerical stability and computational time required. For reason of brevity a more accurate description of the used numerical method is not possible. This Section deals with the numerical implementation of the considered problem.

Governing equations 2 is characterized by remarkable nonlinearities, due to the convection and heat generation, even if composite material properties are assumed as constants. An efficient numerical procedure has been adopted by decoupling heat generation term from the energy equation as follows:

$$\rho_c c_{p,c}\left[\left(\frac{\partial T}{\partial t} + u\frac{\partial T}{\partial x}\right)\right]^{n+1} = \left[k_{x,c}\frac{\partial^2 T}{\partial x^2} + k_{y,c}\frac{\partial^2 T}{\partial y^2} + k_{z,c}\frac{\partial^2 T}{\partial z^2}\right]^{n+1} + \left[\rho_r V_r Q\right]^n, \tag{20}$$

where n is the time step index. Equation 12 is integrated explicitly as follows:

$$\left[\frac{\partial \alpha}{\partial t}\right]^{n+1} = \left[R_r - u\frac{\partial \alpha}{\partial x}\right]^n. \tag{21}$$

In this way the generative term, evaluated in the previous time step, is assumed as a constant and relative non-linearity is avoided.

According A.D.I. method, equation (2) writes:

$$T^{n+1} - T^n = c\delta_x T^{n+1} + r_x\delta_x^2 T^{n+1} + r_y\delta_y^2 T^{n+1} + r_z\delta_z^2 T^{n+1} + q^n, \tag{22}$$

where:

$$\delta_x^2 T = T_{i-1,j,k} - 2T_{i,j,k} + T_{i+1,j,k},$$

$$\delta_y^2 T = T_{i,j-1,k} - 2T_{i,j,k} + T_{i,j+1,k},$$

$$\delta_z^2 T = T_{i,j,k-1} - 2T_{i,j,k} + T_{i,j,k+1}, \tag{23}$$

$$\delta_x T = T_{i,j,k} - T_{i-1,j,k},$$

$$r_x = \frac{k_{x,c}\Delta t}{\rho_c c_{p,c}\Delta x^2}, \; r_y = \frac{k_{y,c}\Delta t}{\rho_c c_{p,c}\Delta y^2}, \; r_z = \frac{k_{z,c}\Delta t}{\rho_c c_{p,c}\Delta z^2},$$

$$c = u\frac{dt}{dx}, \tag{24}$$

$$q = \frac{\rho_r V_r}{\rho_c c_{p,c}}Q.$$

For each time step, the equation 22 is solved using three different sub-step, one for each direction, assuming that the heat generation is equally distributed.

$$\frac{T^{n+1/3}-T^n}{1/3} = c\delta_x T^{n+1/3} + r_x\delta_x^2 T^{n+1/3} + r_y\delta_y^2 T^n + r_z\delta_z^2 T^n + \frac{1}{3}q^n, \tag{25}$$

$$\frac{T^{n+2/3}-T^{n+1/3}}{1/3} = c\delta_x T^{n+1/3} + r_x\delta_x^2 T^{n+1/3} + r_y\delta_y^2 T^{n+2/3} + r_z\delta_z^2 T^n + \frac{1}{3}q^n, \tag{26}$$

$$\frac{T^{n+1}-T^{n+2/3}}{1/3} = c\delta_x T^{n+1/3} + r_x\delta_x^2 T^{n+1/3} + r_y\delta_y^2 T^{n+2/3} + r_z\delta_z^2 T^{n+1} + \frac{1}{3}q^n. \tag{27}$$

For each direction there are three dependent variable to evaluate by Thomas algorithm.
Three-dimensional analysis needs some stability conditions, which imposes the following restrictions:

$$s_x, s_y, s_z \leq 1.5, \tag{28}$$

where:

$$s_x = \frac{k_{x,c}dt}{\rho_c c_{p,c}dx^2}, \; s_y = \frac{k_{y,c}dt}{\rho_c c_{p,c}dy^2}, \; s_z = \frac{k_{z,c}dt}{\rho_c c_{p,c}dz^2}. \tag{29}$$

Convective term imposes some restraints on discretization to obtain the desired accuracy and stability:

$$Pe = \frac{V_r \rho_c c_{p,c} u dx}{k_{x,c}} \leq \frac{1}{1-c} = 2, \tag{30}$$

$$c + 2s_x \leq 1.$$

Equation 21 is solved using the up-wind method and a backward difference (composite part speed is positive) and can be written, with some manipulation, as:

$$\alpha_{i,j,k}^{n+1} = [R_r]^n dt + \left(1 - u\frac{dt}{dx}\right)\alpha_{i,j,k}^n + u\frac{dt}{dx}\alpha_{i-1,j,k}^n. \tag{31}$$

The stability of up-wind method is conditioned by the following condition:

$$c = u\frac{dt}{dx} \leq 1. \tag{32}$$

4 CASE STUDY

The pultrusion of a part with rectangular cross section, proposed in some previous investigations ([1] and [10]), was simulated to validate the present model with experimental data and to compare it with other previous model. Figure 2 shows the geometry of the part and the die for the considered case. Taking into account the symmetry, only a quarter of the problem was modeled.

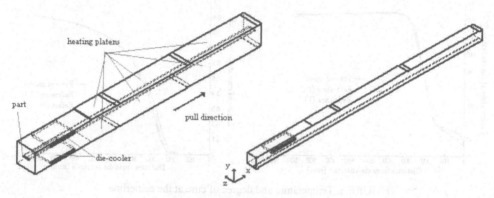

FIGURE 2. Case study

Die length L, width W, and height H are 915 mm, 76.2 mm, and 76.2 mm, respectively. Composite part width w is 25.4 mm and height h is 3.175 mm. Die is heated by six platens, three platens are placed on the top surface and the remaining on the bottom surface. Between the die entrance and the first platen a water cooling channel is placed to avoid resin premature gelation. Die cooler temperature is assumed as 50°C. The part enters into the die at the temperature of 30°C. Pull speed is 200 mm/min. Room temperature is assumed 20°C and convective coefficient is considered as 10 W/mm^2K. The length of the second and the third platens is 270 mm, for the first platen a minor length (100 mm) is considered for the presence of the cooling channel. Consecutive platens are divided by 15 mm and the empty zone is subjected to convective boundary conditions. Platens temperature is assumed as 188°C. Symmetry sections are modelled using adiabatic conditions. The properties of the considered materials and the resin kinetic parameters are listed in tables 1 and 2, respectively.

TABLE 1. Material properties

	ρ [kg/m^3]	c_p [J/kgK]	k_x [W/mK]	k_y, k_z [W/mK]
Epoxy resin	1260	1255	0.21	0.21
Fiberglass	2560	670	11.4	1.04
Lumped (V$_r$=0.361)	2090.7	797.27	0.90	0.56
Die (steel)	7800	473	40	40

TABLE 2. Resin kinetic parameters

H_{tr} [J/kg]	K_0 [s^{-1}]	E [J/mol]	n
324000	192000	60000	1.69

5   RESULTS AND DISCUSSION

Figure 3 shows a comparison between the temperature and the degree of cure in the centerline of the part obtained in the present model and (numerical and experimental) results from references [1] and [10]. Temperature profile has been experimentally evaluated using thermocouples, opportunely placed at die entrance and then pulled with the part.

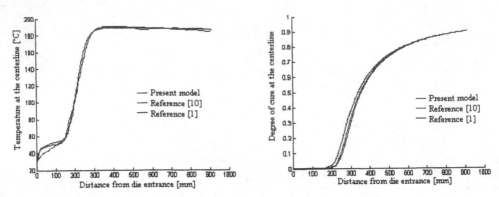

FIGURE 3. Temperature and degree of cure at the centerline

A very good agreement between numerical and experimental data was found.

For the above simulation a mesh constituted of 1101*49*10 (x, y, z direction, respectively) nodes was used, to respect the condition (equation 30) related to the Peclet number and for the reduced part dimension in the y direction. Several simulations have been performed to evaluate the influence of the Peclet number and of the mesh density on the accuracy of the results.

Table 3 lists the different considered value of the Peclet number, the relative space increment in the x direction, indicated as dx, and the calculation time for each simulation. In figure 4 a comparison between the temperature and the degree of cure profiles obtained with different Peclet number is shown. A very good agreement between all profiles was found; negligible differences can be found for Peclet numbers major than 30.

TABLE 3. The effect of the Peclet number on calculation time

PECLET	dx [mm]	CALCULATION TIME [s]
1.82	0.831	8513
10	4.574	1573
20	9.15	884
30	13.45	601
50	22.87	580

The remarkable stability of the proposed model is probably related to the low pull speed considered: a considerable increase of the pull speed could require a relatively lower value for the Peclet number, to obtain the desired accuracy.

In Figure 5 the degree of cure at the exit section is shown for different mesh density, using 12, 20, and 28 nodes in the composite cross section, for the cases a, b, and c, respectively. Negligible differences between the considered meshes are found, probably related to the reduced dimensions of the part. A major relevance of the mesh density was found simulating pultrusion of parts with larger cross sections (not shown for brevity reason).

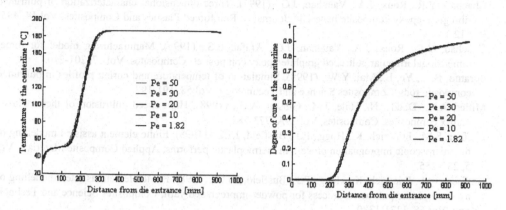

FIGURE 4. Influence of the Peclet number on temperature and degree of cure profiles

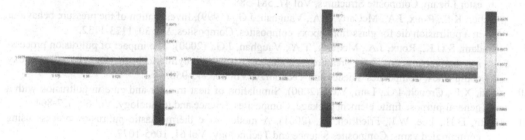

FIGURE 5. Effects of the mesh density on final degree of cure

6 CONCLUSIONS

A three dimensional finite difference model of the thermo-chemical aspects of the pultrusion process has been developed. Numerical modelling is based on the Alternating Direction Implicit (A.D.I.) method and implemented using software MATLAB. Numerical result, i.e. temperature and degree of cure profiles, have been compared with experimental data, and good agreement was found. The proposed procedure results very stable and accurate, and computational time can be considerably reduced using opportune values of the Peclet number, with negligible loss in accuracy. Pultrusion of part of complex cross section can be simulated. The present model is characterized by remarkable flexibility, related to the several materials properties and process parameters considered and can be used as an efficient and economic tool for process analysis and optimization, instead of more expensive experimental investigations.

REFERENCES

1.   Chachad, Y.R., Roux, J.A., Vaughan, J.G., (1995), Three-dimensional characterization of pultruded fibreglass-epoxy composite materials, Journal of Reinforced Plastics and Composites, Vol 14, 495-512.
2.   Chachad, Y.R., Roux, J.A., Vaughan, J.G., Arafat, E.S., (1996), Manufacturing model for three-dimensional irregular pultruded graphite/epoxy composites, Composites, Vol 27, 201-210.
3.   Suratno, B.R., Ye, L., Mai, Y.W., (1998), Simulation of temperature and curing profiles in pultruded composite rods, Composites Science and Technology, Vol 58, 191-197.
4.   Miller, A.H., Dodds, N., Hale, J.M., Gibson, A.G., (1998), High speed pultrusion of thermoplastic matrix composites, Composites, Vol 29A, 773-782.
5.   Haffner, S.M., Friedrich, K., Hogg., P.J., Busfield, J.J.C., (1998), Finite element assisted modelling of the microscopic impregnation process in thermoplastic performs, Applied Composite Materials, Vol 5, 237-255.
6.   Haffner, S.M., Friedrich, K., Hogg., P.J., Busfield, J.J.C., (1998), Finite-element-assisted modelling of a thermoplastic pultrusion process for powder-impregnated yarn, Composites Science and Technology, Vol 58, 1371-1380.
7.   Liu, X.L., Hillier, W., (1999), Heat transfer and cure analysis for the pultrusion of a fibreglass-vinyl ester I beam, Composite Structures, Vol 47, 581-588.
8.   Raper, K.S., Roux, J.A., McCarty, T.A., Vaughan, J.G., (1999), Investigation of the pressure behaviour in a pultrusion die for glass-fibre/epoxy composites, Composites, Vol 30, 1123-1132.
9.   Gadam, S.U.K., Roux, J.A., McCarty, T.A., Vaughan, J.G., (2000), The impact of pultrusion processing parameters on resin pressure rise inside a tapered cylindrical die for glass-fibre/epoxy composites, Composites Science and Techology, Vol 60, 945-958.
10.   Liu, X.L., Crouch, I.G., Lam, Y.C., (2000), Simulation of heat transfer and cure in pultrusion with a general-purpose finite element package, Composites Science and technology, Vol 60, 857-864.
11.   Kim, D.H., Lee, W.I., Friedrich, K., (2001), A model for e thermoplastic pultrusion process using commingled yarns, Composites Science and Technology, Vol 61, 1065-1077.
12.   Joshi, S.C., Lam, Y.C., (2001), Three-dimensional finite-element/nodal-control-volume simulation of the pultrusion process with temperature dependent material properties including resin shrinkage, Composites Science and Technology, Vol 61, 1539-1547.
13.   Liu, X.L., (2001), Numerical modeling on pultrusion of composite I beam, Composites, Vol 32, 663-681.
14.   Li, J., Joshi, S.C., Lam, Y.C., (2002), Curing optimization for pultruded composite sections, Composites Science and Technology, Vol 62, 457-467.
15.   Joshi, S.C., Lam, Y.C., Tun, U.W., (2003), Improved cure optimization in pultrusion with pre-heating and die-cooler temperature, Composites, Vol 34, 1151-1159.
16.   Liu, X.L., (2003), A finite element/nodal volume technique for flow simulation of injection pultrusion, Composites, Vol 34, 649-661.
17.   Britnell D.J., Tucker, N., Smith, G.F., Wong, S.S.F., (2003), Bent-pultrusion-a new method for the manufacture of pultrudate with controlled variation in curvature, Journal of Material Processing Technology, Vol 138, 311-315.
18.   Liu, X.L., (2004), Iterative and transient numerical models for flow simulation of injection pultrusion, Composite Structures, Vol 66, 175-180.

VIRTUAL TRYOUT AND OPTIMIZATION OF THE EXTRUSION PROCESS USING A SHAPE VARIABLES GENERATOR INTEGRATED IN THE CAE PRE-PROCESSING ENVIRONMENT

A. Anglani, A. Del Prete, G. Papadia

Department of Innovation Engineering, University of Lecce, Italy

KEYWORDS: Extrusion, Morphing, Optimization.

ABSTRACT. CAE tools usage to evaluate process performances it has became a matter of fact in cases like: metal forming, foundry, casting and forging. Like in these applications, also for the extrusion processes CAE tools usage has became a convenient opportunity, not only to verify the designed process but also to tune it in a virtual way. In this specific application, it has been evaluated the chance to use an optimization tool in combination with a process solver. The chance to optimize the extrusion process has been investigated using shape design variables for the tool process design, in order to obtain the best extruded profile quality. The applied procedure has shown strength points like: the full integration between the pre-processor and the shape variables generator, without any need to exchange data with the CAD environment during the optimization and weak points, such has the reduced freedom for the shape variation, due to the risk of an excessive distortion of the finite elements which describe the process.

1   INTRODUCTION

The main activities presented in this paper are: definition of an optimization technique and its application of this one to an extrusion die. The CAE tools that have been used are: Altair Hyper-Mesh®, a finite element pre and post processor [1], Altair HyperXtrude® (HX), a finite element based simulation tool for analysis and design of aluminum extrusion dies and process [2], and Altair HyperStudy®, a parametric study and multi-disciplinary optimization tool for robust product design [3]. The CAD-CAE techniques give the change to find the best solution for a production process in terms of: feasibility and product quality. For these reasons it has been implemented a traditional workflow from the CAD to the CAE environment and viceversa; this workflow must be managed by the designer. The chance to introduce a range of CAD alternatives in the CAE analysis and also in the optimization routine gives the possibility: to reduce time, to reduce cost and to improve the quality of the product. It is possible to reach the previous advantages within the usage of morphing-technique. Using morphing it is possible: (1) to change the geometric shape described by a mesh, (2) to alter dimension of a meshed part parametrically, (3) to map an existing mesh onto a new geometry, (4) to create shape changes usable as design variables in a optimization study [4]. In this paper morphing has been used to generate virtual optimization models in which the design variables are the shape changes. These shape changes have influence both on process and product aspects of the considered problem. This work has

Published in: E. Kuljanic (Ed.) *Advanced Manufacturing Systems and Technology*,
CISM Courses and Lectures No. 486, Springer Wien New York, 2005.

been developed in collaboration with To.Ma., a private company specialized in aluminum extrusion.

2 PROCESS SIMULATION AND RESPONSES ANALYSIS

Process simulation has been applied on a test case represented in Figure 1. In Figure 2 is represented the CAD model of the extruded profile and in Figure 3 the cad and real models of the extrusion die.

FIGURE 1. Test case – Extruded Profile

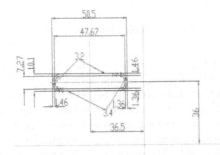

FIGURE 2. CAD of the Extruded Profile

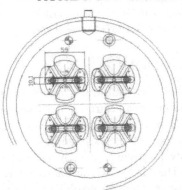

FIGURE 3. Real and CAD model for the extrusion die

The product is obtained by a direct extrusion with a four opens die. In table 1 are presented the fundamental parameters of the extrusion process.

TABLE 1. Parameters of the direct extrusion process

Billet Length	960mm	N° output	4
Billet Weight	64,5kg	Material	AL 6060
Billet Temperature	460°C	N° of profiles for each billet	24
Un-extruded Billet	20mm	Length of single profile	6500mm
Extrusion Velocity	1,4(mm/s)	Extrusion Ratio	43,55
Extruded Profile Velocity	3,6(m/min)		

In Figure 4 it is represented the FEM model of the billet and the extruded profile.

FEM MODEL

Billet   Feeding Zone   Prechamber   Bearing   Profile

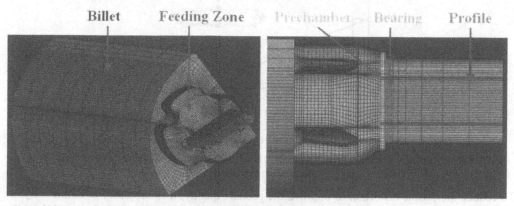

FIGURE 4. FEM model

The discretized model, in combination with boundary conditions set-up in the solver pre-processor (HyperXtrude-Pre), represents the input data to obtain the baseline solution within the solver HX. The results of the simulation are reported in Figure 5 where, the numerical deflection of the profile has been put in evidence.

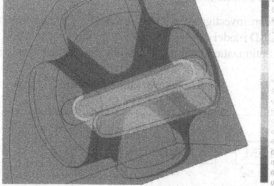

FIGURE 5. Numerical deflection of the test case

From the previous Figure 5 it is a matter of fact that the deflection must be reduced. These deflection is due to the velocity distribution of the extruded profile along the extrusion direction (z) [5]; infact, it can be easily observed (Figure 6) that the portion (x-) is faster than the portion (x+) and also that the portion (y+) is faster than the portion (y-); this non uniform distribution in the output velocity generates the observed deflection.

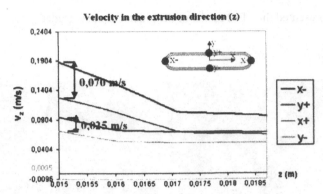

FIGURE 6. Numerical velocities distribution along the different portions of the extruded profile along z (extrusion direction)

To reduce the profile distortion to improve the product quality, it is necessary to define a tryout strategy based on: (1) process parameters (billet temperature and ram velocity) changes [5], or (2) bearing length changes (which is the zone of the die where the extruded profile assumes its final shape) in order to modify the friction conditions and, consequently, the velocity of the extruded profile [6] or (3) feeding zone changes in order to modify the flux of the aluminum inside the die and, so, the velocity of the extruded profile [6]. Each one of the previous points may be considered one at the time or in a combination.

The first option has been investigated without reaching significant results. The second and the third options needed CAD model changes in order to be implemented and, as consequence, a new FEM model for each optimization attempt. The workflow related to this approach is represented in Figure 7:

Traditional working procedure

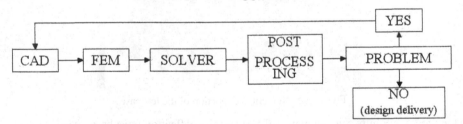

FIGURE 7. The workflow is re-run n times till to the complete resolution of the problems

The only chance to improve the process quality is to find out the right shape modification of the die tools. As consequence, a new CAD model has to be produced as first step to verify the new solution applying the described procedure. This type of analysis presents hided disadvantages like: the impossibility to know, in an initial phase, if the implemented CAD modifications will minimize the profile deflection. At this stage, it is impossible to know how many backwards and forwards cycles are necessary to obtain an optimal solutions. If this goal hasn't been reached, it will be necessary to modify another time the CAD and run once more the procedure or more than one, if necessary. This possible procedure is very time consuming. In order to avoid it, it should be better to create a geometric domain space in which the dimensions of interest can automatically change reaching the optimal value.

3   OPTIMIZATION STRATEGY

The optimization strategy that has been used in this paper is fundamentally based on the chance to use as design variables, geometry changes of the die. Optimization is a very powerful option to down size the risks mentioned above. The effectiveness of the optimization depends from factors like: (1) optimization routines, specifically suitable for the physic problem under test, (2) design variables types, used to explore the design space [7]. About this last point the used pre-processor capabilities are a fundamental aspects in order to define the user's possibility to generate design variables that is, in other words, the capability to define the wanted design space. In this specific case the routine based on the Response Surface Methodology [3] has been used in combination with a shape variable space defined using a morphing technique [4]. In Figure 8 there is a schematic representation of this approach, in which the usage of an optimization software (Altair HyperStudy®) is applied on a design space obtained by the definition of shape design variables each one of them within specific upper and lower bounds using the morphing technique.

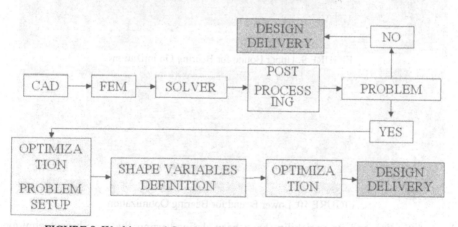

FIGURE 8. Working workflow based on optimization and morphing techniques

The optimization problem has been set up defining as objective function the first term of eq.1: the ratio of the output velocities for which has been imposed a target value

$$(v_+/v_-)=1 \tag{1}$$

where:

v_+ is the average velocity, along the extrusion direction (z), of all nodes with a velocity bigger than v_m

v_- is the average velocity, along the extrusion direction (z), of all nodes with a velocity lower than v_m

v_m is the average velocity of the extruded profile.

In this paper two different optimization approaches have been run: (1) BEARING OPTIMIZATION and (2) FEEDING OPTIMIZATION.

4   BEARING OPTIMIZATION

4.1   PRE-PROCESSING BEARING OPTIMIZATION

To modify the distribution of output velocities it is possible to change the dimension of the bearing, which is the portion of the die by which the aluminum flux reach its definitive profile. First of all, it is necessary to find the portion of the extruded profile that is slower than the others [7]; then it is necessary to reduce the length of the bearing corresponding to this portion of the profile in order to increase its output velocity, by reducing the friction action. Figure 9 shows the initial dimension of the bearing while in Figure 10 it is showed its lower bound configuration.

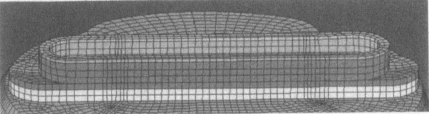

FIGURE 9. Upper Bound for Bearing Optimization

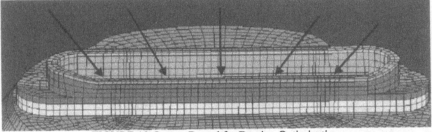

FIGURE 10. Lower Bound for Bearing Optimization

The shape definition and its variability have been defined using the morphing technique and, using the morphing output in combination with the optimization software, it is possible to automatically make the best choice, in terms of possible shapes, for the formulated problem.

4.2   POST-PROCESSING BEARING OPTIMIZATION

At the end of the optimization process (Figure 11) it has been obtained a bearing length different from the one defined in the baseline model (2,5mm instead the initial bearing length of 3,4mm) for which the ratio between average output velocities is equal to one. In Figure 12.1 and 12.2 there is the comparison between the velocities distribution of the baseline model and the velocities distribution of the optimized one. The difference between (x+) and (x-) velocities has been reduced from 0,070(m/s) to 0,035(m/s); at the same time, the difference between (y+) and (y-)

velocities has been reduced from 0,025(m/s) to 0,008(m/s). This new output velocities distribution allows to significantly reduce the extruded profile distortion (Figure 13).

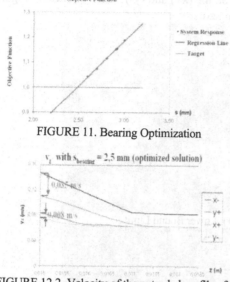

FIGURE 11. Bearing Optimization

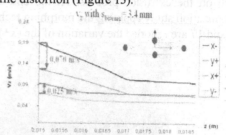

FIGURE 12.1. Velocity of the extruded profile in the initial state

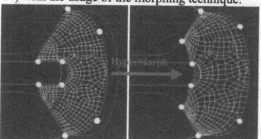

FIGURE 12.2. Velocity of the extruded profile after Bearing Optimization

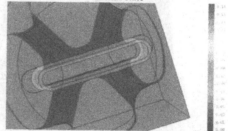

FIGURE 13. Deflection of the extruded profile after Bearing optimization

5 FEEDING OPTIMIZATION

5.1 PRE-PROCESSING FEEDING OPTIMIZATION

The velocities distribution may be also modified changing the feeding zone dimensions (Figure 14) with the usage of the morphing technique.

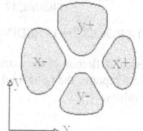

FIGURE 14. Application of morphing technique on feeding zone

FIGURE 15. The four different feeding zones for each extruded profile

First of all it is necessary to identify the area of the feeding zone where the extruded material has the slowest velocity profile; then it is necessary to enlarge the area of the previous zone to increase the output velocity of the material. In the considered test case there are four feeding zones (Figure 15) for each extruded profile; using the same criteria developed for the bearing optimization, these four different feeding zones are indicated with (x+), (x-), (y+) and (y-). A first phase of simulations have been run to identify which, among the four feeding zones, has the major influence on the extrusion process. At the end of this first phase, it has been decided to run an optimization study applying the morphing technique on the (x+) and (y-) feeding zones. In Figure 16 and 17 are reported the variation of the (x+) feeding zone.

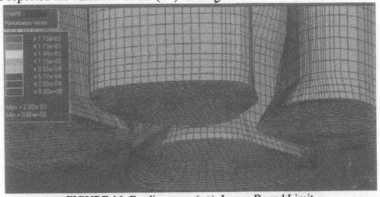

FIGURE 16. Feeding zone (x+). Lower Bound Limit

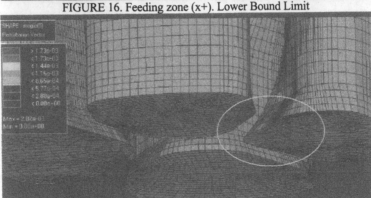

FIGURE 17. Feeding zone (x+). Upper Bound Limit

5.2 POST-PROCESSING FEEDING OPTIMIZATION

The numerical deflection of the extruded profile after the optimization it is reported in Figure 18. In table 2 are reported the optimized areas for the (x+) and (y-) feeding zone, in comparison with the initial values.

TABLE 2.

Feeding Zone	Initial Area (mm^2)	Optimized Area (mm^2)	Delta (mm^2)	Delta (%)
X+	600,06	618,17	18,11	3,13%
Y-	668,89	686,34	17,45	2,6%

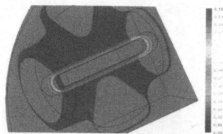

FIGURE 18. Deflection of the extruded profile after Feeding Optimization

The difference between (x+) and (x-) velocities has been reduced from 0,070(m/s) to 0,005(m/s); at the same time the difference between (y+) and (y-) velocities has been reduced from 0,025(m/s) to 0,015(m/s), as it can be seen in Figure 19.

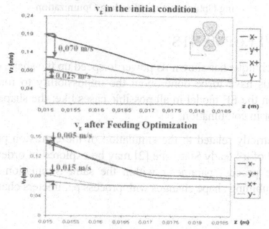

FIGURE 19. Comparison between the z velocity of the extruded profile at the initial condition and after Feeding Optimization

6   CONCLUSIONS

From what has been described is evident like, in reality, the design space (with n dimension where n is the number of the considered variables) is wider than the one really usable which is represented by all the combinations of the shape variables defined which do not compromise the elements quality of the used model. It can be concluded that the result obtained through the Feeding Optimization is better than that one of the Bearing Optimization. The final feedback about the effectiveness of the performed optimization studies may be visualized comparing the different values of the process quality index "Norm of Mesh Displacements" [2]. This last one is the average of the displacements of all the nodes of the free surface of the extruded profile (Figure 20).

The Norm of Mesh Displacements is the lowest for the Feeding Optimization. The used optimization strategy, based on the morphing technique, has presented the following advantages: (1) no time consuming and (2) capable to identify a feasible solution for the right quality of the product. It can be said the usage of this techniques for industrial design applications it is very near to be possible.

Norm of Mesh Displacements

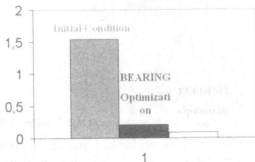

FIGURE 20. Comparison between the Norm of Mesh Displacements of: 1) Initial State, 2) Bearing Optimization, 3) Feeding Optimization

7 FURTHER DEVELOPMENTS

Even if the morphing technique has shown its power to speed up the shape design variable definition, some considerations have to be done about the impossibility to foresee the effect of the shape variables action on the FE model in all possible cases like: the shape variables action considered one at the time or in combination.

Regarding the aspects strictly related to the simulation of the extrusion process, the usage of a Transient option instead of a Steady State one [2] may be explored in order to evaluate the influence of more than one process parameter within the same simulation giving the chance to optimise the process combining shape changes with process parameters changes.

REFERENCES

1. Altair Engineering Inc., (2004), Altair HyperMesh® Users Manual, USA.
2. Altair Engineering Inc., (2004), Altair HyperXtrude® Users Manual, USA.
3. Altair Engineering Inc., (2004), Altair HyperStudy® Users Manual, USA.
4. Altair Engineering Inc., (2004), HyperMorph® Basic Training (Manual), USA.
5. Anglani, A., Del Prete, A., Papadia, G., (2002), Numerical Simulation Applied to Verify andOptimize the Aluminum Extrusion Process, Proceedings of the Annual Conference of ISCS (December 2002).
6. Wright, R.N., (2001), Professional Practice Memorandum- Practical Interpretation of Aluminium Extrusion Press Loads, Aluminum Extrusion Program at Rensselaer PolytechnicInstitute, Troy, New York.
7. Kraft, F.F., (2000), Powers, C., Optimizing extrusion through effective experimentation and analysis, Aluminum Association and the Aluminum Extruders Council

HOT IMPRESSION DIE FORGING PROCESS: AN APPROACH TO FLASH DESIGN FOR TOOL LIFE IMPROVEMENT

R. Di Lorenzo, V. Corona, F. Micari

Dipartimento di Tecnologia Meccanica, Produzione e Ingegneria Gestionale,
Università di Palermo, Italy

KEYWORDS: Hot Forging, Wear, Costs Analysis.

ABSTRACT. In impression die forging the role of the flash geometry is fundamental since a proper design of the flash land strongly influences both the complete die filling and the die wear (i.e. the die life and the related costs). In this paper an integrated approach between numerical simulations and statistical tools was developed with the aim to optimize flash thickness in order to reduce die wear and to minimize material wasting. As wear is regarded, an analytical model depending on sliding velocity, temperature, die hardness and contact pressure was utilized during the numerical simulations of the process in order to reach a wear evaluation for different values of the flash design variables. Thus, it was possible to reach a prediction of tool life for different conditions. Furthermore, a cost analysis was carried out in order to investigate the influence of flash design on tool costs and material wasting costs.

1 INTRODUCTION

The topics of hot impression die forging process design is the determination of the optimal set of operating parameters which allows the achievement of defects free products; in particular the main goals concern the complete filling of the die cavity, the minimization of material loss into the flash and the reduction of the abrasive wear of the dies. Moreover, process design represents a fundamental activity, which has to be carried out taking into account both technological and economical objectives and constraints [1][2].

As technological issues are concerned, in hot forging of parts with complex geometry, the most relevant aspect is related to the achievement of the desired final shape; thus the main objective is the complete filling of die cavity. In fact, when complex components have to produced (for instance rib-web parts) it is necessary to provide an additional volume of material, with respect to the final part volume, in order to guarantee the absence of underfilling defects. In particular, in impression die forging a flash is obtained at the end of the process and the provided surplus of material is wasted into the flash itself. As a consequence one of the objectives in the process design is the minimization of the necessary material surplus since this aspect is related to the reduction of material wasting costs.

The economical analysis of the impression die forging processes leads also to the consideration of die costs. It was demonstrated that die costs are quite relevant in such processes and their reduction is one of the main topics for industrial production. Die costs are strictly related to die life and, as a consequence to die wear. In this way, the proper control of die wear is fundamental in the hot impression die forging design. Several studies were developed in order to investigate die wear

Published in: E. Kuljanic (Ed.) *Advanced Manufacturing Systems and Technology*,
CISM Courses and Lectures No. 486, Springer Wien New York, 2005.

proving its dependence on process parameters such as sliding velocity, die material hardness, contact pressure between die and workpiece and temperature [3][4][5][6][7].

In this context, the analysis of hot impression die forging leads to the following goals: complete die filling has to be reached together with the minimization of die wear and material wasting. In order to reach these objectives proper process parameters have to be determined. Maybe one of the most relevant process parameters to be taken into account is, surely, flash geometry. Flash land geometry was deeply investigated and several studies have highlighted its role in hot forging process mechanics [8]. It was proved that flash design has a relevant influence on final product quality and on costs reduction. In fact, the flash thickness influences both die wear and material wasting. At increasing flash thickness two opposite effects are obtained: the thermo-mechanical stresses on the workpiece decrease, i.e. die wear decreases, while the surplus of material necessary to guarantee die filling increases, i.e. material wasting is higher. As it is evident, flash thickness affects both die costs and material ones.

In this paper the above mentioned issues were taken into account: the hot forging of an AA6061 aluminum alloy rib web component was analyzed with the aim to reach an optimal process design. In particular, flash design was studied and the determination of the most suitable value of flash thickness was carried out.

2   THE INVESTIGATED PROCESS

As mentioned in the introduction, this paper is focused on the optimization of flash thickness in the hot impression die forging of an aluminum alloy axisymmetrical rib web component. The analyzed geometry is shown in Figure 1.

As it can be observed a complex geometry is taken into account with a flash land designed according to the technical literature guidelines; in Figure 1 the final process geometry is represented for a particular case (flash thickness f equal to 2 mm) [9].

As process conditions are concerned, workpiece material was AA6061 aluminum alloy while dies material was AISI H13 steel. In the investigated process the upper die was deeply analyzed since the main wear amount is localized on it. The initial temperature of the upper die was fixed to 150°C while initial workpiece temperature is 470°C. According to industrial data the total forging time was about 0.5 seconds. The initial billet had a cylindrical shape with a diameter equal to 45mm and a shear model for friction with a friction factor of 0,5 was assumed during the numerical simulations of the process [10].

As mentioned, in order to carry out the economical analysis of the process it was necessary to estimate the die life and such issue is strictly related to the evaluation of die wear.

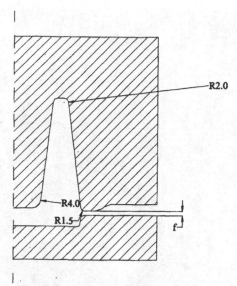

FIGURE 1. The analyzed geometry

In fact, the costs of the dies are related to the number of dies necessary to obtain a fixed production quantity. Once the annual production quanity is fixed, it is necessary to evaluate the number of parts that can be produced with a single die before it goes out of service (i.e. during its useful life). In this way it is possible to calculate the number of dies necessary for the total production and then the dies cost can be calculated.

In order to estimate die life it was necessary to analyze die wear and to take into account the so called die wear tolerance (DWT). The latter parameter indicates, in fact, the amount of wear which corresponds to die failure. In the investigated case the wear tolerance was calculated according to technical guidelines [10] and fixed to 1,3mm.

Therefore it is clear that a proper wear analysis was needed: die wear evolution during the process was investigated by utilizing the wear model proposed by Archard [7] which takes into account the main parameters affecting wear. As shown in the following equation such model can be applied to predict wear (Z) as a function of surface pressure (σ_n), sliding velocity (u) and die hardness (H).

$$Z(r) = \int_0^t k \frac{\sigma_n(r,t) \times u(r,t)}{H} dt \tag{1}$$

In the utilized equation k is a wear coefficient which was suggested by available literature [11].

In order to evaluate wear the utilized wear model was introduced into the numerical code for process simulations. The simulations were carried out by DEFORM 3D numerical environment by modeling the upper die as an elastic object meshed by 40.000 elements [12].

The mesh density was properly defined by taking into account the die regions where stresses are more localized [13].

In Figure 2 the numerical model of the upper die is shown (a quarter of the geometry was taken into account due to symmetry conditions).

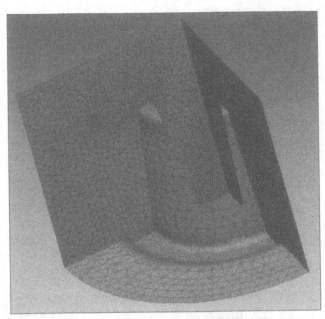

FIGURE 2. The numerical model for upper die

Through the numerical simulations it was possible to estimate the total wear amount for each forging operation and, taking into account the fixed die wear tolerance, the number of dies needed to reach the total production volume was obtained.

As it will be discussed in the following section this number will be utilized for the economical model of the process.

It has to be observed that the above described procedure to estimate die wear was applied for different values of flash thickness.

In fact, the economical model is aimed to optimize flash thickness (f), thus it was necessary to carry out a properly designed numerical simulation campaign which provided wear amount for different values of the design variable (f).

Moreover, through the numerical simulations also the material wasting was calculated for different values of f. In this way, it was also possible to introduce material costs into the developed economical model.

Actually, for each value of flash thickness the complete filling of the die is guaranteed by a certain material surplus. Material surplus ($V\%$) was measured as a additional percentage of the final component volume. It has to be observed that the minimum $V\%$ was determined for each value of f by a trial and error numerical investigation.

Some of the results obtained from the numerical simulations, in terms of die wear and filling, for different combinations of flash thickness and material surplus are summarized in Table 1 where it can be observed that filling is guaranteed only for certain values of material surplus.

Moreover, the opposite effects of flash on material wasting and die wear are shown: at increasing f from 1 to 4 mm die wear decreases but an higher surplus of material is necessary to reach complete filling (5% and 22% respectively).

TABLE 1. Die wear and filling for different values of f and $V\%$.

f [mm]	$V\%$	Wear [10^{-05}mm]	Filling
1	5		No
	7	2,3	Yes
2	8		No
	10	1,9	Yes
3	13		No
	14	1,75	Yes
4	20		No
	22	1,6	Yes

The above considerations prove the necessity to develop an optimization technique able to determine the best compromise between these two effects. Figure 3 shows the final wear distribution on the die in the case $f = 1$mm and $V\% = 7\%$.

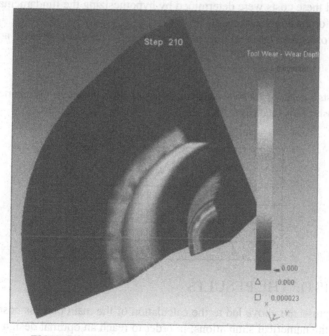

FIGURE 3. Wear distribution for f=1mm and $V\% = 7\%$

3 THE ECONOMICAL MODEL

In order to achieve an effective optimization of the flash thickness, an economical model was developed taking into account the main costs factors, namely material wasting and die costs. Actually, also set-up costs were taken into account since die failure is related also to such cost factor.

The basic hypothesis of the model was to fix an annual production quantity (Q) for the analyzed component equal to 1.000.000 parts per year. Thus, as mentioned, the total number of dies necessary to reach the annual production was determined once the die life was determined through the numerical simulations. Therefore the total die cost (C_{die}) was estimated as follows:

$$C_{die} = \frac{c_{udie} \cdot Z \cdot Q}{DWT} \qquad (2)$$

where:
- c_{udie} is the unit cost of a die;
- Z is the wear accumulated on the die at for each forging operation;
- Q is the annual production quantity;
- DWT is the die wear tolerance for the analyzed die material.

As material wasting is concerned the related cost (C_{mw}) was determined by calculating the total wasted material volume and utilizing material specific weight and unit prize.

Set-up costs (C_{setup}) were calculated taking into account the time necessary for each set-up and the industrial costs related to this time: machine availability, manpower, power supply, industrial indirect costs. All these costs were determined by hypothesizing the fundamental data of the production scenario for the analyzed case.

Finally the total costs taken into account for the optimization can be expressed as:

$$C_{TOT} = C_{die} + C_{mw} + C_{setup} \qquad (3)$$

In Table 2 the costs calculated for four different values of flash thickness are reported, highlighting the relevance of die costs on the total costs.

TABLE 2. Costs for different value of f

f [mm]	C_{die} [€]	C_{TOT} [€]
1	216000	245137
2	180000	223048
3	168000	233912
4	156000	263482

4   ANALYSIS OF THE RESULTS

The cost model presented above led to the calculation of the main process cost factors by means of the results of the numerical simulations; in order to reach an optimal design of flash geometry (flash thickness) the minimization of total costs was needed. In this way, an analytical function which relates the design variable f to the total costs was required and a statistical regression model was utilized to reach such goal. A set of numerical simulations was designed according to statistical tools for design of experiments; in particular the design variable domain was explored by properly varying f from 1mm to 4mm.

Once the simulation data were available, and the total costs were calculated for each value of the design variable, a regression model was applied to determine the analytical function linking the variable f to the total costs. In order to reach this result several statistical investigations were performed with the aim to find out the best performing function shape. The statistical model performance was measured by a correlation coefficient (R^2) which indicates the capability of the chosen regression model to interpret the relation between input and output variables. In this way, a second order polynomial function was chosen as analytical relation between C_{TOT} and f. The application of the regression tool permitted to determine the three coefficient which define the 2nd order polynomial function. Actually, such function provides a correlation coefficient of about 87% proving the effectiveness of the model in describing the variation of the total costs as a function of flash thickness. The total cost function has the following expression:

$$C_{TOT} = 14956 \cdot f^2 - 67433 \cdot f + 301250 \qquad (4)$$

The availability of an analytical function allowed the minimization of total forging costs (by a simple derivative) and the determination of the optimal "economic" value of flash thickness.
The results provided by the minimization led to the conclusion that the best value of flash thickness is equal to 2,2mm. As Figure 4 shows, for f=2,2mm the minimum C_{TOT} is obtained.

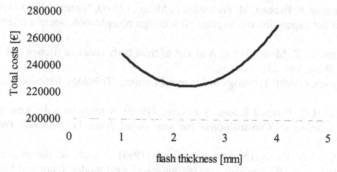

FIGURE 4. C_{TOT} vs. f (minimum costs for f = 2,2mm)

As wear is concerned Figures 5 reports wear distribution on the upper die at the end of the numerical simulation carried out with the optimal value of flash thickness.

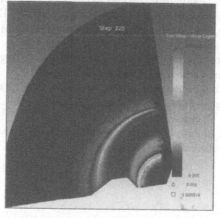

FIGURE 5. Wear distribution on upper die for f = 2,2mm

5 CONCLUSIONS

In the paper a integrated approach between FEM tools, statistical analysis and economical models was presented with the aim to optimize flash thickness in hot impression die forging of rib-web components. The proposed procedure led to the optimal value of flash thickness which guarantees the minimization of forging costs related to material wasting and dies.

A wide numerical investigation was performed in order to analyze wear conditions during the process, providing an effective knowledge base on tool life for the analyzed process. Further developments of the research will concern the application of different optimization technique able to provide optimal solutions through few iterations (i.e. reducing the necessary numerical investigations on the process).

REFERENCES

1. M. Arentoft, P. Henningsen, N. Bay, T. Wanheim, (1994), Simulation of defects in metal forming – an example, Journal of Materials Processing Technology, Vol. 45, 527-532.
2. R. Di Lorenzo, S. Beccari, M. Piacentini, F. Micari, (2004), Numerical and experimental investigation on hot impression die forging: flash design optimization, Journal of Steel GRIPS Vol.2, 153-157
3. S. Stupkiewicz, Z. Mroz, (1999), A model of third body abrasive friction and wear in hot metal forming, Wear, Vol. 231, 124-138.
4. J. H. Beynon, (1998), Tribology in hot metal forming, Tribology International, Vol. 31/1-3, 73-77.
5. J.H. Kanga, I.W. Parkb, J.S. Jaec, S.S. Kang, (1999), A study on a die wear model considering thermal softening: (I) Construction of the wear model, Journal of Materials Processing Technology, Vol. 96, 53-58
6. J.H. Kanga, I.W. Parkb, J.S. Jaec, S.S. Kang, (1999), A study on die wear model considering thermal softening: (II) Application of the suggested wear model, Journal of Materials Processing Technology, Vol. 94, 183-188
7. Ulf Stahlberg, Jonas Hallström, (1999), A comparison between two wear models, Journal of Materials Processing Technology, Vol. 87, 223-229.
8. V. Vazquez, T. Altan, (2000), New concepts in die design-physical and computer modeling applications, Journal of Materials Processing Technology, Vol. 98, 212-223.
9. S. Kalpakjian, (1997), Manufacturing Processes for Engineering Materials 3rd edition, Addison - Wesley Longman Inc., USA.
10. R. Douglas, D. Kuhlmann, (2000), Guidelines for precision hot forging with applications, Journal of Materials Processing Technology, Vol. 98, 182-188.
11. R.G. Snape, S.E. Clift, A.N. Bramley, (1998), Sensitivity of finite element analysis of forging to input parameters, Journal of Materials Processing Technology Vol. 82, 21-26.
12. DEFORM 3D User's Manual.
13. G. Snape, S. Clift, A. Bramley, (2002), Parametric sensitivity analyses for FEA of hot steel forging, Journal of Materials Processing Technology, Vol. 125-126, 353-360.

INNOVATIVE SHEET FORMING PROCESSES: EVALUATION OF THE ECONOMIC CONVENIENCE OF AN ACTUAL CASE

C. Giardini[1], E. Ceretti[2], A. Attanasio[2]

[1] Department of Design and Technology, University of Bergamo, Italy
[2] Department of Mechanical Engineering, University of Brescia, Italy

KEYWORDS: Innovative sheet forming processes, Economic model, Industrial application.

ABSTRACT. In these last few years, the need for sheet parts in low volume batches, has required the development of new production technologies for sheet forming characterized by high flexibility (like the flexible forming process, the sheet die-less forming process or the sheet fluid forming process). These technologies allow new concept products to be used in terms of shape, materials and production volumes. Since these innovative sheet production processes are not widely diffused, there is a lack of knowledge when evaluating their economic convenience even if many factories are developing and introducing economic models. The objective of the present paper is to propose a model able to identify the most convenient technical solution comparing the innovative sheet forming processes with the traditional ones (the economic convenience of hydroforming and sheet incremental forming processes will be compared with the traditional forging, deep drawing and welding production techniques).

1 INTRODUCTION

Low volume batches when producing sheet metal parts together with the need for more and more complex parts, are requiring the development and introduction of new technologies for sheet forming. These latter are flexible forming process, sheet die-less forming process, sheet fluid forming process. The common requirement characterizing these processes is the high flexibility, since these technologies enable new concept products to be used in terms of shape, materials and production volumes.

Compared with other sheet deformation technologies, these processes cover a particular niche in terms of number of produced pieces, even if for certain applications there is no possibility of using traditional processes. Table 1 reports a possible division and comparison between traditional processes and innovative ones as a function of the production volume [1]. Figure 1 shows a qualitative division between different production techniques in terms of costs.

These innovative sheet production processes are not widely diffused at present. In particular there is a lack of knowledge when evaluating their economic convenience even if many companies and researchers are developing and introducing economic models.

The objective of the present paper is to propose a model able to identify the most convenient technical solution comparing the innovative sheet forming processes with the traditional ones (the economic convenience of hydroforming and sheet incremental forming processes will be compared with the traditional forging, deep drawing and welding production techniques).

Published in: E. Kuljanic (Ed.) *Advanced Manufacturing Systems and Technology*,
CISM Courses and Lectures No. 486, Springer Wien New York, 2005.

TABLE 1. Traditional and innovative processes vs. production volumes

VOLUME / LIFE CYCLES	CLASSIFICATION	CURRENT TECHNOLOGY	INNOVATIVE TECHNOLOGIES
1 – 5	Prototypes	Panel beatering	Sheet Incremental Forming
5 – 100	Prototypes	Cast iron / Resin dies	
500 – 10.000	Low volume productions	Cheap dies No automation	
10.000 – 150.000	Medium volume productions	Dies with medium automation	Hydroforming
> 150.000	High volume productions	Dies with high automation	

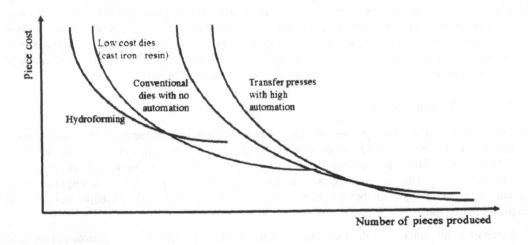

FIGURE 1. Qualitative behaviour of produced piece cost vs. utilised technology

Before starting with the production of a new component some considerations must be drawn independently on the chosen technology. In particular it is necessary to consider:
- Analysis of potentiality / applications,
- Analysis of technological feasibility,
- Industrial verification of the complete cycle,
- Hydroforming presses development (if needed),
- Integrated production system development.

As an example, in the case of hydroforming of automotive components such as fenders, car doors, panels, car roofs, engine hoods (both external sheet and internal frame) the whole process will be realized as a combination of different technologies:
- Part hydroforming (with a hydraulic press),
- Laser trimming (in a laser work cell),
- Part arrangements (in a completion flexible cell),
- Part flanging (in a completion flexible cell).

Figure 2 shows the product/process integrated architecture for Ultraflex. This project is a cooperation between industry, research centers and universities and aims to develop a flexible system for automotive sheet components.

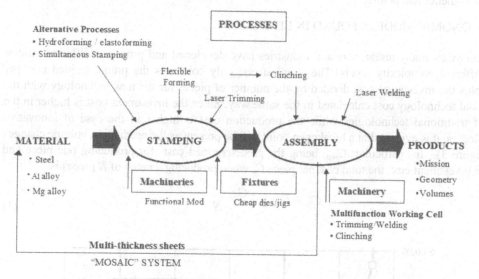

FIGURE 2. The Ultraflex Project – Product / Process integrated architecture

2 THE ECONOMIC MODEL

2.1 PARAMETERS RELEVANT FOR THE ECONOMIC MODEL

The economic model developed by the authors and here presented, aims to find the best solution in terms of minimum part cost as a function of the production volumes. The parameters taken into account for the model implementation are:
- Equipment cost (dies, moulds,…),
- Rough material cost,
- Cycle time,
- Number of working operations to get the final part,
- Machine and man costs per hour,
- Additional costs (programming cost, assembling and finishing operations),
- Market appreciation index.

Among these parameters, the market appreciation index requires some more details. This index, in fact, was introduced in order to consider whether or not the market wants to pay an additional cost to use a new technology. This happens, for example, in niche markets where the technology level is appreciated by the customers (racing motorbikes, cars, peculiar design requirements).

It must be emphasized that the cost can not be the unique element for defining the convenience of using one technology with respect to the others. In fact, also other parameters such as part shaping, roughness, dimensional and geometrical tolerances, thinning, material work hardening or the market needs or requirements must be taken into account.

In order to consider these elements in the economic model the market appreciation index was defined as mentioned before.

2.2 ECONOMIC MODELS FOUND IN LITERATURE

In recent years many researchers and industries have developed and proposed different models with different complexity levels. The simplest one only compares the predetermined cost per piece plus the investment cost divided by the number of pieces for the new technology with the traditional technology cost calculated in the same way. Since the investment cost is higher in the case of traditional technologies while the production cost is higher in the case of innovative technologies, it is evident that a breakeven point will be present as the production volume changes (see Figure 3). In particular, C_{prod} being the predetermined cost for producing one piece and C_{inv} the investment cost, the total cost per piece C_p when producing a batch of N pieces is:

$$C_p = C_{prod} + \frac{C_{inv}}{N}$$

(1)

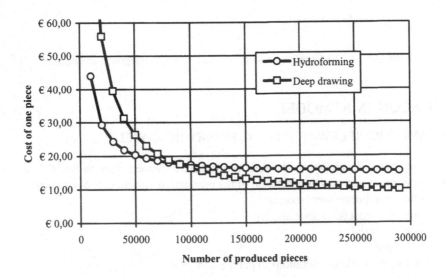

FIGURE 3. Traditional vs. innovative process costs

The limit of this model is that the investment cost is completely applied to this production while, in reality, one machine or piece of equipment can also be used for various productions.

Other models take into account additional costs such as material cost, operation costs composed by equipment, machine, tooling, installation, processing and plant investment, production rate, estimated piece rejection percentage and some correction factors for representing the actual process [2].

In [3] an economic model for the incremental forming operations has been introduced conducting also a breakeven analysis and applying it to an actual case.

In all these models, however, all the cost factors are related to the production and a market appreciation index is not present.

2.3 THE ECONOMIC MODEL DEVELOPED

In the economic model developed by the authors, several parameters have been taken into account. Some of these are "pure cost", some others are "times". In this case the cost can be derived multiplying the time by the cost per hour. In this way, for example, it is possible to take into account the investment cost considering it as a hourly cost of the equipment for the total time necessary to realize the component production.

The cost of the piece can be calculated on the basis of the following assumptions:

- The number of pieces to be produced is N starting form N_r rough pieces whose cost is C_r;
- The production process is subdivided into m phases;
- Each production phase requires a time equal to T_{wi} to be accomplished;
- The total time T_{wi} to complete one phase can be evaluated as the sum of the time required to perform single operations (loading, unloading, forming, welding, ...);
- The scrapping percentage p_{si} for each phase is known;
- The investment cost (hourly) C_{ii} and the equipment and preparation cost C_{pi} are known for each phase; in particular C_{ii} represents the hourly cost due to the investment on machines (like presses or robots) used during the production assuming that these machines are also shared with other productions, while C_{pi} represents the cost due to the preparation of the equipment (die design, die rough material, die production, machine preparation, ...) and is fully applied to the production in question;
- The labour cost (hourly) C_{oi} is known and different for each phase according to the different skills required of the operators;
- The maintenance cost of each station C_{mi} and the $MTBM_i$ (mean time between maintenance) are known;
- Other general costs C_g (fixed cost, occupied space cost, ...) can be also taken into account; these costs are expressed per hour of production;
- The power required by each station Pw_i is known and the unitary cost of the energy C_e is known too;
- The whole system efficiency e is known; this efficiency takes into account on average all the events that can delay or slow down the production (unpredictable events and stops in the production, delay in rough material delivery, ...).

Starting from these hypotheses, we can define other figures, such as:

1) Total production time: $\quad T_p = MAX(T_{wi}) \cdot (N-1) + \sum_{i=1,m} T_{wi}$

To consider the $MAX(T_{wi})$ means taking into account the bottleneck phase within the production process.

2) Number of pieces worked in each phase: $\quad N_i = {N_{i+1}}\Big/{(1 - p_{si})}$

where the number of pieces exiting from the last phase is: $\quad N_{m+1} = N$

3) Number of rough pieces required: $\quad N_r = N\Big/{\prod_{i=1,m}(1 - p_{si})}$

4) Total cost of rough material: $\quad C_{tr} = C_r \cdot N_r$

5) Total maintenance cost: $\quad C_{tm} = \sum_{i=1,m} C_{mi} \cdot {T_{wi} \cdot N_i}\Big/{60 \cdot MTBM_i}$

6) Total investment costs: $\quad C_{ti} = \sum_{i=1,m} C_{ii} \cdot {T_{wi}}\Big/{60} \cdot N_i$

7) Total equipment & setup cost: $\quad C_{tp} = \sum_{i=1,m} C_{pi}$

8) Total general costs: $\quad C_{tg} = C_g {T_p}\Big/{60}$

9) Total energy costs: $\quad C_{te} = C_e \cdot \sum_{i=1,m} Pw_i \cdot {T_{wi}}\Big/{60} \cdot N_i$

10) Total labour costs: $\quad C_{to} = \sum_{i=1,m} C_{oi} \cdot {T_p}\Big/{60}$

11) Total cost per produced piece: $\quad C = \dfrac{C_{tr} + C_{tm} + C_{ti} + C_{tp} + C_{tg} + C_{te} + C_{to}}{N \cdot e} - MAI$

where *MAI* is the market appreciation index, that is the amount of money that the market is willing to pay in order to have a particularly appealing component. This value can also be considered as a percentage discount on the part total cost. In this case the last expression must be rewritten as:

$$C = \dfrac{C_{tr} + C_{tm} + C_{ti} + C_{tp} + C_{tg} + C_{te} + C_{to}}{N \cdot e}\Big/{(1 - MAI)}$$

The consequent economic model has been implemented in an excel datasheet (see Figure 4) where the manager can fill in the needed data field and can obtain the cost per produced piece automatically.

The model can be successfully used for calculating the production cost in the case of traditional forming processes, too, so allowing a direct comparison between different productive solutions.

To calculate the total cost per piece a suitable code has been developed. This code calculates the cost of each piece for four different processes as functions of the production volume. The values so calculated are collected in a table, allowing the manager to directly compare the different production costs as the number of pieces to be produced varies.

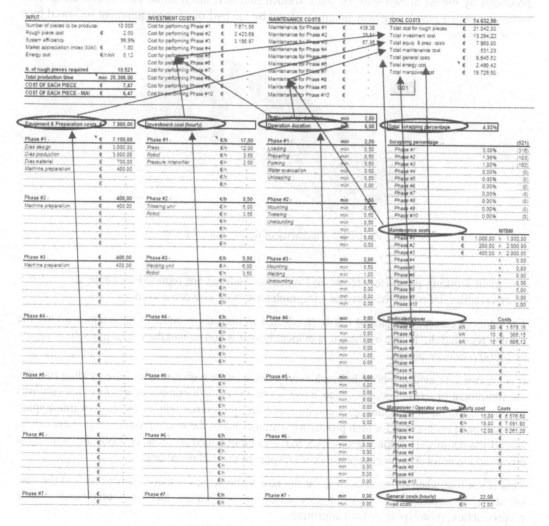

FIGURE 4. Layout of the data sheet

The same data can also be represented by graphs (see Figure 6). In this way it is possible to see how one solution can be more or less convenient than the others and where the breakeven point is for the different technologies.

3   MODEL APPLICATION TO INDUSTRIAL CASE

In order to prove the model ability an actual industrial case has been analyzed considering a part that can be produced either with traditional techniques (stamping and welding) or with innovative ones (hydroforming). In the second case, the cycle time (which can be longer than in traditional processes) can be compensated by reducing the number of post processing operations (assembling, welding, ...). The analyzed part is reported in Figure 5. The piece is actually produced by means of two stamped parts welded together, but it could be realized using THF technique with a punching phase directly realized by the hydroforming equipment followed by a trimming phase.

FIGURE 5. Geometry of the considered piece

The main working parameters (costs and times) considered are reported in Table 2.   All the other variables defined in the previous paragraph have been also considered.

The economic model developed has been applied to both the production techniques giving the results reported in Figure 6 where it is evident how it is convenient the use of hydroforming up to 24000 produced pieces.

It is important to evidence how the break even point is rather high and that, if the welding time for the traditional process increases, it can move up to a very high number of pieces. This confirms the fact that these new technologies can be successfully used not only in the case of prototyping or small batches, but also for high production volumes when the whole production process can be leaned thanks to the complex geometries that can be obtained so reducing both the investment costs and the number of process phases with respect to the traditional ones.

This is not true when the process time for the new technologies is much higher (such in the case of sheet incremental forming): in this case the application field is constrained within preproduction, prototyping or final adjustment.

TABLE 2. The considered parameters for the two technologies compared

Hydroforming technique

Equipment and preparation costs:	Phase #1	*Dies design, Dies production + material + actuators, Machine preparation*
	Phase #2	*Machine preparation, Robot programming*
Investment costs:	Phase #1	*Press, Robot, Pressure intensifier*
	Phase #2	*Trimming robot*
Operation durations:	Phase #1	*Loading, Preparing, Forming, Punching, Water evacuation, Unloading*
	Phase #2	*Loading, Trimming, Unloading*

Conventional technique

Equipment and preparation costs:	Phase #1	*Dies design, Dies production + material, Machine preparation*
	Phase #2	*Dies design, Dies production + material, Machine preparation*
	Phase #3	*Machine preparation, Robot programming, Loading equipment*
Investment costs:	Phase #1	*Press, Robot*
	Phase #2	*Press, Robot*
	Phase #3	*Welding equipment, Robot*
Operation durations:	Phase #1	*Loading, Preparing, Forming, Unloading*
	Phase #2	*Loading, Preparing, Forming, Unloading*
	Phase #3	*Loading, Assembling, Welding, Unloading*

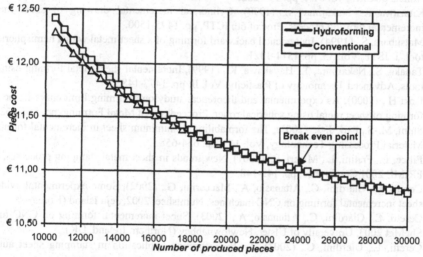

FIGURE 6. Cost curves for the different production solutions

4 CONCLUSIONS

The model described is able to compare the cost of the piece for different forming processes, taking into account different cost elements (fixed, energy, labour, maintenance, set-up, investment, ...). The convenience of using one particular process with respect to others can be identified on the basis of the piece cost. The model has a wide field of application as it can be used for different technologies. In particular an index for taking into account the market interest in using new-technology products (namely, the Market Appreciation Index) is taken into account.

The results show how the use of non conventional plastic deformation processes is convenient when the shape or the requirements of the part to be produced justify the longer cycle time because money can be saved by reducing the total number of steps or furnishing parts with better surface finishing.

It must be remembered that the cost can not be the unique parameter for the definition of the convenience of using one particular technology. Other parameters, such as part shaping, roughness, dimensional and geometrical tolerances, thinning, material work hardening or the market needs or requirements, must be taken into account too.

REFERENCES

1. Gallinaro, G., (2004), Development of innovative projects in the production of sheet components, Int. Conf. on Innovation in Metal Forming, Brescia, Italy.
2. Brun, R., Lai, M., Iammarino, M., (2004), From virtual manufacturing to design for manufacturing, Int. Conf. on Innovation in Metal Forming, Brescia, Italy.
3. Ambrogio, G., Di Lorenzo, R. and Micari, F., (2003), Analysis of the economical effectiveness of incremental forming processes: an industrial case study, VI AITeM Conference, Gaeta, Italy.
4. Filice, L., Fratini, L., Micari, F., (2002), Analysis of material formability in incremental forming, Annals of CIRP, vol 51/1, pp. 199-202.
5. Kitazawa K., Nakajima A., (1999), Cylindrical incremental drawing of sheet metals by incremental forming process, Proc.of 6th ICTP, pp. 1495-1500.
6. Matsubara, S., (1994), Incremental backward forming of a sheet metal with a hemispherical head tool, J. JSTP, Vol. 35, pp. 1311-1321.
7. Tanaka, S., Nakamura, T., Hayakawa, K., (1999), Incremental sheet metal forming using elastic tools, Advanced Technology of Plasticity, Vol. II, pp. 1477-1482.
8. Iseki H., (2000), As experimental and theoretical study on a forming limit curve in incremental forming of sheet metal using spherical roller, Proceedings of Metal Forming, pp. 557-562.
9. Shim, M.-S., Park J.-J., (2001), The formability of aluminum sheet in incremental forming, J. of Material Processing Technology, Vol. 113, pp. 654-658.
10. Filice, L., Fratini, L., Micari, F., (2001), New trends in sheet metal stamping processes, Proc. of PRIME 2001 Conference, pp. 143-148.
11. Ceretti, E., Giardini, C., Attanasio, A., Maccarini, G., (2002), Some experimental evidences in sheet incremental forming on CNC machines, Numisheet 2002, Jeju Island (Korea).
12. Ceretti, E., Giardini, C., Attanasio, A., (2003), Sheet incremental forming on CNC machines, SheMet 2003, University of Ulster, Newtownabbey (Northern Ireland, UK).
13. Ceretti, E., Giardini, C., (2003), An hydroforming application in stamping sheet automotive parts, ESAFORM 2003, Fisciano (I).
14. Giardini, C., Ceretti, E., Attanasio, A., Pasquali, M., (2004), Feasibility limits in sheet incremental forming: experimental and simulative analysis, ESAFORM 2004, Trondheim (N).
15. Giardini, C., Ceretti, E., Attanasio, A., Pasquali, M., (2004), Analysis of the influence of working parameters on final part quality in sheet incremental forming, 3° Int. Conf. on Design and Production of Dies and Molds, Bursa (Turkey).
16. Giardini, C., Ceretti, E., Contri, C., (2004), Analysis of material behavior in sheet incremental forming operations, 8° NUMIFORM 2004, Columbus-Ohio (USA), published on Material Process and Design, American Institute of Physics, vol. 712.

AN ACCURATE-OPTIMIZED MESH FOR THERMO-MECHANICAL ANALYSIS OF LASER FORMING PROCESS

P. Carlone, G. S. Palazzo, M. Puglia

Department of Mechanical Engineering, University of Salerno, Italy

KEYWORDS: Laser Forming, Thermo-Mechanical Analysis, Finite Element Modelling.

ABSTRACT. Laser forming process is used to form metal sheets in more or less complex three-dimensional shapes with good results for small-medium series or prototypes for its remarkable flexibility and precision and for no-dies need. The process is not yet well understood and for this purpose finite element analysis is a good and cheap tool. However numerical analysis can results time expensive for complexity related to this process, modelled as a thermo-mechanical coupled problem, characterized by remarkable geometric and material nonlinearity, to convective-radiative boundary conditions and Gaussian power density distribution in laser beam spot. In this paper a 3D finite element model for laser forming is proposed with an accurate-optimized mesh in order to obtain good accuracy and to reduce computational time. The influence of spot movements simulation is also considered. Numerical results are compared with experimental data.

1 INTRODUCTION

Laser forming is one of the most recent manufacturing applications of laser, in addiction to cutting, drilling, welding, and surface treatment. The process is more and more used to form metal sheets in complex three dimensional shapes for automotive, naval, aerospace, and microelectronic industries. Laser forming process is characterized, with respect to conventional forming processes, by:
- no-die need: the process is tool-less and there is no requirement for external forces;
- high precision for the processed part;
- good fatigue resistance for the processed part, due to compressive residual stresses;
- high flexibility, programmability, possibility of automation and integration with other laser processes, as cutting or welding processes;
- reduced need for part moving, related to remarkable degrees of freedom of the laser head;
- reduced time-cycle and cost for small-medium series or prototypes.

Application of the laser forming process, however, is very limited and nowadays the process set-up is still based on expensive trial and error procedures to obtain the desired shape. A systematic methodology to define heating patterns and to set process parameters to form a sheet metal as desired has not been developed yet.

Laser forming process can be described using three different mechanism: the Temperature Gradient Mechanism (TGM), the Buckling Mechanism (BM), and the Upsetting Mechanism (UM).

The Temperature Gradient Mechanism (see Figure 1) is the most used in laser forming analysis and modelling. According to TGM when a laser beam strikes on the material surface, a different thermal expansion happens through the thickness (thermal gradient is due to the low conductivity

Published in: E. Kuljanic (Ed.) *Advanced Manufacturing Systems and Technology*,
CISM Courses and Lectures No. 486, Springer Wien New York, 2005.

of the material). The top surface expands more then the bottom one and this results in a counterbending. Mechanical properties reduces as the temperature increases; surrounding material opposes to expansion and compressive plastic strains are induced. During the cooling the material contracts and a bending angle, towards laser beam, is induced.

According to the Buckling Mechanism, thermal compressive stresses develop in the sheet as a consequence of the laser heating; large thermo-elastic strains, which result in local thermo-elasto-plastic buckling of the material, are induced. Upsetting Mechanism is used when the heated area is considerably smaller than sheet thickness. Laser heating causes a shortening of the sheet and an increase in thickness with a constant compressive strain. Detailed analysis of the above mechanisms can be found in [1][2][3].

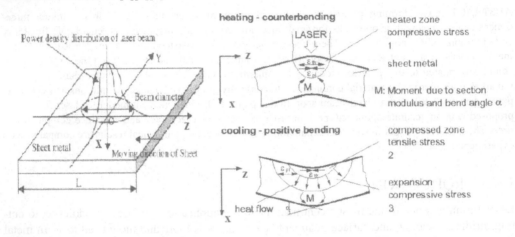

FIGURE 1. Temperature Gradient Mechanism

Analytical modelling has been performed by Kyrsanidi et al. in [7] and by Cheng et al. in [6][8] for thermo-mechanical analysis. Numerical modelling and experimental investigations have been proposed by several researchers, as Kyrsanidi et al. [5], Ji et al. [4][10]. However, power density distribution into the laser beam spot and radiative boundary conditions are not taken into account. A numerical finite element model for the buckling mechanism has been proposed by Hu et al. in [9]. Multi-scan laser forming process has been investigated in [11] by Cheng et al., who considered the effects of superheating, recrystallization, strain hardening, and phase changes on the microstructure and the final shape of the processed part.

Recent researches have been focused on the pulsed laser forming process, in particular on the transient strain field [12][16] and on vibration induced by laser [14]. The influence of preloads on the sheet metal has been investigated by Yanjin et al. in [13]. Several researchers, however, do not take into account effects of the discretization and of an accurate modelling of loads and boundary conditions, whose importance has been evidenced by Zhang et al. in [15][17], in which different approaches to modelling and the minimal discretization requirements are proposed. However, only a specific case, characterized by comparable spot radius and sheet thickness, is taken into account. In this paper a finite element modelling of the laser forming process is proposed, considering a spot radius relatively smaller with respect to material thickness. An opportune mesh has been developed to obtain good accuracy and acceptable computational time. The influence of laser movement discretization has been investigated.

2 NUMERICAL MODELLING

Numerical modelling of the laser forming process has been performed using the finite element code ANSYS.

The process has been modelled as a thermo-mechanical coupled problem. The temperature transient field, due to the heating source and (conductive, convective, and radiative) heat transfer, has been evaluated and then used as a thermal load to evaluate the stress-strain distribution and the final shape of the sheet.

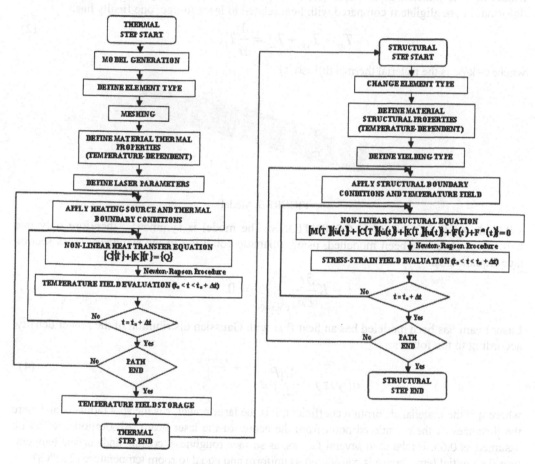

FIGURE 2. Model flow chart

Taking into account the symmetry of the problem, only one half of the sheet has been modelled, using opportune boundary conditions and restraints in the symmetry cross section and applying only one half of the laser beam spot as thermal load.

In Figure 2 the model flow chart is shown.

2.1 THERMAL ANALYSIS

GOVERNING EQUATION. Let us consider a Cartesian axis system as shown in Figure 3. Transient temperature distribution T(x,y,z,t) has been evaluated in according with the non-linear heat transfer equation, written as follows:

$$\frac{\partial}{\partial x}\left(K_x \frac{\partial T}{\partial x}\right) + \frac{\partial}{\partial y}\left(K_y \frac{\partial T}{\partial y}\right) + \frac{\partial}{\partial z}\left(K_z \frac{\partial T}{\partial z}\right) + u_{gen} = \rho c \frac{\partial T}{\partial t}, \tag{1}$$

where K_x, K_y, and K_z are the material conductivities in x, y, and z directions, ρ is the material density, c is the specific heat, and u_{gen} is the heat generation related to material deformation. Modelling the material as isotropic ($K_x = K_y = K_z$) and considering heat generation, due to deformation, negligible if compared with heat related to laser source, one finally has:

$$T_{xx} + T_{yy} + T_{zz} = \frac{1}{a} T_t, \tag{2}$$

where $\alpha = k/\rho c$ is the material thermal diffusivity.

FIGURE 3. Model

INITIAL AND BOUNDARY CONDITIONS. The model is symmetric: therefore only one half of the sheet has been modelled, using adiabatic condition in the symmetry cross section (see Equation 3).

$$-K\frac{\partial T}{\partial n}\bigg|_{x,y,z=0,t} = 0. \tag{3}$$

Laser beam has been modeled has an heat flux with Gaussian distribution of the power density, according to the follows:

$$q(y,z) = \frac{3\eta P}{\pi r^2} e^{\left(-3 \cdot \frac{y^2+z^2}{r^2}\right)}, \tag{4}$$

where η is the material absorption coefficient, P is the laser power, r is the spot radius, x and z are the distances of the considered point from the center of the laser beam. Absorption coefficient, assumed as 0.65, is related to several factors, as surface roughness, coating and surface temperature. The initial temperature is considered as uniform and equal to room temperature (20 [°C]):

$$T(x,y,z,t=0) = T_\infty. \tag{5}$$

Laser beam travels on sheet surface, along y direction, at a constant speed v, then the heat flux q is applied on the elements on which the spot is focused, according to the follows:

$$-K\frac{\partial T}{\partial n}\bigg|_{x=0,y,z,t} = q. \tag{6}$$

On the remaining surface elements convective-radiative boundary conditions are applied, as follows:

$$-K\frac{\partial T}{\partial n}\bigg|_{x,y,z,t} = h_c\left[T(x,y,z,t)-T_\infty\right]+\varepsilon\sigma\left[T(x,y,z,t)^4 - T_\infty^{\ 4}\right] \tag{7}$$

where h_c is the natural convection coefficient, assumed as 10 [W/m²°C], ε is material emissivity, and σ is the Boltzmann constant. Phase changes are not taken into account.

2.2 STRUCTURAL ANALYSIS

ASSUMPTIONS. Material is assumed as isotropic, creep is neglected, considering that a very small amount of material reaches creep temperature, and Bauschinger effect is assumed as negligible. Geometric (large deflections) non-linearity is taken into account. Material nonlinearity and plastic behavior are modelled using a multilinear isotropic hardening option (MISO). Material is described by a temperature dependent multi-linear true-stress true-strain curve, using the Von Mises or equivalent stress (Equation 8) and the yielding criterion.

$$\sigma_e = \left(\frac{1}{2}\left[\left(\sigma_1-\sigma_2\right)^2 +\left(\sigma_2-\sigma_3\right)^2 +\left(\sigma_3-\sigma_1\right)^2\right]\right)^{\frac{1}{2}}. \tag{8}$$

GOVERNING EQUATION. Taking into account geometric and material non-linearity, structural problem formulation results very complex. Equilibrium equation can be achieved using the Virtual Work Principle, written as follows:

$$\int_\Omega \underline{\sigma}:\delta\varepsilon_v dV = \int_{\partial\Omega} \underline{t}\cdot\delta\underline{u}dA + \int_\Omega \underline{b}\cdot\delta\underline{u}dV, \tag{9}$$

where Ω is the considered domain; $\partial\Omega$ is he domain surface; σ is Cauchy tensor; u is the displacement vector; t = σn is the contact force vector; b is the distance force vector; and ε_v is the infinitesimal true strain tensor.

INITIAL AND BOUNDARY CONDITIONS. Neglecting sheet weight, the structural problem is symmetric and then only one half of the sheet can be modelled. Boundary conditions in the symmetric cross section can be written as follows:

$$u_z(z=0)=0,$$
$$\varphi_x(z=0)=0, \tag{10}$$
$$\varphi_y(z=0)=0.$$

Initial sheet temperature is assumed as uniform and equal to room temperature, according with Equation 5. Deformation field is obtained applying a thermal load as follows:

$$T(x,y,z,t)=T^*, \tag{11}$$

where T^* is the temperature distribution previously evaluated.

2.3 MESH

The most recent papers [15][17] on the modelling of laser forming process pointed up the relevance of the spatial discretization and of an accurate simulation of loads and boundary conditions, to obtain good accuracy and reduce computational time. According to the authors, discretization proposed in [17] is a good tool for the analysis of thermal and structural problems in laser forming. However, it results acceptable only when spot diameter to sheet thickness ratio is approximately one. If sheet thickness is three or four times with respect to spot diameter, a different spatial discretization is more opportune. In Figure 4 the proposed mesh is shown.

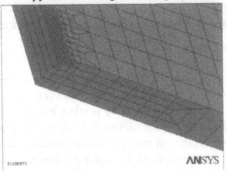

FIGURE 4. Space discretization

This spatial discretization is related to the considerable focalization of the thermal load and the remarkable difference between sheet dimension. A good modelling of the Gaussian distribution of the power density into laser beam spot imposes the use of three elements per radius; as a consequence heating zone is divided into very small elements. An opportune discretization is adopted also along thickness direction, taking into account remarkable thermal and tensional gradients. The use of small elements is not possible for all the sheet, cause convergence is not more achieved (in acceptable computational time) for the excessive number of nodes, elements, and degree of freedom. Relatively larger elements are used away from heating zone along thickness and length directions. The proposed spatial discretization is constituted by 16626 elements and 21470 nodes. Only eight-nodes brick elements are used, avoiding elements degenerations. For the thermal analysis, SOLID70 element (3-D eight-nodes thermal solid element, characterized by thermal material nonlinearities, conduction, convection or heat flux, and radiation capabilities) is used. Each node has one degree of freedom (temperature). For the structural analysis, SOLID45 element (3-D eight-nodes structural solid element, characterized by geometric and material nonlinearities, plasticity, large deflections, and large strain capabilities). Each node has three degrees of freedom (translations in x, y, and z directions). The nonlinear heat transfer equation and the nonlinear transient dynamic structural equation are solved using the Newton-Rapson procedure.

3   EXPERIMENTAL SET-UP

Figure 5 shows the experimental set-up used for the investigation. A CO_2 laser beam has been focused on the material surface by a system of lens. A CNC system controls laser power, travel speed, spot radius (regulating the distance between the focus and the surface), and heating patterns. Temperature distribution on material surface has been measured and recorded by a thermo-camera THERMOVISION and the software CATS201E, and analyzed using the software THERMAGRAM. Acquisition system has been configured using (previously evaluated) material emissivity. Bending angle has been measured in-process using a laser optical displacement sensor.

The material used in this investigation is Fe E355; material surface has been previously coated by an oxide layer to obtain the desired absorption coefficient (0.65). The thickness of the metal sheet was 0.65 mm, width (along heating pattern) was 20 mm and length was 200 mm. An input power of 100 W was used, travel speed was 10 mm/s, and spot radius was 0.12 mm. The final bending angle results equal to 0.097°.

FIGURE 5. Experimental set-up

4 RESULTS AND DISCUSSION

To evaluate the effects of the temporal discretization of the laser beam movement, four simulations have been performed using different space and time increments (see Table 1) to model spot movement at the same travel speed. In this Section are evidenced, for each simulation, the thermal and structural results, computational time employed, and a comparison with experimental data (see Table 2). Numerical and experimental analysis are performed using the same material and process parameters. For the thermal analysis, in simulation 1, convergence was achieved using just one sub-step for each load-step. In all other cases, for the relatively major load discontinuity, convergence resulted more difficult and three sub-steps for each load-step were used; this results in an increase of computational time. Structural analysis resulted more complex and computational time expensive than thermal analysis.

TABLE 1. Simulations

SIMULATION	SPACE INCREMENT	TIME INCREMENT
SIMULATION 1	0.04 [mm]	0.004 [s]
SIMULATION 2	0.08 [mm]	0.008 [s]
SIMULATION 3	0.16 [mm]	0.016 [s]
SIMULATION 4	0.24 [mm]	0.024 [s]

Convergence was achieved using respectively 2, 4, 5, and 8 sub-steps per load-step in simulation 1,2,3, and 4. Figures 6, 7, 8, and 9 illustrate temperature-time profiles obtained by the performed simulations. A remarkable continuity of temperature-time profiles (with major peak temperature) relative to simulation 1, with respect to other simulations, is evidenced (Figure 6). In particular, in Figure 9 is shown the different temperature increase of equally spaced nodes, placed on the heating line, away from sheet edges. The real (and intuitive) situation, as confirmed by infrared vision, is well illustrated in the profile relative to simulation 1, in which each node is interested by the same temperature increase during laser scan.

TABLE 2. Computational time and results

SIMULATION	THERMAL ANALYSIS TIME	MECHANICAL ANALYSIS TIME	FINAL ANGLE	DIFFERENCE WITH EXP. (0.097°)
SIMULATION 1	12600 [s]	75600 [s]	0.095°	2%
SIMULATION 2	14400 [s]	56640 [s]	0.084°	13%
SIMULATION 3	10800 [s]	41400 [s]	0.057°	41%
SIMULATION 4	7200 [s]	46800 [s]	0.039°	59%

Considerably numerical-experimental points of incongruence are found in other profiles, in which different space increments of the spot implies different exposure times of hit elements to laser source (and different density of incident power for some elements). For the same reasons, anomalous temperature profiles, during heating and cooling, can be evidenced for nodes placed along x and z directions (see Figures 7 and 8). In particular, temperature profiles relative to simulations 3 and 4 are characterized by two more temperature peaks, more and more evident as the distance of the considered node from the heating line increases. These points of relative maximum can be related to a different relevance of conductive effects, with respect to laser source. The influence of space and time increments on residual stresses is evidenced in Figure 10. Some differences in the peaks of the curves are found. Magnified images of the central zone (not shown for brevity) evidence some discontinuities for simulations 3 and 4.

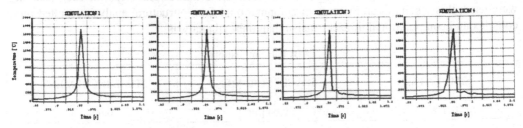

FIGURE 6. Temperature-time profiles on point (x=0; y=10; z=0)

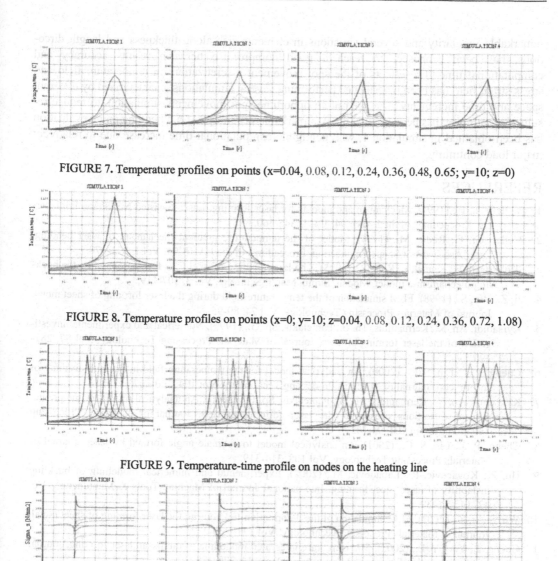

FIGURE 7. Temperature profiles on points (x=0.04, 0.08, 0.12, 0.24, 0.36, 0.48, 0.65; y=10; z=0)

FIGURE 8. Temperature profiles on points (x=0; y=10; z=0.04, 0.08, 0.12, 0.24, 0.36, 0.72, 1.08)

FIGURE 9. Temperature-time profile on nodes on the heating line

FIGURE 10. Residual σ_z on points (x=0.00, 0.08, 0.24, 0.48, 0.65; y=10; z=0)

5 CONCLUSIONS

A three dimensional finite element modelling of the laser forming process has been proposed, using the FE code ANSYS. The process is modelled as a weakly coupled thermo-mechanical problem. Considerable attention has been used to obtain a correct modelling of boundary conditions, loads, and nonlinear material properties. An accurate space discretization, characterized by remarkable regularity and several variations in elements size along thickness and length directions, has been proposed, to reach a good compromise between results accuracy and computational time. The effects of the spot movement discretization on temperature field and stress-strain field has been investigated performing several simulations, characterized by different space and time increments. The need of an opportune spot movements simulation, to reduce results discontinuity, has been pointed up. A remarkable accuracy in final bending angle evaluation has been obtained in simulation 1, with negligible computational time increase due to relatively major load continuity.

REFERENCES

1. Vollertsen, F., (1994), Mechanism and models for laser forming, Laser Assisted Net Shape Engineering, Proceedings of the LANE'94, Vol 1, 345-360.
2. Vollertsen, F., Rodle, M., (1994), Model of the temperature gradient mechanism of laser bending, Laser Assisted Net Shape Engineering, Proceedings of the LANE'94, Vol 1, 371-378.
3. Vollertsen, F., Komel, I., Kals, R., (1995), The laser bending of steel foils for microparts by the buckling mechanism-a model, Modell. Simul. Mater. Sci. Eng., Vol 3, 107-119.
4. Ji, Z., Wu, S., (1998), FEM simulation of the temperature field during the laser forming of sheet metal, Journal of Materials Processing Technology, Vol 74, 89-95.
5. Kyrsanidi, An. K., Kermanidis, Th. B., Pantelakis, Sp. G., (1999), Numerical and experimental investigation of the laser forming process, Journal of Materials Processing Technology, Vol 87, 281-290.
6. Cheng, P. J., Lin, S. C., (2000), An analytical model for the temperature field in the laser forming of sheet metal, Journal of Materials Processing Technology, Vol 104, 260-267.
7. Kyrsanidi, An. K., Kermanidis, Th. B., Pantelakis, Sp. G., (2000), An analytical model for the prediction of distortions caused by the laser forming process, Journal of Materials Processing Technology, Vol 104, 94-102.
8. Cheng, P. J., Lin, S. C., (2001), An analytical model to estimate angle formed by laser, Journal of Materials Processing Technology, Vol 108, 314-319.
9. Hu, Z., Kovacevic, R., Labudovic, M., (2002), Experimental and numerical modelling of buckling instability of laser sheet forming, International Journal of Machine Tools & Manufacture, Vol 42, 1427-1439.
10. Wu, S., Ji, Z., (2002), FEM simulation of the deformation field during the laser forming of sheet metal, Journal of Materials Processing Technology, Vol 121, 269-272.
11. Cheng, J., Yao, Y. L., (2002), Microstructure integrated modelling of multiscan laser forming, Journal of Manufacturing Science and Engineering, Vol 124, 379-388.
12. Lee, K., Lin, J., (2002), Transient deformation of thin metal sheets during pulsed laser forming, Optics & Laser Technology, Vol 34, 639-648.
13. Yanjin, G., Sheng, S., Guoqun, Z., Yiguo, L., (2003), Finite element modelling of laser bending of preloaded sheet metals, Journal of Materials Processing Technology, Vol 142, 400-407.
14. Hsieh, H., Lin, J., (2004), Laser-induced vibration during pulsed laser forming, Optics & Laser Technology, Vol 36, 431-439.
15. Zhang, L., Michaleris, P., (2004), Investigation on Lagrangian and Eulerian finite element methods for modelling the laser forming, Finite Element in Analysis and Design, Vol 40, 383-405.
16. Hsieh, H., Lin, J., (2004), Thermal-mechanical analysis on the transient deformation during pulsed laser forming, International Journal of Machine Tools & Manufacture, 30-34.
17. Zhang, L., Reutzel, E. W., Michaleris, P., (2004), Finite element modelling discretization requirements for the laser forming process, International Journal of Mechanical Sciences, Vol 46, 623-637.

OPTIMIZATION OF AN INDUCTION QUENCHING PROCESS USING THE RESPONSE SURFACE APPROACH

A. Del Taglia, G. Campatelli

Department of Mechanical and Industrial Technologies, University of Firenze, Italy

KEYWORDS: Induction Quenching, Process Optimization, Design of Experiments, Response Surface.

ABSTRACT. The quenching processes are usually, by their nature, very unstable processes, due to the great sensitivity to the product material composition, heating procedure, quenching fluid composition and flow. The paper describes the process set up in order to improve the robustness of an induction quenching process for mass production of an automotive transmission joint. The preliminary step of the study has been the evaluation of the best metric for the process robustness. The experimental part of the study has been carried out in order to investigate which parameters influence the process and which is their relative importance. The factors that could influence the product characteristic has been chosen, among the all, by the application of a process monitoring approach, where historical data were available. The effects and interactions of the remaining factors has been quantified using a Response Surface approach in order to develop a precise process characterization and to allow the definition of optimal process factors tolerances. The process optimization has led to a considerable increase in process robustness and scrap reduction.

1 INTRODUCTION TO ROBUST DESIGN

The overall quality level of a manufacturing process can be enhanced by using Robust Design methodology that is mostly used for product quality improvement. The objective of this methodology is to reduce the variance of the response, thereby making the process less sensitive to the variation of casual factors associated with it. Sensitivity can be reduced by an appropriate choice of the values assigned to the control factors of the process. The definitions formulated by Phadke and Park are useful for introducing the basic concepts of Robust Design. Phadke [1] ascertains that: "The fundamental principle of Robust Design is to improve the quality of a product by minimizing the effect of the causes of variation without eliminating the causes"; Park claims that [2]: "Robust Design is an engineering methodology for optimizing the product and process conditions which are minimally sensitive to the various causes of variation, and which produce high-quality products with low development and manufacturing costs". It must be remembered that the control factors are the variables that the manufacturing engineering is able to control and modify in the process design phase while the noise factors are the variables that, due to economic or technological limitations, cannot be controlled. The operative principle of Robust Design is the definition of the variables directly managed by the process owner (control factors) in such a way that, considered together, they are able to absorb the negative influence of uncontrollable phenomena (generally connected with noise factors).

The scientific literature identifies three phases for Robust Design: Concept Design, Parameter Design and Tolerance Design. The objective of the first phase is to choose the process structure

Published in: E. Kuljanic (Ed.) *Advanced Manufacturing Systems and Technology*,
CISM Courses and Lectures No. 486, Springer Wien New York, 2005.

which is intrinsically most robust. For example, with regard to the human factor, an automated assembly station is more robust, respect to the repeatability of the production, than a manual assembly station. In the second phase appropriate values are chosen for the factors influencing the process so as to ensure an improved level of robustness; this phase involves the choice of specific values for each aspect of the manufacturing process (i.e. for a turning operation the values could be cutting speed, tool geometry, flow of lube, etc.). In the third phase the variance of the factors is reduced to increase process robustness. This phase requires relevant reduction of the control and noise factors variability and often implies investments in new hardware and considerably raises the manufacturing cost. It is carried out only when strictly necessary, i.e. when the previous phases have not reached the objective process performance. .

2   THE METRIC FOR ROBUSTNESS COMPUTATION

To improve the process robustness it is necessary first of all to define a metric for the robustness . The measure proposed by the "father" of Robust Design, G. Taguchi, is the Signal to Noise Ratio (S/N in dB). In our case of induction quenching process each output has a target value so the approach used is "Nominal the Best" as called in Robust Design terminology. S/N compares the mean value of the process output with its standard deviation. This ratio is the argument of a logarithmic function so to use the dB as metric (1).

$$S/N = 10\log_{10}\left(\frac{\mu^2}{\sigma^2}\right) = 20\log_{10}\left(\frac{\mu}{\sigma}\right) \tag{1}$$

The S/N presents however some relevant problems such:
- the lack of relation with the limits of the process specifications;
- it requires an high number of experimental trials to be computed effectively;
- its value is not easily related to the process performance due to the lack of a standard references; it is not possible to evaluate if a specific parameters configuration is the optimal but only to compare which is the best among a group of possibility.

During this study other measures for the calculation of the robustness that could be more useful in industrial cases have been developed. For the selection of other measures of robustness computation is necessary to focus on what is the aim of robustness. As defined earlier the process robustness is the capability of the process to be insensible to the noise factors. In the example reported in figure 1 two functions of the process output (e.g. surface quality) are reported versus the control variable (e.g. cutting speed). A robustness measure would provide a better result for the case 1 than for the case 2 where the sensitivity to the noise factors is represented like a "stiffness" of the process. The noise factors are represented partly as the spread of the control characteristic (cutting speed in the case) and partly with the uncertainty of the process function (not reported in figure for simplicity). Three other measures have been proposed to compute the process robustness, each focusing on a specific aspect. The measure proposed have been:
- Mean value of the process output
- Standard deviation of the process output
- Capability of the process output C_{pk} (figure 1 right)

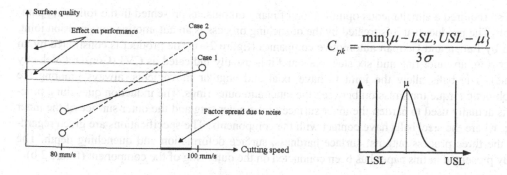

FIGURE 1. Effect of a control factor on the process output (left) and Cpk definition (right)

The first measure could be used in order to optimize the final performance of the process, so to improve the mean performance or to centre the process output distribution respect to the specifications (scrap reduction). The importance of the shifting of the output distribution in the centre of the specification is only partially taken into account by the S/N, and beside in this case the process spread is not considered. The standard deviation is, on the other hand, a measure that takes into account especially the process robustness, defined as reduction of the spread of the process output. This measure however is not able to consider the process output values respect to the specifications limits. This measure is so complementary with the mean of the process. An optimization using the objective to minimize the standard deviation leads to an improvement of the process robustness but, simultaneously, to an increase of process scrap due to the possible change in mean value of the process output. The third measure, the process capability C_{pk}, has the characteristic to summarize both the previous measures, providing an optimal compromise between the optimization of the mean value of the process output and the process spread. It has the objective to reduce the process scraps, it is so strongly related to the range of the process specifications. This measure would probably be the best compromise for the robustness computation in a real industrial case. In the following chapters is presented an application and a comparison of these measure to a real manufacturing process.

3   INTRODUCTION TO INDUCTION QUENCHING PROCESS

The induction quenching process is a surface treatment process that is finalized to an improvement of the surface hardness. The most relevant difference from a traditional quenching is that the surface to be treated is warmed by the current induced by a special inductor. The high frequency assures that the current and the related thermal effect is concentrated on the product surface. The advantages of the inductive quench is that the product is not exposed to a flame (risk of material degradation and inclusion) and the heating of the product is strongly localized. Moreover the induction quenching process grants higher production rate than the traditional approaches due to the higher possible automation of the process. The relevant outputs of the process, and so the objective of the optimization, are the surface hardness, the surface deformation and the quench depth. This case presents, as usual in an industrial case, more objective to be met simultaneously.

It is so required a simultaneous optimization of many outputs, as presented in the following paragraph. The product that it is treated by the quenching process is an automotive transmission joint, used by Italian and German automotive companies (figure 2 a). The product is constituted by an outer ring, an inner ring and six steel balls and it is usually indicated as CVJ (Constant Velocity Joints). The balls allow the joint to have axial and angular movements, always granting the omokinetic torque transmission between the inner and outer rings. The induction quenching process is actually used to harden the inner surface of the outer ring and the outer surface of the inner ring, where the steel balls have contact with the components. The specifications are given regarding the three process outputs: surface hardness, surface deformation and quenching depth. The study presented in this paper has been conducted on the outer ring of the component (figure 2 b).

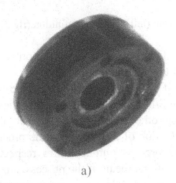
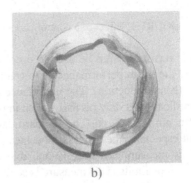

a)   b)

FIGURE 2. Automotive transmission joint

4   THE PROCESS MONITORING

The first step for the process optimization is the evaluation of the adequacy of the manufacturing process to the manufacturing objective in terms of scraps (Concept Design phase). This analysis, carried out thanks to the collaboration with the company's process engineers and a search of the other possible technologies, has proven that the actual process is the best solution for this specific product. The only possibility of economically convenient technology change is represented by the induction single shot quenching [4] that now is actually experimentally used for other products of the company. The single shot quenching uses an inductor wider than the original: its axial dimension is greater than the outer ring height. This inductor is fixed and it does not have any axial movement, so the current induced heats simultaneously the whole ring inner surface. The heating process is so executed in a "single shot" way. The possible advantages, as presented in the scientific literature, are mainly related to the uniformity of the quenching depth obtained and the reduced deformation of the surface. For this specific product economical consideration has led to the choice to maintain the actual technology and to optimize the configuration of the process parameters (Parameter Design phase). So the first step of the process optimization has been the evaluation of which parameters influence, and how much, the process outputs. These relations have to be investigated in order to obtain the optimal set up of these parameters. The first parameters that have been considered as potentially influent, obtained during a brainstorming session with the technical personnel of the company and a study of the scientific literature [5], have been the following:

TABLE 1. Parameters selected

1	Inductor stroke starting point	mm
2	Inductor stroke end point	mm
3	Coolant temperature	°C
4	Quenching oil in cooling fluid (poliquench)	%
5	Quenching fluid flow	Q/Q_{max} %
6	Preheating time	s
7	Inductor speed	V/V_{max} %
8	Quenching time	s
9	Inductor voltage	Volt
10	Inductor type	Nominal

The number of the variables (ten) seems not too high but, if it is considered the need of an experimental test plan to evaluate a mathematical model of the process, this number becomes excessive. With ten factors the experimental combinations to be tested, considering two high/low values for each variable, would be 1024. With a Factorial Fractional [6] test plan the number could be decreased till 128, producing only an approximated linear mathematical model, not sophisticated enough to allow the determination of the process robustness. To reduce considerably the number of the test experiments a process monitoring phase has been carried out. This phase is finalized to evaluate if really every parameter is influent on the process output or if exists an already computable optimal value for any of these parameters. In both cases the parameter could be excluded from the experimental test plan and its value could be "fixed". A couple of examples of the results of the process monitoring are reported in figure 3. The graphs have been obtained through a data collection of the process outputs and characteristics for about 100 working shifts. In the left graph the influence of the inductor voltage has been evaluated in order to obtain a certain quenching depth. The approach used has been a linear regression. The regression (solid line in figure 3) has the aim to evaluate the relation between the value of the inductor voltage and the value of the quenching depth (if any). Similar approach could be used for a series of parameters, such as the choice of the best inductor type. The inductors have been all build by the mechanical shop of the factory, all with lightly different characteristics. The study reported in figure 3 (right) allows to choose the best type of inductor, the one that grant the lower performance spread and an output within the specifications range. This graph has been drawn using the same inductor voltage (330 V) and represents the process output of the normal production during a period of time characterized by the absence of relevant change in the other parameters.

The test to measure the product hardness and quenching depth are destructive and require long time: this implies that the number of tests has to be reduced as much as possible. The process monitoring approach has been applied to most part of the selected parameters (Tab. 1) in order to evaluate the influence of the parameters on the three process outputs. This phase has allowed a screening of the parameters in order to reduce the number to be considered for the experimental test plan; in particular the parameters with a proven low influence on the process outputs have been eliminated.

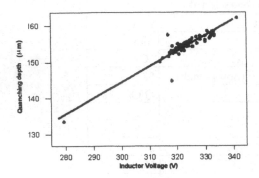 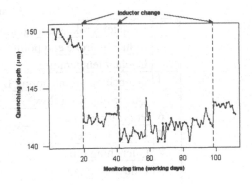

FIGURE 3. Process Monitoring

5 EXPERIMENTAL PLAN: RESPONSE SURFACE APPROACH

The parameters considered for the experimental test plan have been:

TABLE 2. Parameters and levels used for the test plan

Parameter	Metric	Value -2	Value -1	Value 0	Value +1	Value +2
Inductor Stroke	mm	28	29	30	31	32
Coolant temperature	°C	28	32	36	40	44
Quenching oil	%	8,5	9,5	10,5	11,5	12,5
Coolant flow	Q/Q_{max} %	26	31	36	41	46
Inductor speed	V/V_{max} %	51	52	53	54	55

For the experimental test plan a Response Surface [7] approach has been chosen. The reasons of this choice are: it allows to determine a non linear mathematical model, the most useful in order to find how to improve the process robustness, and it allows a consistent reduction in the number of experimental trials needed. In this case the different configurations to be tested are 32 (the test plan chosen has been a Half Central Composite). Considering at least 2 replicas the total number of experimental tests become 64, a number compatible with the time and cost [8] allocated by the company for the optimization of this process. The response surface approach tests each parameter at five different levels. The main levels are, as conventionally indicated, +1 and -1. The values of these two levels are chosen in order to define an adequate value span for the experimental test plan: the values have to be representative of the feasible variation of the parameter, according to the engineering experience of the process. The other values are derived by the previous ones: the first is the middle value between these two and the other are external values, whose distance from the +1 and -1 values depends from the number of factors considered for the test plan. The lower is the number of factors considered the lower is the distance of the external points. In case of five parameters the distance of the external points form the central points is 2 (considering the distance among the central point and one of the first level equal to 1), so the external points assume the conventional value of +2 and -2. To understand the organization of the Response Surface test plan an example of a plan for 3 factors is presented in figure 4 (left). Here the three different type of points characteristic of a response surface test plan could be defined: central point (grey), cube point (from the first levels chosen – white) and axial points (derived from the external points – black).

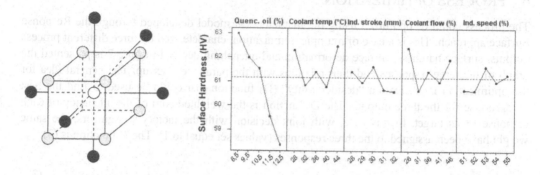

FIGURE 4. Response Surface with 3 factors (left) and influence of the single factors on the Surface Hardness (right)

The output of the response surface has been the following: on one hand it has been possible to quantify the influence of the single factors for each output of the process (figure 4, right) and on the other hand it has been possible to quantify the interaction of the parameters regarding one output through a continuous function (design of a "response surface") that allows an optimization of the parameters values (figure 5).

The mathematical equation of the model developed with the response surface approach is, for the case of 3 different parameters, the following (2):

$$f(x,y,z) = a_{00} + a_{10}x + a_{11}x^2 + a_{12}xy + a_{13}xz + a_{20}y + a_{21}yx + a_{22}y^2 + a_{23}yz + a_{30}z + a_{31}zx + a_{32}zy + a_{33}z^2 \qquad (2)$$

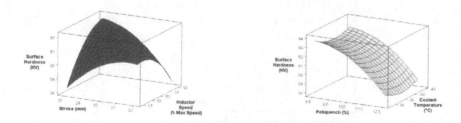

FIGURE 5. Response surfaces of the Surface Hardness

This model could be represented with graphs (response surfaces in figure 5) showing the influence of two parameter on the process output. In this case the other parameters involved are kept fixed at a specific value (usually the optimal value for the parameter –if already found- or the centre value of the experimentation). This allows to understand the interaction among the parameters and to find the optimal value easily.

6   PROCESS OPTIMIZATION

The process could be optimized using the mathematical model developed through the Response Surface approach. This is a case of multiple optimization, characterized by three different process outputs: surface hardness, surface deformation and quenching depth. In figure 7 is presented the graph of the three responses and their relations with the parameters set up. The general idea for the optimization is to create a "desiderability" (D) function that could be used to find the best compromise for the three outputs. The D function is the weighted sum of the adequacy of each response to its target. In this case, with joint decision with the factory's engineers, the same weight have been assigned to the three responses (values set equal to 1). The D function is (3):

$$D = d_1 \cdot w_1 + d_2 \cdot w_2 + d_3 \cdot w_3 \tag{3}$$

where the d_i values represent the adequacy of each response of the i-th response specification and w_i the relative weight of each response. The adequacy of each output has been evaluated using an ad hoc function (figure 6) that allows the normalization of the output in order to perform a correct simultaneous optimization. The adequacy function could assume values from 0 (not adequate, when the process output is out of the specifications) to 1 (fully adequate, when the process is exactly on the output target values). The adequacy of the intermediate cases is evaluate using three different type of assumption. When all the process output values within the specifications have the same importance the adequacy function used is the a), when it is extremely important that the process output is very near to the target the adequacy function used is the c) while in all the other cases the use of a linear function is considered the optimal compromise (function b)).

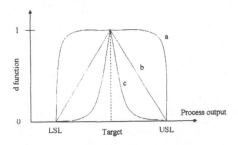

FIGURE 6. Adequacy function and its possible configuration

The optimization of the D function is then carried out with simple iterative methods. The graph reported in figure 7 is a screenshot of the software used for this multiple optimization. In figure are presented the functions that relate the parameters to each output. So this representation allows to evaluate the best configuration regarding the process robustness. To be less sensitive to the parameters variation is necessary that the transfer function reported in figure 7 would prove as horizontal (less "stiff") as possible. Graphically could be chosen the solution that grants the maximum robustness trying to vary the values of the parameters. The optimization process has been carried on with all the four robustness metrics described in paragraph 2: mean, standard deviation, S/N and C_{pk}. The optimization of the standard deviation provides the configuration with the minimum process spread not considering the target values that the output must reach. The optimization using the mean value has as output the configuration that centers the output as near as possible to the centre of the specifications, not considering at all the process spread and so the robustness.

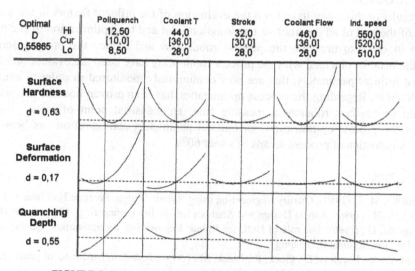

FIGURE 7. Robustness evaluation and multi response optimization

The S/N gives an intermediate configuration that minimize the ratio between the mean value and the standard deviation, improving the process robustness but not centering the process in the specifications. Finally the C_{pk} is able to find the configuration that provides an output as near as possible to the centre of the specifications with the least process spread. The optimization of the process using these four measures has proven how the process capability is able to identify the best configuration from an industrial point of view. In table 3 is reported the comparative analysis of the four optimizations. Obviously in this case the process spread is higher than in the case of the optimization using the goal to minimize the standard deviation or to maximize the S/N but could be proven that the chosen configuration grants the minimum percentage of scraps (reported in ppm – scraps per million of parts), the most important measure for the efficiency of the process. However it is possible to notice how the difference among the process spread of the last three cases is not so relevant while there is a drastic reduction in process scraps using the C_{pk} respect to the other two.

TABLE 3. Optimization using different approaches

Optimization approaches	Parameters					Process outputs	
	Quenching oil (%)	Coolant temp. (°C)	Ind. stroke (mm)	Coolant flow (% Q_{max})	Ind. Speed (% S_{max})	Mean standard deviation	Estimated scraps (ppm)
Mean Value (target)	9,6	34,5	28,9	46	51	0,76	483
Standard Deviation (Min)	10	36	30	36	52	0,36	23
S/N (Max)	9,9	35,8	30	36	53	0,38	24
C_{pk} (Max)	10	36	30	36	55	0,40	1,7

7   CONCLUSIONS

The final results of this study have been the evaluation of the influent factors in the induction quenching of the part of an automotive transmission joint and the optimization of the parameters set up in order to improve the process robustness and reduce the process scraps. In particular the benefits obtained with the process monitoring have been the evaluation of which are the most influent parameters, that are now continuously monitored in order to control the process behaviour. Regarding the process optimization has been proven how the process capability would be the best robustness measure from an industrial point of view. Using this measure for the process optimization, an optimal parameters configuration has been found. This allows a reduction of process scraps of about 60%.

REFERENCES

1.   Phadke, M. S., (1989), Quality Engineering using Robust Design, Prentice Hall International.
2.   Park, S. H., (1996), Robust Design and Analysis for Quality Engineering,, Chapman & Hall.
3.   Taguchi, G., (1996), The role of DOE for Robust Engineering: a commentary, Quality and Reliability International Publishing.
4.   Rudnev, V., Loveless, D., Cook, R., Black, M., (2004), Induction hardening of gears: A review, Heat Treatment of Metals, v 31, n 1, p 11-15.
5.   Rudnev, V.; Loveless, D.; Murray, J.; Escobedo, R., (2000), Intricacies of Induction Tempering for Automotive Industry, ASM Proceedings: Heat Treating, v 2, p 872-878.
6.   Montgomery, D. C., (2000) Design and Analysis of Experiments, Wiley.
7.   Myers, H., Montgomery, D. C., (2002), Response Surface Methodology, Wiley Interscience.
8.   Guerrero Mata, M. P., Colas, R., (1998), Experimental setup for conducting induction quenching, International Automotive Heat Treating Conference, p 264-266.

INFLUENCE OF PROCESS PARAMETERS FOR THIXOTROPIC ALLOYS

A. Barcellona, L. Fratini, D. Palmeri

Dip. di Tecnologia Meccanica, Produzione e Ingegneria Gestionale, Università di Palermo, Italy

KEYWORDS: Thixoforming, Semi-Solid State, Forging.

ABSTRACT With reference to a metallic alloy, the attribute thixotropic is utilized to indicate the behaviour of it in the semi-solid state when its microstructure consists of spheroids in a liquid matrix. Such alloys are characterized by very low values of viscosity under shearing stress in the semi-solid state, while after solidification they show relevant mechanical properties. Actually a structural change from a dendritic structure to a globular one, with the globular grains finely dispersed in a liquid matrix, is observed after particular thermo-mechanical treatments. In the present paper the authors present the results of a wide experimental campaign on the AA 7075 aluminium alloy that shows a large semi-solid window temperature range. In particular, a mechanical device was set up in order to develop a mechanical treatment with the aim to produce thixotropic material to be subsequently forged. The influence of the main operative parameters was taken into account and micrographic analysis and mechanical tests were performed highlighting properties of the obtained structures.

1 INTRODUCTION

Thixoforming process consists in the forming of material when it is in its semisolid state, i.e. at a temperature between solidus and liquidus ones. In these conditions it is possible to combine the fluidity of the liquid state to the formability at the solid state, so combining vantages of the conventional foundry and forging techniques. The process is going to impose on the market because it allows to obtain complex shape products, with high superficial finishing and very low dimensional tolerances by only one forming step. Thus, thixoforming technologies lead to a productivity improvement, due to a minimum pieces discharging, and to a better product quality. Success of a thixoforming process is related to the possibility to obtain a microstructure constituted by globular particles of solid phase surrounded by a continuous film of liquid phase. In such conditions, material shows a thixotropic behaviour; this means that under a shear action, it exhibits a decreasing viscosity in the time, but, when the action ceases, viscosity come back to its initial value. It should be noted that presence of this globular fine and uniform structure determines an increasing of mechanical properties and of the ductility of material. Several methods are today employed and researched in order to obtain such structure; in the present research, material is firstly brought over the fusion temperature and, during the successive cooling, it is subjected to a mechanical action at a fixed temperature inside the semisolid range. Successively, material is cooled at environment temperature. After this first thermo-mechanical treatment, a non-dendritic but still coarse globular structure is obtained. The successive step consists in a re-heating of the material at the same temperature in the semisolid range at which the material was firstly treated, in order to sharpen the obtained structure both in terms of dimensions and spherical shape. Experiments have been conducted

Published in: E. Kuljanic (Ed.) *Advanced Manufacturing Systems and Technology*, CISM Courses and Lectures No. 486, Springer Wien New York, 2005.

on a Al Zn Mg aluminium alloy, that it has firstly brought in its liquid state and successively subjected to different thermo-mechanical cycles. Parameters as cooling rate, thermo-mechanical time treatment, permanence at semisolid temperature time in the re-heating phase have been considered in order to observe the influence of each of them on the obtained globular fine structure. In particular process parameters that lead to a fine and homogenous microstructure have been detected; mean grain dimension between 10 and 15 μm and micro hardness values much higher than those of base material have been obtained. [1], [2], [3], [4], [5].

2   EXPERIMENTAL MATERIALS AND PROCEDURES

Experiments have been performed on the commercial AA 7075 aluminium alloy. Composition, determined by EDAX analysis, is : Al, 6.06% Zn, 2.02% Mg, 1.62% Cu, 0.34% Si. The choice of this alloys depends both by the wide use in the automotive and aerospace fields and the large semisolid temperature range. Liquidus temperature has been found equal to 635°C, while solidus temperature equal to 477°C. Inside this interval, temperature of 610°C has been considered suitable for the mechanical treatment. Such temperature has been computed employing the Scheil equation, choosing a value of solid fraction equal to 60%. This value is the most commonly employed in industrial field because it assure an enough low viscosity to fill the die during the forming step but it allows to avoid the collapse of the billet when it is again brought to the semisolid temperature to be subjected to the forming step[6], [7].

Scheil equation relates the value of the solid fraction at the fusion temperature of the alloy solvent to the process temperature and to the alloy liquidus temperature:

$$f_s = 1 - \left(\frac{T_{mAl} - T}{T_{mAl} - T_{liq}} \right)^{-\frac{1}{1-K}} \tag{1}$$

in which:

f_s is the solid fraction value, defined as the ratio between solid quantity and the total quantity of the alloy existing at the generic T temperature, inside the fusion temperature range;

T_{mAl} is the melting temperature of the solvent, i.e. the pure aluminium, equal to 660°C;

T is the temperature at which the mechanical action is performed;

T_{liq} is the liquidus temperature of the alloy, in the studied case, equal to 635°C;

K is the distribution coefficient, defined as the ratio between solid phase and liquid phase concentration at the chosen semisolid temperature. K is a T function, obtained from the phase diagram, and for AA 7075 it results equal to 0.24.

Experimental test has been performed on cylindrical specimens (60 mm in diameter and height). Each test is performed by heating the material up to the melting temperature for a proper time, and successively by slowly decreasing the temperature down to the chosen semisolid temperature of 610 °C. Specimens are thus maintained at the semisolid temperature for 5 hours, in order to obtain a complete homogenisation, before starting the mechanical agitation.

Mechanical treatment has been performed by employing the tooling. The shear action, needed for the breaking of the dendritic arms and for the obtaining of the globular structure, it has been realised by means of a simple crank mechanism; a long metal billet is going into the electric oven trough a hole and confers to the melting pot an alternate movement.

The employed electric engine has a rotation speed equal to 1400 rpm and has been coupled to a reducer, obtaining a reduction ratio equal to 1/10.

After the mechanical treatment, each specimen has been cooled down to environmental temperature by different cooling rates. Material subjected to this 'first' treatment e is defined as 'precursor' material. Finally, by means of micrographic observations of the final structure, the influence of the two parameters (time length of the mechanical treatment and cooling rate) has been valued. It should be noted that the material has to be still re-heated, before the final result.

Micrographic observations have been performed by grinding and polishing the specimen surfaces and successively by etching them for 5 second in 8% hydrofluoric acid in order to observe the structure.

Classification of the final obtained structures has been performed by means of the evaluation of the following parameters:

- Globular particles mean dimensions;
- Shape Factor SF:

$$SF = \left(\frac{4\pi A}{P^2}\right) \tag{2}$$

defined as the following ratio in which A is the mean grain surface and P its perimeter. Furthermore, for each specimen, hardness Vickers tests, have been executed on the so called 'precursor' material.

After the first treatment step, each specimen has been successively subjected to a re-heating treatment in three steps. The re-heating phase consists in the heating the material up to 350°C for 10 minutes, then up to 570°C for 8 minutes and finally up to the semisolid temperature of 610°C for 5 minutes; the three stages are necessary in order to obtain an all volume specimen uniform temperature in the final stage at 610°C. Finally, for the specimen that has shown the best structure after the re-heating phase, three more tests have been executed, by varying the permanence time at the semisolid temperature in the final stage of the re-heating phase and by employing a water cooling [8].

The same micrographic observations, by computing the above described parameters (mean grain dimensions and Shape Factor), and the micro hardness tests have been performed on the re-heated material.

All the operative parameters and the final results are reported in table 1. Specimen number 1 is considered as the 'base' material; this specimen has been melted and slowly cooled down to environment temperature; a completely dendritic structure was obtained.

TABLE 1. Experimental plane and obtained results

Specimen/ case	OPERATIVE PARAMETERS			RESULTS				
	Mechanical treatment time (min)	Cooling	Re-Heating time (min)	Shape Factor	Mean grain dimension (μm)		Vickers Micro hardness	
Base material								
1	-	Oven (30°C/h)	-	0	(Dendritic structure)		125	
'Precursor' material – Oven cooling								
2 (precursor)	20	Oven	no	0.45	280		98	
3	20	Oven	5	0.5	293		95	
'Precursor' material – Air cooling								
4 (precursor)	8	Air (120°C/h)	No	0.35	150		141	
5	8	Air	5	0.45	152		125	
6 (precursor)	15	Air	No	0.35	156		134	
7	15	Air	5	0.48	158		131	
8 (precursor)	20	Air	No	0.52	480		55	
9	20	Air	5	0.65	501	75	42	-
10 (precursor)	40	Air	No	0.53	590		52	
11	40	Air	5	0.65	600	83	40	80
New stage experiments – Water cooling, new re-heating								
12 (precursor)	20	Water (3°C/sec)	No	0.48	50		151	
13	20	Water	5	0.65	118	9	130	164
14	20	Water	15	0.68	150	12	110	162
15	20	Water	25	0.67	14		161	

3   DISCUSSION ON THE EXPERIMENTAL RESULTS

Reference structure for the 'base' material (specimen 1), is completely dendritic, as shown in figure 1: higher lengths of dendritic arms were found equal to 5-6 mm, and Vickers micro hardness tests shown a value equal to 125 HV.

Specimen 2 has been obtained after a mechanical agitation for 20 minutes and a slow cooling inside the oven; the 'precursor' obtained material shown a non-homogeneous structure with large areas of dendritic structure and narrow zones of rough globular structure (280 μm diameter of grains) with a very low value of SF. Case 3, that is the same for specimen n. 2 but followed by the re-heating phase, still highlighted a non-homogeneous structure in which globular grains were enlarged (293 μm average diameter of grains); just a small improvement of the SF was also observed.

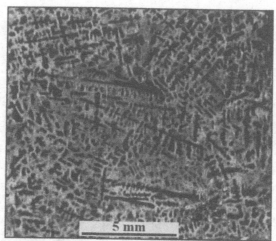

FIGURE 1. Dendritic structure in the 'base' material

Basing on these preliminary results, the successive tests have been performed by considering an air cooling, in order to improve the cooling rate, and assuming 4 different time length of the mechanical treatment at the semisolid temperature, in order to detect the influence of such parameters on the final structure.

Specimen 4 has been mechanically treated for 8 minutes; also in this situation, the obtained structure is resulted non-homogeneous, with large dendritic rosette areas and very narrow globular zones, actually with reducer dimensions (150 μm) with respect to the previous cases but with a still low value of the SF, as illustrated in figure 2.a. Micro hardness value is resulted equal to 141 HV, higher than those of 'base' material. The re-heating treatment (case 5) brought to a light increment of average grain dimensions (152 μm) and determined a decreasing of the hardness value; obtained structure is shown in figure 2.b.

Specimen 6 has been mechanically treated for 15 minutes; the obtained structure, yet showing a larger globular zone, still presents dendritic areas. In the globular section, mean grain dimension is lightly higher than those obtained with the lower time length agitation. The obtained values was 156 μm before the re-heating phase, and 158 μm after it.

Furthermore, SF of the structure, still low, has been improved by the re-heating phase (case 7). As shown in table n.1, hardness values have been obtained lower than those of cases 4 and 5, according to the higher obtained grain dimensions.

Figures 2.c and 2.d show the two structures, respectively.

Specimens 8 and 9 have been mechanically treated for 20 minutes; for the one non subjected to the re-heating step a full globular structure was obtained, characterised by a high SF but still by relevant average dimensions of grain (480-500 μm) and subsequent low hardness values. Nevertheless, obtained structure after the re-heating step is not perfectly homogeneous, in fact some small areas, in which the average grain dimension was equal to 75 μm, were found.

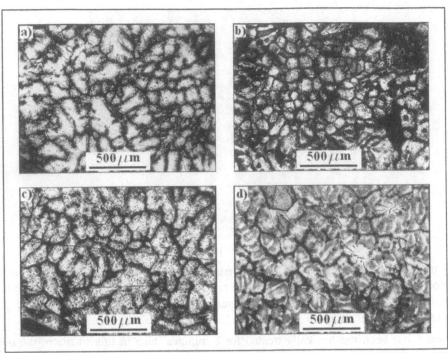

FIGURE 2. a) 8 min MT, air, b) 8 min MT, air, RH, c) 15 min MT, air, d) 15 min MT, air, RH

For the case 10 and 11, time length of the mechanical treatment has been brought up to 40 minutes. For the 'precursor' material, an uniform globular structure with a high value of the shape factor but with too high mean dimensions (590 μm) and a subsequent very low hardness value, was found. For this case, the re-heating treatment has stressed the phenomenon of the non homogeneity, i.e. of the presence of different areas with different average grain dimensions, as already observed in the previous cases 8-9. In this situation, about the 10% of the structure show a globular fine structure (83 μm), and a high value of hardness (80 HV), while the remaining part, still globular, show a globular structure but with a mean grain dimension equal to 600 μm and a low hardness value equal to 40 HV.

Actually, since the 'precursor' material with an uniform globular structure has been obtained only employing a mechanical treatment time length equal to 20 and 40 minutes and also that the structure with the best grain dimension and shape factor was those of case 8, it has been decided to repeat this test only by increasing the cooling rate in order to obtain still a homogeneous globular structure, but with lower dimensions. Case 12 is the same of case 8 but with a higher cooling rate; case 12 is also the 'precursor' of case 13, that has been subjected to the traditional re-heating phase. Obtained structures are shown in figure 3 for the specimen non subjected to the post heating step, and in the figures 4.a and 4.b for the re-heated material, respectively in the two obtained different areas. Obtained structure for specimen 12 resulted globular, with a low shape factor, but with mean grain dimensions lower than those already obtained (50 μm) and a subsequent increasing of the hardness. In the 'traditional 5 minutes' re-heating phase, structure resulted non homogeneous, and two different areas where found:

structure is constituted by large zones in which the original structure has been subjected to an increasing of the mean dimensions, together with an increment of the shape factor and a decrement of the hardness (figure 4); in this area it has been observed that, at the boundary of the rough grains, the formation of an extremely fine structure appears to start (9 μm), with very high values of shape factor and hardness (164 HV). This second area is extended for the 10% of the total area. By considering the above described obtained results, the study was extended in a new stage. As already described, in the first stage, thermal treatment length time and cooling rates have been considered as the independent variables, so as, in the re-heating phase, a permanence time of 5 minutes at 610°C has been assigned to all the specimens. The 'new' stage consists in the variation of the permanence time of the re-heating phase for the specimens that sown the best results of the previous stage. In particular, starting from the best results obtained for the specimen 12, permanence time of the final step in the re-heating phase has been varied and analysed. In order to obtain this extremely fine structure in the total volume of the material, it has been decide to perform on the same case 12 a longer permanence time length of the re-heating phase at the semisolid temperature equal to 15 minutes [9], [10].

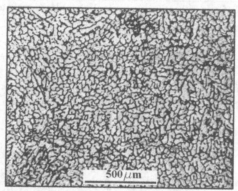

FIGURE 3. 20 min Mechanical Treatment, Water cooling

Still in this situation, the obtained structure, shown in figures 5.a and 5.b, is constitute by a mixed area in which exists together rough and fine grains, the latter characterised by very high values of shape factor and hardness. The extension of this area is resulted higher then the previous case, close to 45%, and for this reason, it has been decided to still increase the permanence length time at the semisolid temperature up to 25 minutes (case 15). The final obtained structure is shown in figure 6. Only one homogeneous fine grain area was finally found, in which grain dimension and hardness values are typical of high mechanical properties [11], [12].

4 FINAL DISCUSSION

The above described observed phenomena, are founded on the consideration that at a semisolid temperature, solid and liquid phase are in chemical equilibrium and solid fraction is constant. Such equilibrium results dynamic: in fact during the permanence at the mechanical treatment temperature, solid phase may locally melt and solidify again into an other zone. This phenomenon determines modifications in shape, dimension and density of nucleating and growing grains.

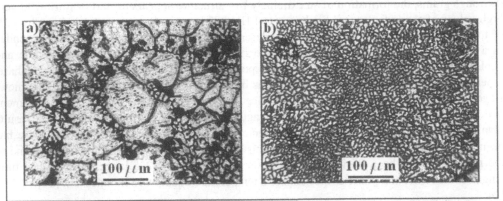

FIGURE 4. 20 min MT, 5 min re-heating, water cooling, a) large grain area, b) fine grain area

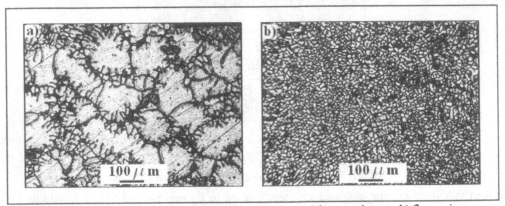

FIGURE 5. 20 min MT, 15 min re-heating, water cooling, a) large grain area, b) fine grain area

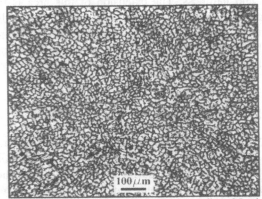

FIGURE 6. 20 min Mechanical Treatment, Water cooling, 25 min re-heating

The mechanical agitation generates a distortion, deformation and fragmentation of dendrites which appear in the first solidification step and for this reason the first globular particles with a low shape factor start to generate itselves. After it, spheroidizing phenomena occurs, depending from the energy reduction at the solid and liquid phase interfaces. From the experimental results it has been obtained that parameters as mechanical agitation time length and cooling rate have a strong influence on dimensions and morphology of final structure.

Observing the results of cases 2 and 3, the slow oven cooling has almost totally annulled the mechanical agitation effect, because the long time permanence at high temperatures determined the re-growing of dendrites, after the end of the mechanical action; furthermore, the few grains that maintained the globular structure grew enormously. For the specimens in which cooling was performed in air, it has been observed that mechanical treatment times lower than 20 minutes resulted not enough to obtain a homogeneous globular structure. Homogeneous structures have been found only when times longer than 20 minutes were utilised. Furthermore, increasing semisolid temperature permanence time to obtain the precursor material, a strong grain dimension increasing has been obtained.

A more drastic water cooling, and mechanical treatment permanence time long enough to obtain 100% globular structure, led to obtain the best structure for the precursor material, both in terms of dimensions, shape factor and hardness; the post re-heating treatment always determined an improving of the shape factor and a low decrement of hardness.

For cases in which a homogeneous globular structure is obtained, the re-heating treatment determined the occurrence of a different structure that resulted extremely fine and show high shape factor and hardness values. Explanation oh the evolution of such phenomenon is schematically illustrated in the figure 7.

FIGURE 7. Microstructure evolution: a), b), c), precursor material; d), e), re-heating

When material is cooled from its melting temperature down to a chosen semisolid temperature, a dendritic solidification starts (a). A sufficient agitation time and not excessively low cooling rates determine the formation of a globular structure that results also stable at environment temperature. Higher cooling rates brings the material in a metastable condition that results determinant for the formation of the final structure in the subsequent re-heating step (b). Initially, by the re-heating treatment, and because of the high temperature permanence, structure tends to increase its mean dimensions and shape factor (c). With the increasing of the permanence time at the semisolid temperature, structure tends to evolve to a stable shape at this temperature and thus tends to come back dendritic. Grain boundaries, more reactive because of their higher energetic content, tend to develop a dendritic shape (d). At this time, the fragmentation dendritic action is not any more developed from the mechanical agitation but it is due to the presence of very near grains that tend to increase themselves, in this way leading to the fragmentation of the growing dendritic arms (d). Finally, because of the slight space between grains, the obtained structure is globular and extremely fine. Furthermore, the occurred globular grains, in order to reduce their energy, tend to increase their shape factor. Operative parameters that permit this phenomenon were above described with particular reference to case 15.

5 SUMMARY AND CONCLUSIONS

To take advantages of thixoforming processes, it is essential to obtain a structure that shows thixotropic behaviour. In this research, the AA 7075 aluminium alloy has been thermo-mechanically treated, and it has been observed that, to obtain a fine structure after the re-heating treatment, it is necessary to achieve a metastable globular structure in the precursor material. Its further permanence at the semisolid temperature in the third stage of the re-heating determines the formation of an initially dendritic structure that 'degenerate' to a globular one due to the action among boundary grains. Final structure is characterised by fine and globular grain, with good shape factor and hardness value.

REFERENCES

1. Barcellona A., Fratini L., Riccobono R., (2003), The thixoforming process: an experimental device, VI AITEM Conference, Gaeta, Italy.
2. Liu D., Atkinson H.V., Kapranos P., Jirattiticharoean W., Jones H., (2003), Microstructural evolution and tensile mechanical properties of thixoformed high performance aluminium alloys, Materials Science and Engineering A361 213-224.
3. Haga T., Suzuki S., (2001), Casting of aluminum alloy ingots for thixoforming using a cooling slope, Journal of Materials Processing technology 118 169-172.
4. Sang-Yong Lee, Se-Il Oh, (2002), Thixoforming characteristics of thermo-mechanically treated AA 6061 alloy for suspension parts of electric vehicles, Journal of Materials Technology 130-131 587-593.
5. Kim N.S., Kang C.G., (2000), an investigation of flow characteristics considering the effect of viscosity variation in the thixoforming process, Journal of materials Processing Technology 103 237-246.
6. Tzimas E., Zavaliangos A. (2000), Evaluation Of Volume Fraction Of Solid In Alloys Formed By Semisolid Processing, Journal Of Materials Science 35 5319–5329.
7. Modigell M., Koke J., (1999), Time-dependent rheological properties of semi-solid metal alloys, Mechanics of time dependent Materials 3 15-30.
8. Sang-Yong Lee, Jung-Hwan Lee, Young-Seon Lee, (2001), Characterization of Al 7075 alloys after cold working and heating in the semi-solid temperature range, Journal of Materials Processing Technology 111 42-47.
9. Dong J., Cui Z., Le Q.C., Lu G.M. (2003), Liquidus semi-continuous casting, reheating and thixoforming of a wrought aluminum alloy 7075, Materials Science and Engineering A345 234-242.
10. Eskin D. G., Suyitno, Katgerman L., (2004), Mechanical properties in the semi-solid state and hot tearing of aluminium alloys, Progress in Materials Science 49 629-711.
11. kleiner S., Beffort O., Wahlen A., Uggowitzer P.J., (2002), Microstructure and mechanical properties of squeeze cast and semi-solid cast Mg-Al alloys, Journal of Light Metals 2 277-280.
12. Kleiner S., Beffort O., Uggowitzer P.J. (2004), Microstructure evolution during reheating of an extruded Mg-Al-Zn alloy into the semisolid state, Scripta Materialia 51 405-410.

A PRELIMINARY STUDY ON A TORQUE SENSOR FOR TOOL CONDITION MONITORING IN MILLING

F. Tognazzi, M. Porta, F. Failli, G. Dini

Department of Mechanical, Nuclear and Production Engineering, University of Pisa, Italy

KEYWORDS: Tool Condition Monitoring, Milling, Cutting Torque Measurement, Inductive Sensors.

ABSTRACT. The measure of cutting forces is one of the most usual methods in Tool Condition Monitoring (TCM). In this paper a cutting torque sensor for tool condition monitoring in milling is proposed. The device is based on inductive sensors that measure the relative rotation, due to the cutting torque, of two slotted disks fitted on the toolholder; a modification of the toolholder has been required to enhance the sensitivity of the device. During a preliminary stage, the measuring system has been designed and the optimization of the toolholder dimensions has been carried out by using both simplified and FEM-based approaches. To check the feasibility of the proposed device, a mechanical simulator installed on a lathe has been realized and its performance has been tested under different cutting conditions. The results show the validity of the method and encourage the use of the modified toolholder as instrument for tool condition monitoring.

1 INTRODUCTION

In modern manufacturing industries, the global competition has generated a more and more intensive policy of costs reduction and product quality improvement, based on better production control and production-time and workforce reduction. For these reasons, process monitoring and automation are key-strategies to be applied.

Since the tool is the weakest part of a machining system, the monitoring of the tool condition is necessary in order to obtain a fully automated cutting process.

The main purpose of a TCM system is the control of the tool damage, that consists of measurement of the tool wear and breakage detection. Various direct and indirect monitoring systems have been proposed and studied in the last twenty years; some typical sensing techniques are based on measurement of cutting forces, axis and spindle motors current/power, vibrations, acoustic emissions and optical methods. Apart from the sensing principle, a sensor for TCM should be easy to use, to mount and to maintain. It also should be characterized by a robust-noise behaviour, a high resolution, low interaction with the process, low cost and it should have the possibility to be positioned close to the cutting zone.

2 STATE OF THE ART OF TCM IN MILLING

The measurement of cutting forces is one of the most employed methods to detect tool wear and tool breakage in milling, and generally in TCM [1,2].

Depending on the mounting position, a distinction can be made between sensors fitted on the workpiece side and those ones fitted on the tool side [3]. The formers are easy to install, but the

Published in: E. Kuljanic (Ed.) *Advanced Manufacturing Systems and Technology*,
CISM Courses and Lectures No. 486, Springer Wien New York, 2005.

signal behavior could not be constant, because the cutting point, and then the signal transmission path, continuously changes, with respect to the sensor position. By mounting the sensor directly on the toolholder, the signal is acquired with a higher amplitude and quality, but for the same workpiece, several sensorized toolholder could be necessary, typically one for each monitored tool.

Some interesting examples of sensorized toolholders for cutting forces monitoring in milling can be found in literature. Aoyama and Inasaki developed a sensor system, based on the magnetostrictive effect, in order to measure the cutting torque [4]. Schul et al. proposed a modular toolholder for monitoring of various cutting processes [5]. Other rotating sensors and commercial rotating dynamometers are nowadays available on the market. Rotating sensors based on piezoelectric transducers are commonly used in research experiments, such as in [6]. Other more economic commercial dynamometers exist, based on strain gauges and mainly used in industrial applications.

Sensorized toolholders are effective instruments for TCM in milling. Since the main drawback of these devices is their high cost, the objective of this work is the design of a low cost, easy to use rotating sensing system.

3   WORKING PRINCIPLE OF THE PROPOSED SYSTEM

A general description of the proposed system is shown in Fig. 1. The main parts are the toolholder, the inductive sensors, a counter board and a computer. Two slotted disks are fitted on the toolholder and two sensors are mounted close to them. Each sensor provides a digital pulse output when the toolholder rotates and the ferromagnetic and non-ferromagnetic zones of the disk pass in front of it.

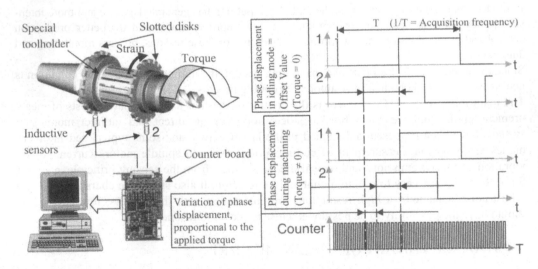

FIGURE 1. The proposed system

The working principle is the following: when the spindle runs in idling mode, the phase displacement between the two signals is constant. During machining, the torque causes the relative rotation of the disks, and the phase displacement changes its value. In order to hold the sensors and to keep them close to the disks, a support positioned under the spindle shaft has been designed. Its structure must be stiff enough to avoid undesirable vibratory phenomena and must allow the automatic tool change. A digital counter, with a frequency of 80MHz, and a computer have to be used to acquire the two signals and measure the phase displacement between them. A rising edge on the signal of the first sensor starts the phase displacement counting, and a rising edge on the signal of the second sensor stops it.

The sensing system is characterized by a low cost, a high level of robustness and it is easy to install on a machine; the sensors are not influenced by cutting fluids and dirtiness, but, since the correct working distance between sensors and disks has to be less than 2mm, chips could fall in that small gap. The problem can be avoided by blowing in some jets of compressed air.

An important advantage of the proposed device is the possibility of measuring the torque_directly on the toolholder, near the cutting point. Furthermore, unlike most rotating dynamometers, there is not need to transmit the signal between rotating and fixed part. Nevertheless, the working principle of the system limits the acquisition frequency f [Hz] as follows:

$$f \leq n \frac{N_s}{60} \tag{1}$$

where n is the spindle speed [rpm] and N_s is the number of slots in each disk.

4 SYSTEM SPECIFICATIONS

As described above, the system detects the machining torque value by the relative rotation of two slotted disks at the ends of the central body of the toolholder. The minimum measurable value of this rotation θ_{res} (i.e. the resolution of the system) depends on the rotational speed of the toolholder n [rpm] and on the counter frequency f_c [Hz] through the expression 2:

$$\theta_{res} = 2\pi \frac{n}{60} \frac{1}{f_c} \tag{2}$$

Considering a maximum rotational speed of 500 rpm and a counter frequency of 80MHz, the minimum relative rotation of disks must be greater than $6.5 \cdot 10^{-7}$ rad.

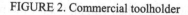

FIGURE 2. Commercial toolholder   FIGURE 3. Modified toolholder

For the whole system, the torque resolution value has been assumed as 0.1 Nm. The commercial toolholder used during the test, under this torque value shows a relative ends rotation of only $9.5 \cdot 10^{-8}$ rad. For this reason the stiffness of the toolholder has been decreased by creating longitudinal slots in the central body and by enlarging the central hole. To obtain a good compromise between sensitivity to the torque and cutting performance, an accurate structural analysis on the modified toolholder has been performed.

In Fig. 2 and Fig. 3 the commercial toolholder and the modified toolholder are shown.

5 TOOLHOLDER DESIGN

As mentioned before the machining of some narrow slots in the body of a commercial toolholder and the widening of its central hole are the solutions adopted to decrease the torsional stiffness without a great reduction of the flexional stiffness. The general aim of this study has been the detection of a suitable configuration of the modified toolholder (number and dimension of the slots) in terms of static and dynamic properties and fatigue life. The modified toolholder has been designed following two steps: simplified analysis and FEM analysis. The first raw results of the simplified analysis have been confirmed and refined by the successive FEM analysis.

5.1 SIMPLIFIED ANALYSIS

In a simplified approach, the structure of the central body has been represented by the structural model shown in Fig. 4.

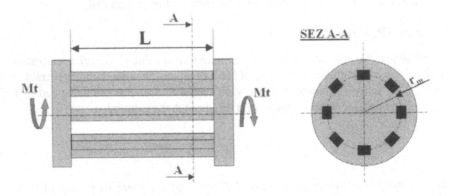

FIGURE 4. Simplified model of the modified toolholder body

Using this approach, the relation between the relative rotation of its ends and the applied torque Mt has been obtained as shown in the following expression:

$$\theta = \frac{1}{12} \frac{Mt\,L^3}{E \cdot I \cdot N \cdot r_m^2} \tag{3}$$

where E is the Young modulus, N is the number of beams, L is their length, I is the moment of inertia of each beam and r_m is the mean distance of the beams from the axis of the whole structure. For each slot, the minimum width value of 3 mm has been chosen. Considering the relative rotation $\theta_{res} = 6.5 \cdot 10^{-7}$ rad, ensuring the resolution of 0.1 Nm, the expression 3 has been iteratively used to obtain acceptable values for N, L, I and r_m. The selected ranges for the different values of parameters are shown in Tab. 1. These values have been the starting values for FEM analysis.

TABLE 1. Parameters range determined by the simplified model of the modified toolholder body

PARAMETER	Range value
N	12-18
L	35-46.5 mm
r_m	15-21 mm
I	120-140 mm^4

5.2 FEM ANALYSIS

Several FEM models with different values of the parameters N, L, I and r_m have been analyzed to investigate on their behavior under heavy cutting conditions.

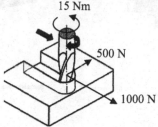

FIGURE 5. Principle of the cutting test with the commercial toolholder

TABLE 2. Parameters and performance of the FEM models

Parameters value	Model 1	Model 2	Model 3	Model 4
N	12	14	16	18
L [mm]	38	40	35	46.5
I [mm^4]	139	132	127	124
r_m [mm]	15	17	16	21
Performances	Model 1	Model 2	Model 3	Model 4
Relative rotation of disks [rad]	$5.2\ 10^{-6}$	$4.2\ 10^{-6}$	$6.9\ 10^{-7}$	$4.1 \cdot 10^{-6}$
Maximum deflection [μm]	93	96	67	73
Max Stresses [MPa]	110	90	70	78
First own frequency [Hz]	834	937	1044	861

The maximum values of cutting forces obtained from tests performed with the commercial toolholder have been used (Fig. 5). The cutting parameters adopted in these tests are: feed 0.36 mm/rev, rotation speed 500 rpm and cutting depth 4 mm.

In Tab.2 the performances of some meaningful considered FEM model are shown.

The selected toolholder has been the 4th model because its performance in term of maximum deflection is very good. Furthermore, the maximum stresses is quite low compared to the material yield strength (900 MPa).

The first own frequency of the selected toolholder is not the highest, but it is high enough to avoid resonance effect (as shown below in Tab. 4). The 3rd model has been discarded, despite to the very low value of deflection, because its relative rotation is just higher than the required minimum relative rotation of disks. So it could generate problems concerning the resolution in torque measurement.

The static stresses and the deflections of the selected modified toolholder are shown in Fig. 6 and Fig. 7.

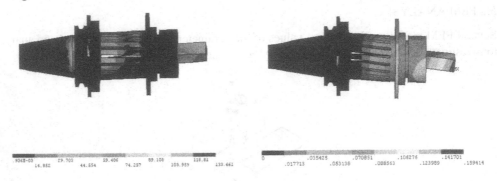

FIGURE 6. Static stresses of the system   FIGURE 7. Deflections of the system

The relative rotation of the disks, the deflections and the maximum stresses of the commercial and of the modified toolholder are visible in Tab. 3.

Tab. 3 shows that the deflection of the modified toolholder increases of 1.7 times compared to the commercial one, while the relative rotation of the disks increases about 43 times.

TABLE 3. System performance

	COMMERCIAL SYSTEM	MODIFIED SYSTEM
DISK RELATIVE ROTATION	$9.5 \cdot 10^{-8}$ rad	$4.1 \cdot 10^{-6}$ rad
DEFLECTION	32 μm	73 μm
MAXIMUM STRESSES	48 MPa	78 MPa

The FEM analysis has been also used for a dynamic study of the modified toolholder. The frequency of the cutting forces in milling depends on the rotational speed and on the teeth number of the tool. To avoid dangerous resonance effects, the system own frequencies must be much greater than the frequencies of the cutting forces. According to the rotational speed range (between 300 and 500 rpm) and to the use of a 3-cutter tool, the cutting frequency ranges in the field of 15-25 Hz. The first 8 own frequencies of the proposed and the commercial systems are shown in Tab. 4.

TABLE 4. System own frequencies

COMMERCIAL TOOLHOLDER		MODIFIED TOOLHOLDER	
Frequency [Hz]	Type	Frequency [Hz]	Type
1490	flexional	861	flexional
1490	flexional	861	flexional
4679	flexional	**968**	**torsional**
4830	flexional	4089	flexional
6043	**torsional**	4116	flexional
8772	flexional	6854	flexional
8800	flexional	7059	flexional
8899	flexional	7114	flexional

As shown in Tab. 5, lower deflection than 73 μm can be obtained by reducing the number of slots N, and increasing I and r_m, but a strong reduction in terms of resolution can be observed. However, according to expression 2, the required relative rotation of the disks that ensures a certain resolution decreases by using a higher counter frequency. So a better behavior of the system in terms of resolution and deflection can be obtained by using a counter frequency higher than 80MHz.

TABLE 5. Relationship between resolution and deflection

N	I	r_m	Deflection	Resolution
18	123.76 mm^4	35 mm	73 μm	0.1 Nm
16	253.30 mm^4	32 mm	58 μm	0.2 Nm
14	544.10 mm^4	28 mm	47 μm	0.3 Nm

6   CUTTING TESTS

During this preliminary stage, the working principle of the sensing system has been simulated and tested by using the experimental facility shown in Fig. 8. The simulator consists of a steel bar whose torsional stiffness has been decreased by 16 axial slots; the bar includes two slotted disks and it is mounted on a lathe.

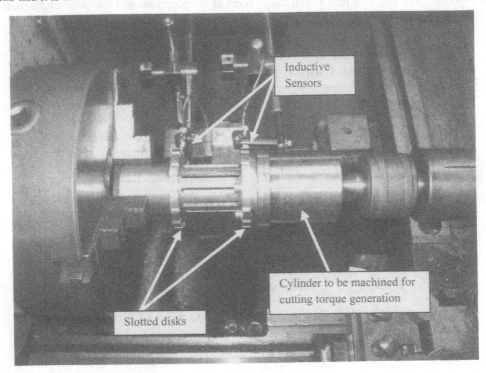

FIGURE 8. The experimental facility used to simulate the working principle of the proposed system

One end of the simulator is clamped in the self-centering chuck; the other end is linked to a cylinder supported by the dead center of the lathe. The cylinder is machined to generate a cutting torque. A proper fixture, placed close to the lathe and isolated from ground vibrations, holds two inductive sensors in front of the two slotted disks of the simulator. The sensors signals are acquired by a counter mounted in a PC, as described in section 3.

The structural and dynamic differences between the simulator and the designed toolholder are not relevant at this evaluation stage.

All the tests described in this section have been carried out under the cutting conditions described in Tab. 6. The torsion value is calculated 16 times per revolution; one degree of torsion corresponds to about 23,400 counter pulses. Since the phase displacement between the two signals in idling mode is normally different from zero, the torsion value presents an initial offset value. This offset has been automatically eliminated by software.

TABLE 6. Cutting conditions

TOOL	Cutting insert: Carbide, P25 grade
CUTTING PARAMETERS	Spindle speed: 570 rpm. Feed: 0.2 mm/rev Lubrication: Dry
WORKPIECE MATERIAL	Not alloyed carbon steel (C: 0.4%)

Fig. 9 shows results obtained in some cutting tests. In Fig. 9a the depth of cut has been step-wise increased; in Fig. 9b the depth of cut has been linearly increased. In this group of tests, the torsion value is smoothed with a moving average filter. In particular, the tests show that even small values of depth of cut can be detected by the system.

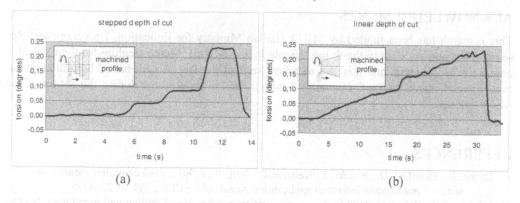

(a)   (b)

FIGURE 9. Cutting tests: a) stepped depth of cut (0.1-0.3-0.7 mm);   b) linearly increased depth of cut (from 0 to 0.7mm)

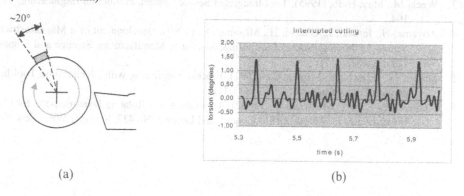

(a)   (b)

FIGURE 10. Interrupted cutting test: a) workpiece (depth of cut: 1mm); b) acquired signal

Tests with interrupted cutting were also carried out, by cutting the profile shown in Fig. 10, where the torsion signal is also reported. These tests show the ability of the system to recognize abrupt variations of the cutting torque.

7 CONCLUSIONS

A new concept of TCM system for milling operations has been proposed. Simplified and FEM approaches have supported the design of the device.
The results obtained from the cutting tests highlight the validity of the working principle. The tested device is able to measure the cutting torque with good resolution and also abrupt variations of the cutting torque can be detected.
Future work will address the development of the proposed system and its implementation on a machining center. Cutting tests will be carried out in milling. Further applications of the system
could involve also other types of rotating tools.

ACKNOWLEDGEMENTS

This research has been funded by MIUR (Italian Ministry for Education, University and Research. The authors wish to thank Ing. Nicola Ricci and Prof. Marco Beghini for their valuable contribute in structural analysis and technical staff of the Department for its helpful cooperation. Special thanks to Prof. Giovanni Tantussi for his suggestions during the set-up of the whole system.

REFERENCES

1. Byrne, G., Dornfeld, D., Inasaki, I., Ketteler, G., Teti, R., (1995), Tool condition monitoring – The status of research and industrial applications, Annals of the CIRP, Vol. 44/2, 541-567.
2. Prickett, P.W., Johns, C., (1999), An overview of approaches to end milling tool monitoring, International journal of machine tool and manufacture, Vol 39, 105-122.
3. Weck, M., May, H. P., (1995), Tool Integrated Sensor System, Production Engineering, Vol II/2, 101-104.
4. Aoyama, H., Inasaki, I., Ohzeki, H., Mashine, A., (1999), Development of a Magnetostrictive Torque Sensor for Milling Process Monitoring, Journal of Manufacturing Science and Engineering, Vol 121, 615-622.
5. Schulz, H., Versch, A., Fiedler, U., (2001), Process Monitoring with Mechatronic Tool holders, Production Engineering, Vol VIII/2, 115-118.
6. Kuljanic, E., Sortino, M., Miani, F., (2002), Application of a Rotating Dynamometer for Cutting Force Measurement in Milling, CISM Courses and Lectures No 437, Springer Wien New York, 15-35.

COMPARISION AND ANALISYS OF IN-PROCESS TOOL CONDITION MONITORING CRITERIONS IN MILLING

S. Garnier, M. Ritou, B. Furet, J. Y. Hascoet

Institut de Recherche en Communications et Cybernetique
de Nantes (IRCCyN), UMR CNRS 6597, Nantes, France.

KEYWORDS: Tool Condition Monitoring, Forces Analysis in Milling, HSM, Tool Breakage.

ABSTRACT. Even with the best process design, incidents may still occur during machining. Considerable damages for the product and the machine-tool, may then be involved by the high feeds currently used. Therefore, process monitoring is suitable to ensure both product quality and process safety. Yet there is a lack of process monitoring solutions for small batch sizes or one-off production, which usually concerns high added value parts. Researchers have proposed various criterions to detect tool breakage, based on the milling force waveform. Our work aimed at estimating the relevancy of these criterions by machining a specific part under various cutting conditions. Criterions are compared and analyzed according to the measured cutting force signals. Then, improvements are suggested in order to increase their efficiency.

1   INTRODUCTION

Expectations in the context of increased productivity, improved part quality and reduced costs, has led to automation. Unmanned manufacturing process has to be completely reliable and the success hinges primarily on the effectiveness of process monitoring and control systems [1-2]. Indeed, the whole parameters of milling operations are not under full control and incidents may still occur; for example, collisions, tool breakages, excessive tool wears, chatters [3]. Therefore, the process has to be monitored to avoid, nevertheless to limit, damages for the product and the machine-tool. This task is done by Tool Condition Monitoring systems (TCM) that have to be reliable. Commercial TCM systems are proposed [4-5]. Most of them are based on the teach-in method: a first part is machined (a trial cut) and a reference signal is stored, then thresholds are placed from part to part of that signal, based on heuristic knowledge [2-3]. But this method is tedious, time consuming and it is not reliable enough [2]. Moreover, as a first part has to be machined, it is incompatible with small batch sizes or one-off production, which usually concerns high added value parts [3]. There is a lack of process monitoring solutions for flexible manufacturing [6].

In this paper, we will focus on the approaches proposed by researchers for tool breakage detection in the context of flexible manufacturing, in milling operations. Figure 1 suggests a classification of those methods. The milling forces waveform has led various authors to use only signal processing methods, such as time series autoregressive models, synchronized averaging or wavelet transform coefficients A4 [7] or D5 [8]. Basic process knowledge is not taken into account and it is harder to distinguish tool breakage from the effects of tool runout or transient cut. Indeed, geometric, kinematic and mechanistic characteristics of the cutting process are, or could be, under control during milling operations [2]. They should be used to improve or simplify

Published in: E. Kuljanic (Ed.) *Advanced Manufacturing Systems and Technology*,
CISM Courses and Lectures No. 486, Springer Wien New York, 2005.

the signal processing and increase the reliability of the monitoring system. Artificial Intelligence includes neural networks and fuzzy logic. Both are often used for sensor data fusion. The latter is rather used for adaptative control. The former is more interesting for tool breakage monitoring but the networks have to be trained with trial cuts and the main problem is the generalization: under other cutting conditions, is the neural network still reliable [2]? Users may have to train again the network to monitor the machining of a new part [9]. Note that hybrid approaches are proposed, combining reasoning capability of fuzzy logic theory with the learning capability of the neural networks [10]. Another approach can be found in literature: the reference signal of the teach-in method is predicted instead of being measured during a trial cut [11]. The average cutting force value per spindle turn is estimated but the gap with the measured average force is important. It is assumed that, in milling, if a tooth is broken, it removes a smaller volume of material than before and the following tooth removes a larger volume [12]. So, should be monitored the cutting force per tooth and not per turn.

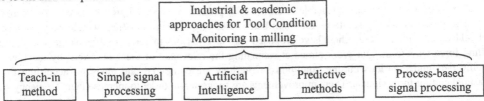

FIGURE 1. Monitoring methods classification

The forces per tooth are considered by the process-based signal processing methods. The criterions are designed taking into account cutting characteristics. During the TCM, only instant data provided by the sensor(s) will allow the system to evaluate the tool condition. Those methods aim at being directly applied, whatever the part to be machined and without training. Therefore this type of monitoring methods is particularly suitable for flexible manufacturing. Altintas et al. [13] synchronized the cutting force measurement and the tooth passing frequency. They introduced two criterions: the first and the second differences of the quasi mean resultant force between adjacent teeth. Altintas [14] also proposed the residual of the first order auto-regressive model of the quasi mean. But it is impossible to distinguish tool breakage from cutter runout [15]. Lee et al. [16] added to the residual a new criterion: the relative variation of the mean force between two consecutive turns, for a given tooth. It introduces the idea of monitoring each tooth individually. But all those criterions are dependent on the cutting conditions changes. Therefore, tool breakage cannot be detected reliably [15]. Kim and Chu [15] proposed the Tool Failure Index (TFI) which is the ratio of the peak-to-valley cutting force between two adjacent teeth divided by its own past average ratio. This average ratio could overcome the cutting condition changes. It's the reason why this criterion has been selected for further experiments. At last, Deyuan et al. [17] proposed two criterions: The peak rate Km is the ratio of the difference to the sum between the peak forces of two adjacent teeth. The relative eccentricity rate Bm is similar to the ratio of the tooth eccentricity to the maximum cutting thickness. The authors precised that those criterions are independent of the cutting conditions. So, the criterions have been selected too.

The more interesting approach for tool breakage detection in flexible manufacturing, seems to be the process-based signal processing. Several criterions have been proposed. In most of cases, a few experiments have been carried out to evaluate the criterion relevancy. They generally consist in machining a straight path under a few cutting conditions [7-8;15-17]. Moreover, the cutting conditions used are far from ones used by industrialists. For example, the spindle speed is usually less than 500 RPM [7-8;14;13;15-17]. In such conditions, it is difficult to conclude about the reliability of a method.

The aim of this paper is to evaluate the relevancy of the TCM criterions kept (TFI, Km and Bm). The influence of the cutting conditions on the criterions will be analyzed analytically. Then a specific part will be designed and toolpaths generated in order to vary the cutting conditions that could influence the criterions. Criterions will be applied on force measurements and the evolution of the criterions will be analyzed thanks to real cutting conditions, calculated from the axis encoders measurements. Finally, improvements about the way of using process-based signal processing criterions for TCM, are suggested.

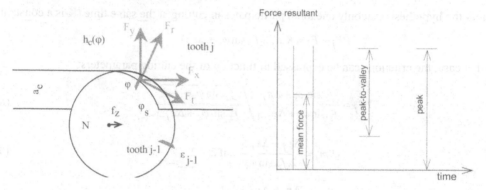

FIGURE 2. Chip thickness and forces in milling

2 CRITERIONS DEFINITION

During milling operations, the cutting forces evolution is periodic, at the tooth passing frequency. F_j and PV_j can represent respectively the peak and the peak-to-valley values of the resultant force applied to the jth tooth for a given spindle turn t. Let Z be the number of teeth. The criterions are defined as follow [15;17]:

$$TFI_j = \frac{PV_j}{PV_{j-1}} \Big/ \frac{\overline{PV_j}}{\overline{PV_{j-1}}} \tag{1}$$

$$Km_j = \frac{F_j - F_{j-1}}{F_j + F_{j-1}} \tag{2}$$

$$Bm_j = \sum_{k=2}^{Z} \left(\frac{F_k}{\overline{F}} - 1 \right) \tag{3}$$

As said earlier, the TFI takes into account the past PV_j values. In equation 1, $\overline{PV_j}$ represents here the mean of PV_j during the ten spindle turns before. Conversely, in equation 3, $\overline{F}$ represents the mean of the peak forces of the Z teeth, at the current spindle turn. So, Deyuan et al.'s criterions depend only on the peak forces at the current turn, whatever happened during the past turns. The three criterions are dimensionless and define one value per turn and per tooth. The underlying influences on the criterions will now be researched. Sabberwal [18] force models, equations 4 and 5, are used; where k_t and k_r are constants, h_c the chip thickness, a_p the depth of cut and f_z the feed per tooth. Note that Cherif at al. [19] specified advances about the force models.

$$F_t = k_t.h_c.a_p \qquad (4)$$

$$F_r = k_r.F_t \qquad (5)$$

The term ε_j defines the tooth radial eccentricity, the shape influence of the cutter, the tool and spindle runout and the amount of cutter chipping. It can be taken into account in the expression of the chip thickness removed by tooth #j [17;20]:

$$h_{cj} = f_z.\sin\varphi + \varepsilon_j - \varepsilon_{j-1} = f_z.\sin\varphi + \Delta\varepsilon_j \qquad (6)$$

Under the hypothesis that only one tooth participates in cutting at the same time (K is a constant),

$$PV_j = F_j = K.a_p.(f_z.\sin\varphi_s + \Delta\varepsilon_j) \qquad (7)$$

In that case, the criterions can be expressed in function of the cutting parameters:

$$TFI_j = \frac{f_z.\sin\varphi_s + \Delta\varepsilon_j}{f_z.\sin\varphi_s + \Delta\varepsilon_{j-1}} \bigg/ \overline{\frac{f_z.\sin\varphi_s + \Delta\varepsilon_j}{f_z.\sin\varphi_s + \Delta\varepsilon_{j-1}}} \qquad (8)$$

$$Km_j = \frac{\Delta\varepsilon_j - \Delta\varepsilon_{j-1}}{2.f_z.\sin\varphi_s} \text{ , if Z = 2} \qquad (9)$$

$$Km_j = \frac{\Delta\varepsilon_j - \Delta\varepsilon_{j-1}}{2.f_z.\sin\varphi_s + \Delta\varepsilon_j + \Delta\varepsilon_{j-1}} \text{ , if Z > 2} \qquad (10)$$

$$Bm_j = \frac{\varepsilon_j}{f_z.\sin\varphi_s} \qquad (11)$$

In case of breakage of the jth tooth, ε_j decreases. Therefore, the criterions should allow tool breakage detection. To do so, the criterion value is compared to a threshold. Kim and Chu proposed to adapt the threshold to predict the magnitude of the tool breakage and they specified that it was possible to distinguish tool breakage from cutter eccentricity and transient cutting. Deyuan et al. fixed both thresholds at 0.8, corresponding to the peak forces $F_j=9.F_{j-1}$ for Km; $F_{j+1}=9.F_j$ for Bm if Z=2 and $F_{j+1}=3.F_j$ for Bm if Z=3. However the influence of the feed per tooth f_z and of the exit angle φ_s can be noticed for each criterion (equations 9 and 10 show that, depending on the number of teeth, the influence of the cutting parameters on Km is different). It could damage the reliability of the tool breakage detections. That's why we have designed a test part and toolpaths so that various feeds Vf and radial width of cut a_e would be encountered during the milling operation.

A pocketing operation has been selected with a zig-zag strategy, allowing up-milling and down-milling. It is composed at the entry of the pocket of common turns (figure 3 zone 1) and in the background of sharp turns (figure 3 zone 2). During the turns, the feedrate should drop. A contouring path finishes the pocket.

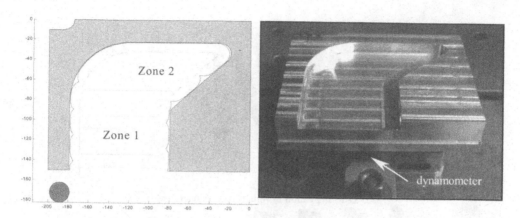

FIGURE 3. Test part and toolpath (from measurements, on the left). Experimental setup (on the right)

3   EXPERIMENTS

The cutting force signals were measured with a 9257A Kistler quartz three-components dynamometer and with a sample frequency set at 64 kHz. The dynamometer and the part were mounted on the table of the SABRE machining center. The X and Y axis position encoders were measured with a sample frequency set at 500 Hz [21]. The part was made of 7075 aluminum alloy. The Taguchi L16 table has been used for the design of experiments. The parameters were the feed per tooth (4 levels: 0.08, 0.12, 0.16 and 0.2 mm/tooth), the radial width of cut (4 levels: 15, 40, 65, 90% of tool diameter) and the tool (2 levels: a 32 mm diameter with 2 inserts and a 20 mm diameter endmill with 3 flutes; the maximum tooth eccentricity is 0.02 mm for both tools). Depth of cut was set to 2.5 mm. Cutting speed was set to 650 m/min, leading to 6 500 and 10 000 RPM spindle speeds and from 1.04 to 6 m/min for the feedrates. Those values correspond to industrial cutting conditions.

The X and Y force components were low-filtered at twice the tooth passing frequency before calculating the force resultant. Then minimum and maximum were researched for each tooth passing to evaluate F_j and PV_j, for each spindle turn. Based on the axis encoder measurements, the feed Vf was calculated as well as width of cut a_e. The edges of the workpiece had been discretised. Then, for each new tool position, intersections with the tool edge led to the entry and exit angles of the teeth and next the width of cut a_e is obtained.

As it could be expected, the machine slows down during the turns (figure 4 a&c). a_e increases during the links between two paths and can reach a full diameter immersion. After a sharp turn, it decreases because the material has yet been removed by the tool (figure 4 b). Note that the a_e setting during toolpath generation, at CAM level, is only respected along the paths (in yellow on figure 4 b).

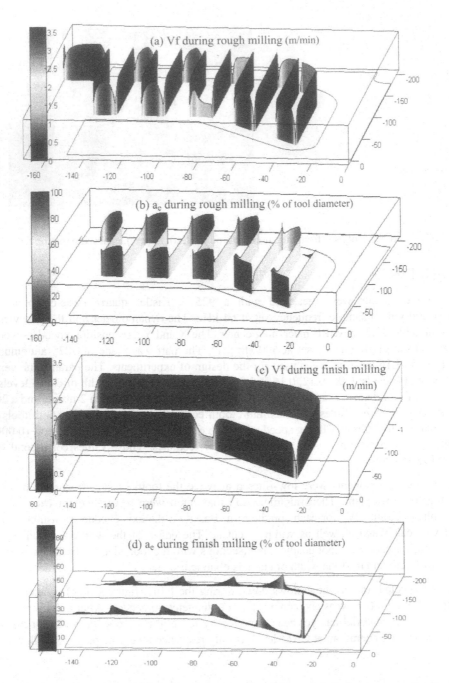

FIGURE 4. Instant value of feedrate (a & c) and radial width of cut (b & d) represented by Z axis
(CAM settings: Vf = 3.6 m/min, a_e= 65%, tool Ø20 mm and 3 teeth)

4 ANALYSIS

4.1 COMMON TURN

Figure 5 corresponds to the machining which links two adjacent paths, in the zone 1 of the figure 3. The axes of the machine slow down when passing the turns because each axis acceleration available is limited [22]. At time 9.5s on figure 5, the first turn is passed and the link toward the next path begins. The tool moves into the material and a full diameter immersion is reached. After the second turn (time 9.76s), a_e drops to the CAM setting (65% of tool diameter).

The criterions have been applied. As the peak forces Fj used by Km and Bm, and the peak-to-valley PVj used by the TFI, the criterions are outlined for each tooth. The grey dashed lines on Km and Bm plots, represent the thresholds proposed by the Deyuan et al. It can be noticed that the criterions values are less than the thresholds ones. Under those changing cutting conditions, the criterions evolution is negligible and a tool breakage could be reliably detected, which is the usual case.

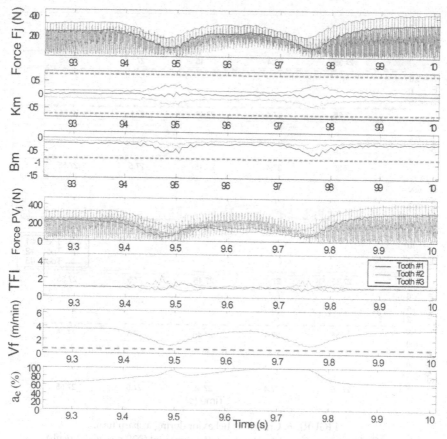

FIGURE 5. Criterions behavior during a common turn.
(CAM settings: Vf = 3.6 m/min, a_e= 65%, tool ⌀20 mm and 3 teeth)

4.2 SHARP TURN

The sharper the turn is, the slower the machine passes it [22]. Figure 6 corresponds to the machining of the corner in the background on the right of the pocket, during the finish path (figure 3 zone 2). During the rough milling, 0.5 mm had been left on the contour, at CAM level. The amount of material left in the corner is important (until 1.5 mm width, figure 3), which involves a peak of the a_e curve. It may be particularly dangerous for the tool condition. Therefore, the reliability of the criterions is important here. Note that two of the three teeth barely participate in the cutting when the cutting conditions are steady. Thus, the Bm criterion always overpasses the threshold because the chip thickness is too less than the cutter eccentricity (eq. 11). Close to time 27.47s, the feedrate decreases so much that two of the teeth have a null chip thickness due to their eccentricity (the corresponding threshold is the grey dashed line on the Vf graph). It also happens during each sharp turn of rough machinings. Km and Bm criterions values overpass the thresholds because of what the fixed values of the thresholds correspond to, in term of peak force (cf. §2).

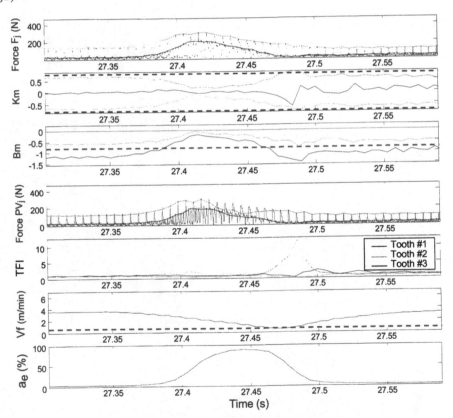

FIGURE 6. Criterions behavior during a sharp turn.
(CAM settings: Vf = 3.6 m/min, a_e= 0.5 mm, tool $\varnothing$20 mm and 3 teeth)

$$TFI_j = \frac{PV_j}{\overline{PV_j}} \bigg/ \frac{PV_{j-1}}{\overline{PV_{j-1}}} = 100\% / 10\% = 10 \tag{12}$$

The TFI can also be expressed as in equation 12. When the cutting efforts of teeth #2 and #3 decrease till nearly zero during the sharp turn, the efforts peak-to-valley values of tooth #1 are constant. So, the ratio PV_1 to mean of PV_1 equals 100%. If the efforts of tooth #3 decrease till 10% of their past value (which is the mean of PV_3), the ratio PV_3 to mean of PV_3 equals 10%. Therefore tooth #1 TFI is important. The value is more than the threshold value proposed by Kim and Chu, even if the threshold value is adapted to the cutting conditions. It results the same when the tool exits the material because, for a while, only the tooth with the smaller eccentricity goes on machining. Those examples clearly demonstrate that, in a few case, the process-based criterions are not reliable. Indeed, cutting conditions can change so that some of the teeth do not participate in the cut. In such a case, it is impossible to distinguish a tool breakage. The criterions are not relevant because this special case has not been taken into account during their design.

5 CONCLUSIONS

There is a lack of reliable solutions for the TCM of flexible productions: TCM criterions based on the cutting mechanistic, seems to be interesting for this task. Indeed, the goal is to distinguish tool breakage from cutter runout or cutting conditions changes. Some of those criterions have been implemented during experiments on a test part, under various industrial cutting conditions. It has been shown that, if in most of cases, the criterions are nearly independent of the cutting conditions. However there are a few transient cuttings where they are very dependent. It happens when some of the teeth no longer participate in the cut, whereas others do. Thus, the criterions are not reliable for tool breakage detection.

In order to increase the reliability, the authors propose to introduce a memory aspect in the TCM criterions. The tool can be characterized independently of the machining operation, to know the tool condition at a given instant. Then, during steady machinings, the current tool condition can be compared with the previous tool condition to detect tool breakage. With such a method, there is no need to catch the instant of the incident. It will be automatically detected during the next steady cut. This involves that, in a first time, there is no need to know how to understand each particular case that can be met. For example, the monitoring would be stopped when the corner (figure 6) is machined and it would start again, 0.1s latterly. In this way, a discrete monitoring during steady cuts could allow a reliable monitoring of any milling operation. Further on, the current tool condition could be compared with the initial tool condition to evaluate the amount of wear on the cutting tool-edges.

REFERENCES

1. Liang, S.Y., Hecker, R.L., Landers, R.G., (2002), Machining process monitoring and control : the state-of-the-art, ASME Proceedings of IMECE, N°32640.
2. O'Donnell, G., Young, P., Kelly, K., Byrne, G., (2001), Towards the improvement of tool condition monitoring systems in the manufacturing environment, Materials Processing Technology, Vol 119/2, 133-139.

3. Furet, B., Garnier, S., (2002), La surveillance automatique de l'usinage à grande vitesse, 2e Assises Machines et Usinage à Grande Vitesse, Lille, 221-230.
4. Byrne, G., Dornfeld, D., Inasaki, I., Ketteler, G., König, W., Teti, R., (1995), Tool Condition Monitoring (TCM) - The Status of Research and Industrial Application, Annals of the CIRP, Vol 44/2, 541-567.
5. Jemielniak, K., (1999), Commercial Tool Condition Monitoring Systems, Advanced Manufacturing Technology, Vol 15/10, 711-721.
6. Klocke, F., Reuber, M., (1999), Process monitoring in mould and die finish milling operations - challenges and approches, Proceedings of the 2nd International Workshop on Intelligent Manufacturing Systems, Leuven, Belgium, 747-756.
7. Lee, B.Y., Tarng, Y.S., (1999), Application of the Discrete Wavelet Transform to the monitoring of tool failure in end milling using the spindle motor current, Advanced Manufacturing Systems and Technology, Vol 15/4, 238-243.
8. Xu, S.X., Zhao, J., Zhan, J.M., Le,G., (2002), Research on a fault monitoring system in free-form surface CNC machining based on wavelet analysis, Materials Processing Technology, Vol 129/3, 588-591.
9. Tarng, Y.S., Chen, M.C., Liu, H.S., (1996), Detection of tool failure in end milling, Materials Processing Technology, Vol 57/1, 55-61.
10. Chen, J.C., (2000), An effective fuzzy-nets training scheme for monitoring tool breakage, Journal of Intelligent Manufacturing, Vol 11/1, 85-101.
11. Saturley, P.V., Spence, A.D., (2000), Integration of Milling Process Simulation with On-Line Monitoring and Control, Advanced Manufacturing Technology, Vol 16/2, 92-99.
12. Prickett, P.W., John, C., (1999), An overview of approaches to end milling tool monitoring, Machine Tools and Manufacture, Vol 39/1, 105-122.
13. Altintas, Y., Yellowley, I., (1989), In-process detection of tool falure in milling using cutting force models, Engineering for Industry, Vol 111, 149-157.
14. Altintas, Y., (1988), In-process detection of tool breakage's using time series monitoring of cutting forces, Machine Tools and Manufacture, Vol 28/2, 157–172.
15. Kim, G.D., Chu, C.N., (2001), In-Process Tool Fracture monitoring in Face Milling Using Spindle Motor Current and Tool Fracture Index, Advanced Manufacturing Systems and Technology, Vol 18/6, 383-389.
16. Lee, J.M., Choi, D.K., Kim, J., Chu, C.N., (1995), Real-time tool breakage monitoring for NC milling process, Annals of the CIRP, Vol 44/1, 59–62.
17. Deyuan, Z., Huntay, H., Dingchang, C., (1995), On-line detection of tool breakages using teletering of cutting forces in milling, Machine Tools and Manufacture, Vol 35/1, 19-27.
18. Sabberwal, A.J.P., (1962), Cutting forces in down milling, International Journal of Machine Tool Design and Research, Vol 2/1, 27-41.
19. Cherif, M., Thomas, H., Furet, B., Hascoet, J.Y., (2004), Generic modelling of milling forces for CAD/CAM applications, Machine Tools and Manufacture, Vol 44/1, 29-37.
20. Kasashima, N., Mori, K., Herrera-Ruiz, G., Taniguchi, N., (1995), Online Failure Detection in Face Milling Using Discrete Wavelet Transform, Annals of the CIRP, Vol 44/1, 483-487.
21. Dugas, A., Terrier, M., Hascoet, J.Y., (2002), Free form surface measurement method and machine qualification for high speed milling, Proceedings of IDMME, 1-10.
22. Dugas, A., Lee, J.J., Hascoet, J.Y., (2002), High speed milling : solid simulation and machine limits, Kluwer Academic Publishers, Netherlands.

DESIGN AND PERFORMANCE ASSESSMENT OF A HSM TOOL HOLDER

A. Del Taglia, G. Campatelli

Department of Mechanics and Industrial Technologies, University of Firenze, Italy

KEYWORDS: Manufacturing Tolerance, Tool Holder Design, Tool Assembly, Manufacturing Cost.

ABSTRACT. High Speed Milling is demanding more and more accuracy of the tool rotation axis and overall tool holder performances. In this paper the design and performance assessment of a thermal shrink fit tool holder is described. The holder is monolithic to improve the stiffness and reduce the tool rotation axis error. The tool is assembled manually, introducing the tool in the pre-heated tool holder; the subsequent thermal shrinkage assures the transmission of the cutting force and torque by friction. The disassembly is critical as the heat flow causes the thermal expansion of both the tool and the tool holder. This retention mechanism needs an accurate study for the choice of the tool holder material and geometry, of the heather geometry and power and of the heating procedure. A mathematical model of the system, based on coupled thermal-stress finite elements, has been set up in order to optimize the design and, consequently, to improve the tool holder expected performances. The tool holder has been manufactured and experimental test has been executed to validate the model and assess the tool holder performances.

1 INTRODUCTION

Machine tools have undergone in the last years huge improvements, especially concerning the following aspects:

- greater and greater static and dynamic stiffness has been achieved, allowing greater cutting force, torque and chatter resistance;
- high speed has been introduced, both for spindles and for feed drives, allowing greater removal rate;
- high power motors are installed, as a consequence of the above said.

This evolution has been driven by the evolution of cutting tools and tools materials that support now high forces and high cutting speed. Up to this time, the necessary element connecting the tool to the machine tool, the tool holder, has not received the same attention; insufficient research has led to not relevant results, and now the tool holder is the weak element, limiting the performances of both the tool and the machine tool [1]. In fact quality (tolerances, surface finish) and economy (machining cost, setup cost, tool life) depend on a chain of elements (machine tool, tool, tool holder, work piece holder) each having comparable importance.

Limiting our attention to the machining operation of end milling, that is widely used in the fields of aeronautical and die manufacturing, we can list the specifications to which the tool holder must comply:

- high gripping force, allowing cutting operations without relative slip between the tool and tool holder;

Published in: E. Kuljanic (Ed.) *Advanced Manufacturing Systems and Technology*,
CISM Courses and Lectures No. 486, Springer Wien New York, 2005.

-   high stiffness allowing great cutting forces and low elastic deformation;
-   sufficient grip force and stiffness at extremely high rotation speed (40,000 rpm and more);
-   good dynamical balancing, in order to avoid vibrations;
-   low inertia to allow faster accelerations and decelerations;
-   high concentricity and alignment between tool axis and spindle axis; such errors cause an unbalance and an uneven distribution of the chip thickness among the mill teethes. In high speed milling, especially using a ball nose mill, the mean chip thickness is so small that even a very small rotation error causes problems; in figure 1 is reported the polar diagram of the cutting forces of a two teethes ball point end mill, cutting a slot and supported by a tool holder with "only" 0.01 mm of concentricity error. It is commonly accepted that an error not exceeding a few μm is recommended for high speed milling.

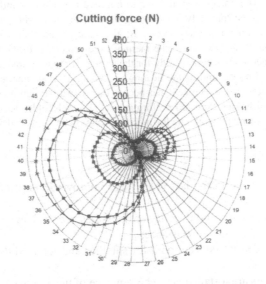

FIGURE 1. Polar diagrams of the tangential, radial, axial components and resultant of the cutting force of a two teethes ball nose end mill cutting a slot, with 10 μm of eccentricity

In recent years the improvements in the milling tool holders has led to the HSK attachment for the tool holder-machine link and the thermal shrink fit attachment for the tool-tool holder link [2, 3]. This attachment is shown in figure 2 d); the same figure 2 reports the elastic collet chuck a), that is the most popular system for connecting cylindrical shank tools, the roll lock tool holder b) and the hydraulic tool holder c). The elastic collet chuck has an intermediate element, that introduces errors of concentricity and of alignment; furthermore the grip force is low, and can become lower at high speed. In the roll lock tool holder, the collet is eliminated, and a unique element is connected to the machine-tool through the conical attachment, and to the tool through a calibrated hole that is shrunk from outside by a sleeve that exerts high radial pressure. This solution gives better results in comparison with the elastic collet, but the weight and the inertia are high and is more expensive. The hydraulic tool holder offers high grip force and rotation accuracy, as the previous, but the stiffness is unsatisfactory and the cost is high too.

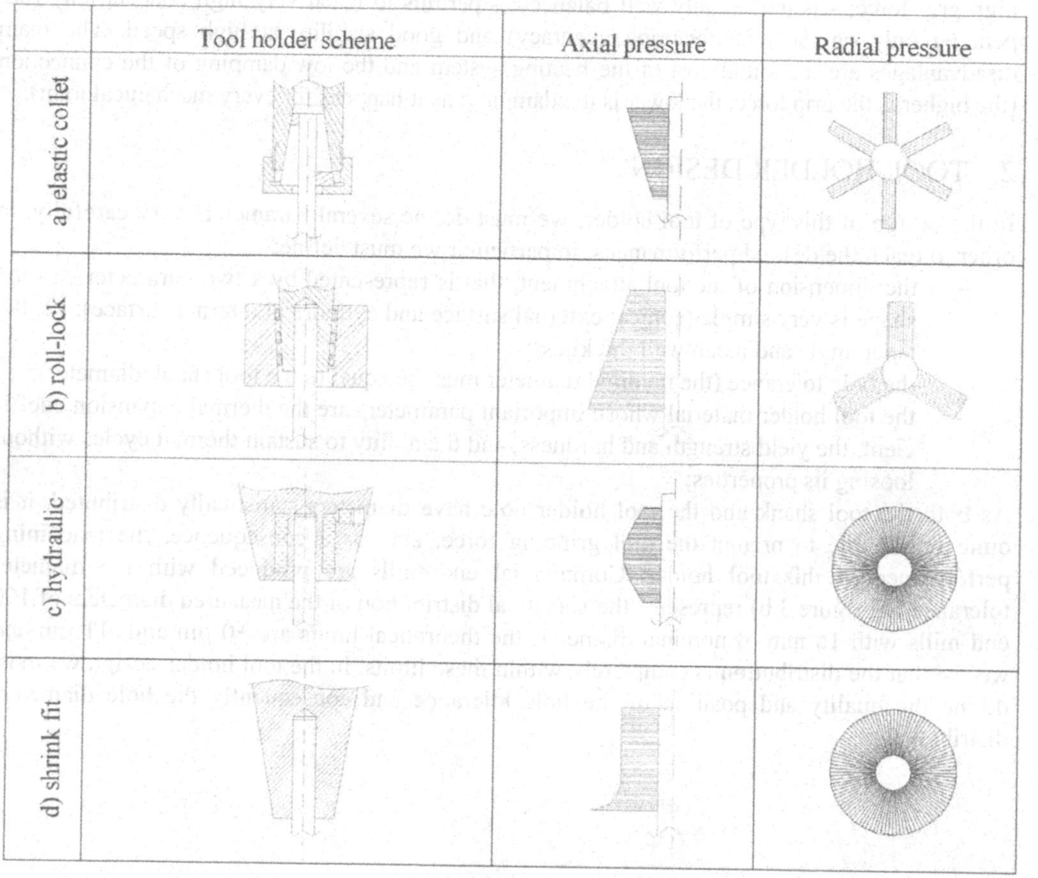

	Tool holder scheme	Axial pressure	Radial pressure
a) elastic collet			
b) roll-lock			
c) hydraulic			
d) shrink fit			

FIGURE 2. Different types of tool holder and pressure distributions

The thermal shrink fit, that is the object of our work, is extremely simple; at the tool side only a hole exists, in which the tool is fitted with interference. Both for assembly and disassembly of the tool, the tool holder is heated in order to cause a thermal expansion. In the assembly phase the tool is at room temperature and is introduced by hand; during the disassembly phase the heating first causes a temperature rise and relative thermal expansion in the tool holder; but after some seconds also the tool's temperature increases and the tool expands too; as time proceeds the temperature tends to become uniform. Disassembly is possible if the material of the tool and of the tool holder have different thermal expansion coefficients (greater in the tool holder and lower in the tool; this is always true using solid carbide end mills); otherwise disassembly is possible in a narrow time span during the heating phase, when a high thermal gradient exists inside the assembly, and the differential expansion of the two elements makes the grip force to decrease or to vanish. This type of connection shows many interesting features: - has high stiffness; - allows high grip force; - is intrinsically well balanced; - permits to reach very high concentricity (depending only on the manufacturing accuracy) and good stability at high speed. The main disadvantages are the initial cost of the heating system and the low damping of the connection (the higher is the grip force, the lower is the damping; as it happens for every mechanical joint).

2   TOOL HOLDER DESIGN

In the design of this type of tool holder, we must define several parameters very carefully, in order to reach the desired performances; in particular we must define:

-   the dimension of the tool attachment, that is represented by a two parameters, as the shape is very simple (conical external surface and cylindrical internal surface); i.e. the taper angle and mean wall thickness;
-   the hole tolerance (the nominal diameter must be equal to the tool shank diameter);
-   the tool holder material whose important parameters are the thermal expansion coefficient, the yield strength and hardness, and the ability to sustain thermal cycles without loosing its properties;

As both the tool shank and the tool holder hole have diameters statistically distributed, it is quite impossible to predict the tool gripping force, and, as a consequence, the machining performance of this tool holder. Commercial end mills are produced with the diameter tolerance h6; figure 3 b) represents the statistical distribution of the measured diameters of 100 end mills with 16 mm of nominal diameter; the theoretical limits are +0 μm and -11 μm and we see that the distribution is completely within these limits. In the tool holder design we must define the quality and position of the hole tolerance and consequently the hole diameters distribution.

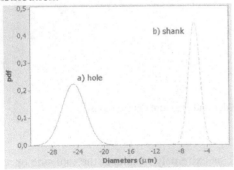 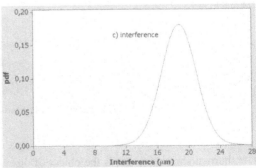

FIGURE 3. Spread of gripping surface and hole diameters

This distribution (for instance distribution a) of figure 3) permits to define the interference distribution (figure 2 c). As the interference controls the grip force, we see that the interference distribution can be defined posing some constraint on the grip force or, better, on the torque and axial force that the tool holder can support without slipping. The constraint has been formulated in this way: - assure a minimum value of torque and axial force (that is values which do not cause slippage using any end mill with diameter within the tolerance limits); - assure values of torque and axial force that cause slip with a given probability, say 5%.

To prevent that too high interference is chosen, also conditions on the maximum tensile stress on the tool holder wall, and on the disassembly condition have been posed. In particular this last conditions has been formulated as the necessity that, during the heating of the tool-tool holder assembly, an instant exists in which the tool slips under its own weight.

3 MATHEMATICAL MODEL OF TOOL-TOOLHOLDER ASSEMBLY

The objective of the mathematical model is to provide an analysis tool that can be used for design optimization; in the model we can change the dimensions, the tolerances, the material, the heating procedure and, with a number of carefully chosen runs of simulation, the best design can be determined.

The diameters that has been taken into account for the simulations are: 6-10-12-16-20-25-32 mm. The geometry of the tool holder used for the simulation is reported in figure 4.

FIGURE 4. Tool holder geometry

The heating of the assembly has been modelled in two ways. The first that considers a constant heat flow from the heater to the tool holder (more suitable for an induction heater) and a second that considers a constant temperature on the tool holder surface (more suitable for a hot air heater). For the simulation it has been considered only the disassembly case for it is the most critical. The disassembly condition is more restrictive than the assembly condition as could be proven that if a tool can be disassembled, it also can be assembled. Moreover, due to the axial symmetry of the assembly, the simulation has been carried out on the simplified model presented in figure 5. The software used for the simulation has been ANSYS®.

Particular attention has been paid to the simulation of the contact between hole and tool gripping surface [4, 5, 6]. The set up of the contact that has showed the best results, in terms of adherence to the real behaviour, is the following:
- Eulerian-Lagrangian (ALE) definition of the elements involved in the contact;
- Augmented Lagrangian Method for contact stiffness;
- constant friction coefficient until the reach of yield stress.

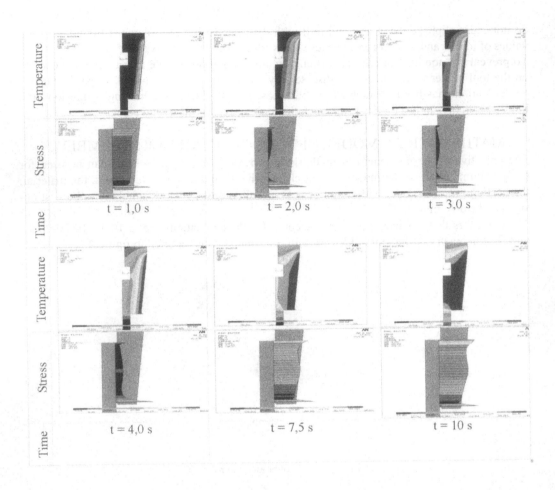

FIGURE 5. Distribution of temperature and radial pressure during disassembly

4   RESULTS OF THE FEM ANALYSIS

In this paragraph the distributions of the stress and temperature of the assembly at various disassembly simulations are reported. Figure 6 reports the temperature distribution versus time in both the tool shank and the tool holder, and the radial pressure distribution at the same values of time.

It is possible to evaluate that after 2 s from the beginning of heating a minimum of radial pressure is reached and the axial slip force can be calculated; if this force is less than the tool weight, the tool is released spontaneously.

Another analysis has been performed in order to verify, during the assembly and disassembly process, that the von Mises stress doesn't exceed the yield stress, to monitor the risk of localized plasticization. Our simulation has proven that no risk of plasticization is present during the assembly/disassembly process. Figure 7 indicates the distribution of von Mises stress in the tool and tool holder in operative conditions.

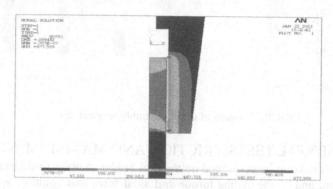

FIGURE 6. Stress distribution of the assembly

Another analysis has been performed in order to determine the temperature that has to be reached in the tool holder to allow the tool disassembly. In this case the temperature depends mainly form three parameters: the design interference, the tool and tool holder materials (thermal dilatation coefficients) and the diameter of the tool. In particular has been considered the adimensional interference, obtained dividing the interference value with the diameter of the tool. The range of value of adimensional interference considered is between 0,001 and 0,003, corresponding to the extreme values of the possible assembly interference.

The simulations have been carried on with two materials for the tool (WC and HSS) and two materials for the tool holder. The tool holder materials are: a tool steel, the UNI X40CrMoV 5 1 1KU that is characterized by an high stress resistance, high wear resistance and high resistance to thermal fatigue, and the other that is a special steel with higher thermal expansion coefficient: the UNI X15CrNiSi25 21 that has lower mechanical characteristics (stress and wear resistance) but a higher thermal fatigue resistance and a higher thermal dilatation coefficient, very useful in case of HSS tools. In fact in this case the thermal coefficient of the original material is very near to that of the tool, generating problems during the disassembly procedure.

The first graph of figure 8 shows how the adimensional interference and the tool diameter influence the needed temperature when the tool is HSS and the tool holder is tool steel. It could be observed that in case of adimensional interference of 0,001 the diameter has a reduced influence. The second graph shows how the material and the interference influence the disassembly temperature for all the possible combinations of the considered materials (in the figure the first tool holder material is called "regular" and the second "special"). From the graph it is possible to derive that only with the special material it is possible to use the greater values of interference: with the first material the value of temperature needed for the disassembly would be too high and must be avoided to prevent material degradation.

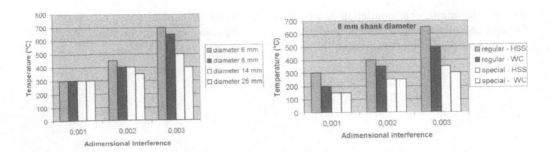

FIGURE 7. Graph of the disassembly temperature

5   EXPERIMENTAL TESTS: FRICTION AND MAXIMUM TORQUE

Experimental tests have to be performed on the tool holder in order to determine the effective friction coefficient and so the cutting torque and axial force that could be applied during the milling process [7]. The experimental test is presented in figure 9. In the tool holder has been mounted a dummy tool, characterized by a carefully determined interference, that has been loaded with a torque, continuously measured by means of a torque meter, until the slippage is reached.

FIGURE 8. Photo of the experimental set-up

A series of experiments has been carried out for the evaluation of the maximum transmissible torque; the expected results has been compared with the numerical simulation results. The graph of the applied torque is reported in figure 10 (left). From the peak value of the torque is possible to find the friction coefficient. The computed value of the coefficient for the used tool is 0,23. Using this value all the simulation give a good accordance with the experimental data. In figure 10 (right) is reported the graph of the maximum transmissible torque for various value of interference and tool diameters.

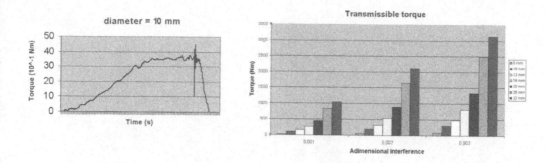

FIGURE 9. Graphs of experimental transmissible torque and maximum torque

6 CONCLUSIONS

The tool holder has been designed and optimized, by the use of a mathematical model of the assembly, for what concerns the torque, axial force and disassembly procedure. In particular the tool holder dimensions, the optimal interference and the best disassembly procedure have been chosen. The temperature, power and time for assembly and disassembly have been determined for the complete series of diameters. Then the tool holder has been manufactured and tested. The tests confirms and validate the model results.

REFERENCES

1. Rivin, E.I., (2000), Tooling Structure Interface between Cutting Edge and Machine Tool, CIRP Annals, v 49/2/2000.
2. Cook, H., (1994), Tool Halder System and Method of Making, U.S. Patent 5,311,654
3. AA.VV., (1997), Shrink-fit toolholding, Cutting Tool Engineering, v 49, n 3, Apr, 5pp.
4. Yang, G.M, Coquille, J.,(2001), Influence of roughness on characteristics of tight interface fit of a shaft and a hub, International Journal of Solids and Structures, v 38, n 42-43, Sep 14, p 7691-7701.
5. Venkateswara, P., Ramamoorthy, B., Radhakrishnan, V., (1995), Role of interacting surfaces in the performance enhancement of interference fits, International Journal of Machine Tools & Manufacture, v 35, n 10, Oct, p 1375.
6. Fontaine, J.F., Siala, I.E., (1998), Optimization of the contact surface shape of a shrinkage fit, Journal of Materials Processing Technology, v 74, n 1-3, Feb, p 96-103.
7. Sackfield, A., Barber, J.R., Hills, D.A., Truman, C.E., (2002), A shrink-fit shaft subject to torsion, European Journal of Mechanics, A/Solids, v 21, n 1, January/February, p 73-84.

A FORCE-TORQUE SENSOR FOR THE APPLICATIONS IN MEDICAL ROBOTICS

A. Biason[1], G. Boschetti[2], A. Gasparetto[1], M. Giovagnoni[1], V. Zanotto[1]

[1] Department of Electrical, Management and Mechanical Engineering, University of Udine, Italy
[2] Department of Innovation in Mechanics and Management, University of Padua, Italy

KEYWORDS: Medical Robotics, Force-Torque Sensor.

ABSTRACT. The paper describes the construction of a force-torque sensor to be employed for applications of medical robotics. The sensor was designed and built in the Mechatronics laboratory of the Dept. of Electrical, Management and Mechanical Engineering of the University of Udine. The aim of the sensor is to measure the values of the forces and torques applied by the external environment to the end effector of a robotic manipulator. This is of fundamental importance when interaction of the robot with the external world must be considered and gauged. In particular, the robot is intended to perform a surgical task, such as the perforation of vertebrae for insertion of peduncular screws. Such an operation is to be done by a master-slave robotic system, where the master robot is connected to the slave by means of a haptic interface. The master sends the command to the slave, which executes the task and feeds back to the master the information of the force and torque values obtained by measuring the interaction of the robot's end effector with the external world. In order to perform this measurement, an adequate sensor has been designed and built. The application of the sensor described in this paper are therefore very important in the medical robotics field, as well as in many other robotic applications.

1   INTRODUCTION

Back pain is one of the major causes of chronic disability. As far as health care is concerned [1], high costs related to back pain treatments increase public expenses. Many surgical protocols to deliver the most effective care to patients are largely employed with relevant success. In order to strengthen the spine, spinal fusion surgery is employed: the surgeon creates a "bridge" of solid bone by linking bones together to assist with alignment and increase strength [2]. Mechanical stability is achieved with pedicle screws, plates and cages fixed to the vertebrae (Figure 1). The pedicle screws used in such operations have a maximum diameter of 5 mm, and a maximum length of 60 mm. Once they are inserted, a little stiffening board, chosen amongst various possible shapes, is fastened to any pair of screws. The tool used to pierce the vertebrae is a dedicated drill, having a maximum weight of 1 kg.

In these kind of surgical tasks, high precision is required, since there exists a not negligible risk of damaging vertebrae, blood vases, nerves and the spinal cord; for example, in a spinal fusion operation, a slightly wrong alignment of the screw with respect to the pre-planned axis could lead to a disability much more serious than the one the operation was supposed to remove.

At the present, spinal fusion remains a standard open procedure. Usually, the patient is turned prone over a curved bed in order to obtain the maximum flexion of the spine. Indeed, the surgeon needs a good access to the vertebrae to make the hole and perform screw insertion with the re-

Published in: E. Kuljanic (Ed.) *Advanced Manufacturing Systems and Technology*,
CISM Courses and Lectures No. 486, Springer Wien New York, 2005.

quired precision (few millimeters). The reason why open procedure is required is that vertebrae displacements occur during intervention because of patient's breathing, patient or table slippage and surgeon-induced movements [3]. As a consequence, the optimum drilling axis moves according to vertebrae motion and percutaneous procedures are not exploitable.

The surgeon can use several electromechanical devices to improve surgical precision; all these mechanisms aim to provide the surgeon with information about the surgical worksite, which cannot be seen at sight [4,5,6]. In this way, the surgeon can guide the drill according to a target shown by a display.

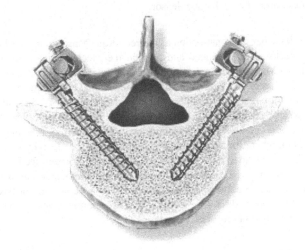

FIGURE 1. Section of a vertebra with two pedicle screws inserted

This paper presents a force transducer carried out at the Mechatronics laboratory of the University of Udine. The transducer is fastened on the wrist of a 6 DoF robot conceived as the slave module of a master-slave system [8] for performing spinal fusion surgical tasks (mainly, the drilling operation). The slave robot is made up of a 3 DoF cartesian robot and a 3 DoF robotic wrist. The robot has been built in the framework of the RIME Project (Robot In Medical Environment) supported by MIUR (Italian Ministry of Education, University and Research). The wrist, that implements the last three degrees of freedom of the robot, will be provided with dedicated tools for surgical operations (such as the drill for pedicle screw insertion into human vertebrae). The robotic wrist is made up of three rotational axes intersecting each other in a single point, so as to allow the orientation of a generic tool in the three-dimensional space and to get the reverse kinematic pose solution in closed form. The operator controls the movements of a surgical drill held by the robot (slave) through a remote haptic device (master). The aim of the torque transducer is to sample the interaction forces between the drill and the patient's vertebra. These force signals are fed back to the haptic device, in such a way the surgeon feels as he/she is contact with the remote patient. Robot specifications are shown in Table 1.

TABLE 1. Robot specifications

Robot Type	Cartesian
Payload	10 kg
X-axis run	700 mm
Y-axis run	600 mm
Z-axis run	500 mm
Repeatability	0.5 mm
Max. Speed	1 m/s
Max. Acceleration	3 m/s2

The payload quoted in Table 1 includes both the tool mass and the mass of the object to be manipulated. The robot overall scheme is shown in Figure 2. The main structure is made of ten aluminum alloy bars, while the three Cartesian axes are provided by means of pre-assembled linear modules, that are a good compromise between costs and accuracy.

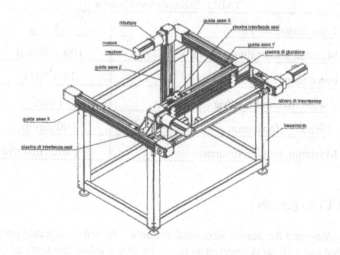

FIGURE 2. Robot architecture scheme

The robot overall dimensions are 1600x1420x2100 mm^3, while the workspace measures 700x600x500 mm^3. A NI Labview software application controls the robot movements through a NI PCI7344 I/O card.

2 THE FORCE/TORQUE TRANSDUCER

In this surgical application, the interaction forces between the drill and the patient's vertebra are less than 70 N on the vertical axis, while the estimated maximum torque around the same axis is 1 Nm. In order to obtain a measure of ±10 V with the electric signal amplification, the deformable arms of the sensor have to resist to a material deformation higher than 400 μm/m. This deformation value is too high if the links dimensions are considered. In order to reduce this deformation value, the strategy to connect the strain gauges by means of a Wheatstone bridge allows to use a material with deformation characteristic near 100μm/m. In fact this method allows to have an electric signal four times larger with the same material deformation.

Another specification involves the mechanical characteristics of the materials used to build the transducer. Particular specifications are requested for the material of which the deformable parts are to be made. The material elasticity has to be compatible with the environment force values that the sensor has to measure. This characteristic has to guarantee that the maximum force value of the environment will not make permanent deformation to the sensor arms. Plastic deformations, in fact, could make the sensor unusable. In order to verify this characteristic, a comparison between maximal force value and material yield stress is needed.

Table 2 shows all the transducer design specifications.

TABLE 2. Sensor specification

Total sensor weight	< 500 g
Maximal sensor diameter value	100-150 mm
Force range values	15 – 70 N
Torque maximum value	1 Nm
Minimal deformation	100 μm/m
Maximum tension on arms	< σ_S arms material

2.1 TRANSDUCER DESIGN

The first step is designing the sensor structural features. The sensor is made up of two concentric rings: the external one will be connected to the wrist of the robot, the internal one directly to the end-effector. These rings are cross linked and need to be built with a material compatible with the specification seen in the previous Section. The starting idea was to make the overall sensor components in aluminum alloy, but this solution raised some problems concerning the linkage of the rings with the links. Moreover the aluminum links were too short to be compatible with the material deformation characteristics.

In order to overcome the linkage problem, the chosen solution was to divide both rings into two part: an upper and a lower one. This strategy allows to realize fixed beam by means of a simple milling. This solution allows inexpensive manufacturing costs and does not require a long mechanical machining work. Choosing an appropriate material for building the links turned out to be a much more complicated problem. The proposed solutions to this problem will be analyzed and discussed in the next Section of the paper.

2.2 TRANSDUCER FEATURES

In order to build a sensor with a diameter less than 110 mm long and meeting all the design requirements, the following characteristics were set:

- Internal ring diameter: 30 mm

- Arms flexible length: 30 mm

- External ring thickness: 10 mm

- Joint depth (on both rings): 5 mm

- Arms section: 8x8 mm

These parameters will be useful to find a material compatible with the specification of the project. Considering that in a cross structure the external forces are symmetrically divided on the four links, the force on a single link will range between 3.75 N and 25 N. This consideration allows to make the force analysis on a single link. It is possible to analyze the arm solicitations by means of hyperstatic structure calculations. Figure 3 represents the structural model.

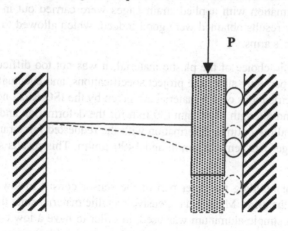

FIGURE 3. Structural model of a single link

Considering the forces of interest in this model, it is easy to find a torque value M_f in the range between 56.25 Nm and 375 Nm. In order to calculate the maximum bending stress, it is important to consider the inertia of a square section $J=s^4/12$, where s is the side length in mm. In this case the inertia parameter results $J = 341.33$ mm^4. Now it is possible to calculate the bending stress by means of the Navier formula: $\sigma_M = (M_f/J)*(s/2)$. Using this formula with the values just calculated, σ_M reaches a value lying in a range between 0.66 N/mm^2 and 4.4 N/mm^2. This surface tension value is very important in order to choose the correct material for the sensor links. As a matter of fact, the links are the deformable members of the sensor, so the material must resist to multiple deformations. We need to use a material capable to resist to the surface tension value just calculated.

Young's modulus is another important parameter to be considered. The value of the Young's modulus E should be known in order to be sure to have a material deformation of 100 µm/m on each link (as said in the specification paragraph). The relation between tension and deformation is given from $E = \sigma/\varepsilon$. Considering the minimum value of σ in order to grant a good deformation, and a minimum value of $\varepsilon = 200$ µm/m, the formula gives the following result: $E < 3300$ N/mm^2. So, any material featuring a Young's module less than 3300 N/mm^2 is suitable to be used for building the force-torque sensor.

2.3 BUILDING THE SENSOR

All the considerations of the previous sections allow to choose the best materials, in particular the material for the flexible links. If only one of the parameters of the last section is not respected, the sensor might not be able to give the requested functionality. A not expensive metallic material with the needed specifications for this sensor could not be found, so plastic materials were considered. The main problem of using a plastic material to build the sensor rises when fixing the strain gages, because such transducers are usually conceived for metallic materials. Some tests of plastic material deformation with applied strain gages were carried out in order to check their compatibility, and the results obtained were good indeed, which allowed to use plastic materials for building the sensor's arms.

Concerning the specific choice of the plastic material, it was not too difficult to find a low cost plastic materials that perfectly suites the project specifications, and eventually simple Hard PVC was chosen. The specifications of this material are given by the ISO R527 norm. The arms with a section of 8x8 mm and a length of 40 mm (30 mm for the deformation and 5 mm for each joint side) of this material allow optimal deformation with the requested force range. The deformation values are in the range between 225 µm/m and 1490 µm/m. This results suite perfectly the requested specification.

The specifications for building the outer part of the sensor consist in low weight and a higher stiffness than that of the arms. Many not expensive metallic material suite these specifications, so in the current project simple aluminum was used, in order to have a low cost malleable material and low cost manufacturing. In order to best suite the sensor with the end-effector of the robot, some changes of the main project have been done: for instance, a circular plate was added to the upper part of the external ring. This plate not only protects the arms of the sensor, but allows the sensor to best suite to every end-effector size.

FIGURE 4. The force – torque sensor built in the laboratory

3 CONCLUSIONS

A low cost force-torque sensor was realized, with interesting specification in terms of structural solution and of used materials. The maximal dimension value is 110 mm and the weight value is less the 410 g. Material and manufacturing solutions allow to maintain a very low production cost. This sensor, still under testing, will be used in spine surgery application. The sensor suites every specification requested in this type of applications.

ACKNOWLEDGEMENTS

The authors would like to thank Mr. Ruben Dal Fabbro and Mr. Elvio Castellarin for their help in designing and building the sensor.

REFERENCES

1. Sciavicco, L., Siciliano, B., (1995), Robotica industriale Modellistica e controllo di Manipolatori, McGraw-Hill.
2. Klearly K., Clifford M., Freedman M., Zeng J., Mun S. K. Watson V. and Henderson F., Technology improvements for image-guided and minimally invasive spine procedures, Transactions on Information Technology in Biomedicine IEEE, Vol. 6, 2002, pp. 249-261.

3. Wider B. H., Future of spine care, CNI Review Vol. 12, 2001, Issue on Spine and Spinal Cord Surgery.
4. Glossop N. and Hu R., Assessment of vertebral body motion during spine surgery, Spine, Vol. 22, 1997, pp. 903-909.
5. Simon D., Intra-operative position sensing and tracking devices, Proceedings of the First Joint CVRMed/MRCAS Conference, 1997, pp. 62-64.
6. Peters T., Image-guided surgery: From x-rays to virtual reality, Accepted for publication in Computer Methods in Biomechanics and Biomedical Engineering, OPA Amsterdam, 2001.
7. De Waal Malefijt J., Image-guided surgery of the spine, MedicaMundi, Vol. 42(1), 1998, pp. 38–43.
8. Santos-Munn J. J., Peshkin M. A., Mirkovic S., Stulberg S. D. and Kienzle III T. C., A stereotactic/robotic system for pedicle screw placement, Proceedings of the Medicine Meets Virtual Reality III Conference, 1995, San Diego, CA.
9. Boschetti G., Gallina P., Rosati G., Rossi A. and Zanotto V., A novel approach to haptic/telerobotic spine surgery, Submitted for Publication to the 12th International Workshop on Robotics in Alpe-Adria-Danube Region RAAD'03, 2003, Cassino, Italy.
10. Carlisle B., Robot Mechanisms. Proceedings of the 2000 IEEE International Conference on Robotics & Automation, 2000, pp. 701-708.
11. Gasparetto A. and Biason A., Design of a robotic wrist for biomedical applications, Proceedings of AMST '02, 2002, pp. 545-552, Udine, Italy.

α-SiAlONs CUTTING TOOL DEVELOPMENT, CHARACTERIZATION AND APPLICATION IN MACHINING OF COMPACTED CAST IRON

[1]J. V.C. Souza, [2]C. A. Kelly, [2]M. R. V. Moreira, [2]C. Santos, [2]M. V. Ribeiro

[1] Departament of Materials and Technology, Faculty of Guaratinguetá Engineering, Brazil
[2] Departament of Materials Engineering, Faculty of Chemical Engineering of Lorena, Brazil
[3] Departament of Mechanical and Aerospatial Engineering, Aerospace Technical Institute, Brazil
[4] Analysis of Rare Materials, Aerospace Technical Center, Brazil

KEYWORDS: SiAlONs Cutting Tools, Machining, Compacted Cast Iron.

ABSTRACT. Recently, scientific and technological development, allowed new materials appearance aiming at to applications in several fields of the engineering, with the agreement between development and environment. Due of this condition, compacted cast iron is detaching in the production of motor blocks to diesel, being limited for its machinability. Therefore, this work had as objective to produze, to characterize and to apply SiAlONs cutting tools in machining of this material. A powder mixture constituted by 82.86 wt % α-Si_3N_4, 10.63 wt % AlN and 6.51 wt % Y_2O_3 was homogenized, uniaxially and cold isostatically pressed, and sintered at 1900°C for 1h. After chacterization, cutting tools were submitted at the different machining conditions: Vc = 200, 350 and 500m/min, f = 0.20 mm/rot and ap = 0.50 mm. Properties obtained in SiAlONs cutting tools together with machining conditions showed higher performance to the found in literature.

1 INTRODUCTION

In last years, engine block to diesel, before manufactured with gray cast iron, it have been substituted for compacted cast iron, aiming at to improve performance of motors to the combustion and of this mode to decrease pollutant discharge, because of raising of the pressure picks in explosion chamber. For that such objectives are reached, there is need to increase to the resistance of material used for manufacture of the explosion chamber. Such condition is possible with use of compacted cast iron, that allows to obtain engine block to diesel with finer walls and for lighter consequence [1-3]. However, it is an extremely abrasive material and difficult machinability, condidtion that limits its application [4].

Due to its good physical and thermomechanical properties, silicon nitride ceramics (Si_3N_4) are promissing materials to be used in machinability of compacted cast iron. Such properties result of its strong covallent character, conferring him low self-diffusion coefficient [5,6], impeding of this mode obtaining of Si_3N_4 dense ceramics for solid phase sinbtering.

Published in: E. Kuljanic (Ed.) *Advanced Manufacturing Systems and Technology*,
CISM Courses and Lectures No. 486, Springer Wien New York, 2005.

However, dense Si_3N_4 ceramics can be obtained for liquid phase sintering by addition of sintering additives small amount, that facilitate the diffusional phenomena, decreasing the porosity of the material and consequently, improving the densification and mechanical properties of the Si_3N_4. Main additives used in the Si_3N_4 sintering are: AlN, Al_2O_3, Y_2O_3, SiO_2, CRE_2O_3 (yttrium and rare earth oxide mixed) or mixtures of these [7].

The Si_3N_4 shows a low resistance to the oxidation and creep if compared at other covalent ceramics, as silicon carbide, SiC. A way to improve such properties is use rich additives in Al, Y and O ions, that it tend to incorporate inside of the crystalline structure of the Si_3N_4, resulting in solid solutions, substitutional and/or interstitial, called SiAlONs. In this material type, during the sintering occurs partial of the α-Si_3N_4 grains in a liquid phase rich in Al, Y and O ions, which Al and O ions replace Si and N, respectively. Whereas, Y ion occupies interstitial void in structure, stabilizing of this way the α phase at high sintering temperatures [8].

Usually, ceramic cut tools are fragile in the cut edges, could happen chipping or fractures. In this case, the use of bevel edege has been recommended. Improvement in the manufacture technical and mechanical properties (hardness and fracture toughness) of the ceramic cut tools manufactred it already allows the use of positive inclination geometry. In Figure 1 [9] are shown tool cutting bit with tip angles more common in the play lateral surface.

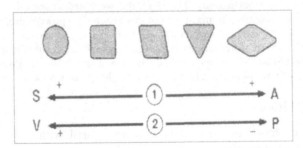

FIGURE 1. Tip angle of the cut tools [9]

This work had as objectives: to develop, to characterize and to apply α-SiAlON cutting tools in machining of the compacted cast iron, evaluating the performance of this tools with relationship to the changes in machining parameters.

2 EXPERIMENTAL PROCEDURE

2.1 MATERIALS

The materials used in this work were: α-Si_3N_4 (99.9 % - H. C. Starck), AlN and Y_2O_3 (Fine grade C – H. C. Starck) and nitrogen (type B50 – Air Liquid Brasil S/A)

2.2 METHOD

The powder batches were prepared in a planetary mill for 3 hours using isopropilic alcohol as vehicle. The suspensions were dried and subsequently sieved. The overall compositions of the powders mixture, as well as its designation are represented in Table 1.

TABLE 1. Composition and green relative density of powder mixture used

Code	Composition (wt %)			Green relative density (% of theoretical)
	α-Si$_3$N$_4$	AlN	Y$_2$O$_3$	
SNAY	82.86	10.63	6.51	59.98 ± 0.20

Green bodies were fabricated by uniaxial pressing under a pressure of 100 MPa and subsequent isostatic pressing under a pressure of 300 MPa. After compactation, samples showed 13.36x13.46x7.5 mm dimensions. The green densities of the compacts was determined geometrically. Prior to sintering, the samples were involved by a 70% Si$_3$N$_4$ + 30% BN as powder bed, then placed in a furnace with a graphite heating element (Thermal Technology Inc. type 1000-4560-FP20) in nitrogen atmosphere. The heating rate employed was 25°C/min up to a maximum sintering temperature at 1900°C, with a holding time for 1 hour. The cooling rate was the same as during heating-up.

The relative density of the sintered samples was determined by the immersion method in distilled water. The weight loss were determined by before and after sintering measurements. The phase analysis were determined for X-ray diffraction with CuKα radiation and scanning speed equal to 0.02°/s.

Samples grinded and polished were submitted to chemical etching in a NaOH:KOH mixture (1:1 at 500°C/10 minutes) to reveal the microstructure. The micrographs of sintered samples were obtained by the use of Scanning Electron Microscopy (SEM).

The hardness was determined by Vicker's indentations under an applied load of 20N for 30 s. For statistical reasons, 20 indentations have been made per sample. The fracture toughness has been determined by the measurement of the crack length created by the Vicker's indentations. The calculation of the fracture toughness values was done by the relation proposed by Evans et al., valid for Palmqvist type cracks [10,11].

After characterisation, samples were grinned in diamond wheel into 13x13x4.8 mm dimensions, with negative bevel of 20° and thickness of bevel equal to 0.8 m, conform ISO 1832 standard.

Some properties and micrograph of compacted cast iron used in the machining tests are listed in Table 2 [2] and Figure 2 [2], respectively. A compacted cast iron bar cylindrical shape with 400 mm length, 120 and 60 mm inner and outside diameter, respectively, has been used.

TABLE 2. Properties of the Ti6Al4V alloy used in this work [2]

Tensile strength [MPa]	Modulus of elasticity [GPa]	Fatigue strength [MPa]	Thermal conductivity [W/(mK)]	Hardness [HB]
500	140	205	35	225

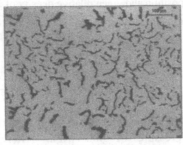

FIGURE 2. Spatial shape of the graphite type of compacted cast iron [2]

Machining tests were realized no fluid cut in an lathe CNC (Romi, mod. Centur 30D). For measure the temperature of piece-cut tool interface was used a infrared radiation pyrometer. The data used in the machining tests are shown in Table 3. Wear analysis of the cut tool were determined for Ra and Ry; and flank wear maximum (VBmax), respectively. Where Ra and Ry are measures realized on machined surface of the test specimen. A flank wear of 0.6 mm (ISO 3685) and abrupt variation of Ry has been used as end of tool life criterion.

TABLE 3. Conditions and cut parameters used in machining tests

Machining conditions	Cut parameters			
	Vc (m/min)	f (mm/rot)	ap (mm)	Lc$_{max}$ (m)
A	200			1658
B	350	0.2	0.5	1076
C	500			1052
D	200		1.0	1446

Obs: Vc= cutting speed, f=feed, ap=cutting depth

After machining tests, surfaces of the compacted cast iron were charcterized for roughness measures (Surftest SJ-201 – Mitutoyo), whereas for to analyze surfaces of the cut tool was used scanning electron microscopy (LEO-1450 VP).

3 RESULTS AND DISCUSSION

3.1 CUTTING TOOLS MANUFACTURED

For to obtain ceramics cutting tools is necessary to combine high relative density and good mechanical properties as: fracture toughness and hardness, that are directly related to the microstructure of material.

National ceramics cutting tools manufacture applied to specific cases it is a reality. But, ceramic cutting tools obtaning with high hardness and good wear resistance usually has been accompanied for higher brittleness, if compared to the high-speed steel. Such condition demands an accurate control of machining parameters, aiming at to determine the cut optimum states.

Aspect of a molded SiAlONs cutting tools and its shaving incidence and exit surfaces, before machining tests, has been shown in Figure 3.

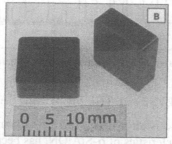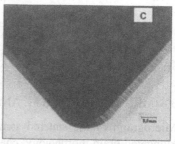

FIGURE 3. Aspects of molded SiAlONs cutting tool and its respectives shaving incidence and exit surfaces, before to the machining tests: (a) produced tool cutting bit, (b) molded tool cutting bit and (c) tool cutting bit with surface finish

3.2 RELATIVE DENSITY AND LINEAR SHRINKAGE

Relative density and linear shrinkage of the sintered samples has been shown in Table 4. Sintering conditions used, aided by the additive system provided the obtaining of a ceramic with high linear shrinkage and relative density. Formed liquid phase amount during sintering is probably another condition that allowed to obtain a ceramic with such properties. Because, it intensified particles rearrangement mechanism and α-Si$_3$N$_4$ grains into α-SiAlONs and β-Si$_3$N$_4$ reprecepitation-solution process.

TABLE 4. Relative density and linear shrikage of the sintered samples

Code	Physical properties	
	Relative density (% of theoretical)	Linear shrinkage (%)
SNAY	97.86	15.26

3.3 PHASE ANALYSIS OF THE SINTERED SAMPLES

X-ray diffraction patterns of the sintered samples are shown in Figure 4. Observing the results shown in this figure, presence of α-SiAlON, β-Si$_3$N$_4$ and Y$_2$Si$_3$N$_4$O$_3$ phases has been noted. However, with predominace of α-SiAlON phase, showing the effectiveness of additive system as stabilization of α-Si$_3$N$_4$ phase and transformation to α-SiAlON.

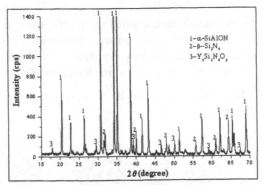

FIGURE 4. X-ray diffraction patterns sample sintered at 1900°C

3.4 MICROSTRUCTURAL ANALYSIS OF THE SINTERED SAMPLES

Micrograph of the sintered sample is shown in Figure 5. A homogeneous microstructure with elongated grain morphology, characteristics of α-SiAlON, has been observed, conform also shown X-ray diffraction patterns.

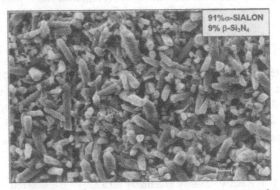

FIGURE 5. Micrograph of the sintered sample at 1900°C

3.5 HARDNESS AND FRACTURE TOUGHNESS

Fracture toughness and hardness of the sintered samples are shown in Table 5. High hardness values are related to predominance of α-SiAlON phase, which shows hardness higher to the β-Si_3N_4. Whereas, fracture toughness results are directly related to the α-SiAlON and β-Si_3N_4 amount, that can activate toughening mechanisms, depending of aspect ratio (length:diameter) of the presents phases.

TABLE 5. Hardness and fracture toughness of the sintered samples

Code	Mechanical properties	
	Hardness (GPa)	Fracture toughness (MPa.m$^{1/2}$)
SNAY	21.36± 0.21	5.28 ± 0.12

3.6 INFLUENCE OF THE MACHINING PARAMETERS IN WEAR OF CUTTING TOOL AND SURFACE OF THE PIECE MACHINED

Influence of the different machining conditions in flank wear of compacted cast iron is shown in Figure 6. In this figure, considering ap=0.5, increase of the flank wear with the cutting speed has been noted. Therefore, the same behavior also has been observed with increase of the cutting depth (ap). These results can be explained by the greater impact frequency in cutting edge. Those phenomenon also has been shown in the work developed for Diniz [12].

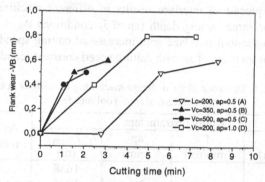

FIGURE 6. Influence of the different machining conditions in flank wear
of compacted cast iron

The influence of cutting length and different machining conditions in the roughness (Ra and Ry) has been noted in Figure 7. Increase of cutting speed and cutting speed contributes for decrease of Ra and Ry, due to the cyclical efforts intensification and temperature abrupt variataions, that facilitate material removal because rise of the friction between test specimen and cutting tool, improving of this mode finish surface. However, for condition D, the roughness (Ra and Ry) practically is maintained constant for cutting length higher to 500 m. In conditions B e C, machining tests has been interrupted for Lc=1052m and Lc=1076 m, respectively, because of wear appearance on surface of the cutting tool. This wear might have been caused by imperfections presents in cutting edge, due to the own preparation process of cutting tool; or same proceeding of the own machining tests.

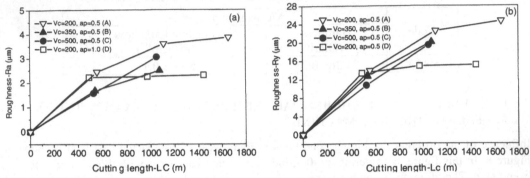

FIGURE 7. Influence of the different machining conditions in roughness: (a) Ra and (b) Ry of compacted cast iron

Avarage temperature of piece-tool interface results in different machining conditions has been observed in Table 6. For same cutting depth (ap=0.5, conditions A, B and C), increase of the avarage temperature of piece-tool interface with increase of cutting speed is noted. However, this behavior also is shown for increase of ap with cutting speed constant (conditions A and D).

TABLE 6. Influence of the different machining conditions in avarage temperature of iece-Tool interface

Machining conditions	Cutting parameters				Average temperature of piece-tool interface (°C)
	V_c (m/min)	f (mm/rot)	ap (mm)	Lc_{max} (m)	
A	200			1658	770
B	350	0.2	0.5	1076	780
C	500			1052	892
D	200		1.0	1446	838

3.7 MICROGRAPHS OF THE CUTTING TOOLS AFTER MACHINING TESTS

Micrographs of the α-SiAlONs cutting tools, after different machining conditions, has been shown in Figure 8. For machining conditions with same cutting depth (ap=0.5), increase of the crater and flank wear has been noted with increase of cutting speed. Similar behavior also has been shown in machining conditions with constant cutting speed, but cutting depth variable.

The condition C showed early wear, with minor cutting time (tc=2,11 min). In this machining condition, crater wear in face and abrasive wear in flank has been noted (See Figure 8c). This behavior can be justified for related problems to the machine-tool or same inclusions in compacted cast iron. Those results are in agreement with them obtained for Leuze, et al [13].

Whereas, condition A that presented minor wear of the cutting tool, proportioning of this mode, cutting time (tc=8.29 min) higher to the found for conditions B and C (condition B, tc=3.08 min; condition C, tc=2.11 min). Crater wear and flank abrasive wear in the cutting edge has been observed in this condition (See Figure 8a). In aggrement with results shown in Figure 8a, can be concluded that for minor cutting speed, the wear can be less intense.

For condition D, cutting tool showed showed a quick and progressive wear, with cutting time varying of 4.88 to 7.23 min, but with wear higher to the recommended by the ISO 3685 standard (VB=0.6 mm). Crater wear, flank abrasive wear in the cutting edge and chipping has been noted in this condition (See Figure D).

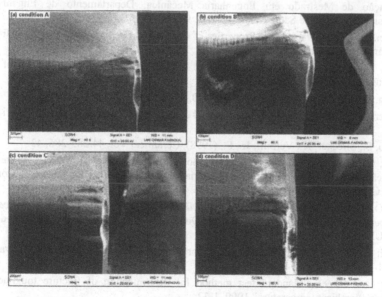

FIGURE 8. Micrographs of the SiAlONs cutting tools obtained of the

4   CONCLUSIONS

Results obtained in this work proved to be possible the structural ceramics obtaining, specifically α-SiAlONs ceramics with good physical and mechnical properties appropriate to the application as cutting tools for compacted cast iron machining, being controlled the machining parameters.

Machining parameters employed led α-SiAlONs cutting tools to an excessive wear, even so with results higher to the found in literature.

Ductile shaving formation during machining tests of compacted cast iron, due to the contact greater time between test specimen and cutting tool had contributed for increase of friction, that added to grindability of the shaving and excessive heat intensified wear of cutting tool.

ACKNOWLEDGEMENTS

The authors would like to express their gratitude to CNPq and CAPES for financial support.

REFERENCES

1.   Guesser, L. W.; Guedes, L. C., (1997), Desenvolvimentos Recentes em Ferros Fundidos Aplicados à Indústria Automobilística, Anais do IX Simpósio de Engenharia Automotiva.
2.   Mocellin, F. *"Avaliação da Usinabilidade do Ferro Fundido Vermicular em Ensaios de Furação."*, dissertação de Mestrado em Engenharia Mecânica, Departamento de Engenharia Mecânica, Universidade Federal de Santa Catarina, 2002, 94 p.
3.   Dawson, S., (1994), Operational Properties of Compacted Graphite Iron: Feedback from ongoing test programmes. SinterCast S.A, Switzerland.
4.   Reuter, U. et al, (1999), Wear Mechanisms in High-Speed Machining, Annals of the Compacted Graphite Iron – Machining Workshop.
5.   Witting, H., (2002), Máquinas e metais, n. 440, 156-165.
6.   Dressler, W., Riedel, R. (1997), International Journal of Refractary & Hards Materials, Vol. 15/1-3, 13-47.
7.   Ribeiro, S., Strecker, K., Vernilli, Jr., F., (1998), Cerâmica, Vol. 44/285-286, 43.
8.   Santos, C., Silva, C.R.M., Strecker, K., Barbosa, M. J. R., Silva, O. M. M., Piorino, N. F., (2004), Avaliação da Resistência à Fluência de Materiais Cerâmicos Utilizados Ensaio de Compressão, Anais do 59° Congresso Anual da ABM.
9.   Silva, O.M.M. *"Processamento e caracterização do nitreto de silício aditivado com carbonato de ítrio e concentrado de terras raras"*, Tese de Doutorado, Faculdade de Engenharia de Guaratinguetá (FEG/UNESP), 2000, 169 p.
10.   ASTM: C1327-99, "Standard test method for vickers indentation hardness of advanced ceramics", 1999, 1-8.
11.   ASTM: C-1421-99, "Standard test method for determination of fracture toughness of advanced ceramics at ambient temperature",1999, 1-32.
12.   Diniz, A. E., Marcondes, F. C., Coppini, N. L., (2000), Tecnologia da Usinagem dos Metais, segunda

edição, Artliber.

13. Leuze, P., (2000), High Productivity Carbide Boring Tools for Roughing and Finishing CGI. Annals of the Compacted Graphite Iron – Machining Workshop.

β-SI₃N₄ CUTTING TOOLS OBTAINING FOR MACHINING OF THE GRAY CAST IRON

[1] J. V.C. Souza, [2] C. A. Kelly, [2] M. R. V. Moreira, [2] M. V. Ribeiro,
[3] M. A. Lanna, [4] O. M. M. Silva

[1] Departament of Materials and Technology, Faculty of Guaratinguetá Engineering, Brazil
[2] Departament of Materials Engineering, Faculty of Chemical Engineering of Lorena, Brazil
[3] Departament of Mechanical and Aerospatial Engineering, Aerospace Technical Institute, Brazil
[4] Analysis of Rare Materials, Aerospace Technical Center, Brazil

KEYWORDS: β-Si₃N₄ Cutting Tools, Machining, Gray Cast Iron.

ABSTRACT. Recently, Si₃N₄ has been used as cutting tools in several alloys machining. Such materials shows SiAlONs and β-Si₃N₄, that it has excellent thermomachanical properties. This work had as objetive to develop and to apply β-Si₃N₄ cutting tools in machining of gray cast iron, starting from a mixture composed by 78.30 wt % α-Si₃N₄, 1.00 wt % Al₂O₃, 14.40 wt % AlN, 3.15 wt % CeO₂ and 3.15 wt % Y₂O₃, that was pressed and sintered at 1900°C. After characterization, cutting tools showed, HV = 20.50 GPa, K_{IC} = 6.45 MPa.m$^{1/2}$, with 95 % β-Si₃N₄ phase. Then, it were submited to machining of gray cast iron using: 1ª step - Vc=180, 240, 300 and 360 m/min, f = 0.12mm/rot and ap = 1.00 mm; and 2ª step – Vc = 300m/min, f = 0.23, 0.33 and 0.40 mm/rot and ap = 1.00 mm. For 7500 m cutting length, results showed excellent performace.

1 INTRODUCTION

Scientific and technological development is allowing the new materials discovery aiming at to specific applications. In some of this applications, material to be used needs to present indispensable properties, as high wear resistance, good impact and compression resistance to high temperatures among others. This conditions usually are found in the machining process of industrialized materials, such as: titanium alloy (aerospace application), compacted cast iron and gray cast iron (automobile application) among others [1, 2].

Gray cast iron, chemically no shows difference significant if compared to the compacted and nodular cast iron [3]. The differences among those alloys is due to morphology types of its graphites, conferring them distinct physical and mechnical properties [4]. Graphite types of the gray cast iron lead that material to a promising application in the manufacture of combustion gas collector, disc brake, engine block to diesel and others, due to its characteristics of thermal conductivity [5].

Turning machining of this material is considered critic mainly in the beginning, that it occurs in impact extreme conditions, causing in the material elastic and plastic deformation, consequently resulting in shearing of the material and shavings formation [6]. Characteristics of chavings obtained during turning machining depends of graphite shape, that in case of the green cast iron present sharp flakes and interconnected creating a crack propagation plan at

Published in: E. Kuljanic (Ed.) *Advanced Manufacturing Systems and Technology*,
CISM Courses and Lectures No. 486, Springer Wien New York, 2005.

front of edge with advance of cutting tools. Material removal occurs due to the attrition of three bodies (tool-shaving-piece) contributing of this way for decreasing of useful life time of the cutting tool. In turning machining of the green cast iron, cutting tool is not in contact with material of piece, because a material portion is removed at front of the cutting tool, creating crater and material extracting of cutting tool, contributing for superficial finish not desired [7].

Recently, ceramic materials, mainly covalent ceramics, as: Cubic Boron Nitride (CBN), silicon nitride (Si_3N_4), and others has been used in machining process of this material, due to its properties [8]. Dense covalent ceramics are difficult of be obtained for solid phase sintering, because of the its low self-diffusion coefficient [9]. However, dense covalent ceramics can be obtained for liquid phase sintering by addition of sintering additives small amount, that facilitate the diffusional phenomena, decreasing the porosity of the material and consequently, improving the densification and mechanical properties. Main additives used in the Si_3N_4 sintering are: AlN, Al_2O_3, Y_2O_3, SiO_2, CRE_2O_3 (yttrium and rare earth oxide mixed) or mixtures of these [10].

The Si_3N_4 shows a low resistance to the oxidation and creep if compared at other covalent ceramics, as silicon carbide, SiC [11]. A way to improve such properties is use rich additives in Al, Y and O ions, that it tend to incorporate inside of the crystalline structure of the Si_3N_4, resulting in solid solutions, substitutional and/or interstitial, called SiAlONs [12]. In this material type, during the sintering occurs partial of the α-Si_3N_4 grains in a liquid phase rich in Al, Y and O ions, which Al and O ions replace Si and N, respectively. Whereas, Y ion occupies interstitial void in structure, stabilizing of this way the α phase at high sintering temperatures [13].

Usually, ceramic cutting tools are fragile in the cutting edges, could happen chipping or fractures. In this case, the use of bevel edge has been recommended. Improvement in the manufacture technical and mechanical properties (hardness and fracture toughness) of the ceramic cutting tools manufactued it already allows the use of positive inclination geometry. In Figure 1 [14] are shown cutting bit with tip angles more common in the exit lateral surface.

The work had as objective to produze and use β-Si_3N_4 cutting tools in machining of the gray cast iron, using different machining parameters.

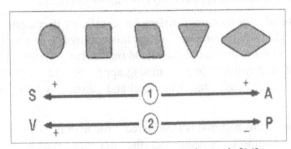

FIGURE 1. Tip angle of the cutting tools [14]

2 EXPERIMENTAL PROCEDURE

2.1 MATERIALS

The materials used in this work were: α-Si$_3$N$_4$ (99.9 % - H. C. Starck), AlN and Y$_2$O$_3$ (Fine grade C – H. C. Starck), Al$_2$O$_3$ (type AS 250 KC – Baikalox), CeO$_2$ (high purity – H. C. Starck – Germany) and nitrogen (type B50 – Air Liquid Brasil S/A).

2.2 METHOD

The powder batch was prepared in a planetary mill for 4 hours using isopropilic alcohol as vehicle. The suspension was dried and subsequently sieved. The overall composition of the powder mixture, as well as its designation are represented in Table 1.

TABLE 1. Composition and green relative density of powder mixture used

Code	Composition (wt %)					Green relative density (% of theoretical)
	α-Si$_3$N$_4$	AlN	Y$_2$O$_3$	Al$_2$O$_3$	CeO$_2$	
SNYC	78.30	14.40	3.15	1.00	3.15	61.02 ± 0.20

Green bodies were fabricated by uniaxial pressing under a pressure of 100 MPa and subsequent isostatic pressing under a pressure of 300 MPa. After compactation, samples showed 16.36x16.36x7.5 mm dimensions. The green densities of the compacts was determined geometrically. Prior to sintering, the samples were involved by a 70% Si$_3$N$_4$ + 30% BN as powder bed, then placed in a furnace with a graphite heating element (Thermal Technology Inc. type 1000-4560-FP20) in nitrogen atmosphere. The heating rate employed was 15°C/min up to a maximum sintering temperature at 1900°C, with a holding time for 1 hour. The cooling rate was 25°C/min.

The relative density of the sintered samples was determined by the Arquimedes method. The weight loss were determined by before and after sintering measurements. The phase analysis were determined for X-ray diffraction with CuKα radiation and scanning speed equal to 0.02°/s.

Samples grinded and polished were submitted to chemical etching, in a NaOH:KOH mixture (1:1 at 500°C/10 minutes) to reveal the microstructure. The micrographs of sintered samples were obtained by the use of Scanning Electron Microscopy (SEM).

The hardness was determined by Vicker's indentations under an applied load of 20N for 30 s. For statistical reasons, 20 indentations have been made for sample. The fracture toughness has been determined by the measurement of the crack length created by the Vicker's indentations. The calculation of the fracture toughness values was done by the relation proposed by Evans et al., valid for Palmqvist type cracks [15,16].

After characterisation, samples were grinned in diamond wheel into 13x13x4.8 mm dimensions, with negative bevel of 20° and thickness of bevel equal to 0.08 m, conform ISO 1832 standard.

Some properties and micrograph of gray cast iron used in the machining tests are listed in Table 2 [17] and Figure 2 [6], respectively. A gray cast iron bar in cylindrical shape into 300 mm length, external diameter equal to 105 mm has been used.

TABLE 2. Properties of the gray cast iron used in this work [17]

Tensile strength [MPa]	Elasticity Modulus [GPa]	Fatigue strength [MPa]	Thermal conductivity [W/(mK)]	Hardness [HB]
235	110	100	48	200

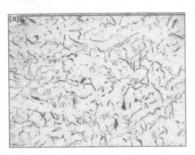

FIGURE 2. Spatial shape of the graphite type of gray cast iron: (a) no etching and (b) chemically etched for Nital 2% [6]

Machining tests were realized no fluid cut in an lathe CNC (Romi, mod. Centur 30D). For measure the temperature of piece-cutting tool interface was used a infrared radiation pyrometer. The data used in the machining tests are shown in Table 3. Wear analysis of the cutting tool were determined for roughness measurements (Ra, Rq and Ry), flank wear maximum (VBmax) and F (N), respectively. Where, Ra and Ry are measures realized on machined surface of the test specimen. Whereas, F (N) is force applied to the test specimen and it depends of the feed (f) and cutting speed (Vc). A flank wear of 0.6 mm (ISO 3685) and abrupt variation of Ry has been used as end of tool life criterion. For each 1500 m were realized measurements with objetive of to monitor behavior of cutting tools during machining tests.

TABLE 3. Conditions and cut parameters used in machining tests

Machining conditions	Cutting parameters			
	Vc (m/min)	f (mm/rot)	ap (mm)	Lc$_{max}$ (m)
1ᵃ step	180			
A				
B	240	0.12		
C	300			
D	360		1.00	7500
2ᵃ step		0.23		
E				
F*	300	0.32		
G		0.40		
H		0.50		

Obs: Vc = cutting speed, f = feed, ap = cutting depth.
* Condition used for wear analysis of the cutting tool.

After machining tests, surfaces of the compacted cast iron were charcterized for roughness measures (Surftest SJ-201 – Mitutoyo), whereas for to analyze surfaces of the cutting tool was used scanning electron microscopy (LEO-1450 VP).

3 RESULTS AND DISCUSSION

3.1 RELATIVE DENSITY AND WEIGHT LOSS

Relative density, linear shrinkage and weight loss results of the sintered samples has been shown in Table 4. High densification and linear shrinkage demonstrate great influence of liquid phase formed on intensification of sintering mechanisms, as particles rearrangement and solution-reprecipitation, main responsible for decrease of porosity and consequent increase of relative density. Those results attend in agreement with obtained in the work developed for Santos, et al [18], that developed Si₃N₄ ceramics additived with AlN/Y₂O₃ and/or AlN/CRE₂O₃.

TABLE 4. Linear shrinkage, relative density and weight loss of the sintered samples

Code	Physical properties		
	Linear shrinkage (%)	Relative density (% of theoretical)	Weight loss (%)
SNYC	16.32 ± 0.12	98.54 ± 0.26	2.66 ± 0.16

3.2 PHASE ANALYSIS OF THE SINTERED SAMPLES

X-ray diffraction results of the sintered samples (Figure 3) indicated only presence of the β-Si₃N₄, showing full α into β-Si₃N₄ transformation. The presence of intergranular phase not has been observed, probably the same has remained in the amorphous form.

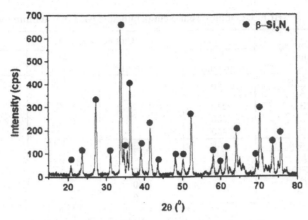

FIGURE 3. X-ray diffraction of the sintered sample

3.3 CUTTING TOOLS MANUFACTURED AND MICROSTRUCTURE OF THE SINTERED SAMPLE

Cutting tool manufactured and micrograph of the sintered sample are shown in Figure 3. Analyzing the result presented in Figure 3b, a heterogeneous microstructure, composed by β-Si_3N_4 eleongated grains with different microstructural thickening has been noted, conform X-ray diffraction results. The same microstructural behavior also has been shown in literature [19].

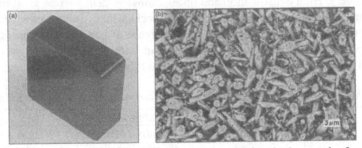

FIGURE 3. Aspects of molded ceramic cutting tool (a) and micrograph of the sintered sample at 1900°C (b)

3.4 HARDNESS AND FRACTURE TOUGHNESS

Fracture toughness and hardness of the sintered samples are shown in Table 5. High hardness values are related to high densification and low weight loss of the sintered samples. Whereas fracture toughness is explained for the presence of a morphology composed by eleongated grains, that improve toughening mechanism, as crack bridging, crack deflection, amomg others [20].

TABLE 5. Hardness and fracture toughness of the sintered samples

Code	Mechanical properties	
	Hardness (GPa)	Fracture toughness (MPa.m$^{1/2}$)
SNYC	20.50± 0.31	6.45 ± 0.22

3.5 INFLUENCE OF THE MACHINING PARAMETERS IN WEAR OF CUTTING TOOL AND SURFACE OF THE PIECE MACHINED

Influence of feed (f) and cuttting speed (Vc) in the force (F) applied for cutting tool on test specimen during machining of gray cast iron is shown in Figure 4. Analyzing the results presents in Figure 4a, increase of machining forces (frictional, passive and feed) has been observed with rise of the f and Vc, but feed force tends stabilizing for feed (f) superior values. This phenomenon are explained by the increase of cut area, requiring of this mode higher force (or power) for to shear test specimen into cutting zone. But, the increase cut force no is directly proportional to the increase of feed, because of decrease in cut specific press [21].

Whereas evaluating Figure 4b, a quick increase of machining force with rise of cutting speed for values until 300 m/min has been shown. Starting from this value, machining force decrease with rise of cut speed. Because, increase of heat caused in shearing zone promotes a decrease in the resistance of the material, allowing of the graphite present in gray cast iron into pasty fluid transformation and of this mode improving the slip of cutting tool on test specimen and protecting the same of possible impacts with surface of cutting tool. Therefore, it can cocluded that increase of the cutting speed improves the performance of cutting tool analyzed during machining of gray cast iron.

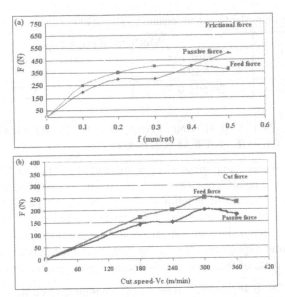

FIGURE 4. Influence of feed (f) and cut speed (Vc) in force (F) applied to the test specimen
during machining of gray cast iron

Figure 5 shows influence of feed (f) and cutting speed (Vc) in the temperature of piece-tool inteface during maching of gray cast iron. Analizyng the results shwon in Figure 5a, increase temperature of piece-tool interface with increase of feed until 0.1 mm/rot has been observed. Whereas, a diminution of this property with increase of feed is shown for values above of 0.1 mm/rot. In machining of the gray cast iron, this behavior is expected, due to the increase of the heat dispersion area, promoted for the increase of feed [22].

In Figure 5b, temperature of piece-tool interface increase with rise of cutting speed. But, for cutting speed upper to 180 m/min, curve temperature versus cutting speed suffers a small decrease in its inclination, tending to stabilize for high cutting speed. Such behavior also has been noted in literature [21].

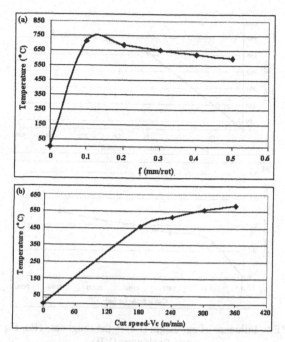

FIGURE 5. Influence of feed (f) and cut speed (Vc) in temperature of piece-cutting tool interface during machining of gray cast iron

Influence of cutting length (Lc) in the roughness (Ra, Rq and Rt) and flank wear (VB) has been presented in Figure 6. Evaluating results in Figure 6a, roughness showed small alterations during machining test, where largest of this property values occurred at the beginning of process. In cut interval among 4500 and 7500 m, roughness stabilized and keeping itself constant, because of the natural accommodation between cutting tool and test specimen, behavior justified for literature [23].

For cutting length until 1500 m, a behavior practically linear of the flank wear has been noted. However, decrease of intensity of the wear for cutting length higher to 1500 m has been shown, pointing out that cutting tools analyzed no didn't reach end of life criterion recommended by ISO 3685/93 standard (0.3 mm) [24].

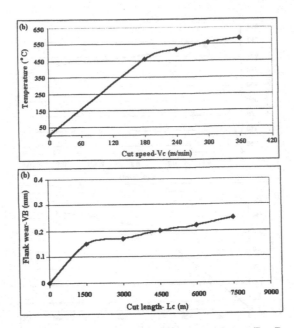

FIGURE 6. Influence of cut length (Lc) in the roughness (Ra, Rq and Rt)
and flank wear (VB)

3.6 MICROGARPHS OF THE CUTTTING TOOLS

The condition used for wear analysis of the cutting tool were found in decrease of feed force, conform results shown in Figure 4, which the same it reduce for Vc=300 m/min (see Figure 4a) and f=0.32 mm/rot (see Figure 4b). However, in Figure 7 are shown micrographs of cutting tool, with Vc=300 m/min, f = 0.32 mm/rot and ap= 1mm (condition F*), modifying cut length to each 1500 m. Analyzing the micrography present in this, increase of crater and flank wera has been noted with increase of cutting length. This behavior probably can be justified for the material exit of the test specimen, causing an attrition between those particles and cut edge. The material diffusion of the test specimen for into cutting tool with increase of the temperature piece-tool interface, might also have contributed for occurrence of such wear.

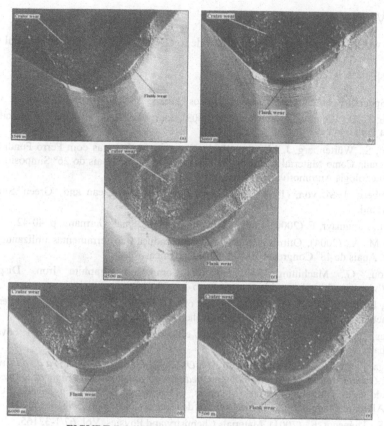

FIGURE 7. Micrographs of the β-Si₃N₄ cutting tool

4 CONCLUSIONS

Obtained results in this work showed to be possible to obtain structural ceramics, specifically Si_3N_4 ceramics with good physical properties (relative density: 98 % of theoretical) and mechanical (hardness: 20 GPa and fracture toughness approximately 6 MPa.m$^{1/2}$), that allowed a excellent performance of the cutting tool during different machining conditions.

The low flank wear found in this work, conform literature and ISO 3685/93 standard (VB=0.3 mm), it assure hereafter application of this cutting tool in more severe conditions during machining of gray cast iron.

The increase of the cutting speed provides a reduction in cutting forces (passive, feed and frictional), however with sensitive increase of the crater and flank wear.

ACKNOWLEDGEMENTS

The authors would like to express their gratitude to CNPq and CAPES for financial support.

REFERENCES

1. ASM. Speciality Handbook: Cast Irons. Estados Unidos: ASM International, 1996, p. 33-267.

2. Ezugwu, E. O.; Bonney, J.; Yamane, Y., (2003), Journal of materials processing and technology, Vol 134, 233 - 253.

3. Dawson, S., Würtenberg, J. M. von, (1993), Experiências Práticas com Ferro Fundido de Grafita Compactada Como Material para Componentes Automotivos, Anais do 26° Simpósio Internacional sobre Tecnologia Automotiva e Automação.

4. Würtemberg, J. M. von, (1994), The Diesel Engine: Lean, Clean and "Green".SinterCast S.A, Switzerland.

5. Hick, H., Langmayr, F, (2000), Engine Tecnology International, Germany, p. 40-42.

6. Lanna, M. A., (2004), Otimização de custos de produção de ferramentas utilizando sinterização normal. Anais do 48° Congresso Brasileiro de Cerâmicas.

7. Georgiou, G., Machining Solutions for Compacted Graphite Iron. Disponível em: <htpp://www.machineshopguide.com> Acesso em: 30/01/2002.

8. Reuter, U.; et al., (2000), The Wear Process of CGI Cutting Machining Developments. Anais do Compacted Graphite Iron – Machining Workshop (Darmstadt).

9. Devezas, T., (1985), Cerâmicos especiais estruturais: 2° Parte: cerâmicos covalentes, ITA Engenharia, vol 6/4.

10. Gonzaga, R. "Influência da substituição da SiO_2 por Al_2O_3 e AlN na mistura de aditivos Y_2O_3/SiO_2 e CTR_2O_3/SiO_2 na microestrutura e propriedades mecânicas de cerâmicas a base de Si_3N_4", Mestrado em Engenharia de Materiais, Departamento de Engenharia de Materiais (DEMAR)/Faculdade de Engenharia Química de Lorena (Faenquil), 1998, 62 p.

11. Baud, S., Thévenot, F., (2001), Materials Chemistry and Physics, Vol. 67/1-3, 165.

12. DUTTA, S., (1982), Journal of the American Ceramic Society, Vol. 68/5, 2.

13. Shin, I.H., Kim, D.J., (2001), Materials Letters, Vol. 47, 329-333.

14. Silva, O.M.M. "Processamento e caracterização do nitreto de silício aditivado com carbonato de ítrio e concentrado de terras raras", Tese de Doutorado, Faculdade de Engenharia de Guaratinguetá (FEG/UNESP), 2000, 169 p.

15. ASTM: C1327-99, "Standard test method for vickers indentation hardness of advanced ceramics", 1999, 1-8.

16. ASTM: C-1421-99, "Standard test method for determination of fracture toughness of advanced ceramics at ambient temperature",1999, 1-32.

17 TECHNICAL ARTICLES, Mechanical Properties of Compacted Graphite Iron. Disponível em: http://www.castingsource.com/tech_art_graphite.asp. Acesso em: 29/08/2000

18. Santos, C., Strecker, K., Ribeiro S., (2000), Sinterização e propriedades mecânicas do nitreto de silício (Si_3N_4) aditivado com misturas de Al_2O_3/Y_2O_3 ou Al_2O_3/CTR_2O_3, Anais do 14° Congresso Brasileiro de Ciência e Engenharia de Materiais (14° CBECIMAT). 2702-2712.

19. Silva, V.A ."Fabricação de ferramentas de corte cerâmico à base de Si_3N_4 dopadas com adição de Y_2O_3, Al_2O_3 e CTR_2O_3, com modificação superficial dos sinterizados por aplicação de diamante-CVD.", Tese de doutorado, Instituto Técnico Aeroespacial (ITA), 1998, 166 p.

20. Santos, C., Baldacim, S. A., Silva, O. M. M., Silva, C. R. M., (2004), Materials Science Engineering A, Vol. 367/1-2, 312-316.

21. Lanna, M. A., Bello, A. A.AL., Souza, J. V. C., (2004), Avaliação das Tensões e Deformações em Ferramentas Cerâmicas de Nitreto de Silício, Anais do 48° Congresso Brasileiro de Cerâmica.

22. Vilella, R. C. *"Metodologia prática visando à otimização das condições de usinagem em células de fabricação"* Dissertação de Mestrado em Engenharia Mecânica, Universidade Estadual de Campinas (UNICAMP), 1998, 105 p.

23. Cunha, E.A., Ribeiro, M.V., (2004), Revista Máquinas e Metais, Vol. 40/457, 132-138.

24. Brandt, G., (1986), Wear, Vol. 112, 39-56.

STAINLESS STEELS
MACHINABILITY ASSESSMENT MODEL

R. Bertelli, R. Cristel, G. Melotti, T. Ceccon

Quality and Research Department, Acciaierie Valbruna S.p.A., Italy

KEYWORDS: Stainless Steel, Machinability, Chemistry and Metallurgical Effect.

ABSTRACT. In this paper a linear mathematical model to evaluate machinability performances of different stainless steel grades is presented. Remarks on the corrosion resistance of enhanced machinability grades of stainless steels are related to the basic metallurgy.

1 INTRODUCTION

Stainless steels yearly production reach 2,2% of the total output of steel in the world today. The growing rate trend for stainless steels is about 3-5%.

These figures give an idea of the importance of stainless steels as an investment material, with an increasing use, based on LCC (Life-Cycle-Cost) considerations. The required composition in order to provide corrosion resistance in stainless steels results in lower machinability than carbon and engineering steels.

See Figure 1 for a qualitative evaluation regarding stainless steels machinability.

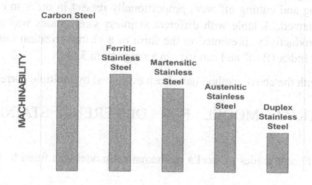

FIGURE 1. Qualitative assessment of various stainless steels families

The reasons why of poor machinability are:

• Mechanical properties of stainless steels show higher elongation to fracture and low yield strength in comparison to tensile strength. These characteristics result in ductile chips.

Published in: E. Kuljanic (Ed.) *Advanced Manufacturing Systems and Technology*,
CISM Courses and Lectures No. 486, Springer Wien New York, 2005.

• Thermal conductivity is 2-3 times lower than in carbon steels. Consequently the heat of cutting edge cannot be absorbed inside worked piece and the higher temperature of the edge decreases tool life.

• Wear coefficient of stainless steels is higher than in carbon steels. The martensitic grades contains a lot of hard carbides (chromium carbides) more abrasive than iron carbides. The austenitic grades show an higher work hardening rate and the resulting effect is an hard layer induced by the tool cutting edge of the worked piece.

The above considerations are related only to the stainless steel bars; tools technology and machining techniques are not involved.

2 BASIC GUIDELINES FOR CONSTRUCTING A MODEL FOR STAINLESS STEELS

Acciaierie Valbruna S.p.A. established continuous relationship with machining works in Europe since 20 years: there is a long job in technical assistance from the steel shop to the customers with established plans for improvements. On the other hand machining works are involved in tool performance's assessment plus machining parameters (cutting speed, feed, cutting fluids…). All this led to a series of improvements well comparable among different works and countries.

In our study experimental data are taken from an evaluation of machinability of different stainless steels grades on the field, i.e. results from production output homogenously collected.

To be more precise because the obtained pieces were quite different in final form a balance concerning the quantity of operations has been made in order to have a single data output based on 8 hour production average number of piece machined. This means that the operations like traversing, drilling, plunging and cutting off were proportionally reasset in order to compare the same volume of chips removed. A table with different stainless steel grades was obtained with homogenous data of productivity, presented in the form of a characterization index. This index is named "Productivity Index (P.I.)" and can vary in a range from 3 to 11.

The data collected with the above method have been evaluated by multiple correlation method.

3 MATHEMATICAL MODEL FOR DIFFERENT STAINLESS STEEL GRADES

In order to assess different grades of steel a mathematical model was fitted by a multiple regression analysis.

In this model steel's analytical content of elements, Yield Strength/Rupture Strenght ratio, Thermal conductivity and Microcleanliness effect have been considered as the independent variables and have been correlated with the Productivity Index (P.I.).

The formula obtained with the correlation method is:

$$P.I. = 11 \cdot (\%C) + 0,34 \cdot (\%Mn) - 1,9 \cdot (\%Si) + 0,03(\%Cr) + 0,08 \cdot (\%Ni)$$

$$- 0,025 \cdot (\%Mo) + 1,7 \cdot (\%S) - 0,24 \cdot (\%Cu) + 2,1 \cdot (\frac{Rp}{Rm}) + 0,025 \cdot (TC) + 1,5 \cdot (MI) \tag{1}$$

Rp/Rm = yield strength – tensile strength ratio
TC = thermal conductivity
MI = microcleanless evaluation by the ASTM E45 method A

The variables limitation in using this formula are:

C ≤ 0,25%
Mn ≤ 3%
Si ≤ 1%
Cr = 11 - 26%
Ni ≤ 20%
Mo ≤ 6%
S ≤ 0,40%
Cu ≤ 3,5%
Rp/Rm ≤ 0,98%

The correspondence between experimental data and calculated output from the model is 77%, with an estimation error of 0,76.

The diagram of Figure 2 shows the correspondence between observed results and model predicted results:

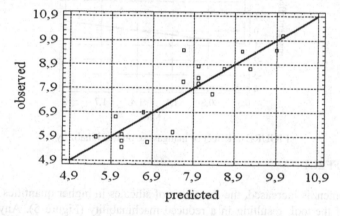

FIGURE 2. Corrispondence between observed and model predicted results

4 SINGLE STEP ANALYSIS OF THE PARAMETERS

The influence of the single elements of the chemistry is shown in the following diagrams.

4.1 CARBON EFFECT

The effect of carbon on machinability is quite directly proportional up to 0,20% (Figure 3). We have no data over 0,2%, but we think a lack of proportionality could happen over 0,2%: this because of increasing in hardness and quantity of non dissolved chromium carbides that would determine an increase tool wear.

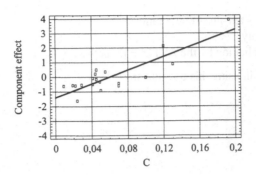

FIGURE 3. Effect of carbon on machinability

4.2 MANGANESE EFFECT

This element has a small positive effect thanks to the increased quantity of manganese sulfides (MnS) that have a lubrication effect on the cutting edge (Figure 4).

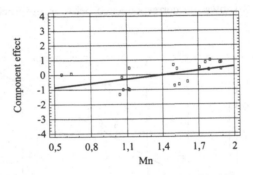

FIGURE 4. Effect of manganese on machinability

4.3 SILICON EFFECT

When silicon content is increased, the formation of silicates in higher quantities determinates an abrasive wear of the tool, resulting in a reduced machinability (Figure 5). Anyway, this small negative effect is compensated by the calcium treatment that envelops the desired increased quantity of inclusions. The aim of silicon-calcium treatment is, in fact, to create complex inclusions having a composition inside the field of anorthites or pseudowallastonite in the ternary diagram $Al_2O_3 - CaO - SiO_2$ (Figure 6). This fabrication technology shows a beneficial effect on the machinability and is mainly applied both austenitic grades and some martensitic grades. In Figure 7 a typical inclusion is analyzed.

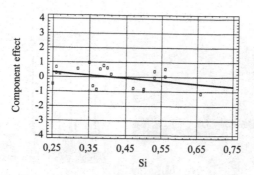

FIGURE 5. Effect of silicon on machinability

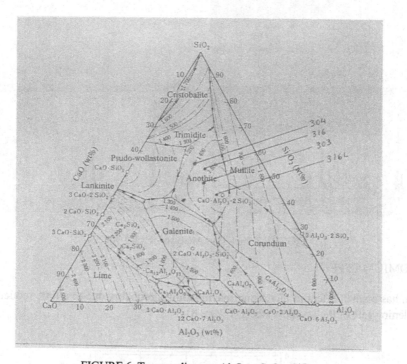

FIGURE 6. Ternary diagram $Al_2O_3 - CaO - SiO_2$

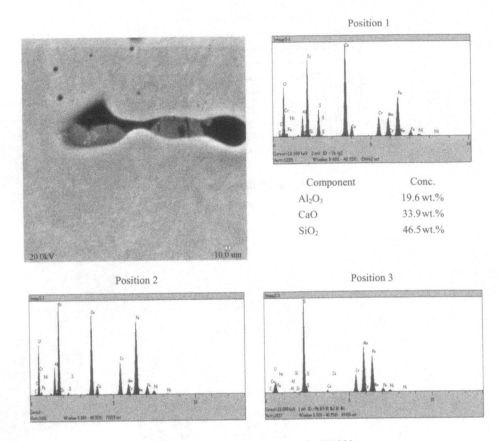

FIGURE 7. Inclusion Analysis of AISI 303

Component	Conc.
Al_2O_3	19.6 wt.%
CaO	33.9 wt.%
SiO_2	46.5 wt.%

4.4 CHROMIUM EFFECT

Chromium has a small effect in increasing machinability, probably due to a little reduction ot the work-hardening rate of the stainless steels (Figure 8).

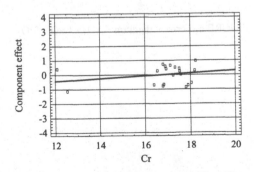

FIGURE 8. Effect of chromium on machinability

4.5 NICKEL EFFECT

Similar to Cr effect because even Nickel decreases the work-hardening rate (Figure 9).

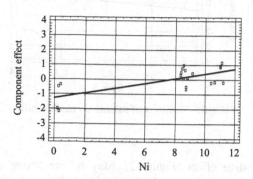

FIGURE 9. Effect of nichel on machinability

4.6 MOLYBDENUM EFFECT

Molybdenum has no effect on work-hardening, but only influence a little the matrix hardness and has very important effect on the corrosion resistance (Figure 10).

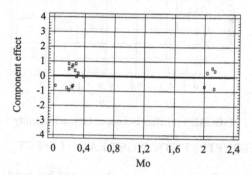

FIGURE 10. Effect of molybdenum on machinability

4.7 SULPHUR EFFECT

As expected Sulphur has a strong positive effect on machinability due to the formation of sulfide inclusions with Manganese, Chromium or Iron (Figure 11). Is important to note that the effect of sulphur is mainly due to the quantity of inclusions and to the final morphology (thickness/length ratio) than to the plain quantity of sulphur. See the discussion on the Cleanliness index below.

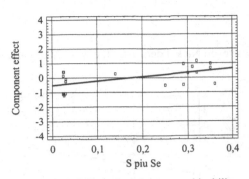

FIGURE 11. Effect of sulphur on machinability

4.8 COPPER EFFECT

Copper has a little negative effect (Figure12). May be the strong effect in reducing work-hardening lead to a softer matrix that results in a sticking-effect as built-up edge or because chip deforms in a plastic way during machining.

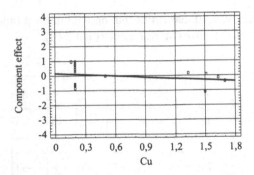

FIGURE 12. Effect of copper on machinability

4.9 YIELD STRENGTH/RUPTURE STRENGHT RATIO EFFECT

The positive effect of this mechanical parameter (Figure 13) is related to the deformation behaviour during tensile test. Much closer are the yield and the rupture strength, more brittle are the chips and two effects happen: small chips, less heat on the cutting edge and less plastic work absorbed by the machining.

It is important to note that the effect of work-hardening obtained by drawing increases the yield strength faster than rupture strength. Drawn bars show different behaviour in machining: see Figure 14 as an example. The ratio Yield Strength/Rupture Strenght take in consideration even the effect of Nitrogen. This element is an interstitial as carbon but is not a carbide former (up to 0.3%). As a consequence its effect can be summarized by the cold working hardening factor inside the Rp/Rm ratio.

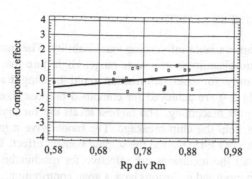

FIGURE 13. Effect of Rp/Rm ratio on machinability

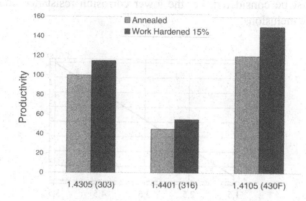

FIGURE 14. Hard-working influence on machinability (annealed 303 as 100)

4.10 THERMAL CONDUCTIVITY

The higher the thermal conductivity the higher the heat transfer in the machined piece and the longer the tool life (Figure 15).

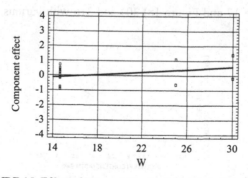

FIGURE 15. Effect of Thermal Conductivity on machinability

4.11   MICROCLEANLINESS EFFECT

This is the most important variable influencing machinability. The modern technology of steel-making can modify micro-cleanliness in a wide range. High pure materials are difficult to machine and are used mainly for cold deformation (cold heading-drawing). The intermediate cleanliness (up the grade 2~) are achieved for combined purpose like limited cold heading or drawing or hot forging plus machining. The highest levels are for high machining productivity because more inclusions help the chip breakage. The model give a greater contribution to the quantity of inclusions (ASTM E45) than to the sulphur content effect. This concept permit a better assessment because all the inclusions are effective for machinability and not only the sulphides. Even the calcium modified inclusions have a great contribution.

An adverse effect must be considered, i.e. the lower corrosion resistance connected with high density of non metallic inclusions.

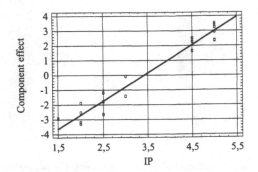

FIGURE 16. Microcleanliness Grade ASTM E45 meth. A

Considering a qualitative test like Strauss test for intercristalline corrosion, there is no problem because the copper – copper sulfate test is not so high sensitive to the quantity of inclusions. The difference can be measured by mean of Boiling Nitric Acid test (Pratice C of ASTM A262 Huey test).

In pictures below (Figure 17 and Figure 18) 304 and 316 with various inclusion content are assessed.

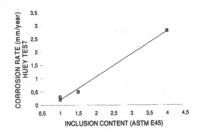

FIGURE 17. Huey test on 1.4306 – 1.4307 (304 - 304L)

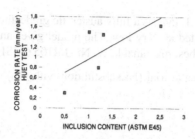

FIGURE 18. Huey test on different 1.4404 grades (316)

5 COMPARISON WITH LITERATURE DATA CONCERNING MACHINABILITY

Quite good correspondence was found between proposed model for estimate machinability and qualitative data reported in Stainless Steel Handbook (Chapter 24 – Peckmer, Bernstein – Table 1).

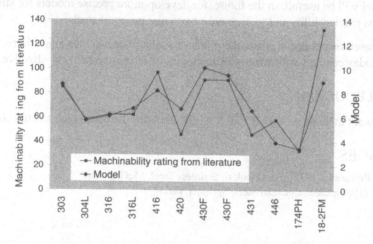

FIGURE 19. Comparison of model with literature data

An useful application of the proposed model can be the evaluation of a stainless steel grade for wich no data are available.

For instance very few data are available in literature about duplex stainless steels. By this model, it is possible to forecast an index of 3,4. This means the machinability of this grade is comparable to a W. Nr. 1.4542 (AISI 630).

Another example is the W. Nr. 1.4567 (AISI 302HQ) used for cold heading, for which there are no data regarding the machinability of the annealed bars. The model gives an evaluation of 2.0, that means a very poor machinability. Actually, with avg. Cu ≅ 3,2%, we have to expect the chips deform without any breakage. Some test in turning were made in our works finding this result as true. Even the low sulphur content (10-20 ppm max) must be considered for this steel grade.

Super austenitic stainless steel 1.4547 is a fully austenitic grade alloyed with 6% Mo and 0,2% N and its machinability is expected as very low. The model gives an assessment of 3,5 that corresponds roughly to 30% of the best machinable W. Nr. 1.4105 (AISI 430F).

For a new duplex in experimental trial (Ni substitution with Mn) our model forecast 3,8, that is quite close to the classic W. Nr. 1.4462.

6 CONCLUSIONS

1 - The proposed model for machinability assessment has quite good correspondence with available data and can be even used for evaluation of new grades machinability.

2 - This model changes a little the weight of some parameters like the sulphur content in the steel that has been found less effective than the quantity of inclusions.

3 - The main driving effect on machinability comes from inclusion content and mechanical properties.

4 - This model will be usefull, in the future, for develop more precise models for single grades in order to assess machinability and to improve a particular group of steels.

5 - The proposed model don't assess the effect of tooling because this is a different field of improvement (today coatings with various technologies have been found very effective).

ACKNOWLEDGEMENTS

The authors want to thank ACCIAIERIE VALBRUNA for the permission to publish this work.

REFERENCES

1.   Pecker, Bernstein, (1977), Handbook of Stainless Steel, McGraw Hill
2.   Ototani, (1986), Calcium Clean Steel, Springer Verlag

EXPERIMENTAL AND MATHEMATICAL MODELING OF CUTTING TOOL WEAR IN MILLING CONDITIONS

V. Gecevska[1], F. Cus[2], M. Kuzinovski[1], U. Zuperl[2]

[1] Faculty of Mechanical Engineering, Ss. Cyril and Methodius University, Skopje, Macedonia
[2] Faculty of Mechanical Engineering, University of Maribor, Slovenia

KEYWORDS: Cutting, Tool Wear, Experimental Modeling.

ABSTRACT. This paper proposes a methodology for experimental and mathematical modeling of cutting tool wear in milling condition with multipass machining operation. The paper describes experimental research, realized in real manufacturing condition for determination of the machinability data indispensable for complex mathematical model of machining process and mathematical modeling of the cutting tool wear phenomena.

1 INTRODUCTION

For increasing of the efficiency and the productivity of machining processes, in the research [3], the methodology for determines the optimal cutting parameters are proposed. The research is realized for the manufacturing of prismatic parts of the machining centers. The maximizing of the technological and the economical effects in manufacturing is used by the optimization of machining process at the machining centers.

The present research proposes that optimal determination of the cutting parameters in multipass machining be addressed as a multi-objective programming mathematical model. In the model, the optimal solution is obtained by using a deterministic method and a genetic algorithm [3]. The optimization method has been performed for the machining process, as object of optimization.

The mathematical model for objective function is defined to describe the object of optimization - machining process and to determine the dependence between cutting conditions. The mathematical model for optimization is considered, by projecting of the optimization function for several machining operation running of the machining centers (for milling, drilling, boring, reaming and threading).

The constraints produce restrictions for cutting parameters and they are projected as boundaries, linear or nonlinear functions [1] [2] [4]. Mathematical equitation of constraints functions is determined to use empirical and analytical relations for machining process and to involve experimental machinability data. The function of constrains are formulated from: (1) cutting tools characteristics and tool wear, (2) cutting tool life in different machining conditions, (3) quality and accuracy of the machining, (4) properties of tool and work piece materials, (5) geometry of the machining work piece, (6) characteristics of the main and idle movements etc. The criteria for optimization result from the economical or technological effects of machining process, as a min machining time, min cost of production or max productivity.

Published in: E. Kuljanic (Ed.) *Advanced Manufacturing Systems and Technology*,
CISM Courses and Lectures No. 486, Springer Wien New York, 2005.

2   PROPOSED METHODOLOGY FOR OPTIMIZATION

In the research [3], created complex mathematical model of machining process are expanded with functions of real values of machining constraints. Functions are determined by parameters with results from real machining conditions, as a:
- Real tool life of the cutting tool,
- Real tool wear of the cutting edges, defined by width of flank wears VB,
- Real engaged power,
- Real produced cutting force.

In the complex mathematical model, these parameters are determined by entry of optimized machining conditions in equitation of constraints. For this reason, in the research it is realized experimental investigation in real machining condition for determination of exponents in Taylor's equitation for milling.

3   EXPERIMENTAL DETERMINATION OF EXPONENTS IN TAYLOR'S EQUITATION FOR MILLING

Experimental research is made for determination of follow functional dependences:
1.   Mathematical model for tool wear as a function of cutting time,
2.   Mathematical model of Taylor's equitation for tool life as a function of cutting parameters.

The tool life is dependant by tool wear of cutting edge. So, the tool life, as a cutting time with the same cutting tool, is a co-proportional with tool wear. In the research, as ISO standards [5] [6], tool wear is verified by width of flank wears VB of cutting edge (Figure 1).

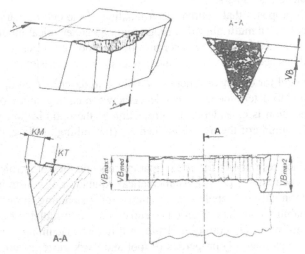

FIGURE 1. Geometrical points of tool wear of cutting edge during the milling [5] [6]

There exist more methods for determination of exponents in equitation for cutting tool life, which based of fundamental and empirical research.

One of them is aproxcimative method for determination of tool wear value of cutting edge by check of VB as a function of cutting time, with ignoring of initial wear. This method is used in created optimization methodology in research [3], for analyzed machining.

Other method base of experiments, where is applicable the methods of plan of experiment, for determination of analytical equitation for experimental conditions. This method is used for determining the equitation of cutting tool wear and tool life in condition of milling.

3.1. EXPERIMENT CONDITION

Determining of value of tool wear during the milling, it is realized experimental investigation on the vertical machining center, type MAZAK VQC-20/50 with n=50-3500[o/min] and power P=10[kW] (Figure 2), in the fabric for produce machine tools-FAM in Skopje, Macedonia.

FIGURE 2. Vertical machining center VQC-20/50

Machining material is steel DIN St70-2, with chemical and mechanical characteristics, according to attest No.775-EN10025 Makstil-Skopje (Table 1).

TABLE 1.

Attest No. 775-EN10025	Chemical characteristics (in %)						
JUS ^0745 DIN St70-2	C	Si	Mn	P	S	Al	N
	0.17	0.30	1.36	0.017	0.013	0.038	0.007
	Mechanical characteristics						
	$R_e[N/mm^2\ min]$	$\sigma_m[N/mm^2]$		KV			HB
				J	^0C		
	379	733		100	0		190

Cutting tool is a face-milling cutter with carbide tips, produced by PP CORUNT (Table 2). Every experiment is made with now cutting edge of carbide tips for recognizing the machining parameters influence of the tool wear of the cutting tool.

TABLE 2.

Milling cutter	Type: R G 62.2-050-10	
Cutting inserts	Type: TPKN 16-03 R	Material: P30
	Stereo metric: $\kappa=90^0$ $\alpha=6^0$ $\gamma=11^0$ $\lambda=5^0$ $\varepsilon=60^0$ a=1.2[mm] b=1[mm]	
Producer:	PP Corun	

3.2. PLANNING OF EXPERIMENT AND MACHINING PARAMETERS

Milling, as a metal cutting process with chip removal, is made by changing the independently variable machining parameters speed V, feed f, depth of cut a, cutting time t, on the two levels (Table 3) by four-factor plan experiment (2^4+4). Experiment planning is made according to coding plan-matrix for defining machining parameters combinations for every single experiment.

TABLE 3.

.1.1 Characteristics of independently variable					
No.	Type	Level-code	Maximal 1	Middle 0	Minimal -1
1.	V [m/min]	x 1	230	151.657	100
2.	s [mm/z]	x 2	0.4	0.200	0.1
3.	a [mm]	x 3	1.8	1.273	0.9
4.	t [min]	x 4	6.0	2.449	1.0

Introducing the time, as an independently variable parameter, enables to decide of:

- Mathematical model for describing tool wear of cutting tool as a function of machining parameters (V,f,a) and cutting time (t):

$$VB_b = C \cdot V^{x_1} \cdot s^{y_1} \cdot a^{z_1} \cdot t^{q_1} \qquad (1)$$

C - constant dependent by machining and cutting materials and machining condition

x, y, z, q - exponents, which done an impact of machining parameters (V,f,a) and cutting time t of the tool wear, explicated by VB.

- Mathematical model of Taylor's equitation for tool life of cutting tool for milling (T=f(V,f,a,VB)), as a function of V,f,a, for criterion VB.

4 RESULTS OF MADE EXPERIMENTS

For determination of mathematical model (4.1) witch will be valid for real machining condition, it is made the experiment in accordance with defined plan experiment (Table 4). On each level of experiment, it is done three results as a number of carbide inserts on the milling cuter.

Result of each experiment is a width of flank wears VB of cutting edge. Detection of results in experimental conditions is made with electronic microscope NICON with accuracy 0.001[mm]. Numerical and statistical preparation of experimental data is made with methodology for selection of adequate mathematical model.

TABLE 4.

No.	Real and coding plan matrix									Values of each edge			Results
	Four factors plan experiment (2^4+4)												
	V [m/min]		s [mm/z]		a [mm]		t [min]			VB_b -1	VB_b -2	VB_b -3	VB_b [mm]
1.	100	-1	0.1	-1	0.9	-1	1	-1		0.080	0.085	0.069	0.078
2.	230	1	0.1	-1	0.9	-1	1	-1		0.124	0.117	0.122	0.121
3.	100	-1	0.4	1	0.9	-1	1	-1		0.089	0.097	0.093	0.093
4.	230	1	0.4	1	0.9	-1	1	-1		0.129	0.149	0.151	0.143
5.	100	-1	0.1	-1	1.8	1	1	-1		0.084	0.097	0.089	0.090
6.	230	1	0.1	-1	1.8	1	1	-1		0.131	0.148	0.136	0.138
7.	100	-1	0.4	1	1.8	1	1	-1		0.092	0.108	0.097	0.099
8.	230	1	0.4	1	1.8	1	1	-1		0.170	0.155	0.149	0.158
9.	100	-1	0.1	-1	0.9	-1	6	1		0.165	0.183	0.179	0.176
10.	230	1	0.1	-1	0.9	-1	6	1		0.255	0.241	0.246	0.247
11.	100	-1	0.4	1	0.9	-1	6	1		0.197	0.199	0.218	0.205
12.	230	1	0.4	1	0.9	-1	6	1		0.297	0.303	0.284	0.294
13.	100	-1	0.1	-1	1.8	1	6	1		0.171	0.185	0.191	0.182
14.	230	1	0.1	-1	1.8	1	6	1		0.251	0.267	0.255	0.258
15.	100	-1	0.4	1	1.8	1	6	1		0.188	0.207	0.191	0.195
16.	230	1	0.4	1	1.8	1	6	1		0.308	0.328	0.312	0.316
17.	151.65	0	0.2	0	1.273	0	2.449	0		0.141	0.157	0.145	0.148
18.	151.65	0	0.2	0	1.273	0	2.449	0		0.164	0.181	0.171	0.172
19.	151.65	0	0.2	0	1.273	0	2.449	0		0.158	0.177	0.165	0.167
20.	151.65	0	0.2	0	1.273	0	2.449	0		0.144	0.168	0.166	0.159

Statistical analysis of condition and results of experimental data is made determination of significant parameters and done exponents and mathematical models for real machining condition, as a:

-tool wears of cutting edge
$$VB_b = 0.0108 \cdot V^{0.494} \cdot s^{0.105} \cdot a^{0.096} \cdot t^{0.400} \tag{2}$$

-Taylor's equitation
$$T = \frac{92.59 \cdot VB_b^{2.5}}{V^{1.235} \cdot s^{0.262} \cdot a^{0.24}} \tag{3}$$

Graphical interpretation of computed values of cutting wear of cutting edge, for determined mathematical model (2), is done of the figure 3. Graphical presentation is done mathematical dependence of cutting wear-VB from cutting time-t, in correlation with speed-V, feed-F and dept of cut-a.

Graphical interpretation of computed values of cutting wear of cutting edge, for determined mathematical model (2), is done of the figure 3. Graphical presentation is done mathematical dependence of cutting wear-VB from cutting time-t, in correlation with speed-V, feed-F and dept of cut-a.

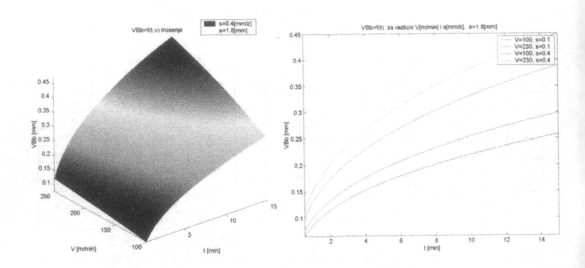

FIGURE 3.a) Graphical presentation of mathematical model for VB during the variable value of speed V[m/mm] and cutting time t[mm]

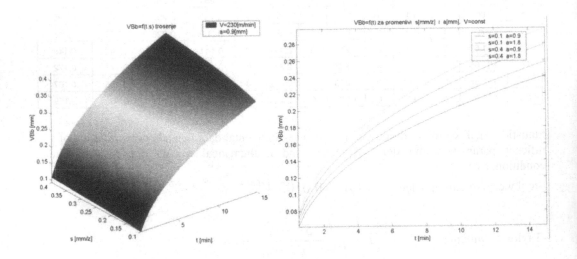

FIGURE 3.b) Graphical presentation of mathematical model for VB during the variable value of feed f[mm/toot] and cutting time t[min]

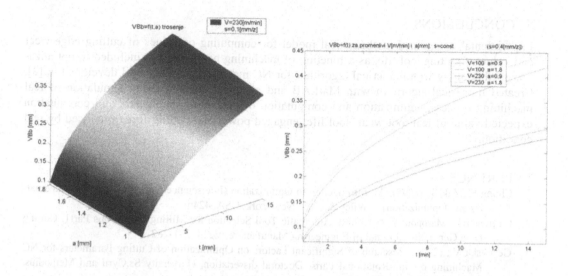

FIGURE 3.c) Graphical presentation of mathematical model for VB during the variable value of dept of cut a[mm] and cutting time t[min]

Graphical presentation of computed values of tool life of the cutting edge for determined mathematical model (3), is done on the figure 4, where is presented tool life as a function of: (1) speed and tool wear T=f(V, VB) and (2) speed and feed T=f(V, s).

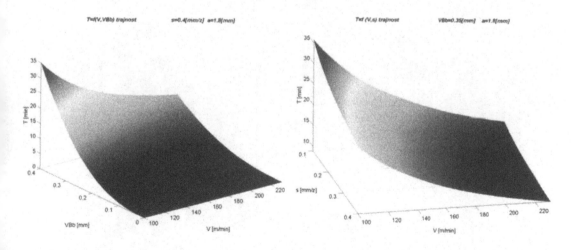

FIGURE 4. Graphical interpretation of mathematical model of tool life of the cutting tool

5   CONCLUSIONS

Experimental determined mathematical model for computing of values of cutting edge wear and value of cutting tool life, as a function of machining parameters, is included in optimization methodology with numerical algorithm for NC machining, created and developed in [3]. Created numerical algorithm with MatLAB and genetic algorithm doing simulation of real machining process, optimization and computation of machining parameters and computation expected value of real tool wear, tool life, engaged power and cutting force produced by real condition.

REFERENCES

1. Chong E.; Zak S., (1995), An Introduction to Optimization (Interscience Series in Discrete Mathematics and Optimization); J.Wiley & Sons, New York, USA, 424p.
2. Carpenter I.; Maropoulos P., (2000), Automatic Tool Selection for Milling Operations Part1. Cutting Data Generation; Journal of Engineering Manufacture, V214, 271-282.
3. Gecevska V., (2002), Research of Significant Factors on Optimization of Cutting Parameters for NC Machining of Non-Rotational Parts, Doctoral dissertation, University Ss.Cyril and Methodius, Macedonia, 227p.
4. Kuljanic, E., Fioretti, M., Beltrame, M. Miani, F., (1998), Milling Titanium Compressor Blades with PCD Cutter, Annals of the CIRP, Vol 47/1, 61-64.
5. ISO 8688-1: Tool life testing in milling – Part 1: Face milling, (1989), 27
6. ISO 8688-2: Tool life testing in milling – Part 2: End milling, 1989, 26

LAMINATED OBJECT MANUFACTURING OF METAL FOIL – PROCESS CHAIN AND SYSTEM TECHNOLOGY

M. Prechtl[1], A. Otto[2], M. Geiger[1,2]

[1] Bavarian Laser Center (BLZ gGmbH), Erlangen, Germany
[2] Chair of Manufacturing Technology, University of Erlangen-Nuremberg, Germany

KEYWORDS: Rapid Tooling, Laser Material Processing, Diffusion Welding.

ABSTRACT. Laminated Object Manufacturing of metal foil as an automated two-step procedure is a novel technology for additive manufacturing of massive three-dimensional parts. It enables for example the production of technical tools like moulds. The great advantage hereby is the high accuracy in combination with a high stability of the resulting tools.

A laser system for the automated stacking of metal foil contours has been developed. The stacking procedure is realised by a combination of layer fixing and contour generation with a laser beam. But the stability of the produced foil stack is insufficient for most kind of application. Therefore a second step to enhance the mechanical properties of the part is necessary. This can for example be realised by high temperature brazing or diffusion welding.

In this paper the laser system for the automated stacking of metal foil contours is described as well as the required preparation of the layer data. Further on some results of investigations on the sub-processes layer fixing, counter generation and final joining are shown.

1 INTRODUCTION

The manufacturing technology "Laminated Object Manufacturing (LOM)" has been commercialised by Helisys, Inc. (Torrance) in 1991. Paper coated with adhesive was used as a basic material which was cut by a laser beam into the desired shape. The paper sheets were stacked and joined by thermal activated gluing in an automated procedure. Nowadays plastic, composites with glass fibres and ceramics could be stacked together using the principle of LOM technology [1]. The produced parts show a relatively high accuracy because the general layer thickness is approximately 0.1 mm. As the parts are produced by joining of layers on the one hand and on the other hand by cutting them the procedure is exactly a hybrid between additive and subtractive manufacturing. However the joining step is dominating so that LOM can be classified as an additive technology. It also should be mentioned that the mechanical and/or thermal properties of such LOM parts are insufficient for applications in many cases. One important example hereby is the manufacturing of functional tools like moulds.

Therefore metal sheet as a LOM base material is a field of research, too [2]. The procedure is quite similar to "paper LOM": After generating the layers by laser beam cutting the contours are glued, bolt or welded together. This technique is a partial automated procedure because only the cutting process is numerically controlled. Produced parts show properties according to the used sheet material and especially the kind of joining process. For example diffusion welding nearly enables a mechanical strength like the base material. But there is also a significant disadvantage

Published in: E. Kuljanic (Ed.) *Advanced Manufacturing Systems and Technology*,
CISM Courses and Lectures No. 486, Springer Wien New York, 2005.

of the sheet metal LOM technique. For the handling of the sheet metal contours e a sufficient self stiffness of the sheet is necessary. Therefore the thickness has to be more than approximately 0.5 mm. As a consequence of the relatively thick layers metal sheet LOM parts suffer from a significant staircase effect. This can be reduced or sometimes avoided by applying first order approximation which means that the cutting angle of the slit is unequal 90° (slanted slits). The surface quality of metal LOM part is often scarce so that in these cases a post processing such as milling, build-up welding or shot peening is necessary.

For a general increase of the accuracy of metal LOM parts the layer thickness has to be reduced where the low self stiffness of the material (foil) is a great challenge. The so called "metal foil LOM technology" combines the advantages of "paper LOM" and "sheet LOM". Therefore the metal foil LOM technology, especially realized by an automated procedure, can be considered as an enhancement of the conventional sheet metal LOM technology.

2   PROCESS CHAIN

The metal foil LOM technology is realised as a two step procedure. First the layers are automatically stacked by the help of a laser beam and afterwards they are joined together in a furnace. The whole procedure is schematically shown in Figure 1 where it can also be seen that the stacking step is even separated in layer fixing and contour generation.

Starting point however is a 3D-CAD model of the part. This model is transferred to a set of layer data including especially the geometric data by a SLICE software. After the data preparation the data of the layers are iteratively treated in the stacking system. Very important is the sequence of the stacking sub-processes. First the current layer is fixed at the layer before and in case of the first layer it is fixed at a changeable metallic base plate. Thus the position of the current layer is

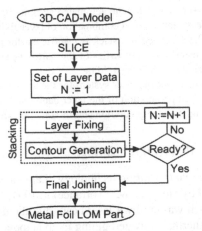

FIGURE 1. Procedure of the metal foil LOM technology (N: layer number)

determined. Afterwards the generation of the layer contour can be done. Segments of the current foil which are not used for the part are removed by hatching and on-line exhausting.

As mentioned the stability of the produced foil stack is mostly insufficient for any kind of application. Therefore the layers are joined together by diffusion welding or high temperature brazing, e. g. with copper. Such produced parts show no pores or inclusions.

The fundamental sub-processes of the metal foil LOM technology are layer fixing, contour generation and final joining. The first two sub-processes are laser processes – for both processes a short pulse solid-state laser (Nd:YAG) is used – whereas the final joining is realised by a thermally activated process in a high temperature furnace. Below some results of experiments on the sub-processes are shown and discussed. Hereby the properties and quality of the resulting metal foil LOM part are of special importance.

2.1 LAYER FIXING

For the fixing of the layers weld spots with a defined distance are made whereas only two foils – the current layer and the layer before – have to be joined. The main demands on the weld spots are bumps as small as possible and a high stability. Small bumps are necessary because wide gaps between the foils can cause high inaccuracy and possibly they can interfere the final joining process. And a high stability of the weld spots is required for the handling of the produced foil stack. At least reproducible spot welds and a process which is not very sensible on fluctuations of the process parameters is aspired.

As described later, the use of a cross-jet reduces the interaction between gas jet and weld pool and enables laser beam spot welding with short pulse length in a wide parameter range. But the produced weld spots show very different quality. It is obvious that the pulse energy is one important parameter because it correlates with the volume of the weld pool. Therefore the maximum bump height $H_{b,max}$ and the maximum shear stress $F_{s,max}$ were investigated as a function of the pulse energy (Figure 2). It has to be mentioned that the pulse length τ and the distance between beam focus and foil surface Δz also influence the quality of the resulting spot weld and they should be chosen to enable a large process window.

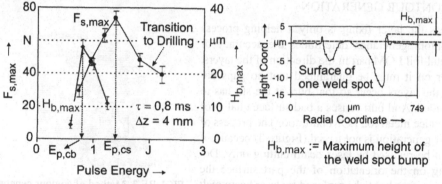

FIGURE 2. Maximum shear stress $F_{s,max}$ and maximum bump height $H_{b,max}$ of weld spots as a function of pulse energy (cross-jet with 3 bar air pressure, focus diameter: 150 μm, wavelength: 1064 nm, foil material: unalloyed steel St2 K50, foil thickness: 100 μm)

The maximum shear stress increases first to a maximum which is reached at the characteristic pulse energy $E_{p,cs}$. For higher pulse energies the melt pool volume get bigger and also the cross sectional area of the weld spot in the contact area of the two metal foils. After increasing the pulse energy further on, the maximum shear stress decreases because more melt evaporates. This means that the welding process turns into a drilling process and therefore the joining between the foils exists only in for of an annulus.

The maximum bump height $H_{b,max}$ shows in principle the same characteristic. To explain the effects for the increase and decrease of the maximum bump height the reasons for the mechanisms of emerging such bumps have to be discussed. Using short laser pulses for spot welding material evaporates in any case. The rebound of the vapour and the vapour pressure itself influences the melt pool so that its surface is deformed. At the same time some melt is displaced to the melt pool border. The displaced melt solidifies in form of border bumps. As a consequence of the deformation of the melt pool the absorption rate changes as well as the dynamic of the melt pool. Two dimensional vibrations of the melt surface can be initiated and can cause bumps in the centre of the weld spot (Figure 2, right side).

By increasing the pulse energy the amount of vapour increases, too. Therefore the deformation of the melt pool surface is bigger and also the border bumps. Additionally the bigger amplitude of the melt pool vibration can cause higher centre bump. The height of bumps which are caused by the vibration mechanism correlates with the processing time. Depending on the pulse length a bump or a pool crater can occur because the moment of solidification is determining for the shape of the weld spot surface. After the characteristic pulse energy $E_{p,cb}$ the vaporisation of the melt partially compensates the development of bumps. The amount of melt is then not enough for producing such high bumps.

But the most important result of this examinations is the difference between the characteristic pulse energies $E_{p,cs}$ and $E_{p,cb}$. For the metal foil LOM Technology weld spots with high stability (shear stress) and also low bumps are required. By using a short pulse laser in combination with a cross-jet this aim was reached. The shear stress increases at pulse energies for which the bump height reduces. When applying the right pulse energy the vaporisation of melt compensates the bump development but it is not sufficient for the laser beam drilling process.

2.2 CONTOUR GENERATION

Whereas the layer fixing is only a helping process the contour generation determines the accuracy of the metal foil LOM part in the direction of the layers. Further on it must be guaranteed that no gaps between the layers occur so that burr at the slit has to be avoided. And burr causes a bad surface quality of the produce metal foil LOM part, too. The process of contour generation is not trivial (Figure 3) because it can not be realised by laser beam cutting only. Depending on the orientation of the part surface the cutting process has to be replaced by laser beam melt ablation. If there is already a foil stack below for the contour of the current layer with no distance to the

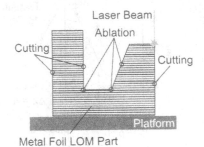

FIGURE 3. Method of contour generation depending on the part surface

foil stack the melt during the laser beam process must be removed towards the laser beam. The big challenge hereby is that the layer before should not be influenced. This can for example be realised by iterative laser beam melt ablation, i. e. the metal foil is divided after more than one irradiation – the required number of irradiations depends on the used parameters – of the same contour.

Generally for the cutting process the width of the slit as well as the cutting angle are of special importance. The slit width correlates with the amount of melt during the cutting process and therefore also with the surface quality of the produced metal foil LOM part because the melt is deposited at the contours of the layers before. Therefore it is necessary that the slit width is as small as possible. In Figure 4 the medial slit width s is shown as a function of pulse lap δ_p which links the laser frequency f with the cutting velocity v and is defined by the equation

$$\delta_p := 1 - \frac{v}{f \cdot D} \tag{1}$$

where D is the diameter of the laser beam at the surface of the metal foil. A decrease of the pulse lap causes a reduction of the slit width as the energy per line (unit: J/mm) is lower. Reaching a characteristic pulse lap the metal foil is not divided only perforated. It should be mentioned that the pulse lap influences also the roughness of the cut.

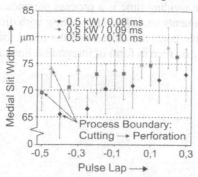 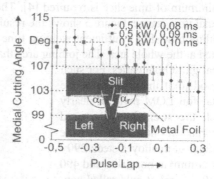

FIGURE 4. Medial slit width and medial cutting angle (material: St2 K50, foil thickness: 100 μm, focus diameter: 50 μm, nozzle diameter: 0,8 mm, Δz = 0 mm, process gas: 5 bar N₂)

Further on the cutting angle is also important for the surface quality of the produced metal foil LOM part especially in correlation with the orientation of the part surface. For vertical surfaces a cutting angle of 90° should be achieved whereas for geometries like a pyramid a cutting angle unequal 90° is better. The cutting angle α is defined as

$$\alpha := \frac{1}{2}(\alpha_l + \alpha_r)$$ (2)

where α_l and α_r are the angles at the left and right side of the slit. It can be seen that the medial cutting angle arises with the pulse lap and is roughly constant after a characteristic pulse lap (Figure. 4). By reducing the pulse lap the influence of the cutting fore on the laser induced temperature field decreases – and also the heat accumulation at the bottom of the cutting fore – so that the peak of the angle is less melted. If the pulse lap is too "high" the metal foil is only divided by a sequence of drill holes and the cutting angle moves to a saturation value.

Summing up a compromise for the slit width and the cutting angle must be made especially for vertical surfaces. The removed melt during the cutting process can greatly deposit at the layers before because the gas jet becomes bigger after passing the slit. One possibility for optimisation is the use of air pressure as process gas. The amount of oxygen causes an exothermal reaction so that more process energy is available. By this additional energy more material can be molten and the cutting angle – it shows than no dependence on the pulse lap as well as on the pulse length – and the roughness of the slit is reduced. The cutting angle is approximately of 107° when using air pressure. And some of the material is burned by the energy of the exothermal reaction so that the amount of deposited melt at the layers before is reduced, too.

2.3 FINAL JOINING

The produced metal foil stack has no stability necessary for any technical application. Therefore the layers are joined by diffusion welding. For the realisation of this process the foil stack is pressed at high temperature in combination with an inert atmosphere for a defined time. Therefore the main process parameters are the surface pressure p_S, the welding temperature T_w and the welding time t_w. Some details concerning the microscopic and atomic mechanisms during the diffusion welding process are shown for example in [3].

Diffusion procedures in solids are atomic procedures with change of functional location for what a minimum of time slice is required [4]. Therefore the welding time is a process parameter with special importance. Figure 5 shows the result of experiments on diffusions welding with different welding time. For a sufficient welding time microscopic pores are closed and the metal foils are joined at the whole area. The joining area than can only be seen in form of integration lines and/or

lineal grain boundaries. For such a good joint the strength of the metal foil LOM part can nearly reach the strength of the used material at normality (tensile strength of unalloyed steel: 290 - 670 N/mm² at normality and 490 - 640 N/mm² at cold rolled condition as foil). But as a consequence of the production layer by layer the strength of the part depends also on the direction of the strain.

It has to be mentioned yet that next to the joining quality be-

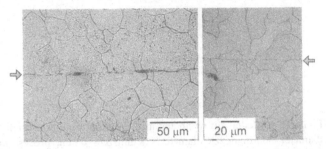

FIGURE 5. Cut through a joined metal foil stack for different welding time (left: 1 h, right: 2 h, surface pressure: 5 N/mm², welding temperature: 900 °C, atmosphere: 95% Ar / 5% H₂, material: St2 K50); Note: The two arrows show the joining area between two metal foils.

tween the layers also the dimensional accuracy of the metal foil LOM part is of great importance. As a result of the applied surface pressure during the diffusion welding process the roughness of each foil surface is levelled. Therefore the height of a metal foil LOM part is lower than the amount of layers corresponds. But the required number of layers for realising a defined height can be calculated whereas a shrinkage-factor has to be investigated by experiment before [5].

3   SYSTEM TECHNOLOGY

For the automated stacking of metal foils contours for an additive manufacturing of three dimensional parts a 3D-CAD-Model of the part is necessary. The solid model design can be done using commercial CAD software like ProE and SolidWorks. This model is first transferred to a set of two dimensional layer data which includes basically the geometric information of the layer. This procedure is realised by a SLICE-software and is quite equal to conventional Rapid Prototyping technologies like LOM. In addition however some information for the sub-process of layer fixing is necessary. This special

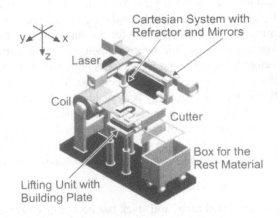

FIGURE 6. Scheme of a machine for automated laser assisted stacking of metal foil contours

LOM software includes the following modules: CAD file import, slicing, post processing, and machining. More details on the SLICE software for metal foil LOM are described in [6].

The produced layer data are than delivered to a machine (Figure 6) where the foil stack is built up iteratively layer by layer. First the metal foil is fed the machine from a coil where the fixing procedure occurs. In the area of the foil where the part is growing some weld spots are made with a laser beam. Afterwards the contour is generated also by a laser beam and the not used material of the inside of the layer is hatched and removed by exhausting. Than the lifting unit moves down for a defined distance depending on the layer thickness and the foil is fed further on. The rest material, i. e. the metal foil without the contours, is cut into pieces which were collected in a box. Now one layer is generated and the procedure occurs again so that the foil stack is produces iteratively.

The basic components of the metal foil stacking system are the beam source, the processing head and the system for measuring the foil thickness. Theses components are specific for the metal foil LOM technology so that in the following the challenges should be discussed.

3.1 USED BEAM SOURCE AND OPTIC

As mentioned above the layer fixing and the contour generation is done by a laser beam process. Therefore it is aimed that both the laser beam welding and the cutting/ablation process can be realised with one beam source. The two possibilities are cutting with a welding laser or welding with a cutting laser.

Cutting with a welding laser is possible as the material is molten by the energy of the laser beam and than removed by a gas jet. But the resulting accuracy is insufficient for the metal foil LOM technology. As a result of the high pulse length (> 1 ms) many of the foil material is molten and the slit shows a big heat affected zone as well as big burr at the flanks. And the resulting slit width is very irregular for constant process parameters because the removing of the molten material is not a continuous process. As the cutting process – in combination with the axes accuracy – however defines the accuracy of the LOM part in layer direction the possibility of cutting with a welding laser is not reasonable.

The laser beam spot welding process is only a helping process used for layer fixing has no influence on the accuracy of the metal foil LOM part. Although a pulse length more than 1 ms is required for laser beam welding [7] enough melt for joining metal foils with a thickness of 100 μm can be produces at a pulse length lower than 1 ms. Therefore this option war chosen and a solid-state laser (1064 nm, $M^2 = 12$) with a pulse length of 80 μs – 1 ms which is generally used for cutting and drilling was integrated into the stacking system.

The relative movement between laser beam and metal foil is realised by a cartesian system with mirrors for the beam lead and a lens for focussing the laser beam. With this system a focus diameter of 50 μm can be realised in combination with an image area of 300 mm x 300 mm. The velocity of the linear induction motors is up to 10 m/min. As the maximal laser frequency is approximately 800 Hz the axe velocity is absolutely sufficient.

3.2 NOZZLE CONFIGURATION

As mentioned above, for the layer fixing weld spots with small bumps are necessary because bumps cause gaps between the metal foils and can therefore influence the process of final joining. Additionally a big process window should be achieved so that the process is not very sensible on fluctuations of the process parameters like the distance between the laser beam focus and the metal foil surface. For the process window the gas jet is of special importance. But a gas jet is needed anyway because the lens must be protected against splashes and steam which can occur during the spot welding process.

Experiments applying a gas jet coaxial to the laser beam have shown that a sufficient process window can only be reached for a distance between laser focus and foil surface more than a characteristic value. This value depends on the distance between the laser focus and the nozzle outlet, the diameter of the nozzle outlet and the pressure of the process gas. The maximum of the velocity of the gas jet reduces after a defined distance to the nozzle outlet and therefore the impulse on the melt pool is lower because the force F_g by the gas jet depends on the gas velocity u (ρ: gas density, R: radius of the gas jet).

$$F_g = \rho u^2 R^2 \pi \tag{3}$$

As this impulse causes a deformation of the melt pool the injected energy is influenced by the gas jet because a multi reflection can happen at a deformed melt pool. For a lower impulse on the melt pool the transition to a laser beam drilling process occurs at higher pulse energy. If the impulse on the melt pool is too high – in the case of a low distance between nozzle outlet and foil surface – a laser beam spot welding process is impossible. But in any case the coaxial gas jet influences the melt pool and therefore causes a small process window and weld spots with generally high bumps.

To avoid any interaction between process gas and melt pool the direction of the gas jet has to be changed. If the gas jet is orthogonal to the laser beam (Figure 7) the protection function of the process gas is ensured and the interaction between gas jet and melt pool reduced. For a sufficient distance of the cross-jet to the metal foil surface there is not influence of the melt pool by the gas.

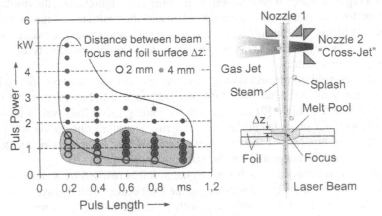

FIGURE 7. Process window of laser beam spot welding using a cross-jet (focus diameter: 150 μm, process gas: 3 bar air pressure, foil material: St2 K50, foil thickness: 100 μm)

If there is no interaction between gas jet and melt pool by applying a cross-jet a clearly bigger process window can be reached. The deformation of the melt pool in this case is only caused by the evaporation (vapour pressure and rebound force) of the material at sufficient injected energy as well as a convection of the melt pool. These effects than determine the transition to the drilling process and consequently the boundary of the process window. In addition to a bigger process window the weld spot quality is increased by using a cross-jet.

3.3 THICKNESS MEASUREMENT

The used foil material is produced by cold rolling and therefore an inaccuracy of thickness can not be avoided. This difference to the ideal foil thickness is of $\Delta t_l = \pm 5$ µm [8]. As a consequence the real height of the foil stack after N layers can differ at

$$(\Delta h)_{max} = \pm N \left| (\Delta t_l)_{max} \right| \tag{4}$$

so that the generated contour do not correspond with the contour of the 3D-CAD-model at this height. Therefore the thickness of the foil is measured during the foil transportation at three positions of the foil by a tactile system. This measure system consists of rollers where the foil is carried through and three push-buttons with an integrated measuring unit. By calculation the average of the three measurements for each layer the real layer thickness is investigated and the real height of the foil stack, too. In combination with the SLICE the tolerance of the foil thickness can be considered. One possibility is that each layer data is generated after producing the physical layer which can be named "on-line SLICE". But there must be enough time between two layers. Another possibility for the combination of SLICE and thickness measurement is the reduction of the (mathematic) layer thickness for the SLICE. That means that the SLICE thickness t_s is only a part of the real layer thickness t_l (dim(t_l) = 1 µm).

$$t_s = \text{int}\left(\frac{1}{n} t_l\right); \quad n \in \{2;3;4;...\} \tag{5}$$

Depending on the real height of the foil stack the layer data for this height – rounded to a value determined by t_s – can be chosen. With this method more data are produced but there is no time for data preparation during the stacking procedure necessary.

4 POTENTIAL OF APPLICATION

The main potential of the metal foil LOM technology is the manufacture of massive moulds which can also show some grooves or something like that. Such technical parts can conventionally only be produced by milling in combination with a high amount of eroding processes. The geometric structure of the part which should be produced is engraved in the massive mould and therefore represented by the casting procedure. Complex parts cause high costs for the mould fabrication in conventional processes. Applying the automated two-step metal foil LOM technology, the production of such parts is done in an automated procedure nearly not needing any manpower which is therefore more economic.

Further on complex cooling systems in moulds can be realised by the metal foil LOM technology. A lot of mould inserts – which can include cavities – and especially sliders and cores often can not be fitted with a suitable cooling system because its manufacture is impossible with mechanical processes because no accessibility exists. Such parts are very common in mould making and show another potential of application of the metal foil LOM technology. Complex cooling systems can be realised by applying an additive procedure, which can produce metallic parts, and expensive erosion processes can be replaced.

5 CONCLUSIONS

The technology of Laminated Object Manufacturing (LOM) is characterised by the generation of contours out of foil or plate which are joined afterwards. For the realisation of metal foil LOM technology a two step process is necessary. First a foil stack is produced in an automated procedure. The layers of the foil stack are then joined in a furnace process.

The stacking procedure itself consists of the two sub-processes layer fixing and contour generation whereas the contour generation occurs after the fixing procedure. Thereby the position of the layers can be determined. The layer fixing is realised by laser beam spot welding. A great challenge for this process is the prevention of great melt bumps in combination with the guarantee of a sufficient weld spot stability. This can be ensured by the use of defined pulse energy. For this pulse energy the weld spot energy shows a maximum whereas the bump height is relatively low.

The process of contour generation depends on the differences between the current contour and the contour before. It must be realised by laser beam cutting and laser beam ablation respectively. These two options can differ in the process parameters like the laser parameters and the rate of feed and especially in the number of irradiations.

Finally a "post-process" is necessary to increase the stability of the produced metal foil stack. This is realised by diffusion welding of the layers in a high temperature press. Using a sufficient welding time the foils are completely joined and no inclusions or gaps in the joining area can be seen. This is a very important condition for high stability of the produced metal foil LOM part which is necessary for the most technical application.

In consequence of the resulting high stability of metal foil LOM parts in combination with the reachable accuracy as well as possible geometric variety the potential of application of this technology is the manufacture of tool or moulds for gravity casting, die casting or injection moulding. Especially the possibility of realising integrated tempering channels makes the technology very important for Rapid Tooling.

REFERENCES

1. Gebhardt, A., (2000), Rapid Prototyping – Werkzeuge für die schnelle Produktentwicklung. Carl Hanser Verlag München Wien.
2. Techel, A., Himmer, T., Gnann, R., (2004), Lamellenwerkzeuge mit konturfolgender Kühlung für Spritzguss- und Schäumwerkzeuge, Anwendertagung für Rapid Technologie, Messe Erfurt.
3. Ortloff, S., (1995), Diffusionsschweißen hochfester Aluminiumlegierungen, Herbert Utz Verlag München, Zugl.: Dissertation, Technische Universität München.
4. Raith, W., (Hrsg.), (1992), Bergmann, Schäfer, Lehrbuch der Experimentalphysik – Band 6: Festkörper, Walter de Gruyter Berlin, New York.
5. Prechtl, M., Niebling, F., Otto, A., (2004), Fertigung von Prototypen und Werkzeugelementen durch iteratives Paketieren und anschließendes Endfügen von Stahlfolien, Tagungsband LÖT 2004, DVS-Verlag Düsseldorf, 169-174.
6. Prechtl, M., Pursche, L., Otto, A., (2004), System Technology and Data Preparation for the Automated Laser Assisted Stacking of Metal Foil Contours, Proceedings of LANE 2004, Meisenbach Verlag Bamberg, 601-610.
7. Meijer, J., Du, K., Gillner, A., Hoffmann, D., Kovalenko, V. S., Masuzawa, T, Ostendorf, A., Poprawe, R., Schulz, W., (2002), Laser Machining by Short and Ultrashort Pulses – State of the Art, Annals of the CIRP, Vol. 51/2.
8. N. N., (2004), Produktspezifikationen Record Metall-Folien GmbH, Mühlheim/Main.

CUSTOMISING A KNOWLEDGE-BASED SYSTEM FOR DESIGN OPTIMISATION IN FUSED DEPOSITION MODELLING RP-TECHNIQUE

C. Bandera[1], I. Cristofolini[2], S. Filippi[1]

[1] Department of Electrical, Management and Mechanical Engineering, University of Udine, Italy
[2] Department of Mechanical and Structural Engineering, University of Trento, Italy

KEYWORDS: Rapid Prototyping, Fused Deposition Modelling, Knowledge-Based Systems.

ABSTRACT. In a Design For Manufacturing context, Rapid Prototyping techniques are some way still considered as "new technologies": the peculiar characteristics of the manufacturing processes are not widely known and may deeply affect the final product functionality. A Knowledge Based System, the Design GuideLines – DGLs, was developed by our Research Group at the University of Udine; it evaluated the products design, in order to verify its feasibility by DMLS (Direct Metal Laser Sintering) Rapid Prototyping technique. During the evaluation process, the DGLs also keep into consideration the aspects relating the verification step, according to the ISO-GPS principles, thus enhancing the completeness of the tool. Aim of this work was to customise the DGLs for design optimisation in FDM (Fused Deposition Modelling), also evidencing the critical aspects and proposing alternative solutions. The contents and structure of the customised version of DGLs are presented in this work.

1 INTRODUCTION

New technologies, advanced materials and information systems are very powerful tools today available for product design and optimisation, but often the designer does not have enough knowledge and/or experience to correctly and conveniently manage them.

The design phase typically begins with the definition of form (in terms of components, connections, configuration and constraints) and materials that can best deliver the required functions. Production technologies, assembly and verification procedures, distribution and recycling phases must then be defined, as much costly and effectively as possible [1].

Rapid Prototyping technologies find their application in this context. Rapid Prototyping (RP) technologies allow construction of physical models starting directly from their CAD representations [2]. Complex objects can be generated without the use of standard technologies, such as NC milling machines, hand finishing, etc., so that costs and times are significantly reduced.

RP activities are preceded by the *design phase*, where the CAD model necessary for the prototyping process is generated. The *prototyping phase* then begins, which can be divided into three main steps:

pre-processing: CAD data preparation for the RP machine (STL conversion, support addition, slicing, etc.) and process parameter setting (*job* generation);

processing: prototype building by the RP machine;

Published in: E. Kuljanic (Ed.) *Advanced Manufacturing Systems and Technology*, CISM Courses and Lectures No. 486, Springer Wien New York, 2005.

post-processing: if necessary, the object, as it comes from the RP machine, is completed using traditional methods, so as to satisfy totally the design requirements (support removal, surface finishing, form and dimension adjustment, etc.).

So far, little attention has been paid to the model design phase within the RP field; emphasis being normally on the development of the technology itself (processes, materials, building strategies, etc.) and on manufacturing parameter optimisation. On the other hand, for a given state of technology, the operations and choices which take place during the design phase are crucial and decisive for the quality of the result and the expectation of success in the use of RP. For this reason, it can be of great use to deepen RP processes from the point of view of model design by developing, collecting, and providing, even at the design stage, the knowledge to take informed decisions and improve the building process.

All these considerations gave rise to this paper. We focused on development of a tool, called Design Guidelines (DGLs), that helps the designer when his/her work is oriented to the use of RP technologies. Up to now, two different releases of the DGLs have been developed; the first derived simply from our experience as designers and RP experts; the second has been heavily influenced and integrated by synergy with some important concepts from the field of Geometric Dimensioning (GPS - Geometric Product Specifications).

In the following paper, we perform a deep analysis of the activities and the results achieved during the various stages of the customization of the DGLs for the FDM technology, thus deriving the objectives for the continuation of our work.

2   THE DGLs LEADING IDEA

The research presented in this work has as the main aim the evaluation of the design phase within RP activities. Due to the wide range of technologies, at the beginning we focused our work on a particular RP process: Direct Metal Laser Sintering (DMLS). This is one of the most promising RP technologies currently available because of its ability to build metal objects, using the same material as foreseen for the final product. This makes it possible to do not only a larger number of tests on the design, but also to apply such methods as *rapid tooling* [3] (i.e. inserts for plastic injection moulding [4]) and *rapid manufacturing*. On the other hand, the DMLS process shows some critical aspects mainly due to the behaviour of metal powders (complex sintering dynamics, residual stress, thermal deformation) [5], so that more study is needed to solve or avoid current limitations even in early activities, during the design phase.

Currently, the most manifest aspect in the relationship between design activities and RP is the need/possibility to modify the 3D model in order to avoid critical situations for the prototyping process and to optimise manufacture in terms of times and quality of results. These operations will be grouped here in the term *re-design*. Modification of the CAD model allows great improvement of RP activities; so our study is mainly linked to this aspect; we are convinced that the existing literature on RP modelling is insufficient, showing a wide margin for improvement. In particular, the research described in this document starts with two considerations:

It is not safe to let the RP operator carry out re-design operations. This is today what happens; but the operator might have little knowledge of the product domain and may not take the best decision in the case of multiple choice; or, even worse, may damage the model function [6]. For this reason, the modifications must remain the responsibility of the designers. Nevertheless, to succeed, designers need specific knowledge [7] of the RP process adopted and of manufacturing problems. A study of *usability* [8] has to be associated to this knowledge, to maximise the effectiveness of use.

Design rules in existing literature [9], which now represent the only help in modifying the 3D model, even if a good starting point, show some limitations that severely reduce their usability during the design phase. These failures concern mainly the absence of a structure to help generation, organisation and use of the rules; the lack of application criteria to guide the modifications; the scarce autonomy due to the fact that classic rules have not been thought out specifically for designers.

The Design Guidelines (DGLs) are based on a set of design rules that complete the existing set and increase the possibilities of their use as an effective guide for the modification of a CAD model during the design phase. Modifications must guarantee not only that the model can be built with DMLS technology, but also that the building process is advantageous when compared to classic technologies. This last fact implies that times and post-processing work must be minimised and that the capabilities of DMLS technology must be exploited.

We follow with a schematic description of the current DGLs release. It includes an introductory description, followed by three elements: the Conceptual diagram, the Knowledge Matrix layout, and the considerations of Pros and Cons. The Conceptual diagram represents the process for which knowledge has been gathered and derived; the Knowledge Matrix (KM) layout describes the structure; the notes at the end of each description attempt to weight the importance of the result reached and list the directions for the following release.

3 CURRENT RELEASE OF THE DGLs

The previous release of the DGLs, not described here, led to a set of important errors, misunderstandings, etc. [10], that must be considered as the starting point of the current release of the DGLs. For this reason, some of them are briefly summarized here.

• Product characterization is one of the most important issues of the whole research. It is an intrinsically difficult task, given the large number of degrees of freedom involved. In the first release of the DGLs our consideration of characterization was extremely simple, merely to test the complete process, from product characterization to product model re-design.

• The location of the domains and their characterization were wrong. The conceptual diagram was too simple and did not represent knowledge generation correctly as occurs in the real world.

• The KM appeared to be too strict; at the same time its structure could lead to considerable redundancy (i.e. when the same technological requirement requires a number of rules, each with several attributes).

• The distinction among the different domains present in the conceptual diagram was lost in the KM. As we shall see below, this is a serious drawback.

• The re-design column corresponded to no element in the conceptual diagram.

Together with the attempt to eliminate these problems/mistakes, during the development of the new release of the DGLs the main effort has been to "widen the view", considering other aspects of the product life cycle. A synergy started with product characterization experts for verification in dimensional, micro, and macro-geometrical terms. All this led to analysis and adoption of the ISO-GPS concepts. This is the reason why the next paragraph shows a brief description of the GPS. A different RP technology was considered in this version, aiming at evidencing that the DGLs can usefully be applied to a wide variety of RP techniques. The Fused Deposition Modelling (FDM) technique was considered, the main characteristics of which are also reported. A description follows of the current release of the DGLs, with a Conceptual diagram, KM layout, and a discussion.

3.1 THE GPS – GEOMETRIC PRODUCT SPECIFICATIONS

The relevant changes in the field of the ISO standardisation will determine a set of rules, which will also markedly influence the designers' way of thinking. As from 1995, a specifically constituted ISO Technical Committee (ISO/TC 213) is working on the harmonisation and development of standards in the field of the geometrical characteristics specification for products, known as GPS (Geometrical Product Specifications).

It is worth underlining two main aspects characterising GPS: firstly, the will to develop standards, which are to be used as "tools" of help during the design phase; then, the need for establishing real links between design and manufacturing and verification. This last implies both that the designer's intents must be unequivocally transmitted to manufacturing and to verification and that the designer must consider manufacturing and verification needs during design. ISO/TC 213 activities are developing in this direction.

The first action of ISO/TC 213, aimed at collecting and harmonising existing standards for the specification and verification of geometrical characteristics of parts, determined the so-called GPS matrix, briefly described below, as presented in *ISO/TR 14638: 1995 – GPS Masterplan* [11].

In the GPS matrix model the concept of *chain of standards* (also referred to as *chain links*) is applied. Referring to a specified geometrical characteristic, each *chain* collects all the related standards, which can be used in the various steps of the production process (from design to verification, also considering manufacturing and metrological aspects). Each single standard in the *chain* affects the other standards, which must necessarily be known to be understood and applied correctly [12].

To reach a precise definition of the specifications, the concept of *operator* was developed. As in ISO/TS 17450-2:2002, an *operator* is defined as "an ordered set of operations", being an *operation* "a specific tool required obtaining features or values of characteristics, their nominal value and their limit(s)". The most interesting idea is to build a system where the specification procedure is parallel to the verification procedure, thus implying a duality in the specification and verification *operators*, as shown in fig. 1 [13].

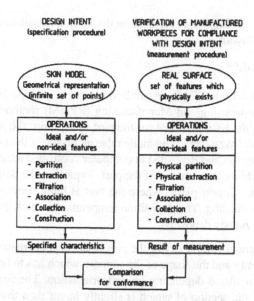

FIGURE 1. Correlation between specification and verification phases

As an example, being the *specification operation* the association of a minimum circumscribed cylinder in the specification of a shaft diameter, the dual *verification operation* will be for example the evaluation of a two-point diameter using a micrometer when verifying the diameter.

To establish this link between the specification and the verification step, the geometrical features defining the components were established. ISO 14660-1:1999 specifies first that geometrical features exist in the "world of specification" (how the work pieces are imagined by the designer), in the "physical world" and in the "world of inspection" (meaning, how the work pieces are represented by means of the results of measurement). The geometrical features are thus precisely defined in each world, as well as the relationships between them, by means of terms as *nominal feature* (design), *real feature* (physical world), *extracted feature* (measurement) and *associated feature* (link between nominal feature and extracted feature) [14].

In this context, the link between DGLs and GPS is immediately evident, from a conceptual point of view, both being basically tools to help the designer in the specification, refinement and communication of product characteristics. As the DGLs is a design tool especially related to a specific manufacturing process, analysing the possibility and opportunity of revising its structure and contents to agree with GPS principles seems of great interest: it could signify establishing the possibility of making a real link between design, manufacturing and verification, enlarging the application domain of the first DGLs version.

As explained below, GPS concepts have been actively adopted for updating the generation and the formalization of knowledge within the DGLs. The most important new elements introduced in the DGLs are the use of a new domain (verification domain), a classification of requirements and rules, and a different KM layout.

3.2 THE FDM TECHNOLOGY

The Fused Deposition Modelling (FDM) system builds parts in multiple thin layers, as is the case with all current Rapid Prototyping and Manufacturing (RP&M) methods. FDM uses spools of thermoplastic filament as the basic material for the part fabrication: the material is heated to just above melting point in a delivery head. The molten thermoplastic is then extruded through a nozzle in the form of a thin ribbon and deposited in computer controlled locations appropriate for the object geometry, thus building the sections of the part. Typically, the delivery head moves in the horizontal plane while the support plane, where the part is built, moves vertically, so that each section is built over the preceding. The deposition temperature is such that the deposing material binds firmly with that previously deposited.

Depending on the geometrical complexity of the part, some support material may be necessary to build the model: the quantity and the shape of the support, which has to be removed from the final part, are automatically calculated depending on part orientation. The first section is in any case built on a support plane, the section of which is slightly larger than that of the model, to allow easy removal of the part from the foam base.

The precision and surface finishing of the parts are affected by the so-called "slicing" (the layering), which depends on the kind of machine used, and can vary typically from 0.33 mm to 0.17 mm. A wide array of thermoplastic materials can be used to build models, among those ABS, polyolefin and polyamide.

The final parts do not need post-processing, except for removal of the support and perhaps grinding for a better surface finish.

Another advantage of the FDM system is that it can be used not only in a laboratory, but also in an office: no high powered lasers are used; the materials are supplied in spool format and present neither special handling nor environmental concerns.

3.3 CONCEPTUAL DIAGRAM OF THE DGLs

Fig. 2 shows the conceptual diagram of the current DGLs release. We have three domains, Design Domain, Manufacturing Domain and Verification Domain: the Manufacturing and Verification domains determine two different sets of requirements. Two different sets of rules (DFM – Design for Manufacturing – rules and DFV – Design for Verification – rules) derive directly from these. The attributes used for describing the product come from both these sets, and the product configuration, once defined in terms of attribute values, determine the generation of the last two sets of "derived rules" (Manufacturing and Verification rules). The meaning of all these elements should be clear considering the content of the KM.

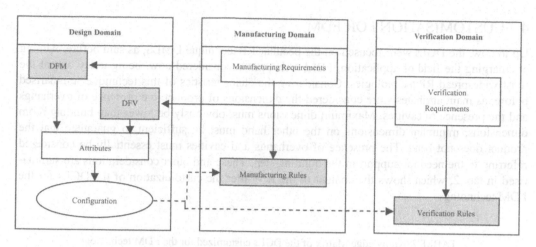

FIGURE 2. Conceptual diagram for the current DGLs release

3.4 KNOWLEDGE MATRIX OF THE DGLs

As regards the KM, here we have many new elements and issues to analyze. First of all, we have to consider that the layout of the matrix (tab. 1) does not match the conceptual diagram. This is because the idea was to bring the knowledge structure closer to the final user than to the technology. The attributes, the most important element from the user's point of view, appear at the left hand side of the matrix. Then there are the Design for Manufacturing and Design for Verification rules connected with the different attributes. To the right there are the Manufacturing and Verification requirements that, in a sense, justify the presence of the rules and the attributes. In the lower part of the KM there are the rules (Manufacturing and Verification) derived during the definition of the product configuration and represented explicitly.

TABLE 1. Knowledge Matrix of the current release of the DGLs

Attribute	Description	DFM and DFV rules	Manufacturing requirements	Verification requirements
Height	Height of the model (linear dimension)	Minimize the height of the model	Minimize building time (building time depends on the height)	
		On the basis of functional requirements and related dimensional tolerances, establish the verification method and indicate it on the drawing with proper symbols		The verification methods and tools have to be chosen considering the characteristics of the features to measure

Manufacturing rules	Verification rules
Use thicker layers	Measure the height according to the indication established on the drawing

4   CUSTOMISATION FOR FDM

Up to now, the DGLs were focused on the peculiar RP technique DMLS, as said before. Aiming at enlarging the field of application, FDM technique is considered now, being today one of the most widespread RP technologies. Considering the characteristics of this technique summarised before, as main attributes were considered the dimensions of product, the presence of overhangs and the presence of cavities. Maximum dimensions must obviously be lower than building room dimensions, minimum dimensions on the other hand must be sufficient to guarantee that the product does not bend. The presence of overhangs and cavities must essentially be considered referring to the need for support in the building step. These and other considerations are summarised in tab. 2, which shows the content of the KM after the customization of the DGLs for the FDM technology.

TABLE 2. Knowledge Matrix of the DGLs customized for the FDM technology

Attribute	Description	DFM and DFV rules	Manufacturing requirements	Verification requirements
Maximum dimensions	Maximum dimensions of the model (linear dimension)	Maximum dimensions of the product must be minor than maximum dimensions of the building room	Building room dimensions	
		Maximum dimensions of the product must be minor than maximum dimensions of the measuring volume		Verification workspace dimensions

Manufacturing rules	Verification rules
Split the model if the maximum dimensions are larger than the dimensions of the building room	Change the orientation of the product in the verification work-space

Attribute	Description	DFM and DFV rules	Manufacturing requirements	Verification requirements
Minimum dimensions	Minimum dimensions of the model (linear dimension)	Minimum dimensions of the product must be greater than minimum dimensions related to the presence of support	Need for support	
		The accessibility must be ensured		Type of probes

Manufacturing rules	Verification rules
Over dimension thin parts	Change the orientation of the product in the verification work-space to ensure the accessibility

Attribute	Description	DFM and DFV rules	Manufacturing requirements	Verification requirements
Overhangs	Presence of overhangs in the model	The presence of overhangs must be evaluated considering the need for support	Need for support	
		The accessibility must be ensured		Type of probes

Manufacturing rules	Verification rules
Change the orientation of the product in the work-space to minimize the quantity of required support	Change the orientation of the product in the verification work-space to ensure the accessibility

Attribute	Description	DFM and DFV rules	Manufacturing requirements	Verification requirements
Cavities	Through holes, blind holes, key-ways...	The dimensions and depth of cavities must be compatible with the need of support	Need for support	
		Dimensions and depth of cavities must be compatible with the characteristics of the probes		Type of probes

Manufacturing rules	Verification rules
Change the orientation of the product in the work-space to get easier the support removal	Change the orientation of the product in the verification work-space to ensure the accessibility

5   CONCLUSIONS

Technologies become more and more complex and sophisticated day by day. Although there is a great effort directed to optimizing human computer interaction, user interfaces, and, generally speaking, usability, great skill is still required to manage them. From this point of view, knowledge-based systems can be of great help.

This paper has described the Design Guidelines (DGLs), a knowledge-based system developed to help the mechanical designer to optimize products and render them compatible with Rapid Prototyping Technologies.

Two releases of the DGLs have been developed to date. Starting from the development of the first DGLs release, the enlargement due to the adoption of the GPS concepts was discussed, both from a conceptual and from a structural point of view.

The customization of the current version for the FDM RP-technology was revealed, giving the necessary information for the generation of an effective database. In the near future the tool will be used in the field and the results of these test cases will be presented.

REFERENCES

1. D.G. Ullman, The mechanical design process, Mc Graw Hill International Editions, 1997

2. Gatto, A., Iuliano, L., Prototipazione rapida: la tecnologia per la competizione globale, Tecniche Nuove, 1998.

3. Jacobs, P. F., Stereolithography & Other Rp&m Technologies: From Rapid Prototyping to Rapid Tooling, Society of Manufacturing Engineers, 1995.

4. Nelson, C., RapidSteel 2.0 Mold Inserts for Plastic Injection Molding, © by DTM Technology.

5. Agarwala, M., Bourell, D., Beaman, J., Marcus, H., Barlow, J., Direct Selective Laser Sintering of Metals, Rapid Prototyping Journal, Vol. 1, No. 1, pp. 26-36, 1995.

6. Kumaran, N., Chittaro, L. (eds.), Reasoning About Function, Artificial Intelligence in Engineering, ELSEVIER, Vol. 12, 1998.

7. Haffey, M., K., D., and Duffy, A., H., B., Knowledge Discovery and Data Mining within a Design Enviroment, In Cugini, U., Wozny, M. (eds.): Knowledge Intensive CAD Volume 4, Proc. Fourth IFIP WG 5.2 Workshop on Knowledge Intensive CAD, pp. 72-87, 2000.

8. Nielsen, J., Usability Engineering, Academic Press, Cambrige, MA, 1993.

9. EOS EOSINT250X user's manual.

10. S. Filippi, C. Bandera, G. Toneatto: Generation and Testing of Guidelines for Effective Rapid Prototyping Activities. Atti della conferenza ADM International Conference on Design Tools and Methods in Industrial Engineering (pp. A2.18-A2.27). Rimini (Italia),2001.

11. ISO/TR 14638:1995(E), Geometrical Product Specification (GPS) – Masterplan, ISO, Switzerland, 1995

12. Bennich, P., "Chains of Standards - A New Concept in GPS Standards", Manufacturing Review, The American Society of Mechanical Engineers, Vol. 7, No. 1, pp.29-38, 1994

13. ISO/TS 17450-2:2002, Geometrical Product Specifications (GPS) — General concepts — Part 2: Basic tenets, specifications, operators and uncertainties, 2002

14. ISO 14660-1:1999, Geometrical Product Specification (GPS) – Geometrical features – Part 1: General terms and definitions, 1999

FILLING BALANCE OPTIMIZATION FOR PLASTICS INJECTION MOLDING

R. Baesso, M. Salvador, G. Lucchetta

DIMEG, University of Padova, Italy

KEYWORDS: Injection Molding, Gate Location, Filling Balance.

ABSTRACT. The quality of an injection molded part is strongly influenced by the filling balance of the mold cavity. Gate location is the principal balance factor and it heavily depends on designer's experience and knowledge. In this paper, a three-step method is proposed which aims at determining the gate location that maximizes the filling balance of a mold cavity. First numerical simulations of the filling process are run changing gate position opportunely. Then third order regression surface method is applied to determine the gate location which minimizes the filling times standard deviation of the mesh nodes located along the cavity boundaries. The proposed method can be used for single gated mold cavities or multiple gated ones where the gates are positioned symmetrically. The results obtained from an industrial case study highlighted the effectiveness of the proposed approach in finding the optimal gate location by a restricted number of simulations.

1 INTRODUCTION

Injection molding of thermoplastics probably represents the most important and diffused process among the plastic forming technologies. The molten plastic flows from the machine nozzle through the sprue and runner system and into the cavities through the gate. In order to manufacture high quality moldings, it is important to identify the optimum gate location for the part, i.e. the position of the gate that creates balanced flow, allowing the extremities of the mold to fill at the same time and at uniform pressure. Besides, when the filling flows are balanced, the maximum injection pressure and clamp force are reduced. For parts with complex shape the best gate location is usually determined by trial and error. In fact, the filling balance is generally difficult to evaluate because no quantitative parameter defines it properly. Therefore the choice of the gate location heavily depends on designers' experience and knowledge.

In recent years, many approaches have been introduced to solve the best gate location problem. Lee and Kim [1] investigated gate locations using the evaluation criteria of warpage, weld lines and Izod impact strength. A local search was used to determine the nodes that optimized the location of the gate. Saxena and Irani [2] proposed a framework for a non-manifold-topology-based environment. A prototype system for gate location design was developed. The criteria for evaluation were based on geometry-related parameters. Lin [3, 4] selected injection location and size of the gate as the major control parameters, and chose the product performance (deformation) as the optimizing parameter. Combining the technologies of abductive networks and the optimization algorithms of annealing simulation, the optimal model for the location and size of the gate was constructed. Zhou et al. [5] proposed to select the best gate location based on an analysis of the plastic parts by reasoning according to an established set of rules. Pandelidis et al. [6, 7]

Published in: E. Kuljanic (Ed.) *Advanced Manufacturing Systems and Technology*, CISM Courses and Lectures No. 486, Springer Wien New York, 2005.

developed a system which can optimize gate location starting from an initial gating strategy. The system controlled and optimized the temperature differential and the number of elements overpacked by means of Moldflow® software for flow analysis.

In these paper it is proposed a three-step method which aims at determining the gate location that maximizes the filling balance of a mold cavity. The proposed method involves to carry out numerical simulations of the filling process by using the software Moldflow Plastics Insight®.

2 THE OPTIMIZATION METHOD

The proposed method aims at determining the best gate location that maximizes the filling balance of a mold cavity. Evaluation of the filling balance is not trivial because there is no quantitative parameter that defines it properly. Therefore it is impossible to objectively compare the filling balance degree of the same mold cavity obtained with different positions of the gate.

When the melt flows that fill a mold cavity are perfectly balanced, the boundary of the cavity is filled at the same time and with uniform pressure. The more different are the filling times of the cavity furthest points, the more unbalanced is the filling phase. Therefore it is proposed to associate the degree of filling unbalance with the variation of filling times among the furthest points of the mold cavity. This variation is well summarized by calculating the standard deviation of filling times for some points located along the cavity boundary. In this way, a quantitative evaluation of the filling balance can be carried out overcoming the limitations of the trial and error method.

The proposed three-step method involves the use of a numerical code for the simulation of the injection molding process. In particular, the software Moldflow Plastics Insight® has been used in this paper, although other similar applications can be employed too. In the first step a restricted number of numerical simulations of the cavity filling process are run. For each of them, while all other settings are maintained fixed, the gate location is opportunely changed within a specific region of the cavity that is supposed to contain the best one. The more numerous and evenly distributed are the analyzed gate locations the better will be the representation of the whole region of the mold cavity under investigation. Unfortunately, this involves also an ever increasing computational time.

In the second step, the standard deviation of filling times is calculated for the nodes of the mesh that are located along the cavity boundaries and for each simulation which was launched in the previous step. This value is used to quantitatively evaluate the filling balance.

In the third step, a third order regression surface is employed to determine the optimal gate location, i.e. the gate position that minimizes the calculated standard deviation. The independent variables of the regression surface are the x and y coordinates of the projection of the analyzed gate locations onto a plane parallel to the parting line of the mold. The dependent variable is the standard deviation of the filling times obtained by the numerical simulations run in the first step. A third order regression surface is proposed instead of a linear one because the former considers both the strong interactions among the independent variables and the curvature effects, even though it is more complex to solve.

Since there is a strong interaction between the variable, a third order is requested to interpolate data obtain from numerical simulations. Application of this technique makes it possible to obtain information about filling balance of mold configurations – in terms of gate location - that where not simulated. In fact, the regression surface allows to extrapolate the value of standard deviation of filling times for mold configurations with a gate location included in the investigation region but not specifically analyzed by the numerical simulations that were run in the first step. Therefore, by finding out the absolute minimum of the regression surface it is possible to determine the coordinates x and y of the gate location which minimizes the standard deviation of the filling times.

The proposed method can be used with single-gated mold cavities or with multiple-gated symmetrical ones whose gates are symmetrically placed.

2.1 THIRD ORDER REGRESSION SURFACE

The regression method is used for fitting a series of data with a curve or a surface [8]. In general, the dependent variable, or response, Y can be related to a number k of independent variables (x_1, x_2,..., x_k). The general model is as follow:

$$Y = \beta_0 + \beta_1 x_1 + \beta_2 x_2 + ... + \beta_k x_k + \varepsilon \tag{1}$$

and it is called multiple linear regression model with k variables. The β_j, ($j = 0, 1,..., k$) parameters are called regression coefficients. ε is the term relative to the random error and it is used to compensate the difference between the value of Y estimated by equation (1) and the real value that variable Y assumes in a specified point $(\overline{x}_1, \overline{x}_2,..., \overline{x}_k)$. In order to estimate values of β_j, the Least Squares Estimation method can be employed.

Often, this method is used as an approximating function too. That is, the real relationship between Y and $(x_1, x_2,..., x_k)$ is not known, however the multiple linear regression model can well approximate it when the value of independent variables are within a certain interval.

Multiple linear regression techniques are often employed to simplify the analysis of higher order models – like the third order model proposed in this paper - by means of the Least Squares Estimation method. The equation of a third order regression surface with two regression variables is as follows:

$$Y = \beta_0 + \beta_1 x_1 + \beta_2 x_2 + \beta_{11} x_1^2 + \beta_{22} x_2^2 + \beta_{12} x_1 x_2 + \beta_{111} x_1^3 + \beta_{222} x_2^3 +$$
$$+ \beta_{112} x_1^2 x_2 + \beta_{122} x_1 x_2^2 + \varepsilon \tag{2}$$

where β_0, β_1, β_2, β_{11}, β_{22}, β_{12}, β_{111}, β_{222}, β_{112}, β_{122} are the regression coefficients; x_1 and x_2 are the regression variables; Y is the dependent variable or response.

The equation (2) can be represented by a multiple linear regression model if it is assumed that:

$$x_3 = x_1^2 \qquad\qquad \beta_3 = \beta_{11}$$
$$x_4 = x_2^2 \qquad\qquad \beta_4 = \beta_{22}$$
$$x_5 = x_1 x_2 \qquad\qquad \beta_5 = \beta_{12}$$

$$x_6 = x_1^3 \qquad\qquad\qquad \beta_6 = \beta_{111}$$

$$x_7 = x_2^3 \qquad\qquad\qquad \beta_7 = \beta_{222}$$

$$x_8 = x_1^2 x_2 \qquad\qquad\qquad \beta_8 = \beta_{112}$$

$$x_9 = x_1 x_2^2 \qquad\qquad\qquad \beta_9 = \beta_{122}$$

In this way equation (2) becomes as follows:

$$Y = \beta_0 + \beta_1 x_1 + \beta_2 x_2 + \beta_3 x_3 + \beta_4 x_4 + \beta_5 x_5 + \beta_6 x_6 + \beta_7 x_7 + \beta_8 x_8 + \beta_9 x_9 + \varepsilon \qquad (3)$$

and can be easily solved by means of the Least Squares Estimation method.

3   AN APPLICATION CASE OF THE OPTIMIZATION PROCEDURE

3.1   FIRST STEP: RUNNIG SIMULATIONS

In order to test the effectiveness of the proposed optimization method an industrial plastic part was studied. Figure 1 shows the geometry of a bed for dogs which is proposed as a case study. The part is made of polypropylene and has a nominal wall thickness of 4.0 mm. The mold cavity is filled by a hot runner system with two gates.

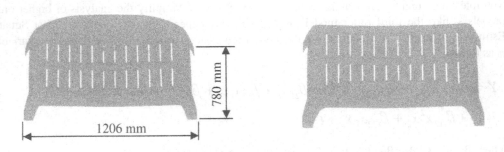

FIGURE 1. Geometry of the analyzed part

As mentioned above, the proposed method can be used with single-gated mold cavities or with multiple-gated symmetrical ones whose gates are symmetrically placed. In the latter case, the cavity mold can be divided into symmetrical fractions that contain only one gate and the analysis can be restricted to one of these fraction. In this way the computational time are reduced without compromising the reliability of the results. The analyzed part belongs to this latter case because it has one symmetry axis and the two gates are symmetrically located. The fraction of the part that was investigated is showed in figure 2.

In the first step of the optimization procedure 10 numerical simulations of the filling phase of the mold cavity were run. For each of them, the gate location was changed within a region of the mold that was supposed to contain the best one according to the evenly distributed points reported in figure 2.

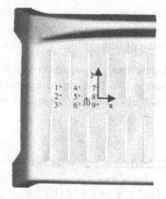

FIGURE 2. Fraction of the analyzed part which was investigated with the 10 gate locations which were analyzed by numerical simulations

3.2 SECOND STEP: STANDARD DEVIATION OF FILLING TIMES

The second step of the proposed method involves calculation of the standard deviation of filling times for the nodes of the mesh that are located along the cavity boundaries and for each simulation which was launched in the previous step.

TABLE 1. Standard deviation of filling times of the boundaries nodes for each gate location analyzed

CASE	x [mm]	y [mm]	$t_{average}$ [s]	σ [s]
1	-160	47	5,091	1,687
2	-160	10	5,224	2,229
3	-160	-27	5,77	3,975
4	-80	47	5,547	1,553
5	-80	10	5,807	3,357
6	-80	-27	6,563	4,421
7	0	47	5,565	1,884
8	0	10	5,695	2,374
9	0	-27	6,135	3,197
10	-40	1	5,936	2,922

Table 1 reports the results obtained for each simulated gate location. x and y are the coordinates of the projection of the analyzed gate locations onto a plane parallel to the parting line of the mold. The origin of the reference coordinate system is shown in figure 2. $T_{average}$ is the average of filling times for the nodes of the mesh that are located along the cavity boundaries.

3.3 THIRD STEP: DETERMINATION OF THE REGRESSION SURFACE

The last step of the gate location optimization method involves finding a mathematical correlation between the independent and dependent variables, i.e. respectively the coordinates x and y of the gate location and the standard deviation of filling times.

The third order regression model described by equation (2) was reduced to a multiple linear regression model as shown in equation (3). In order to estimate the values of the 10 unknown parameters ($\beta_0, \dots, \beta_9$), the Least Squares Estimation method was employed and the following equation was obtained:

$$Y = 3,04 + 8,328 \cdot 10^{-3} x - 6,027 \cdot 10^{-2} y + 2,301 \cdot 10^{-4} x^2 +$$

$$-1,004 \cdot 10^{-3} y^2 + 6,107 \cdot 10^{-4} x \, y + 1,132 \cdot 10^{-6} x^3 + \tag{4}$$

$$+3,751 \cdot 10^{-5} y^3 + 3,292 \cdot 10^{-6} x^2 y - 4,232 \cdot 10^{-6} x \, y^2$$

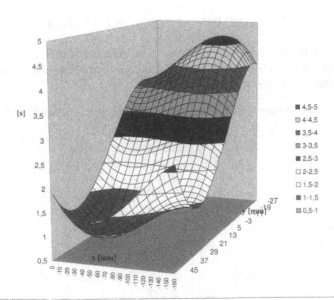

FIGURE 3. 3D representation of the obtained third-order regression surface model

A three dimensional representation of the equation (4) is showed in figure 3. It is worth noticing that the independent variables x and y in the equation (4) were not continuous in their domains because obviously the gate could not be positioned in correspondence of the holes of the part. However, in figure 3, the surface is represented as a continuous function it the x-y domain.

3.4 RESULTS

It was calculated that the absolute minimum of the equation (4) was at x = -60 mm and y = 41 mm and such position was proposed as the best gate location (figure 4a). In order to evaluate the proposed method, several simulations with different gate locations were launched. The obtained results showed that the proposed best gate location was really the best one.

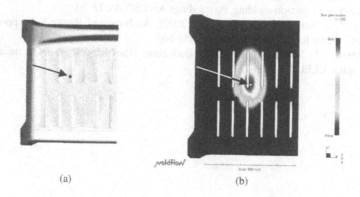

(a) (b)

FIGURE 4. (a) Best gate location calculated and (b)best gate location as resulted from a simulation
with Best Gate Location Module of Moldflow Plastics Insight®

The proposed gate location was then compared with the one obtained by the Best Gate Location module of Moldflow Plastics Insight® (figure 4b). A numerical simulation based on the latter showed a higher standard deviation of the filling times of the boundaries nodes and, furthermore, it would cause a short shot.

4 CONCLUSIONS

In injection molding the gate location selection process relies heavily on the knowledge and experience of mold engineers and is mainly based on a trial-and-error method. In this paper, a new method has been developed to determine the best gate location, i.e. the one that maximizes filling balance by minimizing the standard deviation of the filling times of the cavity boundary nodes. As a practical application of the proposed method, an industrial plastic part was analyzed. The obtained results were compared with the ones obtained from the Best Gate Location module of Moldflow Plastics Insight®. The proposed method has eventually proved to be more accurate in finding the best gate location.

REFERENCES

1. Lee, B. H., Kim, B. H., (1995), Optimization of part wall thickness to reduce warpage of injection-molded parts based on the modified complex method, Polymer Plastics Technology Engineering, 34, 793-811.
2. Saxena, M., Irani, R. K., (1993), An integrated NMT-based CAE environment-part: Application to automated gating plan synthesis for injection molding, Engineering with Computers, 9, 220-230.
3. Lin, J. C., (2001), Optimum gate design of freeform injection mould using the abductive network, International Journal of Advanced Manufacturing Technology, 17, 297-304.
4. Tai, C. C., Lin, J. C., (1999) The optimal position fot the injection gate of a die-casting die, Journal of Material s Processing Technology, 86, 87-100
5. Zhou, Z.Y., Gu, Z. Zh., Shi, J. Y., (200), Research on integrated design techiniques for injection mold runner system, Journal of Computer-Aided Design and Computer Graphics, 12(1), 6-10.
6. Pandelidis, I., Zou, Q., Lingard, T. J., (1998), Optimization of gate location and operational molding conditions for injection molding, Proceedings ANTEC, 46, 18-20.
7. Shi, F., Lou, Z. L., Lu, J. G., Zhang, Y. Q., (2002), An Improved Rough Set Approach to Design of Gating Scheme for Injection Moulding, 662-663.
8. Berti, G., Monti, M., Salmaso, L, (2002), Introduzione alla Metodologia DOE nella Sperimentazione Meccanica, CLEUP, 243-267.

INVESTIGATION ON HEAT TRANSFER IN THE INVESTMENT CASTING

G. Casalino[1], J. Orkas[2], N. Accettura[1]

[1] DIMeG, Mechanic and Operational Engineering Department, Politecnico di Bari, Italy
[2] Foundry Department TKK Helsinki University of Technology, Finland

KEYWORDS: Investment Casting, Heat Transfer, Experimental and Numerical Analysis.

ABSTRACT. Ceramic moulds produce castings with fine details, smooth surface (smoothness 3 mm), and a high degree of dimensional accuracy because the metal contraction can be predicted closely enough to provide castings within excellent tolerances (±0.08 mm on dimensions to 25 mm, ± 0.13 mm on dimensions to 75 mm, and more).
The purpose of this paper is to study the thermal properties of ceramics shell moulds for investment casting, which is also known as "lost wax process". A special attention was given to the analysis of the modes governing the heath transfer in the casting process and their correct mathematical modelling.
Casting temperatures were measured either by a set of thermocouples and a thermo-camera. Experimental data were compared to the result predicted by commercial simulation software.

1 FUNDAMENTALS OF THE INVESTMENT CASTING

Investment casting is also known as the lost wax process. Intricate shapes can be made with high accuracy. In addition, metals that are hard to machine or fabricate are good candidates for this process. It can be used to make parts that cannot be produced by conventional manufacturing techniques, such as turbine blades that have complex shapes, or airplane parts that have to withstand high temperatures [1,2].

The materials used for the slurry are a mixture of plaster of Paris, a binder and powdered silica, a refractory, for low temperature melts. For higher temperature melts, sillimanite an alumina-silicate is used as a refractory, and silica is used as a binder. Depending on the fineness of the finish desired additional coatings of sillimanite and ethyl silicate may be applied. The mould thus produced can be used directly for light castings, or be reinforced by placing it in a larger container and reinforcing it more slurry.

Just before the pour, the mould is pre-heated to about 1000°C (1832 °F) to remove any residues of wax, harden the binder. The pour in the pre-heated mould also ensures that the mould will fill completely. Pouring can be done using gravity, pressure or vacuum conditions. Attention must be paid to mould permeability when using pressure, to allow the air to escape as the pour is done.

The commonly used binders are also siliceous and include colloidal silica, hydrolyzed ethil silicate and sodium silicate. Hybrid binders have also developed, and alumina or zirconia binders are used for some processes [3].

The most common refractories for ceramic shell moulds are siliceous, zircon, and various aluminium silicates composed of mullite and free silica [4].

The types of materials that can be cast are Aluminum alloys, Bronzes, tool steels, stainless steels, Stellite, Hastelloys, and precious metals. Parts made with investment castings often do not require any further machining, because of the close tolerances that can be achieved.

Published in: E. Kuljanic (Ed.) *Advanced Manufacturing Systems and Technology*,
CISM Courses and Lectures No. 486, Springer Wien New York, 2005.

Investment casting can produces large, complex thin-walled engine carcass parts such as diffuser housings and combustors, and generally parts up to 1500 mm across and 600 mm deep. These are replacing sheet metal fabrications, not only because they are more cost effective but because they also provide a high rigidity monolithic component with superior service characteristics. Investment cast integrally bladed turbine wheels for smaller turbine engines, offering large cost savings over mechanical fabrications, have been widely adopted.

Recently, demand has been growing for investment cast blades and fans to be used in land based turbines for power generation. Components are larger and operate at lower temperatures, but the growing demand parallels that for aero engine parts in the 1980s. This sector of the market has in fact sustained airfoil investment casters over the past few years when aircraft demand dropped.

In a different field altogether, there is a long established market for hip replacement prostheses made from cobalt-base superalloys formed by investment casting.

2 HEAT TRANSFER IN CASTING

The heat transfer through the mould is very important during casting and solidification. The formation of an air gap between the casting and the mould has a huge impact on the heat transfer [12]. The relationship between variations in the heat transfer coefficient and the formation of an air gap has been investigated in a number of articles about mould casting. It has been shown that gap formation starts as soon as the solid metal shell is strong enough to withstand the pressure from the liquid metal. Before the formation of a microscopic air gap, heat transfer occurs mainly through conduction. When the air gap starts to grow, the conduction is gradually reduced and the heat transfer can be described by a simple superposition of the radiation and conduction terms. The actual mechanism behind the material shrinkage resulting in a gap between the solidified shell and the inner mould wall has not yet been elucidated. The general explanation is that the shrinkage is caused by thermal stresses in the material due to a temperature gradient [9]. A model has been suggested assuming that the solidification process is greatly affected by the formation and condensation of vacancies.

By its nature, casting is primarily a heat-extraction process. The conversion molten metal into a solid semi-finished shape involves the removal of the following forms of heat:
- superheat from the liquid entering the mould from the tundish.
- the latent heat of fusion at the solidification front, and finally
- the sensible heat (cooling below the solidus temperature) from the solid shell

These heats are extracted by a combination of the following heat-transfer mechanisms:
- convection in the liquid pool
- heat conduction down temperature gradients in the solid shell from the solidification front to the colder outside surface of the cast, and
- external heat transfer by radiation, conduction and convection to surroundings.

Also not less important is the heat transfer before the molten metal is poured into the mould or instance, in the casting of steel, heat transfer is important before the steel enters the mould because control of superheat in the molten steel is vital to the attainment of a predominantly equiaxed structure and good internal quality. Thus, conduction of heat into ladle and tundish linings, the preheat of these vessels, convection of the molten steel and heat losses to the surroundings also play an important role in the investment casting.

Because heat transfer is the major phenomenon occurring in casting, it is also the limiting factor in the operation of a casting machine. The casting speed must be limited to allow sufficient time for the heat of solidification to be extracted from the strand.

Heat transfer not only limits maximum productivity but also profoundly influences cast quality, particularly with respect to the formation of surface and internal cracks. In part, this is because metals expand and contract during periods of heating or cooling. That is, sudden changes in the temperature gradient through the solid shell, resulting from abrupt changes in surface heat extraction, causes differential thermal expansion and the generation of tensile strains. The rate of heat extraction also influences the ability of the shell to withstand the bulging force due to the ferrostatic pressure owing to the effect of temperature on the mechanical properties of the metal.

3 EXPERIMENTAL SET-UP AND MATERIALS

The slurry, which we have used in the experimental part of this project, is a combine between two different coating materials:

First Slurry (Primary Slurry): Bindzil (binder), Zircon_200 (powdwer, first stucco), Wictawet (wetting agent).

Second Slurry (Backup Slurry): Ludox SK-R (binder), Molochite-200 (powder, second and third stucco).

First stucco zircon sand, first backup stucco molochite 50/80 (0.2 mm), remaining backup stucco molochite 16/30 (0.5 -1 mm).

In the experiments of these project Al-Si 10 was used as casting metal, which presents a good pouring properties and a low shrinkage. The he Eutectic composition corresponds to 11,7% in silicon weight, to a temperature of 577°C.

3.1 EXPERIMENTAL DEVICES DESCRIPTION

Figure 1 shows the experimental set-up that was used to study the heat transfer through the ceramic mould. It was made up from the mould, the data logger, and 6 thermocouples.

To minimize the heat contact between the hot aluminium casting and cold thermocouple wire the lasts were covered with ceramic slurry.

Ceramic surface temperature, Te, was measured with 3 type-K thermocouple spot-welded (T2, T4, T6) and other 3 type-K thermocouples were placed at predefined depths within the flux gap to measure flux film temperatures (T1, T2, T3).

The thermocouples were connected through an

FIGURE 1. Experimental set-up

A/D serial board, to a laptop data acquisition system. Temperatures were recorded from each thermocouple every second (0.33 Hz). The assembled apparatus was surrounded by refractory brick, zirconium paper, and Kao-wool to reduce heat losses.

Four experiments were conducted in order to test the influence on the process of several process parameters. The infrared camera and numerical simulation were used. The data coming from the experimental devices and numerical simulation were compared, when possible.

3.2 MOULD PREPARATION

The mould preparation consisted of the following steps.
- Computer aided drawing of the master pattern (Fig 2); the piece has been drawn using Solid Edge 14 and Descartes 3D as support program.
- Producing the master pattern with protoprinting technology [10,11].
- Producing a master mould in plastic from the master pattern.
- Using the model in "red wax" (complex mixture of many compounds including natural or synthetic wax and even water), the positive die was produced mixing two kinds of polyurethane [6,7].
- Producing the wax pattern. For this process "red wax" was used. The red wax melted at 300°C and then was poured in the plastic die. This step was difficult since the wax pattern was fragile and broke down several times when taken away from the plastic mould. A spray "release agent", which covered the plastic mould, eased this the extracting operation [8].

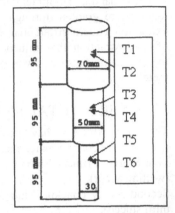

FIGURE 2.Thermocouples positions

- Creating ad-hoc holes to insert the thermocouples inside the mould for surveying the metal temperatures. Straws were attached using a soldering iron.
- Coating the pattern several times, with the investment material, until the shell thickness was reached (3-6 mm).
- Melting the wax pattern to permit it to run out of the mould (650° degrees)
- Heating the mould to give it the right hardness (900° degrees for 1 hour)
- Attaching the thermocouple after that the die was ready. Wrong contacts between thermocouples were avoided (Fig 2).

3.3 FIRST EXPERIMENT

The liquid flux at 730 ° was poured into the apparatus. Then data recorder system was turned on and switched off after 4-5 hours. The resulting cooling curve will be analysed to determine the local solidification time and the effects of modification and cooling rate on the primary crystallization and eutectic growth temperatures.

The temperature histories for the first experiment of all the six thermocouples are given in Figure 3.

It can be seen that the direction changes of the curves are not very clear, for this reason we can concentrate our attention on T5 and T6 thermocouples.

T5 represents the trend of the temperature internal to the die in correspondence of the smallest diameter.

In the T5 experimental cooling curve for Al-Si 10 the formation of primary Al nucleation and Al-Si eutectic reaction occurred at a temperature well below both the liquidus and the eutectic temperatures. The reason for this depends on the nucleation and growth of the crystals in the liquid. In the following discussion, only the expression $dQ/dt = \alpha(V, Cp, \Delta H, \rho, df/dt) \, dT/dt$, (V volume of the sample, Cp heat capacity, dT/dt cooling rate of the liquid "the slope", ΔH heat of solidification, ρ density, df/dt volume fraction of solid formed at a changing temperature) will be used to define the heat transport away from the experimental setup. A striking feature, observed in all experiments, is a sharp spike in the temperature of thermocouples in the first few seconds after pouring the liquid flux into the mould.

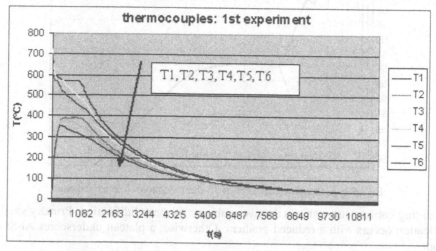

FIGURE 3. Temperature histories of the thermocouples during the 1st experiment

This is due to the reason that when the liquid cools down to the solidification temperature, crystals nucleation begins and there is a reduction in cooling rate and a consequent reduction in the gradient of the cooling curve. The change in shape of the temperature-time curve now depends on the volume fraction of solid formed. The volume fraction of solid formed is related to the number of crystals and their growth rate. Crystals are normally nucleated when the temperature of the liquid reaches the liquidus temperature as well as the eutectic temperatures. In spite of this, the temperature of the liquid will decrease further and initially the curve can be extrapolated to show where no crystals are growing. After a certain time, the slope of the curve begins to decrease, slowly at first and then more rapidly (Fig4). In the early stages of crystal formation, the growing crystals are small (as is heat development) and therefore have no effect on the bulk liquid temperature. The growth rate is also low because of the low supercooling close to the liquidus temperature. The supercooling increases, when the temperature decreases; therefore, the growth rate will increase as the area of the crystal increases. The result is that more heat is given away and the slope of the temperature-time curve begins to decrease. The change in the slope will begin earlier and will proceeds faster. The more crystals there are in the liquid, the earlier the change in slope starts, and the faster it will advance. When the temperature increases, the growth rate of the free crystals decreases again, but they are so large by this time that the solidified volume per unit of time will also be large. As such, the temperature will continue to increase until the growth rate and the growth area of the crystals offset the heat extraction. In conclusion when solidifying hypoeutectic or hypereutectic alloys, (AlSi 10 hypoeutectic alloy) the first solid to form is a single phase, which has a composition different to that of the liquid (Al nucleation). This causes the liquid composition to approach that of the eutectic as cooling occurs. Once the liquid reaches the eutectic temperature it will have the eutectic composition and will freeze at that temperature to form a solid eutectic mixture of two phases.

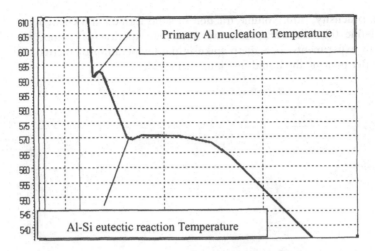

FIGURE 4. Temperature curve of T5 thermocouple (first experiment)

The resulting cooling curve shows the two moments of the solidification. Primary single phase Al nucleation occurs with a reduced gradient. Otherwise, a plateau underscores Al-Si eutectic reaction.

T6 represents the trend of the temperature on the external surface of the die in correspondence of the smallest diameter.

The rate of heat transferred across the ceramic mould, which was measured by that thermocouple, depends mainly on the properties of the mould flux during filling. These properties include phonon and photon conductivity, radiation properties such as the emissivity and absorption coefficient, and the contact thermal resistance where the flux is solid and in correspondence to the mould. Note during the brief initial time in each time the temperature is increased, the interface resistances jumps sharply. It is believed to be due to the heat content of the molten flux and not to a high rate of conduction (convection). When the solidification phase begins the aluminium heat thermal coefficient drops down (from convection to conduction) because of the change in the microstructure of the material and the radiation contribution decreases with the loss of the temperature for this reason the curve stops to increase. There can be seen a period at a constant temperature which depends on the diameter of the cylinder. The bigger the diameter, the wider is the arc of time at constant temperature; it is due to the heat energy of the aluminium mass. Then the temperature starts to decrease and eventually reaches room temperature. The curve has less of a gradient than the one observed from within the mould.

3.4 SECOND EXPERIMENT

In the second experiment the usual cylindrical structure with three diameters and a cup of the same diameter as the middle part of the cylinder, was used. As measuring tools, the usual three thermocouples were placed inside the die and a thermocouple internal to the cup. It can be seen in figure 5 that the three thermocouples inside the cylindrical structure behaved in the same way seen in the other experiments but this time it is important to notice that the cooling curve registered by the thermocouple inside the cup has a totally different trend from the others. After the primary crystallization temperature and solidification temperature the alloy in the cup cooled faster. It is believed to be due to the absence of the radiation phenomena and thermal conductivity that instead existed between the various parts of the cylindrical structure.

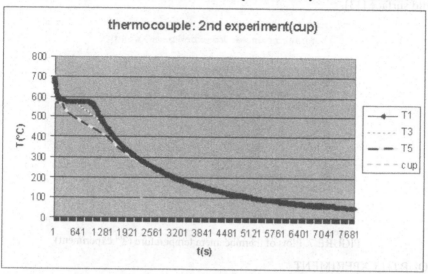

FIGURE 5. Second experiment registered thermocouples temperatures

3.5 THIRD EXPERIMENT

In the third experiment the simple cylindrical structure was used again. As measuring tools, a thermocamera registered the surface heat. Figure 6 shows the hot sample and the positions were the temperatures. Figure 7 the results obtained by the thermocamera the ones used in the first experiment to study the trend of the cooling curves on the external surface of the ceramic mould. It should be noticed that the cooling curve of the X point, even if has the same trend of the three other curves, presented higher temperatures, this is believed to be due to the close position of the X point (Fig 6) to two intersecting surfaces that exchange heat (convection and radiation) between them.

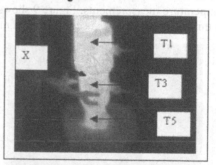

FIGURE 6. Thermocamera view (3rd ex.)

The greater absorption of heat can be explained by radiation effect during the first seconds (before the solidification) and by convective effect in the remaining part of the experiment.

It can be seen that the trend of temperatures obtained by the thermocamera (Fig 7) agreed with those obtained by thermocouples in the first experiment. Otherwise, it must be underscored a big difference between temperatures values obtained from thermocouples from thermo camera. The gap was between 200 °C and 300°C.

Several authors have done many experiments in order to justify the temperature difference, and it has been demonstrated that the explanation is the presence of the "foreign material" in and on the mould. Since the thermocouples wires have different thermal conductivities than the casting and mould material, for this reason there is always some distortion of the true temperature field. The thermocouple errors become more significant for sensor placed close to the mould surface [13].

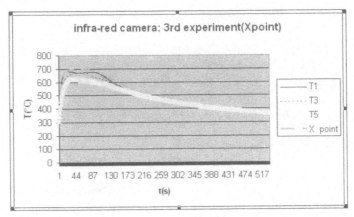

FIGURE 7. Plots of thermocamera temperature (3rd experiment)

3.6 FOURTH EXPERIMENT

In the fourth and last experiment a heat draining differences had been found between experimental measurements and numerical calculations. Two thermocouples were inserted in the mould in two points, to show as good as possible the radiation effect contribution during the solidification and cooling of the casting. Moreover, the thermocamera was use for further temperature measurements. Figure 9 shows the points were temperature was measured.

In this experiment it has used a different mould shape and it was found that (see figure 9)

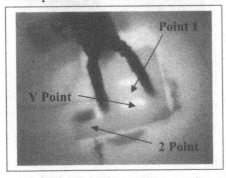

FIGURE 8. Thermocamera picture (4th ex.)

1. the cooling curve concerning the Y, which is internal to the hole, point reached the highest temperature,

2. point 1 had a different distribution of temperatures as regards Point 2, even if they had the same shape (Fig 9).

3. point 1 reached higher temperatures than Point 2, this involved a different cooling of metal in the two considered sections.

In the plot it can also be seen that the point-1 temperature reached 650°C in the moment of the casting (zero in the scale of time). Then the temperature grew up to 700°C since the radiation and convective effects. Then after the solidification phase it starts to cool down to the room temperature.

For the some reasons explained for the point-1, the y-point temperature overcomes that of point-2. Finally the two temperatures converged to the room temperature long after the mould was filled. The explanation is that the different positions of the points influenced the convective effect. Point 1 felt the effect of greater thermal exchange than point-2, because of the highest heat energy presence in the finger 1 area.

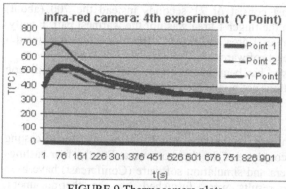

FIGURE 9.Thermocamera plots

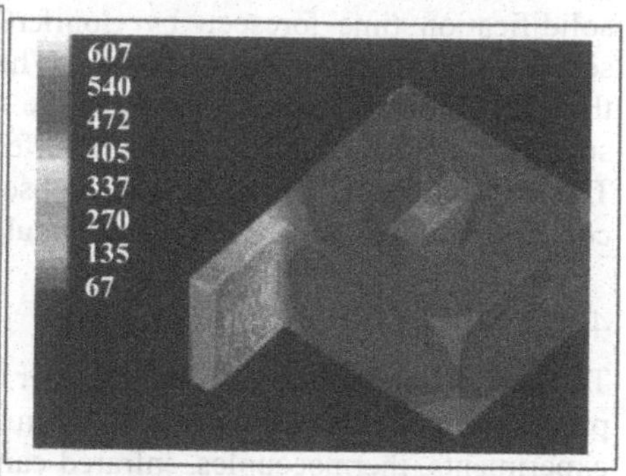

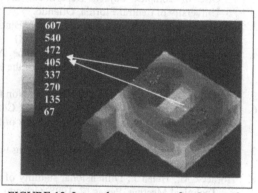

FIGURE 10. Calculated temperature

Finger1area is the hottest part, because of the high rate of energy exchange between the different parts of the mould. Finger-1 area was hotter than the finger-2 area since the last was not surrounded by any other parts of the mould and so it did not have any radiation and convective contribution to heat.

Figure10 represents, for the same piece, the temperature gradient of the mould surface after the solidification obtained by simulation program ConiferCast. ConiferCast®, which is a computer program for simulating casting processes that involve metal flow and solidification. It uses world famous FLOW-3D® fluid simulation software as a calculation engine. The program is based on the fundamental laws of mass, momentum and energy conservation [14].

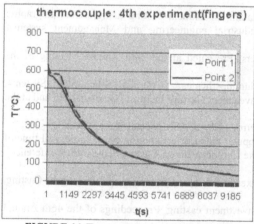

FIGURE 11.Temperature inside the mould

FIGURE 12. Internal temperature after 3 seconds

It can be seen that the finger 2 area was the cold part in agreement with the empiric data but the finger 1 area is not the warmest one. That is believed to be due to the absence in the ConniferCast of a feature that is able to reproduce the radiation effect.

As a verification of the disagreement between software and experimental, it is possible to compare temperatures inside the mould (Fig 12). It can be firstly seen that after about 3 seconds filling time the temperature of the finger 1 area was around 440°degrees, which is very much lower than the one registered by thermocouples inside the mould (fig. 11). Secondly, the solidification time foreseen by ConiferCast consist of 21 seconds instead the 700 (about) seconds registered by thermocouples. Then the finger 1 area was not the warmest part. Finally, the temperature difference, between the finger 1 area and finger 2 thermocouples registered a smaller then that calculated by the Conifercast software.

The explanation for the differences observed has been based on the fact that in the radiation contributions was neglected during calculations not by the thermocouples.

4 CONCLUSIONS

The task of measuring the heat transfer through the ceramic mould in the investment casting process was performed using both thermocouples and the thermocamera. For the four castings experiments, thermocouples, infrared camera and simulation software (Conifercast) have been used. Among those the most reliable results were furnished by the thermocamera. Thermocouples and calculations based on the fundamental laws of mass, momentum and energy conservation were no consistent with the actual thermal fields measured by the thermocamera.

The presence of thermocouple itself affects the temperature field that is to be measured. Comparing the thermocouple curves with infrared camera ones, even if they present the same cooling trend they did not showed the same relation Time –Temperature.

The simulation software needs to be improved by radiation flux equations.

REFERENCES

1. Chua Chee Kai, Rapid Prototyping – Principles and Application in Manufacturing , John Wiley & sons Inc., 1997.
2. Yasser A.Hosni, Jamal Nayfeh, Ravindran Sundaram, Investment casting using stereolithography: case of complex object, Department of industrial engineering and Management systems-University of Central Florida, Orlando.
3. Chester Feagin R., Alumina and Zirconia Binders, Presented at the 29th Annual Meeting of the Investment Casting Institute, October 1981
4. Chester Feagin R., Characteristics of Some Aluminosilicates - Colloidal Silica Shell Systems, Presented at the 26th Annual Meeting of the Investment Casting Institute, October 1978 and at the EICF Conference, June 1978.
5. Metals handbook. Properties and selection: Non ferrous alloys. ASM, Metals Park, Ohio.
6. Jagos M., Macosek I., Pospisil B., Rusin K., Computer aided determination of technological time of solidification of wax patters. Proceedings of the European Invest Casters Federation Conference , 1978.
7. Williams RB. , Update on investment casting waxes, 7th World Conference on Investment Casting, Paper 2, 1998.
8. Horacek M., Helan J., Dimensional stability of investment casting, Proceedings of the 46th Annual Technical Meeting, Paper 17, 1998, Orlando.
9. Campbell J., Castings, Butterworth-Heinemann Ltd, 1993, UK.
10. Leong K.F., Rapid Prototyping : Fundamentals of Stereolithography, McGraw Hill , 1993, New York
11. Terry T. Wohlers, Rapid Prototyping Systems, Wohlers Associates, published in the Proceedings of the First European Rapid Prototyping Convention, June 1993, Paris .
12. Ho K.,Pehlke RD., Mechanisms of heat transfer at a metal-mould interface, AFS Trans 1984.
13. Piwonka S., Woodbury K., Wiest JM., Modelling casting dimensions: effect of wax rheology and interfacial heat transfer, Material and Design 21, 2000, USA.
14. VTT Chemical Technology, Industrial Mathematics, P.O. Box 1404 FIN 02044, Finland.

REVERSE ENGINEERING OF A TURBINE BLADE: COMPARISON BETWEEN TWO DIFFERENT ACQUISITION TECHNIQUES

C. Bandera, S. Filippi, B. Motyl

Department of Electrical, Management and Mechanical Engineering, University of Udine, Italy

KEYWORDS: Reverse Engineering, Methods Comparison, Turbine Blade.

ABSTRACT. Reverse Engineering is a rapidly evolving discipline. Nowadays shape acquisition systems have reached enough capabilities to reproduce complex free form objects. This paper presents a comparison between two different acquisition systems to evaluate their performances and their easy to use capabilities and to evaluate quality and accuracy of the 3D reconstructed models. To achieve this task, a complex free form object, a steam turbine blade, was acquired. After shape acquisition and data elaboration of this particular object an evaluation of the two systems characteristics and of the 3D CAD reconstructed models were done and results are illustrated here.

1 INTRODUCTION

Reverse Engineering (RE) is a rapidly evolving discipline, which covers a multitude of activities. There are several application areas of RE, for example: measurement, quality control, design, virtual reality and reconstruction of artistic and ancient objects. Nowadays, shape acquisition systems have reached enough capabilities to reproduce geometric profiles of complex objects with high accuracy and repeatability. For all these kinds of application a large number of systems, based on different approaches (mechanical, optical, laser based sensors), were developed and used.

The authors propose a comparison between two different acquisition techniques applied for digitizing a complex sample object: a turbine blade. In particular, the aim of this work is to evaluate the system performance and the easy to use capability of the two systems and the accuracy and quality of the 3D reconstructed models in order to find and solve acquisition problems and to fit the best match for complex mechanical components acquisition.

This paper begins with a RE process overview and a shape systems taxonomy. Then the description of the two different acquisition systems used for this study is done. After that, the acquisition of a sample object is presented in order to compare the two different systems, in terms of acquired data volume and quality of acquisition. Lastly, results and outcomes are illustrated.

2 REVERSE ENGINEERING PROCESS: PHASES AND SYSTEMS TAXONOMY

In mechanical design and industrial engineering Reverse Engineering is known as a process that transforms real objects into engineering models and concepts. Typically this process starts with

Published in: E. Kuljanic (Ed.) *Advanced Manufacturing Systems and Technology*,
CISM Courses and Lectures No. 486, Springer Wien New York, 2005.

measuring an existing object in order to deduce its surface and to reproduce its solid form with the advantage of CAD/CAM technologies. The main phases of this process are data acquisition and data elaboration. This two phase can be divided in more detailed sub-phases as: 1) data capture or part digitization; 2) pre-processing; 3) segmentation and surface extraction; 4) CAD model creation [1],[2],[3]. These phases are often overlapping and instead of a sequential execution usually several iterations are required.

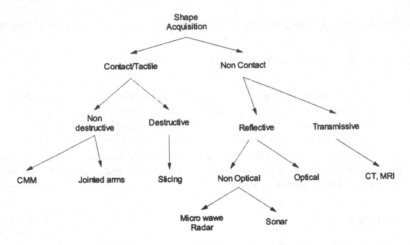

FIGURE 1. Shape acquisition systems taxonomy [4]

There are many different methods for acquiring and digitizing shape data of objects. A first distinction can be done referring to the way of interaction between systems and parts, so there are Contact (or Tactile) and Non Contact methods. Another distinction can be done referring to the mechanism (robotic arms, coordinate measuring machines - CMM) or phenomenon (optical, magnetic, acoustic) used for capturing shape, as shown in Figure 1.

Each method has strengths and weaknesses which require that the data acquisition system be carefully selected for the shape capture functionality desired.

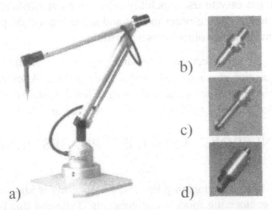

FIGURE 2. a) Robotic arm, b) Conical stylus probe, c) Spherical stylus probe, d) Squared stylus probe

The most popular shape capture systems are CMM machines and robotic arms (as contact methods) and Laser and Optical Systems (as non contact methods). In this work, two different methods were used for acquiring the digital shape of a turbine blade, a robotic arm and a structured light stereo system.

3   DESCRIPTION OF THE COMPARED SHAPE ACQUISITION SYSTEMS

3.1   CONTACT SYSTEM: 5-AXIS ROBOTIC ARM

The contact system used here is a manually driven measurement arm (Figure 2a). It has 5-axis of movement and it is designed to detect profiles and surfaces of a large variety of parts. It works like a CMM, in fact, the arm has a touch probe that can be handily driven by an operator for measuring points on the surface of the object (sensing devices on the joints determine the relative coordinate location of these points).

This system is interfaced with a 3D free form modeling software (Rhinoceros by Mc Neel&Associates) that allows to perform the calibration procedure and digitize and collect the 3D coordinates of the acquired points. Also this modeling software allows preprocessing point clouds and extracting surface patches (if the number of points is limited).

This model of robotic arm can be equipped with different touching probes: a conical stylus with insert in super rapid steel (Figure 2b), a spherical rubin stylus (Figure 2c) and a squared stylus for borders acquisition (Figure 2d). The technical characteristics of the system are summarized in Table 1.

TABLE 1. Technical characteristics of the robotic arm

Robotic Arm model	Baces3D M200/5 AR
Axis	5
Measure diameter	2600 mm
Accuracy	± 0,1 mm
Length precision	± 0,15 mm
Power supply	Universal voltage 100-240V AC 50-60Hz
PC Connection	Serial RS232 with DB9 - DB9 cable. Compatible with USB adapter
Working temperature [°C]	10 - 30

3.2   NON CONTACT SYSTEM: STRUCTURED LIGHT STEREO SYSTEM

The 3D optical digitizer system used here has been developed at Mechanical, Nuclear and Production Engineering department of University of Pisa.

The system is composed of two digital cameras, a multimedia projector (as seen in Figure 3) and it is driven by specific software that allows to manage the hardware and to process the range images acquisition. It is based on an active vision method using coded light in order to acquire complex surfaces. Technical characteristics of this system are summarized in Table 2 [5].

The measurement methodology is based on a light coded approach [6]. A multimedia standard projector is used to generate vertical and parallel, black and white, fringe patterns. The two digital cameras are used to acquire range images of the surface under structured lighting conditions.

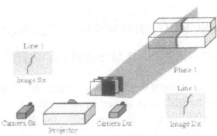

FIGURE 3. The structured light stereo system [6]

This stereo system has to be calibrated by evaluating the intrinsic parameters (focal distances, coordinates of principal points, radial and tangential distortions) and extrinsic parameters (position and orientations with respect to an absolute reference system) of the digital cameras.

TABLE 2. Optical system characteristics

Structured Light Stereo System	DIPN Prototype
Digital camera resolution	Monochrome, 1280x 960 pixel
Digital projector	1024 x 768 pixel
Calibration procedure	Software and plate
Accuracy [mm]	0.02 ÷ 0.04
Lateral resolution [mm]	0.28 ÷ 0.4
Measuring area [mm]	400 x 300
Working distance [mm]	600 ÷ 800
PC interfaces	Dual head video board IEEE 1394 data acquisition board
Acquisition time for one view [sec]	≤20
Accessories	Rotary table
Working temperature [°C]	+5 ÷ +40

In particular, the parameters of each camera are obtained correlating the coordinates of known markers located on a calibrating sample in different positions. An iterative process, specific for this system (developed by its constructors), solves the correlation equations and provides the different parameters. The measurement process is based on correlating data detected in camera images and therefore multimedia projector does not need to be calibrated.

3.3 QUALITATIVE EVALUATION OF THE TWO SYSTEMS

The authors propose a series of general parameters and characteristics, listed in Table 3, to use for a qualitative evaluation and comparison of the two systems analysed before the turbine blade acquisition. These parameters and their meaningfulness in this context have been derived from [7],[8],[9].

TABLE 3. Qualitative evaluation of the two systems

Parameters and characteristics	Robotic arm	Light stereo system
Easy to use	Excellent	Medium
Total time = acquisition time + calibration time	Medium	Sufficient
Time for data elaboration	Medium	Sufficient
Resolution	Sufficient	Excellent
Accuracy	Sufficient	Excellent
Calibration Procedure	Excellent	Medium
Environmental sensitivity	Excellent	Sufficient
Mobile/ non mobile system	Medium	Medium
Software driven acquisition help	Medium	Excellent

4 APPLICATION CONTEXT

To perform a methods and systems comparison, a complex free form object, as a blade of a steam turbine (see Figure 4a) was chosen. This kind of components has a complicate geometry, in particular the turbine airfoil profile curve and the root geometry. In Figure 4b is shown a 2D CAD representation of the acquired blade while in Figure 4c is shown a photo. This kind of turbine blade is a 3DS™ rotating blade of a high pressure (HP) Parsons stage of a steam turbine. It was produced in a local plant, the CBlade spa - Forging & Manufacturing, in Maniago, (PN).

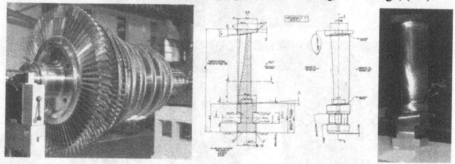

FIGURE 4. a) Steam turbine, b) CAD draw, c) Photo of the blade

5 DATA ACQUISITION

Blade acquisition by robotic arm was carried out at Marmax S.r.l., a spin-off of the University of Udine located inside the Rapid Prototyping Lab. Firstly, the system was connected to a PC and then it was calibrated (there is a specific Rhinoceros plug-in for system calibration and initialization) in order to create a correspondence between real world and CAD modelling virtual environment and to define a reference coordinate system for point data acquisition. In particular, during the blade acquisition, the conical stylus, the spherical and the conical touching probe were used for points digitizing.

To facilitate this operation, the user was suggested to analyse the shape of the target object and to visualize a virtual grid over it, or to draw a real grid just on the surface of the object in order to follow it during the digitization.

Acquisition procedure by the optical system has required more steps, first of all system calibration and object preparation. In fact, for reducing environmental influence and error propagation under acquisition, target objects have to be painted by a white opaque paint as shown in Figure 5a. At the beginning of acquisition the system has to be calibrated for calculate intrinsic and extrinsic parameters. In particular this blade acquisition was taken in several views at the distance of about 700 mm, using a rotary table to facilitate the alignment of the whole acquired data sets (see Figure 5b).

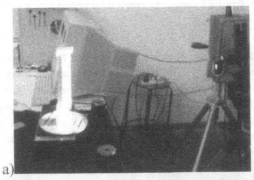

FIGURE 5. a) The optical system during acquisition, b) Optical acquisition example views

6 DATA ELABORATION

After data acquisition, as said previously in the paper, RE process requires a data elaboration step. This phase is the most important step. During this phase, starting from the acquired data, the 3D CAD model of the target object was created (see Figure 6). In this particular case study, data acquired by robotic arm, were first elaborated with a free-form surface modeller as Rhinoceros, and then were optimized with a solid modeller as Solid Edge v15. On the contrary, acquired point clouds by optical structured light system firstly were elaborated by the control software of the system for a preliminary de-noising. Then the data volume was elaborated whit Geomagic Studio, a specific software for Reverse Engineering applications. In particular, this software has tools for point clouds manipulation and for surface reconstruction. Two kinds of models of the blade have been reconstructed, a triangle mesh (in STL format) and a surface-based (iges format). In particular, STL model can be used for Rapid Prototyping techniques while iges model can be used for quality analysis and comparisons between the designed 3D CAD model of the target object and the 3D CAD reconstructed model.

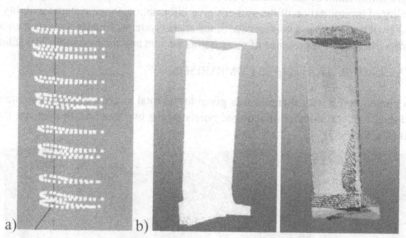

FIGURE 6. a) Robotic Arm point cloud, b) Optical System point clouds

7 SHAPE SYSTEMS COMPARISON

The comparison of the two systems is based on the analysis of some technical characteristics and parameters. The authors divided systems comparison in two steps. One refers to digitization phase characteristics and parameters (see Table 4) and one refers to the 3D CAD models obtained after the data elaboration phase (see Table 5).

TABLE 4. Digitization phase parameters comparison

	Robotic arm	Light stereo system
Measurement method or principle	Contact	Non contact - Optical
Data volume [n° of points]	494	507.851
Total time of acquisition + calibration [hours]	1.5	2
Time for data elaboration [hours]	8	16
Resolution	0.10 mm	0.02 mm
Accuracy	±0.15 mm	±0.04 mm
Calibration Procedure	By plug-in	By software
Calibration Time [minutes]	20	40
Mobile / non mobile system	mobile	mobile
Environmental sensitivity	none	Light, reflection

7.1 DIGITIZATION PHASE COMPARISON

In particular, for the digitization phase, as summarized in Table 4, it was given attention to the acquired data volume as number of points and on the acquisition speed. Another important parameter is the total time of acquisition. Resolution and accuracy, as technical characteristics of the systems give information of the systems performance. Calibration procedure, calibration time and the possibility of transportation of the systems give information about systems usability and about knowledge level required for a normal user of the shape systems. Finally, environmental sensitivity gives information on the presence of measurement errors and their possible prevention.

7.2 DATA ELABORATION PHASE COMPARISON

After data elaboration phase, attention was given to the total amount of acquired data as number of point clouds and total number of acquired points by the two systems, to compare the obtained 3D models (see Figure 7).

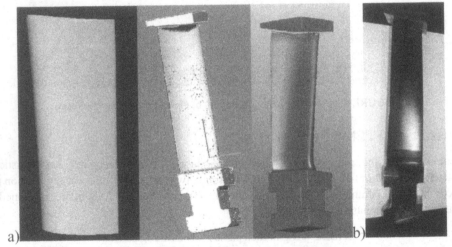

FIGURE 7. a) 3D CAD models of the turbine blade b) Turbine blade photo

Then it was pointed out the difference between the STL models reconstructed in terms of number of triangles and file dimensions (in Kilobytes), and the difference between the surface models reconstructed. Particular attention was given to the de-noising operations and to the dimensional analysis between the reconstructed model and the original CAD model of the blade. These data are summarized in Table 5.

In particular, the dimensional comparison between the reconstructed model from optical acquisition and the original CAD model of the blade, can be visualized like a report where are indicated the maximum and minimum values of the accepted tolerance, maximum and minimum values of deviation and some coloured maps for visualize these data in a graphical way.

TABLE 5. 3D models comparison

	Robotic arm	Light stereo system
N° of partial point clouds	2	8
N° points of total point cloud	641	507.851
N° points of reduced point cloud	-	394.070
N° of triangles in STL model	60.047	185.122
STL file dimensions [KB]	2933	9.040
Kind of Surface model file	iges	iges
Noise elimination	-	yes
Deviation from CAD model	high	low
Manual repair of the model	yes	-

8 RESULTS

The comparison presented in this work is based on the turbine blade acquisition and data elaboration. In particular, it emerged that it is easier, in terms of user training, to perform the acquisition by using the robotic arm than by using the optical system. Furthermore the optical system acquires a more complete data volume than the robotic arm. Another aspect to point out is the total acquisition time; in this case the robotic arm needs less time for total acquisition than the optical system. The calibration procedure for the two systems is software-driven but calibration time is longer for the optical system. Data elaboration in the case of robotic arm is reduced if compared with the optical system where the data amount has to be first de-noised, aligned, merged and finally elaborated. The robotic arm acquisition system seems more reliable for its easy to use capability and for the reduced elaboration procedure, while the volume of acquired data is not as complete as the volume of data acquired by the optical system. To acquire complex free form objects like turbine blades, orthopaedic prosthesis or other mechanical components it is better to have a data set of the whole object surface that allows an exact reconstruction of the physical object. Therefore, it is possible to perform a dimensional analysis and a quality inspection of parts.

9 CONCLUSIONS

In this paper a comparison between two different shape acquisition systems was presented. Two systems were considered: a handily driven 5-axis robotic arm and a stereo optical system based on the projection of a structured light. A steam turbine blade was chosen as a benchmark to compare the capabilities of the two systems for complex free-form surfaces acquisition. The results of the acquisition phase and the elaborated 3D CAD models were described and analyzed in order to point out a qualitative and a quantitative comparison in the case of acquisition of complex free form objects. From the qualitative analysis it emerged that both systems were valid for free form object acquisition for their easy to use capability and their software-driven acquisition aids. On the other hand, from the quantitative analysis, it emerged that the optical system allows acquiring a more complete data volume than the robotic arm. Even if the elaboration of a so complex data volume took a lot of time, the obtained 3D CAD model was more precise. On the contrary, the elaboration of the data acquired by the robotic arm was easier but it required the manual intervention of the CAD designer to draw and define some particulars (i.e. fillets, bend surfaces).

ACKNOWLEDGMENTS

The authors would link to thank prof. S. Barone and ing. A. Razionale of the Mechanical, Nuclear and Production Engineering Department of the University of Pisa, who kindly let us use their shape acquisition system, ing. M. Zanzero of Marmax S.r.l. for the availability of the robotic arm, and ing. G. Visentini of CBlade spa - Forging & Manufacturing, Maniago (PN), who provided the turbine blade.

REFERENCES
1. Várady, T., Martin, R., R., Cox, J., (1997), Reverse engineering of geometric models – an introduction, Computer Aided Design, Vol 29, No.4, 255-268.
2. Bernardini, F., Rushmeier, H., (2002) the 3D Model Acquisition Pipeline, Computer Graphics forum, Vol 21, No 2, 149-172.
3. Motavalli, S., (1998), Review of Reverse Engineering Approaches, Computers industrial Engineering Vol. 35, No 2, 25- 28
4. Curless, B., (2000), Overview of Active vision Techniques, SIGGRAPH 2000 Course on 3D Photography.
5. Barone, S., Curcio, A., Razionale, A., V., (2003) A structured light stereo system for Revere Engineering applications, Proceedings of XIII ADM-XV INGEGRAF- International conference on Tools and Methods evolution in Enginnering Design, Napoli, Italia.
6. Barone, S., Razionale, A., V., (2004), A Reverse Engineering methodology to capture complex shapes, Proceedings of XVI Congreso Internacional de Ingenieria Grafica, Zaragoza, España.
7. Iuliano, L., Minetola, P., Rossino, C., (2004), Reverse engineering in campo artistico-archeologico: confronto tra sistemi ottici per la scansione di sculture, Atti del Convegno Reverse engineering: potenzialità e applicazioni, Modena, Italia.
8. Beraldi, A., J., Gaiani, M., (2003), Valutazione delle prestazioni di sistemi di acquisizione tipo 3D active vision: alcuni risultati, Disegno e Design Digitale- DDD, anno 2, n. 5, http://www.polidesign.net/ddd
9. Broggiato, G., B., Campana, F., Gerbino, S., Martorelli, M. (2002), Confronto tra diverse tecniche di digitalizzazione delle forme per il riverse Engineering, Proceedings of XIV Congreso Internacional de Ingenieria Grafica, Santander, España.

COATING PHOTOELASTICITY APPLIED TO A RAPID PROTOTYPING MODEL OF A HUB FOR HIGH PERFORMANCE CARS

A. Baldini[1], F. Cevolini[2], M. Giacopini[1], M. Piraccini[3], V. Ronco[2], A. Strozzi[1]

[1] University of Modena and Reggio Emilia, Italy
[2] CRP Technology, Modena, Italy
[3] Minardi Team F1, Faenza (Ra), Italy

KEYWORDS: Photoelasticity, Rapid Prototyping, Hub, Stress Concentrations.

ABSTRACT. A photoelastic technique is presented that is based on a photoelastic coating applied along the surface of a model of a mechanical component made with rapid prototyping. The proposed procedure allows more flexible models to be obtained in comparison to metal parts, thus favouring the outset of a higher number of photoelastic fringes. In addition, the proposed approach permits enlarged models to be easily constructed, thus easing the photoelastic readings. The preliminary study here presented addresses a wheel hub for high performance cars, with particular regard to the stress concentrations by the fin fillets.

1 INTRODUCTION

In this paper models obtained with rapid prototyping are proposed to perform a qualitative coating photoelastic analysis of mechanical components, fast and reliable at a time. In particular, a coating procedure with epoxy resins has been developed, that has been applied to the mechanical analysis of a hub for formula one cars. The preliminary results of this non-conventional technique appear to be encouraging, since they have allowed the most stressed zones to be located.

In this work preliminary aims have been fulfilled. In particular:

1) a photoelastic coating technique has been developed, that permits a prototype of a mechanical part geometrically complex to be coated within a reasonable time and with simple tools;

2) a photoelastic analysis of the structural part subjected to assigned loading conditions has been carried out, and a qualitative stress analysis has been effected;

3) a semi-quantitative stress analysis has been performed for the most severely stressed loci.

2 THE PHOTOELASTIC TECHNIQUE

The transmission photoelasticity allows the stresses within a component to be thoroughly evaluated. Anyway, this approach requires a costly model of the component to be made with a

Published in: E. Kuljanic (Ed.) *Advanced Manufacturing Systems and Technology*,
CISM Courses and Lectures No. 486, Springer Wien New York, 2005.

transparent photoelastic resin, so that an alternative photoelastic approach has been proposed, that consists in employing a coating photoelasticity. With this technique the actual component is covered with a thin photoelastic layer, thus avoiding the machining of a model. The photoelastic coating allows the surface stresses, which often dominate, to be estimated.

One problem encountered in employing coating photoelasticity is that the surface strains within the component be high enough to promote a sufficient number of photoelastic fringes within the coating. On the other side, high strains may cause local metal yielding, thus compromising the validity of an elastic analysis. The above commented problems are circumvented in this paper by applying a coating photoelasticity to models obtained with rapid prototyping, that are appreciably more deformable than the actual metal components, thus allowing the onset of a higher number of fringes.

3 THE COMPONENT

The mechanical part analysed in this paper is a front hub of a formula one car, designed by *C.R.P. Technology,* Modena, and made with the titanium alloy Ti-6Al4V Mod., by employing

FIGURE 1. Actual component made of titanium alloy

rapid prototyping and investment casting, Figure 1. The hub supports the wheel system, and it incorporates two roll bearings. Two front suspension beams articulate with the hub, and the steering system interacts with the hub. The hub is stiffened by a series of fins.

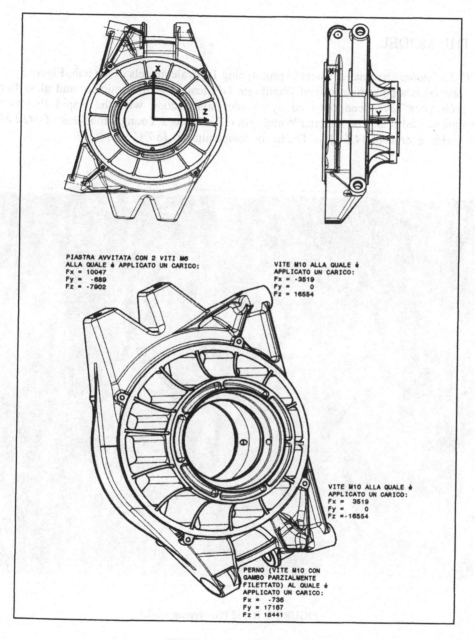

FIGURE 2. Loading conditions

The above perfunctory description aims at explaining the loading conditions adopted in Figure 2. Four forces affect the hub, where the loads F_1 and F_2 represent the action transmitted by the breaking shoes, whereas the forces F_3 and F_4 are produced by the beams contacting the hub.

3 THE MODEL

C.R.P. Technology has made two rapid prototyping 1:1 scale models of the hub, Figure 3. The material is a polyamide named WindForm GF, loaded with aluminium and glass fibres. The prototypes have been obtained by powder sinterization with the Rapid Prototyping technique named SLS. The material WindForm GF exhibits a Young's modulus of *4412 MPa* and a yielding stress of *44.6 MPa*. The fusion temperature is *180°C*.

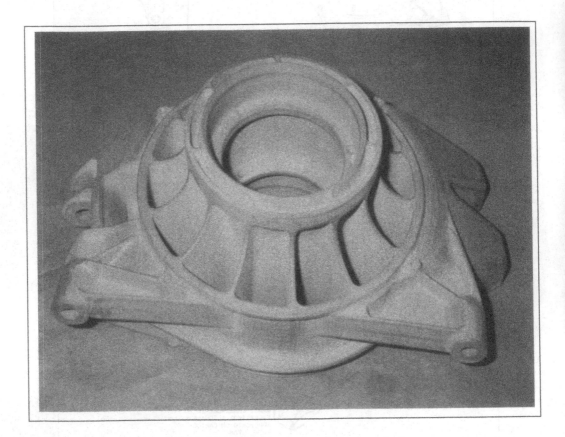

FIGURE 3. Rapid Prototyping model

4   THE PHOTOELASTIC COATING

Since the WindForm GF material is grey and opaque, the component model has been varnished
with a primer and with silver reflective stratum to promote the reflection of the polarised light.
The component has then been covered with a thermosetting, photoelastically active layer.
Experimental tests carried out at the Laboratory of Polymeric Materials, Department of Chemistry, University of Modena and Reggio Emilia have evidenced that the most suitable resin is a vinylister one named DERAKANE MOMENTUM 470-300. The layer is about 1 mm thick.

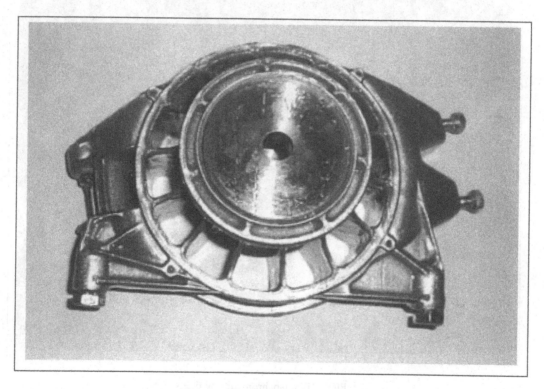

FIGURE 4. Model covered by the photoelastic layer

5   RESULTS

Selected photoelastic fringes are visible in Figures 5 and 6. The fringes are usually more concentrated in the vicinity of fillets along the transition zone between fins and hub, where stress concentrations are deemed to take place, but some highly dense patterns less expectedly occur in flat zones too. Upon measurement of the layer local thickness, it was possible to get a semi-quantitative evaluation of the stress concentrations. A finite element evaluation of the stresses in the neighbourhood of stress raisers requires a highly dense mesh to be employed. In this case of very local stress concentrations, the photoelastic technique is believed to be a valid alternative to the finite element approach.

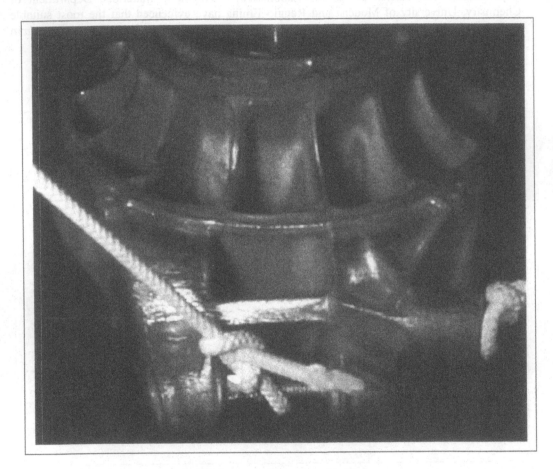

FIGURE 5. Isochromatics by fillets

A drawback of the resin employed for the photoelastic layer is that it exhibits a Young's modulus comparable to the rapid prototyping material. So, the application of the photoelasic layer inevitably causes undesired model stiffening with respect to an uncoated analogue. It will therefore be necessary to develop resins exhibiting a lower Young's modulus.

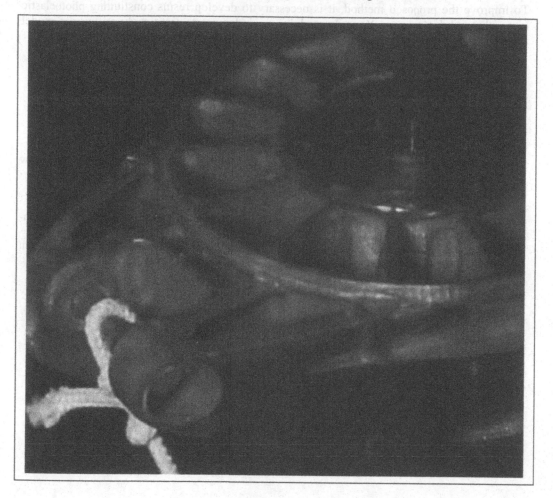

FIGURE 6. Fringes in heavily stressed zones

6   CONCLUSIONS

A non-conventional coating photoelastic technique has been presented, where the model has been made with rapid prototyping. This approach permits a more flexible model than the real component to be obtained, thus easing the obtaining of a higher number of photoelastic fringes. With this approach it is also possible to make an enlarged model of the actual component, thus easing the photoelastic readings. The proposed method has been successfully applied to the mechanical analysis of a hub, where the stress concentrations have been spotted. To improve the proposed method, it is necessary to develop resins constituting photoelastic layer, that exhibit a lower Young's modulus, in order not to fictitiously stiffen the model flexibility. Despite these limits, the results retrieved so far seem encouraging.

REFERENCES

1. Orlov, P., (1980), Fundamentals of Machine Design, MIR, Moscow.
2. Shigley, E., (1986), Mechanical Engineering Design, McGraw-Hill, New York.
3. Juvinall, R.C., Marshek, K.M., (1991), Fundamentals of Machine Component Design, Wiley, New York.
4. Giovannozzi, R., (1965), Costruzione di Macchine, Patron, Bologna.
5. Massa, E., (1975), Costruzione di Macchine, Tamburini, Milano.
6. Hills, D.A., Nowell, D., Sachfield, A., (1993), Mechanics of Elastic Contacts, Butterworth-Heynemann, Oxford.
7. Strozzi, A., Pilati, F., Fabbri, E., (2002), Rapid Prototyping e Fotoelasticita' per ricoprimento, 6th AIMAT National Conference, 8th-11th September 2002, Department of Engineering, Modena.

DESTRUCTIVE DISASSEMBLY OF END-OF-LIFE HOUSEHOLD APPLIANCES: A STRUCTURED ANALYSIS OF CUTTING METHODS OF THE HOUSING

M. Porta, F. Sebastiani, M. Santochi, G. Dini

Department of Mechanical, Nuclear and Production Engineering, University of Pisa, Italy

KEYWORDS: Waste from Electric and Electronic Equipments, Disassembly, Cutting Methods.

ABSTRACT. Materials recycling and components reuse are important key points in reducing the environmental impact of Waste from Electric and Electronic Equipments (WEEE). Disassembly could represent an efficient way for the treatment of end-of-life products, although the choice of suitable tools and methods is necessary for a quick, cheap and safe disassembly.

The removal of the housing is an important step of the treatment process because it gives access to internal components allowing their disassembly.

The aim of this paper is a structured analysis through a QFD like approach of the removal of the washing machine panels by means of some commercial cutting tools. Tests have been performed to collect results on speed, cost and safety of the plasma cutting and other tools. The study shows the performances required in the removal of the domestic appliances housing and confirms that plasma is an excellent method for an economic and safe cutting.

1 INTRODUCTION

The wide diffusion and the short lifetime of consumer goods have increased the quantity of used products to be discarded. The reduction of environmental pollution and the sustainable development impose the recycling of these goods which cannot be wasted. Recycling means recovering materials and/or components from used products to make them available for new products [1].

Some governments are paying attention to recycling and waste recovery, by adopting environmental policies. Those policies include incentives for the use of recycled materials, taxes on virgin materials, manufacturer responsibility for the treatment of end of life products, an additional cost for new products to pay the cost for their treatment, prohibition to discard goods containing dangerous or toxic materials, etc [1].

In Europe the "Directive 2002/96/EC on waste electrical and electronic equipment (WEEE)" has been drawn up with the purposes of reducing the WEEE and of promoting the reusing, the recycling and other forms of recovery of such wastes [2]. In order to reach these goals the Directive introduces the producer responsibility for financing the management of the waste, the set up of convenient facilities for the return of WEEE and the proper marking of electrical and electronic equipment to avoid the wrong disposal of WEEE. Furthermore, to facilitate the correct reuse and recycling of WEEE, producers have to provide treatment information to the disassembly centres for each type of EEE (different components-materials and location of dangerous substances). Regarding the Member States, the Directive suggests to encourage producers to use recycled material and to design EEE in such a way to facilitate their recovery.

Published in: E. Kuljanic (Ed.) *Advanced Manufacturing Systems and Technology*,
CISM Courses and Lectures No. 486, Springer Wien New York, 2005.

The directive also defines the rates of collection, reuse, recovery and recycling of WEEE (4 kg on average per year per inhabitant by December 2006 and new rates by December 2008) [2]. Among the different strategies for the recovery of end-of-life products, disassembly seems to be a good method for the treatment of complex products. Actually disassembly offers, for household appliances such as washmachines and dishwashers, more advantages than other methods, like shredding [1], in terms of purity of separated material, low investments in equipments, recovered subassemblies, etc. On the other hand the main problem of disassembly is the cost of human labour. Actually the great variety of household appliances and their different wear conditions do not permit a significant and profitable automation of disassembly [3]. While products can be "designed for disassembly", existing goods have to be treated according to available technologies. In order to generate profit by the disassembly of present household appliances, methods to characterize and design suitable tools and procedures are needed [1]. Easy and cheap tools and quickly procedures of disassembly are required to reduce the amount of work, to speed up the disassembly process and to reduce labour [4].

2   REMOVAL OF HOUSING IN HOUSEHOLD APPLIANCES

At the Department of Mechanical, Nuclear and Production Engineering of the University of Pisa, a disassembly cell for the treatment of end-of-life washmachines has been designed and realized. The major component of the disassembly cell is an innovative manipulator (patented) which is able to grasp and move any kind of porthole washmachines in the best ergonomic position.

The removal of the housing panels of the domestic appliances is an important step of the disassembly process because it gives good and safe access to internal components, allowing their disassembly. Furthermore this operation must be also cheap to get profit to the treatment for the domestic appliance. So the tool used for the removal of the housing has to be cheap, easy and safe.

As reported in the literature, destructive [5-6-7] or semi-destructive methods and tools [5-8] seem to be the best choice for the removal of the housing of domestic appliances. Actually these destructive tools are often faster than standard tools (screwdriver, pliers, etc.) and can also be unavoidable in case of irreversible connection (e.g. welded points) or part alterations (e.g. oxidation, damages on screw heads, etc.). Moreover, because of many different kind of household appliances (in terms of joining technique, product geometry and unpredictable damage), a high flexibility of tools is required [4].

Many destructive tools are available on the market and most of them are usable for the removal of panels of the housing. In particular, as shown in research projects and disassembly plants [5], plasma seems to be the favourite method, in terms of cost and flexibility, for the cutting of the housing. However no scientific analysis has been done in order to test its performances for this disassembly step.

The aim of this paper is to describe a structured analysis of the performances required for the removal of the household appliances housing and a rational study of the performances given by the plasma method in comparison with other cutting tools.

3 QFD LIKE APPROACH APPLIED TO THE HOUSING CUTTING

"Quality Function Deployment" (QFD) usually provides methods for ensuring quality throughout each stage of the product development process. This is a method for developing a design quality aimed at satisfying the consumer and then translating the consumers' demands into design targets and major quality assurance points to be used throughout the production stage [9]. Akao defines QFD as a method useful in "converting the consumers' demands into quality characteristics and developing a design quality for the finished product by systematically deploying the relationships between the demands and the characteristics, starting with the quality of each functional component and extending the deployment to the quality of each part of the process" [9].

In this paper a QFD like approach is used as a rational method for a structured analysis of the removal process of the sheet metal housing performed with various cutting tools: the plasma, the nibbler and the angle grinder. The House of Performance (Table 5), obtained with the analysis, shows the relationship between the performances required by the removal of the housing in the disassembly of household appliances and the performances of the selected tools.

3.1 REQUIRED PERFORMANCES AND DEPLOYMENT CHART

The "macro" performances required to the tools used for the removal of the housing of the household appliances are the general characteristics needed by the disassembly process: safety and cheapness. These performances have been provided by the literature and verified by many disassembly tests performed in this research project. Actually no other typical procedures of QFD method (interviews, questionnaires etc.) seem to be available. So safety and cheapness could be considered as the voice of customers; in this case customers are considered the people employed in the disassembly field.

First of all the macro performances have been analyzed to explode them into "reworded data". Then these data have been arranged with the "KJ Method of Grouping reworded data" [9-10] as partly shown in Figure 1.

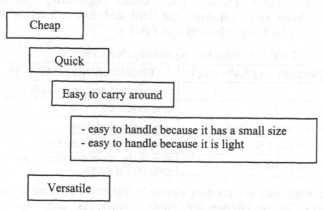

FIGURE 1. KJ Method of grouping reworded data

So a required performance deployment chart has been obtained by arranging the different level items as partly shown in Table 1.

TABLE 1. Required Performance Deployment Chart

1st LEVEL	2nd LEVEL	3rd LEVEL	4th LEVEL
1000 Cheap	1100 Quick	1110 Quick to cut	
		1120 Quick to service	
	1200 Versatile	1210 Easy to carry around	1211 easy to carry around because it is small
			1212 easy to carry around because it is light
			1213 easy to carry around because it has no cables
		1220 Able to work in different positions	

The features required of tools in the cutting of the housing do not have the same importance. The relative importance of various characteristics is based on the third/fourth level details in the required performance deployment chart and has been obtained by the experience of the authors in the disassembly field. This degree of importance has been evaluated using a 1 to 5 scale as shown in Table 2.

TABLE 2. Importance of required characteristics

IMPORTANCE OF CHARACTERISTICS	NUMERIC CONVERSION
Negligible	1
Preferable	2
Important	3
Very important	4
Indispensable	5

3.2 DEPLOYING INTO ENGINEERING CHARACTERISTICS

A number of counterpart characteristics has been found for each individual required performance. These counterpart characteristics, called engineering characteristics, are independent and have been extracted from the third and fourth level of the Requiredd Performance Deployment Chart as partly shown in Table 3.

TABLE 3. Extracting engineering characteristics

REQUIRED PERFORMANCE (3rd–4th LEVEL)	ENGINEERING CHARACTERISTICS
Quick to cut	cutting speed, tool specifications
Quick to service	time to service
Easy to carry around because it is small	size, shape, portability
Easy to carry around because it is light	weight, shape, portability
Able to work in different positions	size, shape, weight, center of gravity, stability, tool specifications

The engineering characteristics are product elements that can be measured to evaluate the required performances. These engineering characteristics, as done for the demanded performances, have been exploded in different levels to make an Engineering Characteristics Deployment Chart as partly shown in Table 4. The evaluation of the importance and the correlation among engineering characteristics of the third level are described in the subsection "Correlations" of paragraph 3.5.

TABLE 4. Engineering Characteristics Deployment Chart

1st LEVEL	2nd LEVEL	3rd LEVEL
Handling	Portability	Dimension
		Weight
	Stability	Center of gravity
Performance	Time required for cutting	Cutting speed

3.3 BUILDING THE MATRIX

The matrix has been made by combining the Required Performance Deployment Chart and the Engineering Characteristic Deployment Chart to build the "House of performance" (partly shown in Table 5), that corresponds to the House of Quality in the traditional use of QFD. As usually done in the QFD method, the symbols in Table 6 have been used to show the degree of correlation between required performances and engineering characteristics. This degree of correlation has been decided by the authors and is based on direct technical experience and on literature. A high degree of correlation means that a little change in the engineering characteristic causes a big change in the satisfaction of the the correlated demanded performance. The numeric conversion of the symbol has been done using the most typical scale of the QFD method [9-10].

TABLE 5. House of performance

Required Performances deployment chart / Engineering Characteristics deployment chart	Absolute Importance performance	Relative Importance performance	size (-)	weight (-)	cutting speed (+)	time set up specification (-)	time for wearing Dpi (-)	Time change cutting element (-)	TOTAL TOOL SCORE (decimal scale)
1110 Quick to cut	5	3.9%	○	○	◎			○	
1120 Quick to service	4					□		◎	
to carry around	3	2.3%				◎			
handling	4	3.1%	○	○					
2400 Cut sheet not dangerous	5	3.9%							
Absolute importance of engineering characteristic			263	295	196	48	38	36	
Relative importance of engineering characteristic			6.9	7.7	5.3	1.2	1	0.9	
Normalized importance of engineering characteristic			6.6	7.4	5.7	1.1	0.9	0.8	
Measuring units			mm³	Kg	m/s	s	s	s	
Plasma cutting — Value obtained			5	5	5	1	1	3	8
Plasma cutting — Normalized value			34.5	37	28.5	1.1	0.9	2.4	
Nibbler — Value obtained			3	3	1	5	5	1	5
Nibbler — Normalized value			20.7	22.	5.7	5.5	4.5	0.8	
Angle grinders — Value obtained			1	1	3	5	3	5	6
Angle grinders — Normalized value			6.6	4.7	17.1	5.5	2.7	4	

TABLE 6. Strength of the relationship

SYMBOL	MEANING OF THE SYMBOL	NUMERIC CONVERSION
◎	Strong correlation	9
○	Average correlation	3
□	Some correlation	1

3.4 MAKING A COMPARISON CHART

In order to compare [9-10] plasma cutting with other competitive tools, a nibbler and an angle grinder have been evaluated for each required performance. The aim of this comparison is to show that plasma cutting is an excellent method, in terms of satisfaction of the demanded performances, in the treatment of end-of-life products.
The evaluation of the engineering characteristics of plasma cutting, nibbler and angle grinder has been obtained by experimental tests, product catalogues and literature.

3.5 ENGINEERING CHARACTERISTICS

EVALUATION OF THE IMPORTANCE

The evaluation of the absolute (w_j) and the relative (w^*_j) importance of engineering characteristics has been obtained by converting the importance of required performance. This conversion was made with the independent scoring method [9-10] and it is based on expressions 1 and 2.

$$w_j = \sum_{i=1}^{n} d_i \cdot r_{ij} \tag{1}$$

$$w^*_j = \frac{w_j}{\sum_{j=1}^{m} w_j} \tag{2}$$

In these expressions d_i is the degree of importance of the required performance and r_{ij} is the strength of the match between the "i^{th}" demanded performance and the "j^{th}" engineering characteristic; n is the number of required performances and m the number of engineering characteristics This strength is described by the numerical conversion of the symbols shown in Table 6.
The engineering characteristic has been marked (as visible in Table 5) with + or with -: the sign +/- means that a greater value of the engineering characteristics gives a greater/lower satisfaction to the correlated demanded performances.

CORRELATIONS

The correlation among engineering characteristics has been displayed in the triangular matrix at the top of the House of Performances. This correlation has been shown with the same symbols used for the strength of the relationship among required performances and engineering characteristics.
In order to make the required performances independent from the number of engineering characteristics, a normalization of the r_{ij} factors is needed.
This goal has been obtained by using Wasserman's normalization [10] as shown in expression 3.

$$r_{i,j}^{norm} = \frac{\sum_{k=1}^{m}\left(r_{ik} \cdot \gamma_{kj}\right)}{\sum_{j=1}^{m}\sum_{k=1}^{m}\left(r_{ij} \cdot \gamma_{jk}\right)} \quad (3)$$

In this expression $\gamma_{kj=}\gamma_{jk}$ represents the strength of the correlation between the engineering characteristics "k^{th}" and "j^{th}", while r_{ik} is the strength of the correlation between the required performance "i^{th}" and engineering characteristic "k^{th}", finally r_{ij} represents the strength of the correlation between the demanded performance "i^{th}" and the engineering characteristic "j^{th}".

4 TOOL PERFORMANCES

4.1 EVALUATION OF ENGINEERING CHARACTERISTICS

According to QFD method, tests to evaluate the engineering characteristics have been set up in the same conditions and using the same devices. Tests were carried out, as much as possible, following the procedure provided by the EN-ISO standards.

Most values of engineering characteristics have been estimated on a specific test-piece, that is a sheet metal of standard dimensions obtained from housings. This standard sheet allows to get values that are not influenced by different household appliances used in tests. On the other hand, some engineering characteristics cannot be evaluated from the test piece: therefore their values have been obtained directly using household appliances as elements to test the tools on. In this paper only the most significant tests are briefly described in paragraph 4.3.

4.2 EVALUATION OF TOOLS

Different values obtained by the tools for each engineering characteristic have been transformed in a 1 to 5 scale. The gap between the greatest and the lowest value of each engineering characteristic has been divided in 3 parts: the lowest value gains 1and the greatest 5 if the engineering characteristic is marked with +; on the contrary the lowest value gains 5 and the greater 1 if the engineering characteristic is marked with -. The other values gain a score from 2 to 4 as explained, as example, in Table 7 (the example is about cutting speed).

TABLE 7. Conversion of engineering values into a numeric scale

	THE GREATEST	GAP BETWEEN TOP AND BOTTOM VALUES			THE LOWEST
VALUE OBTAINED	4.2 mm/s	4.1<v<3.1	3<v<2	1.9<v<1	0.9 mm/s
NUMERIC SCORE	5	4	3	2	1

Then the score assigned to each tool for every engineering characteristic has been multiplied by the "normalized importance of the engineering characteristic" in order to obtain the "normalized value of the engineering characteristic". The sum of all the normalized values is the final score related to the tool. This final score has been transformed in the decimal scale to get an immediate evaluation of tool global performance. In this decimal scale, 10 is the score of a hypothetical tool that obtains the maximum value for all engineering characteristics; on the contrary 1 is the score of a tool that has minimum value in every engineering characteristic.

4.3   EXPERIMENTAL TESTS

Different tests and devices have been set up to evaluate the engineering characteristics for the different tools. In the following paragraph only the most significant tests are reported.

RISK OF CUT FOR WORKERS

The part of the metal sheet remaining on the frame of the housing must not be dangerous for the following steps of disassembly. Consequently the edge of the cut sheet has not to be so sharp to hurt workers. In order to evaluate the risk of cut, a working-clothes, according with EN 340, has been tested following the EN ISO 13997. As described in the EN-ISO procedure, a sample of the protective clothes is attached on a proper convex device. In this standard the contact between the cutting edge and the sample has to be performed at a particular speed and for a specified travel. Thanks to the experimental tests it has been possible to evaluate the normal contact force, between the edge and the textile, that is able to cut completely the working-clothes. This value is assumed as an indicator of the risk of cut because the rip of the textile takes off the protection of the worker's arm and could permits the contact between the cutting edge and skin. Figure 2 shows a simple scheme of test device with the detail of a cutting edge. The cutting edge (1) is fixed to a board (2) able to move at a required speed in one direction (3). A beam (4), connected by a hinge (5) to the wall (6), supports the convex device (7) with the clothes sample (8) and weights (9). The normal contact force between the sample and the cutting edge can be modified by changing the weighs at the extremity of the beam.

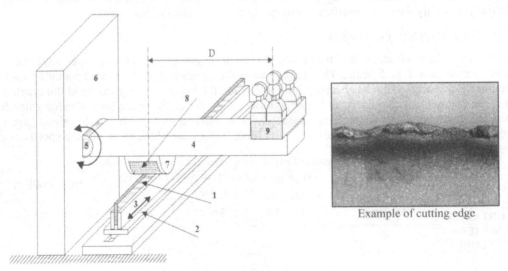

Example of cutting edge

FIGURE 2. Simple scheme of test device

RISK OF FIRE AND BURN FROM SPARKS

The contact between sparks and inflammable material (plastic, rubber, etc) and objects (wirings and rubber water pipes) can cause fire. Furthermore sparks are dangerous because they could burn workers' clothes. As shown in Figure 3, plasma and angle grinder generate a lot of sparks, so their evaluation is an important element for the safety of these tools.

FIGURE 3. Sparks in the plasma cutting

Thanks to a computerized image processing of video recordings of cutting test, the sparks characteristics have been evaluated. The output parameters of tests are the amount of sparks, the distance and the height that they can reach. Every parameter has been considered in terms of pixels percentage filling up every frames: the density of sparks represents the total amount while the other two parameters are evaluated as percentage of sparks that exceed set values of distance/height .

FORCE NEEDED TO CUT THE SHEET METAL

An important parameter to characterize the ergonomic attitude of cutting tools is the force to be operated. Actually a high force is not acceptable for workers in using a tool for a long time, as an entire working shift.

Among the tools considered in tests, plasma does not require any force; on the contrary the nibbler and the angle grinder do. To evaluate the force required by the nibbler and the angle grinder, the procedure visible in Figure 4 has been used. The dynamometer (1) is put between a fixed point (2) and the test piece (3). During tests the dynamometer measures the instantaneous force and stores its highest value that is the assumed as meaningful parameter of test. The figure also shown the angle grinder during the tests.

FIGURE 4. Force test and angle grinder

5 CONCLUSIONS

As mentioned, current household appliances are not yet designed to be easily disassembled. In order to gain an economic advantage from the treatment of these end-of-life products, the

choice of good tools and methods is fundamental. Structured methods of analysis can be useful to select and evaluate suitable tools and procedures for a quick, cheap and safe disassembly.
In this paper the proposed QFD like approach represents a rational method to select the best tool for the removal of the housing of discarded domestic appliances. The performances of three selected tools (plasma, nibbler and angle grinder) are exploited in a House of performances. Among these tools, plasma is clearly the best tool: its total score (8) is higher than nibbler (5) and angle grinder (6) scores. Plasma satisfies very well the performances required by the removal of the housing as shown by its engineering characteristics values.
The values obtain by the tools in each engineering characteristic could also be used to evidence all critical aspects of a tool in order to optimize and redesign it. Actually a low value means that the tool is not good to satisfy the engineering characteristic.

ACKNOWLEDGEMENTS

The authors wish to thank the "Fondazione Cassa di Risparmio di Pisa" for financial support and the technical staff of the Dept. of Mechanical, Nuclear and Production Engineering, Section Production, for its valuable technical support.

REFERENCES

1. Jovane, F., Alting, L., Armillotta, A., Eversheim, W., Feldmann, K., Seliger, L., Roth, N., (1993), A key issue in Product Life Cycle: Disassembly, Annals of the Cirp, Vol 42/2/1993, 651-658.
2. Directive 2002/96/EC of the European Parliament and of the Council of 27 January 2003 on waste electrical and electronic equipment (WEEE), (2003), Official Journal of the European Union.
3. Uhlmann, E., Haertwig, J.P., Seliger, G., Keil, T., (2000), A pilot system for the disassembly of home appliances using new tools and concepts, Proceedings of The Third World Congress on Intelligent Manufacturing Processes & Systems, Cambridge, MA-June 28-30.
4. Santochi, M., Dini, G., Failli, F., (2001), Disassembly for Recycling, Maintenance and Remanufacturing: State of the Art and Perspectives, Proceedings of sixth International Conference on Advanced Manufacturing Systems and Technology (AMST '02), Udine, June 2002, CISM Courses and Lectures NO. 437, Springer Verlag Wien New York, 73-89.
5. Seliger, G., Keil, T., Rebafka, U., Stenzel, A., (2001), Flexible Disassembly tools, Proceedings of the 2001 IEEE International Symphosium on Electronics & the Enviroment, Denver, Colorado, 30-35.
6. Uchiyama, Y., Fujisawa, R., Oda, Y., Hirasawa, E., (1999), Air conditioner and washing machine primary disassembly process, First International Symposium On Environmentally Conscious Design and Inverse Manufacturing 1-3 Feb. 1999, Tokyo, 258-262.
7. Project triennal "Rester Propre", Recerche Socio Tecniques pour le ryciclage de Produits manifactures par Revalorisation, rapport de l'année I :97-98 (in french).
8. Wagner, M., Seliger, G., (1996), Modelling of Geometry-Indipendent End Effector for Flexible disassembly Tools, Proc. of 3rd Int. CIRP Sem. on Life Cycle Engineering, Zurich, Verlag, 219-228.
9. Akao, Y., (1990), Quality Function Deployment: integrating costumer requirement into product design, Yohi Akao Editor.
10. Franceschini, F., (1997), Quality Function Deployment: uno strumento progettuale per coniugare qualità e innovazione, Il sole 24 ore libri (in italian).

QUALITY ASPECTS IN THE SECTOR OF WASTE FROM ELECTRICAL AND ELECTRONIC EQUIPMENT (WEEE)

M. Kljajin

Faculty of Mechanical Engineering in Slavonski Brod, University of Osijek, Croatia

KEYWORDS: Quality, WEEE, Life Cycle Thinking.

ABSTRACT. This paper presents that quality marks do not only refer to the design phase (e.g. recycling equitable design or design for environment) but also to the process chain after the end of use. The configuration of possible QM-systems in WEEE recycling strongly depends on the demands they are driven by the following main recommendations: (1) The envisaged ways of recycling or reusing components will have different demands than a material treatment. (2) Reusing materials can be different in same usage, upgrade or downgrade of materials. (3) The materials that are utilized in products, e.g. the treatment of hazardous or precious materials. (4) The quantity and (5) the loss of quality until the end-of-life and available information about the influences during use period.

1 INTRODUCTION

The electrical and electronic equipment sector is coming under increasing pressure from policy makers to improve the life cycle environmental performance of its products. Policy initiatives aimed at achieving this objective include the Integrated Product Policy (IPP) initiative, the proposed Waste Electrical and Electronic Equipment (WEEE), Restriction of Hazardous Substances (RoHS) Directive, and also proposed the Environment of Electrical and Electronic Equipment (EEE) Directive.

The Integrated Product Policy (IPP) is a European Commission environmental policy initiative aimed at improving the environmental performance of products throughout their life cycle. The Commission has proposed the potential principles, strategies, and instruments it might contain, and the potential roles of various stakeholders in the development and execution of the policy.

According to the Commission, IPP should be based on three key areas, which are: getting the prices right, for example reduced VAT rates on eco-labeled product, stimulating demand for green products, includes instruments to give consumers environmental information on products, and strengthening green production, i.e. promoting eco-design.

According to the Commission, the main objective of the WEEE and RoHS Directives is to reduce the environmental impact of waste equipment and to improve the environmental performance of all economic operators involved in the life cycle of the equipment, particularly operators involved in the treatment phase. The WEEE Directive deals with the collection, transport, treatment, and disposal of waste equipment. It contains detailed provisions on treatment specifications and targets, as well as the allocation of physical and financial responsibilities, which fall not only on producers, but also on municipalities, retailers, and treatment operators.

Published in: E. Kuljanic (Ed.) *Advanced Manufacturing Systems and Technology*,
CISM Courses and Lectures No. 486, Springer Wien New York, 2005.

The RoHS Directive requires the phase out of lead, mercury, cadmium, hexavalent chromium and two brominated flame retardants (PBB and PBDE) from all electrical and electronic equipment. By July 1, 2006, no new electrical and electronic equipment put on the market may contain lead, mercury, cadmium, or hexavalent chromium. Polybrominated biphenyls (PBBs) and polybrominated diphenyl ethers (PBDEs) - two types of flame retardant - are also prohibited. Countries that already have restrictions or prohibitions on the use of these substances in electrical and electronic equipment may keep them in place prior to July 1, 2006.

Quality Management (QM) is the aspect of the overall management function that determines and implements the quality policy. The attainment of desired quality requires the commitment and participation of all members of the organization whereas the responsibility for quality management belongs to top management. Quality management includes strategic planning, allocation of resources and other systematic activities for quality, such as quality planning, operations and evaluations. What is mean in the sector of WEEE and it's recycling? Quality assurance in recycling of WEEE does not only refer to the inspection of single product, but has mainly to cover aspects of the whole material and waste stream. Therefore, product data collection (combined with e.g. statistical methods) is important, in order to get a first impression of the circumstances quality aspects WEEE deals with, and to draw conclusions on the performance of QM-systems. Possible ways for product data collection are: field data collection and series testing.

2   QUALITY ASPECTS IN THE SECTOR OF WEEE

Talking about quality marks it gets obvious that they do not longer only refer to the function of a product, but increasingly to the "loss" for the society after its release and use. In the Taguchi [1][2] idea quality aspects are enlarged to the end-of-life phase of products, especially the harm that used products can produce to environment and society. In order to minimize the unwanted effects and to use the positive potentials end-of-life products possess, quality marks do not only refer to the design phase (e.g. recycling equitable design) but also to the process chain after the end of use.

Quality aspects are often accompanied with standardization. The ISO 8402 defines Quality Management (QM) and QM-Systems, whereas the ISO 9000 till ISO 9004 give rules for the implementation of QM-Systems.

ISO 9000 is a set of standards for quality management systems that is accepted around the world. Currently more than 90 countries have adopted ISO 9000 as national standards. When you purchase a product or service from an organization that is registered to the appropriate ISO 9000 standard, you have important assurances that the quality of what you receive will be as you expect. In addition, with the year 2000 revision of the standard, quality objectives, continual improvement, and monitoring of customer satisfaction provide the customer with increased assurances that their needs and expectations will be met. The standard intended for quality management system assessment and registration is ISO 9001. The standards apply uniformly to organizations of any size or description. Many companies require their suppliers to become registered to ISO 9001 and because of this, registered companies find that their market opportunities have increased. In addition, a company's compliance with ISO 9001 insures that it has a sound quality management system, and that's good business. Registered companies have had dramatic reductions in customer complaints, significant reductions in operating costs and increased demand for their products and services. Other benefits can include better working conditions, increased market share, and increased profits. ISO 9000 registration is rapidly becoming a must for any company that does business in Europe. Many industrial companies require registration by their own suppliers. There is a growing trend toward universal acceptance of ISO 9000 as an international standard.

Beside these, in different countries are several different national quality standards, which all have one thing in common: they are not very concrete concerning treatment of WEEE. The most concrete advice concerning environmental items is given in ISO 9004-1, which in short recommends "to be prosperous, an organization should offer products, which have to fulfill the requirements of the society and regard environmental demands." Unfortunately, until know, there is no valid EC-wide regulation on quality aspects in end-of-life of WEEE product treatment.

ISO 14001 is an emerging standard entitled "Environmental Management Systems - Specification". Although it has no formal relationship to the ISO 9000 family of documents, it is structured much like the ISO 9001 standard. This International Standard specifies requirements for an environmental management system, to enable an organization to formulate a policy and objectives taking into account legislative requirements and information about significant environmental impacts. It applies to those environmental aspects, which the organization can control and over which it can be expected to have an influence. It does not itself state specific environmental performance criteria. This International Standard is applicable to any organization that wishes to (1) implement, maintain and improve an environmental management system; (2) assure itself of its conformance with its stated environmental policy; (3) demonstrate such conformance to others; (4) seek certification/registration of its environmental management system by an external organization; (5) make a self-determination and self-declaration of conformance with this International Standard.

The series is made up of documents related to EMS – environmental management systems (i.e., ISO 14001 and ISO 14004) and documents related to environmental management tools (i.e., all other ISO 14000 series documents). This approach takes the view that establishment and implementation of an organization's EMS is of central importance in determining the organization's environmental policy, objectives, and targets.

Environmental management tools exist to assist the organization in realizing its environmental policy, objectives, and targets. For example: ISO 14001 requires the conduct of EMS audits, and guidance for the conduct of such audits can be found in ISO 14010, ISO 14011 and ISO 14012. ISO 14001 requires an organization to monitor and measure the environmental performance of its activities, products and services in order to continually improve such performance, and ISO 14031 provides guidance for this purpose. ISO 14001 requires that an organization consider the environmental aspects of its products and services.

The ISO 14040 standards assist an organization in the identification and analysis of environmental aspects of products and services. The ISO 14020 standards address guidance on providing information on the environmental aspects of products and services through labels and declarations.

Identifying the environmental aspects of an organization's activities, products and services, and determining their relative significance, are important elements of implementing an EMS or conducting EPE (environmental performance evaluation) in an organization. ISO 14001, ISO 14004 and ISO 14031 provide guidance on identifying significant environmental aspects.

ISO 14040 states in its introduction: "LCA (life cycle assessment) is a technique for assessing the environmental aspects and potential impacts associated (with products and services) ... LCA can assist in identifying opportunities to improve the environmental aspects of (products and services) at various points in their life cycle".

In general, the management of an organization will decide the content and format of any environmental reporting or communicating. However, the organization may find that it has a number of different reporting needs and intended audiences.

ISO 14001 and ISO 14004 provide guidance on reporting and communicating information on the environmental aspects and the EMS of an organization. ISO 14010 and ISO 14011 provide guidance on the preparation, content and distribution of audit reports. As environ-mental declarations and claims can be viewed as ways in which the environmental aspects of products and services are reported or communicated, the ISO 14020 standards provide appropriate guidance.

ISO 14031 provides guidance on reporting and communicating information describing the environmental performance of an organization. The ISO 14040 standards provide guidance on reporting and communicating the results of an LCA study.

All the requirements in this International Standard could to be incorporated into any environmental management system. The extent of the application will depend on such factors as the environmental policy of the organization, the nature of its activities and the conditions in which it operates.

The configuration of possible QM-systems in WEEE recycling strongly depends on the demands they are driven by (1) the envisaged ways of recycling: reusing components will have different demands than a material treatment; (2) reusing materials can be different in same usage, upgrade or downgrade of materials; (3) the materials that are utilized in products, e.g. the treatment of hazardous or precious materials; (4) the quantity, in which materials accrue; (5) the loss of quality, which befalls a product from its production until the end-of-life and available information about the influences during use period.

Quality assurance in recycling of WEEE does not only refer to the inspection of single product, but has mainly to cover aspects of the whole material and waste stream. Therefore, product data collection (combined with e.g. statistical methods) is important, in order to get a first impression of the circumstances quality aspects WEEE deals with, and to draw conclusions on the performance of QM-systems. Possible ways for product data collection are: field data collection and series testing [5].

3 DEMANDS ON QM-SYSTEMS IN ELECTRICAL AND ELECTRONIC COMPONENTS REUSING

Processes aiming at the transfer of products or components in further usage phases are designated as adaptations [6]. Kinds of adaptation are maintenance, repair and remanufacturing. Their objective is the removal of physical changes. Changing requirements can be met by upgrading or downgrading, enlargement or reduction, rearrangement or modernization of electrical and electronic components.

For a successful reuse and re-distribution the product quality has to be specified and monitored, as well as quality parameters for secondary usage have to be defined. However, the origin of products and product components is frequently unknown. The collection of reliability and quality criteria during the product's use phase by means of today available methods is possible, however, only to a limited extent.

Main objective in concept of QM-systems for reusing, are investigating quality marks and defining quality loops that can be used for building up a quality management system for re-usable components.

In case of disassembly, a documented quality management system has to be set up so that the disassembled units and the ones, ready to be recycled, can satisfy the specified demands. The activities relating to this process are as (1) the defining and obtaining all the regulations, processes, controlling equipment, gadgets, the complete disassembling background and expert knowledge that could be necessary in order to attain the requested quality; (2) the choosing the methods of quality regulation and control including the development of technical equipment; (3) the identification (definition) of measuring demands; (4) the definition of acceptability demands relating to all the characteristics and requests including the subjective elements, too; (5) the disassembling and controlling measuring methods and the co-ordination of relating documentation.

Total Quality Management (TQM) can be applied or adapted also to any type of organization in the sector of WEEE. As it is known the TQM processes are divided into four sequential categories: plan, do, check, and act (the PDCA cycle). In the **planning** phase, people define the problem to be addressed, collect relevant data, and ascertain the problem's root cause; in the **doing** phase, people develop and implement a solution, and decide upon a measurement to gauge its effectiveness; in the **checking** phase, people confirm the results through before-and-after data comparison; in the **acting** phase, people document their results, inform others about process changes, and make recommendations for the problem to be addressed in the next PDCA cycle.

4   QUALITY ASPECTS OF SECONDARY MATERIALS COMING FROM WEEE

Those parts of a product classified by the producer as unsuitable for reuse or have been detected as not re-usable by the quality inspection before the re-assembly, have to be processed. The components designated as pollutants that have to be disposed in an environmentally compatible way are excluded from the previously mentioned group. The application of secondary materials from end-of-life products for new products should be labeled and the retention of the quality characteristics should be re-provable as material-specific quantities. Every new electrical and electronic product in the EU must bear a label that (1) verifies that it was put on the market after August 13, 2005, (2) verifies that it will be separately collected, and (3) bears the name of the producer. The EU intends to prepare standards for this label. Producers must provide information to consumers on the collection systems available and on the environmental and health impacts of hazardous substances contained in waste electrical and electronic products. They must also provide information to facilitate the environmentally sound reuse, recycling, and treatment of waste electrical and electronic products. Such information includes the identity of components and materials and the location of dangerous substances inside a product. The EU Eco-label is administered by the European Eco-labeling Board (EUEB) and receives the support of the European Commission, all Member States of the European Union and the European Economic Area (EEA). The Eco-labeling Board includes representatives such as industry, environment protection groups and consumer organizations.

The quality of secondary materials represents an essential second partial aspect of a life-cycle oriented quality management. The main objective of a material recycling is to ensure a quality level of the recyclable materials, which is as constant as possible. A down-cycling of materials that restricts a further use to inferior purposes should be avoided.

At present there are no explicit quality standards or guidelines for recyclable material fractions. Guidelines for the certification of dismantling demand a coarse differentiation of the material fractions, which, however, does not permit any, exact statement about the material quality. The real quality of the material fractions depends primarily on the economical situation on the secondary material market. According to receipts to be achieved for material fractions, the expenditures are individually estimated with regard to the dismounting depth, cleaning and separation of the materials.

In order to come to effective close loop structures, it is important that the term "quality" in WEEE is much more enlarged to the whole life cycle thinking. One essential step in this direction is the implementation of an efficient recycling management, which can only be achieved when end-of-life challenges get to the awareness of producing industry: Quality in end-of-life of WEEE starts in the design phase of electric and electronic equipment. Therefore, the big aim in this field is to integrate recycling management into the quality loop.

The main issues that will have to be covered in future comprise establishing environmental demands as quality mark already in the first design phases, acquisition, storage, processing and transfer of cost effective product data during life time, definition of quality standards and the European Commission wide certifications for reused parts and products, establishment of European Commission unified procedures for documentation of material streams and compliance with safety regulations, and implementation of standards for data recording, storage and transfer during life time.

5   LIFE CYCLE THINKING AS MANAGEMENT AND STRATEGY

A variety of eco-design tools have been developed within companies and by universities and consultants. However, there is often a poor transference of these tools and knowledge between academia and business – due to a lack of appreciation of real-life challenges being faced by **product** developers. The prime focus has been on environmental evaluation tools and especially on Lifecycle Assessment (LCA) [7] [8] [9]. Outside the realm of the experts there is often confusion between LCA and Lifecycle Thinking (LCT) – LCA being a tool and LCT being the "cradle to grave" philosophy of thinking about product development. LCA is also criticized for being too complex, time consuming, costly and imprecise – which means that it often does not get used outside of leadership companies that can afford to employ in-house experts or consultants. There is a big need for simpler LCT based eco-design tools for non-experts and small and medium sized enterprises (SMEs) - particularly as most practicing design engineers will have little or no understanding of environmental issues. In addition, there is a need for a different set of tools dependent on the stage of the product development process one is addressing (see ISO14062). Eco-design requires a multi-disciplinary approach, which means that their needs to be the development of tools focused on the needs of different business functions e.g. tools for marketing written in **marketing language**. Eco-design is an emerging discipline and **sustainable product design** (SPD) is an even newer concept.

Sustainable development can be defined in different ways – some see it as an eco-efficiency issue only, but others see it as covering the "triple bottom line" terms (economic, environmental, social) with social aspects of SPD being very new territory. In addition, with the edges becoming less clear between products, services and product-service-systems (PSS) - perhaps we should be thinking more broadly of **sustainable solutions**. This maybe requires the development of a new discipline of **sustainable solutions developers** who are comfortable with each aspect of the "triple bottom line", are business focused, innovative and have the skills to create and maintain multi-stakeholder partnerships.

At the heart of eco-design is the concept of the product life cycle. It starts with resources taken from nature, goes on to the production of materials and product or component manufacturing processes, the use and maintenance of a product, and concludes at the end-of-life stage ("cradle to grave" approach).

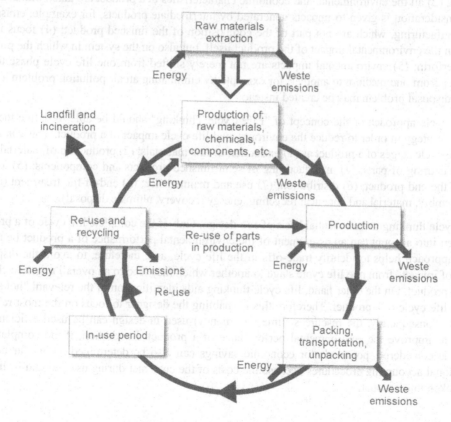

FIGURE 1. The Life Cycle Perspective

"Life cycle thinking" takes into account all the environmental aspects that occur in the complete life cycle of a product. These include energy consumption, materials application, chemical substances, durability, reusability or recyclability, packaging, transport, etc. The complete life cycle includes mining and materials production, production of components and subassemblies, assembly of products, and the reuse and discarding of products. Finding eco-efficient solutions, on the basis of life cycle thinking, is encouraged nowadays by legislators, especially in the European Union (see, for example, the WEEE Directive, the EEE draft directive, and the discussion on the IPP proposal). One way to achieve eco-efficient solutions is by the "producers" responsibility" principle, under which producers are made legally responsible for parts of the life cycle outside their traditional domain of manufacturing; for instance packaging waste, and products discarded by consumers.

Life cycle thinking is a holistic view. Design options should not have a reduced impact at one life-cycle stage at the expense of increasing the impact on the complete life cycle. Therefore trade-offs between design options, as well as between stages of the life cycle, have to be examined carefully.

Consideration of the entire life cycle can help ensure that [10] (1) no materials are arbitrarily excluded; (2) all the environmental and economic characteristics of a product are taken into account; (3) consideration is given to impacts generated by intermediate products, for example, emissions in manufacturing, which are not part of the composition of the finished product; (4) focus is not only on the environmental impact of the product itself, but also on the system in which the product will perform; (5) environmental impacts are not merely shifted from one life cycle phase to another or from one medium to another: for example by eliminating an air pollution problem a solid waste disposal problem may be created instead.

A life cycle approach or the concept of "Life Cycle Thinking" should be able to define the best design strategy in order to reduce the environmental life cycle impact of a product. This can affect the life cycle stages of a product as (1) extraction of raw materials; (2) production of materials; (3) manufacturing of parts; (4) manufacturing of semi-finished products and components; (5) assembly of the end product; (6) distribution; (7) use and maintenance; (8) end-of-life treatment (reuse, disassembly, material and chemical recycling, energy recovery, ultimate disposal).

Life cycle thinking is the essential basis of eco-design. Only if the complete life cycle of a product is taken into account can an assessment of the environmental performance of a product be made. This approach helps to identify trade-offs in the life cycle, and, therefore, to avoid the displacement of impacts from one life cycle stage to another which may lead to an overall negative change in the product. On the other hand, life cycle thinking aids identification of the relevant "hot spots" in the life cycle of a product. Therefore this is enabling the designer to focus on the most relevant issues. Consequently, the resources (time and money) used in design can be used efficiently in order to improve the environmental performance of a product. In addition, if the complete life cycle is considered, potentials for economic savings can also be determined, going far beyond traditional accounting procedures because the costs of the customer during use and end-of-life are also taken into account.

6 CONCLUSIONS

Integrated Product Policy (IPP) still lacks the necessary mechanisms to be operationalised, as well as methodology for setting priorities in conflicting situations; it also lacks a focus on the global context of the initiative, supply networks, services and the social/ethical aspects of sustainable development. There are concerns about the WEEE amendments, which might make the Directive unworkable and prohibitively expensive; these amendments have fostered a lack of trust in European environmental product policy-making. The acceptance of the draft EEE Directive has been mixed, because it still needs clarity, precision and non-conditionality. Some terms need to be defined; while, for example, the mandatory requirement for component and sub-assembly information to be passed down through the supply network has received some positive response. Organizational aspects play an important role in the success of environmental product improvements and environmental product chain management. For eco-design to be successful, it must operate at both a strategic level and an operational level, and be supported by adequate tools and methodologies. Corporate cultures also play an important role in how eco-design is incorporated and managed.

The electronics sector is facing a range of short, medium and longer-term pressures for improved environmental performance that have implications for eco-design e.g. from customers (B2B, government, retailers), legislation and standardization (WEEE, national WEEE, RoHS, EEE, IPP, ISO14062 1, 2). This means that "environment" is becoming a competitive issue in the industry. In parallel, the sector is facing an economic downturn, constant technological change, structural shifts e.g. moves towards outsourcing and contract manufacturing and "knowledge loss" e.g. environmental professionals being made redundant as a result of down-sizing and re-structuring.

In addition, many large companies are moving to become "systems integrators" supplying "black box" services manufactured by other companies. This means that procurement, supply chain, or network management will become an increasing important business process in relation to product development.

The configuration of possible QM-systems in WEEE recycling strongly depends on the demands they are driven by the following recommendations: (1) There is a need to analyze the reasons for poor transference of eco-design knowledge between academia and business. (2) There is a need to examine the implications for the sector of eco-design "knowledge loss" as a result of redundancies due to downsizing and re-structuring, in lieu of the forthcoming implementation of WEEE and RoHS. (3) Universities should stimulate the integration of technical and management considerations of eco-design into engineering, design and business school curricula. (4) Member state governments should develop awareness raising programmes in relation to eco-design and recycling implications of the forthcoming environmental legislation, particularly focused on the needs of SMEs: existing member states and accession countries. (5) Business multipliers should develop eco-design "train the trainer" programmes to be integrated into innovation schemes aimed at SMEs. (6) There should be a wider dissemination of simple eco-design tools designed for non-experts and SMEs. (7) Practical eco-design tools should be developed focused on services and product-service-systems (PSS). (8) Practical eco-design tools should be developed focused on the needs of marketing, procurement; supplies chain management and cross-functional teams. (9) There is a need for a clearer understanding of what sustainable development means for the electronics sector and its implications for product development and subsequently education, and training. (10) The envisaged ways of recycling or reusing components will have different demands than a material treatment. Reusing materials can be different in same usage, upgrade or downgrade of materials. (11) The materials that are utilized in products, e.g. the treatment of hazardous or precious materials. (12) The quantity, in which materials accrue. (13) The loss of quality until the end-of-life and available information about the influences during use period.

ACKNOWLEDGEMENTS

The authors acknowledge the support of the the programme HITRA/TEST – Technological research-developmental project supported by the Ministry of Science and Technology (Republic of Croatia) (Project of modular system for recycling of secondary raw materials, no. TP-02/0120-14).

REFERENCES

1. Taguchi, G., (1986), Introduction to Quality Engineering: Designing Quality into Products and Processes, Asian Productivity Organization, Tokyo - American distribution by UNIPUB/Kraus International Publications, New York.
2. Taguchi, G., Elsayed, E.A., Hsiang, T.C., (1989), Quality Engineering in Production Systems, McGraw-Hill.
3. ..., (2003), Directive 2002/96/EC of the European Parliament and of the Council of 27 January 2003 on Waste Electrical and Electronic Equipment (WEEE), Council of the EU.
4. ..., (2003), Directive 2002/95/EC of the European Parliament and of the Council of 27 January 2003 on the Restriction of the Use of Certain Hazardous Substances in Electrical and Electronic Equipment (RoHS), Council of the EU.
5. ..., EU Framework 5: Competitive and sustainable growth project // Environmental Life Cycle Information Management and Acquisition for Consumer Products - www.elima.org
6. Seliger, G., Grudzien, W., Müller, K., (1998), The acquiring and handling of devaluation, Proceedings of the 5th CIRP International Seminar on Life Cycle Engineering, Stockholm (Sweden), pp 99-108.
7. Cramer, J.M., Stevels, A., (2001), The unpredictable process of implementing eco-efficiency strategies, Chapter 18 in Sustainable Solutions, Martin Charter and U. Tischner, Eds, Greenleaf Publishers Ltd, Sheffield, UK.
8. Frankl, P., Rubik, F., (2000), Life Cycle Assessment in Industry and Business – Adoption Patterns, Applications and Implications, Springer, Berlin, Germany.
9. ..., (1997), ISO 14040 – Environmental Management – Life Cycle Assessment – Principles and Framework, Berlin, Beuth, September.
10. ..., (2001), ISO 14062 – Environmental management – Integrating environmental aspects into product design and development, draft of the ISO/TC 207/WG3, April.

A COMPARISON OF DIFFERENT TECHNOLOGIES FOR AUTOMOTIVE COMPONENT PRODUCTION BY A LIFE CYCLE PERSPECTIVE

M. De Monte, E. Padoano, D. Pozzetto

Department of Mechanical Engineering, University of Trieste, Italy

KEYWORDS: Environmental Impact, LCA, Car Components Production.

ABSTRACT. For several decades now, plastic components have been used more and more in the automotive industry, both for supporting and for aesthetic elements. For such uses, goals like reduction of weight and of production costs, improvement of production plant flexibility, have assumed an ever increasing importance and have been considered important for companies to define their capability on answering to market requirements and challenges. Also the reduction of environmental impacts related to production, use and disposal of these components, was considered as an important goal during recent years. With the aim of investigating some alternative production cycles of plastic components, the paper shows the process and the results of comparison between some different ways to develop the molding and plating processes using the Life Cycle Assessment methodology. Such approach is introductory to the following evaluation of a specific component, a car trim, which has not a complex geometry but that was considered useful to initial targets of a comparison involving a plethora of possible different solutions.

1 INTRODUCTION

The importance of the automotive sector for the Gross National Product is well known for many countries. Such sector, actually, represents one of the most important industrial sectors from many points of view.

From the technical one, experimentation of new materials and production technologies characterize this area, especially to give answers to new design features and to construction requirements, and to reduce environmental impacts without invalidating final product quality.

From the economic one, the complexity of production systems and logistics require a continuous improvement of management policies to answer to globalization challenges.

From resource and from environmental perspectives, the great quantity of final commodities generated by such sector imply large raw material consumption and waste production. These ones, large in amount and variety during the whole life cycle, have to be treated (recycled, reused and disposed) and need designers to put a great attention to production process selection and optimization directly from first steps of new product design [1].

It is also well known that the use of plastic components in the automotive sector presents a growing trend. For such reason, it's important to analyze and improve the study and the knowledge of

Published in: E. Kuljanic (Ed.) *Advanced Manufacturing Systems and Technology*,
CISM Courses and Lectures No. 486, Springer Wien New York, 2005.

plastic materials and related technologies from an environmental point of view, to allow a conscious use of them, gathering improvements without generating problems at any point of product life cycle.

To investigate this matter with respect to the case study of a plastic trim production process, the methodology of Life Cycle Assessment (LCA) was taken into consideration and applied to alternative options. Goals of this study were to verify the quality and quantity of impacts related to different materials and production processes and to compare impacts of new materials and processes with those of materials and processes in use taken into consideration as point of reference.

2   LCA: A METHOD FOR THE EVALUATION OF PRODUCT ENVIRONMENTAL IMPACT

One of the most used and standardized methodologies available to evaluate environmental performances of processes and products is the Life Cycle Assessment procedure (LCA), presented in its guidelines in the ISO 14040 ÷ 43 norms. It may be used to assess environmental performances of products from the very beginning of their production to the end of their use, both for internal uses (e.g. improvement of product performances) and for external uses (e.g. communication or marketing) [2]. LCA is a process "to evaluate environmental burdens related to products, processes or activities, to identify potential impacts on the environment coming from energy or material consumptions, to identify and to evaluate possible product improvements" [3].

The four steps characterizing a general application of LCA, as defined in the UNI EN ISO 14040 norm, are the definition of goal and scope, the inventory analysis, the impact assessment and the final interpretation [4]. Such steps do not describe a static process. All these phases, actually, use a feedback operation to fine-tune initial objectives and to allow the improvement of the quality of final results.

The first phase defines in detail parameters that will have a relevant impact on the whole assessment, such as targets of the evaluation process in terms of its aim and final user of results, system boundaries in terms of elements and processes considered into the analysis, functional units to which all input, output and results refer to, and quality standards for input data in terms of time-window for their collection, geographic area of interest, technologies considered for production processes and data accuracy [5].

The second phase includes data collection on specific processes and materials involved into the specific life cycle [6]. Variables of interest are related to amounts of raw materials and energies, to quantities of emissions and wastes. Data sources to develop such step can be identified into companies interested in the evaluation process and into public sources (databases and researches) available in literature. The first source can be exploited to observe and measure specific process data, the second one, instead, can be exploited for general-purpose ones, also because it's not always suitable and possible develop directly an analysis of all elements.

The quality of collected data is very important for the development of the following part of a LCA study. Results of the whole study, actually, depend directly on the quality of data available at this moment. To reduce problems of data collection, especially taking into account complex life cycles, a splitting activity of complex activities can be suitable. Such operation makes it possible to define elementary processes, not necessarily reproducing physic ones, allowing an easier definition of mass-energy balances.

The third phase exploits the first result of the LCA process, the "Inventory table", to start the final evaluation. Such structured format has, actually, an intrinsic information content that is, unfortunately, suitable only for technicians.

The definition of an exhaustive judgment of environmental performances can be obtained developing the impact assessment phase, which uses some methods ("indicators") to provide a synthetic, quantitative evaluation of impacts.

At the end of the synthesis of results, the final interpretation phase can be developed to ensure the most accurate analysis of them. At this step, some observations can emerge and some feedbacks can be returned to improve both the LCA process and production and design activities toward an environment-friendly perspective.

3   A COMPARISON BETWEEN DIFFERENT TECHNOLOGIES FOR PLASTIC COMPONENT PRODUCTION

To develop the comparison between different materials and production processes for production of plastic parts, it was clear that the selection of a component to be used as element of comparison was needed. Such component had not to be necessarily with a complex form, because this aspect is not a limit for an environmental evaluation taking into consideration mass-energy balances for its development. The target of the study was to compare technologies and not the production processes of a particular component, even if considering a specific component could simplify such objective.

It would be much more important that the component presented aesthetic features, characteristics that are usually required for automotive components, and that some different choices could be proposed for it.

A component that was considered satisfying such requirements and useful to catch final targets of the comparison, was a trim for a dashboard (Figure 1). With respect to this part, the following issues were considered:

- materials:

  - ABS co-polymer;

  - ABS + PC blend;

- molding processes:

  - compact molding;

  - gas aided injection molding (GAIM);

  - bi-injection molding with rotating mould;

- plating processes:

- ▪ standard plating system with electro less nickel;
- ▪ direct plating system;

and, taking into consideration the whole production process for the trim part, the following alternatives were considered:

- trim mono-material, made on ABS, molded with the compact molding process and plated with the standard plating process;
- trim mono-material, made on ABS+PC, molded with the compact molding process and plated with the standard plating process;
- trim mono-material, made on ABS+PC, molded with the GAIM process and plated with the standard plating process,
- trim mono-material, made on ABS+PC, molded with the GAIM process and plated with the direct plating process,
- trim bi-material, structural layer on ABS+PC, aesthetic layer on ABS, molded with a bi-injection process with rotating mould and plated with direct plating process.

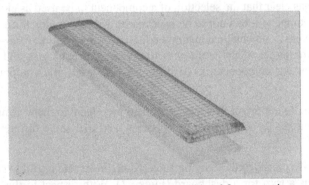

FIGURE 1. The trim, component considered for comparison

In particular, the optimized GAIM process, the rotating bi-injection molding process and the direct plating process without the use of electro-less nickel, can be considered an improvement of the relevant transformation processes.

4   LCA DEVELOPMENT

According to the UNI EN ISO 14040 norm, the evaluation process of considered materials and processes, and of alternative products is presented as follows.

4.1   GOAL AND SCOPE DEFINITION

The aim of the study was to evaluate impacts related to different materials and processes used in the production cycle of plastic components for the automotive sector. In particular, the target was to compare, from the environmental point of view, performances of materials and processes considered both separately and simultaneously with the trim production cycle.

Results were directed to the design team of a company, therefore for an internal use.

In order to make the evaluation uniform between the considered processes and materials, the comparison was developed with respect to the following functional units:

- one kg for polymer materials;

- one m^2 of surface for the plating processes;

- one trim production for the molding processes.

With respect to data used to develop the evaluation, those related to specific processes used by the partners of the company were, if possible, collected or measured directly by them. If data were not available, they were estimated exploiting the basic experience of company's technicians. General-purpose data were, instead, selected from a set of databases taking into account the following two criteria:

- consider a geographic area of interest including the EU Community;

- limit the number of data sources, to guarantee the use of data selected, collected and processed using homogeneous conditions.

The following sources associated with the SimaPro 5.0 SW were used:

- Pré Consultants database;

- Idemat '96 and 2001 databases;

- ETH_ESU '96 database.

In this way, it was possible to guarantee not only the use of appropriate data, but also a uniform quality of them and, together with the direct measurement of specific process variables, an appropriate final quality of results.

The perspective considered into the evaluation is a "from cradle to process" perspective, without the aim of evaluating logistic organizations. For such reason, system boundaries were defined in order to include processes and materials required for component production. Transportation between processing sites and assembly of components were, instead, not considered. Also final disposal after the use was not considered because the aim of the study was to support recycling of materials and the bi-injection molding process could allow such reuse. Recycling, actually, is not related only to product performances, but also to the system capability of collecting, separating and making available secondary materials to production companies.

4.2 INVENTORY ANALYSIS

To develop such phase, the analysis of involved processes and materials was carried out in order to define which parameters could be directly available from the plant and which ones had to be found into general-purpose sources. Such activity was supported by a splitting process of complex production cycles useful to simplify mass-energy balances related to elementary processes. In particular, specific data from the company were used to detail amounts of materials, emissions and energy consumptions of transformation lines. General-purpose databases were, instead, considered for data related to electricity and energies, chemicals, raw materials, etc., for all the elements that can be considered homogeneous, similar, provided by similar but different suppliers.

All such data were the basis for the fulfillment of the Inventory table, starting point for the synthetic results elaborated as follows.

4.3 IMPACT ASSESSMENT

The assessment phase was carried out using the Eco-Indicator '95 method proposed by Leiden University in 1992 [7]. Although this is not one of the most recent methods, it does evaluate impacts with respect to a set of effects like greenhouse effect, ozone depletion, acidification, heavy metals, winter and summer smog, eutrophication and carcinogens, that can be easily understood by technicians and also by non-technicians alike. For such reason, it was considered much more useful for targets of the evaluation under development. With respect to normalization values, the option extrapolating missing data using energy use was considered.

Before taking into consideration the comparison between different production cycles, the comparison between materials and processes was considered. Table 1 resumes impact values (in millipoints – mPt) obtained for the comparison between polymers, molding processes and plating processes respectively. Such results are graphically presented in Figure 2 for polymers, in Figure 3 for molding processes, in Figure 4 for plating processes.

TABLE 1. Impact values for various materials and processes

Impact category	Unit	Total	green-house	ozone layer	acidifi-cation	eutro-phicat.	heavy metals	carci-nogens	winter smog	summ. smog
Unit		mPt	mPt	mPt	mPt	mPt	mPt	mPt	mPt	mPt
ABS Polymer	kg	6,86	0,238	0,653	2,98	0,302	0,306	0,011	1,25	1,12
ABS+PC Blend	kg	5,44	0,406	0,261	2,73	0,443	0,124	0,005	0,732	0,74
Trim Compact Molding	one trim	0,691	0,0954	-0,004	0,304	0,0174	-0,002	-7,3E-5	0,155	0,126
Trim Bi-Injection Rotating	one trim	0,54	0,0796	-0,0076	0,233	0,0091	-0,0036	-0,0001	0,128	0,102
Trim GAIM	one trim	0,456	0,0646	-0,0042	0,199	0,01	-0,002	-6,1E-5	0,104	0,0838
Plating Standard	m²	120	2,89	0,225	36,4	1,53	56,7	0,203	20,5	1,34
Plating Direct	m²	87	2,69	0,219	33,8	0,984	28,7	0,2	19,3	1,17

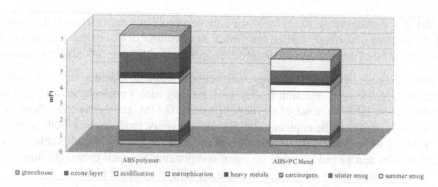

FIGURE 2. Comparison between impacts of polymers of interest for the work

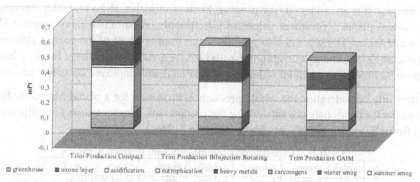

FIGURE 3. Comparison between impacts of molding processes of interest for the work

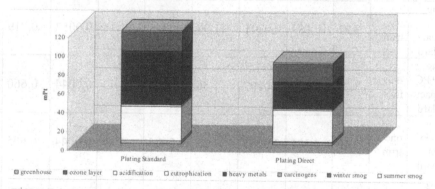

FIGURE 4. Comparison between impacts of plating processes of interest for the work

From the first diagram emerges that the ABS+PC blend has a lower impact than the ABS co-polymer. Such fact is important because the ABS+PC blend presents also better mechanical per-formances, which allow an easier use of it in a sector characterized by very strict technical requirements. Unfortunately, it presents some difficulties when used for components which have to be plated. It demonstrates, basically, some adhesion problems in particular if the molding process of the part is not correctly carried out with a strict control of its process parameters. For such reason it could be of interest to consider production systems allowing a double layer creation.

The second diagram, showing the comparison between molding systems, supports such intuition. The use of a double layer injection system with rotating molds presents, actually, performances which are not so far from those of the better process, GAIM, and better than those of compact molding. Such result can be explained by the contemporary production of both layers using the same clamping force. Such systems, actually, have double or multi-face molds able to rotate into a molding system and to produce one or more components for each cycle. Because of this, the higher energy consumption is balanced by a greater efficiency of the plant.

The third diagram compares the innovative plating process using direct deposition to the commonly used system using electro-less nickel. The new process can allow a greater reduction of impacts because it is an ammonia free process. Such elimination allows a better running of waste water treatment plants, with lower values of all effluents, in particular heavy metals like nickel, into the waste water. The process also guarantees an improvement of the energy efficiency of the plating process (lower energy consumption), and a 30% greater productivity for the plant, due to a shorter process time, with a similar reduction of financial and employee costs for the company.

Taking now into consideration the whole production processes for a plated trim with the different production cycles previously considered for the comparison, impact values (in millipoints – mPt) emerged from the evaluation process are resumed in Table 2.

TABLE 2. Impact values for different trim production cycles

Impact category	Unit	Total	green-house	ozone layer	acidifi-cation	eutro-phicat.	heavy metals	carci-nogens	winter smog	summ. smog
Unit		mPt	mPt	mPt	mPt	mPt	mPt	mPt	mPt	mPt
Drying ABS Compact Standard	one trim	3,92	0,189	0,0741	1,39	0,0843	1,15	0,0052	0,719	0,291
Drying ABS+PC Compact Standard	one trim	3,76	0,208	0,0297	1,36	0,100	1,13	0,0044	0,660	0,248
Drying ABS+PC GAIM Standard	one trim	3,49	0,175	0,0285	1,25	0,0905	1,14	0,0044	0,605	0,202
Drying ABS+PC GAIM Direct	one trim	2,84	0,171	0,0284	1,20	0,0796	0,581	0,0043	0,581	0,199
Drying ABS+PC Bi-inject. Direct	one trim	3,07	0,183	0,0497	1,28	0,0765	0,590	0,0047	0,644	0,247

Such results are graphically presented in Figure 5 with respect to Eco-Indicator categories and in figure 6 with respect to different production phases.

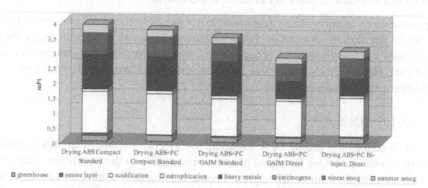

FIGURE 5. Comparison between impacts of trim production with respect to Eco-Indicator categories

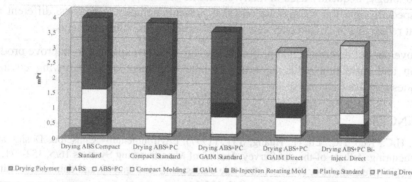

FIGURE 6. Comparison between impacts of trim production with respect to different production phases

4.4 INTERPRETATION

With respect to the whole production process, the results of the evaluation show that the proposed technologies for the molding and the plating phases are able to support the improvement of environmental performances of the selected component. In particular, it is important to underline the great contribution of the innovative plating system to the reduction of impacts of the whole production cycle. Such new process, actually, is able to reduce of about 25% the environmental impact of a process which is critical from the environmental point of view.

Another fact emerging from such evaluation concerns the use of the bi-injection process for the molding phase. Such use, on one hand, allows the implementation of a double material product much more friendly for the plating process (ABS surfaces can be considered easier to be plated than ABS+PC ones) also with scrap reduction. On the other hand, a double layer product can allow the use of recycled materials for the structural layer, materials that cannot be used for the aesthetic one because they do not provide adequate performances and final quality of the surface.

The use of a bi-injection process does not cause, actually, a so great deterioration of environmental performances of the product, and the use of recycled materials for the sub-layer could allow a further reduction of impacts evaluated in the fifth case.

The innovative production systems guarantee performances from the environmental point of view which are in line with the ones of the better technologies available. Besides, proposed processes ensure greater flexibility levels for processes and also for materials that can be used to produce the final part. Because of this, their use can be of interest to improve the whole product quality and environmental friendship.

5   CONCLUSIONS

The paper described the use of LCA methodology to evaluate environmental performances of alternative production cycles for plated plastic components into the automotive context. The aim of the study was to show how such approach could help industry during its management activity, to make conscious environmentally motivated decisions.

The methodology, frequently used in many economic sectors as Decision Support Tool, could be helpful especially into the automotive one that presents complex processes, different materials and a great research work to develop new materials and process improvements.

The improvement of environmental performances can also be exploited to improve product communication and marketing skills, in order to take advantage of the growing environmental consciousness of final customers.

REFERENCES

1. Zhang, H.C., Kuo, T.C., Lu, H., Huang, S.H., (1997), Environmentally Conscious Design and Manufacturing: A State-of-the-Art Survey, Journal of Manufacturing Systems 16/5, 352-371.
2. Vigon, B.W., Jensen, A.A., (1995), Life cycle assessment: data quality and database practitioner survey, Journal of Cleaner Production, 3/3, 135-141.
3. SETAC, (1993), A Conceptual Framework for Life Cycle Impact Assessment, SETAC (Society of Environmental Toxicology and Chemistry), Pensacola, USA.
4. UNI, (1998), UNI EN ISO 14040, October, 31, 1998, Environmental management - Life cycle assessment-Principles and framework (Italian version).
5. De Monte, M., Padoano, E., Pozzetto, D., (2005), Alternative coffee packaging: an analysis from a life cycle point of view, Journal of Food Engineering 66, 405-411.
6. Curran, M.A., (1996), Environmental Life-Cycle Assessment, New York, USA: McGraw-Hill.
7. Goedkoop, M., (1995), The Eco-Indicator '95 Final Report, Pré Consultants, The Nederland.

SIX SIGMA METHODOLOGY: A POSSIBLE SOLUTION TO INCREASE THE PROCESS CAPABILITY IN THE AUTOMOTIVE MARKET

F. Aggogeri, E. Gentili

Department of Mechanical Engineering, University of Brescia, Italy

KEYWORDS: Quality Management, Continuous Improvement, Process Capability.

ABSTRACT. The purpose of this paper is to show the power of the Six Sigma methodology in satisfying the requirements of the automotive market. The company of this project produces cooling air pipes for deluxe vehicles. The problem solving method applied was the DMAIC (Define, Measure, Analyse, Improve, Control). This process review takes into account the increased quality standards of the output; therefore, the real problem was not to increase the system productivity, but to eliminate those process features that the automotive market considers as defects. For example, identifying the CTC (Critical to Customer), the main defect categories were: scraps on the pipes, blows on the devices and contamination of chip. The paper shows all steps of the Six Sigma implementation suggesting the solution to increase the process performance.

1 INTRODUCTION

Six Sigma is a strategic tool to gain efficiency in order to satisfy customers. The introduction of these management philosophies means a wide reorganisation of the company, by adopting a continuous improvement logic and assuring severe changes in business results. The end purpose, in adopting the Six sigma methodology, is to create an output, exactly as desired by the customer, by internally removing all the potential source of defects and reducing no-value activities from the value stream. Six Sigma is an implementation of a measurement system to collect data, analyse results and integrate the information into industrial processes.

Six Sigma means that the process or product will perform with almost no defects, but the methodology goes beyond of this aspect. In fact, the methodology is a better way to manage a business or a department. It "puts the customer first and uses data and facts to drive a better solution" [1]. It could be defined at three distinct levels: metric, for an elevated number of data and statistical concepts, methodology, for a rigorous problem solving method, and philosophy, for a new idea to involve all internal and external resources. In the Six Sigma methodology the communication is an important key for its success. It is fundamental to involve all organisation, showing the philosophy principles, tools and tactics.

Over the last years, Six Sigma is becoming a strong reference for the Industrial Quality and the Business Management. It is not the continuation of Total Quality Management. In fact even if the methodologies have the same goals and similar application tools, Six Sigma and Total Quality Management were born in different places and have their own distinct managerial features in a problem solving approach.

Published in: E. Kuljanic (Ed.) *Advanced Manufacturing Systems and Technology*,
CISM Courses and Lectures No. 486, Springer Wien New York, 2005.

The Six Sigma methodology is characterised by two different problem solving methods: the DMAIC (Define, Measure, Analyse, Improve, Control) and the DFSS (Design For Six Sigma). It is possible to use the DMAIC when the processes or the system are already existing, while it is preferable to apply the DFSS in optimising a design process [2].

Defining, measuring, analysing, improving and controlling the activities of the organisations can increase the efficiency that is gained by improving equipment usage, improving management methods and implementing strategic plans and goals [3].

Business must satisfy each of customer needs, produce quality products at lowest cost possible and sustain acceptable profitability. "The DMAIC roadmap is not only useful for problem troubleshooting; it also works well as a checklist when doing a project" [4]. The methodology leaves a large field to interpret its adaptability in the different contexts; it delivers a clear structure that could be used and decomposed in many solutions. Its flexibility, if applied correctly, enables a significant improvement and saving also in service companies.

Sigma defines the performance level of a process, described by a normal distribution level, to satisfy the customer needs. Supposing that a process is centered on the tolerance interval, the level of sigma denotes the distance between the process average and the nearest specification limit. Three sigma processes represent a highly unsatisfactory performance [5], because they generate 2,700 defects per million parts. The methodology goal is "to produce at least 99.999966% "quality" at "the process step" or part level within assembly" [6]. This means no more than 3.4 defects per million parts or process steps if the process has a shift of as much as 1.5 in the long run. It is important to remember that another Six Sigma metric is the capability index, Cp and Cpk. A Six Sigma quality level assures index values for Cp and Cpk requirement of 2.0 and 1.5 respectively.

2   THE KOMO PROJECT

The purpose of this case study, namely KOMO project, is to show the power of the Six Sigma methodology in satisfying the requirements of the automotive market. The company of this project produces air cooling pipes for deluxe vehicles, Figure 1.

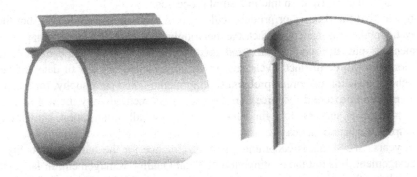

FIGURE 1. The project product, an air cooling pipe for deluxe vehicles

This process review takes into account the increased quality standards of the output; therefore, an analysis of defect causes was necessary to eliminate those process features that the automotive market considers as defects.

In the project it was very important to collect input information on customer needs and expectations, translate the VOC input into meaningful terms and define requirements for the process, and product.

The main goal of this Six Sigma project was to increase the performance level of the whole manufacturing system.

In modern organisations processes are fragmented in different departments. Therefore, the team should map the primary process and alternative paths, providing a context picture. In this way, information and data are the inputs for the other Six Sigma steps. To understand the complexity of the KOMO system, a process mapping was necessary. The production stream of the product is described: the extruded pipes arrive at warehouse from the supplier, and then they are transported to the cutting process. In this position an operator cuts the pipes in devices of 400 mm and eliminates the produced chips.

The devices are worked by five machine tools producing on the piece all features required by the customer. Subsequently the devices, having been washed in an industrial washing machine, are sent to a DC (distribution centre), that then delivers the vehicle constructors.

In order to verify customer's requirements, a 100 % sampling was implemented in the distribution centre, using a visual inspection.

The applied problem solving method was the DMAIC (Define, Measure, Analyse, Improve, Control), but in certain project aspects there are references to DFSS (Design for Six Sigma).

In order to deploy this project the top management had constituted a team, consisting of the production manager, the control and assurance quality leader, two expert operators and continuous improvement consultants. It was essential to make an effective use of the main human resources of the company.

In the first part of the study it was necessary to collect information and data concerning the manufacturing system.

The tools used to collect information on Voice of the Customer "include many simple and sophisticated market research methods, requirement analysis concepts, and newer technologies, such as data warehouses and data mining" [7].

This activity was divided into two steps:

-   to assess the technical reports of the Distribution Centre, classifying the main defect categories and deploying an economic analysis on the Cost of Poor Quality and on the lost gains
-   the implementation of an internal visual inspection to check the impact of the Six Sigma actions in order to understand if there was an effective improvement of the system performance level.

Considering this information it was possible to plan the main milestones for the improvement actions.

The first step of the project was to identify the CTCs (Critical to Customer), assessing the technical reports given by the DC inspection.

The collection of these data and information allowed us to constitute a defect list, as follows:

-   Blows on device

-   Off centring error
-   Internal rugosity
-   Extrusion bubbles
-   Scraps on device

that represented the CTCs (Critical to Customer).

Figure 2 shows a Pareto Diagram that identifies which defect category had a strong impact on the result.

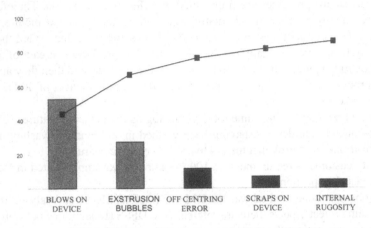

FIGURE 2. Pareto Diagram on main defect categories

It is possible to note that the main category is blows on device, followed by extrusion bubbles. These customer needs had to be correlated with the Critical to Quality (CTQs) characteristics, using the Quality Function Deployment, developed by the Six Sigma team.

This tool can be used to link the Voice of the Customer directly to internal process. It involves the entire company in the design and control activity, reporting the Critical to Customer information and comparing it with the Process and the Product [8].

The CTCs are in the rows while the CTQs, the features of the process or product giving value to customer satisfaction are in the columns, Table 1.

This first qualitatively analysis showed that the Six Sigma team had to focus on the washing process and in particular on the industrial washing machine with reference to the extremity zone of the piece. The last part of the define phase consisted in an economic analysis of the KOMO project. The production cost of a piece is 4 € and the transportation fee of the devices rejected by the DC is 0.5 € / piece. Considering a production volume of 60,000,000 pieces / year, it is possible to quantify the COPQs measuring the performance level of the KOMO system, as follows.

These customer needs had to be correlated with the Critical to Quality (CTQs) characteristics, using the Quality Function Deployment, developed by the Six Sigma team.

3   THE SYSTEM MEASUREMENT

Measures are important because they help in creating baselines and targets for improvement, and provide a common language and focus for a cross-functional group [9].

In fact, the measurement enables us to show how the reduction of defect and the elimination of the variability gives a link to a correct collection of data and information.

Analysing the Distribution Centre reports on defects, it was possible to calculate the KOMO system yield. Considering a production volume of 60,000,000 pieces / year, it was 98.5% defining 900,000 defect pieces / year.

In order to check the impact of the improvement actions, the Six Sigma team implemented a visual inspection (100 % sampling) after the washing process, using the more expert operators of the company as inspectors.

Nevertheless, in the first analysed batch they rejected a lot of pieces as scraps, without classifying them in a particular defect category. This aspect highlighted that an inspection training was necessary to improve the performance level of the measurement system.

TABLE 1. The QFD for KOMO project

	Extremity	Cutting	Washing	Machine Tool	Priority Level
BLOWS	10	5	10	4	7
BUBBLES	1	1	1	1	5
OFF CENTRING	1	1	1	8	3
RUGOSITY	1	1	1	1	1
SCRAPS DEVICE	8	10	8	5	3
TOTAL	103	74	103	73	

4 THE ANALYSIS OF THE KOMO MANUFACTURING SYSTEM

The analysis phase was divided into two steps: manufacturing processes, considering the machinery and all internal features of the system, and materials, involving the supplier processes and external resources.

To begin the Six Sigma team considered blows on device as the main defect category, because it had a strong impact on the final result. In order to assess the scraps, the team studied those piece zones in which there were a lot of blows, identifying the extremity of the device as a critical area on which the team should focus for immediate improvement, as shown in the qualitatively analysis (QFD).

A global review of the system enabled us to identify the washing machinery as a defect source. In fact the machinery basket could damage the device when the operator inserts the piece in the machinery, as shown in Figure 3.

The brainstorming analysis focused on the other defect categories, for example it was possible to note that the scraps on device could be caused by the cutting process, in particular by the maintenance of the cutter. The bad clamping of the piece in the machine tools could give rise to the off centring errors.

To identify the source of the extrusion bubbles, the Six Sigma team involved the pipe supplier, imposing on it a global review of its processes.

This step was deployed by the marketing managers, defining the main operative milestones to reach the performances requested by the automotive market.

It was essential to study the performance level of the inspection process; in fact in the first measurements the operators classified many pieces in the unknown defect category. Thus a training for operators to define the real defect classes was necessary.

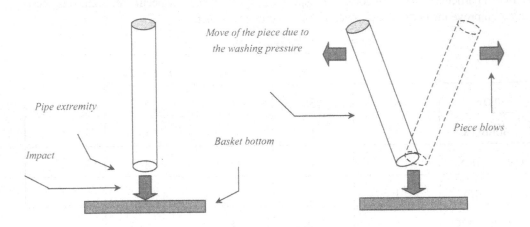

FIGURE 3. Washing simulation before Six Sigma implementation

5 THE IMPROVEMENT AND CONTROL PHASES

The first step of the improvement phase was the deployment of a new training to show the operators the different typologies of defects. This action reduced strongly the "unknown" category. Before the Six Sigma application the "unknown" voice was 13% in the defect classification and 2.3 % in methodology implementation, as shown in Table 2.

In order to eliminate the blows on the device, the Six Sigma team studied a new method to wash pieces, by changing the basket cover. This choice enabled us to protect the pieces during the washing, eliminating the blows on the device.

TABLE 2. The improvement of the inspection process

	Before Six Sigma	After training	Last measurement
Unknown category	13%	2.3%	1.5%

The maintenance person used rubber to cover the basket that soften the blows due to water pressure during the washing.

Figure 4 shows the washing basket before the Six Sigma analysis (a) and after the implementation of the improvement actions.

Another important step was to involve the pipe supplier, in order to eliminate the extrusion bubbles. The Six Sigma Team imposed a global review of the process, suggesting a deployment of the DOE (Design of Experiments).

Using The Design of Experiment, it was possible to identify which process variables had a strong influence of the final result.

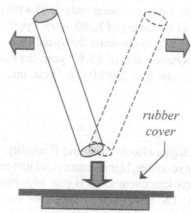

rubber cover

FIGURE 4. Washing process after the implementation of the cover

The team suggested checking the extrusion speed and temperature. At present the analysis of these data are in the process of the being worked: nevertheless the first results of this study are shown in the last visual inspection measurements. In implementing Six Sigma the company obtained these results:

- the KOMO system yield was 98.50 %
- at present, the results of the last measurements shows that the new yield is 99.45 %

The Six Sigma methodology provides a problem solving method to design product and process, utilising tools, training and measurements.

The design purpose is to meet customer expectations at Six Sigma quality levels. The name of this problem solving method is Design for Six Sigma and this can be integrated in the DMAIC.

As just stated the framework deploys five phases: define, measure, analyse, design and verify (DMADV).

In the KOMO project the team used prevalently the DMAIC, but in some situations there was the possibility to design new elements or activities of the system.

For example, during the analysis phase the team assessed the possibility to move the washing process to the centre of the layout reducing the cycle time and avoiding the blows and scraps due to internal transportation. This action is not easy to implement.

Nevertheless the team is working to fulfil this expectation. In collaboration with the Distribution Centre (DC), the company is analysing the possibility to eliminate the visual inspection in the DC.

Before the Six Sigma implementation the more expert operators were on the washing process in order to avoid the scraps on the pieces. Now it is possible to use them in the company's internal inspection process, reducing the costs due to rejected batches.

The main problem of Design for Six Sigma is the implementation time. In the KOMO project a high quality standard was required in a short time.

Before the Six Sigma implementation the KOMO system yield was 98.5 % with a COPQ of 3,600,000 € / year (considering a production volume of 60,000,000 pieces / year and a production cost of 4 € / unit).

In implementing the methodology, the new process yield is 99.45% (330,000 defects / year), with a COPQs of 1,320,000 € / year and the saving of 2,280,000 € / year.

It is possible to consider the possibility to eliminate the visual inspection in the Distribution Centre. Supposing an average transportation cost of 0.5 € / year, the company saving is 165,000 € / year. Therefore the total company saving is 2,445,000 € / year, underling that there are no investments in the project.

6 CONCLUSIONS

This project underlines the Six Sigma effectiveness and flexibility. The methodology assures an increased customer loyalty, more revenues, higher returns and increased earnings.

Six Sigma is more than a process improvement. It is part of a planned and monitored business strategy steered toward success. The major key to obtain its successful implementation is the alignment of the organisation's visions, values and systems.

It is also fundamental to identify the sources of resistance to Six Sigma and planning a strategy to overcome that resistance. The application of Six Sigma confirms that this approach is essential to satisfy the customers need requirements reducing costs.

This paper intends to show the particular method to approach industrial problems, independently of their nature enabling the company to guarantee the satisfaction of customer needs at the lowest possible cost.

In the project there are no significant investments in fact the team took advantage only of resources that already existed in the company.

Six Sigma should not replace existing organisational initiatives, but instead create an infrastructure that offers a tactical approach to determinate the best solution for a given situation [10]. In this way, the reduction of the COPQs is assured in short time.

The Authors can conclude that Six Sigma is an application to increase the business performances and to improve management operations. It enables employees to deliver the greatest value to customers and owners.

The power of the methodology is the possibility to quantify the results of its application as company savings or profitability: this aspect is often neglected in other improvement techniques.

The organisational mistake is often to think that the methodology is only a statistical technique. Six Sigma is more than a statistical method, it is a part of business strategy that involves all company stakeholders.

ACKNOWLEDGMENTS

Special thanks to Mrs. Mary Flynn who checked the manuscript.

REFERENCES

1. Pande, P., Holpp, L., (2002), What is Six Sigma?, McGraw Hill, New York, 2-3.
2. Yang, K., El-Haik, B., (2003), Design For Six Sigma, McGraw Hill, New York.
3. Adams, C., Gupta, P., Wilson, C., (2003), Six Sigma Deployment, Butterworth Heinemann, Amsterdam.
4. Brussee, W., (2004), Statistics for Six Sigma Made Easy!, McGraw-Hill, New York, 13.
5. Brue, G., (2003), Six Sigma for Manager, McGraw-Hill, New York, p.4.
6. Basu, R., Wright, J. N., (2004), Quality Beyond Six Sigma, Butterworth Heinemann, New York, 36.
7. Pande, P., Holpp, L., (2002), What is Six Sigma?, McGraw Hill, New York, 56.
8. Pyzdek, T., (2003) The Six Sigma Handbook, New York, McGraw-Hill.
9. Stamatis, D.H., (2002), Six Sigma and Beyond Foundations of Excellent Performance, Volume I, St. Lucie Press, New York, 120.
10. Breyfogle, F.W III, Meadows, B., (2000), The Six Sigma Implementation Process, Smarter Solution Inc, www.smartersolution.com.

DETECTING CHANGES IN AUTOREGRESSIVE PROCESSES WITH A RECURRENT NEURAL NETWORK FOR MANUFACTURING QUALITY MONITORING

M. Pacella[1], Q. Semeraro[2], A. Anglani[1]

[1] Dipartimento di Ingegneria dell'Innovazione, Università degli Studi di Lecce, Italy
[2] Dipartimento di Meccanica, Politecnico di Milano, Italy

KEYWORDS: Manufacturing, ARMA Models, Neural Networks.

ABSTRACT. The traditional use of control charts assumes the independence of data. It is widely recognized that many processes are autocorrelated thus violating the main assumption of independence. As a result, there is a need for a broader approach to quality monitoring when data are time-dependent or autocorrelated. The aim of this work is to present a new procedure for manufacturing process quality control in the case of serially correlated data. In particular, a recurrent neural network is introduced for quality control problem. Performance comparisons between the neural-based algorithm and control charts are also presented in the paper in order to validate the proposed approach. The simulation results indicate that the neural-based procedure is quite effective as it achieves improved performance over control charts.

1 INTRODUCTION

Control charts are effective for detecting the presence of an assignable cause of variation, which is potentially upsetting the natural functioning of a manufacturing process, and thus for making possible quality improvement [1]. Inherent in the construction of control charts is the assumption that the sampled process is a normal distribution whose observations are independent and identically distributed (IID). Many processes, such as those found in chemical manufacturing or refinery operations, have been shown to have autocorrelated observations, thus violating the main assumption of independence. With the growing of automation in manufacturing, process quality characteristics are being measured at higher rates and, as a result, the data are more likely to be time-dependent (i.e. serially autocorrelated). Autocorrelation, which violates the independence assumption of standard control charts, is known to have an adverse effect on the average run length (ARL) performance.

A common approach to solve this problem is to filter out autocorrelation by means of Auto-Regressive Integrated Moving-Average (ARIMA) time-series [2]. If the time-series model is accurate enough, the residuals (i.e. the prediction errors) are uncorrelated, and common control charts can be applied to them. The main limitation of residual-based control charts, also called Special Cause Chart (SCC), is that a time series model has to be identified before residuals can be obtained. Time series modeling of the process data, however, is not always simple and the potential effect of a modeling error may render the developed control chart useless or unpredictable. The robustness of the time series modeling approach depends on the sufficiency of the data used for model fitting and the adequacy of the model.

Published in: E. Kuljanic (Ed.) *Advanced Manufacturing Systems and Technology*, CISM Courses and Lectures No. 486, Springer Wien New York, 2005.

On the other hand, the ARMAST chart [3] applies the autoregressive moving average statistic (ARMA) directly to the autocorrelated data without identifying the process parameters. A control chart, implemented to monitor this statistic, is shown to outperform SCCs particularly in the case of small mean-shift of stationary time series (the focus of this work). A special case of ARMAST chart is the EWMAST procedure that applies the exponentially weighted moving average (EWMA) statistic directly to the output of a stationary process. The ARMAST and EWMAST approaches are alleviated from the need for building a time series model. However, selecting proper charting parameters can be cumbersome, and it becomes more complicated when higher orders autoregressive model are considered. This problem points to the need for other flexible methods, such as the one proposed in this paper.

In recent years, witnessing the emerging power of artificial neural networks (ANN) and their significant successes in many applications, there is a great interest in ANN models for quality monitoring, also for detecting manufacturing process shifts in the presence of autocorrelation [4,5]. Among various models of ANN, recurrent neural networks, exemplified by the Elman's neural network (ENN), are the most suitable for quality monitoring in the case of autoregressive data since they exhibit dynamic behaviors.

In this paper, we present an innovative yet simple neural network model for the problem of process-monitoring in the case of autoregressive data. In particular, the simplest Elman's recurrent neural network is proposed which has been shown to be useful in detecting changes of autoregressive processes. The proposed neural network approach applies the ENN algorithm directly to process data and, unlike the aforementioned control charts, it does not require any assumptions about time series model. Performance comparisons between the neural-based algorithm and three benchmark control charts (SCC, EWMAST and ARMAST) are also presented in five simulated test cases. The adaptability of the neural-network approach is demonstrated through the flexible design of the training data set. A network can be trained with historic data, or data generated from a properly selected set of underlying model structures and parameter values if historic data are not available.

The paper is structured as follows. Sections 2 and 3 give an overview of ENN and of the process model. Sections 4 and 5 describe the proposed neural network architecture and the training procedure. In section 6 the performance are analyzed and compared to those of benchmark control charts. Finally brief conclusions are given.

2   RECURRENT NEURAL NETWORKS

In recent years, attention has been devoted to ANN algorithms for solution to difficult real-world problems, including manufacturing control applications and quality monitoring. It has been proved that a useful architecture for quality monitoring is the multilayer perceptron (a feed forward neural network). Indeed, it has been widely applied in the literature [6]. Apart from a feed forward ANN, there have been a number of alternative types, e.g. adaptive resonance theory (ART) and recurrent neural networks (RNNs), which can be usefully exploited to solve many practical problems, including quality monitoring [7,8,9].

RNN [11] is similar to a multilayer perceptron in that all connections feed forward. The difference is that RNN has one or more feedback loops that can originate from hidden or output neurons not only during current time step, but also during some number of previous time steps. The recurrency allows the network to remember cues from the recent past but does not significantly complicate the structure and training of the whole network. Thus, a RNN has superior dynamic characteristics over static neural networks, such as feed forward neural networks, for sequential processing and time-series modeling. RNNs are biased towards learning patterns which occur in temporal order, thus a minor drawback is that they are less prone to learning random correlations which do not occur in temporal order.

A simple way of constructing a recurrent network is to add feedback loops to the multilayer perceptron as proposed by Elman [11]. Elman adds a set of context units that buffer the output of the hidden layer units. The context units feed again into all of the hidden layer units of the network, which operates in discrete time steps, in the same way as the input units. In this way the output of the hidden units is applied in the next time steps to the hidden units along side with new input. At each time step the input to the network consists only of the current value of the time series. Contrary to the normal feed forward networks, training data for the time series has to be presented to the network in sequence. At a given time the output of the context units is dependent on the sequence of values previously presented to the network.

3 PROCESS MODEL

Process monitoring consists in using a procedure that at regular intervals checks the desired stable state of the process. Usually, the process is analyzed to verify a constant mean (assumed equal to zero) with some natural variation (in-control state of the process). The use of a control system can lead to the elimination of special causes of variation pointed to by changes in the process mean (out-of-control state of the process). Let $\{X_t\}_{t=1,2,...}$ be the random time series of the quality characteristic measurements. Suppose that data obey the following model:

$$X_t = Z_t + s \tag{1}$$

Where t is an index of time (or part number), $\{Z_t\}_{t=1,2,...}$ is a time series of autocorrelated natural deviations with zero mean, and s is a shift due to assignable causes. The aim of quality monitoring is to test the null hypothesis $H_0 : s = 0$ (in-control state of the process) against the alternative hypothesis $H_0 : s \neq 0$ (out-of-control state of the process). An important example of autocorrelated data is that arising from a stationary time series. In such a case, the general linear model for the time series $\{Z_t\}_{t=1,2,...}$ is a weighted sum of present and past values of a stochastic sequence $\{\varepsilon_t\}_{t=1,2,...}$ which are IID with zero mean and common variance σ_ε^2. Stationary ARMA(1,1) models are special cases of this general linear model, they are defined as follows.

$$Z_t = \phi Z_{t-1} - \theta \varepsilon_{t-1} + \varepsilon_t \qquad \varepsilon_t \, \square \, NID(0, \sigma_\varepsilon) \tag{2}$$

Where the parameters ϕ (i.e. the autoregressive coefficient) and θ (i.e. the moving-average coefficient) must satisfy the condition $|\phi| < 1$ and $|\theta| < 1$ as it has been assumed that the process is both stationary and reversible. Without loss of generality, the variance of the time series $\{\varepsilon_t\}_{t=1,2,...}$ is assumed equal to:

$$\sigma_\varepsilon^2 = \frac{1-\phi^2}{1+\theta^2-2\phi\theta} \tag{3}$$

Thus $\{X_t\}_{t=1,2,...}$ is a time series of mean equal to $E[X_t]=s$ and variance equal to $Var[X_t]=1$.

4   NEURAL NETWORK DESIGN

Generally, it is difficult to obtain *a priori* information on the structure of a network (i.e. the number of nodes in each level and the connections) to do a given task. This necessitates the training of various sized networks by trail and error. This difficulty is attributed to excess degrees of freedom of a neural network. Additionally, the topology of the neural network directly affects two of the most important factors of neural network training: generalization and training time.

Simulations have shown that larger networks tend to over-fit the training data, producing a poor generalization, while a too small neural network may not be even able to learn the training samples. In general, a large neural network requires more computational time than a smaller one. Moreover, a smaller network may be more desirable because of model understanding. Currently there are no formal methods to directly adapt or select the network structure: the most common approach is the trial-and-error method. The neural network is trained with different sizes and the smallest structure that learns the training examples is selected. It is worth noting that smaller networks achieve good generalizations, but take longer to train.

In this work, the ENN with the simplest multi-layer structure (one input node, one hidden layer neuron, and one output neuron) is used as sequential quality monitoring system. The Matlab/Simulink© software has been used to simulate the performance of the proposed neural network. The hidden layer neuron or node uses a nonlinear activation function (*tanh*) while the output layer neuron uses a linear activation function (*purelin*). Let x_t and y_t be the mono-dimensional input and output value of the neural network respectively, the mathematical model of the ENN-based sequential controller can be summarized as follows.

$$\begin{cases} y_t = b_2 + w_3 z_t \\ z_t = \tanh\left[b_1 + w_1 x_t + w_2 z_{t-1}\right] \end{cases} \tag{4}$$

Where b_1 and b_2 are the biases of the hidden and output layer respectively. w_1 is the weight from the input to hidden layer. w_2 is the recurrent weight from the hidden output layer. w_3 is the weight from the hidden to output layer. The activation function adopted in this work is

$$\tanh(x) = \frac{1-e^{2x}}{1+e^{2x}} \tag{5}$$

The internal structure of the designed neural network is shown in figure 1, where circles represent the input and output nodes. The *tansig* function of figure 1 is mathematically equivalent to *tanh*. It differs in that it runs faster than the Matlab implementation of *tanh*. This function is a useful trade-off for neural networks, where speed is essential and the exact shape of the transfer function is not. The input of the network is the absolute value of the process output at instant t ($x_t = |X_t|$) while the output y_t is a number ranging between 0 and 1.

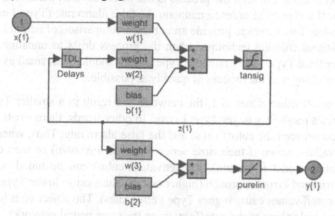

FIGURE 1. The proposed neural network architecture

5 TRAINING

The network is trained in the conventional supervised manner by minimizing the Mean Squared Error function (MSE) with respect to the network weights. The training goal has been fixed to: $MSE < 0.001$. The network is trained with a sequence of two temporal patterns. The first is obtained by Monte Carlo simulation based on the model of equations (1-2) where $s = 0.0$ (i.e. the natural process data with training target $y_t = 0$); the second is obtained by shifting the same sequence by $s = 3.5$ (i.e. the unnatural process output with training target $y_t = 1$). Consequently, an output value close to 1 would indicate that the process mean has changed.

Optimization of the weights has been achieved using an extension of the back propagation (BP) scheme, which is based on the gradient descent principle. In our work, BP learning rules with momentum and adaptive learning rate are used to adjust the weights and biases of networks to minimize the sum-squared error of the network. This is done by continually changing the value of the network weights and biases in the direction of steepest descent with respect to the error. The BP with momentum method decreases BP's sensitivity to small details in the error surface. This helps the training process to avoid being stuck in shallow minima. Training time has been decreased by the use of an adaptive learning rate, which attempts to keep the learning rate step size as large as possible while keeping learning stable. The adaptive gradient rule was tuned automatically by the software Matlab during training. For each of the test cases, the CPU time required for network training averaged 20 minutes on a 1300-MHz machine.

6 PERFORMANCE EVALUATION

In this application, the desired output of neural network is either 0 if no change has been detected, 1 otherwise. Due to random noise, and to different values of mean-shift, the output of neural network is actually a real number ranging between 0 and 1. Hence, an activation cutoff value is defined: if the network's output is greater than such a cutoff, the procedure concludes that a change has been detected. The choice of a cutoff value will usually be based on the considerations of errors of Type I (some action is taken although the process is in-control, i.e. false alarm) and Type II (no action is taken although the process is out-of-control). In particular, in our approach the activation cutoff is chosen in order to maintain the false alarm rate (Type I error) about equal to a predefined value. This serves to provide an unbiased comparison of neural network perform-ance to any traditional charting technique when the process drifts to unnatural states (Type II error rates). Under fixed Type I error rate, the Type II error should be as small as possible in order to signal any mean changes in the process as quickly as possible.

Note that as the cutoff value closer to 1, the network will result in a smaller Type I error while smaller cutoff values result in a larger Type I error. In other words, there exists a monotonically decreasing relation between the cutoff value and the false alarm rate. Thus, when natural process data are either available (even if their time series model is unknown) or they can be simulated (their autocorrelation model is known), the activation cutoff can be tuned off-line through a straightforward 'trial-and-error' approach (higher cutoff values cause lower Type I error rates and *vice versa* lower cutoff values cause higher Type I error rates). The objective is to assess the Type I error for different trial values of the cutoff (by using the same neural network).

In our work, computer simulations were conducted to illustrate the validity of the proposed con-trol procedure. The training and testing data were generated as described by equation (2). In order to compare directly the neural network performance to those of three charting techniques, the values of ϕ and θ used, respectively the AR and MA parameter listed in equation (5), have been selected from those exploited in the literature ([3] – table 4).

$$(\phi, \theta) = (0.475, 0); (0.95, 0); (0.475, -0.9); (0.95, 0.45); (0.95, -0.9) \qquad (5)$$

Training and test data sets were developed for each combination of mean shift ($s = 0$ or $s = 3.5$) size and correlation coefficients. The data sets included one input and one output value, with the input representing the process parameters at time t, and the output value denoting the absence or presence of a shift in mean (0 for non-shifted, 1 for shifted). The training data set included exam-ples of both the non-shifted data pattern and the shifted data patterns.

As noted earlier, a neural network must be trained with sufficient examples in order to be general-ized. However, no evident improvement in performance was attained in our simulations by extending the training set beyond (approximately) 1000 examples. Therefore, one thousand data vectors (500 examples of non-shifted data, 500 of shifted data) of random numbers with a normal distribution of mean s and standard deviation equal to one were utilized for training and testing. Clearly, the training data were not used in the testing process. The weights and the biases of the trained ENN-based control system are given in table 1.

The Average Run Length (ARL), computed from independent simulation runs, is used as the performance criterion used for conventional control charts. ARL is defined as the number of observations needed until an out-of-control signal is released. A good monitoring scheme should have long ARL for an in-control process and short ARL for an out-of-control process. An ARL of 370 for an in-control process is used as a reference for performance evaluation throughout this study. As previously mentioned, appropriate cutoff values have been chosen in each of the five test cases in order to obtain for the in-control process ARLs comparable to the reference value of 370. Table 2 summarizes the ARL (column 'Mean') and the standard deviation of run lengths (column 'StDev') derived from the networks trained in this study.

TABLE 1. Weights and biases (rows) of the trained ENN for each ARMA model (columns)

	$(0.475, 0)$	$(0.95, 0)$	$(0.475, -0.9)$	$(0.95, 0.45)$	$(0.95, -0.9)$
w_1	1.1071	0.5773	0.9643	0.4751	0.7286
w_2	2.3018	1.8271	2.0058	1.8008	1.7207
w_3	0.5047	0.5147	0.5064	0.5165	0.5128
b_1	-1.4230	-0.7241	-1.4451	-0.6160	-1.1682
b_2	0.4963	0.4945	0.4961	0.4909	0.4957

TABLE 2. Simulation results and comparisons for five ARMA models

	Process Parameters			Recurrent Neural Network			Control Charts Jiang, Tsui and Woodall (2000)		
Shift	AR(1)	MA(1)	Cutoff	Mean	StDev	SE Mean	ARMAST	EWMAST	SCC
0.0	0.475	0.0	0.9785	387.19	36.72	5.19	370	370	370
0.5				58.95	9.10	1.29	65.6	83.3	253
1.0				12.10	1.50	0.21	20.3	22.4	118
2.0				2.79	0.23	0.03	6.61	6.17	22.6
3.0				1.55	0.10	0.01	3.67	3.40	4.20
0.0	0.95	0.0	0.7810	388.92	33.64	4.76	370	370	370
0.5				6.28	0.31	0.04	226	237	331
1.0				3.22	0.12	0.02	102	108	139
2.0				1.96	0.04	0.01	25.8	25.7	1.08
3.0				1.07	0.04	0.01	8.65	8.30	1.00
0.0	0.475	-0.90	0.9435	384.01	38.01	5.38	380	370	370
0.5				65.38	9.90	1.40	84.7	105	109
1.0				13.44	1.97	0.28	25.4	29.8	22.8
2.0				2.69	0.24	0.03	7.94	7.68	2.79
3.0				1.43	0.10	0.01	4.29	4.02	1.01
0.0	0.95	0.45	0.8350	387.55	39.67	5.61	378	370	370
0.5				6.96	0.26	0.04	224	226	350
1.0				3.44	0.11	0.02	95.4	97.5	275
2.0				1.93	0.05	0.01	23.6	21.9	43.5
3.0				1.34	0.07	0.01	5.14	7.15	1.30
0.0	0.95	-0.90	0.7460	380.08	38.11	5.39	370	370	370
0.5				6.04	0.27	0.04	42.8	240	42.8
1.0				3.07	0.08	0.01	1.00	110	1.00
2.0				1.97	0.04	0.01	1.00	26.4	1.00
3.0				1.01	0.01	0.00	1.00	8.48	1.00

The column 'SE Mean' indicates the standard deviation of the ARL. Also listed in the same table are the benchmark ARLs of the SCC, EWMAST, and ARMAST charts [3]. The batch-mean procedure was exploited in order to guarantee that simulation measures can be assumed independent and normal-distributed (50 batches of 100 runs in the case of shift equal to 0 and 50 bathes of 40 runs in the case of a shift greater than 0). Figure 2 shows the box plots of the run length obtained from the proposed monitoring schema compared to the best performance among ARMAST, EWMAST and SCC control charts. Similar graphs are depicted on figure 3 in the case of two AR processes.

The comparison, although limited, seems to indicate clearly that the performance of the proposed neural-network based monitoring scheme is superior to that of other statistics-based control charts in almost all instances except when large shifts occur under a high level of autocorrelation. Also worth noticing in table 2 is that the improvement in ARL is quite drastic for $\phi - \theta$ equal to (0.95, 0), (0.475, -0.9) and (0.95, 0.45) cases (i.e. for high values of the AR and/or of the MA parameter) when the shift magnitude is low to medium (i.e. equal to 0.5 and 1.0).

The performance of the proposed monitoring system are compared to those obtained from a feed-forward neural network based approach (the Extended-Delta-Bar-Delta BP neural network), which is recently appeared in the literature [4]. Again, the comparison indicates that the performance of the proposed recurrent neural-network based monitoring scheme is superior to that of the other non-recurrent neural network in all instances except when small shifts occur under a slight autocorrelation. This result seems to support our idea that a recurrent neural network can outperform feed-forward neural networks in the case of autocorrelated processes.

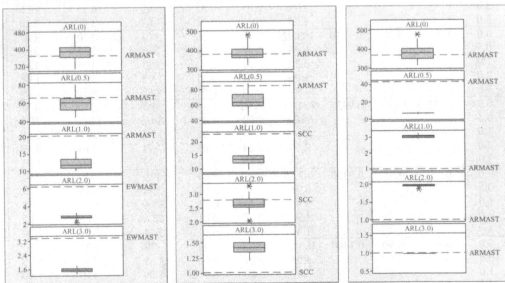

FIGURE 2. Box plots for the proposed ENN's ARLs compared to best control chart performances. From left to right ARMA(1,1) equal to (0.475,0); (0.475, -0.9); (0.95, -0.9)

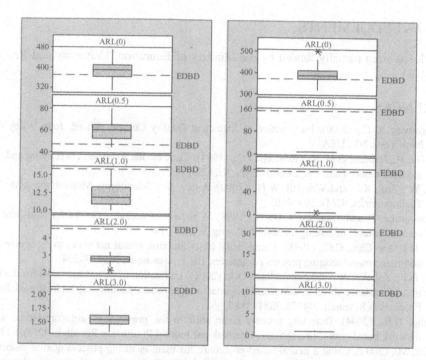

FIGURE 3. Box plots for the proposed ENN's ARLs compared to EDBD neural network performances (Hwarng 2004). From left to right ARMA(1,1) equal to (0.475,0); (0.95,0)

7 CONCLUSIONS

With the growing of automation in manufacturing, process quality characteristics are being measured at higher rates and, as a result, data generated by manufacturing operations are more likely to be autocorrelated. The presence of autocorrelation can significantly increase the generation of false out-of-control signals in traditional control charts. A neural network method to identify shifts in process parameters in the presence of autocorrelation was developed to allow the benefits of SPC to be gained in manufacturing processes where significant autocorrelation exists. The simplest Elman's recurrent neural network has been proposed which has been shown to be useful in detecting changes of autoregressive processes. Its advantages are ease of training and accuracy for detection. The comparative study shows that the performance of the proposed neural-network based monitoring scheme achieves improved performance over statistics-based control charts in several instances. Given the simplicity of this network topology, this approach is proved from simulation experiments, to be a feasible alternative for quality monitoring in the case of autocorrelated process data. Further study may include a rigorous statistical analysis for the neural network algorithm.

ACKNOWLEDGEMENTS

This work has been partially funded by the Ministry of Education, University and Research of Italy (MIUR).

REFERENCES

1. Montgomery, D.C., (2000). Introduction to Statistical Quality Control, 4th ed. John Wiley & Sons, New York, NY, USA.
2. Box, G.E.P., Jenkins, G.M. and Reinsel, G.C., (1994). Time Series Analysis: Forecasting and Control. 3rd ed. Prentice Hall, Englewood Cliffs, NJ, USA.
3. Jiang, W., Tsui, K.L. and Woodall, W.H., (2000). A new SPC Monitoring Method: the ARMA Chart. Technometrics, 42(4), 399–410.
4. Zorriassantine, F. and Tannock, J. D. T., (1998). A review of neural networks for statistical process control. Journal of Intelligent Manufacturing, 9, 209–224.
5. Cook, D.F. and Chiu, C.C., (1998). Using radial basis function neural networks to recognize shifts in correlated manufacturing process parameters. IIE Transactions, 30, 227–234.
6. Cook, D.F., Zobel, C.W. and Nottingham, Q.J., (2001). Utilization of neural networks for the recognition of variance shifts in correlated manufacturing process parameters, International Journal of Production Research, 39(17), 3881–3887.
7. Hwarng, H.B., (2004). Detecting process mean shift in the presence of autocorrelation: a neural-network based monitoring scheme, International Journal Production Research, 42(3), 573–595.
8. Pacella, M., (2003). Using a neural-based procedure for manufacturing process quality monitoring in the case of serially correlated data. Proc. 6th A.I.Te.M. Conference 'Enhancing the science of manufacturing', Gaeta, Italy, September 8th – 10th.
9. Pacella, M., Semeraro, Q. and Anglani, A., (2003). Manufacturing process quality control by means of a Fuzzy ART neural algorithm. Proc. 6th International Conference on Artificial Neural Networks and Genetic Algorithms (ICANNGA03), Roanne, France, April, 23rd–25th.
10. Pacella, M., Semeraro, Q. and Anglani, A., (2004). Manufacturing quality control by means of a Fuzzy ART network trained on natural process data. Engineering Applications of Artificial Intelligence, 17, 83–96.
11. Tsoi, A.C. and Back, A., (1997). Discrete time recurrent neural network architectures: A unifying review. Neurocomputing, 15, 183–223.

A NEW FMEA APPROACH BASED ON AVAILABILITY AND COSTS

G. D'Urso[1], D. Stancheris[1], N. Valsecchi[1], G Maccarini[1], A. Bugini[2]

[1] Dipartimento di Progettazione e Tecnologie, Università di Bergamo, Italy
[2] Dipartimento di Ingegneria Meccanica, Università di Brescia, Italy

KEYWORDS: FMEA, Machine Tools, Operating Failures.

ABSTRACT. The paper reports on a new FMEA (Failure Mode and Effects Analysis) technique for the evaluation of operating failures on machine tools. The traditional FMEA methodology focuses the analysis on failure problems and usually does not take into account other parameters, such as availability and costs. This new technique was setup for predicting operating failures during the design process of machine tools, so reducing costs and time to market. This approach is based on a new index depending on machine tool availability and customer costs related to failure time. For the validation and the evaluation of the method, an industrial case was studied; data collected from this analysis proved to be in good agreement with failures detected by the customer. The adopted method showed a higher reliability compared with traditional FMEA results.

1 INTRODUCTION

The FMEA methodology (Failure Mode and Effects Analysis) is an operative technique used to define, identify and eliminate known or potential failures and problems from products or services before they reach the customer. According to the needs for continuous improvement and to the worldwide competition between providers of goods and services, modern quality systems consider the FMEA approach as an essential aspect for the compliance to the modern certification procedures. If we consider, as an example, the VISION 2000 normative, it is possible to notice how the concepts of quality and intervention procedures are nowadays more concentrated to the planning (instead of acting), to the prevention (instead of controlling and correction) and to the continuous improvement; this approach may reduce costs and time to market so giving to the customers a better image of the company.

The FMEA methodology, introduced into industrial sector by Henry Ford in 1972 can be defined as a precise and systematic analysis of all the possible operating failure modes for the evaluation of strength and weakness aspects of an item and for the prediction of possible problems before they happen [1, 2]. The most important aim is to assure that all the feasible actions for controlling quality and reliability have been accomplished, minimizing in this way the probability of failure occurrence; moreover, the identification of critical aspects during design activities, when the flexibility for adjustments is higher, is useful to reduce development costs and time to market [3]. These aspects become more and more important if we consider the changeover from the push economy to the pull economy during the last years and the awareness that aspects like market fragmentation, high quality, flexibility, short production cycles, innovation and high personalization of products are now essential to be competitive on the market [4, 5].

Published in: E. Kuljanic (Ed.) *Advanced Manufacturing Systems and Technology*, CISM Courses and Lectures No. 486, Springer Wien New York, 2005.

It is important to remark that FMEA techniques are very helpful for controlling and improving reliability but they could be not so accurate on the evaluation of costs related to failures. In effect, not only reliability should be taken into account but the proper balance between reliability, costs and availability. The traditional FMEA approach, even if useful for the definition of the intervention priority (it gives a detailed evaluation of failure risks), can not be employed as reference tool for taking decisions about availability and costs. Basing on this assumption, a new approach for the evaluation of critical aspects during the design process of a specific product using a relation between RPN (the traditional FMEA index based on occurrence, severity and detection), availability and cost variables could be largely advantageous.

Aim of this work is the set up of a new method to estimate operating failures taking into account not only the reliability aspect but the two variables mentioned above as well. A new index based on machine tool availability and customer costs related to failure time was defined and discussed; furthermore, for the validation and the evaluation of the method, an industrial case concerning design problems of a portal-type milling machine was studied.

2   THE ANALYSIS OBJECT: THE MILLING MACHINE

2.1 PORTAL-TYPE MILLING MACHINE DESCRIPTION

Object of this analysis is the operating unit of a 5-axis portal-type milling machine specifically designed and manufactured basing on customer's inputs and needs (see figure 1). This unit is composed by seven basic devices: ball screw feed system, support system with hydraulic slideways, hydraulic balancing system, positioning system (linear and rotary encoder), two brushless engines, two-range epyciclic reduction gearbox and C axis.

FIGURE 1. The milling machine

It is important to give a more accurate description of the two devices that will result critical after FMEA analysis: reduction gearbox and C axis. The first device (see figure 2) is a two-range epyciclic reduction gearbox composed by three parts that transfer the rotation movement to the spindle axis.

FIGURE 2. Two-range epyciclic reduction gearbox

With regard to the C axis it is important to observe that such device is not provided with a spindle and it transmits only the rotary movement of the principal servomotor to the coupling point with other accessories. This technical solution is related to the high flexibility required by the customer and to the machine tool operative purpose. The C axis working mode depends on a particular device (namely ringspann) based on some overlapped rings composed by Belleville springs; the pressure of an hydraulic piston straightens the ring and joint the C axis to the rotation shaft.

3   THE FMEA ANALYSIS

3.1   FMEA IMPLEMENTATION AND RESULTS

The FMEA analysis, executed with traditional methods was implemented on the 7 basic devices described in the previous paragraph. A risk priority number (RPN) was associated to each failure source and the same severity index was assigned to each failure mode. Corrective actions were only trigged for RPN values higher than 100 in order to modify detection and occurrence. No actions were adopted on the severity index since adjustment of this aspects would require a new design of the whole device. Analytical results concerning the two devices that showed design problems (the reduction gearbox and the C axis showed a RPN higher than 100) are reported in Table 1. Occurrence (O) is the frequency of the failure, severity (S) is the seriousness of the failure, detection (D) is the ability to detect the failure before it reaches the customer and RPN* is the risk priority number after corrective actions are taken. The critical topic in the first failure cause is not connected to the cause itself but to the leak of a checking procedure; for this reason the D index resulted to be equal to 10.

TABLE 1. FMEA results

No	CAUSES	DEVICE	FAILURE	S	O	D	RPN	RPN*
1	Design and computation mistakes	Reduction Gearbox	Inability to work with a specific speed, power and torque	8	3	10	240	72
2	Ringspann operating defects	C Axis	Low positioning accuracy Low repeatability accuracy	8	10	5	400	72
3	Hydraulic collector leak caused by ringspann high working pressure	C Axis	Low positioning accuracy Low repeatability accuracy Low hydraulic pressure	8	10	3	240	72
4	Electric collector breakage	C Axis	Short-circuit	8	2	10	160	48
5	Non uniform thermal changes	C Axis	Low positioning accuracy Low static and dynamic stiffness Engine short life Noisy device	8	3	5	120	48

Corrective actions adopted to solve this problem (alternative computations, comparison with similar designs and test procedures) reduced the D index to 3 and consequently the RPN value to 72. With regard to the C axis, design problems are especially about ringspann, hydraulic collector and electric collector. Basing on these statement, a whole redesign of the device was done and the adoption of a more traditional solution adjusted the RPN values respectively to 72, 72, 48. It is important to remark that not only redesign could be adopted, but other solutions based on internal test and co-design activities (with the components supplier) can be devised as well. Redesign was preferred in order to minimize the intervention time; anyway, the solution based on corrective actions should be preferred in a middle-long term. Finally, the fifth failure cause, concerning non uniform thermal changes on the operating unit external surfaces, is due to thermal nonsymmetry of the device and may result in bending during vertical movements. This problem, widespread on several machine tools categories, was controlled by employing some special mechanical and hydraulic components, so dropping the RPN index to 48.

These failure causes, as previously discussed, were compared to the technical service feedback based on service reports; results are showed in table 2. Basing on these results it is possible to confirm a good capability of FMEA methodology in predicting operating failures during the design process of machine tools. In fact, all reported failures can be correlated with the set of failure causes evaluated by FMEA (see table 1).

TABLE 2. Correspondence between FMEA results and failures claimed by the customer

CUSTOMER PROBLEM DESCRIPTION	CORRESPONDANCE TO FAILURE CAUSES No.
Chain breakage and consequent C axis short-circuit	5
Positioning divergence using different accessories	1
Problems during C axis positioning	1
Spindle gears breakage	2
Leak on the RAM	4
Swinging problems on the RAM	3

3.2 EVALUATION OF FMEA ECONOMIC EFFECTIVENESS

Once the technical effectiveness of the FMEA methodology applied to this study was proved some economic aspects of the problem were taken into account. The possible reduction in costs obtained as a result of the FMEA setup during the machine tool design was estimated. The analysis was carried out by comparing the forecasted costs related to the setup of an FMEA procedure and the effective costs for the company when failures (described in the previous paragraph) are detected once the machine tool is already delivered to the customer. The extra cost for FMEA (taking into account the labor cost of an inter-functional design team of six people) was estimated to be equal to € 47,000. The failure costs calculated basing on redesign costs, costs for substitutive items (purchase or production) and labor cost for dismounting, supply and fitting of replacement parts resulted to be equal to € 76,800. The difference in costs obtained by a successful set up of an FMEA procedure is equal to € 29,800, so confirming the economic effectiveness of FMEA techniques in machine tools design.

4 THE NEW APPROACH

4.1 BASIC DESCRIPTION

Since companies are today more and more customer oriented, the need for a revision of production methods, including FMEA, is essential in order to pay more attention to the final user and to be competitive on the market [6]. In particular, "classical FMEA" may assign low values of RPN to failures that, although very frequent, are very easy to detect; this approach is clearly unrealistic: it underestimates the possible effects of high costs related to frequent but easily detectable failures. Basing on these assumptions, it may result useful a new problem approach focused not only on the traditional RPN index but on availability and costs as well. The traditional FMEA analysis was then modified as described below [7,8]. The more critical failure sources, previously found out, were taken into account and the corresponding RPN indexes were recalculated by setting the detection parameter to 1. Adopting this method [9], the dependence of the new index (RPN_N) by detection and the presence of some possible incongruities were removed. A new cost variable based on the estimation of hourly cost of the machine tool incurred by the customer was adopted. An assumption of failure times related to each failure cause was carried out. Starting from availability and costs parameters a new index I was finally obtained.

4.2 EVALUATION OF MACHINE TOOL COSTS FOR THE CUSTOMER

The hourly cost of machine tool was obtained by evaluating fixed costs, operational costs, dismissing costs and effective machine tool working time (hours per year). Taking into account the large number of data required, their variability and dependence by external factors, some assumptions were required.

TABLE 3. Hourly cost

YEAR	MACHINE TOOL AVAILABILITY (hour per year)	HOURLY COST
1	5700	€ 151.43
2	5640	€ 148.27
3	5580	€ 145.04
4	5520	€ 141.74
5	5460	€ 138.36

Results of this analysis are reported in table 3; the hourly cost for the first year (it is the period of time in which the machine tool builder warranty is still active) is the more significant parameter considered in the next steps.

4.3 EVALUATION OF FAILURE TIME

The machine tool considered for this analysis is expected to work 24 hours per day with an average availability for the first year equal to 0.95. The failure time for each cause was estimated using Eq.1:

$$T_s = DS \cdot 24 \cdot 0.95 \tag{1}$$

where DS represents the days of stop necessary to repair the failure; the whole time takes into account dismounting, redesign and production of new items (when necessary), supply and fitting of replacement parts. Basing on these data, for each failure cause, failure time was estimated and reported in table 4.

TABLE 4. Failure time

CAUSE No.	CAUSES OF FAILURE	Days of stop	FAILURE TIME (hours)
1	Design and computation mistakes	180	4104
2	Ringspann operating defects	125	2850
3	Hydraulic collector leak	45	1026
4	Electric collector breakage	15	342
5	Non uniform thermal changes	75	1710

4.4 NEW EXPERIMETNAL INDEX

The new experimental index was obtained by the combination of availability and cost indexes. The availability of a device can be defined as the ratio between effective and nominal machine running time (see Eq. 2); T_w is the nominal running time (machine tool availability), T_s is the failure time related to a specific failure cause. T_w was calculated as described in Eq. 3, where HY is the hours of work per year (2000 hours/year), S is the number of ship for each day (the value was set equal to 3) and 0.95 is the availability for the first year. The cost index was calculated assuming the worst failure condition for the customer (items can not be fabricated using other internal resources and an outsourcing support is required) and assuming a 50% increase in production costs related to the contribution margin.

$$\eta = (T_w - T_s)/T_w \qquad (2)$$

$$T_w = HY \cdot S \cdot 0.95 \qquad (3)$$

Basing on these hypothesis, the second index was calculated using Eq. 4; C_h is the hourly cost for each year as reported in table 3, $(1.5 \cdot T_s \cdot C_h)$ is the cost incurred by the customer for outsourcing production during the machine tool failure time, $(T_w - T_s) \cdot C_h$ represents the costs incurred by the customer in any case during failure time, $(T_w \cdot C_h)$ is the cost for the customer per year with no failure occurrence. The final index I was obtained using the relation 5.

$$C = [(1.5 \cdot T_s \cdot C_h) + (T_w - T_s) \cdot C_h]/(T_w \cdot C_h) \qquad (4)$$

$$I = (C / \eta) \tag{5}$$

Results obtained using this new approach are reported in table 5. It is important to remark that failure causes no. 1 and 2 result now the most critical, in agreement with the feedback received from the customer. The experimental approach demonstrates that causes 1 and 2 are the most critical even if availability and cost indexes are considered separately. No limits for the I index were set by now, since the estimation of a critical value for the definition of possible failure causes would require a more specific and large analysis on several machine tools.

TABLE 5. New index values

No	CAUSES OF FAILURE	DEVICE	η	C	I
1	Design and computation mistakes	Reduction Gear	0.280	1.360	4,857
2	Ringspann operating defects	C Axis	0.508	1.246	2,453
3	Hydraulic collector leak	C Axis	0.820	1.090	1,329
4	Electric collector breakage	C Axis	0.940	1.030	1,096
5	Non uniform thermal changes	C Axis	0.700	1.150	1,643

A comparison between experimental index I, RPN_N index and results gathered using the traditional FMEA methodology is reported in figure 3 (values are normalized).

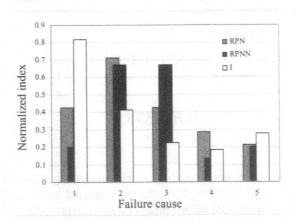

FIGURE 3. Indexes comparison

5 CONCLUSIONS

The effectiveness of FMEA analysis in terms of both technical and economical aspects was proved in this study. In particular, RPN indexes resulted in good agreement with the most critical failures found out by the feedback and costs for the set up of an FMEA procedure resulted lower than failure costs. The new methodology for the evaluation of failures introduced in this work and resumed in the index I showed a good failure prediction capability and a better correspondence with customer needs and problems. This new approach partially disagrees with the FMEA results since the customer orientation of this methodology changed the priority of failure modes previously detected. The tools adopted in this study resulted useful for an efficient and economical design of machine tools. The most important advantages of this technique could be:

- reduction of number and cost of plants and devices modification

- reduction in design and development time

- reduction in cost for production rejects and assistance after sale

- lower risk in terms of builder responsibility

- higher customer satisfaction

A wider generalization of this approach to the whole machine tool sector would require other accurate analysis regarding different companies. Results obtained are anyway a first step for proving how the FMEA can be applied successfully not only in the automotive sector. Future developments will regard a more specific and large analysis on several machine tools in order to estimate and set some critical values for the index I. Besides, a more specific correlation between RPN and I index would be largely desirable.

ACKNOWLEDGMENTS

This work was made possible thanks to Innse-Berardi Spa – BRESCIA (Italy).

This work was made possible thanks to MIUR ex 60% funds.

REFERENCES

1. Stamatis, D.H., (1995), Failure Mode and Effects Analysis: FMEA from Theory to Execution, ASQ, Milwaukee.
2. Chrysler Corporation, Ford Motor Company, General Motors Corporation, (1995), Potential Failure Mode and Effects Analysis (FMEA): Reference Manual.
3. Bassi, A., Surace, F., (1997), Metodologie dell'affidabilità previsionale in progettazione, S.IN.TE.SI AB, Milano.
4. Sciuccati, F., Tanaka, M., (1994), Riprogettare il sistema di produzione. Quality, Cost, Delivery: i tre pilastri della competitività. Il Sole 24 ore libri, Milano.
5. Bianchi, F., Koudate, A., Shimizu, F., (1996), Dall'idea al cliente, come sviluppare nuovi prodotti in minor tempo con le tecniche dell' High Speed Management, Il Sole 24 ore libri, Milano.
6. Guinta Lawrence R.,(1993), The QFD book the team approach to solving problems and satisfying customers through QFD, Amacom, New York.

7. Kmenta, S., Ishii, K., (1998), Advanced FMEA Using Meta Behaviour Modelling for Concurrent Design of Products and Controls, Proceedings of the 1998 ASME Design For Manufacturing Conference, Atlanta USA.
8. Kmenta, S.,. Ishii, K., (2000), Scenario–based FMEA: A Life Cycle Cost Perspective, Proceedings of the 2000 ASME Reliability, Stress Analysis and Failure Prevention Conference, Baltimore, USA.
9. Locatelli, E., Valsecchi, N., Maccarini, G., Bugini, A., (2002), Failure Scenario FMEA: Theoretical and Applicative Aspects, Proceedings of the 6th International Conference on Advanced Manufacturing Systems and Technology, Springer Wien New York, 657-664.

RELIABILITY ANALYSIS IN PRODUCT DEVELOPMENT

F. Galetto

Politecnico di Torino, Dipartimento Sistemi di Produzione ed Economia dell'Azienda

KEYWORDS: Quality, Reliability, Quality Methods in Product Development.

ABSTRACT. In 1999 the author met managers of a "certified" company developing a new engine; he saw use of the "Duane Method" for predicting the in-service MTBF by elaborating the test reliability data. The paper shows the relationship between the Reliability R(t), the "mean number of failures" M(t) and the predicted MTBF using both the "Integral Theory of Reliability" and the "Duane Method"; the difference between the two methods enhances their respective real applicability. Later he saw "fuzzy Theory" used to evaluate product Quality: is that scientific? Since the absence of the scientific approach before, during and after the experiments typically results in relatively uninformative output of questionable general validity, in order to make sound decisions, managers have to be aware of the consequences of their decisions. The same is valid for Lecturers at Universities. This means that professors MUST teach, in a scientific way, Quality ideas on Quality. Unfortunately several people with very little competence, knowledge and experience are in charge of teaching. Professors need Quality metamorphosis.

1 INTRODUCTION

To make Quality of products and services, knowledge of Quality ideas and Quality tools for achieving Quality are absolutely needed, for any person involved in any Company management (Universities…). To find and use the Quality tools for Quality achievement, education of Managers on Quality is essential. Unfortunately too many (I insist, too many) managers [and not only managers ...] know very little about Quality ideas and Methods.

One can see the level of "disquality" (the opposite of Quality) generated by the so-called "practitioners" by analysing the new ISO documents on the Quality Systems: ISO 9001:2000 - "Quality Management Systems-Requirements"; ISO 9004:2000 - "Quality Management Systems - Guidelines"; ISO 9000 - "Quality Management Systems: Fundamentals and Vocabulary". The new ISO documents on the Quality Systems still lack correct ideas of Quality and of Quality Management for PREVENTION: as a matter of fact, they use very badly the ideas on Improvement and Prevention. Some points are worse than before, in spite of the fact that various documents were sent (by F. Galetto) to the Committee Secretaries on December 1998. "ISO-drugged managers" do not give due consideration to the fact that Company Management, as said by Deming (in his very good book Out of the Crisis), must be aware that "Experience alone, without theory, teaches management nothing about what to do to improve quality and competitive position, nor how to do it.), ... understanding of quality requires education. There is no substitute for knowledge. It si a hazard to copy. It is necessary to understand the theory of what one wishes to do or to make..... hundreds of people are learning what is wrong. .... I make this statement on the basis of experience, seeing every day the devastating effects of incompetent teaching and faulty applications. Again, teaching of beginners should be done by a master, not by a hack". Managers have to learn Logic, Design of Experiments and Statistical

Published in: E. Kuljanic (Ed.) *Advanced Manufacturing Systems and Technology*, CISM Courses and Lectures No. 486, Springer Wien New York, 2005.

Thinking to draw good decisions. Professors have to, as well. Quality is their number one objective: they have to learn Quality methods as well, using Intellectual Honesty.

2 QUALITY AND RELIABILITY DEFINITIONS

In the author's opinion, the first step to Quality achievement, through problem prevention, is to define logically what Quality is. It is very important defining correctly what Quality means, because Quality is a serious and difficult business; in order to provide a practical and managerial definition, since 1985 F. Galetto was proposing the following one: Quality is the set of characteristics of a system that makes it able to satisfy the NEEDS of the Customer, of the User and of the Society. This definition highlights the importance of the needs of the three actors: the Customer, the User and the Society.

Prevention is the fundamental idea present in this definition: you possibly satisfy the needs only by preventing the occurrence of any problem against the needs.

Teachers must teach this to students (future Managers); to teach Quality, professors have to understand what Quality entails. Too many professors think like prof. Montgomery(1996): "Quality is inversely proportional to variability", "We prefer a modern definition of quality", "Quality Improvement is the reduction of variability in processes and products", "Note that this definition implies that if variability decreases, the quality of the product increases.".

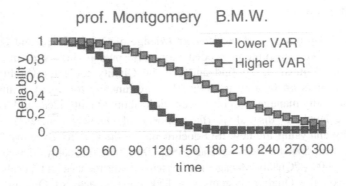

FIGURE 1. Proof that an item with more variability has better Quality (hence Montgomery ...)

Unfortunately such professors really do know very little about Quality practice: Reliability is very important for "Customer/User/Society needs satisfaction". [Quality Tetrahedron, fig. 2] Let's consider a customer buying a car; would you think he will buy a better reliability car or a worse reliability car? Look at the fig. 1: the more reliable item has lower Quality according to Prof. Montgomery ideas!!! It is the same for ISO!!! As a matter of fact, one can find in the paragraph 8.5.2 "corrective action" that the "identification of causes of deviations" generate the improvement: the cause of deviation ("from what?", in fig. 1) is its better reliability: you have to lower the reliability if you want reduce the "deviation"!!! I don't know if there were professors in the committee ISO/TC 176 ... Let's T be the random variable (r.v.) "time to failure", Reliability is $R(t)=P(T>t)$, a function depending on the duration of the interval $0^- t$. The expected value $E(T)$ called MTTF (Mean Time To Failure) is the area under the curve

R(t), does not depend on t. The "failure rate" $h(t)=f(t)/R(t)$ generally depends on t. IF and only IF $h(t)=\lambda$, a constant not depending on t, $MTTF=1/\lambda$, and $\lambda=1/MTTF$ and $R(t)=\exp(-\lambda t)$: the item is always "as good as new". In any other case the failure rate is $h(t)\neq 1/MTTF$, and hence $MTTF \neq 1/h(t)$. For the Weibull distribution, $R(t)=\exp[-(t/\eta)^{\beta}]$, ($\eta$ characteristic life, β shape parameter); $h(t)=(\beta/\eta)(t/\eta)^{\beta-1}$, and $MTTF \neq 1/h(t)$: many professors do not know that.

Defining $H(t) = \int_0^t h(s)ds$, we have $H(t)=\lambda t$ (exponential r.v.), and $H(t)=(t/\eta)^{\beta}$ (Weibull). R(t) is a function related to the first failure of a system. To deal with the various subsequent failures of a repaired system F. Galetto defined the "extended reliability $R_k(t)$" as $P[T_k>t]$ probability of the r.v. T_k "time to the k-th failure". Letting N(t) be the r.v. "number of failures occurred up to time t", the function $M(t)=E[N(t)]=\sum[1-R_j(t)]$ is the "mean number of failures till time t":
$$M(t) = F(t)+ \int_0^t f(s)M(t-s)ds,$$ if the system is renewed at any failure ("renewal equation").

If the system is repaired (but not renewed) at any failure and its reliability after repair is the same as just before failure $M(t) = H(t)$ ["as bad as old" process] (reliability does not improve). It is evident from fig. 1 that the item with higher MTTF is better than the other one and will experience less failures M(t), BUT has lower Quality, according to prof. Montgomery!!!

It is very amazing to note that Montgomery book, which has so silly a definition and other wrong concepts, is suggested to students: what kind of Logic can prove that a wrong idea, even if professed by thousands people, becomes a good idea? Why there are so many professors waffling about Quality and related methods and with little knowledge of Reliability, Design of Experiments (DOE), Statistics, Quality Management, ...? Managers have to learn Logic, DoE and Statistical Thinking to draw good decisions. Professors have to, as well. Quality is their number one objective: they must learn Quality methods as well, with Intellectual Honesty. Prevention, Quality, reliability, process control, corrective actions, QFD, DOE, analysis of variance, statistical tests of hypotheses, confidence intervals, aliasing and entanglement are concepts so important for Customers/Users/Society that professors should know them correctly and teach them correctly. Management and Professors must understand that they have a new job: learning how to use Quality methods for Quality.

Therefore schools must teach with Quality Quality ideas on Quality Management and not follow silly ideas just because many people are saying that they are good ideas: a disquality idea does not become a Quality idea just because thousands of people say it is good.

IF I am right, the fact that the document ISO 9004:2000 got approval (almost one hundred of Nations) as it is, does not make it a "Quality document". Students, as future managers, must know NOW that: managers must understand and learn Quality ideas because too many companies are well behind the desired level of Quality management practices. For F. Galetto, Manager is the person who achieves the Company goals, economically, through other people, recognise existing problems, prevents future problems, states priorities, dealing with their conflicts, makes decisions thinking to their consequences, with rational and scientific method, using thinking capability and knowledge of people. These Managers are very rare ....

In many books (e.g. Montgomery, Taguchi, ...) the NEEDS are left out!!! On the contrary, the Galetto definition insists on the needs; ten characteristics are defined to satisfy the needs (Quality Tetrahedron, fig. 2): one can see very important items like Safety, Reliability, Maintainability, Durability. ISO 9004:2000 uses quite a lot the three terms: needs, expectations and requirements (and "customer satisfaction"); one statement is "The needs and expectations of all interested parties should be translated into requirements for the processes inside the organization."; please note that the requirements are referred only to processes!!! You can fulfil the requirements of a process with complete disquality of the product provided by that process!!! (there are plenty of cases in F. Galetto references)

Quality Tetrahedron

FIGURE 2. The Quality Tetrahedron

Various "practitioners" suggest the use of a panacea for quality planning (many times considered equivalent to "prevention" by incompetent people): QFD, Quality Function Deployment. QFD is not mentioned in the ISO 9004:2000, that states: "Planning, Deployment, Checking, Improvement", are the "actions needed to achieve the continual improvement of performance of the organization." This PDCI cycle is very different from the PDCA of the same ISO? QFD, like ISO 9004:2000, is full of "needs and expectations" (and "customer satisfaction"), that are put in the "customer voice room" of the HoQ (house of quality):we note explicitly that Prevention is always absent in the QFD literature, and this absence is evident from the fist two rooms of the "house of quality": Quality does not live there.

I think that the following statement of the Nobel prize M. Gell-Mann is very suitable for describing the "reality of QFD": "Once that such a misunderstanding has taken place in the publication, it tends to become perpetual, because the various authors simply copy one each other." ("The Quark and the Jaguar: Adventures in the Simple and the Complex", 1994])

One more example in the reliability field is from Akao book; in chapter 7 "Quality Deployment and Reliability Deployment" you can find: "We define reliability as the ability of a product to perform its functions long after it has been purchased." "When we speak of the reliability of a product, we are really talking about the life of each of the product's basic functions." "Although a product has a number of quality features, the life of a product really means the life of the product's main function." "The lifespan of the various quality features will differ ...." "... the life of the product's quality characteristics." "... product life can be measured by success ratio, survival probability which can be measured by mean time to failure (MTTF). [!?!?]" "... decreasing function [!?!?] which can be measured by mean time between failures (MTBF). [!?!?]" "Reliability means guaranteeing that the basic functional quality features of a product will last over a certain time period." "When deploying reliability, we need to identify quality elements for those characteristics whose quality could be difficult to maintain over the desired timespan." The author of that chapter 7 knows very little about reliability (he does not know neither the basic definitions, nor the Mathematics for reliability calculations: is it better for "preventive maintenance" (and reliability, when the Weibull is adequate) a product with

MTTF=100 hours [and β=2.5] or a product with MTTF=200 hours [and β=0.7] ?

Duane was not better in 1964; and are not better the managers I saw using his method in 1999: the method was provided to the Company by "quality" [sic] Consultants!!! Duane derived an empirical relationship, based on a silly "Duane postulate": he defined the "instantaneous cumulative $MTBF_c$= total time (t_c) accumulated by the tested items/total number $M(t_c)$ of failures by time t_c" and found that $MTBF_c= MTBF_0 (t_c/t_0)^\alpha$, where α ranged between 0.2 and 0.4, correlated with the effort on reliability improvement; t_0 is the "total time" at the start of the monitoring period, at which we know that the MTBF is $MTBF_0$. Later we will see how bad is this postulate.

Management must learn that solving problems is essential but it is not enough: they must prevent future problems and take preventive actions: Safety, Reliability, Durability, Maintainability, Ecology, Economy can be tackled rightly only through good methods and preventive actions; the PDCA cycle is useless for prevention; it very useful for improvement. Several of the Quality characteristics [in the Quality Tetrahedron] need prevention; reliability is one of the most important: very rarely failures can be attributed to blue collar workers. Failures arise from lack of prevention, and prevention is a fundamental aspect and responsibility of Management. The same happens for safety, durability, maintainability, ecology, economy, ... The Schumacher accident at Silverstone Formula 1 race (11 July 1999) is a good example of ...

Quality Improvement is important but is very different from Problem Prevention which needs quite different methods and gets much better results: problem prevention is always looking for "Customer/User/Society Needs Satisfaction" and hence looking for better ways of providing attractive Quality to the benefit of Customers/User/Society. (fig. 3)

Too many "ISO-drugged managers" and too many Companies [very often "certified firm" as seen on many trucks running in the italian highways and buildings nearby) are still lacking Quality Management on Quality: the most huge problem against the Quality. To overcome this paramount drawback there is a MUST: Quality Education on Quality for Managers.

But, as said by Deming, education on Quality matters have to be provided by masters, not by hacks: "anyone that engages teaching by hacks deserves to be rooked".

Design of product/service/educational system is the most important phase for Quality. This is evident if one looks at the Development Cycle (fig. 3) of any product/service. From fig. 3, it is clear that Prevention is quite different from Correction: they are related to different problems, carried out at different times and tackled with different methods; Prevention is carried out during the development of product and service, while Correction and Improvement are carried out after production and use of the product/service.

To achieve Quality, Quality tools on Quality are essential: Management need to grow-up their knowledge on methods, especially those needed for Prevention.

Only recently western nations have recognised that education and training are essential for Quality achievement, but in some way they are not making quality decisions: they use blindly methods imported from Japan, just because they are Japanese methods [QFD and Taguchi Methods]. ISO 9001 never states explicitly that the methods used for Quality must be "methods imbedded of Quality"; perhaps this is an "implied need" according the definition 2.1 "Quality" of the standard ISO 8402. ISO 9004:2000 is not better!

If a Manager is the person defined previously, it is evident that he must learn the way to become and to be like that, through serious learning, based on sound education provided by Quality teachers on Quality.

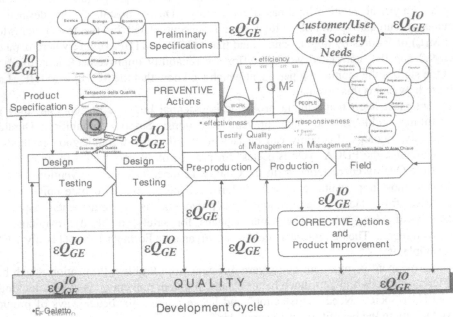

FIGURE 3. The Development Cycle for making Quality product and services

Quality is "needs satisfaction". "Needs satisfaction" is so important and so badly understood that in a translation of one of my papers, where I was insisting that there is a BIG difference between "Customer needs satisfaction and "customer satisfaction" the statement "Customers/Users/Society needs satisfaction must be converted from a slogan to real practice ..." was translated in a foreign language as "The need of satisfaction of Customers/Users/Society must be converted from a slogan to real practice ..."giving the opposite meaning to my own ideas!!! This distinction is not understood, as well, by the document ISO 9004:2000! Unfortunately, that is the Culture of Quality that is professed by the professors of "Total Quality"!!!! Total Quality is a statement with no sense at all.

Such professors think that "certification (or accreditation) of curricula" [using the "credits"] will provide and assure Quality of the courses. Many Universities are wasting a lot of money following those silly ideas [that have been wasting a lot of money, for long-time, in many Companies]!!! TQM and TQM Masters have recently entered higher education, but did Quality enter? If you look at those course programs, you can see that they generally lack some very important subjects like Reliability (system and testing, FMECA), Design Of Experiments, and Prevention; but there is something worse: if you look at the course teaching, you can see that they teach wrong ideas to students in very important fields like testing hypotheses, data analysis, improvements tools, confusion of Problem Solving with Prevention, ... You get Reliability mainly through prevention of problems: for more than 15 years the Ferrari race team did not have enough Reliability due to lack of preventive actions. Looking at the

Development Cycle (fig. 3), one can see that preventive actions are taken during the phases of design, testing and preproduction, well before the product is sold; when the product is in the field a Company is involved in correcting the in-service problems and improving the product behaviour: in this area it is important to prevent again some new future problems (different from the ones that have been corrected). Since "prevention is better than cure", we name "prevented [from deficiencies] product" a product on which "preventive actions" have been taken.

3   QUALITY AND QUALITY METHODS: DUANE VERSUS RIT

Quality of Quality Methods is often disregarded by the "so-called quality professionals" or "quality practitioners". Interested people can find many cases. Many cases analysed show that facts and figures are useless, if not dangerous, without a sound theory. [F. Galetto] It is important to note that most of the people waffling about "total quality", QFD, Bayes methods and Taguchi Methods generally do not distinguish between "planning" (like in the QFD) and Prevention. The interested reader can find many cases, if he uses his Intellectual Honesty.

As seen before the "Duane method" predicts either t_c or α or $MTBF_c$ from t_0 and $MTBF_0$. For example let's use the following data; we know that the starting MTBF is $417=MTBF_0$ after $1500h=t_0$ of development testing and we want $\alpha=0.3$; if our goal is $MTBF_c=700h$ we need an additional testing effort of 6933h.

To understand why the "Duane method" makes no sense, it is important to discuss a bit some ideas on the reliability tests. The number of failures $G(t)$ experienced [up to time t] during a reliability test depends on the items reliability; for Weibull distribution it depends on both the parameters η and β. Using the Integral Theory of Reliability [RIT] we can describe a reliability test through a "system related to the reliability test", where the state "i" of the system is such that $G(t)=i$, and $b_{i,i+1}(s|r)ds$ [called "kernel of the stochastic process"] is the probability that the system makes the transition $i \Rightarrow i+1$ (i.e. it experiences the (i+1)-th failure) in the interval $s \overline{\ } s+ds$, given that it entered the state i at the time instant r;

$$R_i(t|r) = 1 - W_i(t|r) + \int_r b_{i,i+1}(s|r)R_{i+1}(t|s)ds \quad \text{where } i = 0, \dots g \text{ e } b_{g,g+1}(s|r) = 0$$

(fundamental system of the Integral Theory of Estimators)

the functions $b_{i,i+1}(s|r)$ provide $1-W_i(t|r)$, the probability that the system wait in the state i (i.e. it does not experience the next (i+1)-th failure) for the whole interval $r \overline{\ } t$, given that it entered the state i at the time instant r; $R_i(t|r)$ is the probability that the system is not in the state g. $R_0(t|0)$ is the probability that system experienced less that g failures: $G(t)<g$.

The previous integral system allows one to find the estimates and their confidence limits (lower and upper) of the parameters of the failure probability density function.

In the special case of the exponential distribution, $h(t)=\lambda$ (constant failure rate), the estimate of the MTTF is the ratio of "total time on test divided by the total number of failures".

For the Weibull distribution the failure rate is $h(t)=(\beta/\eta)(t/\eta)^{\beta-1}$ [and therefore MTTF $\neq$ $1/h(t)$]; hence the estimate of the MTTF is NOT the ratio "total time on test divided by the total number of failures", but a funtion of the estimates of the parameters η and β.

Based on test data, and the consequent parameters estimation, a Company during the development can "predict" the reliability of the system and the mean number of failures M(t).

A way of predicting reliability (a very important component of Quality) is the use of the theory of stochastic processes.

Many times a "Renewal Process" is adequate for reliability predictions; in that case, one can compute the "mean number of failures M(t)" by solving the following integral equation $M(t) = F(t) + \int_0^t f(s)M(t-s)ds$ where F(t) and f(t) are, respectively, the distribution and the probability density functions of the random variable "time to failure". The integral equation is easily solved by the Laplace Transforms, or by numerical methods.

Look at the following "transition matrix". The integral equation can be suitably transformed into a differential equation If and only If the stochastic process (that rules the phenomenon) can be modelled by a Homogeneous Markov Process (with added states if necessary). The proof can be found in F. Galetto. Therefore is quite silly to state, as done by prof. P. Erto at Pescara (1992) Conference of the Italian Statistical Society (SIS), that "you use a life model RP [Renewal Process] like in the following transition matrix, where t_i are the times of failures and (of renewal) and z(t) is the failure rate" to describe the system behaviour: [you must know Reliability to understand the error]

TABLE 1. Renewal [according P. Erto]

Time				t_1	t_2	t_3			
	State	0	1	2	3	4		
	0	$1-z(t)\Delta t$	$z(t)\Delta t$					
T_1	1		$1-z(t-t_1)\Delta$	$z(t-t_1)\Delta t$				
T_2	2			$1-z(t-t_2)\Delta t$	$Z(t-t_2)\Delta t$			
T_3	3				$1-z(t-t_3)\Delta t$	$z(t-t_3)\Delta t$	......	
.....	....	....	....	......	....	....	......	

How can one compute correctly the various reliability characteristics, if he uses wrong formulas? How can you decide correctly and managerially on reliability, if you use wrong ideas? How much will that cost? [cost of disquality]

I hope that Erto's students use Logic, as my students do, so avoiding such mistakes.

Many other times a "Non-Homogenous Poisson Process (NHPP)" is adequate for reliability predictions; in that case the "failure intensity m(t)" must be assessed; from that, one can compute the "mean number of failures M(t)" by computing the integral $M(t) = \int_0^t m(s)ds$; IF one can assume an intensity $m(t)=(\beta/\eta)(t/\eta)^{\beta-1}$ [that is assuming that the failure rate h(t) of the Weibull distribution can be used as such even after the repair of the system] then M(t)=H(t).

Back to "Duane postulate": he defined MTBF=1/h(t) of the Weibull [actually he did not mention this!], i.e. $MTBF=t^{1-\beta}\eta^\beta/\beta$: putting $\alpha=1-\beta$ and $MTBF_0/t_0^{1-\beta}$ =const, we get $MTBF_c= MTBF_0 (t_c/t_0)^\alpha$. The definition MTBF=1/h(t) is NONSENSE from a Logic, Scientific and Statistical point of view; therefore is going to waste the Company money (cost of disquality!!!): what king of "quality" manager is the one that is so incompetent to buy rubbish by consultants? Does the "Company CERTIFICATION" (the company that bought that method is certified, according to ISO 9002 Standard) give any proof of its Quality?

Following Deming ideas, the Quality Manager can estimate the reliability growth by using statistical sound methods.

Let's use the suffix 0 for any item "before" fixing a problem during development and 1 for the item "after" the problem has been fixed. One can estimate if $MTTF_1 > MTTF_0$ by using and analysing Scientifically the test data, before and after the fix, and the fundamental system of the Integral Theory of Estimators. He can also predict the "mean number of failures in field use" $M(t)$ through the "statistical analysis of the observed $M_0(t)$, $M_1(t)$, $M_2(t)$, ..., $M_n(t)$, ..." after any fix; but a good manager can also predict the "mean number of failures after the i-th fix" $M_{i+1}(t)$ either through the "statistical analysis of the observed $M_0(t)$, $M_1(t)$, $M_2(t)$, ..., $M_i(t)$", ... or the probability calculus of $M(t)$ using stochastic processes. F. Galetto proved that, for any component of a system, there is an asymptote to the Mean Number of Failures curve $M(t)$, which is the limit, for $t \to \infty$ of the solution of the integral equation $Y(t) = F(t) - \int_0^t R(s)ds / MTTF + \int_0^t f(s)Y(t-s)ds$; since any system can be considered as the "superposition" of many components, and the number of failures of the system is the sum of the number of failures of each subsystem, we derive that any system has an asymptote to its curve $M(t)$; the asymptote depends on the failure rate and on the MTTF of the system. At the asymptote the system experiences a steady state failure intensity that is related to the "in-field" failure rate and MTTF. Therefore we can compare the "predicted in-field reliability" by comparing the asymptotes, before the "preventive actions" and after the "preventive actions". The intercept with the axis of the ordinate provides another important figure for measuring the transient period still needed for development before we can rely on the "predicted in-field reliability", after we have made the preventive actions. Fig. 4 shows the idea. The "prevented product" (the one on which preventive actions are made, has lower slope and intercept of the asymptote (better Reliability, MTTF and MTBF) than the "previous product". It is easily seen that there is no need of using wrong, and completely NOT scientific, ideas like the "Duane postulate" to estimate the effect of prevention. ($l*t$ is the line computed from an estimated "constant failure rate" l [λ in the formulae].

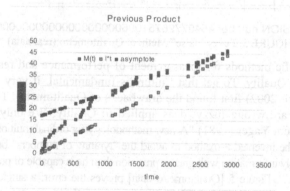

FIGURE 4. M(t) computed from real data

4 QUALITY AND QUALITY METHODS: FUZZY THEORY

Let's consider the case of some "new" methods for Quality evaluation; they were developed because "there is now a strong need for proper evaluation tools", [Franceschini, Romano, Rossetto, 1998-99-2000]. They say the "Qualitometro I method was devised (1998) ... in order to evaluate and check on-line service quality". Later it was presented and discussed "a new proposal for data processing that enhances elaboration capabilities of Qualitometro I. This new procedure, named Qualitometro II, is able to manage information given by customers on linguistic scales, without any arbitrary and artificial conversion of collected data. Collecting and treating data by means of the Qualitometro II eases this process providing a method for performing elaboration closer to customers fuzzy thoughts. … Qualitometro II method can be interpreted as a Group Decision Support Tool for service quality design/redesign … able to handle information expressed on linguistic scales, without any artificial numeric scalarization." Hence they introduce a "new instrument that can fulfil the formal properties of a linguistic scale and allow for the expression of the variety in the decisional logic of the evaluator. … The fuzzy operator that is used allows for this flexibility in the decision logic." (underlinement is due to F. Galetto)

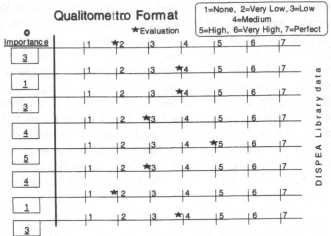

DECISION number 4540772675109600000000000000000000000
FIGURE 5. Fuzzy "false" Method: Qualitometro (real data)

Effective and scientific methods for measurement of performance and results, is the absolute condition to achieve Quality. To get that Culture is fundamental. It very interesting to notice that a student L. Perri (2002) first found the drawbacks of Qualitometro. There is not space for showing how much are wrong fuzzy ideas applied to Quality. We only mention that those wrong ideas come from Yager (1981) "A new methodology for ordinal multiobjective decision based on fuzzy sets", where he invented a method to avoid the "tyranny of numbers" because "… forcing the decision maker to supply information with greater precision than he is capable of providing. This may lead to incorrect answers …". Figure 5 [Qualitometro data] proves the error: a single number provides the decision function (using "prime numbers")! The referees of those papers could not find what students can find. This panacea idea was before used in the paper "Design for Quality: selecting a product's technical features" (Franceschini F., Rossetto S., 1997, Quality Engineering), where the two "authors thank Dr. D. Schaub for her helpful comments and suggestions". Was Dr. D. Schaub the referee? The referee of the paper could not find what students can find.

5 CONCLUSIONS

In this paper we stressed again that Prevention with "future consequences of present decisions" [the futurity] is essential in order to provide the Customers/Users/Society integrally the Quality they need. In Companies, management need management metamorphosis. In Universities, professors need Quality management metamorphosis. because students that are going to be the future managers must know that facts and figures are useless, if not dangerous, without a sound theory. (F. Galetto). Managers have to learn Logic, Quality, Reliability, Design of Experiments and Statistical Thinking to draw good decisions. Professors as well. Quality is their number one objective: they have to learn Quality methods as well, using Intellectual Honesty. [Quality Tetralogy]

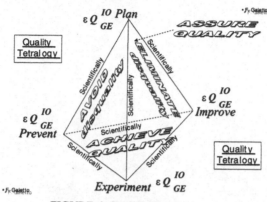

FIGURE 6. Quality Tetralogy

Let's conclude with Deming statements in § 1 and ST John "And there are also many other things, the which, if they should be written everyone, I suppose that even the world itself could not contain the books that should be written." and ST Mark "A prophet is not without honour, but in his own country, and among his own kin, and in his own house." The personal conclusion is left to the Intellectually Honest reader to whom is offered the Quality Tetralogy: Prevent, Experiment, Improve, Plan, SCIENTIFICALLY to avoid disquality, to eliminate disquality, to achieve Quality, to assure Quality, using Intellectual Honesty, as shown in the figure. Brain is the most important asset: let's not forget it.

The author acknowledges the support offered by QuASAR (Moro 8, 20090 Buccinasco, Italy).

REFERENCES

1. J.T. DUANE (1964), Learning Curve Approach to Relibility Monitoring, IEEE Trans., A-2, 563-566
2. W.E. DEMING (1986), Out of the crisis, Cambridge Press
3. W.E. DEMING (1997), The new economics for industry, government, education, Cambridge
4. FRANCESCHINI F., ROSSETTO S. (1999-2000) Qualitometro II Method, Quality Engineering
5. GALETTO F. (1973), Numero dei guasti di un sistema ...., Atti del VIII Congr. AICQ, Napoli
6. GALETTO F. (1984) Affidabilità, Volume 1: Teoria e Metodi di Calcolo, CLEUP, Padova.
7. GALETTO F. (1985) Affidabilità, Volume 2: Prove di Affidabilità, CLEUP, Padova
8. GALETTO F. (1999) GIQA, the Golden Integral Quality Approach, TQM, Vol. 10, No. 1
9. GALETTO F. (1999) Quality Methods for Design of Experiments, 5th AMST 99, Udine
10. GALETTO F. (1999) Quality Function Deployment, Managerial Concerns, AITEM99, Brescia
11. GALETTO F. (2000 4th ed.) Qualità. Alcuni metodi statistici da Manager, CLUT, Torino

12. GALETTO F. (2004) Gestione Manageriale della Affidabilità, CLUT, Torino
13. MONTGOMERY D.C. (1996), Introduction to Statistical Quality Control, Wiley & Sons (wrong …)

NON – CONTACT VOLUMETRIC MEASUREMENT FOR SPINE EVALUATION AND PRODUCTION OF SPINE ORTHOTICS

Karlo Obrovac[1], Toma Udiljak[2], Jadranka Vuković Obrovac[1]

[1] Corkit d.o.o., Zagreb, Croatia
[2] Faculty of Mechanical Engineering and Naval Architecture, University of Zagreb, Croatia

KEYWORDS: Stereophotogrammetry, Spine Deformities, Human Body Scanning.

ABSTRACT. Today it is possible to measure human body with completely non-invasive devices [1,2,3,4] but there is no widely used method for quick, accurate and non-invasive evaluation of spine.[5,6,7,8].Another problem is the fact that such patient should be transported in some specialized center where the appropriate therapeutic solution could be obtained, and where there are technical and medical resources. This work presents the idea, already supported with published work of other authors, of using 3D optical scanning system for evaluating of spine deformities and asymmetry. Considering the fact that commercially available devices for 3D scanning are very expensive, the possibility to scan human body with cameras from several directions and make the reconstruction of body shape seems very promising and competitive. The reconstructed model can be measured with the computer aid which gives tool for quick and quality follow up of shape change through time, and comparison before and after applied therapy. The system could also offer a possibility to design and produce high quality spine orthotics, or highly comfortable and individually adjusted spine supports.

1 INTRODUCTION

Measurement of outer human boundaries is an old discipline. There is almost no human activity where such of measurement isn't important in some aspect. It is difficult to imagine modern furniture, automobile, textile, shoe-making, orthotic and prosthetic industry without having proper information about human body shape and dimensions.[3,4,5,6,8]

Old-fashioned way of measurement of human body includes impression of the whole body or its parts into the foam or warping with plaster bandages. Such procedures have a lot of disadvantages and require skilled technician, while it is inaccurate, time demanding and not always well tolerated by the examined person.

Today there are numerous ways of use human body topography measurements . Here, we primairily think about diagnostics and navigation in surgical procedures. Such applications show tremendous progress and we believe that will change characteristics of classic medical branches.[7,8,10,11,12,13]

During our work with measurement device for the human body digitalization we have thought about its possible use for evaluation of spine deformities.

Published in: E. Kuljanic (Ed.) *Advanced Manufacturing Systems and Technology*,
CISM Courses and Lectures No. 486, Springer Wien New York, 2005.

2 MEASUREMENT OPTIONS

Modern 3D scanning technologies bridge lot of practical issues. Although there are lot of technologies built for this purpose, two principles surpasses others – laser scanning and stereophotogrammetric devices [1,2,3]. Laser scanning is widely used in all kind of engineering applications. Results are highly accurate and with use of specific hardware it may be quite fast. Disadvantage relies in fact that the price of an average laser scanning 3D system often discourage companies from its implementation. Also, there are some more limitations for use of this system for direct body scanning but it beats any other system in accuracy.

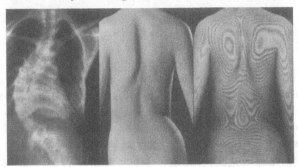

FIGURE 1. Spine deformities and contemporary evaluation methods

Stereophotogrammetric systems are based on the binocular vision.[2,14,15,16,17] Measurements are conducted with typical use of two calibrated cameras (it also may be one or more) pointed to same object. There are some different ways of calculating depth from the images but the most common is recognizing of corresponding points on both images and 3D reconstruction of the scene, through geometric calculations from point positions in the images and known parameters such as distance between the cameras and intrinsic camera parameters.[16] This technique certainly requires appropriate conditions for imaging but allows for fast and accurate scanning. The accuracy of this system may be improved by projection of different patterns onto the object what improves matching of the corresponding points and lessen the number falsely recognized ones.

FIGURE.2. Equipment applied for stereophotogrammetry

3 THE PROBLEM

Spine deformities are seen in the offices of general practitioner, orthopedic surgeon and physical therapist. Often stay undetected or inadequately evaluated what can lead to numerous complications. Reasons for that are, among others, inability for adequate mass screening in schools and lack of the experience for medical evaluation of the problem.

Currently there is no widely used method for quick, accurate and non-invasive evaluation of spine based on a mobile, simple to operate device. Prior to diagnosis transportation of patients to some specialized center where the quality diagnostic and therapeutic solution could be obtained is cost prohibitive.[12,13,14,15,16,17]

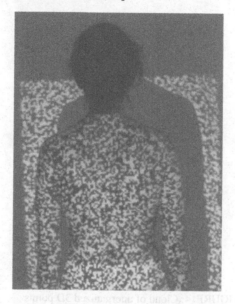

FIGURE.3. Random pattern

Even more important is an issue of invasiveness of commonly used diagnostic techniques. With the help of X-rays measurement conclusion about the size of spine deformity could be done but is not recommendable for patient to do it frequently.

4 METHODOLOGY

When consider spine deformities, it is possible to scan human back with cameras from the several directions and make 3D reconstruction of the given part.

Reconstructed model can be measured with the computer aid (measurement of angles and distances in space), what gives us tool for quick and proper follow up of shape change through time, and comparison before and after applied therapy. One of the advantage of this method is the possibility for the results to be forwarded in digital form via network to the highly specialized centers where professionals could give their evaluation based on given data.

4.1. DEVELOPED AND APPLIED TECHNIQUE

Digital stereophotogrammetric system has been developed based on two (even one) regular (off the shelf) digital cameras that undergo calibration procedure. Using it, and with some additional inexpensive hardware (that is widely available), imaging of human body can be taken. The system can scan body in upright and laying positions, in inclination and declination, and in other needed positions too.

FIGURE 4. Cloud of unorganized 3D points

Patient is aligned in wanted position what is controlled by laser device with projection of vertical and horizontal lines. Then random generated pattern is projected onto the body surface by regular slide projector. Examinee is asked to stand still during the procedure. Imaging is done from two positions with some conditions fulfilled – paralleled optical axes and registration of measurements settings (camera focal length, distance between cameras, etc.). Pictures are processed, with well defined parameters, by specially developed software in user friendly package, providing solutions for camera calibration, recognition of corresponding points in two (or more) pictures, 3D surface restoration from given (unorganized) points and tools for measurement of angles and distances. The results are in form of clouds of unorganized points in space, and in that form they are further processed.

Besides mentioned software modules, modules for data transfer via network, and database for data archiving have been developed. Database with its environment and functions provides data storage, data comparison, and saving of all needed data related to certain patient, condition or so.

5 THE RESULTS

In comparison with measurements given from other systems (laser and tactile device measurements) the registered measurement accuracy was 0.25 mm, which is satisfactory for that purpose. Results are visible in hyper realistic 3D environment with possibility of free object positioning and rotation as well as it's increasing and decreasing in size.

FIGURE 5. Rendered surface

Software with it's characteristics and tools allows for measurements and comparisons which could be difficult to conduct in reality. Initial results encourage further work. All done measurements show differences in symmetry, shape and size of measured human body part when results are compared with referent ones. But, because it is early phase in realization of the project, measurements are done with small number of patients and there are no deep analysis done, which will be done after results that are expected from measurements done in specialized medical centers that are expected to use such devices. Pilot machining of comfortable spine orthotics have been done and, according to patient testimony, proofs to comply well with body shape. In further steps obtained results will be subject of objective and quamtitative verification.

6 CONCLUSIONS

Our project has shown other possibilities for use of stereophotogrammetry in spine deformities evaluation. Compared with commercial available systems it is cheap, portable, reliable and easy to use. Clinical testing through medical (and other) specialties, where such system is applicable, is the further step. Up to date survey showed its quality already and we expect it will find its place in clinical work in view of price within reach, reliability, speed and simplicity in near future.

REFERENCES

1. Pommert, J. K. et al., (1996), "Three Dimensional Imaging In Medicine: Method And Applications", Computer Integrated Surgery (Eds R.H. Taylor et al.), Ch. 9, 155-174

2. J. Spencer, P.M.James, "Three-dimensional facial growth studied by optical surface scanning", British Journal of Orthodontics, Vol. 27, No. 1, 31-38., 2000.

3. J.Norton, N.Donaldson, L. Dekker, "3D whole body scanning to determine mass properties of legs", J Biomech.,35(1):81-86.,2002.

4. Yu, C.Y., Lo, Y.H., Chiou, W.K., (2003), "The 3D scanner for measuring body surface area: a simplified calculation in the Chinese adult", Appl Ergon. 34(3):273-278

5. R.B.Winter, MB, E.Lonstein, "To brace or not to brace : The true value of school screening", Spine. 22, 12, 1283-1284 .,1997.

6. N Chockalingam, PH Dangerfield, G Giakas, T Cochrane and JC Dorgan, "Computer assisted assessment of spine deformities", European Spine Journal 11(4), 353-357.,2002.

7. PD Masso, GE Gorton III, "Quantifying changes in standing body segment alignment following spinal instrumentation and fusion idiopathic scoliosis using and optoelectronic measurement system", Spine; 25: 457-62.,2000.Spine; 19: 236-248. ,1994.

8. Theologis, TN., Fairbank, JC., Turner-Smith, AR., Pantazopoulos, T. (1997), "Early detection of progression in adolescent idiopathic scoliosis by measurement of changes in back shape with integrated shape imaging system scanner", Spine, 22, 1223-1228

9. M.J. Thali, M. Braun, J. Wirth, P. Vock, R. Dirnhofer, " 3D surface and body documentation in forensic medicine: 3-D/CAD Photogrammetry merged with 3D radiological scanning", J Forensic Sci. ,48(6):1356-1365., 2003.

10. Aubin, CE., Dansereau, J., Petit, Y., Parent, F., DeGuise, J., (1996), "Validation of a 3D reconstruction technique for the validation of scoliotic spine wedging", First Scientific meeting of the The International Research Society of Spinal Deformities, Stockholm

11. Stokes, IA., (1994), "Three-dimensional terminology of spinal deformity: A report presented to the Scoliosis Research Society by the Scoliosis Research Society Working Group on 3-D terminology of spinal deformity", Spine 15;19(2):236-248

12. S.Delorme, P.Violas, Dansereau J., J.A. de Guise, C.É.Aubin, H.Labelle, "Preoperative and early postoperative three-dimensional changes of the rib cage after posterior instrumentation in adolescent idiopathic scoliosis", Eur. Spine J., 10 : 101-106., 2000.

13. Stokes, IA., Armstrong, JG., Moreland, MS., (1998), "Deformity and back surface asymmetry in idiopathic scoliosis", I. orthop.res - 6, 129-137

14. Baillard, C., Zisserman, A., (1999), "Automatic Reconstruction of Piecewise Planar Models from Multiple Views", Proceedings of CVPR '99, 559-565., Fort Collins,CO,USA

15. Reid, G.T., Rixon, R.C., Messer, H.J., (1984), "Absolute and Comparative Measurements of Three Dimensional Shape by Phase Measuring Topography", Optics and Laser Technology, 16, 315-319

16. Jericevic, Z., Wiese, B., Bryan, J., Smith, L.C., (1989), "Validation of an Imaging System", Meth. Cell Biol., 30:48-83

17. N. Ayache and F. Lustman, " Fast and Reliable Passive Trinocular Stereovision", Proceedings of ICCV '87,422-427., London, UK,1987.

INFLUENCE OF THE MEASURING PROCESS ON THE VERIFICATION OF GEAR GEOMETRICAL SPECIFICATIONS

G. Casalino, V. De Totero

[1]DIMeG, Mechanic and Operational Engineering Department, Politecnico di Bari, Italy.

KEYWORDS: ISO-TS Geometrical Specification, Measure Uncertainty, Gear.

ABSTRACT. Measurement principle, method, procedure, and conditions can heavily affect the measuring uncertainty and in turn the measuring reliability.

Moreover, the use of data coming straight from measures taken in the working place is rising for decisional process and statistical process control. The losses from mistaken measures and high uncertainty can hit the economy of production. Therefore it is evident that a correct design of the measuring principle, method and procedure as much as the measuring conditions are required.

In this paper a total quality approach is presented for the design of the verification for the product specification of a gear for automotive application. Measurements were taken from the working place. A Doe experiment was conducted for the measure uncertainty minimisation. The approach was borrowed from EN ISO 1253 specifications.

1 INTRODUCTION

The problem of geometrical specifications verification of workpieces can be tackled only after the design process of the product has been conceived. Based on the functional requisites, which are pre-set by the marketing, the designer will establish some characteristics of the product on which it is opportune to define the tolerances (T), which are defined as the interval between the upper specification limit (USL) and the lower specification limit (LSL).

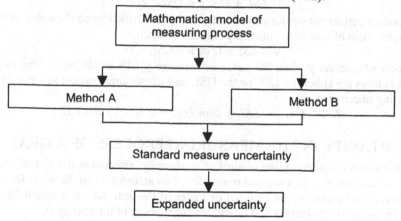

FIGURE 1. GUM flow chart for uncertainty estimation

To effect the geometrical process specifications of a workpiece it is necessary to consider the measure uncertainty. When controlling and industrial testing (typically the verification of

Published in: E. Kuljanic (Ed.) *Advanced Manufacturing Systems and Technology*, CISM Courses and Lectures No. 486, Springer Wien New York, 2005.

conformity of pieces produced in series) the evaluation of the measurement uncertainty can be based on the application of the method of the ISO standard [1], which requires measuring system and devices that guarantee a pre-determined uncertainty (figure 1).

The uncertainty is defined as the parameter, associated with a stated value or a relation that characterizes the dispersion of the values, which could reasonably be attributed to the stated value or relation. This parameter is associated with the result of a measurement, that characterizes the dispersion of the values that could reasonably be attributed to the measurand [2].

The expanded uncertainty is a quantity defining an interval about the result of a measurement that may be expected to encompass a large fraction of the distribution of values that could reasonably be attributed to the measurand.

The evaluation of the uncertainty through the method A is based on the statistic analysis of the observations series.

The B-type evaluation differentiates from the evaluation type A because of the formality with which the result is gotten is not "statistic type " but it is based on " previous experiences" on the measuring operation.

The standard ISO 14253-1 establishes that the tolerances zones are the interval of values of a characteristic of the product for which the product can be declared conforming. The specification limits (lower LSL and upper USL) used in the project setup are conceived not considering the uncertainty of the measuring process by means of that the product conformity to the specifications will be verified.

The standard establish that, when the result of a measurement is compared with the zone of specification of the characteristic of the inspected product, it can happen one of the following three options (where y is the output of the measuring process):

- condition of certain conformity: condition that is verified when the value of measure y falls in the zone of sure conformity, which happens when

$$LSL + U < y < USL - U$$

- condition of certain not-conformity: condition that is verified when the value of measure y falls in the zone of sure not conformity and that is when

$$y < LSL - U \text{ or } y > USL + U$$

- condition of ambiguity when the value of measure y falls inside one of the two intervals whose centres are either the LSL or the USL and whose amplitude is as large as the actual measuring uncertainty

$$LSL - U < y < LSL + U \text{ or } USL - U < y < USL + U$$

2 UNCERTAINTY IN THE MEASURING PROCESS OF A GEAR

We enter the worth considering, the control of conformity, effected at the end of the operation of grinding of the rear gear of a manual transaxle for an automotive application. In such control the characteristic of quality is the external diameter of the gear, which is called M_{dk}. This is a functional requisite for the correct and noisy free functioning of the rear gear.

The tolerances for the external diameter (d) were $\pm 0,039 \ mm$.

It is important to define the relationship between tolerances and functionality of the product as a consequence it is possible to establish how much that characteristic of quality is critical for the functionality of the final product. Nevertheless, narrow tolerances without correlation with the functionality can create increases of costs of production and control, without advantage for the product [3].

In the verification phase of production of the product specifications the control department has to activate a system of measure to declare the conformity or less to the tolerances prescribed by the design. This system of measure will introduce an uncertainty of measure U of which it is necessary to keep in mind reducing the zone of tolerance to a zone of sure conformity. Obviously if it were possible to measuring without uncertainty, all the pieces whose characteristics are in the assigned tolerances would be accepted; but surely elevated costs of measure should sustained. On the other hand, if the uncertainty were as large as T/2, which is (LSL-USL)/2, it would be impossible to decide whether to accept the pieces, which would fall in the ambiguity zone [4].

The process of selecting the right measuring uncertainty starts with defining the measuring process itself. The first step is to recognize in the budget of the uncertainties the contributors to the uncertainty of the process of measure according to the ISO/TS 14253/2 standard. Then each contribution has to be estimated and the uncertainty measured [5].

2.1 UNCERTAINTY CONTRIBUTERS

In the experiment the measurements were done using a simple touch probe that has a lodge for the gear and two ball-ending arms for measuring the external diameter. The M_{dk} measurement consists in the average of several diameters taken at different angles (figures 2).

The operations to be completed before positioning the piece on the lodge are:

1. let the piece cooling down for 10÷15 minutes;
2. cleaning the piece either with a cloth or compressed air;
3. positioning the piece in the lodge correctly.

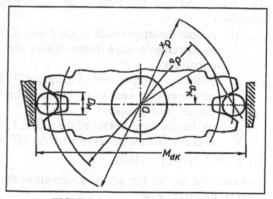

FIGURE 2. M_{dk} measuring layout

Then it was necessary to verify, through a designed experiment which, among these factors, produce an effect on the uncertainty, determining the magnitude and the type of effect. Finally the uncertainty of the measuring process can influence the final decision on the conformity of the piece (figure 3).

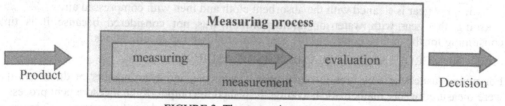

FIGURE 3. The measuring process

The experimental design container four factors that are:

A. piece temperature;
B. piece cleaning;
C. numerousness of diameters measured;
D. duration of each measure.

Each of these factors had three levels which produced a factorial plan 34 composed by 81 tests. Five repetitions for each test were done.

Once built the factorial plan the following step was the execution of all the 405 tests (divided in three equal groups and performed in three different days) then the analysis of the results followed, which permitted to individualize the factors that influence the uncertainty of the process of measure. Choice, then, the optimal procedure, and the combination of the factors that minimizes the same uncertainty completed the investigation.

2.1.1 PIECE TEMPERATURE

The thermal expansion of the metals pushes to choose the temperature of the piece as a possible factor of influence for the uncertainty of the measuring procedure. At the end of the grinding operation the gear's temperature rises up to about 36°C that can causes a consistent increase of the diameter M_{dk}, distorting the result of the measurement. Therefore, after grinding operation, three waiting times were selected so that the diameters were measured at different temperatures.

The times corresponded defined the temperature at which the measurements were taken, respectively:

1. so that the temperature of the piece, T, is contained in the interval 30°C≤T<36°C, the measurements were taken within the fifteenth minute from the end of grinding operation;
2. so that the temperature of the piece, T, is contained in the interval 28°C≤T<30°C, the measurements were taken between the fifteenth minute and the 45th minute from the end of grinding operation;
3. so that the temperature of the piece, T, is contained in the interval 26°C≤T<28°C, the measurements were taken after the 45th minute from the end of grinding operation.

2.1.2 PIECE CLEANING

Towards the end of the grinding operation the machine rotates the piece to get rid of the residual oil on the gear.

Nevertheless, the gear is still dirty and soaked with oil. This condition can alter the measurement to give a non-truthful result, that's why this factor is considered inside the factorial plan.

The three different levels correspond three different cleaning conditions:

1. the gear is measured as clean as it is after the rotation;
2. the gear is measured after been cleaned with an absorbent cloth;
3. the gear is cleaned with the absorbent cloth and then with compressed air.

Washing the gear with water or liquid solution was not considered because it is time consuming for the on-line control of the gear conformity to the specifications.

2.1.3 NUMEROUSNESS OF DIAMETERS MEASURED

For numerousness of measures for each gear it was considered the number of diameters that were measured whose mathematical average furnished the output of the measurement process.

Every of the three different levels corresponded to a precise number of measures withdrawn by a gear, which was rotated in the probe's lodge to the correct position for ball-ended arms. The angles between two successive rotations were (see figure 4)

1. 120° for the first level, which required three measurements;
2. 90° for the second level, which required four measurements;
3. 60° for the first level, which required six measurements.

FIGURE 4. Position of measurements for each level

2.1.4 DURATION OF EACH MEASURE

To get a single measurement the gear was positioned in the lodge and the probe's arms were released so the balls touched the profile of the gear teeth.

Whenever the gear is positioned and arms released it took some seconds to the probe to register the measurement through the electronic acquisition system. The operator can set time for acquisition and it has showed to have some influence on the precision of the measurement.

Therefore three different levels for the duration of the acquisition of measurement were:

1. 5 seconds
2. 10 seconds
3. 15 seconds

Table 1 shows the experimental plan at-a-glance.

TABLE 1. Factorial plan for uncertainty

		FACTORS			
		A **temperature**	**B** **cleaning**	**C** **number of** **measurements**	**D** **duration of test**
	1	30°C≤T<36°C	absent	3	5 sec
LEVELS	**2**	28°C≤T<30°C	rough	4	10 sec
	3	26°C≤T<28°C	fine	6	15 sec

3 DOE EVALUATION OF CONTRIBUTIONS

The analysis of the variance is used for identifying and quantifying controllable causes in random process.

The sample mean square error of the measurement, which is the squared root of the variance, was calculated for each one of the 81 treatments.

The $\overline{M_{dk}}$ was the average on 5 repetitions of the measure.

The graphics of mains effects plots shows that only the numerousness of measured diameters is important for the uncertainty of the measuring process (factor C in the mains effects plots in figure 5). The p-values of interactions between factors in the analysis of variance were lower than 0,05.

Therefore they were neglected in the successive optimization process.

Besides the analysis of the variance signaled four outliers observations, which could be caused by an operator's error either during measuring or recording the data after an accurate control. The values were discharged. [6].

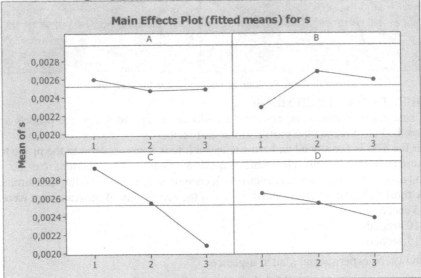

FIGURE 5. Plots of main effects

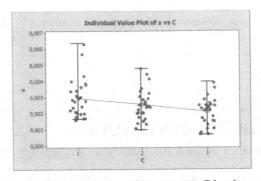

FIGURE 6. Mean squared error versus C levels

The lower uncertainty associated to the procedure of measure has been that to which corresponded the 3rd level of the factor C, and that is, a numerousness of measured diameters as large as six.

Figure 6 shows the mean square error for each treatment against the C factor levels.

4 UNCERTAINTY MINIMISATION

Uncertainty calculation requires some important sources [7].

From the variance analysis the only factor that produced a principal effect on the uncertainty of measure of the procedure was the number of measures for each gear. The level of this factor that minimised the uncertainty of the measuring process was 6 with a 60° angle between each diameter.

Once fixed that level, the levels of the other factors, which formed the measuring treatments, were chosen on the basis of practical considerations. In practice, none particular recommendations were done for the temperature, which is upper bounded by that of the grinding machine. The cleaning was made by the absorbent cloth. The duration of test was fixed to 5 seconds.

The uncertainty related to the above-depicted treatments was evaluated by 75 further measurements.

The variance of the sample was calculated with the following formula based on the normalized distribution and applied to the M_{dk}.

$$s^2\left(\bar{z}\right) = \frac{s^2(z_i)}{n} = \frac{1}{n(n-1)}\sum_{i=1}^{n}\left(z_i - \bar{z}\right)^2 \tag{1}$$

$$s^2\left(M_{dk}\right) = \frac{1}{(75-1)}\sum_{i=1}^{75}\left(M_{dk\,i} - \overline{M_{dk}}\right)^2 = 0,0000869504 mm^2 \tag{2}$$

The variance of the average of the M_{dk} is

$$s^2\left(\overline{M_{dk}}\right) = \frac{s^2\left(M_{dk\,i}\right)}{75} = \frac{0,0086950404}{75} = 0,0000011593 mm^2 \tag{3}$$

Finally an estimation of the measuring process uncertainty is

$$u = s\left(\overline{M_{dk}}\right) = \sqrt{s\left(\overline{M_{dk}}\right)^2} = 0,0010767259 mm \tag{4}$$

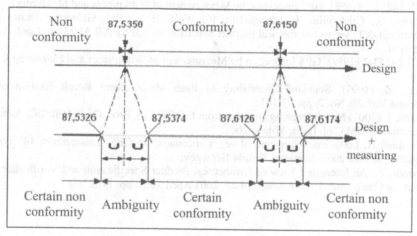

FIGURE 7. Comparison between specifications and actual intervals of conformity

Now the composed uncertainty can be calculated considering the contribution of the procedure but also caring the contributions of the instrument and the operator contributions, u_{probe} = 0,00048935 mm and u_{man} =0,00010038 mm, which were estimated before this investigation. Both are evaluations of category A. Therefore the composed uncertainty is calculated reducing the contributions to normal distribution 1s:

$$u_{pp} = \sqrt{u_{probe}^2 + u_{man}^2 + u_{method}^2} = 0,0011869618mm \qquad (5)$$

The expanded uncertainty, which is a 95% level of confidence interval, is

$$U = 2 \cdot u_{pp} = 2 \cdot 0,0011869618 = 0,0023739235mm \qquad (6)$$

In figure 7, the intervals of certain verification, ambiguity and certain non verification of conformity is represented assuming the expanded uncertainty as the measuring process uncertainty.

The design specifications interval, upper interval, is compared with the actual conformity interval that is smaller due to the measuring uncertainty [8].

5 CONCLUSIONS

In this article, after having completed the evaluation of the uncertainty of measure of the system for the verification of conformity of the external diameter of a gear, some of the causes of variability of the measuring process were selected and investigated.

The influence of these causes on the uncertainty of measure was evaluated by means of the calculation following the ISO standard suggestions. It was demonstrated by comparison that the uncertainty of measure could alter the process of verification of the geometric dimensions of the piece even if the value of the uncertainty is minimised.

Therefore it is necessary to consider the measuring process as a process itself. The latter should be designed contemporary to the product and its manufacturing process.

REFERENCES

1. ISO/TR 14638:1995 (1995). Guide to the Expression of Uncertainty in Measurement.
2. ISO/TS 14253-1 (1998). GPS Inspection by Measurement of Workpieces and Measuring Equipment.
3. Concheri, G., Cristofolini, I., Meneghello, R., Wolf, G. ((2001). Geometric Dimensioning and Tolerancing (GD&T) versus Geometrical Product Specification (GPS). XII ADM Int. Conf. Rimini, italy, 5th – 7th Sept.
4. ISO/TS 14253-2 (1999). GPS Inspection by Measurement of Workpieces and Measuring Equipment – Part 2.
5. Godec., Z. (1997). Standard Uncertainty in Each Measurement Result Explicit or Implicit-Measurement Vol. 20, No. 2, pp. 97-101.
6. Dovmark, J. (2001) New interesting concepts from ISO/TC 213, Proc. Of INTERSEC, Annual meeting of Associazione CMM Club Italia, Milano 2001.
7. G. Casalino, A. Ludovico. Estimation of target uncertainty in GPS measurement. 14th DAAAM Int. Sym. 22-25th October 2003. Sarajevo, Bosnia-Herzegovina.
8. Srinivasan,V., An Integrated View of Geometrical Product Specification and Verification, 7th CIRP Inter. Sem. on Comp. Aided Tolerancing, 24-25, 2001 April 2001, pp. 7-16.

SIX SIGMA – THE METHODOLOGY FOR ACHEIVING TOTAL BUSINESS EXCELLENCE

V. Majstorovic[1], N. Stefanovic[2]

[1] Laboratory for Production Metrology and TQM, Mechanical Engineering Faculty, University of Belgrade, Serbia and Montenegro
[2] Information Systems Division, Zastava Automobiles, Kragujevac, Serbia and Montenegro

KEYWORDS: Six Sigma, Quality, Business Excellence.

ABSTRACT. The involvement of improvement teams and the use of quality tools, that improved business performance could be achieved only through better planning, capable processes and the involvement of people. Six Sigma (6σ) is a business-driven, multi-faceted approach to process improvement, reduced costs, and increased profits. With a fundamental principle to improve customer satisfaction by reducing defects, its ultimate performance target is virtually defect-free processes and products. Six Sigma is a revolutionary business process geared toward dramatically reducing organizational inefficiencies that translates into bottom-line profitability. This paper shows how to transform Six Sigma to address today's most crucial business challenges: the challenge to execute and the challenge to maximize value. It also discusses practical issues related to its implementation, project management, techniques, and relations with other standards, concepts and tools for achieving synergistic effect.

1 INTRODUCTION

There are three key drivers of business performance excellence. They are a focus on delivering value to customers, a focus on internal operational processes, and a commitment to learning as an entity and to staff learning. We have also found the following key organizational ingredients leading to business performance excellence [1]:

➢ The organizations have senior leadership that is committed to the organization and its staff. These leaders are visionary.
➢ World-class organizations are focused on understanding customers, building relationships with customers, and satisfying customers.
➢ Organizations of the highest caliber project what their customers' needs are both in the present and in the future.
➢ High-functioning organizations have defined their key processes and understand them well.
➢ These organizations exhibit strong financial performance as compared to their competitors, even during economic downturns

The Ultimate Six Sigma [2] process has several reach-out objectives for industry. The objectives are to:

➢ Develop techniques/disciplines that can truly impact corporate profitability, not just fool around its edges.

Published in: E. Kuljanic (Ed.) *Advanced Manufacturing Systems and Technology*,
CISM Courses and Lectures No. 486, Springer Wien New York, 2005.

> Implement a practical, how-to guide to propel a company to quickly achieve the benefits (discussed in the next section) of the Ultimate Six Sigma.
> Develop a comprehensive infrastructure that goes well beyond the narrow confines of quality (the small Q) to encompass all areas of business excellence (the Big Q).
> Maximize all stakeholder loyalty—customer loyalty, employee loyalty, supplier loyalty, distributor/dealer loyalty, and investor loyalty.
> Maximize business results: profits, return on investment, asset turns, inventory turns, and sales/value-added per employee. Furthermore, go beyond just the financials.

2 A DEFINITION OF SIX SIGMA

Companies exist to be profitable. Profitable companies provide jobs and pay taxes that benefit the community, state, and country where they make their products or provide their services. Making a profit is based on having customers who want your product or service. Wanting your product or service is just the beginning. Every customer has requirements regarding the product or service. Requirements are those characteristics about customer's experience that determine whether they are happy or not. If an organization is meeting their requirements, it is being effective. If their requirements are not being met it is being ineffective. Effectiveness through meeting (and preferably exceeding) requirements is only half the battle. Because to be a profitable business, an organization must also be efficient. Efficiency relates to the amount of resources consumed in being effective. Efficiency can be measured in time, cost, labour, or value. For example, if an organization has to hire more people working on the production line to fulfil production plan, hire more people in order to distribute orders on time, and pay extra people to improve customer support quality, they quickly will recognize that the cost of being totally focused on effectiveness without efficiency will result in an unprofitable situation. Since businesses exist to make a profit, being focused on the customer without also being focused on efficiency will not be a good business decision.

Six Sigma, at its basic level, is attempting to improve both effectiveness and efficiency at the same time. Six Sigma is a measure of customer satisfaction that is near perfection. Most companies are at the two to three sigma level of performance–that means between 308,538 and 66,807 customer dissatisfaction occurrences per million customer contacts.

With other quality approaches, management played little if any role other than approval of bringing in external consultants to train the workforce. With Six Sigma, the work begins with management. First, executives create the Process Management system. Before work is done that affects the average worker, management has already spent several months working on identifying and measuring the processes of their organization.

A process is defined as the series of steps and activities that take inputs provided by suppliers, add value and provide outputs for their customers. Six Sigma as a management philosophy instructs management to begin identifying the 20 or 30 most important processes in their business. Next management measures the current sigma performance of each of these processes. Many, if not all, of the processes will be operating at two to three sigma performance. Some processes may even be lower than two sigma. Once management has identified their processes and personally been involved in measurement of their current performance, they then identify the lowest performing processes that have the most direct impact on the company's business objectives. Business objectives are the five to seven most important goals a company establishes each year. Sometimes they are financially stated (like profits) but there are others like customer satisfaction or employee satisfaction.

Once the processes having the worst performance with the greatest impact to the business objectives are identified, project teams are formed. That's where the individual worker comes in. They will become part of a five to seven person team that will have the responsibility of improving the performance of the worst performing processes. These teams usually exist for four to six months. They are taught a series of tools and concepts to help them use their skills to improve sigma performance to achieve greater effectiveness and efficiency.

2.1 AREA OF APPLICATION

The Six Sigma methodology uses a specific problem-solving approach and Six Sigma tools to improve processes and products. This methodology is data-driven, with a goal of reducing unacceptable products or events. The real-world application of Six Sigma in most companies is to make a product that satisfies the customer and minimizes supplier losses to the point that it is not cost-effective to pursue tighter quality. The Six Sigma philosophy can be applied to a wide range of business domains:

Manufacturing Managers, Engineers, and Technicians: Implementing Six Sigma gives manufacturing and engineering teams a common language and approach to problem solving. No matter what skills people currently possess, the use of Six Sigma makes those skills more effective.

Sourcing: Six Sigma is equally valuable when applied to suppliers. The joint use of these tools makes your vendors an extension of your company. Suppliers are eager to participate, since they know that the Six Sigma process will both improve their product and strengthen the bond with their customer. These are almost always win-win situations.

Design Engineers: Production problems are best solved in the design stage. Six Sigma uses data and customer input to assist in designing products and production equipment that are more likely to be problem-free.

Marketing and Sales: Being able to demonstrate how your company uses Six Sigma tools to improve and control processes is a powerful marketing and sales tool. Many leading companies use Six Sigma and a degree of prestige and perceived technical prowess comes with incorporating it. In addition, the tools are an assist in spotting significant changes in demand or sales.

Accounting, Software Development, Insurance, etc.: Although most of the initial Six Sigma applications have been in manufacturing, there is a growing awareness that these techniques work equally well in reducing costs or errors in other fields. These techniques can be used to compare people, processes, companies, events, etc. to spot significant differences or trends. Use the various customer input tools to benefit from the knowledge of everyone affected and to get maximum buy-in.

3 SIX SIGMA FRAMEWORK

Management strategies, such as TQC, TQM, and Six Sigma, are distinguished from each other by their underlying rationale and framework. As far as the corporate framework of Six Sigma is concerned, it embodies the five elements of top-level management commitment, training schemes, project team activities, measurement system and stakeholder involvement. At the core of the framework is a formalized improvement strategy with the following five steps: define, measure, analyse, improve and control (DMAIC).

3.1 TOP-LEVEL MANAGEMENT COMMITMENT AND STAKEHOLDER INVOLVEMENT

The easiest way to implement Six Sigma into an organization is to have complete commitment from top management. This commitment would include company-wide communications explaining the process and its goals, with some explanation of the reasons why the company was going to invest the time and energy into implementing the Six Sigma methodology. This buy-in demonstrates to the whole company that management believes in this methodology. This way the required investment in people and training will happen, along with everyone's active participation. When incorporating Six Sigma, many companies start with outside consultants/instructors and then transition to in-house people as trainers.

There are numerous pragmatic ways for the CEO (chief executive officer) to manifest his commitment: setting the vision and long-term or short-term goal for Six Sigma, allocate appropriate resources in order to implement such Six Sigma programs as training schemes, project team activities and measurement system, check the progress of the Six Sigma program to determine whether there are any problems which might hinder its success, and hold a Six Sigma presentation seminar regularly, say twice a year, in which the results of the project team are presented and good results rewarded financially.

Stakeholder involvement means that the hearts and minds of employees, suppliers, customers, owners and even society should be involved in the improvement methodology of Six Sigma for a company. In order to meet the goal set for improvements in process performance and to complete the improvement projects of a Six Sigma initiative, top-level management commitment is simply not enough. The company needs active support and direct involvement from stakeholders.

3.2 SIX SIGMA TRAINING

Company training is not often regarded as important when resources are required elsewhere to meet higher priorities. This changes with Six Sigma. Very structured and detailed training is necessary to provide the skills and knowledge that are needed to successfully implement the work ethic of Six Sigma. Without a focused and fully committed training effort, Six Sigma may fail in its early stages of development. So, who needs to be trained? According to Greg Bruce [3], the "key players" in the total Six Sigma effort are:

➢ An executive leader — who is committed to Six Sigma and who promotes it
➢ Champions — who remove barriers
➢ Master Black Belts — who work as trainers, mentors, and guides

> Black Belts — who work and manage the Six Sigma projects full-time
> Green Belts — who assist Black Belts on a part-time basis

Generally speaking, the number of Black Belts in some organizations represents roughly one percent of the workforce. The requirements for full-time Six Sigma employees will vary for the size and type of organization. At the start-up, one may consider a Project Champion, a Master Black Belt working with perhaps 15 Black Belts that lead 15 teams, and approximately 150 Green Belts to assist the Black Belts, a ratio of 1:15.

Black Belts working with teams are expected to complete 5-8 projects per year. For every 10-15 Black Belts, a Master Black Belt is devoted full time for support and training activities. Green Belts assist in research, form and facilitate teams, and under the leadership of Black Belts, complete their own projects. Project Champions provide the resources and funding for the teams. They also remove any roadblocks that teams may encounter.

3.3 MEASUREMENT SYSTEM

A Six Sigma company should provide a pragmatic system for measuring performance of processes using a sigma level, ppm (parts-per-million) or DPMO (defects per million opportunities). The measurement system reveals poor process performance and provides early indications of problems to come. There are two types of characteristics: continuous and discrete. Both types can be included in the measurement system. Continuous characteristics may take any measured value on a continuous scale, which provides continuous data. In continuous data, normally the means and variances of the CTQ characteristics are measured for the processes and products.

3.4 DMAIC PROCESS

Many would remember Shewhart's Plan-Do-Check-Act (PDCA) cycle as part of the toolbox for TQM. A more refined process is used within Six Sigma's DMAIC roadmap, which consists of the Define, Measure, Analyze, Improve, and Control stages. Simply put, Six Sigma teams use this roadmap to complete every quality or process improvement project.

There are five high-level steps in the application of Six Sigma tactics. As can be seen in Figure 1, the first step is Define. In the Define step, the project team is formed, a charter is created, customers, their needs and requirements are determined and verified, and, finally, a high-level map of the current process is created.

The second step of the application of Six Sigma tactics is Measure. It is in this second step that the current sigma performance is calculated, sometimes at a more detailed level than occurred at the strategic level of Six Sigma.

The third step in applying Six Sigma tactics is Analysis. During this step, the team analyzes data and the process itself, finally leading to determining the root causes of the poor sigma performance of the process.

The fourth step of applying Six Sigma tactics is Improve. In this step, the team generates and selects a set of solutions to improve sigma performance.

The fifth and last step is Control. Here a set of tools and techniques are applied to the newly improved process so that the improved sigma performance holds up over time.

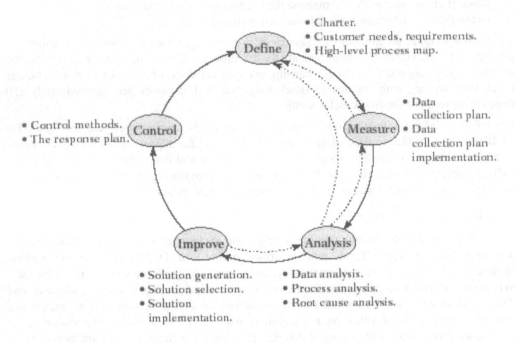

FIGURE 1. DMAIC improvement methodology

3.5 PROJECT MANAGEMENT PLAN AND TEAM ACTIVITIES

Wilemon and Baker [4] observed, after studying a multitude of projects, that there seems to be no single panacea in the field of project management; some factors work well in one environment while other factors work well in other environments. It is commonly agreed that the success of any project requires commitment by the management, planning, resources and formal reviews.

The suggested flow of the project team activities in transactional/service processes is DMARIC [5]. At each step, the actions shown in Table 1 are recommended.

TABLE 1. Suggested actions in each step of DMARIC project team activities

Step	Action
Definition (D)	1. Define the scope and surrounding conditions of the project. 2. Identify critical customer requirements and CTQy's. 3. Check the competitiveness of the CTQy's by benchmarking. 4. Describe the business impact of the project.
Measurement (M)	1. Identify the project metrics for the CTQy's. 2. Measure the project metrics, and start compiling them in time series format by reflecting the long-term variabilities. 3. Address financial measurement issues of project.
Analysis (A)	1. Create a process flowchart/process map of the current process at a level of detail that can give insight into what should be done differently. 2. Create a cause-and-effect diagram or matrix to identify input variables, CTQx's, that can affect the process output, CTQy. 3. Rank importance of input variables using a Pareto diagram. 4. Conduct correlation, regression and analysis of variance studies to gain insight into how input variables can impact output variables.
Redesign (R)	1. Consider using DOEs to assess the impact of process change considerations within a process. 2. Consider changing work standards or process flow to improve process quality or productivity. 3. Determine optimum operating windows of input variables from DOEs and other tools.
Implement (I)	1. Set up the best work standards or process flow. 2. Test whether the optimum operating windows of input variables are suitable, and implement them. 3. Verify process improvements, stability, and performance using runcharts.
Control (C)	1. Update control plan. Implement control charts to check important output and input variables. 2. Create a final project report stating the benefits of the project. 3. Make the project report available to others within the organization. 4. Monitor results at the end of 3 and 6 months after project completion to ensure that project improvements are maintained.

4 SIX SIGMA IMPLEMENTATION

Some companies set up Six Sigma as a separate organization, which then services the rest of the company. As a separate organization, the Six Sigma people work in parallel with the various groups already in place, identifying and implementing Six Sigma projects in addition to whatever projects the groups have already defined. The advantage of this approach is Six Sigma can be implemented with fewer people trained initially and the effect of the methodology can be more readily tracked. The downside of this approach is that the separate Six Sigma organization is often looked at as a group of prima donnas, with their own set of agendas. This causes some resentment among others in the organization and stifles cooperation. It also discourages the input of experts into Six Sigma projects, since many of these experts feel threatened. There can also be some feeling that current ideas are "stolen" and then labelled as Six Sigma.

An alternate approach is to incorporate Six Sigma as part of the organization, not as a separate entity [6]. Six Sigma then becomes an integral part of everyone's job, using a relatively few highly trained Six Sigma people as reference instructors. This is a somewhat more difficult way to implement Six Sigma, because of the large number of people to be trained, but a common Six Sigma language and philosophy will then permeate the organization. As the Six Sigma methodology unfolds in the coming chapters, it will be seen that Six Sigma is helpful to everyone in the organization and therefore it should become an integral part of everyone's job.

When a company intends to introduce Six Sigma for its new management strategy, we would like to recommend the following seven-step procedures:

1. Top-level management commitment for Six Sigma is first and foremost. The CEO of the corporation or business unit should genuinely accept Six Sigma as the management strategy. Then organize a Six Sigma team and set up the long-term Six Sigma vision for the company.

2. Start Six Sigma education for Champions first. Then start the education for WBs, GBs, BBs and MBBs in sequence. Every employee of the company should take the WB education first and then some of the WBs receive the GB education, and finally some of the GBs receive the BB education. However, usually MBB education is practiced in professional organizations.

3. Choose the area in which Six Sigma will be first introduced.

4. Deploy CTQs for all processes concerned. The most important is the company's deployment of big CTQy from the standpoint of customer satisfaction. Appoint BBs as full-time project leaders and ask them to solve some important CTQ problems.

5. Strengthen the infrastructure for Six Sigma, including measurement systems, statistical process control (SPC), knowledge management (KM), database management

system (DBMS) and so on.

6. Designate a Six Sigma day each month, and have the progress of Six Sigma reviewed by top-level management.

7. Evaluate the company's Six Sigma performance from the customers' viewpoint, benchmark the best company in the world, and revise the Six Sigma roadmap if necessary. Go to step 1 for further improvement.

5   SIX SIGMA AND OTHER MANAGEMENT INITIATIVES

5.1   TQM AND SIX SIGMA

While Six Sigma is definitely succeeding in creating some impressive results and culture changes in some influential organizations, it is certainly not yet a widespread success. Total Quality Management (TQM) seems less visible in many businesses than it was in the early 1990s. However, many companies are still engaged in improvement efforts based on the principles and tools of TQM.

Ronald Snee [7] points out that although some people believe it is nothing new, Six Sigma is unique in its approach and deployment. Snee goes on to claim that the following eight characteristics account for Six Sigma's increasing bottom-line (net income or profit) success and popularity with executives:

➢   Bottom-line results expected and delivered

> Senior management leadership
> A disciplined approach (DMAIC)
> Rapid (3–6 months) project completion
> Clearly defined measures of success
> Infrastructure roles for Six Sigma practitioners and leadership
> Focus on customers and processes
> A sound statistical approach to improvement

Other quality initiatives including TQM have laid claim to a subset of these characteristics, but only Six Sigma attributes its success to the simultaneous application of all eight.

5.2 ISO 9000 SERIES AND SIX SIGMA

The ISO 9001:2000 standard specifies requirements for a quality management system for which third-party certification is possible, whereas ISO 9004:2000 provides guidelines for a comprehensive quality management system and performance improvement through Self-Assessment.

The authors believe that Six Sigma is needed regardless of whether a company is compliant with the ISO 9000 series. The two initiatives are not mutually exclusive and the objectives in applying them are different. A Six Sigma program is applied in organizations based on its top-line and bottom- line rationales. The primary objective for applying the ISO 9000 series standards is to demonstrate the company's capability to consistently provide conforming products and/or services.

The ISO 9000 series standards have from their early days been regarded and practiced by industry as a minimum set of requirements for doing business. The new ISO 9000:2000 standards do not represent a significant change to this perspective. Six Sigma on the other hand, aims at world-class performance, based on a pragmatic framework for continuous improvement.

5.3 LEAN MANUFACTURING AND SIX SIGMA

Lean evaluates the entire operation of a factory and restructures the manufacturing method to reduce wasteful activities like waiting, transportation, material hand-offs, inventory, and over-production. Lean and Six Sigma are promoted as different approaches and different thought processes. Yet, upon close inspection, both approaches attack the same enemy and behave like two links within a chain – that is, they are dependent on each other for success. They both battle variation, but from two different points of view. The integration of Lean and Six Sigma takes two powerful problem-solving techniques and bundles them into a powerful package. The two approaches should be viewed as complements to each other rather than as equivalents of or replacements for each other [8].

Lean and Six Sigma, working together, represent a formidable weapon in the fight against process variation. Six Sigma methodology uses problem-solving techniques to determine how systems and processes operate and how to reduce variation in processes. In a system that combines the two philosophies, Lean creates the standard and Six Sigma investigates and resolves any variation from the standard. In addition, the techniques of Six Sigma should be applied within an organization's processes to reduce defects, which can be a very important prerequisite to the success of a Lean project.

5.4 SIX SIGMA AND CMMI

In general, CMMI (Capability Maturity Model Integration) identifies what activities are expected in a process. Six Sigma helps make those activities more effective and efficient. So, for example, an organization will use CMMI to make sure it is implementing activities such as estimating and risk management to improve overall project planning. Six Sigma can be used to continually improve the accuracy of estimating and risk management. CMMI-based process improvement efforts often have little impact on business results such as fewer defects, lower cost, and higher efficiency, whereas Six Sigma activities are totally driven to yield measurable business results in these areas. Six Sigma can provide specific tools for implementing the CMMI practices in DAR (Decision Analysis and Restoration), CAR, and QPM [9]. Six Sigma is weak in terms of establishing an improvement infrastructure whereas this is one of CMMI's strengths through OPF, OPD, OT, and the Generic Practices (GPs).

6 CONCLUSIONS

The establishment of a Six Sigma work ethic within organization will indeed become a strong driving force. It will significantly increase customer satisfaction and act as an enabler for an organization to reach world-class status among competitors. Furthermore, it will also:
- ➤ Promote a common language and understanding of teamwork in the quality arena within the organization
- ➤ Tie directly into existing TQM, ISO-9000, or Lean Manufacturing activities
- ➤ Assist in cost and cycle time reduction, defect rework and waste elimination
- ➤ Attack variation at the supplier, process, product, and service levels
- ➤ Support the Malcolm Baldrige National Quality Award criteria

An essential organizational benefit realized from the establishment and ongoing implementation challenges of Six Sigma quality is the strategic alignment and the improved level of communications and teamwork among organizational units and the customer. Used in all business process, and together with other quality management initiatives, Six Sigma philosophy can assist organizations in their quest for business excellence.

REFERENCES

1. Calingo R., (2002), The Quest for Global Competitiveness Through National Quality and Business Excellence Awards, APO, Japan.
2. Bhote K., (2003), The Power of Ultimate Six Sigma, Amacom, USA.
3. Bruce, G. (2000). Six Sigma for Managers. New York: McGraw-Hill, p. 80.
4. Wilemon, D.L. and Baker, B.N. (1983). Some Major Research Findings Regarding the Human Element in Project Management. In Cleland, D.I. and King, R.W. (Eds) Project Management Handbook. New York, Van Nostrand Reinhold Co.
5. Sung H. P., (2003), Six Sigma for Quality and Productivity Promotion, APO, Japan.
6. Brussee W., (2004), Statistics for Six Sigma Made Easy, McGraw-Hill, USA.
7. Snee, R. (1999), Why Should Statisticians Pay Attention to Six Sigma?: An Examination for Their Role in the Six Sigma Methodology, Quality Progress, 32(9), pp. 100-103.
8. Pyzdek T. (2000), Six Sigma and Lean Production, Quality Digest, p. 14.
9. West M., (2004), Real Process Improvement Using the CMMI, Auerbach Publications, USA.

AN INTELLIGENT SYSTEM FOR AUTOMATIC EXTRACTION OF MANUFACTURING FEATURES FROM CAD SOLID MODELS USING A SIMPLE AND LOGICAL APPROACH

A. S. Deshpande

Department of Industrial and Production Engineering, Gogte Institute of Technology,
Belgaum 590 008, India

KEYWORDS: CAD, Product Model, Feature.

ABSTRACT. Features often form the basis of knowledge about various design and manufacturing tasks like process planning, fixture design, inspection, assembly planning etc. Automatic Feature-s recognition is likely to be an essential requirement for future integrated design and manufacturing systems and in the development of fully automated process planning systems. Majority of the current design and manufacturing data used on shop floors is associated with CAD models. Planning for Product development starts with a 3 D Solid CAD model where as manufacturing drawings are represented in 2 D. Any basic manufacturing process expects the representation of the finished product in terms of an Engineering model. Solid models have replaced the 2D models owing to the trend of the industry towards automation. Any CAD model, when associated with the manufacturing attributes, can be termed as a product model in a general sense. An overview of research on extraction of feature-s from Solid models indicates that, researchers are not comfortable with the drawing interfaces such as IGES, SAT etc. due to dynamic and complex nature of coding. Also, most of the work in this category is based on complicated algorithms. In the perspective of these facts, the present work makes use of the simple and logical approach with DXF interface, which has been so far attempted, only for 2 D CAD Models. DXF has been a widely used interface for wire frame entities, due to clarity and well defined structure. An innovative theory viz.3 Segment theory has been used in for interpreting the features. Further the machining attributes such as surface finish, tolerance associated with the extracted features is used for selecting and sequencing the processes. The decisions are supported by an knowledge based expert system based on a database .The software has been developed in C and an interactive front end has been developed using Visual Basic V6. The concept and the logic used in the system is very much convincing. This would be probably one of the earliest work having used DXF interface for extracting the features from Solid Models successfully.

1 GENERAL BACKGROUND AND INTRODUCTION

Since the invention of computer graphics in the year 1950, CAD models are being used extensively for engineering applications [1]. The majority of CAD systems have been used for few manufacturing activities such as NC cutter path generation etc. are based on 2D or 3D wire frame models.

In real life, many engineering objects have free form surfaces. This is the basis of surface modelling. Various techniques[2] such as Coon's patch, Bezier's surface, B-spline surface and NURB's surface have been developed to model these free form surfaces.

The powerful solid model is a latecomer in the part representation area. Computation of volumetric properties such as volume, weight and moment of inertia and topological properties

Published in: E. Kuljanic (Ed.) *Advanced Manufacturing Systems and Technology*,
CISM Courses and Lectures No. 486, Springer Wien New York, 2005.

such as connectivity relationships , are possible with these models. Various techniques[2] used for building a solid model are CSG, B-rep, Sweeping, Spatial Occupancy Enumeration, Cell Decomposition, and Primitive Instancing etc.

The concept of Product model is gaining more importance in the current research in CAD modelling. A Product model can be considered as a Solid model with the manufacturing and technological attributes defined on it. Thus, a product model becomes more suitable for Product development and process planning.

Manufacturing or CAM modules expect the complete geometrical as well as technological information from CAD models in a form acceptable to them. An overall review indicates that, very few CAD modellers satisfy this major requirement in the CIM environment.

CAPP systems usually serve as a link between CAD and CAM. However, this is a partial link, because most of the existing CAD drafting systems do not provide part feature information, which is the essential data for CAPP. The clear and complete information of geometrical and technological aspects of the features of the CAD models is important for CAPP decisions.

Features are the characteristics of the models having some manufacturing significance; for example, slots, holes, pockets, etc. Each feature can be associated with certain amount of manufacturing knowledge. Every manufacturing operation is associated with certain technological attributes, which are represented by a feature. Features guide in judging the overall shape and size of a given component. Features are conveniently classified as either Geometric form features or Manufacturing features.

A geometric feature can be defined as 'a portion of the part boundary, which comprises a set of connected faces having certain recognizable manufacturing characteristics'.

Manufacturing features A manufacturing feature is one that gives technological information, associated with the manufacturing operations and tools. These include the dimensions of the model in different axes of representation. These also incorporate the linear and geometric tolerances. Manufacturing features also include geometric tolerances of surface finish, cylindricity, etc.

Feature extraction plays an important role [3] in manufacturing and design systems as well. Feature extraction acts as a bridge between CAD and CAM, this is because the entities incorporated in a component during the design stage are recognized by a feature extraction system before the part is actually forwarded for the various manufacturing operations.

Systems for feature extraction provide Feature based user interface for the designing processes and in case of manufacturing, they help in determining the appropriate operations in process planning.

Thus, one can infer from the above diagram that feature extraction forms the foundation part of the entire structure of a Computer Aided Process Planning system.

The current trends in research in CAPP emphasize on Automatic Process Planning approach i.e. Generation of Process Plan / CNC Part program directly from CAD models without any assistance of human decisions.

Feature Extraction research is mainly based on the theory that, a representation of the drawing can be generated in some neutral interfaces with the formats such as DXF, IGES and the emerging international standard STEP etc. These sequential file formats have been defined to permit the transfer of Product data between different CAD systems.

On overview indicates that the concept of Automatic Feature-s Extraction from CAD Models has gained attention only in the last decade. Both, 2-D and 3-D CAD models have been considered by researchers. Meeran & Pratt (CAD V25 1993)[6] and Deshpande, et.al.(1996) [3] have provided a successful basic setup for extracting the features from 2-D CAD models using DXF interface. Few other researchers also have preferred DXF interface effectively for feature extraction task mainly due to its simple, easy and clear structure.

However, IGES interface has been used extensively in the feature extraction from 3 D CAD model.[4].The main difficulties reported with IGES are the frequent revisions , wordy and lengthy structure etc. Hence, a totally innovative approach , has been used successfully in the present work based on DXF interface.

2 METHODOLOGY AND OUTLINE OF PRESENT WORK

2.1 STEPS AND BASIC APPROACH

A simple and conceptual approach as per the following steps is used in the present work.
1. Developing Solid model using CAD Modeller .
2. Exploding the Solid model in to Surface model.
3. Further exploding the Surface model in to Wire frame model.
4. Creating the DXF interface for the model.
5. Analysing the DXF file for identification of the features.
6. Selecting and sequencing the manufacturing processes for the extracted features.

AutoCAD has been used in the present work. However, any CAD modeller which is capable of generating DXF can also be used for modelling. In almost all the reported work on Automatic Feature-s Extraction, IGES interface has been used for 3-D & surface models, whereas DXF for 2-D models. The innovative part of the present work is that, DXF model has been successfully used for 3-D Solid models.

'Feature analysis' part is analyzcd by grouping the extracted features into two categories viz. 'Prismatic' and 'Rotational'. The input for this analysis is the DXF file generated in the CAD modeller. The features considered in the present analysis are:

Prismatic features	Rotational features
✦ Rectangular slot. ✦ Dovetail slot. ✦ Rectangular pocket.	✦ Plain Hole. ✦ Taper Hole. ✦ Tapped Hole. ✦ Counter Bored Hole. ✦ Counter Sunk Hole.

An unique and exhaustive method has been used in the present work, to identify the prismatic and rotational features separately. This works on the basis of linking the basic entities supplied by the drafting package namely: Lines, Circles and Arcs.

2.2 FEATURE RECOGNITION

PRISMATIC FEATURES

Here, we assume that each feature is made up of three basic segments: S1, S2 and S3. This theory is proposed as the *3 segment theory*. Thus, for every prismatic feature, the algorithm works as follows:

FEATURE:	SEGMENTS:			OTHER REMARKS:
	S1	S2	S3	
RECTANGULAR SLOT	Yes	Yes	Yes	–
DOVETAIL SLOT	Yes	Yes	Yes	S1 and S3 are inclined lines.
RECTANGULAR POCKET	Yes	Yes	Yes	S1 and S3 are joined by a HORIZONTAL line on the TOPMOST surface.
STEP	Yes	Yes	No	–

The algorithm of 3 – segment theory opens the DXF file and analyses it with the help of various functions of the program developed. The function *Identify()* is used to recognize all the basic entities like lines, circles, and arcs. This forms the foundation for the 3-segment algorithm.

The algorithm begins with the execution of several user-defined functions to identify *rectangular features* like slots (rectangular, dovetail, 'v', etc.), pockets, steps and *rotational features* like holes (plain, taper, counterbored, countersunk, etc.).

The most vital part of the working is the function *FindProfile()*. This function identifies the feature forming line segments for the rectangular features.

ROTATIONAL FEATURE

In this part, the identification is relatively easier, as the task is limited to establishing of relations between the identified circles and the lines. AutoCAD provides a set of four parallel generators for every cylinder. Thus, the endpoints of the generators meet the circles on their circumferences.

The *tapered holes* are also identified in a similar way, only for the exception that the diameters of the two limiting circles have different values. Other features that can be recognized on similar lines are *counterbored* and *countersunk* holes. The *tapped hole* cannot be identified directly. However, the tapping specifications can be obtained from the user if he desires to provide an internal threading.

After interpreting all the existing *rectangular features*, three subsequent functions are used to recognize the *rotational features*. The function *Display()* facilitates the printing of the details of each of the feature forming entities in the rotational category.

The function *FindHole()* determines the details of holes, like diameter, depth, etc. Again, the working of this function too goes on similar lines as the former. *First*, the feature forming entities are identified (here, lines and circles). *Secondly*, relations are built between the *hole forming* circles (called the *complimentary circles*) and the corresponding line segments, that join these pair of circles at their peripheries. The other two functions: *FindCBHole()* and *FindCSHole()* recognize the *counterbored* and *countersunk* holes respectively.

All these details (features and their dimensions) are saved in a text file as the output.

Once the features are identified, further job is delegated to the GUI of the Windows environment. Here, the user is provided with a facility to give dimensional as well as geometric tolerances. These input values are then compared with the standard database values to arrive finally at the sequence of operations to be recommended. This work becomes much easier in the *user-friendly*, Visual Basic. An extensive knowledge base of Process capabilities is kept ready to backup the decision making process .

The above mentioned text file then forms the basis for further interpretations in Visual Basic. The text file is read line by line, and the user is provided with interactive forms, depending on the type of features. Further interaction with the user is done through a number of dialogue boxes to input the information regarding manufacturing attributes such as tolerances, surface finish etc. and selection and sequencing of processes is executed.

3   RESULTS AND CONCLUSIONS

The sample outputs are shown in the following figures. Considering the complexities and dynamic nature of the domain, feature extraction has been attempted only for few simple features. However, the logic and concept can be extended to combined and complex features too. The system has been quite successful in extracting Prismatic and Rotational features from 3 D Solid CAD models using DXF interface, which is expected to be an innovative and simple approach.

3.1 SAMPLE OUTPUT 1 [PRISMATIC FEATURE]

➢ The Drawing in AutoCAD:

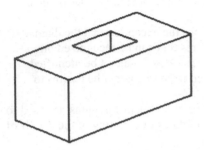

Component feature: Rectangular Pocket
Dimensions: 20x20x30 mm
File name: Pocket1.DWG

The DXF file of this model, Pocket1.DXF is created, which is subsequently processed by the C program, through Visual Basic. Following are the forms that appear in this context:

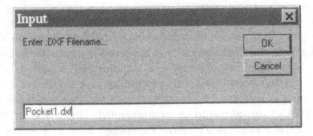

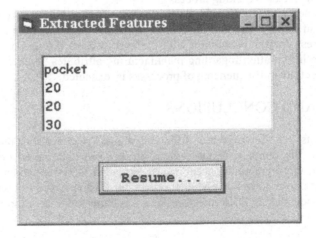

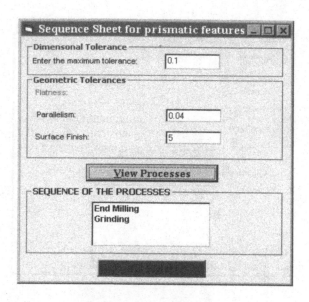

3.2 SAMPLE OUTPUT 2 [ROTATIONAL FEATURE]

> The drawing in AutoCAD:

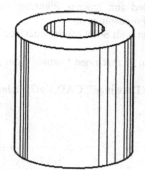

Component feature: Hole
Dimensions: 20 dia. X 100 mm
File name: Hole1.DXF

The procedure followed for a rotational feature is the same as the procedure for a prismatic feature. Therefore, the initial forms for input of DXF file and the extracted features are similar. The differentiating factor is that for a rotational feature the *Sequence Sheet* input and output forms are necessarily different. The following forms appear in this case:

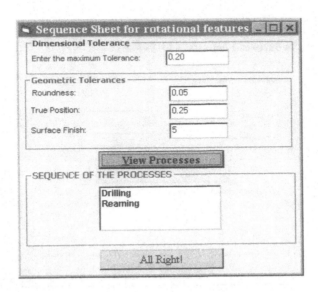

REFERENCES

1. Chang, T.C., "Expert Process Planning for Manufacturing", Addison – Wesley USA, 1990.
2. Chang, T.C., and Wysk, R.A., "An Introduction to Automated Process Planning System", Prentice Hall New Jersy, 1985.
3. Deshpande A.S., Kustagi V.K., Balaji Rao., "Computer Aided for Process Planning for manufacturing: Need of an Integrated and Intelligent Approach",Proceedings of
4. Joshi S., Vissa, N., Chang, T.C., "Expert Process Planning System with Solid Model Interface", Int. J. Prod. Res. Vol. 26/5/1988, PP 863-885.
5. Joshi S., and Chang T.C., "Graph based heuristic for recognition of machined features from a 3D solid model", Computer Aided Design. 20(2):PP58-66.
6. Meeran, S.,Pratt,M.J.,"Automated Feature Recognition from 2-D Drawings", CAD,Vol25/1,Jan 1993, PP 7-17.

ANALYSIS OF THE DEPENDENCE BETWEEN MANUFACTURING AND DESIGN SPECIFICATION

A. Rivière[1], A. Clément[2], P. Serré[1]

[1] Laboratoire d'Ingénierie des Systèmes Mécaniques et des Matériaux, 3 rue Fernand Hainaut, 93407 Saint-Ouen Cedex, France
[2] Dassault Systèmes, 9 quai Marcel Dassault, 92156 Suresnes Cedex, France

KEYWORDS: Metric Tensor, Geometrical Specification, Dimensioning Specification, Tolerancing Specification.

ABSTRACT. The general objective of this paper is to analyse the dependency relations that may exist between manufacturing and design specifications of a geometric object. This object will be modelled on the basis of the TTRS concept and vector modelling synthesised by the metric tensor of a set of vectors. The first problem that we propose to solve therefore is the formal expression of these relations. After that, the dependency will be analysed and illustrated considering a 3D object.

1 INTRODUCTION

ISO dimensioning and tolerancing standards for mechanical parts are not geometric specification standards but standards that specify the metrological verification procedures for a part that already exists, with all the attendant advantages and disadvantages of this restriction. Proof of this assertion can be found in the actual wording of the basic principle, called the "independence principle" expressed as follows:

Every dimensional or geometric requirement specified on a drawing must be individually (independently) complied with, unless a relation is specified. (ISO 8015)

The strong point of standardised, zone-based ISO tolerancing is its transactional nature: it enables exchanges to be made between a project manager and subcontractors with the utmost safety. It constitutes a major technological advance over the previous state of the art situation. Its main weakness resides in its non-applicability to dimensioning and tolerancing at the preliminary project stage (conceptual design) of a mechanical system. The possibility of specifying the parameters of a mechanical part in a non-nominal state by indicating a variation interval for certain of these parameters (parametric tolerancing) is not totally ruled out. On the other hand, it is impossible to obtain an instance of this part in its final state subsequent to parametric differentiation. Due to this inadequacy, justified mistrust has arisen from an industrial standpoint regarding the use of 3D dimension chains resulting from a parametric differential. Hence the current desire by manufacturers to make sure of the existence of the final object for any combination of tolerances. The explicit values of the dimensions of the final object – based on parametric tolerances – actually result from a computation which calls for extensive expertise, except in simple cases, and which does not necessarily imply the existence of the object.

Published in: E. Kuljanic (Ed.) *Advanced Manufacturing Systems and Technology*, CISM Courses and Lectures No. 486, Springer Wien New York, 2005.

In order to explain the problem in greater detail from a mathematical point of view, you simply need to imagine a certain nominal dimension $f_j\left(x^1, x^2, \ldots, x^n\right)$, dependent on n parameters, the variation of which must be specified by the designer. All current parametric technology is based on analysis of the differential form: $df_j = \dfrac{\partial f_j}{\partial x^i} . dx^i$.[Chase et al., 1997], [Laperrière et al., 2003] and [Serré et al., 2003]. The designer is not only obliged to specify all the dx^i with values that depend on the manufacturing process (which he does not know), for values of coefficients of influence (partial local derivatives) that he is not even aware of, but, above all, nothing goes to prove the existence of the object after a variation of this kind, however small it may be. In other words, there is no biunivocal relation in manufacturing reality between the nominal state and evolution of a system. On a mathematical level, this stems from a possible change in rank of the Jacobian $\left\|\dfrac{\partial f_j}{\partial x^i}\right\|$ in the neighbourhood of the nominal point. The "nominal value" and, therefore, this Jacobian do not have any precise meaning for an instance made of the object. This is the normal case for mechanical parts where the relative arrangements of points, straight lines and planes are specific (for example, in a combustion engine, the cylinder axis intersects the crankshaft axis at right angles).

When a designer has spent days or even weeks constructing a complex object using CAD, he cannot devote the same time to exploring the dimension variations of the same object. In an endeavour to meet this need for manufacturing security at the design stage, the normal differential parametric approach is completely abandoned in this study in order to adopt the concept of perturbation of the initial state. It is a method of investigation of the final real form of a geometric object with a known nominal form when it is subjected to independent, finished variations (small or otherwise).

NB. This perturbation concept is a generalisation of the "small displacement torsor", which represented a variation of the position of a solid body [Ballot et al., 2001]; here a perturbation is finished or infinitely small and, above all, the object is subjected to deformation.

2   TENSOR MODELLING OF GEOMETRIC OBJECTS

In analytical geometry, every element of which a complex geometric object is composed is in a unique position relation with the Cartesian reference point (the complexity is $O(n)$). There is no direct topological connection between 2 adjoining elements, except in the eyes of the designer. The extraordinary advantage of the Cartesian reference point is the small number of independent relations to be processed. The major disadvantage is, in point of fact, that there is no relation between 2 adjoining objects, it is up to the user to "propagate" any modifications. It is simple for the machine but difficult for the designer. Therefore, the problem is not properly posed!

In declarative geometry, a complex geometric object is perceived as a system of specifications between geometric objects. Each object is potentially in relation with all the others (the complexity is $O(n^2)$). The vector modelling based on the TTRS concept, already presented throughout precedent seminars [Clément *et al.*, 1999], [Serré *et al.*, 2001] and in a PhD [Duong, 2003], and the tensor representation will enable the complexity to be reduced to $O(3n)$, that is to say, almost to the level of analytic geometry, at the same time explicitly preserving the relations between objects.

Historically speaking, the inventor of the application of tensor computation to technology was G. Kron, who used the application for electrical networks and rotating machinery. In particular, he represented Kirchhoff's laws (mesh and node laws) using n-dimension, rank 2 tensors, the latter being the electrical currents of active or inactive components of this network.

In physics we normally only use rank 3, 3-Dimension metric tensors since these geometric objects need to be made in our Euclidian 3-Dimension space. G. Kron [Kron, 1942], however, the first to discover that this relation of equality between the dimension and rank of the metric tensor is not necessarily subject to certain manipulation restrictions. This tensor concept can be extended to any number (n) of vectors forming a geometric object; however, for it to be possible in $\mathbb{R}^3$ space, the metric tensor must be restricted to rank 3 and each 3*3 sub-tensor must be defined as positive. Moreover, not all the usual mathematical operations are valid on these tensors. Inversion, in particular, is an impossible transformation whereas transposition is still valid.

Numerical construction of the metric tensor associated with a valid geometric object in our 3D space composed of hundreds of vectors presents no difficultly. However, the terms of this tensor are not independent and are composed of thousands of implicit relations: as a result, the slightest variation of one of them invalidates the object. From a mathematical standpoint, it seems that expressing all these terms according to a certain number of independent parameters would be all it takes to vary the object in a straightforward manner. Unfortunately, this is very difficult in practice for a number of reasons, the main one being that you cannot explicitly determine all the Euclidian geometry theorems that apply to the object under consideration and it is therefore almost impossible to know all the constraints linking the geometric elements.

3 NEW MODELLING: PERTURBATION OF A KNOWN OBJECT

The solution adopted is not to use the geometric parameters of the object but the parameters of a Ω perturbation of the object and analyse the existence of the "perturbed" object. The perturbation model then becomes independent from the specification model.

Definition of the Ω perturbation

The ΔE perturbation of the sheaf of vectors E of an object forming a rank 3 sub-space

$$E_{init} = \left(\vec{e_1}, \vec{e_2}, \vec{e_3}, \ldots, \vec{e_n} \right)$$

will create a valid sheaf of vectors for the $\mathbb{R}^3$ space, written as follows:

$$\Delta E_{init} = \left(\overrightarrow{\Delta e_1}, \overrightarrow{\Delta e_2}, \overrightarrow{\Delta e_3}, \ldots, \overrightarrow{\Delta e_n} \right)$$

if, and only if, each perturbation is a linear combination of primitive vectors.

Expressed as:

$$\overrightarrow{\Delta e_i} = \sum_k \omega^k . \vec{e_k}$$

The $\vec{e}_k$ vectors are not independent but any linear combination gives a valid and unique vector Δe_i

Globally, the Ω perturbation of all these vectors will be expressed as:

$$\begin{pmatrix} \Delta \vec{e}_1 \\ \Delta \vec{e}_2 \\ \vdots \\ \Delta \vec{e}_n \end{pmatrix} = \Omega \cdot \begin{pmatrix} \vec{e}_1 \\ \vec{e}_2 \\ \vdots \\ \vec{e}_n \end{pmatrix}$$

The final object will thus be composed of the list of vectors:

$$E_{final} = E_{init} + \Delta E_{init} \quad \ldots\ldots\ldots\text{Eq1}$$

The perturbation defined in this way will give a unique vectoral object, valid only if all the coefficients ω_i^k are real numbers.

3.1 COMPUTATION OF THE METRIC TENSOR OF THE FINAL OBJECT

The initial metric tensor is written: $G_{init} = E_{init} \otimes {}^t E_{init}$

The new E_{final} vectors represent the vectors of the final object.

We deduce the expression of the final tensor from the equation (Eq1) via the tensor product

$$G_{final} = E_{final} \otimes {}^t E_{final}$$

i.e. by substituting the value of E_{final}

$$G_{final} = \left(E_{init} + \Delta E_{init} \right) \otimes {}^t \left(E_{init} + \Delta E_{init} \right)$$

then by developing the equation

$$G_{final} = \left(E_{init} + \Omega \cdot E_{init} \right) \otimes {}^t \left(E_{init} + \Omega \cdot E_{init} \right)$$

i.e. by expressing the unit matrix of dimension n as I_n:

$$G_{final} = \left(I_n + \Omega \right) \cdot \left(E_{init} \otimes {}^t E_{init} \right) \cdot {}^t \left(I_n + \Omega \right)$$

Finally

$$G_{final} = \left(I_n + \Omega \right) \cdot G_{init} \cdot {}^t \left(I_n + \Omega \right) \quad \text{Eq2}$$

This is the basic formula for the dimensional variations of a geometric object. This tensor equation mathematically defines a differentiable parametric manifold, the ω_i^k parameters of which are real numbers. In this way, a topological connection is made between 3 models: the G_{init} nominal model, the Ω perturbation model and the final specified model, G_{final}.

NB. The factor $\left(I_n + \Omega \right)$ demonstrates that a Ω perturbation has definitely been added to unit I_n since, for a nil Ω perturbation, we find $G_{final} = G_{init}$

3.2 FUNDAMENTAL PROPERTY

The theorem for the product of determinants immediately shows that if the rank of G_{init} is 3, then G_{final} is also rank 3, even if rank r of Ω is $3 < r < n$.

In other words, the Eq2 formula always gives a valid object for the 3D space, irrespective of the real Ω perturbation.

With this tensor equation, we are faced with the usual 2 problems, as follows: one, called the **"direct problem"** which, since Ω and G_{init} are known, consists of calculating G_{final}, the other, called the **"reverse problem"**, which since G_{final} is partially known, consists of determining Ω, followed by the complete G_{final}, and then returning to the direct problem.

3.3 ANALYSIS OF THE PROPERTIES OF THE Ω MATRIX

The ω_i^k coefficients of the Ω matrix are similar to the $\dfrac{\partial f_j}{\partial x^i}$ coefficients of influence used in the differential computation. Nonetheless they are very different because they are independent from one another and represent a veritable specification of the direction of the perturbation in the n-dimension space. We will call them "directional coefficients of influence".

By assumption, the sheaf of vectors $E_{init} = \left(\vec{e_1}, \vec{e_2}, \vec{e_3}, \ldots, \vec{e_n} \right)$ forms a rank 3 sub-space and it does not apparently appear useful to define the variations of a vector of this sub-space in the $\mathbb{R}^n$ space. However, this method of proceeding introduces major advantages to the direct and reverse problems.

3.4 DIRECT PROBLEM

For example, let us consider the perturbation of the vector $\vec{e_i}$:

$$\Delta \vec{e_i} = \omega_i^1 . \vec{e_1} + \omega_i^2 . \vec{e_2} + \omega_i^3 . \vec{e_3} + \omega_i^4 . \vec{e_4} + \ldots\ldots + \omega_i^n . \vec{e_n}$$

The designer specifies the desired perturbation by giving values to the ω_i^k coefficients of the Ω matrix. This specification method gives him all the flexibility required to "sculpt" the final object since he has $\dfrac{n!}{6 \cdot (n-3)!}$ independent ways of indicating it for each vector.

For example:

$\left(\omega_i^1, 0, \omega_i^3, 0, \omega_i^5, 0, 0, \ldots, 0 \right)$ specifies that the $\vec{e_i}$ vector is subjected to a variation in 3D space, the perturbations of which are imposed on the triplet: $\left(\vec{e_1}, \vec{e_3}, \vec{e_5} \right)$.

NB. It is obviously possible to specify more than 3 components, provided they are real numbers.

The designer can thus specify a variation by adding perturbations in relation to numerous successive references. The final vector itself will naturally have a position resulting from the combination of these perturbations in the 3D space.

3.5 REVERSE PROBLEM

This substantially important advantage in the preliminary project stage thus brings greater security to the resolution of the reverse problem. In fact, over and above 3, the additional ω_i^k variables allow vector variation redundancy to be ensured by introducing free parameters – in principle serving no purpose – but which will help the solver find solutions in specific cases of poor conditioning. In IT, the limited accuracy of floating numbers very often changes the rank of a matrix during a computation, which numerically becomes rank n although it is theoretically rank 3, for example. It then turns out to be numerically advantageous to carry out intermediary computations in an n-dimension, rank n Ω space to avoid breaking the calculation chain. The actual existence of a 3-D space vector with these n dependent but coherent components must then be ascertained.

NB. Experiments have broadly confirmed this result in all cases. Moreover, it speeds up convergence of the resolution process.

3.6 ALGEBRAIC EXPRESSION OF CONSTRAINTS

Specification of the metric tensor $\left(G_{final} \right)_{i,j} = SPEC_{i,j}$
Given the list of angle specifications

$$SPEC_{i,j} = \left(\left(I_n + \Omega \right) \cdot G_{init} \cdot {}^t \left(I_n + \Omega \right) \right)_{i,j} \qquad \text{Eq3}$$

where G_{init} is known and Ω undetermined.

Specification of the possible closure constraint, where ΔL represents the perturbation of the lengths of vectors of the loop under consideration:

$$\Delta L \cdot \left(I_n + \Omega \right) \cdot G_{init} \cdot {}^t \left(I_n + \Omega \right) \cdot {}^t \Delta L = 0 \qquad \text{Eq4}$$

Since the problem is still underconstrained, we will seek the Ω matrix that verifies all the Eq3 and Eq4 type specifications, at the same time minimising the function $\left(\Omega^2 + \Delta L^2 \right)$. The problem is properly posed: search for the minimum of a convex function subjected to convex constraints. Any off-the-shelf software such as Matlab, automatically provides a valid response, i.e. a certain value for the Ω matrix. The solution is unique if, and only if, the function to be minimised and the constraints are convex, otherwise a local minimum will be obtained which, in any case, is of interest to the designer.

This valid response offers two possibilities:

The constraints Eq3 and Eq4 are accurately verified: this is the solution sought. The G_{final} tensor is then recalculated using the basic Eq2 formula.

One or several constraints are not verified; however, the designer is shown the G_{final} object obtained by applying the basic Eq2 formula. This view will enable the designer to understand the modifications he needs to make to the specifications.

4 CASE STUDY

To illustrate the approach proposed, the study of the perturbation of the geometric object presented in FIGURE 1, is examined. This object is composed of seven planes, called n1, n2 ,.. n7, and a cylinder called l1.

Without losing its general nature, the study only covers angle specifications and, as a result, the generation of vectoral closure equations is not presented in the following explanation.

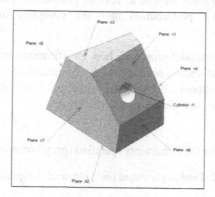

FIGURE 1: Visualisation of the geometric object studied

4.1 INTERNAL REPRESENTATION OF THE OBJECT

Tensor modelling applied to the initial object enables the angle specification model to be obtained, which is the initial metric tensor, called G_{init}. This defines a vectoral space for the sheaf of vectors E useful in the representation of the specifications covered in this paragraph.

Here, $E = \left(\overrightarrow{n_1}, \overrightarrow{n_2}, \overrightarrow{n_3}, \overrightarrow{n_4}, \overrightarrow{n_5}, \overrightarrow{n_6}, \overrightarrow{n_7}, \overrightarrow{l_1}, \right)$ with the following convention: the normal to plane n_i is called $\overrightarrow{n_i}$ and the director vector of the axis of the l_1 cylinder is called $\overrightarrow{l_1}$.

The values of the angles between the vectors, measured on the initial object, are noted in the table below.

TABLE 1: Initial values of angles (in °)

angle	$\overrightarrow{n_1}$	$\overrightarrow{n_2}$	$\overrightarrow{n_3}$	$\overrightarrow{n_4}$	$\overrightarrow{n_5}$	$\overrightarrow{n_6}$	$\overrightarrow{n_7}$	$\overrightarrow{l_1}$
$\overline{n_1}$	0°	40°	40°	83.3239°	124.2760°	45°	77.5901°	180°
$\overline{n_2}$	sym.	0°	0°	85°	85°	85°	85°	140°
$\overline{n_3}$	sym.	sym.	0°	85°	85°	85°	85°	140°
$\overline{n_4}$	sym.	sym.	sym.	0°	83.7013°	85.1644°	159.5953°	96.6761°
$\overline{n_5}$	sym.	sym.	sym.	sym.	0°	165.6258°	113.0501°	55.7240°
$\overline{n_6}$	sym.	sym.	sym.	sym.	sym.	0°	76.2796°	135°

| $\overline{n}_7$ | sym. | sym. | sym. | sym. | sym. | sym. | 0° | 102.4099° |
| $\overline{l}_1$ | sym. | sym. | sym. | sym. | sym. | sym. | sym. | 0° |

4.2 PROBLEM POSED

Based on an initial object of a known form, the designer wishes to obtain a new object (called the final object) which meets certain angle constraints.

This scenario corresponds to what we call a "reverse problem". The method of resolution consists of determining the Ω perturbation then the complete metric tensor, with G_{final}, representing the final object.

To show the genericity of the model proposed, two specifications are studied. For each of them we have noted the numerical values obtained after computation in one table. A second table shows the angle differences obtained between the initial and final values and, finally, two images show the initial object and the final object after perturbation.

Specification 1:

Any value has been chosen for the constraints specified (grey boxes in TABLE 2).

TABLE 2: Final angle values (in °) obtained for specification 1

0°	40.0098°	44.8648°	85.3239°	125.5864°	47°	79.5901°	178°
sym.	0°	4.8599°	86°	87°	87°	84.9059°	141°
sym.	sym.	0°	86°	82.2412°	91.8491°	86.0000°	136.1401°
sym.	sym.	sym.	0°	77.9964°	86.1644°	164.4357°	92.9796°
sym.	sym.	sym.	sym.	0°	163.4088°	114.0501°	54.9089°
sym.	sym.	sym.	sym.	sym.	0°	80.7734°	131.8911°
sym.	sym.	sym.	sym.	sym.	sym.	0°	102.0285°
sym.	sym.	sym.	sym.	sym.	sym.	sym.	0°

Despite the low angle differences specified (all between 1° and 2°), we can see in TABLE 3 that certain differences can be large (see the boxes with bold borders).

TABLE 3: Angle differences (in °) for specification 1

0°	0.0098°	2°	2°	1.4648°	2°	2°	-2°
sym.	0°	1.9922°	1°	2°	2°	0.5336°	1°
sym.	sym.	0°	1°	0.0442°	3.9880°	1°	-0.9922°
sym.	sym.	sym.	0°	-4.8966°	1°	4.5680°	-3.6965°
sym.	sym.	sym.	sym.	0°	-1.4640°	1°	-0.9365°
sym.	sym.	sym.	sym.	sym.	0°	3.7706°	-3.1089°
sym.	sym.	sym.	sym.	sym.	sym.	0°	-0.4021°
sym.	sym.	sym.	sym.	sym.	sym.	sym.	0°

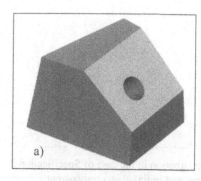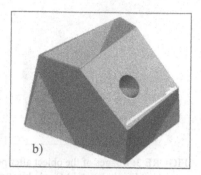

FIGURE 2: Images of the object after perturbation in the case of Specification 1
a) Final object, b) Final object in grey and initial object in white

Specification 2:

The angle constraints specified (grey boxes in TABLE 4) are placed on the first two diagonals, i.e. the angular position of a vector is defined in relation to the two vectors that precede it.

TABLE 4: Final angle values (in °) obtained for Specification 2

0°	41°	42°	128°	91.1804°	85.4806°	81.2767°	137.4902°
sym.	0°	1°	87°	87.0895°	81.8917°	119.8227°	128.7360°
sym.	Sym.	0°	86°	87°	81.8504°	120.7479°	128.2132°
sym.	Sym.	sym.	0°	84.7013°	87.1644°	145.9142°	68.2537°
sym.	Sym.	sym.	sym.	0°	166.6258°	115.0501°	48.0625°
sym.	Sym.	sym.	sym.	sym.	0°	77.2796°	137°
sym.	Sym.	sym.	sym.	sym.	sym.	0°	103.4099°
sym.	Sym.	sym.	sym.	sym.	sym.	sym.	0°

As already noted, despite the low angle differences specified, certain differences may be large. In this instance, they are actually very large (see the boxes bordered in bold in TABLE 5) and the form of the final object is no longer the same as the initial object.

TABLE 5: Angle differences (in °) for Specification 2

0°	1°	2°	44.6761°	-33.0956°	40.4806°	3.6866°	-42.5098°
sym.	0°	1°	2°	2.0895°	-3.1083°	34.8227°	-11.2640°
sym.	Sym.	0	1°	2°	-3.1496°	35.7479°	-11.7868°
sym.	Sym.	sym.	0°	1°	2°	-13.6811°	-28.4224°
sym.	Sym.	sym.	sym.	0°	1°	2°	-7.6615°
sym.	Sym.	sym.	sym.	sym.	0°	1°	2°
sym.	Sym.	sym.	sym.	sym.	sym.	0°	1°
sym.	Sym.	sym.	sym.	sym.	sym.	sym.	0°

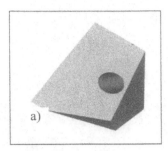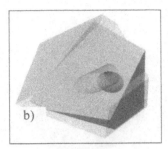

FIGURE 3: Images of the object after perturbation in the case of Specification 2
a) Final object, b) Final object grey and initial object transparent

5   CONCLUSION

We have initially shown that, if necessary, the dimension variations, specified or executed for a 3D geometric object are dependent, and have then demonstrated the basic formula for dimensional variations of a geometric object. There is a wide variety of domains of application for this formulation, for example: the calculation of the tolerancing transfert between manufacturing and design specifications, the geometric tolerancing of mechanical assemblies with the development of analysis and tolerancing synthesis tools, the analysis and synthesis of complex engineering problems by associating the geometry equations presented in this article with other engineering equations describing the globally specified problem.

REFERENCES

1. Ballot, E., Bourdet, P., Thiébaut, F., (2001), Determination of Relative Situations of Parts for Tolerance Computation, In: 7th CIRP International Seminar on Computer Aided Tolerancing, Cachan (FRANCE).
2. Chase, K.W., Gao, J., Magleby, S.P., (1997), Tolerance Analysis of 2-D and 3-D Mechanical Assemblies with Small Kinematic Adjustments, In: Advanced Tolerancing Techniques, John Wiley, pp. 103-137.
3. Clément, A., Rivière, A., Serré, P., (1999) Global Consistency of Dimensioning and Tolerancing, Keynote paper of CIRP Computer Aided Tolerancing, Enschede, The Nederlands.
4. Duong, A.N., (2003), Spécification, analyse et résolution de problèmes géométriques 2D et 3D modélisés par contraintes, Thèse de Doctorat, Ecole Centrale de Paris.
5. Kron, G., (1942) A short course in tensor analysis for electrical engineers, Wiley, New York; Chapman & Hall, London.
6. Laperrière, L., Ghie, W., Desrochers, A., (2003) Projection of Torsors: a Necessary Step for Tolerance Analysis Using the Unified Jacobian-Torsor Model, In: 8th CIRP International Seminar on Computer Aided Tolerancing, Charlotte, USA.
7. Serré P., Rivière A., Clément A., (2001), Analysis of functional geometrical specification, In: 7th CIRP International Seminar on Computer Aided Tolerancing, Cachan, FRANCE.
8. Serré, P., Rivière, A. Clément, A., (2003), The clearance effect for assemblability of over-constrained mechanisms, In: 8th CIRP International Seminar on Computer Aided Tolerancing, Charlotte, USA.

AEROBIC LIVING-ORGANISM ANALOGY:
A CONCEPTUAL MODEL

M. Dassisti

Department of Management and Mechanical Engineering, Polytechnic of Bari, Italy

KEYWORDS: Decision Making, Conceptual Model, Analogical Reasoning.

ABSTRACT. The most part of scientific discoveries of human being borrow ideas and inspiration from nature. This point gives the rationale of the conceptual model here presented, which funds on the basic assumption that it is possible to put the analogy between manufacturing systems, as perceived by decision maker, and aerobic living organisms. The aerobic living-organism (ALO) conceptual model consists of a set of axioms and criteria that allow one to translate his own view of a real manufacturing system into a schemata of an aerobic living-organism, with given features and characteristics. Starting from this schemata it is then possible to build an ALO simulation model, to better study the manufacturing system's behaviour under different operating scenarios. Criticalities of the ALO conceptual model are also discussed.

1 INTRODUCTION

Analogy is one of the most powerful techniques to support decision making known so far: «analogy pervades all our thinking, our everyday speech and our trivial conclusions as well as artistic ways of expression and the highest scientific achievements» [1]. Analogy means bringing new hints and ideas into the decision-making process by following a mapping criteria between two different reasoning domains [2]. Even unexpected conclusions might be found whenever similarities are found into the behavioural rules. The question, most often, is to determine if and how an analogy might be successful in supporting decisional efforts [3]: to the extreme, an analogy can be misleading if principles and rules behind the actors of the reasoning domain do not share the same functional criteria with the actors corresponding in the domain selected. This might occur because different levels of analogy can hold; say, for instance, analogy into the perceived aspect (physical analogy), into the basic functioning (functional analogy) or even into the operating rules (behavioural analogy).

Any decision making process resemble to a complex pattern of "if-then" logical constructs, with an high degree of interdependency between precedent and subsequent statements. The higher the complexity of the system the stronger becomes the decisional task to be performed. This is the case, for instance, for most of the manufacturing systems, where decisions are driven in a sort of "deterministic myopia" : these are driven on the basis of few principal variables, based on some indicators and on the assumption that most of the behaviour of systems can be , in some way, linear or harnessed with a first-order logic. As an example, to improve the throughput of a manufacturing system, the most frequent decision taken is to intervene on the production capacity: the background variables of the systems -as well as all the side effects of this action- are often neglected. Building on this example, a simple increasing of capacity might bring to strong pressing

Published in: E. Kuljanic (Ed.) *Advanced Manufacturing Systems and Technology*,
CISM Courses and Lectures No. 486, Springer Wien New York, 2005.

onto the human resources (technical training, overloading, etc.) and, at the same time, to an increase of the intermediate-buffer levels which, in turn, in the sum of effects may result into a decrease of the same throughput. The higher the complexity of the system, the stronger is the effect of this "deterministic myopia" way of reasoning.

The most known and better performing complex systems known so far are aerobic living organisms. The main feature that characterises complex systems is essentially the strong dependencies between its components. Each individual component contributes to the system output and thus the resulting causal relations are neither linear nor additive. In such a context, it is not possible to extrapolate each component from the system to which it belongs and forecast its behaviour alone. One of the most interesting lessons learned form living organisms is that it seems to exist a "intelligent vision" in taking their decisions, which seems to be a sort of retaliation of the previously mentioned "deterministic myopia". This is mainly the reason of the scientific interest in studying a number of living organism and their organisation, and to learn from these. In some sense, it seems to be an "intrinsic" intelligence in the systems themselves more than an extrinsic high-level intelligence governing everything.

The analogical reasoning about living organisms dated back to the beginning of the history of human being, while several trans-disciplinary studies of analogies of complex systems with ethological systems or ecological systems as well can be found in the recent scientific literature (see, e.g. [4], [5]). These latter try to find out techniques and solutions for decision support on the basis of principles and rules derived from natural complex systems; say, for instance, the dominance order within animal societies or the theory of self-organization of biological systems ([6], [7]).

If the scope of the approach is to determine what rules make these complex systems behave in a effective and robust way -in order to borrow solutions from the "superior mind" that conceived it- the idea discussed in this paper is if it is feasible to use the analogy of a manufacturing system with a living organisms, that are the most perfect and coherent complex system existing in nature. The attractive analogy between nature and industry, based on the similarity of natural functions and certain industrial activities been rarely explored so far as such [5], in particular exploiting the analogies between metabolic functions to support analysis of manufacturing systems, apart the number of non-scientific metaphoric approaches or the managerial ones ([8]).

The non-trivial question in systems crafted by humans is how to apply principles of this *intrinsic* intelligence to systems build with that kind of *extrinsic* intelligence.

The basic assumption on which relies this paper is that it is possible, for a decision maker, to use an analogy reasoning with aerobic living-organisms (ALO) to analyze manufacturing systems and also to drive sound decision concerning its design or management. The conceptual model presented will provide axioms and criteria to reason about a true manufacturing system, i.e. a coordinated set of human and material resources, and as it was an aerobic living-organism, thus facilitating the intuition of its behaviour and criticalities.

2   THE AEROBIC LIVING-ORGANISM AXIOMS

This idea of ALO analogy here discussed is a viable alternative to forecast characteristic behaviour of manufacturing systems by the decision-maker: it consists in making structured the metaphoric translation of the manufacturing systems into an aerobic living-organism schemata, so as to give to it a scientific validity. It is important to stress the difference in the reasoning domains in this latter case, since the living organism is a coherent, and in some sense hierarchically organized, set of components, while an manufacturing systems does not necessarily share this organizational shape; that means an higher degree of freedom in the relations between its components is allowed. In this sense, the analogy proposed might result inappropriate if a physical analogical correspondence would be imposed.

On the other hand, it is here sustained that basing the analogy on the correspondence of the overall behaviour –similarly to the use of entropy in physics - this metaphoric approach may work well to infer conclusion on manufacturing system behaviour, based on analogies with the aerobic living-organism.

The basic assumption of ALO analogy is, in fact, that as the final goal of living organism is to live and to allow its species to survive, the same is for a manufacturing system, where the same term survival assumes in this latter case a slightly different meaning. The basic assumption for the approach to held is that the manufacturing system can evolve as well, but that its features do not vary in time in the period of analysis (stationary condition of the observations).

The conceptual model is funded on the following basic axioms:

BEHAVIOURAL ANALOGY: ENTROPIC AXIOM - since (*manufacturing systems as a whole strives to survive and grow like aerobic living-organisms do*) then (*it is always possible to find out a behavioural analogy between transformations of manufacturing systems as a whole and aerobic living organism transformations*)

To explicit it better, analogously behaving manufacturing systems and aerobic living-organism both tend to determine the same dynamical transformations of the living states **S** in time:

$$S_E(T) \; \Theta \rightarrow S_E(T+1) \cong S_{ALO}(T) \; \Theta \rightarrow S_{ALO}(T+1) \qquad (1)$$

Where the symbol $\Theta \rightarrow$ indicates a function that induces transformations of the system state at any given period of time T.

FUNCTIONAL ANALOGY: DECOMPOSITION AXIOM - descending from the entropic axiom (*it is always possible to find at least one kind of functional analogy between components of an manufacturing systems with fundamental parts of an aerobic living-organism on the basis of their influence on the overall behaviour*).

To explicit it better: since (*interacting parts of the aerobic living-organism determine the transformation function that dynamically changes living states*) then (*it always possible to define manufacturing systems components such as they interact in a similar fashion to the ALO's parts, so as the manufacturing system as a whole behaves similarly to the aerobic living-organism*).

These statements essentially gives the rationale of the mapping process shown in Figure 1, that will be commented throughout the paper.

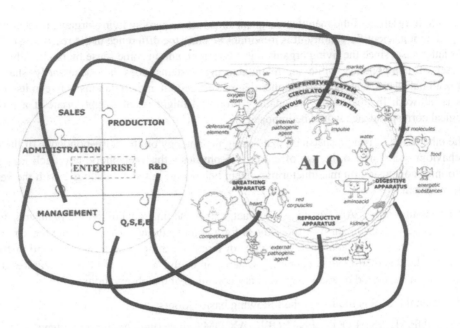

FIGURE 1. Mapping the manufacturing system onto the aerobic living-organism schemata

According to the above, the problem become to find out the mapping criteria, and thus the correspondences between parts of a manufacturing systems and the components of the aerobic living-organism schemata. The criteria here adopted to realize a one-to-one mapping of parts rely on the equivalence of the global functional characteristics of parts, where all parts of the ALO responds to a common goal: the survival of its species. The basic scopes and the related functions of the apparatuses (breathing, digesting, reproducing) and systems (nervous, circulatory and defensive) of the aerobic living organism can all be seen as contributing to its survival, i.e. having a positive effect on the survival of its species in time. In the same way, some actors of the external environment (competitor, external pathogenic agents), or internal pathogenic agents can be characterized with respect to their negative potential to survival. Bearing in mind this analogical model, it is easy to translate the manufacturing systems into an ALO schemata, by explicating the one-to-one correspondences of its components, since their scopes and functions will respond to the survival criterion.

3   AEROBIC LIVING ORGANISM SCHEMATA

According to the above, the ALO conceptual model need to define criteria to clearly identify the one-to-one functional correspondences of the basic components of the manufacturing system with the aerobic living-organism schemata.

The schemata in Figure 1 consists of internal components and elements of the external environment. These latter will represent exogenous factors which regulate the life of the organism, while internal elements determine its life. Either internal or external components are here characterized by scopes and functions defined with respect to the final goal of the ALO survival.

Throughout the paper reference will be made to a manufacturing systems producing material goods; it is evident that similarly the example can be extended to a service manufacturing system.

3.1 ALO SCHEMATA: EXTERNAL ELEMENTS

In the external environment there are several actors as in Table 1.

TABLE 1. Featuring external elements of the ALO schemata (sub-components in brackets)

ELEMENTS	SCOPE	FUNCTIONS
AIR (Oxygen atom)	To assure a vital environment	To surround ALO and brings oxygen atoms to ALO
OXYGEN ATOM	To assure vital functions take place	To support exothermal reactions useful for the life of the ALO
FOOD (molecules of food)	To assure vital energy to ALO	To bring molecules of food that can be reduced into molecules of food.
FOOD MOLECULES	To bring reserve of vital energy and principal structural components	To contain energetic substances (sugars and lipids) and plasticizers (aminoacid) and water molecules and can release this when required.
EXTERNAL PATHOGENIC AGENT (viruses, bacteria)	To strive for its own survive	To subtract vital energy and components from ALO to increase in number and their dimension.
WATER (water molecule)	To assure vital functions and reactions	Water conveys molecules of water
COMPETITOR	Strive to its own survival	To subtract all the primary factors for survival (air, food, water). It may fight also other ALO or help them for opportunistic reasons.

3.2 ALO SCHEMATA: INTERNAL COMPONENTS

The aerobic living-organism schemata is conceived as a set of interconnected apparatuses coordinated by appropriate systems (circulatory, nervous, defensive): respiratory, digestive, reproductive (see Figure 1 and Figure 2). Several elements should be considered in this schemata as in Table 2.

TABLE 2. Featuring internal components of the ALO schemata (sub-components in brackets)

COMPONENTS	SCOPE	FUNCTIONS
ENERGETIC SUBSTANCES (sugar, lipid)	To assure vital energy at different times	Contained into the food molecules, sugars bring vital energy immediately available while lipids allow to accumulate it
STRUCTURAL SUBSTANCES (aminoacid)	To assure fundamental components	Contained into the food molecules, aminoacid are the bricks of ALO and assure the same existence of parts and their structural consistence.
WATER MOLECULES	To assure vital functions and reactions	Water molecules convey substances in sites where are necessary and support energetic reactions
INTERNAL PATHOGENIC AGENT (scraps)	To reproduce itself by using vital substances or even structural components within its ALO environment.	Internal pathogenic agents tend to replicate themselves and thus to damage the ALO. They result from ill-functioning or errors of the same ALO metabolism (cancer; auto-immunological illness, etc.) and they produce scraps.
SCRAPS	To occupy useful room for fundamental parts of the ALO.	Scraps may be accumulated into given sites of ALO, and to do so they subtract room to other vital components. Sometime they may consist of an excess of vital substances (sugar, lipids, water, etc.).
EXAUST	To bring away unutilised substances resulted from vital reactions.	Useless or excess substance are segregated and be expelled by the ALO.
RED CORPUSCLES	To bring oxygen molecules.	Red corpuscles capture oxygen molecules and release them when requested for vital reactions
IMPULSE	To bring information	Information is an electric impulse with given amplitude and frequency.
BLOOD PLATELET	To repair critical injuries or to assure ordinary maintenance	It repairs injuries by providing appropriate substances and procedures.
BREATHING APPARATUS (Air conducts; Lungs)	To introduce air into the ALO and to capture oxygen molecules	The apparatus pumps fresh air into lungs; extracts oxygen molecules and introduces them into the circulatory apparatus, then expels exhausted air.
DIGESTIVE APPARATUS (Masticatory apparatus, Stomach; Liver; Intestine)	To introduce food into the ALO, to transform it into food molecules, to decompose these into energetic and structuring substances, to eliminate exhausts.	Food is brought into by the masticatory apparatus, there processed and transformed into food molecules. The stomach transforms, thanks to the liver, food molecules into energetic and structural substances and introduces them into the circulatory apparatus. Intestine brings out the exhausts.
REPRODUCTIVE APPARATUS (Cerebral cortex; reproductive system)	To ameliorate continuously the state of health and even to produce new ALO.	The apparatus performs activities which bring the ALO to maximize its healthiness and its capability to reproduce itself.
CIRCULATORY SYSTEM (Heart;	To interconnect all the apparatuses bringing all the	It brings oxygen from lungs and carbon dioxide to them. It brings energetic and structural ele-

Arterial passageway; kidneys; urinary system; blood molecules)	fundamental elements throughout the ALO	ments from stomach. It conveys defensive and maintenance actors in all sites. It brings also exhausts.
BLOOD MOLECULES	To transport physical elements in all ALO sites.	Under the stimuli of heart, blood molecules brings energetic and structural substances (sugar, lipids, aminoacid, water)
NERVOUS SYSTEM (Brain vegetative; Nerves)	To coordinate and control basal functions	It controls and coordinates the vital processes through impulses based also on external stimuli.
DEFENSIVE SYSTEM (defensive elements; lymphatic system)	To defend organism from external pathogenic elements	It constantly produces and also conveys defensive elements (white corpuscles and blood platelet) for ordinary control or to react to critical external injuries
DEFENSIVE ELEMENTS (white corpuscles, blood platelet)	To recognize and eliminate external pathogenic elements	Defensive elements are constantly trained to recognise and eliminate external pathogenic elements. To assure ordinary and extraordinary maintenance

4 ANALOGICAL CRITERIA FOR MAPPING AN ALO MODEL

As already stated, the aerobic living-organism conceptual model funds on the entropic axiom and the decomposition axiom, as well as onto the ALO schemata to translate a given real manufacturing system into an analogous ALO manufacturing system model. That means, once the entropic axioms holds, a set of criteria can be defined to correctly identify the one-to-one analogical correspondence of components of a given manufacturing system with and the ALO schemata on the basis of the functional analogies recognised in Table 1 and Table 2. This will allow to build an ALO model of the manufacturing system to be used to support decisions with the help of the confidence in the well-assessed knowledge of the living-organism behaviour.

Here below the analogical criteria for mapping the manufacturing system are explained, that are based on declaring analogous those parts/components having the same SCOPE and the same FUCTIONS with respect to the common aim of survival.

TABLE 3. Analogies of external elements of the manufacturing system with the ALO schemata

ALO ELEMENTS	ANALOGY TO REAL MANUFACTURING SYSTEM	Analogy rationale
AIR	Order	The higher the availability of air the higher the probabilities of survival.
OXYGEN ATOM	Minimum reward per order allowable.	Each atom should correspond to the minimum amount of reward allowable for an order to be remunerable. The higher the density of oxygen atom in the air the higher the remunerability for that orders.
FOOD	Supply lot required to carry out a given production.	The higher the availability of food, the higher the probabilities of survival

FOOD MOLECULES	Materials contained into the supply lot: raw materials, supply materials and spare components.	Food molecules contain several different types of internal components that are used by ALO. The variety of these is important as much as their availability.
EXTERNAL PATHOGENIC AGENT	Errors deriving from external environment (suppliers, customers, competitors, etc.)	Growth or reduction rate of viruses depends from the state of healthiness of the ALO, while their number and aggressiveness depends from the operating environment.
WATER	Monetary circulation	Water represent a monetary resource available to the customers
COMPETITOR	Enemies present in the ALO living environment.	Competitors lowers the volume of air available as well as food by competing with our living organism. The aggressiveness of competitors change inversely with the abundance of living elements.

TABLE 4. Analogies of internal components of the manufacturing system with the ALO schemata

ALO COMPONENTS	REAL MANUFACTURING SYSTEM	Analogy rationale
ENERGETIC SUBSTANCES	Sugars corresponds to earnings coming from selling finished products, while lipids are represented by the value of products on-stock.	Contained into the food molecules, sugars represent the semi-finished or finished products that are immediately sold, therefore bringing fresh energy. Lipids represents those part of raw material, semi-finished and finished products on stock, which can be sold later to bring new energy.
STRUCTURAL SUBSTANCES	Supply materials products functional to production	Aminoacid are those materials necessary assure the regular functioning of ALO..
WATER MOLECULES	Raw materials and semi-finished products functional to production	Water molecules represents the monetary reserves, which can used under certain circumstances (investments; etc.), and give the consistence of the manufacturing system.
INTERNAL PATHOGENIC AGENT	Worst practices, wrong procedures, disorganisation or internal inefficiencies.	Internal pathogenic agents tend to be repeated and damage the ALO. They also produce scraps.
SCRAPS	Scraps or reworked elements, as well as monetary waste.	Scraps accumulated subtract room to vital functions. Sometime they may consist of an excess of stocks.
EXHAUST	Exhausted materials	Residual substance are segregated and expelled.
RED CORPUSCLES	Administrative officers or any professional operator which manages resources.	Brings vital resources (either monetary or physical) and release them when requested for vital reactions
IMPULSE	Information or knowledge required to control or coordinate	Information is an electric impulse with given amplitude and frequency.
BLOOD PLATELET	Maintenance operators	It repair injuries by providing appropriate sub-

		stances and procedures.
BREATHING APPARATUS	Sales department	It is the interface with the external environment that provides to get fresh orders. The higher the number of operators and their competence, the stronger the breathing potentialities under certain energetic condition.
DIGESTIVE APPARATUS	Production function in response to customer's orders.	Masticatory apparatus is the acceptance of material, the stomach the manufacturing settings that process products and liver the production engineering. The intestine is the disposal function of exhausted materials. The higher the number of operators and their competence, the stronger the production potentialities under certain energetic condition.
REPRODUCTIVE APPARATUS	Brain is the applied research functions. The reproductive system is the development function.	Both subcomponents works for assuring a continuous improvement in the capability of production as well as for expanding the same production in other fields. The higher the number of operators and their competence, the stronger the innovation potentialities under certain energetic condition.
CIRCULATORY SYSTEM	Hart is the administrative function.	The administrative function provides distribute resources. Logistics brings all physical components with a given frequency. The higher the number of operators (blood molecules), the stronger the transportation potentialities under certain energetic condition.
BLOOD MOLECULES	The resources of logistical apparatus.	The higher the number of operators, the stronger the transportation potentialities under given energetic conditions.
NERVOUS SYSTEM	Management function and all the related operators. Nerves correspond to the communication system.	Impulses are information used to manage all the other functions. Basic knowledge can be also accumulated into the brain. The higher the number of operators (cerebral cells), the stronger the information processing potentialities up to a threshold condition.
DEFENSIVE SYSTEM	Quality, Safety, Environment and Ethical management	Continuous revision of integrated procedures allow to repair external and internal pathogenic agents.
DEFENSIVE ELEMENTS	White corpuscles are quality, safety and environmental operators. Blood platelet are maintenance operators.	Defensive elements constantly are trained to recognise and eliminate external pathogenic elements, to assure ordinary and extraordinary maintenance

5 VALIDATION CRITERIA OF THE CONCEPTUAL MODEL

The most appropriate use of the analogies identified in Table 3 and Table 4 for supporting decision making is to build an appropriate simulation tool; the analogy rationale, in fact, represents the principles to which a given model should respond. The appropriate characterisation of components features and their functional interactions –summarised in Figure 2- will permit to observe the real manufacturing system resembling to a living organism striving for survival in different environments. In that Figure 2, the links represents interactions between an element source and the corresponding recipient element: the direction of the causal links indicated the recipient of the effect. As a result, a link with a positive sign in parenthesis can be interpreted as follows: *the higher the (SOURCE) the higher the (RECIPIENT)*.

It is important to stress that the design criteria of the simulation model, implementing the ALO conceptual model, should be good enough to derive appropriate global "life-state" indicators to monitor the overall performance as a function of all the governing laws regulating each apparatus, system or elements. Also the concept of health requires a very clear definition, since several possible interpretation can be given, up to the comprehensive, multiscale and dynamic measure proposed by [9].

Two possible global performance indicators can be recognised: namely life state and life potential. Life state can be defined as the instantaneous value of the internal energy, i.e. the total energy. The life potential (i.e. the healthiness) will be represented by the set of values of the critical variables that represent the manufacturing system capability to survive as a function of the external conditions. The combinations of these two global parameters will reflect synthetically the overall status of the system, i.e. its good health or illness, up to the extreme of risk of death, in any given period of time *T*.

FIGURE 2. Correlation loops among components of the aerobic living-organism schemata

Descending from this, in order to validate initially the ALO simulation model, two different kind of validations need to be performed: a congruence test and a sensitivity analysis. The former means to analyse several well assessed conditions (i.e. *combination of external elements features*)

and see if the overall expected behaviour of the simulated system is logically consistent to what is naturally expected for an equivalent aerobic living-organism. The latter consists in a set of experiments where the rate of variation of given output variables is analysed as a function of the gradients of variations of some controlled variables of the model (some combination of *the internal components features*). By simply relying on the profound knowledge of the ALO behaviour with respect to the survival the decision-maker can derive sound conclusions of the validity of the simulation model per se.

6 DISCUSSION

The conceptual model presented in this paper for decision making apparently seems to contrast with the recent theories of self-organisation, where the attempt is to show that complex collective behaviour may emerge from interactions between individuals. Despite the pre-assigned role and scheme of interactions between parts of the ALO conceptual model, the appropriate selection of components features and their relationships might bring to global non-linear behaviour, thus in line with the above mentioned studies.

This characterisation remains the critical point for the success of the conceptual model with respect to its practical significance in supporting the decision maker in driving sound decisions. It is also important to select an appropriate code to translate the ALO schemata of the manufacturing system, able to represent all its components and relationships: agent based modelling concepts can be a viable solution to this question.

The dynamic aspect of life is, finally, a questionable matter for the use of the conceptual model, whether it is correct to state that manufacturing systems are doomed to follow a life-cycle curve.

AKNOWLEDGEMENTS

The author is deeply indebted with Mrs. Angela Tursi for the precious help in preparing figures.

REFERENCES

1. Iding, M.K., (1997), How analogies foster learning from science texts, Instructional Science, Vol 25, 233-253.
2. Harding, A.L., (2000), Coleridge, natural history, and the 'Analogy of Being', History of European Ideas, Vol 26, 143-158.
3. Samuelson, L., (2001), Analogies, Adaptation, and Anomalies, Journal of Economic Theory, Vol 97, 320-366.
4. Costanza, R., King, j., (1998), The first decade of Ecological Economics, TENTH ANNIVERSARY SURVEY ARTICLE, Ecological Economics, Vol.28, 1-9.
5. Ayres, R.U., (2004), On the life cycle metaphor: where ecology and economics diverge, Ecological Economics, Vol.48, 425-438.
6. Bonabeau, E., Theraulaz, G., Deneubourg, J-L., (1999), Dominance Orders in Animal Societies: The Self-organization Hypothesis Revisited, Bulletin of Mathematical Biology, Vol 61, 727-757.
7. Kube, R.C., Bonabeau, E., (2000), Cooperative transport by ants and robots, Robotics and Autonomous Systems, Vol 30, 85-101.
8. Bordia, R., Kronenberg, E., Neely, D., (2005), Innovation'sOrgDNA, BoozAllen Hamilton Inc., 20050009/4/05, USA.

9.   Costanza, R., Mageau, M., (1999), What is a healthy ecosystem?, Aquatic Ecology, Vol.33, 105-115.

MANUFACTURING SYSTEM ANALYSIS VIA AEROBIC LIVING-ORGANISM ANALOGY: A CASE STUDY

M. Dassisti

Department of Management and Mechanical Engineering, Polytechnic of Bari, Italy

KEYWORDS: Living Organism Analogy, Multi-Agent Simulation, Manufacturing System Analysis.

ABSTRACT. The paper presents an application of the aerobic living-organism (ALO) conceptual model developed by the author in a previous paper, to perform an analysis of an Italian small enterprise producing and assembling components for industrial and civil applications. Starting from the ALO enterprise schemata an agent-based simulation model was built using the Starlogo® environment to support re-engineering decisions. The application presented proved its potential usefulness into practical decisional tasks. Criticalities of the implementation of the ALO conceptual model are also discussed, based on the experience gained into the specific application, driving some general hints for future researches.

1 INTRODUCTION

In a previous paper the hypothesis that it is possible to analyze enterprises by referring to the analogy with aerobic living-organisms (ALO) has been presented [1]. The related conceptual model has been discussed to provide all the criteria to put the basis for building an analogies of enterprises with an aerobic living organism. The rationale of the model presented was to facilitate the decision maker the intuition of the model behaviour and thus allow him to derive sound conclusions.

Starting from a aerobic living-organism schemata, the ALO analogy allows to build a simulation model of the real enterprise whose actors are distinguished into external elements and internal components as better specified later in this paper. By appropriately assigning functions to these actors the simulation model can capture embedded functioning mechanisms and thus represent the overall behaviour of the enterprise in a synthetic manner. Since the approach relies on the analogy with the most complex systems known so far – namely the aerobic living organisms- the most appropriate approach for building the simulation model can be the agent-based one. This seems in fact to be the most promising developed so far for capturing the complexity of systems with its bottom-up approach. Despite the apparent prescriptive conception of the ALO conceptual model, it strongly relies on the appropriate characterisation of actors in the model to represent the overall behaviour as a result of complex interactions and combination of variables. External components and internal apparatuses and systems are, in fact, simply tied by functional relationships, which coordinate directions and efforts of those actors which are appointed to convey information, materials or moneys.

Simulation is one of the most well assessed tool to support decision whenever future behaviour is not known; by predicting the effect of some decision projected into the future time is sometime the only way to found an answer to decisional matters. But it is evident how this approach is a

Published in: E. Kuljanic (Ed.) *Advanced Manufacturing Systems and Technology*,
CISM Courses and Lectures No. 486, Springer Wien New York, 2005.

sort of "blind search" into a wide space of potential actions. Despite techniques for accelerating the search process, still it is affected by the before mentioned "deterministic myopia" of the extrinsic governing intelligence. It would be better to find out a self-regulating system, which has an adequate degree of flexibility to cut the decisional space into few potential decision on the basis of parallel signals (constraints) coming from the set of coordinated sensors forming the intrinsic self-regulating intelligence.

In this sense, agent based modelling (ABM) techniques well adapt to represent the ALO schemata, where agents are components with given features, whose interactions will result into an overall behaviour dependent from the components features and also from the interactions with other enterprises. Agents are being advocated as a next generation model for engineering complex, distributed systems. ABM simulation are such that is only necessary to specify the capabilities of each agent by defining one or more tasks it could perform and the related behavioural rules. Agent-based computing represents a new synthesis both for Artificial Intelligence (AI) [2]. A number of agent based simulation environment have been developed so far are for several applications: [3] provides a comprehensive overview of general programming languages and toolkits suitable for agent-based computational economics (ACE) and complex adaptive systems (CAS) modelling. To enable students to build their own CAS models the StarLogo programming language was build at MIT and environment specifically designed to support simulation design, construction and testing [4]. Recently, a number of researchers have attempted to apply agent technology to concurrent engineering, manufacturing enterprise integration, supply chain management, manufacturing scheduling and control [5].

2 THE INDUSTRIAL TEST CASE

The company analysed in the paper (Tubinsud S.r.l.) is an Italian company build in 1969, producing and assembling piping and industrial equipments for industrial and civil applications. It is organised as a job-shop and its survival is strongly dependent on orders issued to it by big companies in competition with several other enterprises. It had 32 permanent employees, with a sunk capital of about 1.242.000,00[€] and a net profit in the year of analysis of about 7.236,00[€]. On the basis of a statistical analysis in the recent years it resulted that there are an average of 260 orders/year. From an initial market analysis, taking also into account the strong recurrence to electronic auction, it resulted that the average number of strong competitors was 3, with a market share of about 30%. The company has a certified quality management system (according to the ISO 9000/2000) and a certified environmental management system (according to ISO 14001). The official organisation consists of the following roles: one sole administrator; one general manager (GM) with three staff members -quality manager, environmental manager and IT responsible-; one managing director with three staff members; sales department; engineering department; production engineering department with one director and two staff members –supply warehouse management- and a number of flat teams assigned to orders, managed by a team responsible.

3 MAPPING ASSUMPTION FOR BUILDING THE SIMULATION MODEL

According to the ALO conceptual model approach, building the enterprise simulation model means to translate external elements and internal components to reproduce the scopes and functions so as their interaction results into a coherent behaviour.

It is important to distinguish between local variables and global ones. The formers are those defining the features of elements of the ALO model, while the latter's refer to the overall manufacturing system and determine the operating scenario or the perception of its global performances.

Set-up variables, signed with a § symbol, are those that need to be assigned by the decision maker before any simulation run to initialise it.

3.1 GLOBAL ALO PERFORMANCE INDICATORS

In order to capture the essence of the healthiness of the enterprise studies it is necessary to clearly define a system of life-state performance indicators. More generally, on the basis of the entropic axiom, it would be interesting to derive a state $(S(T))$ indicator which better allow to represent the transformation function $(S_E(T) \; \Theta \rightarrow S_E(T+1))$, which depends form the overall behaviour of the system under analysis.

Descending from this, two main global state indicators were recognised: the "life potential" (or "healthiness") and the "life state" (or "total energy") of the system. The system state would then be a function of these two indicators, that varies over time (T):

$$S (T) = f(LP(T), LS(T))$$ (1)

No particular hypothesis were here made on the nature of the function f indicated in (1).

It is interesting to explicit the nature of these two global indicators. Life potential is a synonymous of healthiness because the higher the life potential the better is the healthiness, i.e., the system capability to survive and reproduce. In the ALO analogy, healthiness is a direct function of the number and state of (lipids) and (blood molecules) and (blood platelet) and (red corpuscles)

On the other hand, Total Energy is synonymous life state since it is a good state that where there is a lot of energy to perform vital activities. Life state is a function of is a direct function of the number of (molecules of oxygen) and (water molecules) and inverse function of efficiency of all the apparatuses and systems, which consumes energy.

It is important to say that all the reasoning about local and global variables holds up to a threshold below or under which the first-order logical inference does not hold any more. For instance, it is true that the life state is better when energy is higher but, above a given threshold value, energy may become a danger to the same life for the efforts required to manage it correctly. The same reasoning holds for the life potential.

3.2 EXTERNAL ALO VARIABLES

Among global variable there are the environmental ones. Environment represents the external causes and the operating scenario: the decision maker can study the behaviour and tendency of the system by simply varying initial value of the related controlled variable. Table 1 summarise all these variables and their theoretical relationships with all the other variables. The number of competitors as well as the amount of food interact reciprocally during the simulation, determining a dynamical operating scenarios.

TABLE 1. Implementation hypothesis of external elements from the ALO schemata

ALO ELEMENTS [enterprise analogy]	CHARACTERISTIC VARIABLE (§ set-up variable name)	THEORETICAL RELATIONSHIPS
AIR [=Orders]	§ Volume of air in the environment in a given period of observation and its law of variation (Vair) Number of orders got by the enterprise in a given period (No)	$V_{air}(Nco+1, T, AGco)$ $No(Nco, healthiness)$
OXYGEN ATOM [Minimum net profit per order allowable]	§ N_{O2} = Number of oxygen atom for a specific order or its average § P_{O2} = net profit per atom	N_{O2} (order type, market) P_{O2} (order type; healthiness)
FOOD [Supply lot]	Total quantity of material necessary per order (Qf) § Availability of lots (Af)	$Q_f (N_{O2})$ $A_f (Nco, t)$
FOOD MOLECULES [Supply components]	Percentage of raw material to be used for an order (Ns), § Percentage of raw material to be put on-stock (Nl), § Percentage of supply material (Na) § Availability of raw material (As), raw material to be put on-stock (Al), supply material (Aa)	$N_{s,l,a}(order type)$ $A_{s,l,a}(Nco, T, AGco)$
EXTERNAL PATHOGENIC AGENT [External source of errors]	§ Number of viruses and bacteria (Nvb) Grow/reduction rate (ΔNvb) § Energy consumption (Eepa), i.e. pathogenic potentiality of viruses and bacteria	$N_{vb}(market, suppliers)$ $E_{epa}(healtiness; market; suppliers)$ ΔNvb (healtiness; market, T)
WATER [market monetary reserve]	Amount of fresh money obtainable from orders (TW)	$TW(Nco, AGco, market)$
COMPETITOR [Competing enterprises]	§ Number of competitors (Nco) § Aggressiveness (AGco); i.e. number of orders subtracted	$Nco (Vair, AGco, market)$ $AGco(Vair, AGco, market)$

Market is an important global external variable, which influences prices of products and thus remunerability of orders, represented in the ALO schemata as the density of oxygen atoms. Market will be represented indirectly into the local variables

3.3 INTERNAL ALO VARIABLES

Descending form the ALO schemata presented in a previous paper, to implement into a simulation model it is necessary to recognise the set of local (or internal) variables. In this paragraph the ideas already presented are expanded in the following Table 1 and Table 2, where a set of local characteristic variables are presented per each element or component. Theoretical relationships between these and between global variables or indicators are reported as well.

TABLE 2. Implementation hypothesis of internal components derived from the ALO schemata

ALO COMPONENTS	CHARACTERISTIC VARIABLE	RELATIONSHIPS
ENERGETIC SUBSTANCES [sugar= direct raw material; lipid=raw material on-stock]	Amount of finished products on stock (Nli) Amount of finished products manufactured (Nsu) § Net profit of finished products (Psu) § Net profit of finished products on-stock (Eli)	$N_{su,li}$ *(Qf, order type, ηdi)* P_{su}*(order type)* P_{li}*(order type, market, T)*
STRUCTURAL SUBSTANCES [amino-acid= supply material]	§ Amount of supply material available (Qa) § Amount of money available for payments or investments (Qw)	Q_a *(healthiness, T)* Q_w *(healthiness, T, market, Nco, AGco)*
INTERNAL PATHOGENIC AGENT [wrong procedures]	§ Number of pathogenic agents (Nipa) § Energy consumption (Eipa)), i.e. pathogenic potentiality of internal pathogenic agents	N_{ipa} *(healthiness, T)* E_{ipa}*(healthiness)*
SCRAPS [internal errors]	Number of internal errors (N_{er})	N_{er}*(healthiness, T)*
WATER MOLECULES [Cash reserve]	Amount of monetary reserve in the manufacturing system (W)	$W(N_{o2},healthiness)$
EXHAUST [exhausted materials]	Amount of exhaust produced (N_{ex}).	N_{ex}*(healthiness, Vair)*
RED CORPUSCLES [administrative officers]	§ Number of administrative officers or operators (Nrc) §Efficacy (Hrc), i.e capability to bring a given amount of resources when required §Energy consumption (Erc)	$Nrc=(Vair)$ Hrc*(healthiness, T)* E_{rc}*(healthiness)*
IMPULSE [control information]	§Average amount of information supported (Qi) §Energy consumption (Ei)	Q_i*(healthiness)* E_i*(healthiness)*
BLOOD PLATELET [maintenance operator]	§Number of maintenance operators (N_{bp}) §Efficacy of corrective actions (H_{bp}) §Threshold value of efficacy of corrective actions below which platelet are not able to destroy pathogenic agents. (THbp)	N_{bp}*(healthiness, Vair)* H_{bp}*(healthiness)* TH_{bp}*(healthiness)*
BREATHING APPARATUS [Sales de-	§Breathing capacity (Pa), i.e. maximum amount of air inspired versus the total	Pa*(healthiness, Nco, AGco, market)*

	amount of air available §Total amount of energy consumption (TEbr) §Efficiency (η_{br})	$TE_{br}(healthiness)$ $\eta_{br}(healthiness,)$
partment]		
DIGESTIVE APPARATUS [production function]	§Processing capacity = production volume (Qdi) §Total amount of energy consumption (TEdi) §Manufacturing efficiency (η_{di})	$Qdi(healthiness, Vair)$ $TE_{di}(healthiness)$ $\eta_{di}(healthiness)$
REPRODUCTIVE APPARATUS [Research & development]	Number or innovations or new prototypes (Nre) § Total amount of energy consumption (TEre) §Efficiency (η_{re})	$Nre(healthiness, W)$ $TEre(healthiness, W)$ $\eta_{re}(healthiness)$
CIRCULATORY SYSTEM [logistic]	Volume of material conveyed (Vtr) §Total amount of energy consumption (TEtr) §Efficiency (η_{tr})	$Vtr(healthiness, Vair, Q_s, Q_l, Q_a, Q_w)$ $TEtr(healthiness)$ $\eta_{tr}(healthiness)$
BLOOD MOLECULES [logistic system]	§Number of logistic operators (Nbm) §Transport capacity (Cbm)	$N_{bmi}(healthiness, Vair)$ $C_{bm}(healthiness)$
NERVOUS SYSTEM [communicaton system]	§Frequency of contact or meetings (Fns) in the period of observation T §Total amount of energy consumption (TEns) §Efficiency (η_{ns})	$F_{ns}(healthiness, Vair)$ $TEns(healthiness)$ $\eta_{ns}(healthiness)$
DEFENSIVE SYSTEM [Quality, Safety, Environmental and Ethical management system]	§Total amount of energy consumption (TEde) §Efficiency (η_{de}), function of the number of new procedures and error elimination	$TEde(healthiness)$ $\eta_{de}(healthiness)$
DEFENSIVE ELEMENTS [Quality, safety, environmental and ethical operators]	§Number of operators [Nde] §Energy consumption (Ede)	$N_{de}(healthiness, W)$ $E_{de}(healthiness)$

4   ENTERPRISE SIMULATION MODEL

Descending form the reasoning presented in the previous paragraph and the industrial case presented, it was possible to put the basis for implementing a simulation model. The following Table 3 and Table 4 give the outcome of the design activity performed.

TABLE 3. Modelling assumption for the enterprise analysed (*=initial values; RND=random)

ALO ELEMENTS	MODELLING ASSUMPTION	Parameters values
AIR [=Orders]	$V_{io} = (1-AGco)/100 * Nco *RND[0, Vo\_average]$ if Nco>0 $V_{io} = RND[0, Vo\_average]$ for Nco=0 normalised to the maximum number of order/month	$RND[0, 22]$ $Vo\_average =22$ order/month

	(30)	
OXYGEN ATOM [= minimum profit per order]	Ni_{O2}=const P_{O2}= average value = (total net profit/year) /(total number of orders/year)	Ni_{O2}=1 P_{O2}=2/10
FOOD [=total amount of material per order]	Q_f=Value of material purchased/total value of the order= constant	Q_f=34/50
FOOD MOLECULES	Not specified	====
EXTERNAL PATHOGENIC AGENT [inefficiencies]	Nvb = constant in the observation period; Inefficiencies are calculated as the average number of supplier's delays and customer's delays and then normalised to $Eepa$= $\sum (Pr_{vb}*AGepa)/ \sum (Pr_{vb})$ where Pr_{vb} are the probability of occurrence and $AGepa$ is calculated as the average % incidence of increase in the total leadtime and the increase of cost per order and normalised.	Nvb=2/10 $Eepa$=8/10
WATER [cash reserve]	Not included	==
COMPETITOR [competing enterprises in electronic auctions]	$AGco$ = percentage of the number of orders subtracted in each electronic auction $Nco(t)$= $N*co$+ RND [0, Nco_{MAX}]	$AGco$ = 3/10 $N*co$=3=const

TABLE 4. Modelling assumption for the enterprise analysed

ALO COMPONENTS	MODELLING ASSUMPTION	Parameters values
ENERGETIC SUBSTANCES [sugar= assembly to satisfy orders] [lipids = subcomponents]	Amount of finished products on stock and to order is proportional to the total amount of food Q_f*No Net profit of finished products (Psu) is assigned constant in each simulation run Net profit of finished products on-stock (Pli) is assigned constant in each simulation run	Nsu= 0.40*Q_f*No Nli=0.50*Q_f*No Psu= 4/10 Pli =5/10
WATER [cash reserve]	Total value of sunk capital + net profit of the preceding financial year normalised to the average values of the enterprise belonging to the same sector.	W=5/10
STRUCTURAL SUBSTANCES	Not specified	====
INTERNAL PATHOGENIC AGENT [errors and delays]	$Nipa$ = constant in the observation period; Inefficiencies are of three types: manufacturing errors, production delays and cost estimation errors $Eipa$= $\sum (Pr_{vb}*AGipa)/ \sum (Pr_{vb})$ where Pr_{vb} are the probability of occurrence and $AGipa$ is calculated as the average % incidence of increase in the total leadtime and the increase of cost per order	$Nipa$ = 3/10 $Eipa$=3/10
SCRAPS [wrong subcomponents	Ner = constant % of the total amount of products per order in the observation period. Inefficiencies	Ner = 1/10 * Q_f*No

manufactured or assembled]	*are calculated as the average number of manufacturing errors, production delays and cost estimation errors in the reference period of time T*	
EXAUST	*Not included*	====
RED CORPUSCLES	*Not included*	====
IMPULSE [information flows between departments]	*Average amount of information exchanged, evaluate in three possible levels (high=10, medium=6 and low=2) (Qi)* *Energy consumption = Weighted Average of production flows on the basis of an annual observation*	$Qi=3/10$ $Ei=8/10$
BLOOD PLATELET [corrective actions]	*Number of maintenance operators (Nbp)* *A threshold value exists below which platelet are not able to destroy pathogenic agents (Th$_{bp}$)* *The efficacy (H$_{bp}$) increases of the amount of the pathogenic potentiality of viruses and bacteria destroyed*	$N_{bp}=6/10$ $Th_{bp}=29/350$ $H_{bp}=5/10$
BREATHING APPARATUS [sales department]	*Total amount of energy consumption (TEbr)* *Efficiency (η_{br})*	$\eta_{br}=2/10$ $TEbr=1/10$
DIGESTIVE APPARATUS [production and assembly department]	*Total amount of energy consumption (TEdi)* *Manufacturing efficiency (η_{di})*	$TE_{di}=3/10$ $\eta_{di}=9/10$
REPRODUCTIVE APPARATUS [quality engineer]	*Presence = 0 = absence (corresponding to absence of innovation, i.e. reset of impulses to initial state) or =1, presence (corresponding to a reset of living functions to initial condition in case of strong injuries).*	*Presence = 1, coincident with the quality department*
CIRCULATORY SYSTEM [administrative office]	*Total amount of energy consumption (TEtr)*	$TE_{tr}=2/10$
BLOOD MOLECULES	*Considered as red corpuscles*	==
NERVOUS SYSTEM	*Fns = constant in the reference period T.* *Total amount of energy consumption (TEns)*	$Fns = 30 / year$ $TE_{ns}=2/10$
DEFENSIVE SYSTEM	*Total amount of energy consumption (TEde)*	$TE_{de}=3/10$
DEFENSIVE ELEMENTS	*Coincident with blood platelet*	==

5 VALIDATION FOR A TEST CASE

The ALO simulation model was built using the Starlogo® environment [6]. The relative simplicity of this environment and its ABM features allowed quite easy the translation into a model of the modelling assumptions presented in Table 3 and Table 4. Figure 1 presents a program session after a simulation run. The model run on a simple PC in few minutes of real time simulation.

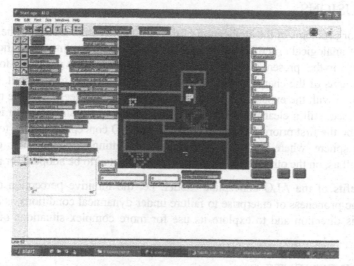

FIGURE 1.   Aerobic living-organism simulation program in Starlogo®

To validate initially the ALO simulation model built, either congruence tests and sensitivity analysis were performed; these are not reported here for the sake of brevity.

To give a the flavour of the usefulness to the managers f the company considered, set of experiment were made over a period of T=170 unit of simulated time, performing 15 replications: the average values of the output are reported in the following Table 5. Three different scenarios were testes: the normal one, corresponding to data put in Table 3 and Table 4; a strong condition – Nbp=7/10 and Hbp=7/10- and, finally, ,a weak condition –where the Vo_average=11.

TABLE 5. Output of the simulation under the conditions state in Table 3 and Table 4

Elements	Normal	Strong	Weak
Life state indicator	272	331	-35
Life potential indicator	305	317	-19
Blood Platelet	3	7	3
Lipids	322	322	0
Red corpuscles	28	27	18
Viruses and bacteria	30	24	11

Despite not clearly significant of the real operating conditions (i.e. *combination of external elements features*) the overall expected behaviour of the simulated system appears logically consistent to what is naturally expected for that enterprise.

This statement derives from the profound knowledge of the ALO behaviour with respect to the survival the decision-maker can derive sound conclusions of the validity of the simulation model per se.

6   CONCLUSIONS

One of the major problems of the application of the ALO conceptual model is the verification of the goodness of analogical criteria allowing to map a real enterprise into a significant simulation model. As shown in the present application, several assumption were necessary to derive quantitative measurements of the characteristic variables of the simulation model built. Despite results derived are in line with the expected outcome and with the overall feeling on the true state of the enterprise analysed, still a clear proof of the validity of the approach presented is lacking. This aspect need to be the first priority of future research on ALO conceptual model, to bring it out of the metaphoric sphere where it actually lies. Parameters setting can be another critical issue to this point as well as, on the other hand, the correct level of detail to be adopted for modelling.

The final benefits of the ALO conceptual model, i.e. the intuitive perception of the decision maker about the proneness of enterprise to failure under dynamical conditions, is a good point to continue in this direction and to explore its use for more complex situations of interoperating enterprises.

AKNOWLEDGEMENTS

The author is indebted with ing. Ivan Passiatore for his precious support in discussing the contents of the present paper. This work is partially supported by the Commission of the European Communities under the sixth framework programme (INTEROP Network of Excellence, Contract N° 508011, http://www.interop-noe.org) .

REFERENCES

1. Dassisti, M., (2005), AEROBIC LIVING-ORGANISM ANALOGY: A CONCEPTUAL MODEL, Proc. Int conf, on Advance Systems and Manufacturing Technologies, Udine, (to appear).
2. Jenning, N.R., (1999), On agent-based software engineering, Artificial Intelligence, Vol. 117, 277–296.
3. Tesfatsion, L., (16 January 2005), General Software and Toolkits - Agent-Based Computational Economics (ACE) and Complex Adaptive Systems (CAS), URL http://www.econ.iastate.edu/tesfatsi/; Department of Economics; Iowa State University; Iowa.
4. Klopfer, E., (2003), Technologies to support the creation of complex systems models—using StarLogo software with students, BioSystems, Vol. 71, 111–122.
5. Shen, W., Maturana, F., Norrie, D.H., (2000), , MetaMorph II: an agent-based architecture for distributed intelligent design and manufacturing, Journal of Intelligent Manufacturing, Vol. 11, 237-251.
6. Klopfer, E., Colella, V., Resnick, M., (2002), New paths on a StarLogo adventure, Computers & Graphics, Vol. 26, 615–622.

AUTHORS INDEX

AUTHORS INDEX

Printed in the United States
By Bookmasters

Printed in the United States
By Bookmasters